LINEAR STATISTICAL MODELS

LINEAR STATISTICAL MODELS

AN APPLIED APPROACH

Bruce L. Bowerman
Richard T. O'Connell
MIAMI UNIVERSITY, OHIO
David A. Dickey
NORTH CAROLINA STATE UNIVERSITY

Duxbury Press
Boston

PWS PUBLISHERS

Prindle, Weber & Schmidt • Duxbury Press • PWS Engineering • Breton Publishers •
Statler Office Building • 20 Park Plaza • Boston, Massachusetts 02116

PWS Publishers is a division of Wadsworth, Inc.

Library of Congress Cataloging-in-Publication Data

Bowerman, Bruce L.
Linear statistical models, an applied approach.

Bibliography: p.
Includes index.
1. Regression analysis. 2. Linear models (Statistics)
I. O'Connell, Richard T. II. Dickey, David A.
III. Title.
QA278.2.B687 1986 519.5'36 85-20418
ISBN 0-87150-904-0

ISBN 0-87150-904-0

Printed in the United States of America

86 87 88 89 90 — 10 9 8 7 6 5 4 3 2 1

Sponsoring Editor: Michael Payne
Production Coordinator: Ellie Connolly
Manuscript Editor: Alice Cheyer
Interior and Cover Design: Helane Manditch-Prottas
Cover Photo: Carol Lee/The Picture Cube
Interior Illustration: ANCO/Boston, Inc.
Typesetting: Syntax International
Cover Printing: New England Book Components
Printing and Binding: Halliday Lithograph

PREFACE

LINEAR STATISTICAL MODELS is designed as a textbook for courses in applied regression analysis and applied linear statistical models. It is appropriate for advanced (junior and senior level) undergraduates and graduate students in business, engineering, and the sciences (including mathematics, statistics, operations research, and computer science). The required mathematical and statistical background for this book is college level algebra and basic statistics. The text does employ elementary matrix algebra, which is clearly explained in Chapter 4, prior to our discussion of regression analysis.

The text consists of 10 chapters. Chapter 1 serves as an introduction, and Chapters 2 and 3 discuss basic statistics and probability distributions. Chapter 4 explains the needed matrix algebra, and Chapter 5 initiates our formal discussion of regression analysis by presenting the simple linear regression model and by introducing multiple regression models (discussions of quadratic models and interaction terms are included). We present geometrical interpretations of these models, define and calculate the least squares point estimates of their parameters, and show how to use these models to carry out point estimation of the mean value of the dependent variable and point prediction of a future (individual) value of the dependent variable.

In Chapter 6 we cover statistical inference in regression analysis. Although we present a brief discussion of confidence intervals for a population mean in optional Section 3.4, the discussion of statistical inference in Chapter 6 is totally self-contained. (Optional sections are indicated with an asterisk in the Table of Contents.) One of the strengths of this book is the careful consideration in Chapter 6 of repeated sampling. We use this idea to interpret the meaning of the "level of confidence" (as it pertains to confidence and prediction intervals) and to explain the meaning of *t*-tests in regression analysis. In Chapter 7 we discuss building an appropriate regression model. Included are sections on coefficients of determination and correlation, multicollinearity, the C_p statistic, comparing regression models on the basis of various criteria, stepwise procedures, *F*-tests, and dummy variables. The book offers flexibility to instructors who wish to discuss coefficients of determination and correlation and the *F*-test for the overall regression model before discussing individual *t*-tests: They can proceed to Section 7.1 after studying Section 6.2 and then return to Section 6.3 to study the remainder of Chapter 6.

In Chapters 8, 9, and 10 we cover more advanced topics in regression analysis. Chapter 8 gives a complete discussion of analyzing data from designed experiments (completely randomized experiments, factorial experiments, and randomized block experiments). We present both the regression approach and the analysis of variance approach to analyzing the experimental situations discussed in Chapter 8. However, since the analysis of variance approach is less general, we make its study optional. Chapter 8 also includes discussion of statistical inferences for linear combinations of regression parameters and explains how to compute Scheffé simultaneous confidence intervals. Chapter 9 gives a detailed presentation of the meaning of the assumptions behind regression analysis. The assumptions of correct functional form, constant variance, independence, and normality are fully explained, and methods (including residual plots) used to detect violations of these assumptions are presented. Outlying and influential observations are also discussed. Chapter 10 contains a comprehensive treatment of remedies that can be employed when the regression assumptions are violated. Included are discussions of transformations to achieve linearity, transformation techniques for handling unequal variance situations (including 0–1 dependent variables), time series models, and handling autocorrelated errors by the Cochran-Orcutt technique. A strength of the text is the abundance of thoroughly explained examples in Chapters 8, 9, and 10. Note that instructors who wish to teach a regression course (without analysis of variance situations and experimental design considerations) can omit Chapter 8, because Chapters 9 and 10 can be read independently.

Our experience indicates that students who have had one or two statistics courses have forgotten many of the concepts. We have therefore spent considerable time making the material in this book self-contained and readable. The material has been extensively class-tested, and many students have told us it is so clearly written that they can use this book for study on their own. In addition, a large number of challenging exercises accompany the chapters. These exercises include data sets that can be analyzed beyond the context of the exercises themselves.

Many people have contributed to this book. We would like to thank the many reviewers of the book including Harry G. Costis, California State University at Fresno; James Horrell, University of Oklahoma; H. W. Lilliefors, George Washington University; Glen Milligan, Ohio State University; and Michael Rabbit, Central Missouri State University. We would also like to thank the fine people at Duxbury Press. In addition, we want to thank Alice Cheyer, copy editor, and two of our students, Peggy Sexton and Carrie Clark, who contributed to the book by running many computer programs and helping to develop the solutions manual that accompanies the text. Finally, our wives, Drena, Jean, and Barbara, deserve special thanks for their patience and understanding during the writing process.

Bruce L. Bowerman
Richard T. O'Connell
David A. Dickey

CONTENTS

1

AN INTRODUCTION TO REGRESSION ANALYSIS

In this chapter we introduce regression analysis and describe what regression analysis is and how it can be used.

Regression analysis is a statistical methodology concerned with relating a variable of interest, which is called the **dependent variable** and denoted by the symbol y, to a set of independent variables, which are denoted by the symbols $x_1, x_2, \ldots, x_p$. The objective is to build a **regression model** that will enable us to adequately *describe, predict,* and *control* the dependent variable on the basis of the independent variables.

For example, in Chapters 5 through 8 we build a regression model relating the dependent variable

y = the demand for Fresh liquid laundry detergent

to the independent variables

x_1 = the price of Fresh

x_2 = the average industry price of competitors' similar detergents

x_3 = the advertising expenditures made to promote Fresh

x_4 = the type of advertising campaign used to promote Fresh

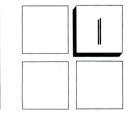

Here, each of the independent variables x_1, x_2, and x_3 is a **quantitative independent variable**, which we define to be an independent variable that assumes numerical values corresponding to the points on a line. In contrast, the independent variable x_4 is a **qualitative independent variable**, which we define to be an independent variable that is not quantitative. Whereas the levels of a quantitative independent variable are numerical, the levels of a qualitative independent variable can be defined by describing them. For example, we would define the levels of

$$x_4 = \text{the type of advertising campaign used to promote Fresh}$$

by describing the different advertising campaigns. Having constructed an appropriate regression model relating y to x_1, x_2, x_3, and x_4, we would then use this model

1. *To describe* the relationships between y and x_1, x_2, x_3, and x_4. For example, we would describe the effect that increasing advertising expenditure has on the demand for Fresh and whether this effect is dependent upon the price of Fresh.
2. *To predict* future demands for Fresh on the basis of future values of the independent variables x_1, x_2, x_3, and x_4.
3. *To control* future demands for Fresh by controlling the price of Fresh and by controlling the advertising expenditures and type of advertising campaigns used to promote Fresh. Note that since we cannot control the price of competitors' similar detergents or their advertising expenditures used to promote these detergents, we cannot perfectly control (or, for that matter, predict) future demands for Fresh.

The following examples illustrate a few of the many other useful applications of regression analysis.

EXAMPLE 1.1

The chief engineer of the physical plant at a university might use regression analysis to predict future fuel consumption (in tons of coal) at the university on the basis of future atmospheric temperatures and wind velocities, so that adequate supplies of fuel can be ordered (see Chapters 5, 6, and 7).

EXAMPLE 1.2

An investor might use regression analysis to relate the rate of return on a particular stock to the rate of return on the overall stock market, so that she might use predicted changes in the overall market to determine whether she should purchase, hold, or sell the particular stock (see Chapter 6).

EXAMPLE 1.3

The manufacturer of an unleaded gasoline might use regression analysis to find the combination of additives that will yield an unleaded gasoline that is reasonably priced and gives high mileage (see Chapter 8).

EXAMPLE 1.4

A farming cooperative might use regression analysis to predict its quarterly gas bills on the basis of trends in natural gas prices and the season of the year (see Chapter 10).

To conclude this section, we should note that although in this book we sometimes speak of the effect of an independent variable upon a dependent variable, regression analysis cannot be used to *prove* that a change in an independent variable *causes* a change in the dependent variable. Rather, regression analysis can only be used to establish that the two variables "move together." For example, although regression analysis can be used to establish that as liquor sales have increased over the years, college professors' salaries have also increased, this does not prove that increased liquor sales increase professors' salaries. Rather, both variables are influenced by a third variable—long-run growth in the national economy.

2

POPULATIONS AND SAMPLES

This chapter presents some basic concepts of statistics that we need to know in order to study regression analysis. We begin in Section 2.1 by discussing finite populations. We see that populations are described by what we call parameters, and we define four such parameters—the population mean, range, variance, and standard deviation. Section 2.2 extends our discussion to infinite populations. Next we present (in Section 2.3) some elementary concepts concerning probability. In Section 2.4 we find that since the true values of population parameters are unknown, we must compute estimates of these parameters. In order to compute such estimates, we must randomly select a sample from the population of interest. Section 2.4 describes how this is done and also shows how sample statistics (which are descriptive measures of samples) can be used as point estimates of population parameters.

2.1 THE MEAN, VARIANCE, AND STANDARD DEVIATION OF A FINITE POPULATION

Frequently we seek information about a collection of objects, or **elements**.

> We define a **population** to be the entire collection of elements about which information is desired.

A population may contain a finite number of elements, in which case we call the population finite. We denote by the symbol N the number of elements in a finite population. If there is no finite limit to the number of elements that could potentially exist in a population, we say that the population has an infinite number of elements and call the population infinite. In this section we discuss finite populations, and in the next section we discuss infinite populations.

If a population consists of a finite number of N elements, we can arbitrarily number these elements and denote them by the symbols $EL_1, EL_2, \ldots, EL_N$. We are often interested in studying properties of some numerical characteristic of these elements. Here each element in the population possesses a particular value of the numerical characteristic under study. In this book a value of a numerical characteristic will always be a number on the real line—that is, a number between negative infinity and positive infinity. For $h = 1, 2, \ldots, N$, let y_{EL_h} denote the value of the numerical characteristic possessed by the hth element, EL_h, and call y_{EL_h} the hth element value of the numerical characteristic. Thus, for the population of elements $EL_1, EL_2, \ldots, EL_N$, there is a corresponding population of element values $y_{EL_1}, y_{EL_2}, \ldots, y_{EL_N}$.

EXAMPLE 2.1

Table 2.1 lists a numerical characteristic that might be of interest for each of the indicated finite populations.

TABLE 2.1

Finite Population of Elements	Numerical Characteristic of an Element
The countries on the planet Earth	Number of inhabitants below poverty level in a country
The states in the U.S.A.	Number of family farms in a state
The cities with a population of at least 5,000 in the state of Nebraska	Percentage of workers unemployed in a city
The inhabitants of Cleveland, Ohio	Yearly income (in dollars) of an inhabitant
The members of the United States Senate	Number of votes missed during 1986 by a Senator
The teams in the National Football League	Number of games won last year by a team

EXAMPLE 2.2

National Motors Company, Inc., is an automobile manufacturer that produces a model called the Gas-Mizer. All Gas-Mizers are equipped with a popular option package consisting of (1) a four-cylinder, 200-hp engine, (2) a standard transmission, and (3) a factory-installed air conditioning system. National Motors wishes to study the gasoline mileage (measured in miles per gallon, or mpg) obtained by the population of all Gas-Mizers produced this year.

Here we are considering Gas-Mizers that would be driven roughly 50,000 miles in normal weather conditions by a driver possessing normal driving habits, with approximately 60 percent of these miles being city driving and approximately 40 percent being highway driving. Assume that National Motors will produce a population of $N = 100{,}000$ Gas-Mizers this year, and for $h = 1, 2, \ldots, 100{,}000$, let GM_h denote the hth Gas-Mizer that will be produced. Furthermore, let y_{GM_h} = the number of miles per gallon averaged by the hth Gas-Mizer when it is driven 50,000 miles. We will measure y_{GM_h} to the nearest tenth of a mile per gallon and will refer to y_{GM_h} as the hth Gas-Mizer mileage. Thus, for the population of Gas-Mizers $GM_1, GM_2, \ldots, GM_{100{,}000}$, there is a corresponding population of Gas-Mizer mileages $y_{GM_1}, y_{GM_2}, \ldots, y_{GM_{100,000}}$. Although the 100,000 Gas-Mizers are alike in that they have a four-cylinder, 200-hp engine, a standard transmission, and a factory-installed air conditioning system, and each will be driven roughly 50,000 miles under the previously described conditions, the Gas-Mizers differ with respect to other characteristics affecting gasoline mileage. For example, although National Motors has a quality-control program, the performance of the Gas-Mizer assembly process and the quality of materials used in the production of Gas-Mizers will vary over time. In addition, the actual conditions under which the Gas-Mizers are driven will be somewhat different. Thus, since the Gas-Mizers $GM_1, GM_2, \ldots, GM_{100,000}$ differ with respect to various characteristics that affect gasoline mileage, the corresponding Gas-Mizer mileages $y_{GM_1}, y_{GM_2}, \ldots, y_{GM_{100,000}}$ are different from each other.

We now define a **parameter** to be a descriptive measure of a population and we define four parameters: the population mean, range, variance, and standard deviation, which are denoted by μ, RNG, σ^2, and σ, respectively.

FOR A FINITE POPULATION OF ELEMENT VALUES $y_{EL_1}, y_{EL_2}, \ldots, y_{EL_N}$:

1. The *population mean* is

$$\mu = \frac{\sum_{h=1}^{N} y_{EL_h}}{N} = \frac{y_{EL_1} + y_{EL_2} + \cdots + y_{EL_N}}{N}$$

2. The *population range, RNG,* is the difference between the largest element value and the smallest element value in the population.
3. The *population variance* is

$$\sigma^2 = \frac{\sum_{h=1}^{N} (y_{EL_h} - \mu)^2}{N}$$

$$= \frac{(y_{EL_1} - \mu)^2 + (y_{EL_2} - \mu)^2 + \cdots + (y_{EL_N} - \mu)^2}{N}$$

4. The *population standard deviation* is

$$\sigma = \sqrt{\sigma^2} = \sqrt{\frac{\sum_{h=1}^{N} (y_{EL_h} - \mu)^2}{N}}$$

Whereas the population mean, which is the average of the N element values $y_{EL_1}, y_{EL_2}, \ldots, y_{EL_N}$, is a measure of the central tendency of these element values, the population range, variance, and standard deviation are measures of the spread, or variation, of the element values. For example, the population variance

$$\sigma^2 = \frac{\sum_{h=1}^{N} (y_{EL_h} - \mu)^2}{N}$$

is the average of the N squared deviations of the individual element values from the mean, μ, of all of the element values. To see that the population variance measures the spread of the element values, consider the following argument. First, suppose that the element values are spread far apart. In this situation, many element values will be far away from the mean, μ. This means that many of the deviations

$$\begin{array}{c} y_{EL_1} - \mu \\ y_{EL_2} - \mu \\ \vdots \\ y_{EL_N} - \mu \end{array}$$

will be large (positive or negative), that many of the squared deviations will be large, that the sum of the squared deviations will be large, and that the average of the squared deviations—the population variance—will be relatively large. On the other hand, if the element values are clustered close together, many element values will be close to the mean, μ. This means that many of the preceding deviations will be small (positive or negative), that many of the squared deviations will be small, that the sum of squared deviations will be small, and that the average of the squared deviations—the population variance—will be small. Thus, we conclude that the greater the spread of the element values, the larger is the population variance.

Before presenting an example, we should mention that one might be tempted at this point to look for some "hidden meaning" behind (or practical interpretation of) the population variance. This temptation is natural, since one might wonder why we are discussing this somewhat strange measure of population spread. However, do not look for a hidden meaning! For now, all that should be understood is (1) the population variance is the average of the squared deviations of the individual element values from the average, μ, of all of the element values, and (2) the reason we are studying the population variance, σ^2, is that the population standard deviation σ, which is the positive square root of the population variance, will be found to be important in later sections of this book.

EXAMPLE 2.3

Reconsider the Gas-Mizer problem of Example 2.2. Suppose we wish to study the population of the first $N = 3$ Gas-Mizers produced by National Motors. Assume that each of these Gas-Mizers is driven 50,000 miles under the previously described driving conditions and that the following gasoline mileages (in miles per gallon) are recorded: $y_{GM_1} = 32.4$, $y_{GM_2} = 30.6$, $y_{GM_3} = 31.8$. For this population

1. The mean is

$$\mu = \frac{\sum_{h=1}^{3} y_{GM_h}}{3} = \frac{y_{GM_1} + y_{GM_2} + y_{GM_3}}{3}$$

$$= \frac{32.4 + 30.6 + 31.8}{3}$$

$$= \frac{94.8}{3}$$

$$= 31.6$$

2. The range is

$$RNG = 32.4 - 30.6 = 1.8$$

3. The variance is

$$\sigma^2 = \frac{\sum_{h=1}^{3} (y_{GM_h} - \mu)^2}{3} = \frac{(y_{GM_1} - \mu)^2 + (y_{GM_2} - \mu)^2 + (y_{GM_3} - \mu)^2}{3}$$

$$= \frac{(32.4 - 31.6)^2 + (30.6 - 31.6)^2 + (31.8 - 31.6)^2}{3}$$

$$= \frac{(.8)^2 + (-1)^2 + (.2)^2}{3}$$

$$= \frac{.64 + 1 + .04}{3}$$

$$= \frac{1.68}{3}$$

$$= .56$$

4. The standard deviation is

$$\sigma = \sqrt{\sigma^2} = \sqrt{.56} = .7483$$

In Example 2.3 we can calculate μ, RNG, σ^2, and σ because the population under consideration contains only three element values. In many situations the population under consideration is so large (it contains so many element values) that no person (or organization, etc.) could afford (because of the time and expense involved) to compute the parameters of the population. In such a situation, in order to more fully explain the meanings of the statistical results in this book, we will sometimes assume that there is a "supernatural power" who does know *the true values of the population parameters.*

EXAMPLE 2.4

Since National Motors wishes to study the gasoline mileage obtained by its entire fleet of Gas-Mizers, it is not really interested in the population of the first three Gas-Mizers it produces. Therefore, consider the population of gasoline mileages that would be obtained by the 100,000 Gas-Mizers that National Motors produced this year. The mean of this population is

$$\mu = \frac{\sum_{h=1}^{100,000} y_{GM_h}}{100,000} = \frac{y_{GM_1} + y_{GM_2} + \cdots + y_{GM_{100,000}}}{100,000}$$

In order to prove that federal gasoline mileage standards for this year are met by the Gas-Mizer, National Motors would like to determine the population mean, μ. However, in order to calculate μ, the manufacturer would have to know $y_{GM_1}, y_{GM_2}, \ldots, y_{GM_{100,000}}$, which are the mileages that would be obtained by all 100,000 Gas-Mizers produced this year. Since National Motors must of course have some Gas-Mizers left to sell after its study of Gas-Mizer gasoline mileage has been completed, the company cannot do the testing required to determine all 100,000 mileages, and hence it will never know the true value of the population mean, μ. However, we assume that (a supernatural power knows that) the true value of μ is 31.5 mpg. Moreover, assume that (this supernatural power knows that) the true value of the population variance is

$$\begin{aligned}\sigma^2 &= \frac{\sum_{h=1}^{100,000} (y_{GM_h} - \mu)^2}{100,000} \\ &= \frac{(y_{GM_1} - \mu)^2 + (y_{GM_2} - \mu)^2 + \cdots + (y_{GM_{100,000}} - \mu)^2}{100,000} \\ &= \frac{(32.4 - 31.5)^2 + (30.6 - 31.5)^2 + \cdots + (31.3 - 31.5)^2}{100,000} \\ &= .64\end{aligned}$$

and thus that the true value of the population standard deviation is

$$\sigma = \sqrt{\sigma^2} = \sqrt{.64} = .8$$

If (as in the Gas-Mizer problem) the value of a population parameter is unknown to us, and if information concerning the parameter is desired, our only recourse is to take a **sample**, or a subset of elements, from the population of interest. **Statistical inference** is the science of using the information contained in a sample to make a generalization about a population. One type of statistical inference is **statistical estimation**, which is the science of using the information contained in a sample (1) to find an estimate of an unknown population parameter, and (2) to place a reasonable bound on how far the estimate might deviate from the unknown population parameter (that is, place a reasonable bound on how wrong the estimate might be). We begin our formal presentation of sampling and statistical estimation in Section 2.4.

2.2 INFINITE POPULATIONS

In the Gas-Mizer problem so far, we have thought of the population of "this year's Gas-Mizers" as a finite population containing the 100,000 Gas-Mizers that National Motors produced this year. However, it might also be useful to think of the population of "this year's Gas-Mizers" as an infinite population containing all Gas-Mizers that National Motors could *potentially* produce this year. We refer to this population as infinite because, although National Motors will not actually produce an infinite number of Gas-Mizers, the company *could* always (with the needed resources and productive capacity) produce another Gas-Mizer. So we might say there is no finite limit to the number of Gas-Mizers that National Motors could potentially produce. Of course, this argument is idealized. There is some finite limit to the number of Gas-Mizers that could potentially be produced, because National Motors has a limited amount of resources that can be used to produce Gas-Mizers. However, since this limit is probably very large, and since it will be convenient to do so in later chapters, we can use our idealized argument and consider the population of all Gas-Mizers that National Motors could produce this year to be infinite.

In general, an infinite population is an idealized representation of a very large population—a population so large that *ideally* there is no finite limit to the number of elements that could potentially exist in the population. As in the case of a finite population of elements, if we are interested in studying some numerical characteristic (for example, gasoline mileage) of an infinite population of elements (for example, Gas-Mizers), then there is a corresponding infinite population of element values (for example, Gas-Mizer mileages). However, since there is no finite limit to the number of element values that could potentially exist in the infinite population, precise definitions of the population mean, μ, and variance, σ^2, require the use of advanced mathematical notation. Rather than giving these definitions, suffice it to say that we can think of the mean of an infinite population as the average of the infinite number of element values that could potentially exist in the population, and we can think of the variance of an infinite population as the average of the infinite number of squared deviations of the individual element values from the mean, μ, of all of the element values. Here, as in the case of finite populations, the population mean is a measure of central tendency of the infinite population, while the population variance and the population standard deviation (which is the positive square root of the population variance) are measures of spread, or variation, of the infinite population. (The larger the population variance, the greater is the variation displayed by the element values in the population.)

EXAMPLE 2.5

In Example 2.4 we stated that the true values of the mean and variance of the *finite* population of the 100,000 gasoline mileages of the Gas-Mizers that National Motors produced this year are, respectively, $\mu = 31.5$ and $\sigma^2 = .64$. In this example we assume that the true values of the mean and variance of the *infinite* population of gasoline mileages of the Gas-Mizers that National Motors could potentially produce are also, respectively, $\mu = 31.5$ and $\sigma^2 = .64$. Realistically, the mean and variance of the finite population would not necessarily equal the mean and variance of the infinite population. However, if we remember that we are measuring gasoline mileages to the nearest one-tenth of a mile per gallon, it might be possible that (with rounding) these means and variances would indeed be equal. We assume that this is the case in the Gas-Mizer problem.

2.3 PROBABILITY

The concept of probability is employed in describing populations and in using sample information to make statistical inferences about populations. To begin our discussion of probability, we first define an **experiment** to be any process of observation that has an uncertain outcome. An **event** is an experimental outcome that may or may not occur. The **probability** of an event is a number that measures the chance, or likelihood, that the event will occur when the experiment is performed. If the symbol A denotes an event that may or may not occur when an experiment, denoted by EXP, is performed, we denote the probability that the event A will occur by the symbol $P(A)$. If the experiment is performed n_{EXP} times, and the event A occurs n_A of these n_{EXP} times, then the proportion of the time event A has occurred is

$$\frac{n_A}{n_{EXP}}$$

The probability that the event A will occur is often interpreted in the following way. Suppose the experiment EXP is performed repeatedly, and suppose that after each repetition the ratio n_A/n_{EXP} is calculated. Now consider repeating the experiment EXP a number of times approaching infinity, and consider the sequence of numbers obtained by calculating the ratio n_A/n_{EXP} after each repetition. The limit of this sequence is interpreted to be the probability of the event A. Stated mathematically,

$$P(A) = \lim_{n_{EXP} \to \infty} \frac{n_A}{n_{EXP}}$$

For example, when we say that the probability of a head appearing when we toss a fair coin is .5, we mean that if we tossed the coin a number of times approaching infinity, the proportion of heads obtained would approach one-half. That is, suppose we toss a fair coin a number of times approaching infinity and obtain the following results (where H denotes a head and T denotes a tail): H, T, H, H, T, H, T, T, H, H, ... Defining the event A as "a head appears," Table 2.2 shows n_A (the number of repetitions on which a head has appeared), n_{EXP} (the number of times the coin has been tossed), and the ratio n_A/n_{EXP} (the proportion of the time a head has

TABLE 2.2
Calculation of the Ratio n_A/n_{EXP} for Repeated Coin Tosses

Repetition	Outcome	Number of Heads (n_A)	Number of Repetitions (n_{EXP})	$\frac{n_A}{n_{EXP}}$
1	H	1	1	$\frac{1}{1}$
2	T	1	2	$\frac{1}{2}$
3	H	2	3	$\frac{2}{3}$
4	H	3	4	$\frac{3}{4}$
5	T	3	5	$\frac{3}{5}$
6	H	4	6	$\frac{4}{6}$
7	T	4	7	$\frac{4}{7}$
8	T	4	8	$\frac{4}{8}$
9	H	5	9	$\frac{5}{9}$
10	H	6	10	$\frac{6}{10}$
⋮	⋮	⋮	⋮	⋮

appeared) for each repetition. Thus, we obtain the following sequence of ratios n_A/n_{EXP}:

$$\frac{1}{1}, \frac{1}{2}, \frac{2}{3}, \frac{3}{4}, \frac{3}{5}, \frac{4}{6}, \frac{4}{7}, \frac{4}{8}, \frac{5}{9}, \frac{6}{10}, \ldots$$

So when we say that the probability of a head appearing is .5, we are saying that the limit of this sequence as n_{EXP} approaches infinity is .5.

Of course, in practice we cannot perform an experiment a number of times approaching infinity. So from a practical standpoint the probability of an event is roughly equal to the proportion of the time the event would occur if the experiment were performed a very large number of times. Consequently, one way we can estimate the probability of an event is to perform the related experiment a very large number of times and estimate the probability to be the proportion of times the event occurs during the repetitions of the experiment.

If we wish to estimate the probability of an event when we cannot perform the related experiment a very large number of times, we might estimate this probability using previous experience with similar situations and intuitive judgment. For example, a company president might estimate the probability of success for a one-time business venture to be .7 if, based on his knowledge of the success of previous similar ventures, on the opinions of company personnel involved with the venture, and on other pertinent information, he believes that there is a 70 percent chance the venture will be successful. If we can neither perform the related experiment a very large number of times nor use subjective judgment, there are many other methods available to estimate the probability of an event. One such method—the use of continuous probability distributions—is discussed in Section 3.1.

Finally, note that since $P(A)$, the probability of the event A, is a long-run proportion, and since any proportion is greater than or equal to zero and less than or equal to 1, it follows that $P(A)$ is greater than or equal to zero and less than or equal to 1. If $P(A) = 0$, this means that the event A cannot occur, whereas if $P(A) = 1$, this means that the event A is sure to occur.

2.4 RANDOM SAMPLES AND SAMPLE STATISTICS

The calculation of many population parameters—such as a population mean or population standard deviation—requires knowledge of all the element values in the (finite or infinite) population of element values. If we do not know all the element values and thus cannot calculate a population parameter of interest, the only recourse is to select randomly a sample (or subset) of n element values from the population. Then, the information contained in the sample can be used to make statistical inferences concerning the population parameter (for example, the parameter can be estimated).

We can randomly select a sample of n element values from a population of element values by first randomly selecting a sample of n elements from the population of elements. This is done by making n selections from the population of elements in such a way that on any particular selection *each element remaining in this popula-*

tion on that selection is given the same probability, or chance, of being selected. Here, we can randomly select the sample with or without replacement. If we **sample with replacement**, we place the element selected on a particular selection back into the population and thus give this element a chance to be selected on any succeeding selection. In such a case, all the elements in the population remain for each and every selection. If we **sample without replacement**, we do not place the element chosen on a particular selection back into the population, and thus we do not give this element a chance to be selected on any succeeding selection. In this case, the elements remaining in the population for a particular selection are all the elements in the population except for the elements that have previously been selected. *It is best to sample without replacement.* The reason is that if on a particular selection we randomly select an element from the population that is unrepresentative of the population—that is, an element considerably different from the other elements in the population, which might throw off our estimates of population parameters—then, if we sample with replacement, we might select the unrepresentative element again, which would result in our estimates of population parameters being thrown off even more drastically. On the other hand, if we sample without replacement, the unrepresentative element cannot be selected again. We assume in this book that all sampling is done without replacement.

If we have randomly selected a sample of n elements from the population of elements, the values of the numerical characteristic of interest possessed by the n elements make up the randomly selected sample of n element values. Specifically, for $i = 1, 2, \ldots, n$, we define the ***i*th observed element**, denoted by OEL_i, to be the element that we actually observe when we make the ith random selection from the population of elements, and the ***i*th observed element value**, denoted by y_i, to be the value of the numerical characteristic under study possessed by OEL_i, the ith observed element.

Then it follows that y_i is the element value that we actually observe when we make the ith random selection from the population of element values and that the set $SPL = \{y_1, y_2, \ldots, y_n\}$, which we call the **observed sample**, is our randomly selected sample of n element values.

We now define a **sample statistic** to be a descriptive measure of the observed sample. Very often we use sample statistics as point estimates of population parameters.

> A **point estimate** of a population parameter is a single number used as an estimate, or guess, of the population parameter.

Following are definitions of three sample statistics that are used as point estimates of the population mean, the population variance, and the population standard deviation.

POINT ESTIMATES OF THE POPULATION MEAN, VARIANCE, AND STANDARD DEVIATION:

Assume the observed sample

$SPL = \{y_1, y_2, \ldots, y_n\}$

has been randomly selected from a finite or infinite population of element values.

1. The *sample mean* is defined by the equation

$$\bar{y} = \frac{\sum_{i=1}^{n} y_i}{n}$$

and is used as a point estimate of the population mean, μ.

2. The *sample variance* is defined by the equation

$$s^2 = \frac{\sum_{i=1}^{n} (y_i - \bar{y})^2}{n - 1}$$

and is used as a point estimate of the population variance, σ^2.

3. The *sample standard deviation* is defined by the equation

$$s = \sqrt{s^2} = \sqrt{\frac{\sum_{i=1}^{n} (y_i - \bar{y})^2}{n - 1}}$$

and is used as a point estimate of the population standard deviation, σ.

The rationale behind the use of these sample statistics as point estimates is that each sample statistic is exactly, or nearly, the sample counterpart of the corresponding population parameter. Here, we define the **sample counterpart** of a population parameter to be the same function of the n element values in the observed sample $SPL = \{y_1, y_2, \ldots, y_n\}$ that the population parameter is of the element values in the population. For example, the sample mean

$$\bar{y} = \frac{\sum_{i=1}^{n} y_i}{n}$$

is the sample counterpart of the population mean

$$\mu = \frac{\sum_{h=1}^{N} y_{EL_h}}{N}$$

However, although

$$\frac{\sum_{i=1}^{n} (y_i - \bar{y})^2}{n}$$

is the sample counterpart of the population variance

$$\sigma^2 = \frac{\sum_{h=1}^{N} (y_{EL_h} - \mu)^2}{N}$$

we use the sample variance

$$s^2 = \frac{\sum_{i=1}^{n} (y_i - \bar{y})^2}{n - 1}$$

as the point estimate of σ^2, because it can be shown that dividing by $n - 1$ rather than n makes s^2 a better estimate (in some senses) of σ^2.

EXAMPLE 2.6

In order to estimate the mean, μ, and standard deviation, σ, of the infinite population of all Gas-Mizer mileages, National Motors will randomly select a sample of $n = 5$ Gas-Mizer mileages from this population. In order to do this, the company will first randomly select a sample of five Gas-Mizers from the infinite population of all Gas-Mizers by randomly selecting five Gas-Mizers from a finite subpopulation of 1,000 Gas-Mizers, which National Motors has produced and which, we assume, is representative of the infinite population of all Gas-Mizers.

For $i = 1, 2, 3, 4, 5$, we define the ith observed Gas-Mizer, denoted by OGM_i, to be the Gas-Mizer selected by National Motors when it makes the ith random selection from the infinite population of all Gas-Mizers, and the ith observed Gas-Mizer mileage, denoted by y_i, to be the number of miles per gallon averaged by OGM_i when it is driven 50,000 miles under the previously described conditions. It follows that y_i is the Gas-Mizer mileage that National Motors will actually observe when it makes the ith random selection from the infinite population of all Gas-Mizer mileages, and that the set $SPL = \{y_1, y_2, y_3, y_4, y_5\}$ is National Motors' randomly selected sample of $n = 5$ Gas-Mizer mileages.

Suppose that when National Motors has employed the above procedure, it has randomly selected the following five Gas-Mizers:

$$OGM_1 = GM_{604}$$

$$OGM_2 = GM_{372}$$

$$OGM_3 = GM_{17}$$

$$OGM_4 = GM_{896}$$

$$OGM_5 = GM_{522}$$

and thus it has randomly selected the following five Gas-Mizer mileages:

$$y_1 = y_{GM_{604}} = 30.7$$

$$y_2 = y_{GM_{372}} = 31.8$$

$$y_3 = y_{GM_{17}} = 30.2$$

$$y_4 = y_{GM_{896}} = 32.0$$

$$y_5 = y_{GM_{522}} = 31.3$$

Therefore, since

$$\begin{aligned} SPL &= \{y_1, y_2, y_3, y_4, y_5\} \\ &= \{30.7, 31.8, 30.2, 32.0, 31.3\} \end{aligned}$$

is the observed sample,

1. The sample mean is

$$\bar{y} = \frac{\sum_{i=1}^{5} y_i}{5} = \frac{30.7 + 31.8 + 30.2 + 32.0 + 31.3}{5} = \frac{156}{5} = 31.2$$

   and is the point estimate of the population mean, μ.

2. The sample variance is

$$\begin{aligned} s^2 &= \frac{\sum_{i=1}^{5} (y_i - \bar{y})^2}{5 - 1} \\ &= \frac{(y_1 - \bar{y})^2 + (y_2 - \bar{y})^2 + (y_3 - \bar{y})^2 + (y_4 - \bar{y})^2 + (y_5 - \bar{y})^2}{4} \\ &= [(30.7 - 31.2)^2 + (31.8 - 31.2)^2 + (30.2 - 31.2)^2 \\ &\quad + (32.0 - 31.2)^2 + (31.3 - 31.2)^2] \div 4 \\ &= \frac{(-.5)^2 + (.6)^2 + (-1)^2 + (.8)^2 + (.1)^2}{4} \\ &= \frac{2.26}{4} \\ &= .565 \end{aligned}$$

   and is the point estimate of the population variance, σ^2.

3. The sample standard deviation is

$$s = \sqrt{s^2} = \sqrt{.565} = .7517$$

   and is the point estimate of the population standard deviation, σ.

Since $\bar{y}$, s^2, and s are calculated from a randomly selected sample of five Gas-Mizer mileages, these point estimates do not equal (unless National Motors is very lucky) the respective population parameters μ, σ^2, and σ (which would be calculated using all the potential Gas-Mizer mileages). For example, recall that the true value of the mean gasoline mileage, μ, is 31.5 mpg. Thus, the sample mean

$$\bar{y} = 31.2$$

which is the point estimate of the population mean,

$$\mu = 31.5$$

is .3 mpg smaller than μ. We call the difference

$$\bar{y} - \mu = 31.2 - 31.5 = -.3$$

the error of estimation obtained when estimating μ by $\bar{y}$. National Motors does not know the size of this error of estimation, because it does not know the true value of μ. However, we will see in Chapter 6 (and in optional Section 3.4) that the science of statistics can utilize the sample standard deviation, s (along with other quantities), to provide National Motors with a bound on the error of estimation that tells National Motors the farthest that $\bar{y}$ might be from μ.

3

PROBABILITY DISTRIBUTIONS

In order to compute probabilities we can often use continuous probability distributions. These distributions are discussed in Section 3.1. Then in Sections 3.2 and 3.3 we study three continuous probability distributions that will be important in our presentation of regression analysis—the normal, *t*-, and *F*-distributions. We complete this chapter with optional Section 3.4, which discusses using probability distributions to find confidence intervals for population means.

3.1 PROBABILITIES CONCERNING A RANDOMLY SELECTED ELEMENT VALUE, AND CONTINUOUS PROBABILITY DISTRIBUTIONS

3.1.1 Probabilities Concerning a Randomly Selected Element Value

Before we randomly select the observed element value, y, from a population of element values, y can potentially be any of the element values in the population of interest. Thus, we can consider calculating probabilities of events concerning the values y might attain when the random selection is actually made. Specifically, if a and b are numbers on the real line, and a is less than b, and if we denote the interval of all numbers on the real line that are greater than or equal to a and less than or equal to b by the symbol $[a, b]$, then we often wish to find

$P(y$ will be in $[a, b]) =$ the probability that y, an element value that will be randomly selected from the population of interest, will be in the interval $[a, b]$

This probability can be written more simply as

$$P(a \leq y \leq b)$$

and can be interpreted as being equal to the proportion of element values in the population that are greater than or equal to a and less than or equal to b.

We now present an example which demonstrates that if we consider any population of element values having a finite mean, μ, and a finite standard deviation, σ, then a high proportion or a high percentage of the element values in the population are within (plus or minus) 2 standard deviations of the mean of the population and thus lie in the interval

$$[\mu - 2\sigma, \mu + 2\sigma]$$

The exact proportion of the element values that lie in this interval depends upon the exact nature of the population.

EXAMPLE 3.1

Recall that the population of all Gas-Mizer mileages has mean $\mu = 31.5$ and standard deviation $\sigma = .8$. Here we denote by y the gasoline mileage obtained by a Gas-Mizer that will be randomly selected (purchased) from the population of all Gas-Mizers that were (or could potentially be) produced by National Motors this year.

Suppose that we wish to find the probability that y will be in the interval

$$\begin{aligned}[\mu - 2\sigma, \mu + 2\sigma] &= [31.5 - 2(.8), 31.5 + 2(.8)] \\ &= [31.5 - 1.6, 31.5 + 1.6] \\ &= [29.9, 33.1]\end{aligned}$$

First, if we consider the finite population of the 100,000 Gas-Mizers that National Motors produced this year, then in order to find

$$P(29.9 \leq y \leq 33.1)$$

we would have to know $N_{[29.9,33.1]}$, which we define to be the number of Gas-Mizer mileages in the population of all such mileages that are in the interval [29.9, 33.1]. Of course, National Motors would not know $N_{[29.9,33.1]}$, because it cannot afford to test each Gas-Mizer for gasoline mileage. However, for purposes of illustration, we assume that $N_{[29.9,33.1]}$ equals 95,440. Hence,

$$\begin{aligned}P(\mu - 2\sigma \leq y \leq \mu + 2\sigma) &= P(y \text{ will be in } [\mu - 2\sigma, \mu + 2\sigma]) \\ &= P(y \text{ will be in } [29.9, 33.1]) \\ &= \text{the proportion of Gas-Mizer mileages in the population of all Gas-Mizer mileages that are greater than or equal to 29.9 mpg and less than or equal to 33.1 mpg} \qquad (1) \\ &= \frac{N_{[29.9,33.1]}}{N} = \frac{95{,}440}{100{,}000} = .9544 \qquad (2)\end{aligned}$$

Next, if we consider the infinite population of all Gas-Mizers that could potentially be produced by National Motors, and if $P(y \text{ will be in } [29.9, 33.1]) = .9544$, then (1) above is an intuitive interpretation of $P(y \text{ will be in } [29.9, 33.1])$. However, (2) above involves a mathematical operation that is not legitimate (because the total number of Gas-Mizer mileages is infinite).

To summarize, the statement

$$P(y \text{ will be in } [\mu - 2\sigma, \mu + 2\sigma]) = P(29.9 \leq y \leq 33.1) = .9544$$

says that 95.44 percent of all Gas-Mizer mileages are within (plus or minus) 2 standard deviations of the mean μ of all Gas-Mizer mileages and thus lie in the interval $[\mu - 2\sigma, \mu + 2\sigma] = [29.9, 33.1]$.

3.1.2 Continuous Probability Distributions

We can often use what is called a continuous probability distribution to find

$$P(y \text{ will be in } [a, b])$$

Specifically, continuous probability distributions determine probabilities by assigning probabilities to intervals of numbers on the real line. To understand this idea, consider a continuous function of the numbers on the real line. In general, we denote this function by $f(v)$. Consider the continuous curve that results when $f(v)$ is graphed against the numbers on the real line. If $[a, b]$ denotes an interval of numbers on the real line, we denote the *area under the curve* $f(v)$ *corresponding to the interval* $[a, b]$ by

$$A[[a, b]; f(v)]$$

A hypothetical curve $f(v)$ and the area $A[[a, b]; f(v)]$ are illustrated in Figure 3.1.

> The curve $f(v)$ is the **continuous probability distribution of a randomly selected element value** y if
>
> $$P(y \text{ will be in } [a, b]) = A[[a, b]; f(v)]$$
>
> that is, if the probability that y will be in the interval $[a, b]$ is the area under the curve $f(v)$ corresponding to the interval $[a, b]$.

For example, if the curve $f(v)$ illustrated in Figure 3.1 is the continuous probability distribution of y, then, as shown in the figure,

$$P(a \le y \le b) = A[[a, b]; f(v)] = .68$$

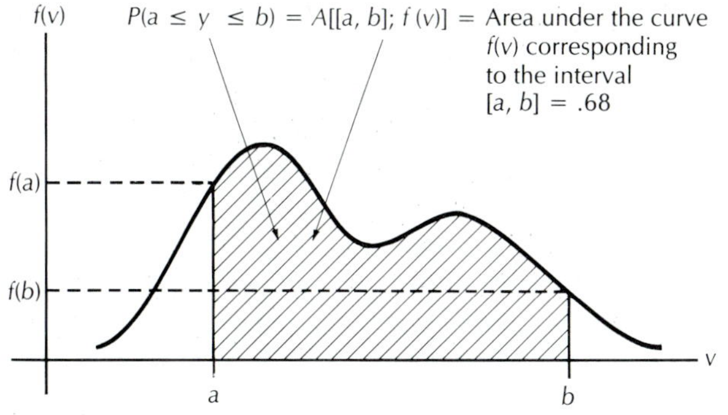

FIGURE 3.1
A Hypothetical Curve $f(v)$

which says that 68 percent of the element values in the population from which y will be randomly selected are greater than or equal to a and less than or equal to b.

For some continuous curves, areas such as the one shown in Figure 3.1 can be found by using integral calculus. However, areas under the important continuous curves studied in this book can be found by using statistical tables.

If the curve $f(v)$ is the continuous probability distribution of y, we say that y, or the population of element values from which y will be randomly selected, *is distributed according to the continuous probability distribution, or continuous probability curve,* $f(v)$. Since the probability that y will be in the interval $[a, b]$ is given by the area under the curve $f(v)$ corresponding to the interval $[a, b]$, the height of the curve $f(v)$ at a given point on the real line represents the relative probability, or chance, that y will be in a small interval of numbers around the given point. Thus, for example, if the continuous probability distribution $f(v)$ of y is as shown in Figure 3.1, it follows (since $f(a)$, the height of the curve at the point a, is greater than $f(b)$, the height of the curve at the point b) that it is more probable that y will be in a small interval of numbers around a than it is probable that y will be in a small interval of numbers around b. Said another way, the height of the curve $f(v)$ at a given point on the real line represents the relative proportion of element values in the population that are in a small interval of numbers around the given point.

Before looking at some continuous probability curves that have been found useful in practice, we must consider two general properties that are satisfied by a continuous probability distribution $f(v)$ of y:

1. The first property is that for any number v on the real line, $f(v) \geq 0$. Intuitively, this property must be satisfied by $f(v)$, because the height of the probability curve $f(v)$ at the point v represents the relative probability that y will be in a small interval of numbers around the point v, and because any probability must be greater than or equal to zero.
2. The second property is that the total area under a continuous probability curve equals 1. This property holds because the total area under a continuous probability curve equals the probability that y will fall between $-\infty$ and ∞, and y is sure to fall between $-\infty$ and ∞.

The shape of the continuous probability distribution illustrated in Figure 3.1 is only one of many possible shapes. In fact, very few populations of element values are distributed according to a continuous probability distribution having the shape of the curve in Figure 3.1. Two common shapes displayed by continuous probability curves used in practice are shown in Figure 3.2. The continuous probability curve illustrated in Figure 3.2a is symmetrical, while the continuous probability curve depicted in Figure 3.2b is skewed to the right. Since the curve of Figure 3.2a is symmetrical, $f(\mu + \epsilon)$, the height of the curve at the point $\mu + \epsilon$, equals $f(\mu - \epsilon)$, the height of the curve at the point $\mu - \epsilon$. Notice that whereas the population mean μ is not located under the highest point of the skewed curve in Figure 3.2b, μ is located under the highest point of the symmetrical curve in Figure 3.2a.

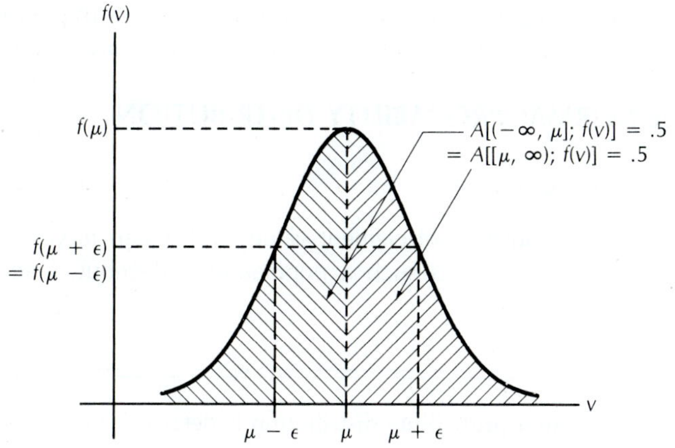

(a) A symmetrical continuous probability distribution

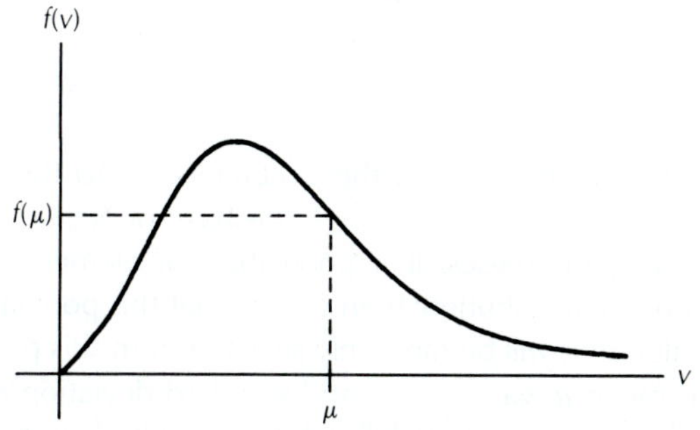

FIGURE 3.2 Some Common Shapes of Continuous Probability Distributions

(b) A continuous probability distribution that is skewed to the right

Again consider Figure 3.2a. We define

$$A[(-\infty, \mu]; f(v)]$$

to be the area under the curve $f(v)$ corresponding to the interval $(-\infty, \mu]$, which is the interval of numbers on the real line that are less than or equal to μ. Similarly, we define

$$A[[\mu, \infty); f(v)]$$

to be the area under the curve $f(v)$ corresponding to the interval $[\mu, \infty)$, which is the interval of numbers on the real line that are greater than or equal to μ. Then, by the symmetry of the curve in Figure 3.2a it follows that (for this curve)

$$A[(-\infty, \mu]; f(v)] = A[[\mu, \infty); f(v)]$$

Since the total area under the curve $f(v)$ equals 1, these equal areas must each equal .5. Thus, if a population is described by a symmetrical continuous probability curve, then 50 percent of the element values are less than or equal to the population mean μ, and 50 percent of the element values are greater than or equal to μ.

3.2 THE NORMAL PROBABILITY DISTRIBUTION

3.2.1 The Normal Curve

Consider a population of element values with mean μ, variance σ^2, and standard deviation σ. Sometimes such a population is distributed according to a normal probability distribution.

> The **normal probability distribution** is defined by the probability curve
>
> $$f(v) = \frac{1}{\sigma\sqrt{2\pi}} \exp\left(-\frac{(v-\mu)^2}{2\sigma^2}\right) \qquad \text{for } -\infty < v < \infty$$

Here $\pi = 3.14159\ldots$ is the ratio of the circumference to the diameter of a circle, and exp denotes taking $e = 2.71828\ldots$, the base of Naperian logarithms, to the power specified in parentheses. If a population of element values is distributed according to a normal distribution, then we say that this population (or equivalently, y, an element value that will be randomly selected from this population) is **normally distributed** with mean μ, variance σ^2, and standard deviation σ.

We denote the normal probability distribution by $N(\mu, \sigma)$. This means that the shape of the curve, called the **normal (probability) curve**, that results when

$$N(\mu, \sigma) = \frac{1}{\sigma\sqrt{2\pi}} \exp\left(-\frac{(v-\mu)^2}{2\sigma^2}\right) \qquad \text{for } -\infty < v < \infty$$

is graphed depends on the mean, μ, and the standard deviation, σ, of the population of element values. Note that we omit the v in the symbol $N(\mu, \sigma)$ because it is a convention to do so. The normal probability curve is illustrated in Figure 3.3 and is often described as a bell-shaped curve. The following are several important properties of the normal curve.

1. The normal curve is centered at the population mean μ.
2. The mean μ corresponds to the highest point on the normal curve.
3. The normal curve is symmetrical around the population mean.
4. Since the normal curve is a probability distribution, the total area under the curve is equal to 1.
5. Since the normal curve is symmetrical, the area under the normal curve above the mean equals the area under the normal curve below the mean, and each of these areas equals .5 (see Figure 3.3).

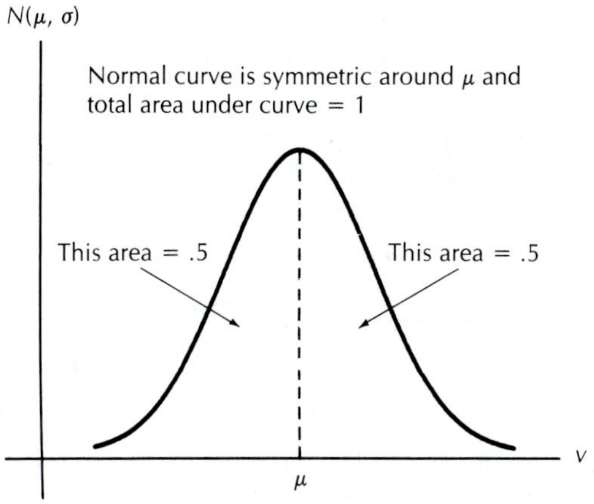

FIGURE 3.3
The Normal Probability Curve

Before continuing, we should make several other comments concerning the normal curve. First, since the normal curve is centered at the population mean μ, the mean μ positions the normal curve on the real line. This is illustrated in Figure 3.4, which shows two normal curves with different means μ_1 and μ_2 (where $\mu_1 > \mu_2$) and with the same standard deviation σ. Second, the variance σ^2 (or the standard deviation σ) of the normal distribution measures the spread of the normal curve. This is illustrated in Figure 3.5, which shows two normal curves with the same mean μ but different standard deviations σ_1 and σ_2, where $\sigma_1 > \sigma_2$. Since the first normal curve, with mean μ and standard deviation σ_1, has a larger standard deviation than the second normal curve, with mean μ and standard deviation σ_2, the first normal curve is more spread out than the second normal curve.

If y is normally distributed with mean μ and standard deviation σ, then

$$P(a \leq y \leq b) = A[[a, b]; N(\mu, \sigma)]$$

That is, the probability that y will be greater than or equal to a and less than or equal to b equals the area under the normal curve with mean μ and standard deviation σ corresponding to the interval $[a, b]$. Such an area is illustrated in Figure 3.6. In

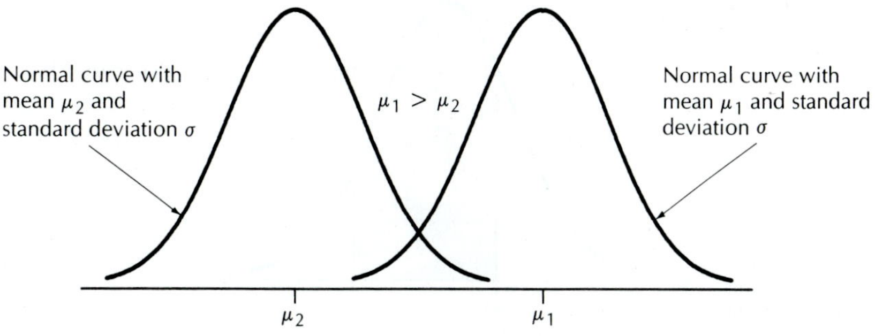

FIGURE 3.4
Two Normal Curves With Different Means and Equal Standard Deviations

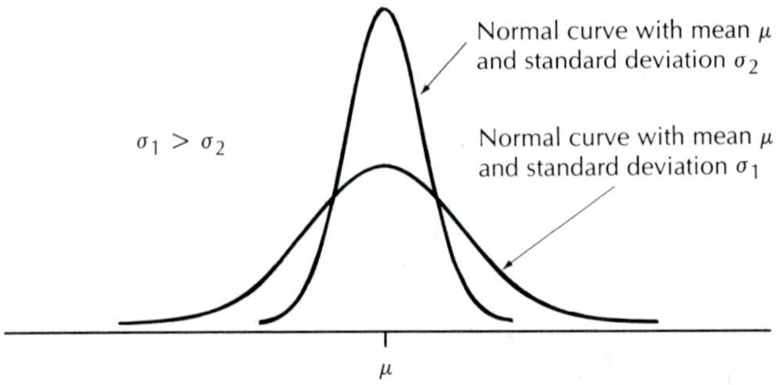

FIGURE 3.5
Two Normal Curves With the Same Mean and Different Standard Deviations

order to find areas under a normal curve, we can use a statistical table called a **normal table**. A normal table is presented as Table E-1 in Appendix E. In Section 3.2.2 we briefly discuss how to find areas under a normal curve.

Although we will not need to find many areas under a normal curve in this book, there are three important areas under the normal curve that we wish to emphasize (see Figure 3.7). If y is normally distributed with mean μ and standard deviation σ, it can be shown (using a normal table) that

1. $$P(\mu - \sigma \leq y \leq \mu + \sigma) = A[[\mu - \sigma, \mu + \sigma]; N(\mu, \sigma)] = .6826$$

   This means that 68.26 percent of the element values in the population are within (plus or minus) 1 standard deviation of the population mean.

2. $$P(\mu - 2\sigma \leq y \leq \mu + 2\sigma) = A[[\mu - 2\sigma, \mu + 2\sigma]; N(\mu, \sigma)] = .9544$$

   This means that 95.44 percent of the element values in the population are within (plus or minus) 2 standard deviations of the population mean.

3. $$P(\mu - 3\sigma \leq y \leq \mu + 3\sigma) = A[[\mu - 3\sigma, \mu + 3\sigma]; N(\mu, \sigma)] = .9973$$

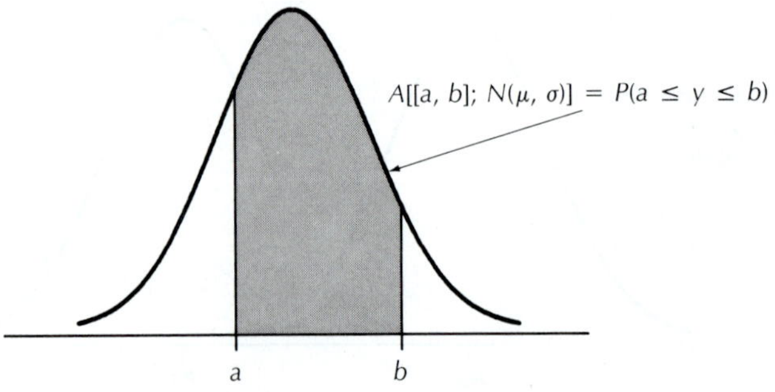

FIGURE 3.6
An Area Under a Normal Curve Corresponding to the Interval [a, b]

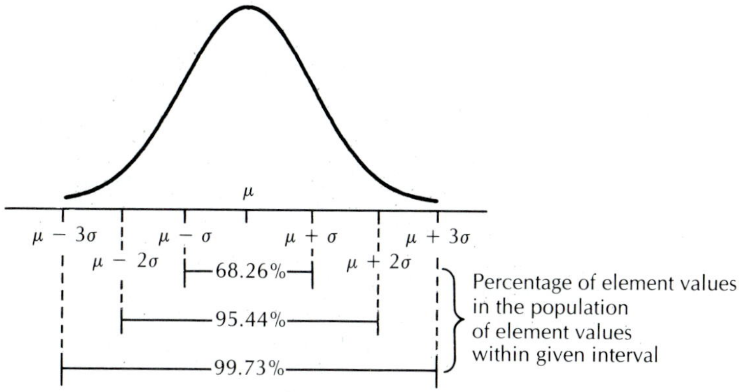

FIGURE 3.7 Three Important Percentages Concerning a Normally Distributed Population With Mean μ and Standard Deviation σ

This means that 99.73 percent of the element values in the population are within (plus or minus) 3 standard deviations of the population mean.

EXAMPLE 3.2

In the Gas-Mizer problem, assume that (a supernatural power knows that) the population of all Gas-Mizer mileages is normally distributed with mean $\mu = 31.5$ and standard deviation $\sigma = .8$.

1. Then 68.26 percent of all Gas-Mizer mileages lie in the interval

$$\begin{aligned}[\mu - \sigma, \mu + \sigma] &= [31.5 - .8, 31.5 + .8] \\ &= [30.7, 32.3]\end{aligned}$$

2. Also, 95.44 percent of all Gas-Mizer mileages lie in the interval

$$\begin{aligned}[\mu - 2\sigma, \mu + 2\sigma] &= [31.5 - 2(.8), 31.5 + 2(.8)] \\ &= [31.5 - 1.6, 31.5 + 1.6] \\ &= [29.9, 33.1]\end{aligned}$$

3. And 99.73 percent of all Gas-Mizer mileages lie in the interval

$$\begin{aligned}[\mu - 3\sigma, \mu + 3\sigma] &= [31.5 - 3(.8), 31.5 + 3(.8)] \\ &= [31.5 - 2.4, 31.5 + 2.4] \\ &= [29.1, 33.9]\end{aligned}$$

Since National Motors does not know the true values of μ and σ, it cannot calculate these intervals. However, using the sample mean $\bar{y} = 31.2$ as the point estimate of μ, and using the sample standard deviation $s = .7517$ as the point estimate of σ (see Example 2.6 for the calculations of $\bar{y}$ and s), National Motors can *estimate* that if the population of all Gas-Mizer mileages is normally distributed, then 68.26 percent of the Gas-Mizer mileages in this population lie in the interval

$$[31.2 - 1(.7517), 31.2 + 1(.7517)] = [30.4483, 31.9517]$$

Also, we estimate that 95.44 percent of the mileages lie in the interval

$$[31.2 - 2(.7517), 31.2 + 2(.7517)] = [29.6966, 32.7034]$$

And we estimate that 99.73 percent of the mileages lie in the interval

$$[31.2 - 3(.7517), 31.2 + 3(.7517)] = [28.9449, 33.4551]$$

The results obtained in this example depend upon the assumption that the population of all Gas-Mizer mileages is normally distributed. If we wish to verify the validity of this assumption, a sample of only $n = 5$ Gas-Mizer mileages is not sufficient. We would need to select a larger sample of mileages.

Table 3.1 lists a larger sample of $n = 49$ Gas-Mizer mileages, which we assume National Motors has randomly selected from the preceding population. Table 3.2 groups the 49 mileages into a **frequency distribution** having six intervals. Here, we have chosen the number of intervals by using a general rule that the number of

TABLE 3.1 An Observed Sample of $n = 49$ Gas-Mizer Mileages

$y_1 = 30.8$	$y_{11} = 30.9$	$y_{21} = 32.0$	$y_{31} = 32.3$	$y_{41} = 32.6$
$y_2 = 31.7$	$y_{12} = 30.4$	$y_{22} = 31.4$	$y_{32} = 32.7$	$y_{42} = 31.4$
$y_3 = 30.1$	$y_{13} = 32.5$	$y_{23} = 30.8$	$y_{33} = 31.2$	$y_{43} = 31.8$
$y_4 = 31.6$	$y_{14} = 30.3$	$y_{24} = 32.8$	$y_{34} = 30.6$	$y_{44} = 31.9$
$y_5 = 32.1$	$y_{15} = 31.3$	$y_{25} = 30.6$	$y_{35} = 31.7$	$y_{45} = 32.8$
$y_6 = 33.3$	$y_{16} = 32.1$	$y_{26} = 31.5$	$y_{36} = 31.4$	$y_{46} = 31.5$
$y_7 = 31.3$	$y_{17} = 32.5$	$y_{27} = 32.4$	$y_{37} = 32.2$	$y_{47} = 31.6$
$y_8 = 31.0$	$y_{18} = 31.8$	$y_{28} = 31.0$	$y_{38} = 31.5$	$y_{48} = 32.2$
$y_9 = 32.0$	$y_{19} = 30.4$	$y_{29} = 29.8$	$y_{39} = 31.7$	$y_{49} = 32.0$
$y_{10} = 32.4$	$y_{20} = 30.5$	$y_{30} = 31.1$	$y_{40} = 30.6$	

$$\bar{y} = \frac{\sum_{i=1}^{49} y_i}{49} = \frac{1546.1}{49} = 31.553061 \approx 31.6$$

$$s^2 = \frac{\sum_{i=1}^{49} (y_i - \bar{y})^2}{48} = \frac{30.666}{48} = .638875 \approx .64$$

$$s = \sqrt{s^2} = \sqrt{.638875} = .799 \approx .8$$

TABLE 3.2
A Frequency Distribution of the $n = 49$ Gas-Mizer Mileages

Interval	Frequency
[29.8, 30.3]	3
[30.4, 30.9]	9
[31.0, 31.5]	12
[31.6, 32.1]	13
[32.2, 32.7]	9
[32.8, 33.3]	3

intervals should be the smallest integer K such that $2^K > n$ (where n is the number of observations that we wish to group into a frequency distribution). Since $n = 49$ in the Gas-Mizer problem, and since $2^5 = 32 < 49$ and $2^6 = 64 > 49$, it follows that we should use $K = 6$ intervals. The first interval, [29.8, 30.3], is then formed by adding .5 to 29.8, the smallest mileage in Table 3.1, which yields an interval containing six measurement values—29.8, 29.9, 30.0, 30.1, 30.2, and 30.3—measured to the nearest one-tenth of a mile per gallon. The decision to include six measurement values in the first interval (and in the other intervals) is based on calculating (see Table 3.1)

$$\frac{[\text{Largest mileage} - \text{Smallest mileage}]}{K} = \frac{33.3 - 29.8}{6} = \frac{3.5}{6} \approx .6$$

which means that in order to include the smallest measurement and the largest measurement in the $K = 6$ classes, each class should contain six measurement values. The five other intervals are formed in exactly the same way (see Table 3.2). The frequency distribution of Table 3.2 is depicted graphically in the form of a histogram in Figure 3.8. We see that this histogram looks reasonably bell-shaped and symmetrical. Since it is customary to look for pronounced rather than subtle departures from the normality assumption, the histogram in Figure 3.8 does not provide much evidence to reject this assumption.

Finally, note that since the sample mean $\bar{y} = 31.6$ (calculated from the $n = 49$ observed mileages in Table 3.1) has been calculated by using more mileages than

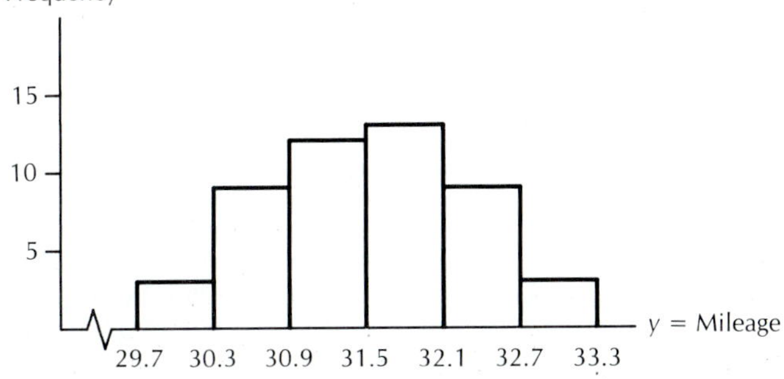

FIGURE 3.8
A Histogram of the $n = 49$ Gas-Mizer Mileages

$\bar{y} = 31.2$ (calculated from the $n = 5$ observed mileages in Example 2.6), National Motors would expect $\bar{y} = 31.6$ (the point estimate of μ based on the larger sample) to be closer to the population mean, μ, than $\bar{y} = 31.2$ (the point estimate of μ based on the smaller sample). This, however, is not guaranteed. Since the true value of μ is 31.5, this expectation would in fact be correct.

3.2.2 *z* Values and Finding Normal Probabilities

If y is an element value that is randomly selected from a normally distributed population with mean μ and standard deviation σ, then we define the z value corresponding to y to be

$$z = \frac{y - \mu}{\sigma} = \text{the number of standard deviations that } y \text{ is from the mean } \mu$$

Since there is a z value corresponding to each element value in the population of element values, there is a population of z values corresponding to the population of element values. Thus, we can assume that when y is randomly selected from the population of element values,

$$z = \frac{y - \mu}{\sigma}$$

is randomly selected from the corresponding population of z values. Moreover, it can be proved that if y (or the population of element values) is normally distributed, with mean μ and standard deviation σ, then z (or the population of z values) is also normally distributed, with mean zero and standard deviation 1. Note that a normal distribution with mean zero and standard deviation 1 is often referred to as a **standard normal distribution**.

If we subtract μ from the inequality

$$a \leq y \leq b$$

and divide by σ, we obtain the following inequalities:

$$\frac{a - \mu}{\sigma} \leq \frac{y - \mu}{\sigma} \leq \frac{b - \mu}{\sigma}$$

or

$$\frac{a - \mu}{\sigma} \leq z \leq \frac{b - \mu}{\sigma}$$

or

$$z_a \leq z \leq z_b$$

where $z_a = (a - \mu)/\sigma$ is the z value corresponding to the number a, and $z_b = (b - \mu)/\sigma$ is the z value corresponding to the number b. We see that

$$P(a \leq y \leq b) = P(z_a \leq z \leq z_b)$$

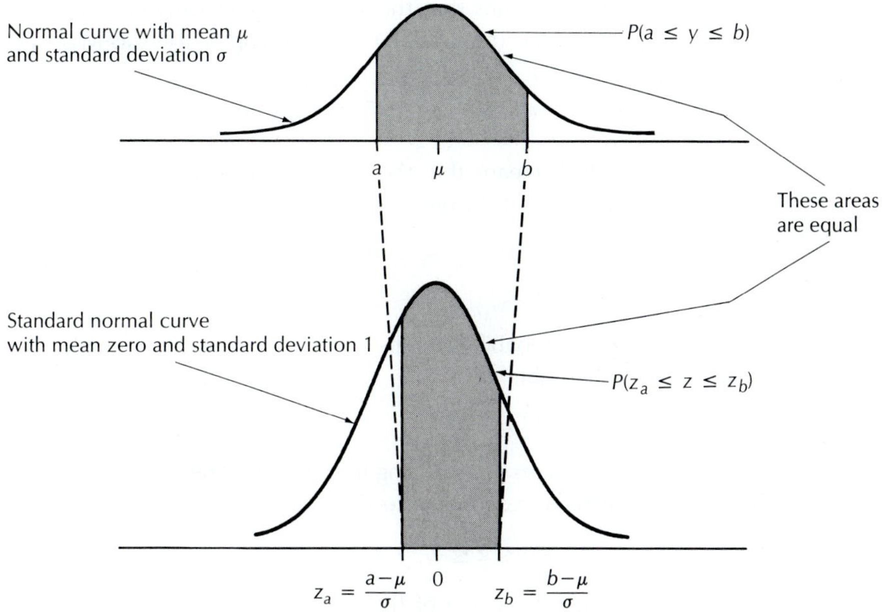

FIGURE 3.9
$P(a \leq y \leq b) =$
$P(z_a \leq z \leq z_b)$

Thus, in order to find the probability

$$P(a \leq y \leq b)$$

we can calculate the z values corresponding to a and b and then find

$$P(z_a \leq z \leq z_b)$$

which is the area under the standard normal curve corresponding to the interval $[z_a, z_b]$. This is illustrated in Figure 3.9.

EXAMPLE 3.3

In the Gas-Mizer problem, the population of all Gas-Mizer mileages is normally distributed with mean $\mu = 31.5$ and standard deviation $\sigma = .8$. We now show that, as stated in Example 3.2, 95.44 percent of all Gas-Mizer mileages lie in the interval [29.9, 33.1]. That is, we show that

$$P(29.9 \leq y \leq 33.1) = .9544$$

The z value corresponding to the Gas-Mizer mileage 29.9 mpg is

$$z_{29.9} = \frac{29.9 - \mu}{\sigma} = \frac{29.9 - 31.5}{.8} = \frac{-1.6}{.8} = -2$$

which means that the Gas-Mizer mileage 29.9 is 2 standard deviations below $\mu = 31.5$. The z value corresponding to the Gas-Mizer mileage 33.1 mpg is

$$z_{33.1} = \frac{33.1 - \mu}{\sigma} = \frac{33.1 - 31.5}{.8} = \frac{1.6}{.8} = 2$$

which means that the Gas-Mizer mileage 33.1 is 2 standard deviations above $\mu = 31.5$. It follows that

$$\begin{aligned} P(29.9 \leq y \leq 33.1) &= P(z_{29.9} \leq z \leq z_{33.1}) \\ &= P(-2 \leq z \leq 2) \end{aligned}$$

This probability can be found using the normal table, Table E-1, in Appendix E. That table gives

$$P(0 \leq z \leq z_c)$$

for values of z_c ranging from 0.00 to 3.09. Looking at this table, we see that the area corresponding to $z_c = 2$ is .4772. Thus,

$$P(0 \leq z \leq 2) = .4772$$

Since the curve of the standard normal distribution is symmetrical about its mean, it follows that

$$\begin{aligned} P(-2 \leq z \leq 2) &= P(-2 \leq z \leq 0) + P(0 \leq z \leq 2) \\ &= .4772 + .4772 = 2(.4772) = .9544 \end{aligned}$$

Thus,

$$P(29.9 \leq y \leq 33.1) = P(-2 \leq z \leq 2) = .9544$$

Before ending this section, we must discuss one more point concerning the normal probability curve. In order to perform the calculations involved in the statistical inference procedures we will present, we sometimes need to find the z value so that the area under the standard normal curve to the right of this z value is γ. Denoting this z value as $z_{[\gamma]}$, we refer to $z_{[\gamma]}$ as *the point on the scale of the standard normal curve so that the area under this curve to the right of* $z_{[\gamma]}$ *is* γ. This point is illustrated in Figure 3.10, which also tells us that $z_{[\gamma]}$ is *the point on the scale of the standard normal curve so that the area under this curve between zero and* $z_{[\gamma]}$ *is* $.5 - \gamma$. (This follows because the area under the standard normal curve to the right of zero is .5). This fact allows us to easily find $z_{[\gamma]}$. For example, if we wish to find $z_{[.025]}$, which is the point on the scale of the standard normal curve so that the area under this curve to the right of $z_{[.025]}$ is .025, then we note that $z_{[.025]}$ is the point on the scale of the standard normal curve so that the area under this curve between zero and $z_{[.025]}$ is $.5 - .025 = .475$. Looking at Table E-1 in Appendix E, we see that the z value corresponding to an area of .4750 is 1.96. It follows that $z_{[.025]}$ equals 1.96.

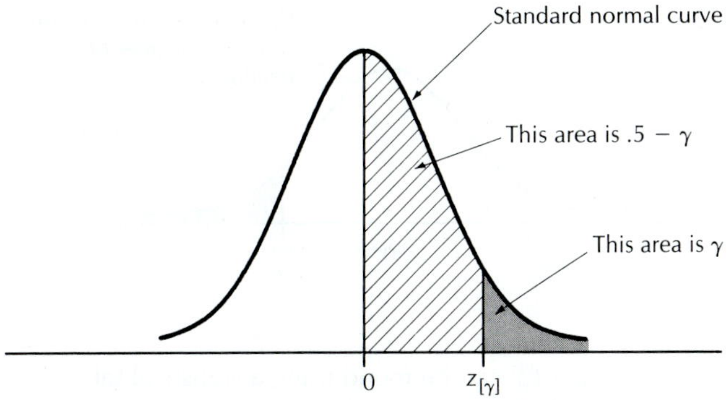

FIGURE 3.10

$z_{[\gamma]}$ = The Point on the Scale of the Standard Normal Curve So That the Area Under This Curve to the Right of $z_{[\gamma]}$ Is γ

3.3 THE *t*-DISTRIBUTION AND THE *F*-DISTRIBUTION

3.3.1 The *t*-Distribution

Sometimes a population of element values, or equivalently, y, an element value randomly selected from this population, is distributed according to what is called a ***t*-distribution**. The probability curve of the *t*-distribution has the following properties:

1. The probability curve of the *t*-distribution has the appearance of a normal curve—that is, it is symmetrical and bell-shaped.
2. The probability curve of the *t*-distribution is symmetrical about zero, which is the mean of the *t*-distribution.
3. The standard deviation, σ, of the *t*-distribution is always greater than 1.
4. The exact spread, or standard deviation, σ, of the *t*-distribution depends on a parameter that is called **the number of degrees of freedom** of the *t*-distribution.
5. As the number of degrees of freedom approaches infinity, the standard deviation, σ, of the *t*-distribution approaches 1.
6. As the number of degrees of freedom approaches infinity, the probability curve of the *t*-distribution approaches (that is, becomes shaped more and more like) the probability curve of a standard normal distribution.

In order to carry out the calculations involved in regression statistical inference procedures we must know how to find the point on the scale of the *t*-distribution having a given number of degrees of freedom so that the area under this curve to the right of this point is γ. Such a point is illustrated in Figure 3.11 and is denoted $t_{[\gamma]}^{(df)}$, where the superscript (df) refers to the parameter (the number of degrees of freedom) that describes the *t*-distribution and the subscript $[\gamma]$ refers to the size of the area under the curve of this *t*-distribution to the right of the point $t_{[\gamma]}^{(df)}$. In general, we will refer to the point $t_{[\gamma]}^{(df)}$ as *the point on the scale of the t-distribution having df degrees of freedom so that the area under this curve* (or equivalently, *in the tail of this t-distribution*) *to the right of* $t_{[\gamma]}^{(df)}$ *is* γ, or more simply, as the ***t*-point** $t_{[\gamma]}^{(df)}$.

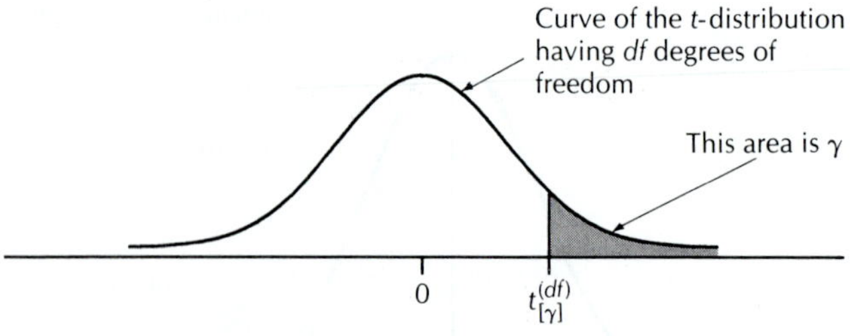

FIGURE 3.11 $t_{[\gamma]}^{(df)}$ = The Point on the Scale of the *t*-Distribution Having *df* Degrees of Freedom So That the Area Under This Curve to the Right of $t_{[\gamma]}^{(df)}$ Is γ

The *t*-point $t_{[\gamma]}^{(df)}$ can be found using a statistical table called a ***t*-table**. A *t*-table is shown in Table E-2 of Appendix E. This table lists values of $t_{[\gamma]}^{(df)}$ for values of γ (areas under the curve to the right of $t_{[\gamma]}^{(df)}$) of .10, .05, .025, .01, and .005. The values of $t_{[\gamma]}^{(df)}$ are tabulated according to the number of degrees of freedom, *df*, of the *t*-distribution. Thus, to find the point $t_{[\gamma]}^{(df)}$, we simply scan across the top of the *t*-table to find the column corresponding to the appropriate value of γ (the area under the curve to the right of $t_{[\gamma]}^{(df)}$) and scan down the left-hand side to find the row corresponding to the proper number of degrees of freedom, *df*. The value in the column corresponding to γ and in the row corresponding to *df* is $t_{[\gamma]}^{(df)}$. Thus, for example, $t_{[.025]}^{(11)}$, which is the point on the scale of the *t*-distribution having 11 degrees of freedom so that the area under this curve to the right of $t_{[.025]}^{(11)} = .025$, is $t_{[.025]}^{(11)} = 2.201$. Notice that the *t*-table lists values of $t_{[\gamma]}^{(df)}$ for degrees of freedom from 1 to 29 and ∞. The reason that values of *df* greater than 29 are not listed is that (as stated earlier in Property 6) as the number of degrees of freedom approaches infinity, the probability curve of the *t*-distribution approaches the probability curve of a standard normal distribution. Thus, when the number of degrees of freedom is large, the value $t_{[\gamma]}^{(df)}$ is very close to the value $z_{[\gamma]}$, the point on the scale of the standard normal curve so that the area under this curve to the right of $z_{[\gamma]}$ is γ. Values of $z_{[\gamma]}$ for values of γ equal to .10, .05, .025, .01, and .005 are given in the *t*-table in the row corresponding to ∞. Generally, if the number of degrees of freedom is 30 or more, it is sufficient to use the value $z_{[\gamma]}$ for $t_{[\gamma]}^{(df)}$, since $z_{[\gamma]}$ will be very close to $t_{[\gamma]}^{(df)}$.

3.3.2 The *F*-Distribution

Sometimes we find that a population of element values, or equivalently, *y*, an element value randomly selected from this population, is distributed according to what is called an ***F*-distribution**. The probability curve of the *F*-distribution has the following properties:

1. The probability curve $f(v)$ of the *F*-distribution is positive (above the real line) for all values of $v > 0$.
2. The probability curve of the *F*-distribution is skewed to the right.
3. The exact form of the probability curve of the *F*-distribution depends on two parameters, called the numerator degrees of freedom (denoted r_1) and the denominator degrees of freedom (denoted r_2).

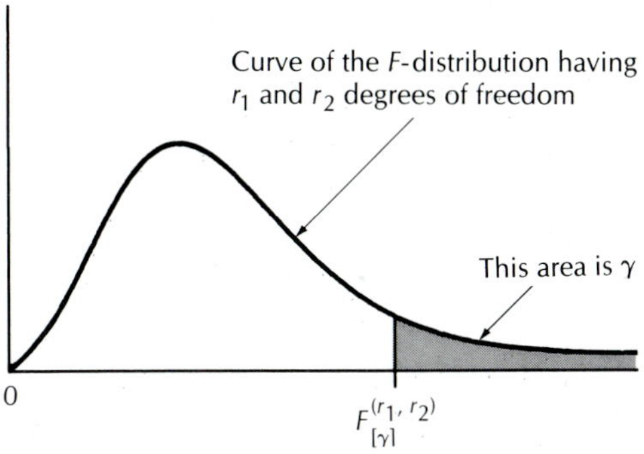

FIGURE 3.12 $F_{[\gamma]}^{(r_1,r_2)}$ = The Point on the Scale of the F-Distribution Having r_1 and r_2 Degrees of Freedom So That the Area Under This Curve to the Right of $F_{[\gamma]}^{(r_1,r_2)}$ is γ

In order to perform manipulations discussed in later chapters, we must know how to find the point on the scale of the F-distribution having r_1 and r_2 degrees of freedom so that the area under this curve to the right of this point is γ. Such a point is illustrated in Figure 3.12 and is denoted as $F_{[\gamma]}^{(r_1,r_2)}$, where the superscript (r_1, r_2) refers to the parameters (the numerator and denominator degrees of freedom) that describe the F-distribution and the subscript $[\gamma]$ refers to the size of the area under the curve of this F-distribution to the right of the point $F_{[\gamma]}^{(r_1,r_2)}$. In general, we will refer to the point $F_{[\gamma]}^{(r_1,r_2)}$ as *the point on the scale of the F-distribution having r_1 and r_2 degrees of freedom so that the area under this curve to the right of $F_{[\gamma]}^{(r_1,r_2)}$ is γ*. The ***F*-point** $F_{[\gamma]}^{(r_1,r_2)}$ can be found using a statistical table called an ***F*-table**. An F-table is shown in Table E-3 of Appendix E. In this table γ (the area under the curve of the F-distribution to the right of $F_{[\gamma]}^{(r_1,r_2)}$) is .05. That is, this table lists values of $F_{[.05]}^{(r_1,r_2)}$ tabulated according to values of the parameters r_1 and r_2. So in order to find the point $F_{[.05]}^{(r_1,r_2)}$ we scan across the top of the F-table to find the column corresponding to the parameter r_1 and we scan down the side to find the row corresponding to the parameter r_2. The value in the column corresponding to r_1 and in the row corresponding to r_2 is $F_{[.05]}^{(r_1,r_2)}$. Thus, for example, $F_{[.05]}^{(8,10)}$, which is the point on the scale of the F-distribution having $r_1 = 8$ and $r_2 = 10$ degrees of freedom so that the area under this curve to the right of $F_{[.05]}^{(8,10)} = .05$, is $F_{[.05]}^{(8,10)} = 3.07$. Although Table E-3 tabulates values of $F_{[\gamma]}^{(r_1,r_2)}$ for the value $\gamma = .05$, tables for other values of γ (other areas under the curve to the right of $F_{[\gamma]}^{(r_1,r_2)}$) are also available (for example, see Table E-4 in Appendix E).

*3.4 AN APPLICATION OF PROBABILITY DISTRIBUTIONS: CONFIDENCE INTERVALS

3.4.1 Confidence Intervals for a Population Mean

We first present an intuitive discussion of confidence intervals and then a precise discussion of the statistical meaning and derivation of these intervals. Recall that if we randomly select a sample of n element values

$$SPL = \{y_1, y_2, \ldots, y_n\}$$

from a population of element values, then (as discussed in Section 2.4) we use the sample mean and the sample standard deviation

$$\bar{y} = \frac{\sum_{i=1}^{n} y_i}{n} \qquad \text{and} \qquad s = \sqrt{\frac{\sum_{i=1}^{n} (y_i - \bar{y})^2}{n-1}}$$

as the point estimates of the population mean, μ, and population standard deviation, σ, respectively. We call the difference between $\bar{y}$ and μ

$$\bar{y} - \mu$$

the *error of estimation obtained when estimating* μ *by* $\bar{y}$. Intuitively, this error of estimation expresses how far the point estimate $\bar{y}$ is from μ. Since we do not know the true value of μ, we cannot know the true size of the error of estimation (that is, exactly how far $\bar{y}$ is from μ). However, the science of statistics can provide us with a quantity called the *100(1 − α)% bound on the error of estimation obtained when estimating* μ *by* $\bar{y}$. This bound, which is denoted by $BE_{\bar{y}}[100(1-\alpha)]$, can be computed so that if certain assumptions are satisfied, we are $100(1-\alpha)\%$ confident (for example, 95% confident if $\alpha = .05$) that the absolute value of the error of estimation obtained when estimating μ by $\bar{y}$ is less than or equal to the bound

$$BE_{\bar{y}}[100(1-\alpha)]$$

Said mathematically, we are $100(1-\alpha)\%$ confident that

$$|\bar{y} - \mu| \leq BE_{\bar{y}}[100(1-\alpha)]$$

This algebraic statement is equivalent to the algebraic statement

$$\mu - BE_{\bar{y}}[100(1-\alpha)] \leq \bar{y} \leq \mu + BE_{\bar{y}}[100(1-\alpha)]$$

In other words, we are $100(1-\alpha)\%$ confident that the point estimate $\bar{y}$ is within $BE_{\bar{y}}[100(1-\alpha)]$ units of the true value of μ. Now, the algebraic statement

$$\mu - BE_{\bar{y}}[100(1-\alpha)] \leq \bar{y} \leq \mu + BE_{\bar{y}}[100(1-\alpha)]$$

can be rearranged to give the algebraic statement

$$\bar{y} - BE_{\bar{y}}[100(1-\alpha)] \leq \mu \leq \bar{y} + BE_{\bar{y}}[100(1-\alpha)]$$

and we are $100(1-\alpha)\%$ confident that μ is contained in the interval

$$[\bar{y} - BE_{\bar{y}}[100(1-\alpha)], \bar{y} + BE_{\bar{y}}[100(1-\alpha)]]$$

We call this interval a *100(1 − α)% confidence interval for* μ: We are $100(1-\alpha)\%$ confident that μ is greater than or equal to the lower bound

$$\bar{y} - BE_{\bar{y}}[100(1-\alpha)]$$

and less than or equal to the upper bound

$$\bar{y} + BE_{\bar{y}}[100(1-\alpha)]$$

of this confidence interval. Since we do not know the true value of μ, we are not absolutely certain (not 100% confident) that μ is contained in the $100(1-\alpha)\%$ confidence interval for μ. Exactly what we mean when we say that we are $100(1-\alpha)\%$ confident that μ is contained in the interval will be discussed later. For now, suffice it to say that if $100(1-\alpha)\%$ is chosen to be a very high level of confidence (such as 95% confidence), then $100(1-\alpha)\%$ confident means very confident. The following is a formula for a $100(1-\alpha)\%$ confidence interval for μ.

A CONFIDENCE INTERVAL FOR A POPULATION MEAN:

Assume that $\bar{y}$ is the mean and s is the standard deviation of

$$SPL = \{y_1, y_2, \ldots, y_n\}$$

which has been randomly selected from a normally distributed population of element values with mean μ and standard deviation σ. Then, a $100(1-\alpha)\%$ confidence interval for μ is

$$[\bar{y} - BE_{\bar{y}}[100(1-\alpha)], \bar{y} + BE_{\bar{y}}[100(1-\alpha)]]$$

where the *$100(1-\alpha)\%$ bound on the error of estimation obtained when estimating μ by $\bar{y}$* is

$$BE_{\bar{y}}[100(1-\alpha)] = t_{[\alpha/2]}^{(n-1)}\left(\frac{s}{\sqrt{n}}\right)$$

Here, $t_{[\alpha/2]}^{(n-1)}$ is the point on the scale of the t-distribution having $n-1$ degrees of freedom so that the area in the tail of this t-distribution to the right of $t_{[\alpha/2]}^{(n-1)}$ is $\alpha/2$.

Regarding this result, first, if $n-1$ degrees of freedom are 30 *df* or more, then it is sufficient to use the value $z_{[\alpha/2]}$ for $t_{[\alpha/2]}^{(n-1)}$ (see Section 3.3.1). Here, $z_{[\alpha/2]}$ is the point on the scale of the standard normal curve so that the area under this curve to the right of $z_{[\alpha/2]}$ is $\alpha/2$ (see Section 3.2.2).

Second, it has been shown that the previous confidence interval formula approximately holds for many populations that are not normally distributed. In particular, these formulas approximately hold for a population described by a probability curve that is mound-shaped (even if this curve is somewhat skewed to the right or left). Recall that the normality assumption (or mound-shaped assumption) can be checked by using sample data to construct a histogram.

Third, although the preceding confidence interval formula and all the other confidence interval formulas discussed in this chapter apply to populations that are infinite (for example, a normally distributed population is infinite), it can be shown that these formulas also approximately apply to many finite (mound-shaped) populations. (However, better formulas, which yield shorter confidence intervals, often exist for finite populations.) We will assume (unless otherwise stated) that for any example discussed in this book the formula that we use is appropriate.

EXAMPLE 3.4

In Example 2.6 we saw that $\bar{y} = 31.2$ mpg is the mean and $s = .7517$ mpg is the standard deviation of

$$\begin{aligned} SPL &= \{y_1, y_2, y_3, y_4, y_5\} \\ &= \{30.7, 31.8, 30.2, 32.0, 31.3\} \end{aligned}$$

which National Motors has randomly selected from the infinite population of all Gas-Mizer mileages. It follows that $\bar{y} = 31.2$ is the point estimate of μ, the mean of all Gas-Mizer mileages, and that the $100(1 - \alpha)\%$ bound on the error of estimation obtained when estimating μ by $\bar{y}$ is

$$\begin{aligned} BE_{\bar{y}}[100(1 - \alpha)] &= t_{[\alpha/2]}^{(n-1)}\left(\frac{s}{\sqrt{n}}\right) \\ &= t_{[\alpha/2]}^{(5-1)}\left(\frac{.7517}{\sqrt{5}}\right) \\ &= t_{[\alpha/2]}^{(4)}(.3362) \end{aligned}$$

If we wish to compute the 95% bound (and thus the 95% confidence interval for μ), then $100(1 - \alpha)\% = 95\%$, which implies that $\alpha = .05$. Thus, to calculate the 95% bound, we need to find

$$t_{[\alpha/2]}^{(4)} = t_{[.05/2]}^{(4)} = t_{[.025]}^{(4)}$$

which is the point on the scale of the t-distribution having 4 degrees of freedom so that the area in the tail of this t-distribution to the right of $t_{[.025]}^{(4)}$ is .025. Looking up $t_{[.025]}^{(4)}$ in Table E-2 in Appendix E, we find that

$$t_{[.025]}^{(4)} = 2.776$$

This implies that the 95% bound on the error of estimation obtained when estimating μ by $\bar{y}$ is

$$\begin{aligned} BE_{\bar{y}}[95] &= t_{[.025]}^{(4)}(.3362) \\ &= 2.776(.3362) \\ &= .9333 \end{aligned}$$

and that the 95% confidence interval for μ is

$$\begin{aligned}[\bar{y} \pm BE_{\bar{y}}[95]] &= [31.2 \pm .9333] \\ &= [30.3, 32.1]\end{aligned}$$

This interval says that National Motors is 95% confident that μ, the mean gasoline mileage that would be achieved by the fleet of all Gas-Mizers that could potentially be produced is at least 30.3 mpg and no more than 32.1 mpg. Since (a supernatural power knows that) the true value of μ is 31.5 mpg, μ is contained in the 95% confidence interval for μ, [30.3, 32.1]. National Motors is not absolutely certain but rather is 95% confident that μ is contained in this interval. However, the company can use this interval as follows: Suppose that new federal gasoline mileage standards state that in order to avoid paying a heavy fine, National Motors must convince the government that μ is at least 30 mpg. Since the 95% confidence interval for μ makes National Motors 95% confident that μ is greater than or equal to the lower bound 30.3, and since this lower bound is itself greater than 30 mpg, the manufacturer can be at least 95% confident that μ is greater than 30 mpg. Thus, National Motors can report to the federal government that there is strong evidence that the Gas-Mizer not only meets but exceeds current gasoline mileage standards.

If we wish to calculate the 99% confidence interval for μ, then

$$100(1 - \alpha)\% = 99\%$$

which implies that $\alpha = .01$. Thus, to calculate the 99% confidence interval, we need to find

$$t^{(4)}_{[\alpha/2]} = t^{(4)}_{[.01/2]} = t^{(4)}_{[.005]}$$

which equals 4.604 (see Table E-2 in Appendix E). This implies that the 99% bound on the error of estimation when estimating μ by $\bar{y}$ is

$$\begin{aligned}BE_{\bar{y}}[99] &= t^{(4)}_{[.005]}(.3362) \\ &= 4.604(.3362) \\ &= 1.5479\end{aligned}$$

and thus that the 99% confidence interval for μ is

$$\begin{aligned}[\bar{y} \pm BE_{\bar{y}}[99]] &= [31.2 \pm 1.5479] \\ &= [29.7, 32.7]\end{aligned}$$

which is longer than [30.3, 32.1], the 95% confidence interval for μ. Hence, increasing the level of confidence from 95% to 99% (1) has the advantage of making us more confident that μ is contained in our confidence interval for μ, but (2) has the disadvantage of increasing the length of our confidence interval, which results in a less precise guess of the true value of μ. With respect to (2), note that since 29.7, the lower bound of the 99% confidence interval for μ, [29.7, 32.7], is not greater than or equal to 30 (the federal mileage standard), National Motors cannot on the basis of this confidence interval be 99% confident that μ is at least 30. However, since National Motors is at least 95% confident that μ is greater than 30, there is substantial evidence that the Gas-Mizer not only meets but exceeds current gasoline mileage standards.

Federal mileage standards also state that if National Motors wishes to avoid paying a heavy fine, three years from now μ must be at least 33 mpg. Since the 95% confidence interval for μ, [30.3, 32.1], makes National Motors 95% confident that μ is less than or equal to the upper bound 32.1, and since this upper bound is itself less than 33 mpg, National Motors can be at least 95% confident that μ is less than 33 mpg. Thus, National Motors is very confident that the current model of the Gas-Mizer will not meet federal gasoline mileage standards three years from now; the company should probably begin a research and development project to improve the mileage of the Gas-Mizer.

Finally, the Gas-Mizer's main competitor, the Gomega, is advertised to obtain a mean gasoline mileage of 31 mpg. National Motors would like to claim in a new advertising campaign that the Gas-Mizer achieves better gasoline mileage than the Gomega (that is, that μ is greater than 31 mpg). However, since 30.3, the lower bound of the 95% confidence interval for μ, is not greater than 31 mpg, National Motors cannot, on the basis of this confidence interval, be 95% confident that μ is greater than 31 mpg. However, the company is still (subjectively) convinced that μ really is greater than 31 mpg. Consequently, in order to attempt to justify this claim, National Motors will need to obtain a more precise estimate of μ by randomly selecting a larger sample. We discuss later in this section how National Motors can choose the size of this sample. We will use this larger sample to justify the company's claim in Example 3.6.

To complete this example, we will discuss the meaning of 95% confidence as it pertains to the 95% confidence interval for μ, [30.3, 32.1], which we have calculated by using the observed sample

$$\begin{aligned} SPL &= \{y_1, y_2, y_3, y_4, y_5\} \\ &= \{30.7, 31.8, 30.2, 32.0, 31.3\} \end{aligned}$$

Figure 3.13 depicts

1. Three possible samples—which we have arbitrarily numbered SPL_1, SPL_2, and SPL_3—from the *population of all possible samples of five Gas-Mizer mileages* that could have been randomly selected from the infinite population of all Gas-Mizer mileages. Note that the population of all Gas-Mizer mileages is illustrated as normally distributed with mean μ and standard deviation σ, which (a supernatural power knows) equal 31.5 and .8, respectively. Also, note that SPL_1 is the sample that National Motors has actually observed.
2. The sample means yielded by SPL_1, SPL_2, and SPL_3, which are denoted $(\bar{y})_{SPL_1}$, $(\bar{y})_{SPL_2}$, and $(\bar{y})_{SPL_3}$.

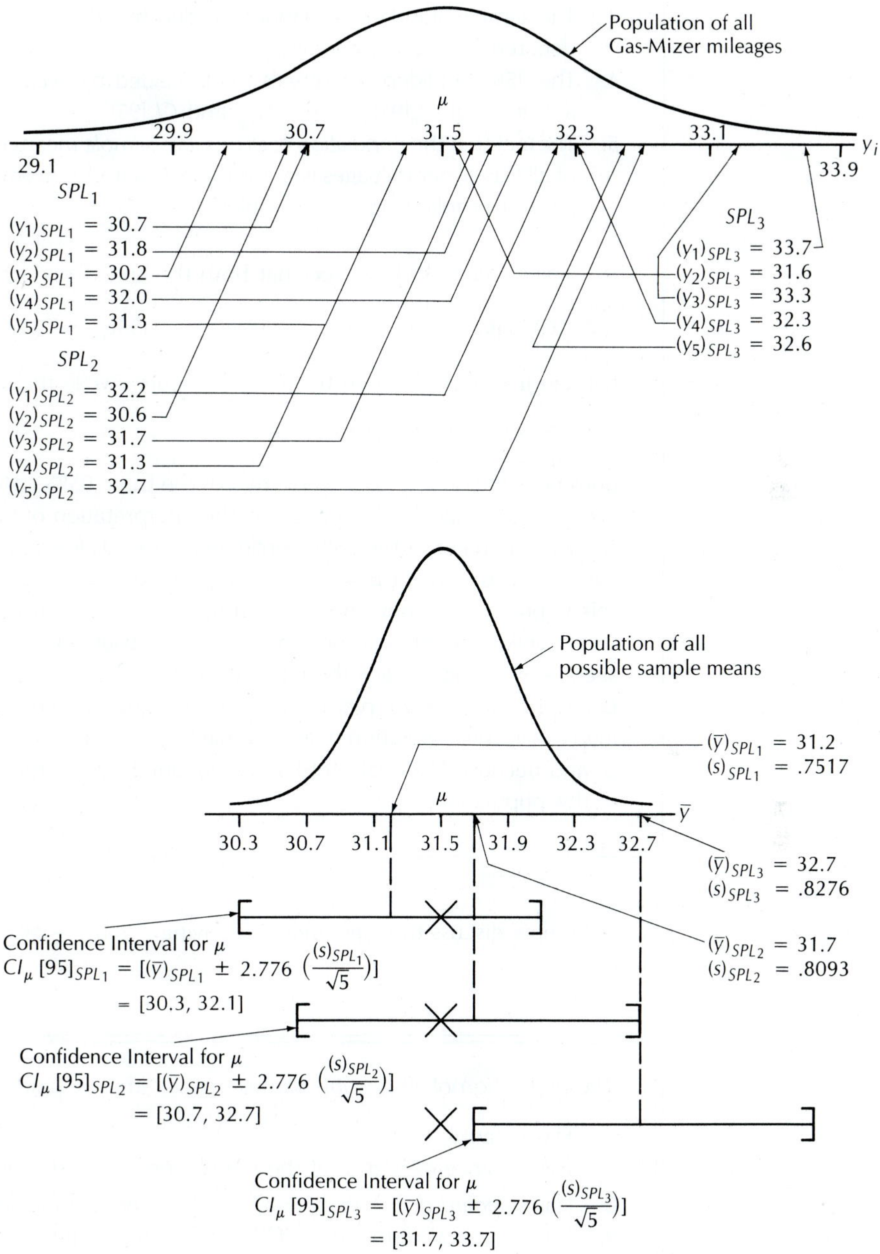

FIGURE 3.13
An Illustration of the Meaning of a 95% Confidence Interval in the Gas-Mizer Problem

3. The sample standard deviations yielded by SPL_1, SPL_2, and SPL_3, which are denoted $(s)_{SPL_1}$, $(s)_{SPL_2}$, and $(s)_{SPL_3}$.
4. The 95% confidence intervals for μ yielded by SPL_1, SPL_2, and SPL_3, which are denoted $CI_\mu[95]_{SPL_1}$, $CI_\mu[95]_{SPL_2}$, and $CI_\mu[95]_{SPL_3}$.
5. The fact that the *population of all possible sample means* has (if the population of all Gas-Mizer mileages is normally distributed) a normal distribution (more about this following this example).

Examining Figure 3.13, we see that both the confidence intervals

$$CI_\mu[95]_{SPL_1} = [30.3, 32.1] \quad \text{and} \quad CI_\mu[95]_{SPL_2} = [30.7, 32.7]$$

contain $\mu = 31.5$ (denoted by $\times$ in the figure), while the confidence interval

$$CI_\mu[95]_{SPL_3} = [31.7, 33.7]$$

does not contain $\mu = 31.5$. Thus, two of the three 95% confidence intervals for μ depicted in Figure 3.13 contain μ. The interpretation of 95% confidence here is that 95 percent of the 95% confidence intervals for μ *in the population of all such intervals* contain $\mu = 31.5$, while 5 percent of the confidence intervals in this population do not contain μ. Thus, when National Motors computes the 95% confidence interval for μ by using the observed sample (which, we have seen, is SPL_1) and obtains the interval [30.3, 32.1], National Motors can be 95% confident that this interval contains μ, because 95 percent of the confidence intervals in the population of all possible 95% confidence intervals for μ contain μ, and because National Motors has obtained one of the confidence intervals in this population.

We now discuss the logic behind deriving a $100(1 - \alpha)\%$ confidence interval for μ.

The population of all possible sample means (that is, point estimates of μ)

1. Has *mean* $\mu_{\bar{y}} = \mu$.
2. Has *variance* $\sigma^2_{\bar{y}} = \sigma^2/n$ (if the population sampled is infinite).
3. Has *standard deviation* $\sigma_{\bar{y}} = \sigma/\sqrt{n}$ (if the population sampled is infinite).
4. Has a *normal distribution* (if the population sampled has a normal distribution).

See Appendix A for a proof of (1), (2), and (3).

The mean of the population of all possible sample means, $\mu_{\bar{y}}$, equals μ, the population mean. For this reason, we say that when we use the observed sample

$$SPL = \{y_1, y_2, \ldots, y_n\}$$

to calculate the sample mean

$$\bar{y} = \frac{\sum_{i=1}^{n} y_i}{n}$$

which is the point estimate of μ, we are using an **unbiased** estimation procedure. This unbiasedness property tells us that although the sample mean $\bar{y}$ that we calculate (from the observed sample) probably does not equal μ, the average of all the different possible sample means that we could have calculated (from all the different possible samples) *is* equal to μ.

We note that $\sigma_{\bar{y}}^2$ and $\sigma_{\bar{y}}$, the variance and standard deviation of the population of all possible sample means, measure the variation, or spread, of the sample means in the population of all possible sample means. Here, a large variance $\sigma_{\bar{y}}^2$ (or a large standard deviation $\sigma_{\bar{y}}$) indicates that the possible sample means (possible point estimates of μ) are widely dispersed around μ, while a small variance $\sigma_{\bar{y}}^2$ (or a small standard deviation $\sigma_{\bar{y}}$) indicates that the possible sample means are clustered closely around μ.

EXAMPLE 3.5

For the Gas-Mizer problem, recall that (a supernatural being knows that) the true values of μ, σ^2, and σ, the mean, variance, and standard deviation of the infinite population of all Gas-Mizer mileages, are 31.5, .64, and .8, respectively. So (a supernatural power knows that) the mean, variance, and standard deviation of the infinite population of all possible sample means (that would be calculated using all possible samples of size $n = 5$) are, respectively,

$$\mu_{\bar{y}} = \mu = 31.5$$

$$\sigma_{\bar{y}}^2 = \frac{\sigma^2}{n} = \frac{.64}{5} = .128$$

$$\sigma_{\bar{y}} = \frac{\sigma}{\sqrt{n}} = \frac{.8}{\sqrt{5}} = .358$$

Since $\sigma = .8$ and $\sigma_{\bar{y}} = .358$, we see that the sample means are more closely clustered around μ than are the original Gas-Mizer mileages. Moreover, assuming that the population of all Gas-Mizer mileages is normally distributed, the population of all possible sample means is also normally distributed (with mean $\mu_{\bar{y}} = \mu = 31.5$ and standard deviation $\sigma_{\bar{y}} = \sigma/\sqrt{n} = .358$) (see Figure 3.1). Thus, as discussed in Section 3.2.1, 68.26%, 95.44%, and 99.73% of all possible sample means lie in, respectively, the intervals

$$[\mu_{\bar{y}} \pm \sigma_{\bar{y}}] = [31.5 \pm .358] = [31.142, 31.858]$$

$$[\mu_{\bar{y}} \pm 2\sigma_{\bar{y}}] = [31.5 \pm 2(.358)] = [31.5 \pm .716] = [30.784, 32.216]$$

$$[\mu_{\bar{y}} \pm 3\sigma_{\bar{y}}] = [31.5 \pm 3(.358)] = [31.5 \pm 1.074] = [30.426, 32.574]$$

If the population that is sampled is normally distributed, the population of all possible sample means is normally distributed (with mean $\mu_{\bar{y}} = \mu$ and standard deviation $\sigma_{\bar{y}} = \sigma/\sqrt{n}$). Therefore, if we define the $z_{[\bar{y},\mu]}$ *statistic* to be

$$z_{[\bar{y},\mu]} = \frac{\bar{y} - \mu_{\bar{y}}}{\sigma_{\bar{y}}} = \frac{\bar{y} - \mu}{\sigma/\sqrt{n}}$$

the population of all possible $z_{[\bar{y},\mu]}$ statistics has a standard normal distribution. Since the $z_{[\bar{y},\mu]}$ statistic involves the unknown population mean μ, we might be tempted to use it to derive a 100(1 — α)% confidence interval for μ. However, since the $z_{[\bar{y},\mu]}$ statistic also involves the unknown population standard deviation σ, we are led to estimate σ by the sample standard deviation s and thus to define the $t_{[\bar{y},\mu]}$ *statistic*:

A CONFIDENCE INTERVAL FOR μ BASED ON THE *t*-DISTRIBUTION:

Assume that the population sampled has a normal distribution. Noting that the point estimate of $\sigma_{\bar{y}} = \sigma/\sqrt{n}$ is $s_{\bar{y}} = s/\sqrt{n}$, which we call the *standard error of the estimate* $\bar{y}$, and defining the $t_{[\bar{y},\mu]}$ *statistic* to be

$$t_{[\bar{y},\mu]} = \frac{\bar{y} - \mu}{s_{\bar{y}}} = \frac{\bar{y} - \mu}{s/\sqrt{n}}$$

the population of all possible $t_{[\bar{y},\mu]}$ statistics has a *t*-distribution with $n - 1$ degrees of freedom. This implies that a *100(1 — α)% confidence interval for μ* is

$$\left[\bar{y} \pm t_{[\alpha/2]}^{(n-1)}\left(\frac{s}{\sqrt{n}}\right)\right]$$

In order to derive this 100(1 — α)% confidence interval for μ, consider the *t*-point $t_{[\alpha/2]}^{(n-1)}$ and Figure 3.14. The fact that the population of all possible $t_{[\bar{y},\mu]}$ statistics has a *t*-distribution with $n - 1$ degrees of freedom implies (as shown in

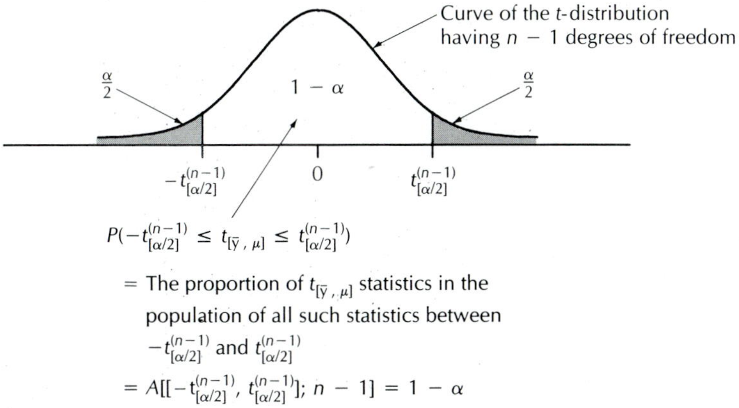

FIGURE 3.14

If the Population
of All Possible $t_{[\bar{y}, \mu]}$
Statistics Has a
t-Distribution With
$n - 1$ Degrees of
Freedom, Then
$P(-t_{[\alpha/2]}^{(n-1)} \le t_{[\bar{y}, \mu]} \le t_{[\alpha/2]}^{(n-1)})$
$= 1 - \alpha$

$P(-t_{[\alpha/2]}^{(n-1)} \le t_{[\bar{y}, \mu]} \le t_{[\alpha/2]}^{(n-1)})$

= The proportion of $t_{[\bar{y}, \mu]}$ statistics in the
population of all such statistics between
$-t_{[\alpha/2]}^{(n-1)}$ and $t_{[\alpha/2]}^{(n-1)}$

$= A[[-t_{[\alpha/2]}^{(n-1)}, t_{[\alpha/2]}^{(n-1)}]; n - 1] = 1 - \alpha$

Figure 3.14) that the probability

$$P(-t_{[\alpha/2]}^{(n-1)} \le t_{[\bar{y}, \mu]} \le t_{[\alpha/2]}^{(n-1)}) = P\left(-t_{[\alpha/2]}^{(n-1)} \le \frac{\bar{y} - \mu}{s/\sqrt{n}} \le t_{[\alpha/2]}^{(n-1)}\right)$$

$$= A[[-t_{[\alpha/2]}^{(n-1)}, t_{[\alpha/2]}^{(n-1)}]; n - 1] = 1 - \alpha$$

which is the area under the curve of the t-distribution having $n - 1$ degrees of free-
dom between $-t_{[\alpha/2]}^{(n-1)}$ and $t_{[\alpha/2]}^{(n-1)}$. Multiplying the inequality in the probability statement

$$P\left(-t_{[\alpha/2]}^{(n-1)} \le \frac{\bar{y} - \mu}{s/\sqrt{n}} \le t_{[\alpha/2]}^{(n-1)}\right) = 1 - \alpha$$

by $s/\sqrt{n}$ (which is positive), we see that this probability statement is equivalent to

$$P\left(-t_{[\alpha/2]}^{(n-1)}\left(\frac{s}{\sqrt{n}}\right) \le \bar{y} - \mu \le t_{[\alpha/2]}^{(n-1)}\left(\frac{s}{\sqrt{n}}\right)\right) = 1 - \alpha$$

which implies (subtracting $\bar{y}$ through the above inequality) that

$$P\left(-\bar{y} - t_{[\alpha/2]}^{(n-1)}\left(\frac{s}{\sqrt{n}}\right) \le -\mu \le -\bar{y} + t_{[\alpha/2]}^{(n-1)}\left(\frac{s}{\sqrt{n}}\right)\right) = 1 - \alpha$$

which in turn implies (multiplying the above inequality by -1) that

$$P\left(\bar{y} + t_{[\alpha/2]}^{(n-1)}\left(\frac{s}{\sqrt{n}}\right) \ge \mu \ge \bar{y} - t_{[\alpha/2]}^{(n-1)}\left(\frac{s}{\sqrt{n}}\right)\right) = 1 - \alpha$$

This probability statement is equivalent to

$$P\left(\bar{y} - t_{[\alpha/2]}^{(n-1)}\left(\frac{s}{\sqrt{n}}\right) \le \mu \le \bar{y} + t_{[\alpha/2]}^{(n-1)}\left(\frac{s}{\sqrt{n}}\right)\right) = 1 - \alpha$$

This probability statement says that the proportion of confidence intervals containing the population mean μ in the population of all possible $100(1-\alpha)\%$ confidence intervals for μ is equal to $1-\alpha$. That is, if we compute a $100(1-\alpha)\%$ confidence interval for μ by using the formula

$$\left[\bar{y} \pm t_{[\alpha/2]}^{(n-1)}\left(\frac{s}{\sqrt{n}}\right)\right]$$

then $100(1-\alpha)\%$ (for example, 95%) of the confidence intervals in the population of all possible $100(1-\alpha)\%$ confidence intervals for μ contain μ, and $100(\alpha)\%$ (for example, 5%) of the confidence intervals in this population do not contain μ. We illustrated this fact in Example 3.4 and Figure 3.13.

The preceding confidence interval, which is based on the t-distribution, assumes that the population sampled is normally distributed (or, at least mound-shaped). We would like to have a confidence interval that is valid no matter what probability distribution describes the population sampled.

A CONFIDENCE INTERVAL FOR μ BASED ON THE NORMAL DISTRIBUTION:

The **Central Limit Theorem** states that if the sample size n is large (say, at least 30), then the population of all possible sample means approximately has a normal distribution (with mean $\mu_{\bar{y}} = \mu$ and standard deviation $\sigma_{\bar{y}} = \sigma/\sqrt{n}$), no matter what probability distribution describes the population sampled (see Figure 3.15). Therefore, if *n is large*, and if we define the $z_{[\bar{y},\mu]}$ statistic to be

$$z_{[\bar{y},\mu]} = \frac{\bar{y}-\mu_{\bar{y}}}{\sigma_{\bar{y}}} = \frac{\bar{y}-\mu}{\sigma/\sqrt{n}}$$

then *the population of all possible $z_{[\bar{y},\mu]}$ statistics approximately has a standard normal distribution*. This implies that

$$\left[\bar{y} \pm z_{[\alpha/2]}\left(\frac{\sigma}{\sqrt{n}}\right)\right] \quad \text{and} \quad \left[\bar{y} \pm z_{[\alpha/2]}\left(\frac{s}{\sqrt{n}}\right)\right]$$

are *approximately correct $100(1-\alpha)\%$ confidence intervals for μ, no matter what probability distribution describes the population sampled*. Here, the second interval follows from the first by approximating σ by s.

In order to derive the first of these $100(1-\alpha)\%$ confidence intervals for μ, consider the z value $z_{[\alpha/2]}$ and Figure 3.16. The fact that the population of all possible

$z_{[\bar{y},\mu]}$ statistics approximately has a standard normal distribution implies (as shown in Figure 3.16) that

$$P(-z_{[\alpha/2]} \leq z_{[\bar{y},\mu]} \leq z_{[\alpha/2]}) = P\left(-z_{[\alpha/2]} \leq \frac{\bar{y}-\mu}{\sigma/\sqrt{n}} \leq z_{[\alpha/2]}\right)$$

$$\approx A[[-z_{[\alpha/2]}, z_{[\alpha/2]}]; N(0, 1)] = 1 - \alpha$$

which is the area under the curve of the standard normal distribution between $-z_{[\alpha/2]}$ and $z_{[\alpha/2]}$. Using algebraic manipulations analogous to those carried out

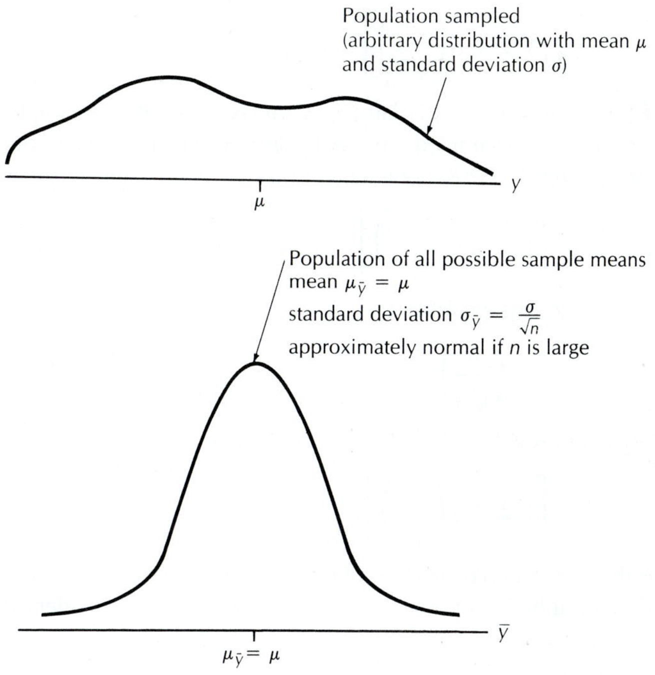

FIGURE 3.15
The Central Limit Theorem

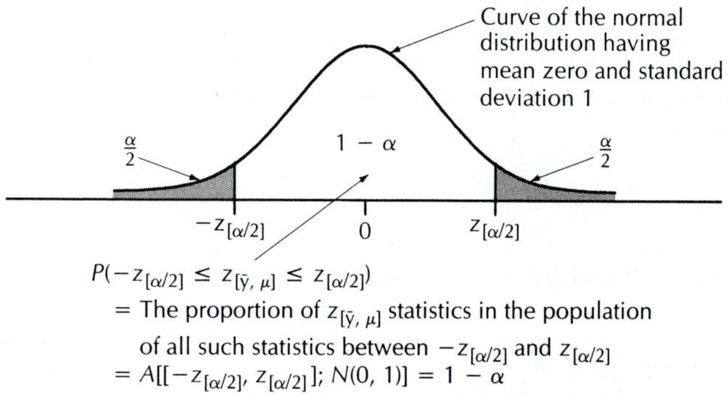

FIGURE 3.16
If the Population of All Possible $z_{[\bar{y},\mu]}$ Statistics Has a Standard Normal Distribution, Then $P(-z_{[\alpha/2]} \leq z_{[\bar{y},\mu]} \leq z_{[\alpha/2]}) = 1 - \alpha$

in deriving the confidence interval for μ based on the t-distribution, we find that

$$P\left(\bar{y} - z_{[\alpha/2]}\left(\frac{\sigma}{\sqrt{n}}\right) \leq \mu \leq \bar{y} + z_{[\alpha/2]}\left(\frac{\sigma}{\sqrt{n}}\right)\right) \approx 1 - \alpha$$

This implies that $[\bar{y} \pm z_{[\alpha/2]}(\sigma/\sqrt{n})]$ is an approximately correct $100(1 - \alpha)\%$ confidence interval for μ.

To summarize our discussion of confidence intervals for a population mean μ, we can say that assuming we do not know the true value of the population standard deviation σ, we should use the $100(1 - \alpha)\%$ confidence interval for μ based on the normal distribution

$$\left[\bar{y} \pm z_{[\alpha/2]}\left(\frac{s}{\sqrt{n}}\right)\right]$$

if the sample size, n, is large (say, at least 30). If the sample size, n, is small and the population sampled is normally distributed (or at least mound-shaped), we should use the $100(1 - \alpha)\%$ confidence interval for μ based on the t-distribution

$$\left[\bar{y} \pm t_{[\alpha/2]}^{(n-1)}\left(\frac{s}{\sqrt{n}}\right)\right]$$

Next, the quantity

$$z_{[\alpha/2]}\left(\frac{\sigma}{\sqrt{n}}\right)$$

in the interval

$$\left[\bar{y} \pm z_{[\alpha/2]}\left(\frac{\sigma}{\sqrt{n}}\right)\right]$$

is the *large sample $100(1 - \alpha)\%$ bound on the error of estimation obtained when estimating μ by $\bar{y}$.* Consequently, we are $100(1 - \alpha)\%$ confident that the sample mean, $\bar{y}$, is within

$$z_{[\alpha/2]}\left(\frac{\sigma}{\sqrt{n}}\right)$$

units of the population mean, μ. Suppose we wish to determine the sample size, n, so that we are $100(1 - \alpha)\%$ confident that $\bar{y}$ is within B units of μ. By setting the bound

$$z_{[\alpha/2]}\left(\frac{\sigma}{\sqrt{n}}\right)$$

equal to B and solving for n, we can determine the necessary sample size. We do this as follows:

$$z_{[\alpha/2]}\left(\frac{\sigma}{\sqrt{n}}\right) = B$$

which implies that

$$z_{[\alpha/2]}\sigma = \sqrt{n}\,B$$

which implies that

$$n = \left[\frac{z_{[\alpha/2]}\sigma}{B}\right]^2$$

To summarize,

> A sample of size
>
> $$n = \left[\frac{z_{[\alpha/2]}\sigma}{B}\right]^2$$
>
> makes us 100(1 − α)% confident that the sample mean, $\bar{y}$, is within B units of the population mean, μ.

Since this formula involves the population standard deviation, σ, which is probably unknown to us, we must often find an estimate of σ. The following are three ways to obtain this estimate:

1. A theorem called Chebyshev's Theorem implies that

   $$\sigma \approx \frac{R}{4}$$

   where R is the range of the element values in the population to be sampled. Therefore, if we know the range R, we can obtain a rough estimate of σ by dividing R by 4.
2. Sometimes theory or knowledge concerning a population similar to the population to be sampled allows us to obtain a rough estimate of σ.
3. We can calculate the standard deviation, denoted by s_p, of a preliminary sample of n_p element values randomly selected from the population to be sampled. If $n_p - 1$ is at least 30, we calculate n by the formula

   $$n = \left[\frac{z_{[\alpha/2]}s_p}{B}\right]^2$$

   If $n_p - 1$ is less than 30, we calculate n by the formula

   $$n = \left[\frac{t_{[\alpha/2]}^{(n_p-1)}s_p}{B}\right]^2$$

EXAMPLE 3.6

Suppose that National Motors wishes to determine the sample size, n, so that it is 95% confident that $\bar{y}$ is within $B = .3$ mpg of μ, the mean of the infinite population of all Gas-Mizer mileages. The company can regard

$$\begin{aligned} SPL &= \{y_1, y_2, y_3, y_4, y_5\} \\ &= \{30.7, 31.8, 30.2, 32.0, 31.3\} \end{aligned}$$

(discussed in Example 3.4) as a preliminary sample, which has a standard deviation of .7517. Then, since

$$t_{[.025]}^{(n_p-1)} = t_{[.025]}^{(5-1)} = 2.776$$

National Motors should randomly select a sample of size

$$n = \left[\frac{t_{[\alpha/2]}^{(n_p-1)} s_p}{B}\right]^2 = \left[\frac{(2.776)(.7517)}{.3}\right]^2 = 48.38, \text{ or } 49 \quad \text{(rounding up)}$$

Note that 48.38 has been rounded up to 49 in order to guarantee that our confidence interval is at least as precise as specified by setting $B = .3$ mpg (rounding down would result in a less precise interval). Assume that when National Motors randomly selects this sample, it obtains the sample of $n = 49$ Gas-Mizer mileages listed in Table 3.1, which has mean $\bar{y} = 31.5531$ and standard deviation $s = .799$. It follows that $\bar{y} = 31.5531$ is the point estimate of the population mean, μ. Moreover, since $n = 49$ is large, a 95% confidence interval for μ is (see Table E-1 in Appendix E)

$$\begin{aligned} \left[\bar{y} \pm z_{[.025]}\left(\frac{s}{\sqrt{n}}\right)\right] &= \left[31.5531 \pm 1.96\left(\frac{.799}{\sqrt{49}}\right)\right] \\ &= [31.5531 \pm .2237] \\ &= [31.3, 31.8] \end{aligned}$$

Recall from Example 3.4 that National Motors would like to claim in its advertisements that μ is greater than 31 mpg, the mean gasoline mileage claimed for the Gas-Mizer's main competitor, the Gomega. Since this 95% confidence interval for μ has a lower bound of 31.3, National Motors can be very confident that μ is greater than 31 mpg. This convinces the company that it can legitimately claim (in a new advertising campaign) that μ is greater than 31 mpg.

3.4.2 Confidence Intervals for Population Proportions

Sometimes we wish to estimate the proportion of elements in a population that possess a particular characteristic (for example, the proportion of beer drinkers who prefer Brand X Beer to all other brands). Denoting this population proportion by the symbol p, a reasonable point estimate of p is $\hat{p}$, which denotes the proportion of elements possessing the characteristic of interest in a sample of elements randomly selected from the population. In addition to finding this point estimate of p, we often wish to find a confidence interval for p.

CONFIDENCE INTERVALS FOR *p* BASED ON THE NORMAL DISTRIBUTION:

The population of all possible sample proportions (that is, point estimates of p)

1. Has *mean* $\mu_{\hat{p}} = p$.
2. Has *variance* $\sigma^2_{\hat{p}} = p(1 - p)/n$ (if the population sampled is infinite).
3. Has *standard deviation* $\sigma_{\hat{p}} = \sqrt{p(1 - p)/n}$ (if the population sampled is infinite).
4. *Approximately has a normal distribution* (if the *sample size, n, is large*). Here, n should be at least 30 *and* large enough so that the interval $[p \pm 3\sqrt{p(1 - p)/n}]$ does not contain zero or 1. Note that (since p is unknown) p must be estimated in order to use this rule of thumb.

Then, if n is large, and if we define the $z_{[\hat{p},p]}$ statistic to be

$$z_{[\hat{p},p]} = \frac{\hat{p} - p}{\sigma_{\hat{p}}} = \frac{\hat{p} - p}{\sqrt{p(1 - p)/n}}$$

the population of all possible $z_{[\hat{p},p]}$ statistics approximately has a standard normal distribution. This implies that

$$\left[\hat{p} \pm z_{[\alpha/2]}\sqrt{\frac{p(1 - p)}{n}}\right] \quad \text{and} \quad \left[\hat{p} \pm z_{[\alpha/2]}\sqrt{\frac{\hat{p}(1 - \hat{p})}{n - 1}}\right]$$

are *approximately correct 100(1 − α)% confidence intervals for p*. Here, the second interval follows from the first by estimating $p(1 - p)$ by $\hat{p}(1 - \hat{p})(n/(n - 1))$, where, it can be shown, multiplying $\hat{p}(1 - \hat{p})$ by $n/(n - 1)$ provides a better point estimate of $p(1 - p)$.

EXAMPLE 3.7

Antibiotics occasionally cause nausea as a side effect when administered to patients. Scientists at Pharmco, Inc., a large drug manufacturer, have developed a new antibiotic, Phe-Mycin. The company wishes to estimate p, the proportion of patients who would experience nausea as a side effect of taking Phe-Mycin. Suppose that a sample of $n = 200$ randomly selected patients are treated with Phe-Mycin, and that 35 of these 200 patients experience nausea as a side effect. Then

1. The point estimate of the population proportion p is the sample proportion

$$\hat{p} = \frac{35}{200} = .175$$

   which says that Pharmco *estimates* that 17.5 percent of all patients would experience nausea as a side effect of taking Phe-Mycin.

2. An approximate 95% confidence interval for p is

$$\left[\hat{p} \pm z_{[.025]}\sqrt{\frac{\hat{p}(1-\hat{p})}{n-1}}\right] = \left[.175 \pm 1.96\sqrt{\frac{.175(.825)}{200-1}}\right]$$
$$= [.175 \pm .053]$$
$$= [.122, .228]$$

   which says that Pharmco is 95% confident that at least 12.2 percent and at most 22.8 percent of all patients would experience nausea as a side effect of taking Phe-Mycin. Note that use of this interval is valid because $n = 200$ and n is large enough to ensure that the interval

$$\left[p \pm 3\sqrt{\frac{p(1-p)}{n}}\right] \approx \left[\hat{p} \pm 3\sqrt{\frac{\hat{p}(1-\hat{p})}{n-1}}\right]$$
$$= \left[.175 \pm 3\sqrt{\frac{.175(.825)}{200-1}}\right] = [.175 \pm .081]$$
$$= [.094, .256]$$

does not contain zero or 1.

Now, assume that before this sample (for which $\hat{p} = .175$) was selected, some of the scientists at Pharmco felt (because of theoretical considerations) that the true percentage of all patients who would experience nausea is less than or equal to 10 percent (that is, $p \leq .10$), and other scientists felt that this percentage is greater than 10 percent (that is, $p > .10$). Since the previously calculated 95% confidence interval for p has a lower bound of .122, Pharmco can be very confident that p is greater than .10. However, this interval is fairly long and does not provide a very precise estimate of p. We discuss how to obtain a more precise estimate of p (that is, how to obtain a shorter confidence interval for p) by taking a larger sample later in this section.

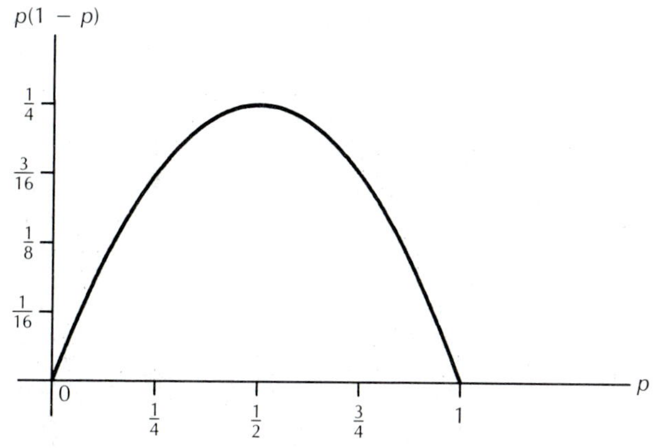

FIGURE 3.17
The Graph of $p(1-p)$ Versus p

Figure 3.17, the graph of $p(1-p)$ versus p, indicates that $p(1-p)$ equals 1/4 when p equals 1/2 and is never any larger than 1/4 for values of p between zero and 1. Therefore, the expression $\sqrt{p(1-p)/n}$, which is employed in the formula

$$\left[\hat{p} \pm z_{[\alpha/2]}\sqrt{\frac{p(1-p)}{n}}\right]$$

cannot be any larger than $\sqrt{1/4n}$. Thus, a conservative confidence interval can be calculated to be

$$\left[\hat{p} \pm z_{[\alpha/2]}\sqrt{\frac{1}{4n}}\right]$$

However, when the unknown population proportion is likely to be substantially different from 1/2, this conservative interval would be substantially longer than (less precise than) the interval

$$\left[\hat{p} \pm z_{[\alpha/2]}\sqrt{\frac{\hat{p}(1-\hat{p})}{n-1}}\right]$$

Regarding the choice of an appropriate sample size, first note that the quantity

$$z_{[\alpha/2]}\sqrt{\frac{p(1-p)}{n}}$$

in the interval

$$\left[\hat{p} \pm z_{[\alpha/2]}\sqrt{\frac{p(1-p)}{n}}\right]$$

is the large sample $100(1-\alpha)\%$ bound on the error of estimation obtained when estimating p by $\hat{p}$. Consequently, we are $100(1-\alpha)\%$ confident that the sample proportion $\hat{p}$ is within

$$z_{[\alpha/2]}\sqrt{\frac{p(1-p)}{n}}$$

of the population proportion p. Suppose we wish to determine the sample size, n, so that we are $100(1-\alpha)\%$ confident that $\hat{p}$ is within B units of p. By setting the bound

$$z_{[\alpha/2]}\sqrt{\frac{p(1-p)}{n}}$$

equal to B and solving for n, we can determine the necessary sample size. We do this as follows:

$$z_{[\alpha/2]}\sqrt{\frac{p(1-p)}{n}} = B$$

which implies that

$$z_{[\alpha/2]}\sqrt{p(1-p)} = B\sqrt{n}$$

which implies that

$$n = p(1-p)\left(\frac{z_{[\alpha/2]}}{B}\right)^2$$

A SAMPLE OF SIZE

$$n = p(1-p)\left(\frac{z_{[\alpha/2]}}{B}\right)^2$$

makes us $100(1-\alpha)\%$ confident that the sample proportion $\hat{p}$ is within B units of the population proportion p.

Since this formula involves $p(1-p)$, which is probably unknown, we should determine the largest value that $p(1-p)$ could reasonably be by using theoretical knowledge or previous sample information concerning the true value of p, and also Figure 3.17, the graph of $p(1-p)$ versus p. In particular, Figure 3.17 indicates (1) if p is less than 1/2, the largest reasonable value for $p(1-p)$ corresponds to the largest reasonable value for p, and (2) if p is greater than 1/2, the largest reasonable value for $p(1-p)$ corresponds to the smallest reasonable value for p.

If p could be near 1/2, then we should set $p(1-p)$ equal to 1/4 and calculate n by the equation

$$n = \frac{1}{4}\left(\frac{z_{[\alpha/2]}}{B}\right)^2$$

Since Figure 3.17 indicates that $p(1-p)$ can never be any larger than 1/4, this equation will always yield an n that is large enough to yield the needed precision, but if p is substantially different from 1/2, the equation will yield an n that is larger than needed.

EXAMPLE 3.8

Assume that Pharmco wishes to determine the sample size, n, so as to be 99% confident that $\hat{p}$ is within $B = .02$ of p, the proportion of all patients who would experience nausea as a side effect of taking Phe-Mycin.

1. The equality $100(1-\alpha)\% = 99\%$ implies that $\alpha = .01$ and thus that

$$z_{[\alpha/2]} = z_{[.01/2]} = z_{[.005]} = 2.58$$

2. The confidence interval computed for p in Example 3.7, [.122, .228], indicates that the largest reasonable value for p is .23 and thus that the largest reasonable value for $p(1-p)$ is $.23(1-.23) = .1771$ (see Figure 3.17).

Therefore, Pharmco should randomly select a sample of

$$n = p(1-p)\left(\frac{z_{[\alpha/2]}}{B}\right)^2 = .1771\left(\frac{2.58}{.02}\right)^2 = 2947 \text{ patients}$$

3 EXERCISES

3.1 Assume that the population of all Gas-Mizer mileages is normally distributed with mean $\mu = 31.5$ and standard deviation $\sigma = .8$, and let y denote a Gas-Mizer mileage randomly selected from this population. Find the following probabilities:

a. $P(30.7 \leq y \leq 32.3)$; b. $P(29.1 \leq y \leq 33.9)$;
c. $P(29.5 \leq y \leq 32.3)$; d. $P(31.0 \leq y \leq 31.3)$;
e. $P(y \leq 29.5)$; f. $P(y \geq 29.5)$;
g. $P(y \geq 33.4)$; h. $P(y \leq 33.4)$.

3.2 Using Table E-1 in Appendix E, find

a. $z_{[.05]}$; b. $z_{[.02]}$; c. $z_{[.01]}$; d. $z_{[.005]}$.

3.3 Using Table E-2 in Appendix E, find
a. $t_{[.05]}^{(7)}$; b. $t_{[.01]}^{(7)}$; c. $t_{[.005]}^{(7)}$.

3.4 Using Table E-3 in Appendix E, find
a. $F_{[.05]}^{(2,5)}$; b. $F_{[.05]}^{(5,2)}$.

3.5 Assuming that the population of all Gas-Mizer mileages is normally distributed, use the following sample of $n = 5$ Gas-Mizer mileages

$$SPL = \{32.3, 30.5, 31.7, 31.4, 32.6\}$$

to find 90%, 95%, 98%, and 99% confidence intervals for μ, the population mean Gas-Mizer mileage.

3.6 Using the three samples in Figure 3.13, calculate three 99% confidence intervals for μ. What percentage of these three confidence intervals contain μ $(=31.5)$? Discuss the meaning of 99% *confidence*.

3.7 Zenex Radio Corporation has developed a new way to assemble an electrical component used in the manufacture of radios. The company wishes to determine whether μ, the mean assembly time of this component using the new method, is less than 20 minutes, which is known to be the mean assembly time of the component using the current method.

Suppose that Zenex Radio randomly selects a sample of $n = 6$ employees, thoroughly trains each employee to use the new assembly method, has each employee assemble one component using the new method, records the assembly times, and calculates the mean and standard deviation of the sample of $n = 6$ assembly times to be $\bar{y} = 14.29$ minutes and $s = 2.19$ minutes.

Assuming that the population of all assembly times has a normal distribution, calculate a 99% confidence interval for μ. Using this confidence interval, can Zenex Radio be at least 99% confident that μ is less than 20 minutes? Justify your answer.

3.8 Referring to Exercise 3.7, determine the sample size, n, so that Zenex Radio is 99% confident that $\bar{y}$ is within $B = 2$ minutes of μ (regard the sample in Exercise 3.7 as a preliminary sample).

3.9 National Motors has equipped the Gas-Mizer with a new disc-brake system. We define the stopping distance for a Gas-Mizer to be the distance (in feet) required to bring the Gas-Mizer to a complete stop from a speed of 55 mph under normal driving conditions using this new brake system. Therefore, corresponding to the infinite population of all Gas-Mizers, there is an infinite population of all possible Gas-Mizer stopping distances, the mean of which we denote by μ. Intuitively, μ is the average of all the stopping distances that would be achieved by the population of all Gas-Mizers that could potentially be produced. The Gas-Mizer's main competitor, the Gomega, is advertised to achieve a mean stopping distance of 60 ft. National Motors would like to claim in a new advertising campaign that the Gas-Mizer achieves a shorter mean stopping distance than the Gomega (a μ less than 60 ft).

Suppose that National Motors randomly selects a sample of $n = 64$ Gas-Mizers, records the stopping distances of each of these Gas-Mizers, and calculates the mean and standard deviation of the sample of $n = 64$ Gas-Mizer stopping distances to be $\bar{y} = 58.22$ ft and $s = 6.13$ ft.

Calculate 90%, 95%, 98%, and 99% confidence intervals for μ. Using the 95% interval, can National Motors be at least 95% confident that μ is less than 60 ft? Using the 98% interval, can National Motors be at least 98% confident that μ is less than 60 ft?

3.10 Referring to Exercise 3.9, determine the sample size n so that National Motors is 98% confident that $\bar{y}$ is within 1 ft of μ (regard the sample in Exercise 3.9 as a preliminary sample).

3.11 It can be shown that if μ and σ^2 are the mean and variance of a finite population of N element values, and if n is the size of a sample randomly selected from this population, then the population of all possible sample means (that is, point estimates of μ) has mean $\mu_{\bar{y}} = \mu$, variance $\sigma^2_{\bar{y}} = (\sigma^2/n)((N - n)/(N - 1))$, and standard deviation $\sigma_{\bar{y}} = (\sigma/\sqrt{n})\sqrt{(N - n)/(N - 1)}$. Moreover, if the sample size, n, is large, the population of all possible sample means approximately has a normal distribution, and thus, defining

$$z_{[\bar{y},\mu]} = \frac{\bar{y} - \mu}{(\sigma/\sqrt{n})\sqrt{(N - n)/(N - 1)}}$$

the population of all possible $z_{[\bar{y},\mu]}$ statistics approximately has a standard normal distribution.

a. Use the preceding information to show that if n is large, then

$$\left[\bar{y} - z_{[\alpha/2]}\left(\frac{\sigma}{\sqrt{n}}\right)\sqrt{\frac{N-n}{N-1}},\ \bar{y} + z_{[\alpha/2]}\left(\frac{\sigma}{\sqrt{n}}\right)\sqrt{\frac{N-n}{N-1}}\right]$$

is an approximately correct $100(1 - \alpha)\%$ confidence interval for μ.

b. It can be shown that

$$s\sqrt{\frac{N-1}{N}} = \sqrt{\frac{\sum_{i=1}^{n} (y_i - \bar{y})^2}{n-1}}\sqrt{\frac{N-1}{N}}$$

is the appropriate point estimate of σ when N is finite. Use this fact and the confidence interval in part (a) to show that

$$\left[\bar{y} - z_{[\alpha/2]}\left(\frac{s}{\sqrt{n}}\right)\sqrt{\frac{N-n}{N}},\ \bar{y} + z_{[\alpha/2]}\left(\frac{s}{\sqrt{n}}\right)\sqrt{\frac{N-n}{N}}\right]$$

is an approximately correct $100(1 - \alpha)\%$ confidence interval for μ.

c. Assume that Dynamics, Inc., employs 900 scientists. Because of lack of secretaries and laboratory assistants, the company is concerned that these scientists must spend too much time performing "trivial tasks." A sample of $n = 70$ scientists is randomly selected, and each scientist is asked to record the number of hours he or she spends doing "trivial tasks" during the upcoming week. When the results are summarized, a sample mean of $\bar{y} = 9.75$ hours and a sample standard deviation of $s = 2.14$ hours are obtained. Use the formula in part (b) to calculate a 95% confidence interval for μ, the *mean* number of hours spent last week doing "trivial tasks" per scientist by all 900 scientists. Interpret the results.

d. Note that if we let τ denote the *total* number of hours spent last week doing "trivial tasks" by all 900 scientists, then $\mu = \tau/N$, and thus $\tau = N\mu$. It follows that a point estimate of $\tau = N\mu$ is $\hat{\tau} = N\bar{y}$ and that an approximately correct $100(1 - \alpha)\%$ confidence interval for τ is

$$\left[N\bar{y} - z_{[\alpha/2]}N\left(\frac{s}{\sqrt{n}}\right)\sqrt{\frac{N-n}{N}},\ N\bar{y} + z_{[\alpha/2]}N\left(\frac{s}{\sqrt{n}}\right)\sqrt{\frac{N-n}{N}}\right]$$

which has been obtained by multiplying through the formula in part (b) by N. Use these formulas and the sample information in part (c) to calculate a point estimate of, and a 95% confidence interval for, τ. Interpret the results.

3.12 A new product is to be test marketed in the Toledo, Ohio, metropolitan area, and advertising is to be based on a balanced mixture of TV, radio, newspaper, and magazine ads. Suppose we wish to estimate p, the proportion of all consumers in the Toledo metropolitan area who are aware of the product two months after the advertising campaign has begun.

a. If we wish to determine the sample size n so that we are 95% confident that the point estimate $\hat{p}$ is within $B = .04$ of p, verify that the necessary sample size, n, is 601 consumers (assume that the true value of p could be near 1/2).

b. Suppose that when we randomly select the sample of 601 consumers, 475 out of the 601 are aware of the product being test marketed. Using this sample information, find a point estimate of, and a 95% confidence interval for, p. Interpret the results.

4

MATRIX ALGEBRA

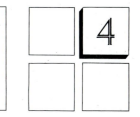

Before beginning our discussion of regression analysis, it is necessary to present some introductory concepts in a branch of mathematics known as matrix algebra.

4.1 MATRICES AND VECTORS

A **matrix** is a rectangular array of numbers (called elements) that is composed of rows and columns. An example of a matrix is

$$\mathbf{A} = \begin{bmatrix} 1 & 5 & 3 & 10 \\ 12 & 6 & 7 & 4 \\ 9 & 2 & 11 & 8 \end{bmatrix}$$

The notation **A** is used to indicate that we are referring to a matrix rather than a number.

The **dimension** of a matrix is determined by the number of rows and columns in the matrix. Since the matrix **A** has 3 rows and 4 columns, this matrix is said to have dimension 3 by 4 (commonly written 3×4). In general, a matrix with m rows and n columns is said to have dimension $m \times n$. As another example, the matrix

$$\mathbf{X} = \begin{bmatrix} 1 & 0 & 0 \\ 1 & 1 & 0 \\ 1 & 2 & 0 \\ 1 & 0 & 1 \\ 1 & 1 & 1 \\ 1 & 2 & 1 \\ 1 & 0 & 2 \\ 1 & 1 & 2 \\ 1 & 2 & 2 \end{bmatrix}$$

has dimension 9×3, since it has 9 rows and 3 columns.

In general, a matrix with dimension $m \times n$ can be represented as

$$\mathbf{A}_{m \times n} = \begin{bmatrix} a_{11} & a_{12} & \cdot\ \cdot & a_{1j} & \cdot & a_{1n} \\ a_{21} & a_{22} & \cdot\ \cdot & a_{2j} & \cdot & a_{2n} \\ \vdots & \vdots & & \vdots & & \vdots \\ a_{i1} & a_{i2} & \cdot\ \cdot & a_{ij} & \cdot & a_{in} \\ \vdots & \vdots & & \vdots & & \vdots \\ a_{m1} & a_{m2} & \cdot\ \cdot & a_{mj} & \cdot & a_{mn} \end{bmatrix}$$

where a_{ij} is the number in the matrix in row i and column j, and the subscript $m \times n$ indicates the dimension of **A**.

A matrix that consists of one column is a **column vector**, for example,

$$\mathbf{B}_{4\times 1} = \begin{bmatrix} 5 \\ 3 \\ 4 \\ 1 \end{bmatrix} \quad \text{and} \quad \mathbf{C}_{3\times 1} = \begin{bmatrix} 101 \\ 73 \\ 51 \end{bmatrix}$$

A matrix that consists of one row is a **row vector**, for example,

$$\mathbf{E}'_{1\times 4} = [10 \quad 7 \quad 6 \quad 12]$$

$$\mathbf{F}'_{1\times 6} = [1 \quad 2 \quad 7 \quad 11 \quad 5 \quad 8]$$

Note that the prime mark (′) is used to distinguish a row vector from a column vector.

4.2 THE TRANSPOSE OF A MATRIX

> The **transpose** of a matrix is formed by interchanging the rows and columns of the matrix.

For example, consider the matrix

$$\mathbf{A}_{2\times 3} = \begin{bmatrix} 5 & 6 & 7 \\ 3 & 2 & 1 \end{bmatrix}$$

The transpose of **A**, which is denoted **A**′, is

$$\mathbf{A}'_{3\times 2} = \begin{bmatrix} 5 & 3 \\ 6 & 2 \\ 7 & 1 \end{bmatrix}$$

Thus, the first row of **A** is the first column of **A**′, and the second row of **A** is the second column of **A**′.

Notice that the transpose of the column vector

$$\mathbf{E}_{4\times 1} = \begin{bmatrix} 10 \\ 7 \\ 6 \\ 12 \end{bmatrix}$$

is the row vector

$$\mathbf{E}'_{1\times 4} = [10 \quad 7 \quad 6 \quad 12]$$

As a last example, consider the matrix

$$\mathbf{X}_{9\times 3} = \begin{bmatrix} 1 & 0 & 0 \\ 1 & 1 & 0 \\ 1 & 2 & 0 \\ 1 & 0 & 1 \\ 1 & 1 & 1 \\ 1 & 2 & 1 \\ 1 & 0 & 2 \\ 1 & 1 & 2 \\ 1 & 2 & 2 \end{bmatrix}$$

The transpose of **X** is

$$\mathbf{X}'_{3\times 9} = \begin{bmatrix} 1 & 1 & 1 & 1 & 1 & 1 & 1 & 1 & 1 \\ 0 & 1 & 2 & 0 & 1 & 2 & 0 & 1 & 2 \\ 0 & 0 & 0 & 1 & 1 & 1 & 2 & 2 & 2 \end{bmatrix}$$

4.3 SUMS AND DIFFERENCES OF MATRICES

Consider two matrices, **A** and **B**, which have the same dimensions.

> The **sum** of **A** and **B** is a matrix obtained by adding the corresponding elements of **A** and **B**.

For example, for the matrices

$$\mathbf{A}_{2\times 3} = \begin{bmatrix} 1 & 4 & 2 \\ 5 & 3 & 2 \end{bmatrix} \quad \text{and} \quad \mathbf{B}_{2\times 3} = \begin{bmatrix} 7 & 0 & 4 \\ 3 & 1 & 5 \end{bmatrix}$$

the sum is

$$\mathbf{C}_{2\times 3} = \mathbf{A}_{2\times 3} + \mathbf{B}_{2\times 3} = \begin{bmatrix} 1+7 & 4+0 & 2+4 \\ 5+3 & 3+1 & 2+5 \end{bmatrix} = \begin{bmatrix} 8 & 4 & 6 \\ 8 & 4 & 7 \end{bmatrix}$$

In general, if **A** and **B** have the same dimensions and $\mathbf{C} = \mathbf{A} + \mathbf{B}$,

$$c_{ij} = a_{ij} + b_{ij}$$

where

c_{ij} = the number in **C** in row i and column j

a_{ij} = the number in **A** in row i and column j

b_{ij} = the number in **B** in row i and column j

Again consider two matrices, **A** and **B**, which have the same dimensions.

The **difference** of **A** and **B** is a matrix obtained by subtracting the corresponding elements of **A** and **B**.

For example, for the matrices

$$\mathbf{A}_{2\times 3} = \begin{bmatrix} 1 & 4 & 2 \\ 5 & 3 & 2 \end{bmatrix} \quad \text{and} \quad \mathbf{B}_{2\times 3} = \begin{bmatrix} 7 & 0 & 4 \\ 3 & 1 & 5 \end{bmatrix}$$

the difference is

$$\mathbf{D}_{2\times 3} = \mathbf{A}_{2\times 3} - \mathbf{B}_{2\times 3} = \begin{bmatrix} 1-7 & 4-0 & 2-4 \\ 5-3 & 3-1 & 2-5 \end{bmatrix} = \begin{bmatrix} -6 & 4 & -2 \\ 2 & 2 & -3 \end{bmatrix}$$

In general, if **A** and **B** have the same dimensions and **D** = **A** − **B**,

$$d_{ij} = a_{ij} - b_{ij}$$

where

d_{ij} = the number in **D** in row i and column j

a_{ij} = the number in **A** in row i and column j

b_{ij} = the number in **B** in row i and column j

4.4 MATRIX MULTIPLICATION

We now consider *multiplication of a matrix by a number.*

The **product of a number λ and a matrix A** is a matrix obtained by multiplying each element of **A** by the number λ.

For example, multiplying the following matrix

$$\mathbf{Z}_{2\times 3} = \begin{bmatrix} 1 & 4 & 7 \\ 3 & 2 & 3 \end{bmatrix}$$

by $\lambda = 5$, we get

$$5\mathbf{Z}_{2\times 3} = 5\begin{bmatrix} 1 & 4 & 7 \\ 3 & 2 & 3 \end{bmatrix} = \begin{bmatrix} 5(1) & 5(4) & 5(7) \\ 5(3) & 5(2) & 5(3) \end{bmatrix} = \begin{bmatrix} 5 & 20 & 35 \\ 15 & 10 & 15 \end{bmatrix}$$

In general, if λ is a number, **A** is a matrix, and $\mathbf{E} = \lambda\mathbf{A}$,

$$e_{ij} = \lambda a_{ij}$$

where

e_{ij} = the number in **E** in row *i* and column *j*

a_{ij} = the number in **A** in row *i* and column *j*

We next consider *multiplication of a matrix by a matrix.*

> Consider two matrices **A** and **B** where the number of *columns* in **A** is equal to the number of *rows* in **B**. Then the **product of the two matrices A and B** is a matrix calculated so that the element in row *i* and column *j* of the product is obtained by multiplying the elements in row *i* of matrix **A** by the corresponding elements in column *j* of matrix **B** and adding the resulting products.

For example, consider the following matrices:

$$\mathbf{A}_{2\times 2} = \begin{bmatrix} 4 & 3 \\ 2 & 2 \end{bmatrix} \quad \text{and} \quad \mathbf{B}_{2\times 2} = \begin{bmatrix} 2 & 1 \\ 3 & 5 \end{bmatrix}$$

Suppose we wish to find the product **AB**. The number in row 1 and column 1 of the product is obtained by multiplying the elements in row 1 of **A** by the corresponding elements in column 1 of **B** and adding these products. We obtain

$$4(2) + 3(3) = 8 + 9 = 17$$

The number in row 1 and column 2 of the product is obtained by multiplying the elements in row 1 of **A** by the corresponding elements in column 2 of **B** and adding these products. We obtain

$$4(1) + 3(5) = 4 + 15 = 19$$

The number in row 2 and column 1 of the product is obtained by multiplying the elements in row 2 of **A** by the corresponding elements in column 1 of **B** and adding these products. We obtain

$$2(2) + 2(3) = 4 + 6 = 10$$

The number in row 2 and column 2 of the product is obtained by multiplying the elements in row 2 of **A** by the corresponding elements in column 2 of **B** and adding these products. We obtain

$$2(1) + 2(5) = 2 + 10 = 12$$

Thus the product **AB** is as follows:

$$\mathbf{A}_{2\times 2}\mathbf{B}_{2\times 2} = \begin{bmatrix} 4 & 3 \\ 2 & 2 \end{bmatrix}\begin{bmatrix} 2 & 1 \\ 3 & 5 \end{bmatrix} = \begin{bmatrix} 4(2) + 3(3) & 4(1) + 3(5) \\ 2(2) + 2(3) & 2(1) + 2(5) \end{bmatrix}$$
$$= \begin{bmatrix} 17 & 19 \\ 10 & 12 \end{bmatrix}$$

In general, we can multiply a matrix **A** with m rows and r columns by a matrix **B** with r rows and n columns and obtain a matrix **C** with m rows and n columns. Moreover, c_{ij}, the number in the product in row i and column j, is obtained by multiplying the elements in row i of **A** by the corresponding elements in column j of **B** and adding the resulting products. Note that the number of columns in **A** must equal the number of rows in **B** in order for this multiplication procedure to be defined.

The multiplication procedure is illustrated here:

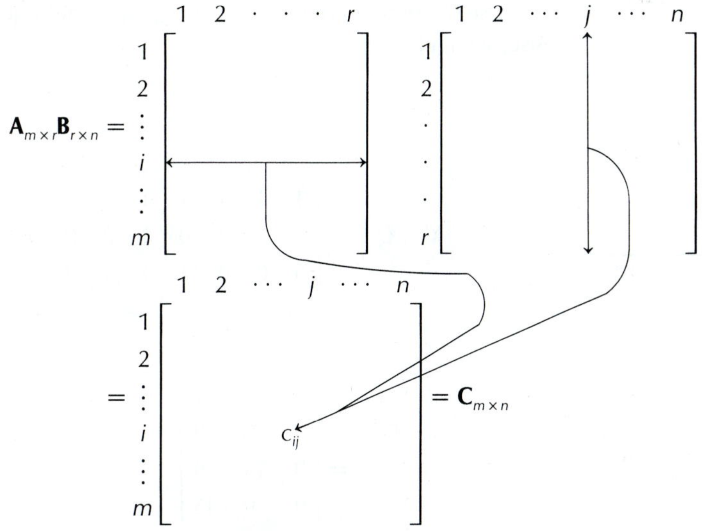

We now present several more examples. Consider the following matrices:

$$\mathbf{W}_{3\times 2} = \begin{bmatrix} 1 & 6 \\ 2 & 5 \\ 3 & 4 \end{bmatrix} \qquad \mathbf{U}_{2\times 2} = \begin{bmatrix} 2 & 2 \\ 1 & 3 \end{bmatrix}$$

$$\mathbf{X}_{9\times 3} = \begin{bmatrix} 1 & 0 & 0 \\ 1 & 1 & 0 \\ 1 & 2 & 0 \\ 1 & 0 & 1 \\ 1 & 1 & 1 \\ 1 & 2 & 1 \\ 1 & 0 & 2 \\ 1 & 1 & 2 \\ 1 & 2 & 2 \end{bmatrix} \qquad \mathbf{y}_{9\times 1} = \begin{bmatrix} 18 \\ 21 \\ 20 \\ 20 \\ 22 \\ 20 \\ 19 \\ 21 \\ 20 \end{bmatrix}$$

Then we find that

$$\mathbf{W}_{3\times 2}\mathbf{U}_{2\times 2} = \begin{bmatrix} 1 & 6 \\ 2 & 5 \\ 3 & 4 \end{bmatrix} \begin{bmatrix} 2 & 2 \\ 1 & 3 \end{bmatrix} = \begin{bmatrix} 8 & 20 \\ 9 & 19 \\ 10 & 18 \end{bmatrix}$$

but that

$$\mathbf{U}_{2\times 2}\mathbf{W}_{3\times 2} = \begin{bmatrix} 2 & 2 \\ 1 & 3 \end{bmatrix} \begin{bmatrix} 1 & 6 \\ 2 & 5 \\ 3 & 4 \end{bmatrix} \quad \text{does not exist}$$

because the number of columns in $\mathbf{U}_{2\times 2}$ does not equal the number of rows in $\mathbf{W}_{3\times 2}$. Also, we find that

$$\mathbf{X}'_{3\times 9}\mathbf{X}_{9\times 3} = \begin{bmatrix} 1 & 1 & 1 & 1 & 1 & 1 & 1 & 1 & 1 \\ 0 & 1 & 2 & 0 & 1 & 2 & 0 & 1 & 2 \\ 0 & 0 & 0 & 1 & 1 & 1 & 2 & 2 & 2 \end{bmatrix} \begin{bmatrix} 1 & 0 & 0 \\ 1 & 1 & 0 \\ 1 & 2 & 0 \\ 1 & 0 & 1 \\ 1 & 1 & 1 \\ 1 & 2 & 1 \\ 1 & 0 & 2 \\ 1 & 1 & 2 \\ 1 & 2 & 2 \end{bmatrix}$$

$$= \begin{bmatrix} 9 & 9 & 9 \\ 9 & 15 & 9 \\ 9 & 9 & 15 \end{bmatrix}$$

and that

$$\mathbf{X}'_{3\times 9}\mathbf{y}_{9\times 1} = \begin{bmatrix} 1 & 1 & 1 & 1 & 1 & 1 & 1 & 1 & 1 \\ 0 & 1 & 2 & 0 & 1 & 2 & 0 & 1 & 2 \\ 0 & 0 & 0 & 1 & 1 & 1 & 2 & 2 & 2 \end{bmatrix} \begin{bmatrix} 18 \\ 21 \\ 20 \\ 20 \\ 22 \\ 20 \\ 19 \\ 21 \\ 20 \end{bmatrix} = \begin{bmatrix} 181 \\ 184 \\ 182 \end{bmatrix}$$

As a last example, consider the matrices

$$\mathbf{A}_{2\times 2} = \begin{bmatrix} 1 & 1 \\ 2 & 2 \end{bmatrix} \quad \text{and} \quad \mathbf{B}_{2\times 2} = \begin{bmatrix} 0 & 1 \\ 1 & 0 \end{bmatrix}$$

Then we find that

$$\mathbf{A}_{2\times 2}\mathbf{B}_{2\times 2} = \begin{bmatrix} 1 & 1 \\ 2 & 2 \end{bmatrix}\begin{bmatrix} 0 & 1 \\ 1 & 0 \end{bmatrix} = \begin{bmatrix} 1 & 1 \\ 2 & 2 \end{bmatrix}$$

In this case **B** is said to be premultiplied by **A**, or **A** is said to be postmultiplied by **B**. Now consider

$$\mathbf{B}_{2\times 2}\mathbf{A}_{2\times 2} = \begin{bmatrix} 0 & 1 \\ 1 & 0 \end{bmatrix}\begin{bmatrix} 1 & 1 \\ 2 & 2 \end{bmatrix} = \begin{bmatrix} 2 & 2 \\ 1 & 1 \end{bmatrix}$$

Here **A** is said to be premultiplied by **B**, or **B** is said to be postmultiplied by **A**. Note that in this case **AB** is not equal to **BA**. In general, if **A** and **B** are matrices, then $\mathbf{AB} \neq \mathbf{BA}$.

4.5 THE IDENTITY MATRIX

> A matrix in which the number of rows is equal to the number of columns is called a **square matrix**.

For example, the following matrices are square matrices:

$$\mathbf{A}_{3\times 3} = \begin{bmatrix} 3 & 4 & 1 \\ 6 & 10 & 2 \\ 3 & 1 & 5 \end{bmatrix} \quad \mathbf{B}_{2\times 2} = \begin{bmatrix} 1 & 6 \\ 2 & 3 \end{bmatrix}$$

A square matrix in which the numbers on the main diagonal (the diagonal which runs from upper left to lower right) are 1's and in which all numbers off this diagonal are zeros is called an **identity matrix** and is denoted **I**.

The following are examples of identity matrices:

$$\mathbf{I}_{2\times 2} = \begin{bmatrix} 1 & 0 \\ 0 & 1 \end{bmatrix} \qquad \mathbf{I}_{3\times 3} = \begin{bmatrix} 1 & 0 & 0 \\ 0 & 1 & 0 \\ 0 & 0 & 1 \end{bmatrix}$$

Such a matrix is called an identity matrix because premultiplication or postmultiplication of a square $n \times n$ matrix **A** by the $n \times n$ identity matrix **I** leaves the matrix **A** unchanged. For example, if

$$\mathbf{A}_{3\times 3} = \begin{bmatrix} 2 & 1 & 3 \\ 4 & 1 & 2 \\ 2 & 2 & 1 \end{bmatrix}$$

then we see that

$$\mathbf{I}_{3\times 3}\mathbf{A}_{3\times 3} = \begin{bmatrix} 1 & 0 & 0 \\ 0 & 1 & 0 \\ 0 & 0 & 1 \end{bmatrix}\begin{bmatrix} 2 & 1 & 3 \\ 4 & 1 & 2 \\ 2 & 2 & 1 \end{bmatrix} = \begin{bmatrix} 2 & 1 & 3 \\ 4 & 1 & 2 \\ 2 & 2 & 1 \end{bmatrix}$$

and also that

$$\mathbf{A}_{3\times 3}\mathbf{I}_{3\times 3} = \begin{bmatrix} 2 & 1 & 3 \\ 4 & 1 & 2 \\ 2 & 2 & 1 \end{bmatrix}\begin{bmatrix} 1 & 0 & 0 \\ 0 & 1 & 0 \\ 0 & 0 & 1 \end{bmatrix} = \begin{bmatrix} 2 & 1 & 3 \\ 4 & 1 & 2 \\ 2 & 2 & 1 \end{bmatrix}$$

In general, if $\mathbf{A}_{m\times n}$ is not a square matrix, then

$$\mathbf{A}_{m\times n}\mathbf{I}_n = \mathbf{A}_{m\times n} \qquad \text{and} \qquad \mathbf{I}_m\mathbf{A}_{m\times n} = \mathbf{A}_{m\times n}$$

4.6 LINEAR DEPENDENCE AND LINEAR INDEPENDENCE

We now discuss two concepts known as linear independence and linear dependence. Consider the following matrix **A**:

$$\mathbf{A}_{3\times 3} = \begin{bmatrix} 1 & 4 & 2 \\ 3 & 2 & 6 \\ 2 & 1 & 4 \end{bmatrix}$$

Notice that the third column in this matrix is a multiple of the first column in the matrix. In particular, the third column is simply the first column multiplied by 2. That is,

$$\begin{bmatrix} 2 \\ 6 \\ 4 \end{bmatrix} = 2 \begin{bmatrix} 1 \\ 3 \\ 2 \end{bmatrix}$$

In a situation like this, when one column in a matrix **A** is a multiple of another column in matrix **A**, the columns of **A** are said to be linearly dependent. More generally,

> If one of the columns of a matrix **A** can be written as a linear combination of some of the other columns in **A**, then the columns of **A** are said to be **linearly dependent**.

As an example, consider the matrix

$$\mathbf{A}_{3\times 4} = \begin{bmatrix} 1 & 3 & 10 & 4 \\ 5 & 2 & 5 & 1 \\ 2 & 2 & 7 & 3 \end{bmatrix}$$

In this case, column 3 is the sum of column 4 plus 2 times column 2. That is,

$$\begin{bmatrix} 4 \\ 1 \\ 3 \end{bmatrix} + 2 \begin{bmatrix} 3 \\ 2 \\ 2 \end{bmatrix} = \begin{bmatrix} 4 \\ 1 \\ 3 \end{bmatrix} + \begin{bmatrix} 6 \\ 4 \\ 4 \end{bmatrix} = \begin{bmatrix} 10 \\ 5 \\ 7 \end{bmatrix}$$

Thus the columns of **A** are linearly dependent, because column 3 can be expressed as a linear combination of columns 2 and 4.

> If none of the columns in a matrix **A** can be written as a linear combination of other columns in **A**, then the columns of **A** are **linearly independent**. The maximum number of linearly independent columns in a matrix **A** is called the **rank** of the matrix. When the rank of a matrix **A** is equal to the number of columns in **A**, the matrix **A** is said to be of **full rank**.

4.7 THE INVERSE OF A MATRIX

Now consider a square matrix $\mathbf{A}_{n\times n}$, which is of full rank.

> The **inverse** of the matrix **A** is another matrix, denoted $\mathbf{A}^{-1}$, which satisfies the condition
>
> $$\mathbf{A}\mathbf{A}^{-1} = \mathbf{A}^{-1}\mathbf{A} = \mathbf{I}_{n\times n}$$
>
> where $\mathbf{I}_{n\times n}$ is the identity matrix with dimension $n \times n$.

It should be emphasized that $\mathbf{A}^{-1}$ exists if, and only if, **A** is a square matrix of full rank. As an example, consider the matrix

$$\mathbf{A}_{3\times 3} = \begin{bmatrix} 9 & 9 & 9 \\ 9 & 15 & 9 \\ 9 & 9 & 15 \end{bmatrix}$$

The inverse of **A**,

$$\mathbf{A}^{-1}_{3\times 3} = \begin{bmatrix} \frac{4}{9} & -\frac{1}{6} & -\frac{1}{6} \\ -\frac{1}{6} & \frac{1}{6} & 0 \\ -\frac{1}{6} & 0 & \frac{1}{6} \end{bmatrix}$$

since

$$\mathbf{A}_{3\times 3}\mathbf{A}^{-1}_{3\times 3} = \begin{bmatrix} 9 & 9 & 9 \\ 9 & 15 & 9 \\ 9 & 9 & 15 \end{bmatrix}\begin{bmatrix} \frac{4}{9} & -\frac{1}{6} & -\frac{1}{6} \\ -\frac{1}{6} & \frac{1}{6} & 0 \\ -\frac{1}{6} & 0 & \frac{1}{6} \end{bmatrix} = \begin{bmatrix} 1 & 0 & 0 \\ 0 & 1 & 0 \\ 0 & 0 & 1 \end{bmatrix}$$

and

$$\mathbf{A}^{-1}_{3\times 3}\mathbf{A}_{3\times 3} = \begin{bmatrix} \frac{4}{9} & -\frac{1}{6} & -\frac{1}{6} \\ -\frac{1}{6} & \frac{1}{6} & 0 \\ -\frac{1}{6} & 0 & \frac{1}{6} \end{bmatrix}\begin{bmatrix} 9 & 9 & 9 \\ 9 & 15 & 9 \\ 9 & 9 & 15 \end{bmatrix} = \begin{bmatrix} 1 & 0 & 0 \\ 0 & 1 & 0 \\ 0 & 0 & 1 \end{bmatrix}$$

It can be shown that $\mathbf{A}^{-1}\mathbf{A} = \mathbf{I}_{n\times n}$ if, and only if, $\mathbf{A}\mathbf{A}^{-1} = \mathbf{I}_{n\times n}$.

Although we will not discuss them here, general formulas exist that allow the calculation of matrix inverses. Also, computer programs are often used to calculate matrix inverses.

4.8 SOME REGRESSION-RELATED MATRIX CALCULATIONS

We now demonstrate how to perform two types of matrix calculations that we will encounter in later chapters of this book. The first calculation is of the form $(\mathbf{X}'\mathbf{X})^{-1}\mathbf{X}'\mathbf{y}$ where $\mathbf{X}$ is a matrix and $\mathbf{y}$ is a column vector. Suppose that

$$\mathbf{y} = \begin{bmatrix} 18 \\ 21 \\ 20 \\ 20 \\ 22 \\ 20 \\ 19 \\ 21 \\ 20 \end{bmatrix} \quad \text{and} \quad \mathbf{X} = \begin{bmatrix} 1 & 0 & 0 \\ 1 & 1 & 0 \\ 1 & 2 & 0 \\ 1 & 0 & 1 \\ 1 & 1 & 1 \\ 1 & 2 & 1 \\ 1 & 0 & 2 \\ 1 & 1 & 2 \\ 1 & 2 & 2 \end{bmatrix}$$

To compute $(\mathbf{X}'\mathbf{X})^{-1}\mathbf{X}'\mathbf{y}$ we make the following calculations:

$$\mathbf{X}' = \begin{bmatrix} 1 & 1 & 1 & 1 & 1 & 1 & 1 & 1 & 1 \\ 0 & 1 & 2 & 0 & 1 & 2 & 0 & 1 & 2 \\ 0 & 0 & 0 & 1 & 1 & 1 & 2 & 2 & 2 \end{bmatrix}$$

$$\mathbf{X}'\mathbf{X} = \begin{bmatrix} 1 & 1 & 1 & 1 & 1 & 1 & 1 & 1 & 1 \\ 0 & 1 & 2 & 0 & 1 & 2 & 0 & 1 & 2 \\ 0 & 0 & 0 & 1 & 1 & 1 & 2 & 2 & 2 \end{bmatrix} \begin{bmatrix} 1 & 0 & 0 \\ 1 & 1 & 0 \\ 1 & 2 & 0 \\ 1 & 0 & 1 \\ 1 & 1 & 1 \\ 1 & 2 & 1 \\ 1 & 0 & 2 \\ 1 & 1 & 2 \\ 1 & 2 & 2 \end{bmatrix}$$

$$= \begin{bmatrix} 9 & 9 & 9 \\ 9 & 15 & 9 \\ 9 & 9 & 15 \end{bmatrix}$$

$$(\mathbf{X}'\mathbf{X})^{-1} = \begin{bmatrix} \frac{4}{9} & -\frac{1}{6} & -\frac{1}{6} \\ -\frac{1}{6} & \frac{1}{6} & 0 \\ -\frac{1}{6} & 0 & \frac{1}{6} \end{bmatrix} \quad \text{since}$$

$$(\mathbf{X}'\mathbf{X})(\mathbf{X}'\mathbf{X})^{-1} = \begin{bmatrix} 9 & 9 & 9 \\ 9 & 15 & 9 \\ 9 & 9 & 15 \end{bmatrix} \begin{bmatrix} \frac{4}{9} & -\frac{1}{6} & -\frac{1}{6} \\ -\frac{1}{6} & \frac{1}{6} & 0 \\ -\frac{1}{6} & 0 & \frac{1}{6} \end{bmatrix}$$

$$= \begin{bmatrix} 1 & 0 & 0 \\ 0 & 1 & 0 \\ 0 & 0 & 1 \end{bmatrix}$$

Here assume that $(\mathbf{X}'\mathbf{X})^{-1}$ has been calculated by computer.

$$\mathbf{X}'\mathbf{y} = \begin{bmatrix} 1 & 1 & 1 & 1 & 1 & 1 & 1 & 1 & 1 \\ 0 & 1 & 2 & 0 & 1 & 2 & 0 & 1 & 2 \\ 0 & 0 & 0 & 1 & 1 & 1 & 2 & 2 & 2 \end{bmatrix} \begin{bmatrix} 18 \\ 21 \\ 20 \\ 20 \\ 22 \\ 20 \\ 19 \\ 21 \\ 20 \end{bmatrix} = \begin{bmatrix} 181 \\ 184 \\ 182 \end{bmatrix}$$

Finally, we find that

$$(\mathbf{X}'\mathbf{X})^{-1}\mathbf{X}'\mathbf{y} = \begin{bmatrix} \frac{4}{9} & -\frac{1}{6} & -\frac{1}{6} \\ -\frac{1}{6} & \frac{1}{6} & 0 \\ -\frac{1}{6} & 0 & \frac{1}{6} \end{bmatrix} \begin{bmatrix} 181 \\ 184 \\ 182 \end{bmatrix}$$

$$= \begin{bmatrix} 19.44 \\ .50 \\ .167 \end{bmatrix}$$

The second type of matrix calculation we will encounter is of the form $\mathbf{x}_0'(\mathbf{X}'\mathbf{X})^{-1}\mathbf{x}_0$, where $\mathbf{x}_0$ is a column vector and $\mathbf{X}$ is a matrix.

We illustrate this type of calculation using

$$\mathbf{x}_0 = \begin{bmatrix} 1 \\ 2 \\ 1 \end{bmatrix} \quad \text{and} \quad \mathbf{X} = \begin{bmatrix} 1 & 0 & 0 \\ 1 & 1 & 0 \\ 1 & 2 & 0 \\ 1 & 0 & 1 \\ 1 & 1 & 1 \\ 1 & 2 & 1 \\ 1 & 0 & 2 \\ 1 & 1 & 2 \\ 1 & 2 & 2 \end{bmatrix}$$

As shown previously,

$$(\mathbf{X}'\mathbf{X})^{-1} = \begin{bmatrix} \frac{4}{9} & -\frac{1}{6} & -\frac{1}{6} \\ -\frac{1}{6} & \frac{1}{6} & 0 \\ -\frac{1}{6} & 0 & \frac{1}{6} \end{bmatrix}$$

Then we find that

$$\mathbf{x}_0'(\mathbf{X}'\mathbf{X})^{-1}\mathbf{x}_0 = [1 \quad 2 \quad 1] \begin{bmatrix} \frac{4}{9} & -\frac{1}{6} & -\frac{1}{6} \\ -\frac{1}{6} & \frac{1}{6} & 0 \\ -\frac{1}{6} & 0 & \frac{1}{6} \end{bmatrix} \begin{bmatrix} 1 \\ 2 \\ 1 \end{bmatrix}$$

$$= [-\tfrac{1}{18} \quad \tfrac{1}{6} \quad 0] \begin{bmatrix} 1 \\ 2 \\ 1 \end{bmatrix} = \tfrac{5}{18}$$

4 EXERCISES

4.1 Let

$$\mathbf{A} = \begin{bmatrix} 1 & 2 \\ 3 & 1 \\ 2 & 2 \end{bmatrix}$$

a. Calculate $\mathbf{A}'$. b. Calculate $\mathbf{A}'\mathbf{A}$.

4.2 Let

$$\mathbf{A} = \begin{bmatrix} 1 & 3 & 1 \\ 2 & 1 & 1 \\ 1 & 3 & 3 \end{bmatrix} \quad \text{and} \quad \mathbf{B} = \begin{bmatrix} 0 & .6 & -.2 \\ .5 & -.2 & -.1 \\ -.5 & 0 & .5 \end{bmatrix}$$

a. Calculate $\mathbf{A} + \mathbf{B}$. b. Calculate $\mathbf{AB}$.
c. Calculate $\mathbf{BA}$. d. How are $\mathbf{A}$ and $\mathbf{B}$ related?

4.3 Let

$$\mathbf{A} = \begin{bmatrix} .02 & 0 & 0 & 0 & 0 \\ 0 & .01 & 0 & 0 & 0 \\ 0 & 0 & .004 & 0 & 0 \\ 0 & 0 & 0 & .005 & 0 \\ 0 & 0 & 0 & 0 & .002 \end{bmatrix} \quad \mathbf{c} = \begin{bmatrix} 1000 \\ 300 \\ 500 \\ 80 \\ 250 \end{bmatrix} \quad \mathbf{x} = \begin{bmatrix} 1 \\ 2 \\ 5 \\ 4 \\ 10 \end{bmatrix}$$

a. Calculate $\mathbf{c} + \mathbf{x}$. b. Calculate $\mathbf{Ac}$. c. Calculate $(\mathbf{Ac})'$.
d. Calculate $(\mathbf{Ac})'\mathbf{c}$. e. Calculate $\mathbf{x}'$. f. Calculate $\mathbf{x}'\mathbf{Ax}$.

5

SOME USEFUL REGRESSION MODELS

In this chapter our goal is to answer two important questions: (1) How do we express a regression model that relates a dependent variable y to a set of independent variables $x_1, x_2, \ldots, x_p$? and (2) What are some particular regression models that are useful in making statistical inferences?

We begin in Section 5.1 by presenting a general regression problem. In Section 5.2 we discuss the simple linear regression model, and in Section 5.3 we discuss a general regression model called the linear regression model. These models involve regression parameters.

In Section 5.4 we explore the least squares point estimates of these regression parameters, and in Section 5.5 we show how to use the least squares point estimates to find a point estimate of the future mean value of the dependent variable and a point prediction of the future (individual) value of the dependent variable. We conclude the chapter with Section 5.6, which discusses the quadratic regression model, and Section 5.7, which explains how to use interaction terms in regression models.

5.1 THE DEPENDENT VARIABLE, THE INDEPENDENT VARIABLES, AND THE OBSERVED DATA IN A GENERAL PROBLEM

Suppose we wish to study a variable of interest, which we call the *dependent variable* and denote by the symbol y. Also suppose that the dependent variable depends upon additional variables, which we call *independent variables*. There are usually a great number of independent variables affecting a dependent variable. In some instances, there may be virtually no limit to the number of independent variables affecting a dependent variable. Therefore, in order to study the dependent variable, we limit our attention to the independent variables that "significantly affect" the dependent variable. Although what we mean by this is somewhat arbitrary, we subsequently study several ways to determine which independent variables "significantly affect" the dependent variable. A reasonable procedure, then, is (1) to use knowledge of the problem under study to choose intuitively which independent variables might significantly affect the dependent variable, and (2) to use certain statistical techniques to determine whether the independent variables chosen do indeed significantly affect the dependent variable. Let p denote the number of independent variables that, in a particular problem, we intuitively believe significantly affect the dependent variable. We denote these independent variables by the symbols $x_1, x_2, \ldots, x_p$. Here, x_1 is called the first independent variable, x_2 is called the second independent variable, and so on, with x_p being called the pth independent variable.

For many problems we are in the present but must consider a future situation. In general, we use the subscript 0 (zero) to refer to quantities relating to the future situation. Thus, we denote the values of the p independent variables $x_1, x_2, \ldots, x_p$ that we will observe in the future by $x_{01}, x_{02}, \ldots, x_{0p}$. Moreover, we let y_0, *the future (individual) value of the dependent variable*, denote the value of the dependent variable that will occur in the future situation, when the future values of the p independent variables will be $x_{01}, x_{02}, \ldots, x_{0p}$. We often need to predict the future value y_0.

EXAMPLE 5.1

Suppose we are responsible for acquiring fuel for the physical plant at State University. In particular, we must be sure that buildings at State University will be adequately heated during the winter months. To help in planning fuel purchases, we wish to predict the amount of fuel (in tons of coal) that will be used by the university to heat buildings in future weeks. Our experience indicates that (1) weekly fuel consumption substantially depends on the average hourly temperature (in degrees Fahrenheit) during the week, and (2) weekly fuel consumption also depends on factors other than average hourly temperature that contribute to an overall "chill factor" at State University. Some of these factors (besides average hourly temperature) are

1. Wind velocity (in miles per hour) during the week,
2. "Cloud cover" during the week,
3. Variations in temperature, wind velocity, and cloud cover during the week (perhaps caused by the movement of weather fronts).

In this chapter we use regression analysis to predict the dependent variable, weekly fuel consumption. We show that regression analysis can be used to find a **prediction equation**, which predicts the dependent variable on the basis of the independent variables. In order to use regression analysis to predict the dependent variable (fuel consumption) for a future week, we must obtain predictions of the corresponding values of the independent variables for the future week. Therefore, we have decided to employ a professional weather forecasting service, Weather Forecasts, Inc., to predict the independent variables that our regression model will use to predict weekly fuel consumption.

Weather Forecasts believes that it can accurately predict average hourly temperature for a future week but feels that it would be difficult and expensive to define in quantitative terms and to predict the future values of each of the independent variables—wind velocity, "cloud cover," and variations in temperature, wind velocity, and cloud cover. Hence, to simplify the situation, Weather Forecasts suggests defining a *chill index*, which for a given average hourly temperature would express the combined effects of all the major factors other than average hourly temperature on the general chill factor at State University during a week. In other words, Weather Forecasts would develop a chill index by taking into account such considerations as the fact that a strong wind and a very cold temperature combine to increase the chill factor more than an equally strong wind and a warmer temperature.

With our approval, Weather Forecasts decides to express the chill index as an integer between 0 and 30. A weekly chill index near zero will indicate that, given the average hourly temperature during the week, all major factors other than average hourly temperature will increase only minimally the chill factor at State University. A weekly chill index near 30 will mean that, for the given average hourly temperature during the week, all major factors other than average hourly temperature will maximally increase the weekly chill factor at State University. In general, then, for a particular week, the larger the increase in the chill factor that will be produced (for a given average hourly temperature) by all major factors other than average hourly temperature, the larger the chill index.

To summarize, we have decided to predict the dependent variable, weekly fuel consumption (y), by using (predictions of) the independent variables, average hourly temperature (x_1) and the chill index (x_2). To make our discussion more concrete, we refer to the time at which we are making our prediction as *the present* and to the week for which our prediction is being made as *the future week*. The rest of our discussion concerning this prediction problem will be stated as if we are in the present and are making a prediction for the future week.

Suppose that Weather Forecasts predicts that the average hourly temperature during the future week will be $x_{01} = 40.0$ and that the chill index during the future week will be $x_{02} = 10$. Although it is possible that these predictions might be somewhat in error, we will assume that Weather Forecasts is so accurate that for all practical purposes the predictions are precisely correct.

The most straightforward approach that might be used to predict y_0, the amount of fuel (in tons of coal) that will be used to heat buildings at State University during the future week, would be to observe n weeks that are identical to the future week because each of these n weeks has an average hourly temperature of $x_{01} = 40.0$ and a chill index of $x_{02} = 10$. Then, we could record the fuel consumptions

$$y_1, y_2, \ldots, y_n$$

for each of the n weeks, and the average of these n fuel consumptions

$$\bar{y} = \frac{\sum_{i=1}^{n} y_i}{n}$$

would be the prediction of y_0, the future fuel consumption. The problem with this approach is that since we cannot control the weather, we cannot force n weeks with an average hourly temperature of $x_{01} = 40.0$ and a chill index of $x_{02} = 10$ to occur. Furthermore, suppose that the future week is next week. Then, we could not wait for n weeks with an average hourly temperature of $x_{01} = 40.0$ and a chill index of $x_{02} = 10$ to occur before making our prediction. So, in order to obtain a prediction of y_0, our only recourse is to use historical

TABLE 5.1 The Fuel Consumption Data Listed in Time Order

The tth Week	Average Hourly Temperature (°F) (x_{t1})	Chill Index (x_{t2})	Weekly Fuel Consumption (Tons) (y_t)
1st week	$x_{11} = 28.0$	$x_{12} = 18$	$y_1 = 12.4$
2d week	$x_{21} = 32.5$	$x_{22} = 24$	$y_2 = 12.4$
3rd week	$x_{31} = 28.0$	$x_{32} = 14$	$y_3 = 11.7$
4th week	$x_{41} = 39.0$	$x_{42} = 22$	$y_4 = 10.8$
5th week	$x_{51} = 57.8$	$x_{52} = 16$	$y_5 = 9.5$
6th week	$x_{61} = 45.9$	$x_{62} = 8$	$y_6 = 9.4$
7th week	$x_{71} = 58.1$	$x_{72} = 1$	$y_7 = 8.0$
8th week	$x_{81} = 62.5$	$x_{82} = 0$	$y_8 = 7.5$

data, or **observed data,** on average hourly temperature, the chill index, and weekly fuel consumption.

Suppose that since we have been interested in constructing a fuel consumption prediction equation for some time, we have already gathered data concerning these variables. These data, given in Table 5.1, consist of the average hourly temperature (in degrees Fahrenheit), the chill index, and the weekly fuel consumption (in tons of coal) at State University for each of the $n = 8$ weeks prior to the current week. We will call the data in Table 5.1, which are listed in the time order in which they have been observed, the historical fuel consumption data, or the observed fuel consumption data. In Table 5.1 the letter t denotes the time order of a previously observed week, with $t = 1$ referring to the first observed week, or the week observed in the most distant past, and $t = 8$ referring to the eighth observed week, or the week observed most recently. Moreover, x_{t1} denotes the average hourly temperature (x_1) that has occurred in the tth observed week, x_{t2} denotes the chill index (x_2) that has occurred in the tth observed week, and y_t denotes the fuel consumption (y) that has occurred in the tth observed week.

The techniques of regression analysis do not require that we list the historical fuel consumption data in the time order in which they were observed. We could, in fact, arrange these data in any order. For example, Table 5.2 lists the data of Table 5.1 in order of increasing average hourly temperature during the observed weeks. Thus, in Table 5.2, the subscript i, where $i = 1, 2, \ldots, 8$, denotes order, in terms of the level of average hourly temperature of a previously observed week. Here $i = 1$ refers to the week having the lowest average hourly temperature, $i = 2$ refers to the week having the second lowest average hourly temperature, and $i = 8$ refers to the week having the highest average hourly temperature. For $i = 1, 2, \ldots, 8$ we refer to the observed week having the ith lowest average hourly temperature as *the* i*th observed week,*

TABLE 5.2 The Fuel Consumption Data Listed in Order of Increasing Average Hourly Temperature

The *i*th Observed Week (OWK_i)	The *i*th Observed Average Hourly Temperature (°F) (x_{i1})	The *i*th Observed Chill Index (x_{i2})	The *i*th Observed Fuel Consumption (Tons) (y_i)
OWK_1	$x_{11} = 28.0$	$x_{12} = 18$	$y_1 = 12.4$
OWK_2	$x_{21} = 28.0$	$x_{22} = 14$	$y_2 = 11.7$
OWK_3	$x_{31} = 32.5$	$x_{32} = 24$	$y_3 = 12.4$
OWK_4	$x_{41} = 39.0$	$x_{42} = 22$	$y_4 = 10.8$
OWK_5	$x_{51} = 45.9$	$x_{52} = 8$	$y_5 = 9.4$
OWK_6	$x_{61} = 57.8$	$x_{62} = 16$	$y_6 = 9.5$
OWK_7	$x_{71} = 58.1$	$x_{72} = 1$	$y_7 = 8.0$
OWK_8	$x_{81} = 62.5$	$x_{82} = 0$	$y_8 = 7.5$

and denote this week by the symbol OWK_i. Thus,

1. The first observed week, denoted by the symbol OWK_1, is the observed week having the lowest average hourly temperature.
2. The second observed week (OWK_2) is the observed week having the second lowest average hourly temperature.

   ⋮

8. The eighth observed week (OWK_8) is the observed week having the highest average hourly temperature.

Furthermore, let x_{i1} denote the average hourly temperature (x_1) that has occurred in OWK_i (the *i*th observed week), x_{i2} denote the chill index (x_2) that has occurred in OWK_i, and y_i denote the fuel consumption (y) that has occurred in OWK_i. From now on, we will refer to the historical fuel consumption data as they are listed in Table 5.2, and we will also use the subscript i as it is used in Table 5.2.

The historical data in Table 5.2 are consistent with our belief that the lower the average hourly temperature during a week, the higher the fuel consumption during the week (other factors being equal), and the higher the chill index for the week, the higher the fuel consumption during the week (other factors being equal). For example, noticing that observed week 1 was colder than observed week 5 in the sense of having a lower average hourly temperature ($x_{11} = 28.0$ is lower than $x_{51} = 45.9$) and a higher chill index ($x_{12} = 18$ is higher than $x_{52} = 8$), we see that the fuel consumption in observed week 1 was higher than the fuel consumption in observed week 5 ($y_1 = 12.4$ is higher than $y_5 = 9.4$). We show later in this chapter that we can use the historical data in Table 5.2 to find the prediction equation

$$\hat{y} = 13.109 - .0900x_1 + .0825x_2$$

which expresses $\hat{y}$, the prediction of y (weekly fuel consumption), as a function of x_1 (average hourly temperature during the week) and x_2 (the chill index during the week). The coefficient $-.0900$, the number multiplied by x_1, is called the *regression coefficient that relates* x_1 *to* $\hat{y}$, the prediction of y. Since the regression coefficient $-.0900$ is negative, it says that the higher x_1, the average hourly temperature, is, the lower is $\hat{y}$, the prediction of y (other factors being equal). This is consistent with the fact that higher values of x_1 would be associated with lower values of y. The coefficient .0825, the number multiplied by x_2, is called the *regression coefficient that relates* x_2 *to* $\hat{y}$, the prediction of y. Since the regression coefficient .0825 is positive, it says that the higher x_2, the chill index, is, the higher is $\hat{y}$, the prediction of y (other factors being equal). This is consistent with the fact that higher values of x_2 would be associated with higher values of y. Later in this chapter we discuss in more detail the meanings of these regression coefficients. For now, we wish to show how to use the prediction equation

$$\hat{y} = 13.109 - .0900x_1 + .0825x_2$$

which utilizes the regression coefficients, to predict a future fuel consumption.

Since Weather Forecasts has told us that the average hourly temperature in the future week will be $x_{01} = 40.0$ and the chill index in the future week will be $x_{02} = 10$, the preceding prediction equation yields the prediction

$$\begin{aligned}\hat{y}_0 &= 13.109 - .0900x_{01} + .0825x_{02} \\ &= 13.109 - .0900(40.0) + .0825(10) \\ &= 10.333 \text{ tons of coal}\end{aligned}$$

Since the prediction $\hat{y}_0 = 10.333$ is a single number, it is a **point prediction** of y_0, the future fuel consumption. Note that this point prediction is based solely on x_{01}, the average hourly temperature in the future week, and x_{02}, the chill index in the future week. Based on intuition and on some statistical information to be subsequently discussed, we find it reasonable to believe that a point prediction, $\hat{y}_0$, based solely on x_{01} and x_{02} will be accurate. Nevertheless, the future fuel consumption, y_0, does not depend solely on x_{01} and x_{02} but also depends on other factors. For example, y_0 probably also depends upon the average hourly thermostat setting during the week. Since the average hourly thermostat settings at State University do vary a little, but (because of state regulations) do not vary greatly from week to week, average hourly thermostat setting probably does not greatly affect weekly fuel consumption. If this is the case, and if no other factors besides x_{01} and x_{02} greatly affect weekly fuel consumption, then basing the point prediction $\hat{y}_0$ solely on x_{01} and x_{02} should yield an accurate prediction. To summarize, since the point prediction $\hat{y}_0 = 10.333$ is based solely on x_{01} and x_{02}, and since y_0, the future fuel consumption, depends not only on x_{01} and x_{02} but also on other factors (hopefully, only to a minor extent), it follows that $\hat{y}_0 = 10.333$ is probably not a perfectly accurate prediction of y_0, the future fuel consumption. Since regression analysis is a statistical technique, it can tell us the farthest that the point prediction $\hat{y}_0 = 10.333$ might reasonably be from y_0, the future fuel consumption. How this is done is explained in Chapter 6.

As illustrated in the preceding example, we sometimes use observed data on the independent variables and the dependent variable to develop a prediction equation that can be used to predict y_0, the future value of the dependent variable, on the basis of $x_{01}, x_{02}, \ldots, x_{0p}$, the future values of the independent variables. For this purpose we assume that we have observed n situations prior to the present and that we have observed the values of the independent variables $x_1, x_2, \ldots, x_p$ and the value of the dependent variable y that have occurred during each of these n situations.

Let the subscript i, where $i = 1, 2, \ldots, n$, denote the order in terms of some criterion (such as time or the level of one of the independent variables) of a previously observed situation. For $i = 1, 2, \ldots, n$, we refer to the observed situation which is ith in this order as *the ith observed situation*, denoted by OST_i. Thus,

1. The first observed situation, denoted by the symbol OST_1, is the observed situation which is first in this order.
2. The second observed situation, denoted by the symbol OST_2, is the observed situation which is second in this order.

   ⋮

n. The nth observed situation, denoted by the symbol OST_n, is the observed situation which is nth (or last) in this order.

Furthermore, we let $x_{i1}, x_{i2}, \ldots, x_{ip}$ denote the (measurements of the) values of the p independent variables $x_1, x_2, \ldots, x_p$ that have occurred in OST_i (the ith observed situation), and we let y_i denote the (measurement of the) value of the dependent variable that has occurred in OST_i. In Table 5.3 we present the general notation

TABLE 5.3
The Observed Data

The ith Observed Situation (OST_i)	The Observed Values of the p Independent Variables (x_{i1}	x_{i2}	$\cdots$	x_{ip})	The ith Observed Value of the Dependent Variable (y_i)
OST_1	x_{11}	x_{12}	$\cdots$	x_{1p}	y_1
OST_2	x_{21}	x_{22}	$\cdots$	x_{2p}	y_2
⋮	⋮	⋮		⋮	⋮
OST_n	x_{n1}	x_{n2}	$\cdots$	x_{np}	y_n

describing the values of the p independent variables $x_1, x_2, \ldots, x_p$ and the values of the dependent variable y that we assume have occurred in the n observed situations.

5.2 THE SIMPLE LINEAR REGRESSION MODEL

We can obtain a prediction equation that can be used to predict the dependent variable y on the basis of the independent variables $x_1, x_2, \ldots, x_p$ by using the observed data in Table 5.3 to develop a *regression model* relating y to $x_1, x_2, \ldots, x_p$. In this section we study the *simple linear regression model.*

EXAMPLE 5.2

Consider the fuel consumption problem and the eight observed fuel consumptions listed in Table 5.2. In Example 5.1 we said that we could use the historical data in Table 5.2 to develop a prediction equation to predict y (weekly fuel consumption) on the basis of x_1 (average hourly temperature) and x_2 (the chill index).

In this example we consider a simpler problem—developing a prediction equation to predict y on the basis of x_1. Therefore, we will ignore the observed chill indices in Table 5.2 and consider Table 5.4, which lists the observed average hourly temperatures and the observed fuel consumptions. Our strategy here will be to construct a regression model relating y to x_1 and then to employ this model to develop our prediction equation. We will accomplish this in several steps.

TABLE 5.4 The Observed Average Hourly Temperatures and the Observed Fuel Consumptions

The *i*th Observed Week (OWK_i)	The *i*th Observed Average Hourly Temperature (°F) (x_{i1})	The *i*th Observed Fuel Consumption (Tons) (y_i)
OWK_1	$x_{11} = 28.0$	$y_1 = 12.4$
OWK_2	$x_{21} = 28.0$	$y_2 = 11.7$
OWK_3	$x_{31} = 32.5$	$y_3 = 12.4$
OWK_4	$x_{41} = 39.0$	$y_4 = 10.8$
OWK_5	$x_{51} = 45.9$	$y_5 = 9.4$
OWK_6	$x_{61} = 57.8$	$y_6 = 9.5$
OWK_7	$x_{71} = 58.1$	$y_7 = 8.0$
OWK_8	$x_{81} = 62.5$	$y_8 = 7.5$

Step 1. Expressing an Observation as the Sum of a Mean and an Error Term.

The *i*th observed fuel consumption, y_i, (1) is assumed to have been randomly selected from *the ith historical population of potential weekly fuel consumptions*, and (2) can be expressed as the sum of a mean and an *error term*. To make our presentation as concrete as possible, we discuss $y_5 = 9.4$, the fifth observed fuel consumption. It should be realized, however, that there is nothing special about the fifth observed fuel consumption. We could discuss any one of the eight observed fuel consumptions.

We begin by considering OWK_5, the fifth observed week. Theoretically speaking, associated with OWK_5 (in which the average hourly temperature was $x_{51} = 45.9$) there is an infinite population of potential weeks identical to OWK_5 in that during each of these potential weeks the average hourly temperature can be assumed to be $x_{51} = 45.9$. We call this population the *fifth historical population of potential weeks*, and we define a *unique characteristic* of any potential week in this population to be any characteristic of that week other than the average hourly temperature of $x_{51} = 45.9$ that affects weekly fuel consumption. For example, the average hourly wind velocity and the average hourly thermostat setting at State University are unique characteristics of any potential week. Now, corresponding to each potential week in the fifth historical population of potential weeks, there is a fuel consumption that we would have observed at State University if that potential week had occurred. Thus, corresponding to the fifth historical population of potential weeks, there is the *fifth historical population of potential weekly fuel consumptions*, which is the infinite population of potential weekly fuel consumptions that we could have observed when the average hourly temperature was $x_{51} = 45.9$. Thus, the unique characteristics (such as the average hourly wind velocities and the average hourly thermostat settings) of the potential weeks will cause at least some of the corresponding potential weekly fuel consumptions to be different from each other. For example, although the average hourly temperature is $x_{51} = 45.9$ for any two potential weeks, perhaps there would be a lower average hourly wind velocity and *thus a smaller fuel consumption* in one such potential week than in another such potential week.

We assume that $y_5 = 9.4$ is the fuel consumption we have actually observed when a *random selection* has been made from the fifth historical population of potential weekly fuel consumptions. Recall that in the Gas-Mizer problem of Chapter 2, since National Motors can produce as many Gas-Mizers as it desires with a four-cylinder engine, a standard transmission, and factory-installed air conditioning, National Motors can produce a finite subpopulation of the (potentially) infinite population of all such Gas-Mizers. Thus, National Motors can make a random selection from the finite subpopulation, which (we have previously stated) is equivalent to making a random selection from the infinite population of all possible Gas-Mizers. However, since we cannot control the weather, we cannot directly make a random selection from the infinite population of potential weekly fuel consumptions that we could observe when the average hourly temperature is $x_{51} = 45.9$. Nevertheless, it can be shown that if it is reasonable to believe that certain assumptions hold, then it is reasonable to believe that our assumption concerning the random selection of y_5 is valid. (See Chapter 6, Section 6.2, for the assumptions.)

Next, we note that we can express y_5 in the form

$$y_5 = \mu_5 + \epsilon_5$$

Here we assume the following:

1. μ_5 is the *mean* of the fifth historical population of potential weekly fuel consumptions. Intuitively, μ_5, which we call the *fifth mean fuel consumption,* is the average of all the potential weekly fuel consumptions that we could have observed when the average hourly temperature was $x_{51} = 45.9$. We do not know the true value of μ_5, because we do not know and cannot average all these potential weekly fuel consumptions.
2. ϵ_5, which we call the *fifth error term,* describes the effects on y_5 of all factors that have occurred in OWK_5, the fifth observed week, other than the average hourly temperature of $x_{51} = 45.9$. For example, as previously mentioned, the average hourly wind velocity and the average hourly thermostat setting are two such other factors. Since the equation $y_5 = \mu_5 + \epsilon_5$ implies that $\epsilon_5 = y_5 - \mu_5$, we see that ϵ_5 measures the distance between y_5 and μ_5, and thus measures how far y_5 deviates from μ_5. It should be noted that although we have observed $y_5 = 9.4$, since we do not know the true value of μ_5, we do not know the true value of ϵ_5. Hence, we say that we *cannot observe* the true value of ϵ_5. This is why we call ϵ_5 the *fifth error term* but not the fifth *observed* error term.

Now, generalizing the preceding discussion concerning y_5, and considering all eight observed fuel consumptions in Table 5.4, we assume that for $i = 1, 2, \ldots, 8$, *the ith observed fuel consumption,* y_i, is the fuel consumption that we have actually observed when a *random selection* has been made from *the ith historical population of potential weekly fuel consumptions,* which is the infinite population of potential weekly fuel consumptions that we could have observed when the average hourly temperature was x_{i1}. Moreover, we can express y_i in the form

$$y_i = \mu_i + \epsilon_i$$

Here we assume the following:

1. μ_i is the *mean* of the *i*th historical population of potential weekly fuel consumptions. Intuitively, μ_i, which we call *the ith mean fuel consumption,* is the average of all of the potential weekly fuel consumptions that we could have observed when the average hourly temperature was x_{i1}. This mean represents the effect on y_i of the average hourly temperature x_{i1}.

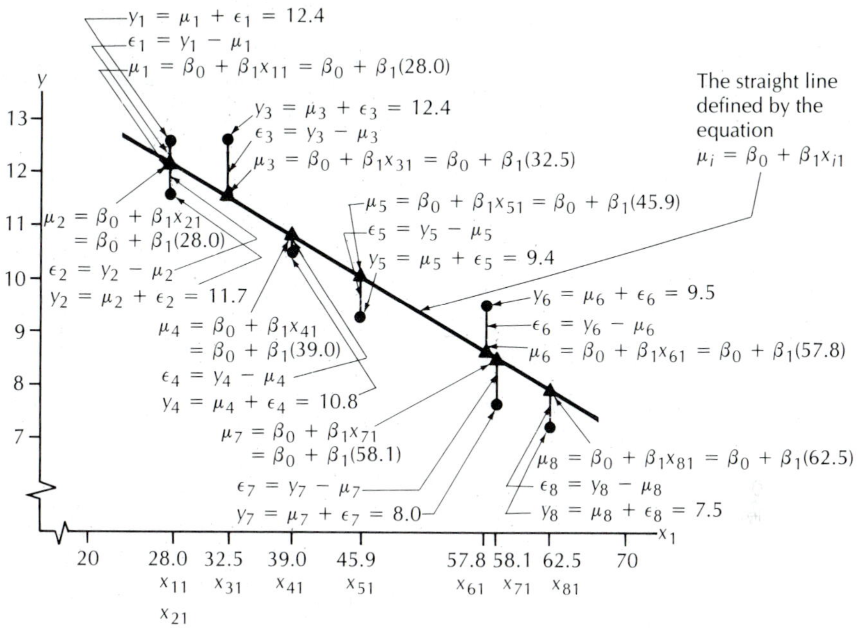

FIGURE 5.1 The Simple Linear Regression Model Relating y (Weekly Fuel Consumption) to x_1 (Average Hourly Temperature)

2. ϵ_i, which we call *the ith error term*, describes the effects on y_i of all factors that have occurred in OWK_i, the ith observed week, other than the average hourly temperature of x_{i1}.

Step 2. Constructing a Regression Model Relating y to x_1.

Having shown that we can express the eight observed fuel consumptions in the form

$$y_1 = \mu_1 + \epsilon_1$$
$$y_2 = \mu_2 + \epsilon_2$$
$$\vdots \qquad \vdots$$
$$y_8 = \mu_8 + \epsilon_8$$

we next plot these y values against the increasing values of x_1, average hourly temperature (see Figure 5.1). We see that the y values (dots in the figure) tend to decrease in a straight line fashion as the values of x_1 increase. Although the eight points

$$(x_{11}, y_1)$$
$$(x_{21}, y_2)$$
$$\vdots$$
$$(x_{81}, y_8)$$

representing the eight combinations of the eight observed average hourly temperatures and the eight observed fuel consumptions, do not exactly lie on a straight line, we can define a *regression model relating the y's to the* x_1*'s* by assuming that the eight points

$$(x_{11}, \mu_1)$$
$$(x_{21}, \mu_2)$$
$$\vdots$$
$$(x_{81}, \mu_8)$$

representing the eight combinations of the eight observed average hourly temperatures and the eight *mean* fuel consumptions (triangles in the figure) *do lie on a straight line*. We denote this straight line by the equation

$$\mu_i = \beta_0 + \beta_1 x_{i1}$$

Here, μ_i is the ith mean fuel consumption, x_{i1} is the ith observed average hourly temperature, β_0 is the ***y*-intercept** of the straight line, and β_1 is the **slope** of the straight line. To interpret the meaning of the *y*-intercept, β_0, assume that $x_{i1} = 0$. Then

$$\begin{aligned}\mu_i &= \beta_0 + \beta_1 x_{i1} \\ &= \beta_0 + \beta_1(0) = \beta_0\end{aligned}$$

So, β_0 is the mean weekly fuel consumption for all potential weeks having an average hourly temperature of 0°F (we later discuss the fact that this interpretation of β_0 is of dubious practical value). In order to interpret the meaning of the slope, β_1, consider two different weeks. Suppose that for the first week average hourly temperature is c. The mean weekly fuel consumption for all such potential weeks, denoted by $\mu_{(c)}$, is

$$\mu_{(c)} = \beta_0 + \beta_1(c)$$

For the second week, suppose that the average hourly temperature is $c + 1$. The mean weekly fuel consumption for all such potential weeks, denoted by $\mu_{(c+1)}$, is

$$\mu_{(c+1)} = \beta_0 + \beta_1(c + 1)$$

The difference between the mean weekly fuel consumptions $\mu_{(c+1)}$ and $\mu_{(c)}$ is

$$\begin{aligned}\mu_{(c+1)} - \mu_{(c)} &= [\beta_0 + \beta_1(c + 1)] - [\beta_0 + \beta_1(c)] \\ &= \beta_1\end{aligned}$$

Thus, the slope, β_1, is the change in the mean weekly fuel consumption that is associated with a 1-degree increase in average hourly temperature. Note that since we do not know the true values of β_0 and β_1, the position of the straight line depicted in Figure 5.1 is only hypothetical. However, we do know that β_1 must be a negative number, because this would say that the mean weekly fuel consumption decreases as the average hourly temperature increases (as shown in Figure 5.1).

Since we are assuming that the ith mean fuel consumption, μ_i, equals $\beta_0 + \beta_1 x_{i1}$, we can describe the ith observed fuel consumption

$$y_i = \mu_i + \epsilon_i$$

by using the equation

$$\begin{aligned} y_i &= \mu_i + \epsilon_i \\ &= \beta_0 + \beta_1 x_{i1} + \epsilon_i \end{aligned}$$

We refer to this equation as the **simple linear** (or **straight line**) **regression model** relating y_i to x_{i1}, and we call β_0 and β_1 the parameters of the model. As illustrated in Figure 5.1, this model says that the eight error terms

$$\epsilon_1, \epsilon_2, \ldots, \epsilon_8$$

cause the eight observed fuel consumptions (dots in figure)

$$\begin{aligned} y_1 &= \mu_1 + \epsilon_1 = \beta_0 + \beta_1 x_{11} + \epsilon_1 \\ y_2 &= \mu_2 + \epsilon_2 = \beta_0 + \beta_1 x_{21} + \epsilon_2 \\ &\vdots \qquad\qquad \vdots \\ y_8 &= \mu_8 + \epsilon_8 = \beta_0 + \beta_1 x_{81} + \epsilon_8 \end{aligned}$$

to deviate from the eight mean fuel consumptions (triangles in figure)

$$\begin{aligned} \mu_1 &= \beta_0 + \beta_1 x_{11} \\ \mu_2 &= \beta_0 + \beta_1 x_{21} \\ &\vdots \qquad \vdots \\ \mu_8 &= \beta_0 + \beta_1 x_{81} \end{aligned}$$

which exactly lie on the straight line defined by the equation

$$\mu_i = \beta_0 + \beta_1 x_{i1}$$

To simplify notation, we sometimes omit the subscript i (which refers to the particular observed fuel consumption we are considering) and write the simple linear regression model as

$$\begin{aligned} y &= \mu + \epsilon \\ &= \beta_0 + \beta_1 x_1 + \epsilon \end{aligned}$$

TABLE 5.5 The Calculation of the Point Estimates b_0 and b_1 of the Parameters in the Fuel Consumption Model $y_i = \mu_i + \epsilon_i = \beta_0 + \beta_1 x_{i1} + \epsilon_i$

y_i	x_{i1}	x_{i1}^2	$y_i x_{i1} = x_{i1} y_i$
12.4	28.0	$(28.0)^2$	(12.4)(28.0)
11.7	28.0	$(28.0)^2$	(11.7)(28.0)
12.4	32.5	$(32.5)^2$	(12.4)(32.5)
10.8	39.0	$(39.0)^2$	(10.8)(39.0)
9.4	45.9	$(45.9)^2$	(9.4)(45.9)
9.5	57.8	$(57.8)^2$	(9.5)(57.8)
8.0	58.1	$(58.1)^2$	(8.0)(58.1)
7.5	62.5	$(62.5)^2$	(7.5)(62.5)

$$\sum_{i=1}^{8} y_i = 81.7 \qquad \sum_{i=1}^{8} x_{i1} = 351.8$$

$$\sum_{i=1}^{8} x_{i1}^2 = 16{,}874.76 \qquad \sum_{i=1}^{8} x_{i1} y_i = 3413.11$$

$$\bar{y} = \frac{81.7}{8} = 10.21 \qquad \bar{x}_1 = \frac{351.8}{8} = 43.98$$

$$b_1 = \frac{8\sum_{i=1}^{8} x_{i1}y_i - \left(\sum_{i=1}^{8} x_{i1}\right)\left(\sum_{i=1}^{8} y_i\right)}{8\sum_{i=1}^{8} x_{i1}^2 - \left(\sum_{i=1}^{8} x_{i1}\right)^2}$$

$$= \frac{8(3413.11) - (351.8)(81.7)}{8(16{,}874.76) - (351.8)^2} = -.1279$$

$$b_0 = \bar{y} - b_1\bar{x}_1 = 10.21 - (-.1279)(43.98)$$

$$= 15.84$$

Step 3. Calculating Least Squares Point Estimates of the Regression Parameters.

Although we do not know the true values of the parameters β_0 and β_1, we can use the observed average hourly temperatures and the observed fuel consumptions in Table 5.4 to calculate point estimates b_0 and b_1 of these parameters by utilizing the formulas presented in Table 5.5. Since Table 5.5 indicates that $b_0 = 15.84$ is the point estimate of β_0, we estimate that the mean weekly fuel consumption for all potential weeks having an average hourly temperature of 0°F is 15.84 tons of coal. Since Table 5.5 indicates that $b_1 = -.1279$ is the point estimate of β_1, we estimate that mean weekly fuel consumption decreases (since b_1 is negative) by .1279 tons of coal when average hourly temperature increases by 1 degree.

We now discuss why we use the formulas in Table 5.5 to calculate the point estimates b_0 and b_1. This can be understood by using b_0 and b_1 to predict, for $i = 1, 2, \ldots, 8$, y_i, the ith observed fuel consumption, on the basis of x_{i1}, the ith observed average hourly temperature.

Note that the ith observed fuel consumption

$$\begin{aligned} y_i &= \mu_i + \epsilon_i \\ &= \beta_0 + \beta_1 x_{i1} + \epsilon_i \end{aligned}$$

is the sum of the ith mean fuel consumption, $\mu_i = \beta_0 + \beta_1 x_{i1}$, and the ith error term, ϵ_i. Denoting the point estimate of μ_i by $\hat{y}_i$, it seems intuitive to calculate $\hat{y}_i$ by replacing the parameters β_0 and β_1 with their respective point estimates b_0 and b_1. Thus, the point estimate of μ_i is

$$\hat{y}_i = b_0 + b_1 x_{i1}$$

In order to see how to predict the ith error term, ϵ_i, remember that we do not know the true value of $\epsilon_i = y_i - \mu_i$, since we do not know the true value of μ_i. Under several assumptions discussed in Chapter 6—assumptions that imply that ϵ_i has a 50 percent chance of being positive (which is true if y_i is greater than μ_i) and a 50 percent chance of being negative (which is true if y_i is less than μ_i)—it is customary to predict ϵ_i to be zero. Doing this implies that

$$\hat{y}_i = b_0 + b_1 x_{i1}$$

the point estimate of the ith mean fuel consumption, $\mu_i = \beta_0 + \beta_1 x_{i1}$, is also the point prediction of the ith observed fuel consumption,

$$\begin{aligned} y_i &= \mu_i + \epsilon_i \\ &= \beta_0 + \beta_1 x_{i1} + \epsilon_i \end{aligned}$$

Next, we define, for $i = 1, 2, \ldots, 8$, *the* i*th residual* to be $e_i = y_i - \hat{y}_i$, which is the difference between y_i, the ith observed fuel consumption, and $\hat{y}_i$, *the* i*th predicted fuel consumption*, and we define the **sum of squared residuals** to be

$$\begin{aligned} SSE &= \sum_{i=1}^{8} e_i^2 = \sum_{i=1}^{8} (y_i - \hat{y}_i)^2 \\ &= \sum_{i=1}^{8} (y_i - (b_0 + b_1 x_{i1}))^2 \end{aligned}$$

Intuitively, if any particular values of b_0 and b_1 are good point estimates of β_0 and β_1, they will, for $i = 1, 2, \ldots, 8$, make the ith predicted fuel consumption,

$$\hat{y}_i = b_0 + b_1 x_{i1}$$

TABLE 5.6 The Values of *SSE* Given by the Least Squares Point Estimates $b_0 = 15.84$ and $b_1 = -.1279$ and by the Point Estimates $b_0 = 16.22$ and $b_1 = -.1152$

Predictions Using the Least Squares Point Estimates $b_0 = 15.84$ and $b_1 = -.1279$

The *i*th Observed Week (OWK_i)	The *i*th Observed Average Hourly Temperature (°F) (x_{i1})	The *i*th Observed Fuel Consumption (Tons) (y_i)	The *i*th Predicted Fuel Consumption ($\hat{y}_i = b_0 + b_1x_{i1} = 15.84 - .1279x_{i1}$)	The *i*th Residual ($e_i = y_i - \hat{y}_i$)
OWK_1	28.0	12.4	12.2560	.1440
OWK_2	28.0	11.7	12.2560	−.5560
OWK_3	32.5	12.4	11.6804	.7196
OWK_4	39.0	10.8	10.8489	−.0489
OWK_5	45.9	9.4	9.9663	−.5663
OWK_6	57.8	9.5	8.4440	1.0560
OWK_7	58.1	8.0	8.4056	−.4056
OWK_8	62.5	7.5	7.8428	−.3428

$$SSE = \sum_{i=1}^{8} e_i^2 = 2.57$$

Predictions Using the Point Estimates $b_0 = 16.22$ and $b_1 = -.1152$

The *i*th Observed Week (OWK_i)	The *i*th Observed Average Hourly Temperature (°F) (x_{i1})	The *i*th Observed Fuel Consumption (Tons) (y_i)	The *i*th Predicted Fuel Consumption ($\hat{y}_i = b_0 + b_1x_{i1} = 16.22 - .1152x_{i1}$)	The *i*th Residual ($e_i = y_i - \hat{y}_i$)
OWK_1	28.0	12.4	12.9944	−.5944
OWK_2	28.0	11.7	12.9944	−1.2944
OWK_3	32.5	12.4	12.4760	−.0760
OWK_4	39.0	10.8	11.7272	−.9272
OWK_5	45.9	9.4	10.9323	−1.5323
OWK_6	57.8	9.5	9.5614	−.0614
OWK_7	58.1	8.0	9.5269	−1.5269
OWK_8	62.5	7.5	9.0200	−1.5200

$$SSE = \sum_{i=1}^{8} e_i^2 = 9.8877$$

fairly close to the ith observed fuel consumption,

$$y_i = \beta_0 + \beta_1 x_{i1} + \epsilon_i$$

Thus, the ith residual, $e_i = y_i - \hat{y}_i$, will be fairly small, and the sum of squared residuals, SSE, will be fairly small.

We use the formulas in Table 5.5 to calculate the point estimates b_0 and b_1 of the parameters β_0 and β_1 because it can be proved (see Appendix B) that the point estimates b_0 and b_1 given by these formulas give a value of SSE that is smaller than would be given by any other values of b_0 and b_1. Since the point estimates b_0 and b_1 given by the Table 5.5 formulas minimize SSE, we call these point estimates the **least squares point estimates**.

In order to illustrate the fact that the least squares point estimates $b_0 = 15.84$ and $b_1 = -.1279$ minimize

$$SSE = \sum_{i=1}^{8} (y_i - \hat{y}_i)^2$$

$$= \sum_{i=1}^{8} (y_i - (b_0 + b_1 x_{i1}))^2$$

consider Table 5.6. Comparing the upper and lower parts of the table, we see that $SSE = 2.57$, the value given by the least squares point estimates $b_0 = 15.84$ and $-.1279$, is less than $SSE = 9.8877$, the value given by $b_0 = 16.22$ and $b_1 = -.1152$, which are not the least squares point estimates. Hence, using one set of values of b_0 and b_1 different from the least squares point estimates, we have illustrated that the least squares point estimates give a value of SSE smaller than would be given by any other values of b_0 and b_1. (There is nothing special about the values 16.22 and $-.1152$. We could choose *any* values of b_0 and b_1 different from the least squares point estimates and obtain a similar result. See Appendix B for proof of this.)

Step 4. Writing the Prediction Equation for Predicting y on the Basis of x_1. Since

$$\hat{y}_i = b_0 + b_1 x_{i1}$$
$$= 15.84 - .1279 x_{i1}$$

is the point prediction of y_i, the ith observed fuel consumption, we call the equation

$$\hat{y} = b_0 + b_1 x_1$$
$$= 15.84 - .1279 x_1$$

a prediction equation. Examining Figure 5.2, we see that whereas the eight predicted fuel consumptions $\hat{y}_1, \hat{y}_2, \ldots, \hat{y}_8$ (squares in the figure) lie exactly on the straight line defined by this prediction equation, the eight observed fuel consumptions $y_1, y_2, \ldots, y_8$ (dots in the figure) fluctuate around the straight line

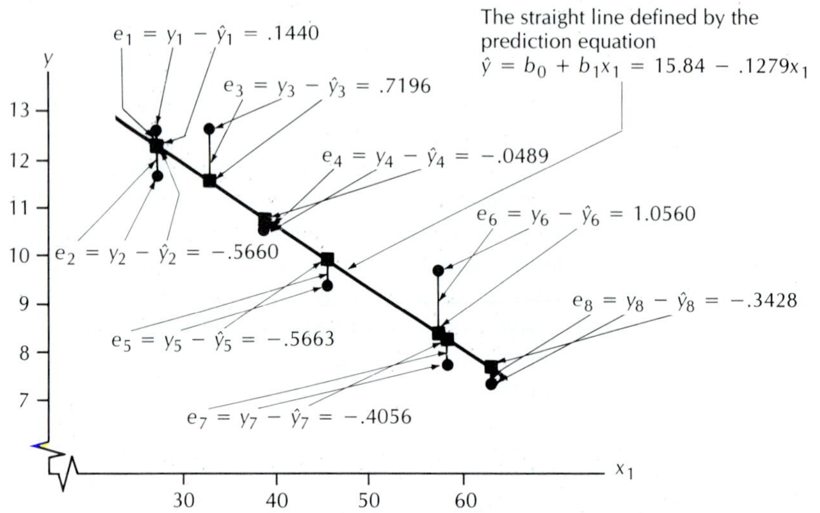

FIGURE 5.2
The Eight Residuals Given by the Prediction Equation $\hat{y} = b_0 + b_1x_1 = 15.84 - .1279x_1$

defined by this prediction equation. Moreover, the eight line segments drawn between the squares and dots are the eight distances between the predicted fuel consumptions and the observed fuel consumptions and thus represent the sizes of the eight residuals $e_1, e_2, \ldots, e_8$. It follows that we can interpret

$$
\begin{aligned}
SSE &= \sum_{i=1}^{8} e_i^2 \\
&= \sum_{i=1}^{8} (y_i - \hat{y}_i)^2 \\
&= \sum_{i=1}^{8} (y_i - (b_0 + b_1x_{i1}))^2 \\
&= \sum_{i=1}^{8} (y_i - (15.84 - .1279x_{i1}))^2 \\
&= (.1440)^2 + (-.5560)^2 + \cdots + (-.3428)^2 = 2.57
\end{aligned}
$$

which is the sum of squared residuals given by the least squares point estimates, as the sum of squared distances between the observed fuel consumptions and the predicted fuel consumptions calculated by using the straight line prediction equation

$$
\begin{aligned}
\hat{y} &= b_0 + b_1x_1 \\
&= 15.84 - .1279x_1
\end{aligned}
$$

Since the least squares point estimates minimize *SSE,* we can interpret them as positioning the straight line prediction equation so that the sum of squared distances between the observed fuel consumptions and the predicted fuel consumptions given by *this* prediction equation is smaller than the sum of squared

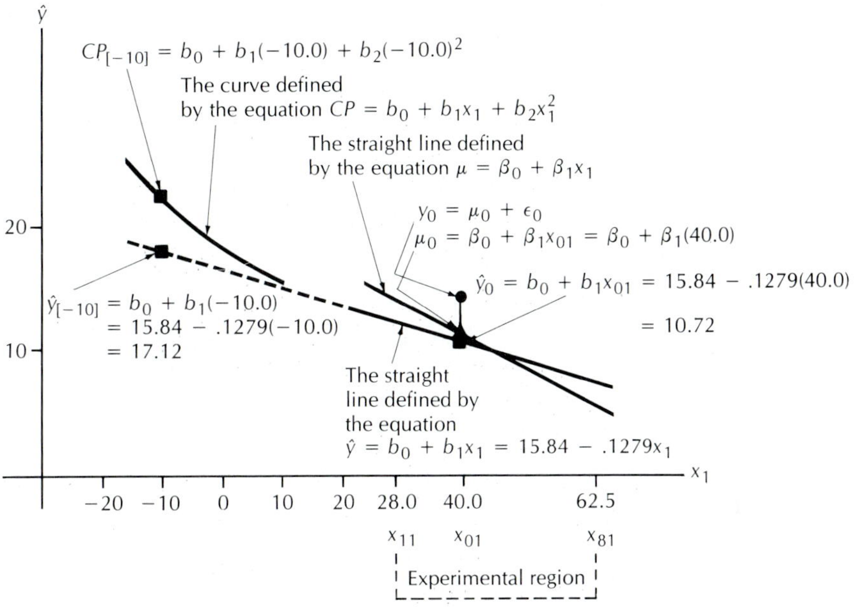

FIGURE 5.3
A Comparison of the Straight Lines Defined by the Equations $\mu = \beta_0 + \beta_1 x_1$ and $\hat{y} = b_0 + b_1 x_1$

distances between the observed fuel consumptions and the predicted fuel consumptions that would be given by any other straight line prediction equation that uses any other values of b_0 and b_1 than the least squares values. In this sense, then, we can say that the prediction equation which employs the least squares point estimates b_0 and b_1 is the best straight line prediction equation that can be drawn through the points representing the combinations of the observed average hourly temperatures and the observed fuel consumptions.

Next, comparing the straight lines in Figures 5.1 and 5.2, we see that the straight line in Figure 5.2, which connects the eight predicted fuel consumptions $\hat{y}_1, \hat{y}_2, \ldots, \hat{y}_8$ and which is defined by the prediction equation

$$\hat{y} = b_0 + b_1 x_1$$
$$= 15.84 - .1279x_1$$

is the *estimate* of the straight line in Figure 5.1, which connects the eight mean fuel consumptions $\mu_1, \mu_2, \ldots, \mu_8$ and which is defined by the equation

$$\mu = \beta_0 + \beta_1 x_1$$

Furthermore, as illustrated in Figure 5.3, since the least squares point estimates $b_0 = 15.84$ and $b_1 = -.1279$ differ (unless we are extremely lucky) from the true values of the parameters β_0 and β_1, the straight line defined by the prediction equation

$$\hat{y} = b_0 + b_1 x_1$$
$$= 15.84 - .1279x_1$$

will be different from the straight line defined by the equation

$$\mu = \beta_0 + \beta_1 x_1$$

Note, however, that the relative positions of the straight lines in Figure 5.3 are only hypothetical, because we do not know the true values of the *y*-intercept, β_0, and the slope, β_1.

Step 5. Predicting for a Future Week.

Now, suppose that we wish to predict y_0, the amount of fuel (in tons of coal) that will be used to heat buildings at State University during the future week, when the average hourly temperature will be $x_{01} = 40.0$. We can express this future (individual) fuel consumption by the equation

$$y_0 = \mu_0 + \epsilon_0$$

Here we assume the following:

1. y_0 is the fuel consumption we will actually observe when a random selection is made from the *future population of potential weekly fuel consumptions,* which is the infinite population of potential weekly fuel consumptions that we could observe when the average hourly temperature will be $x_{01} = 40.0$.
2. μ_0 is the *mean* of the future population of potential weekly fuel consumptions. Intuitively, μ_0, which we call the *future mean fuel consumption,* is the average of all the potential weekly fuel consumptions that we could observe when the average hourly temperature will be $x_{01} = 40.0$.
3. ϵ_0, which we call the *future error term,* describes the effects on y_0 of all factors that will occur in the future week other than the average hourly temperature of $x_{01} = 40.0$.

Since we have concluded that the straight line defined by the equation

$$\mu = \beta_0 + \beta_1 x_1$$

relates the eight mean fuel consumptions $\mu_1, \mu_2, \ldots, \mu_8$ to the eight observed average hourly temperatures $x_{11}, x_{21}, \ldots, x_{81}$, and since, as illustrated in Figure 5.3, $x_{01} = 40.0$, the future average hourly temperature, is in the **experimental region** (which we define to be the range of the eight observed average hourly temperatures—28.0°F to 62.5°F), then it intuitively follows that this straight line (or equation) also relates the future mean fuel consumption μ_0 to $x_{01} = 40.0$. That is, since $x_{01} = 40.0$ is in the experimental region, it seems reasonable to conclude that (1) the equation

$$\begin{aligned}\mu_0 &= \beta_0 + \beta_1 x_{01} \\ &= \beta_0 + \beta_1(40.0)\end{aligned}$$

relates μ_0, the future mean fuel consumption, to $x_{01} = 40.0$, and thus (2) the future (individual) fuel consumption

$$y_0 = \mu_0 + \epsilon_0$$

can be described by the equation

$$\begin{aligned} y_0 &= \mu_0 + \epsilon_0 \\ &= \beta_0 + \beta_1 x_{01} + \epsilon_0 \\ &= \beta_0 + \beta_1(40.0) + \epsilon_0 \end{aligned}$$

Thus,

$$\begin{aligned} \hat{y}_0 &= b_0 + b_1 x_{01} \\ &= 15.84 - .1279(40.0) \\ &= 10.72 \text{ tons of coal} \end{aligned}$$

is the point estimate of μ_0 and is the point prediction of y_0. The logic behind this estimation and prediction is that, if we replace the parameters β_0 and β_1 by their respective least squares point estimates b_0 and b_1, then $\hat{y}_0$ is the obvious point estimate of μ_0. Also, since we predict the future error term, ϵ_0, to be zero, $\hat{y}_0$ is also the point prediction of $y_0 = \mu_0 + \epsilon_0$.

Note that although the relative positions of $\hat{y}_0 = 10.72$, $\mu_0 = \beta_0 + \beta_1 x_{01} = \beta_0 + \beta_1(40.0)$, and $y_0 = \mu_0 + \epsilon_0$ in Figure 5.3 are only hypothetical (since we do not know the true values of β_0 and β_1, or the value of the future error term ϵ_0), Figure 5.3 does illustrate the fact that (unless we are extremely fortunate) $\hat{y}_0$ will differ from both μ_0 and $y_0 = \mu_0 + \epsilon_0$. In Chapter 6 we discuss how to calculate a bound for the farthest $\hat{y}_0$ might be from μ_0, and a bound for the farthest $\hat{y}_0$ might be from y_0.

To conclude this example, we note that it would be dangerous to use the straight line prediction equation

$$\begin{aligned} \hat{y} &= b_0 + b_1 x_1 \\ &= 15.84 - .1279 x_1 \end{aligned}$$

to predict a future value of y (weekly fuel consumption) on the basis of a future value of x_1 (average hourly temperature) that is far outside the experimental region. For example, -10.0 (in Figure 5.3) is a future value of x_1 that is far outside the experimental region. Thus, while Figure 5.1 indicates that, for values of x_1 in the experimental region (that is, between 28.0 and 62.5), the observed values of y tend to decrease in a straight line fashion as the values of x_1 increase, there are no observed data to tell us whether the observed values of y tend to decrease in a straight line fashion as the values of x_1 increase from, say, -20.0 to 20.0 (a range outside the experimental region). If, for example, we observe values of x_1 between -20.0 and 20.0, and if the corresponding observed values of y tend to fluctuate around the *curve* illustrated in Figure 5.3, then instead of using a straight line prediction equation, we should use a curved prediction equation such as

$$CP = b_0 + b_1x_1 + b_2x_1^2$$

to predict a future value of y on the basis of a future value of $x_1 = -10.0$. We discuss the exact nature of a curved prediction equation in Section 5.6. For now, suffice it to say that b_0, b_1, and b_2 are least squares point estimates that would be computed from the observed data and that would relate CP, the prediction of y (weekly fuel consumption), to x_1 (average hourly temperature) and to x_1^2 (the square of average hourly temperature). If we use the prediction equation

$$CP = b_0 + b_1x_1 + b_2x_1^2$$

to make a prediction of a future value of y occurring during a future week having an average hourly temperature equal to -10.0, and denote this prediction using the symbol $CP_{[-10]}$, it follows that the prediction is (as illustrated in Figure 5.3)

$$CP_{[-10]} = b_0 + b_1(-10.0) + b_2(-10.0)^2$$

To see why it is dangerous to use the straight line prediction equation

$$\begin{aligned}\hat{y} &= b_0 + b_1x_1 \\ &= 15.84 - .1279x_1\end{aligned}$$

to predict a future value of y on the basis of a future value of x_1 that is far outside the experimental region, note from Figure 5.3 that if this straight line prediction equation were extrapolated below 28.0 (the lower bound of the experimental region of 28.0 to 62.5) we would obtain the prediction

$$\begin{aligned}\hat{y}_{[-10]} &= b_0 + b_1(-10.0) \\ &= 15.84 - .1279(-10.0) \\ &= 17.12\end{aligned}$$

which (as illustrated in Figure 5.3) is far below the prediction

$$CP_{[-10]} = b_0 + b_1(-10.0) + b_2(-10.0)^2$$

given by the correct curved prediction equation

$$CP = b_0 + b_1x_1 + b_2x_1^2$$

Finally, recall that we have interpreted the y-intercept β_0 of the regression model

$$\begin{aligned}y &= \mu + \epsilon \\ &= \beta_0 + \beta_1x_1 + \epsilon\end{aligned}$$

to be the mean weekly fuel consumption for all potential weeks having an average hourly temperature of 0°F. However, since zero degrees is not in the experimental region, we do not know whether this regression model describes weeks having an average hourly temperature of 0°F, and thus this interpretation of β_0 is of dubious practical value.

As illustrated in the preceding example, we sometimes need to use the simple linear (or straight line) regression model to relate a dependent variable y to a single independent variable x. Assume, then, that we have observed n situations

$$OST_1, OST_2, \ldots, OST_n$$

and the n values of the independent variable x

$$x_1, x_2, \ldots, x_n$$

and the n values of the dependent variable y

$$y_1, y_2, \ldots, y_n$$

that have occurred during these n situations. In the following box we summarize some key results pertaining to the simple linear regression model.

The *simple linear (or straight line) regression model* is

$$\begin{aligned} y_i &= \mu_i + \epsilon_i \\ &= \beta_0 + \beta_1 x_i + \epsilon_i \end{aligned}$$

1. *The ith mean value of the dependent variable,* $\mu_i = \beta_0 + \beta_1 x_i$, is the average of all the potential values of the dependent variable that we could have observed when the value of the independent variable, x, was x_i. This mean describes the effect on y_i of the value x_i of the independent variable x.
2. *The ith error term,* ϵ_i, describes the effects on y_i of all factors that have occurred in OST_i, the ith observed situation, other than the value x_i of the independent variable x.
3. *The y-intercept* β_0 is the average of all the potential values of the dependent variable that we could observe when the value of the independent variable equals zero.
4. *The slope* β_1 is the change in the mean value of the dependent variable that is associated with a one-unit increase in the value of the independent variable. If β_1 is positive, the mean value of the dependent variable increases as the value of the independent variable increases (see Figure 5.4a). If β_1 is

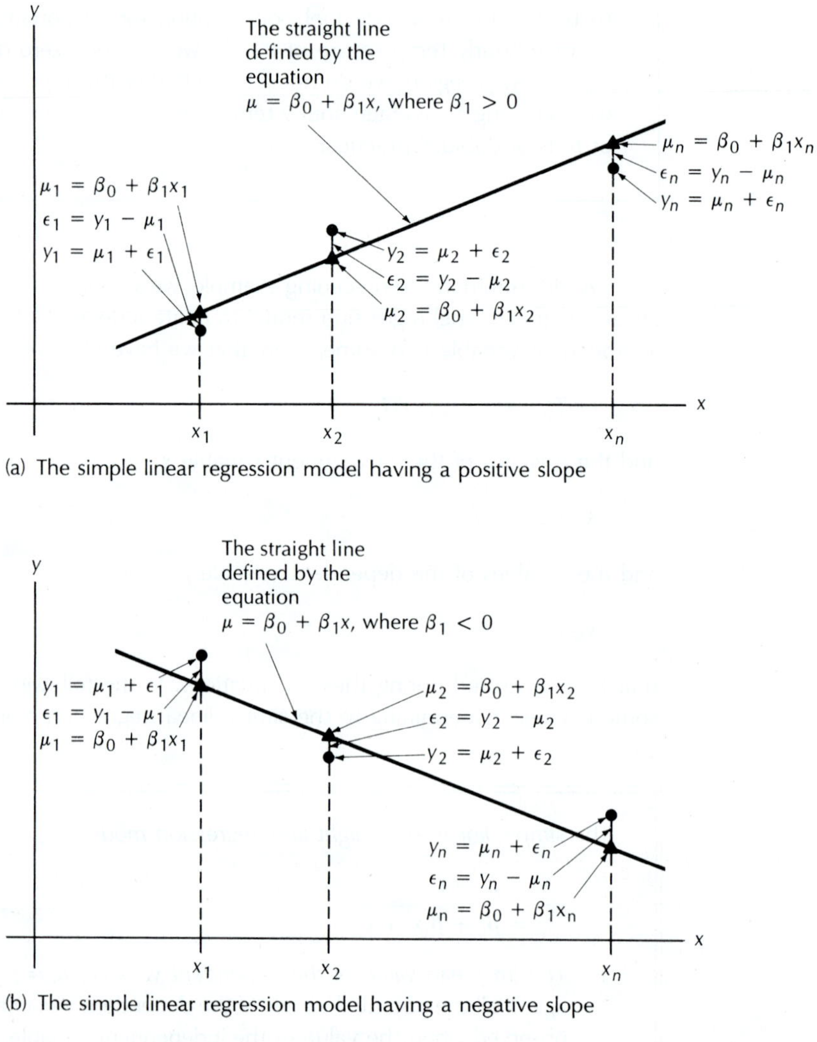

FIGURE 5.4
The Simple Linear Regression Model $y_i = \mu_i + \epsilon_i = \beta_0 + \beta_1 x_i + \epsilon_i$

negative, the mean value of the dependent variable decreases as the value of the independent variable increases (see Figure 5.4b).

5. *The least squares point estimates* b_0 and b_1 of the parameters β_0 and β_1 are the values of b_0 and b_1 that minimize the sum of squared residuals

$$SSE = \sum_{i=1}^{n} e_i^2 = \sum_{i=1}^{n} (y_i - \hat{y}_i)^2$$

$$= \sum_{i=1}^{n} (y_i - (b_0 + b_1 x_i))^2$$

and are calculated by using the equations

$$b_1 = \frac{n\sum_{i=1}^{n} x_i y_i - \left(\sum_{i=1}^{n} x_i\right)\left(\sum_{i=1}^{n} y_i\right)}{n\sum_{i=1}^{n} x_i^2 - \left(\sum_{i=1}^{n} x_i\right)^2}$$

$$b_0 = \bar{y} - b_1\bar{x}$$

where

$$\bar{y} = \frac{\sum_{i=1}^{n} y_i}{n} \quad \text{and} \quad \bar{x} = \frac{\sum_{i=1}^{n} x_i}{n}$$

6. If x_0, the future value of the independent variable x, is in the experimental region (the range of the n observed values $x_1, x_2, \ldots, x_n$), then

$$\hat{y}_0 = b_0 + b_1 x_0$$

is the *point estimate of the future mean value of the dependent variable*

$$\mu_0 = \beta_0 + \beta_1 x_0$$

and the *point prediction of the future (individual) value of the dependent variable*

$$\begin{aligned} y_0 &= \mu_0 + \epsilon_0 \\ &= \beta_0 + \beta_1 x_0 + \epsilon_0 \end{aligned}$$

EXAMPLE 5.3

Again consider the fuel consumption problem and the eight observed fuel consumptions listed in Table 5.2. In this example we define a regression model relating y (weekly fuel consumption) to x_2 (the chill index). Therefore, we ignore the observed average hourly temperatures in Table 5.2 and consider Table 5.7, in which we have arranged the values of x_2 in increasing order, along with the corresponding values of y. Since Figure 5.5 indicates that the y values (dots in the figure) tend to increase in a straight line fashion as the values of x_2 increase, we conclude that the simple linear regression model having a positive slope ($\beta_1 > 0$)

$$\begin{aligned} y_i &= \mu_i + \epsilon_i \\ &= \beta_0 + \beta_1 x_{i2} + \epsilon_i \end{aligned}$$

TABLE 5.7 Values of x_2, the Chill Index, Arranged in Increasing Order

The *i*th Observed Chill Index (x_{i2})	The *i*th Observed Fuel Consumption (Tons) (y_i)
0	7.5
1	8.0
8	9.4
14	11.7
16	9.5
18	12.4
22	10.8
24	12.4

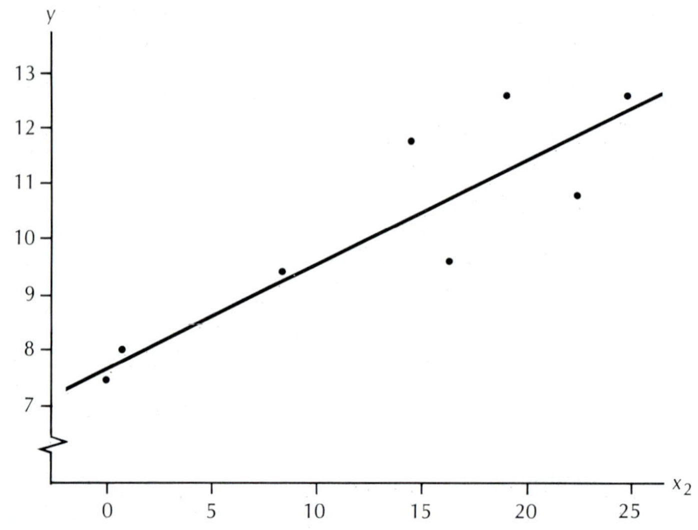

FIGURE 5.5 Plot of the Observed Values of y (Weekly Fuel Consumption) Against Increasing Values of x_2 (the Chill Index)

is a reasonable model relating y_i to x_{i2}. Recall that in Example 5.2 we concluded that the simple linear regression model having a negative slope ($\beta_1 < 0$)

$$\begin{aligned} y_i &= \mu_i + \epsilon_i \\ &= \beta_0 + \beta_1 x_{i1} + \epsilon_i \end{aligned}$$

is a reasonable model relating y_i to x_{i1}. In Section 5.3 we combine the two models to arrive at a single regression model relating y_i to x_{i1} and x_{i2}.

EXAMPLE 5.4

Comp-U-Systems, a new computer manufacturer, sells and services the Comp-U-Systems Microcomputer. As part of its standard purchase contract, Comp-U-Systems agrees to perform regular service on its microcomputer. Because of its lack of experience in servicing computers, Comp-U-Systems wishes to obtain information concerning the time it takes to perform the required service. To this end, Comp-U-Systems has collected data for its first eleven service calls. These data are presented in Table 5.8, which lists the number of microcomputers serviced (x) and the time (in minutes) required to perform the needed service (y) for each of these service calls. Here, for $i = 1, 2, \ldots, 11$,

x_i = the number of microcomputers serviced on the ith service call

y_i = the number of minutes required to perform service on the ith service call

In Figure 5.6 the observed values of y are plotted against the increasing values of x. We see that as the values of x increase, the observed values of y tend to

TABLE 5.8 Comp-U-Systems Service Call Data

Service Call (i)	Number of Microcomputers Serviced (x_i)	Number of Minutes Required on Call (y_i)
1	4	109
2	2	58
3	5	138
4	7	189
5	1	37
6	3	82
7	4	103
8	5	134
9	2	68
10	4	112
11	6	154

increase in a straight line fashion. This implies that the simple linear regression model having a positive slope ($\beta_1 > 0$)

$$\begin{aligned} y &= \mu + \epsilon \\ &= \beta_0 + \beta_1 x + \epsilon \end{aligned}$$

might be an appropriate regression model relating y to x.

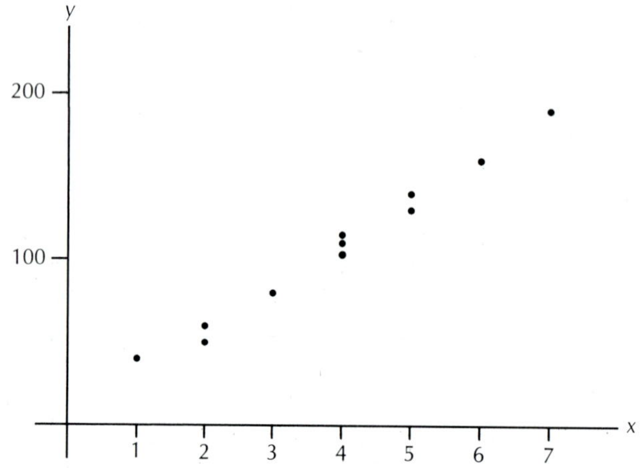

FIGURE 5.6
Plot of Comp-U-Systems Service Call Data

By using the data in Table 5.8, we can calculate the least squares point estimates b_0 and b_1 of β_0 and β_1 to be

$$b_1 = \frac{11 \sum_{i=1}^{11} x_i y_i - \left(\sum_{i=1}^{11} x_i\right)\left(\sum_{i=1}^{11} y_i\right)}{11 \sum_{i=1}^{11} x_i^2 - \left(\sum_{i=1}^{11} x_i\right)^2} = 24.6022$$

$$b_0 = \bar{y} - b_1\bar{x} = 11.4641$$

Since β_1, the slope of the regression model, is the change in the mean service time (in minutes) associated with each additional microcomputer serviced on a call, the least squares point estimate $b_1 = 24.6022$ of β_1 says that we estimate that the mean service time increases by 24.6022 minutes for each additional microcomputer serviced. Furthermore, if we consider the infinite population of all potential calls to service four microcomputers, it follows that

$$\begin{aligned} \hat{y}_0 &= b_0 + b_1 x_0 \\ &= 11.4641 + 24.6022(4) \\ &= 109.873 \text{ minutes} \end{aligned}$$

is the point estimate of

$$\begin{aligned} \mu_0 &= \beta_0 + \beta_1 x_0 \\ &= \beta_0 + \beta_1(4) \end{aligned}$$

the mean time required to service four microcomputers, and is the point prediction of

$$\begin{aligned} y_0 &= \mu_0 + \epsilon_0 \\ &= \beta_0 + \beta_1(4) + \epsilon_0 \end{aligned}$$

the (individual) service time (in minutes) required to service four microcomputers. Other point estimates and predictions of mean and individual service times can be found in a similar fashion.

5.3 THE LINEAR REGRESSION MODEL

In this section we extend the discussion of the previous section to the *linear regression model* relating the dependent variable y to the independent variables $x_1, x_2, \ldots, x_p$.

EXAMPLE 5.5

Consider the fuel consumption problem and the eight observed fuel consumptions listed in Table 5.2.

Part 1. Constructing a Regression Model Relating y to x_1 and x_2.

Figure 5.1 indicates that the observed values of y (weekly fuel consumption) tend to decrease in a straight line fashion as the values of x_1 (average hourly temperature) increase, and thus implies that the simple linear regression model having a negative slope ($\beta_1 < 0$)

$$\begin{aligned} y_i &= \mu_i + \epsilon_i \\ &= \beta_0 + \beta_1 x_{i1} + \epsilon_i \end{aligned}$$

relates y_i to x_{i1}. Furthermore, Figure 5.5 indicates that the observed values of y tend to increase in a straight line fashion as the values of x_2 (the chill index) increase, and thus implies that the simple linear regression model having a positive slope ($\beta_1 > 0$)

$$\begin{aligned} y_i &= \mu_i + \epsilon_i \\ &= \beta_0 + \beta_1 x_{i2} + \epsilon_i \end{aligned}$$

relates y_i to x_{i2}. Thus, it seems reasonable to combine these two models to form the regression model

$$\begin{aligned} y_i &= \mu_i + \epsilon_i \\ &= \beta_0 + \beta_1 x_{i1} + \beta_2 x_{i2} + \epsilon_i \end{aligned}$$

which relates y_i, to x_{i1} and x_{i2}. To simplify notation, we sometimes omit the subscript i (which refers to the particular observed fuel consumption we are considering) and write this model as

$$\begin{aligned} y &= \mu + \epsilon \\ &= \beta_0 + \beta_1 x_1 + \beta_2 x_2 + \epsilon \end{aligned}$$

TABLE 5.9 The Eight Observed Fuel Consumptions Expressed in the Form $y_i = \mu_i + \epsilon_i$

The *i*th Observed Week (OWK_i)	The *i*th Observed Average Hourly Temperature (°F) (x_{i1})	The *i*th Observed Chill Index (x_{i2})	The *i*th Historical Population of Potential Weekly Fuel Consumptions	The *i*th Mean Fuel Consumption (μ_i)	The *i*th Observed Fuel Consumption (Tons) ($y_i = \mu_i + \epsilon_i$)
OWK_1	$x_{11} = 28.0$	$x_{12} = 18$	First historical population	$\mu_1 = 12.042$	$y_1 = \mu_1 + \epsilon_1$ $= 12.042 + .358 = 12.4$
OWK_2	$x_{21} = 28.0$	$x_{22} = 14$	Second historical population	$\mu_2 = 11.698$	$y_2 = \mu_2 + \epsilon_2$ $= 11.698 + .002 = 11.7$
OWK_3	$x_{31} = 32.5$	$x_{32} = 24$	Third historical population	$\mu_3 = 12.166$	$y_3 = \mu_3 + \epsilon_3$ $= 12.166 + .234 = 12.4$
OWK_4	$x_{41} = 39.0$	$x_{42} = 22$	Fourth historical population	$\mu_4 = 11.429$	$y_4 = \mu_4 + \epsilon_4$ $= 11.429 + (-.629) = 10.8$
OWK_5	$x_{51} = 45.9$	$x_{52} = 8$	Fifth historical population	$\mu_5 = 9.625$	$y_5 = \mu_5 + \epsilon_5$ $= 9.625 + (-.225) = 9.4$
OWK_6	$x_{61} = 57.8$	$x_{62} = 16$	Sixth historical population	$\mu_6 = 9.277$	$y_6 = \mu_6 + \epsilon_6$ $= 9.277 + .223 = 9.5$
OWK_7	$x_{71} = 58.1$	$x_{72} = 1$	Seventh historical population	$\mu_7 = 7.961$	$y_7 = \mu_7 + \epsilon_7$ $= 7.961 + .039 = 8.0$
OWK_8	$x_{81} = 62.5$	$x_{82} = 0$	Eighth historical population	$\mu_8 = 7.492$	$y_8 = \mu_8 + \epsilon_8$ $= 7.492 + .008 = 7.5$

This regression model assumes the following:

1. y_i, the ith observed fuel consumption, is the fuel consumption that we have actually observed when a *random selection* has been made from *the ith historical population of potential weekly fuel consumptions,* which is the infinite population of potential weekly fuel consumptions that we could have observed when the average hourly temperature was x_{i1} and the chill index was x_{i2}. Examining Table 5.9, which lists the eight historical populations of potential weekly fuel consumptions, we see that, for example, $y_5 = 9.4$, the fifth observed fuel consumption, is the fuel consumption that we have actually observed when a random selection has been made from the *fifth historical population of potential weekly fuel consumptions,* which is the infinite population of potential weekly fuel consumptions that we could have observed when the average hourly temperature was $x_{51} = 45.9$ and the chill index was $x_{52} = 8$. Again referring to Table 5.9, we let

$$\begin{aligned} SPL &= \{y_1, y_2, y_3, y_4, y_5, y_6, y_7, y_8\} \\ &= \{12.4, 11.7, 12.4, 10.8, 9.4, 9.5, 8.0, 7.5\} \end{aligned}$$

which we call the *observed sample,* denote the set containing the eight observed fuel consumptions. Since we assume that for $i = 1, 2, \ldots, 8$, the ith observed fuel consumption, y_i, has been randomly selected from the ith historical population of potential weekly fuel consumptions, we say that this observed sample has been randomly selected from the eight historical populations of potential weekly fuel consumptions.
2. $\mu_i = \beta_0 + \beta_1 x_{i1} + \beta_2 x_{i2}$ is the *mean* of the ith historical population of potential weekly fuel consumptions. Intuitively, μ_i, which we call *the ith mean fuel consumption,* is the average of all of the potential weekly fuel consumptions that we could have observed when the average hourly temperature was x_{i1} and the chill index was x_{i2}. We do not know the true value of μ_i, because we do not know and cannot average all these potential weekly fuel consumptions. The mean μ_i describes the effect on y_i of the values x_{i1} and x_{i2} of the independent variables x_1 and x_2. As an example, the fifth mean fuel consumption

$$\begin{aligned} \mu_5 &= \beta_0 + \beta_1 x_{51} + \beta_2 x_{52} \\ &= \beta_0 + \beta_1(45.9) + \beta_2(8) \end{aligned}$$

is intuitively the average of all the potential weekly fuel consumptions that we could have observed when the average hourly temperature was $x_{51} = 45.9$ and the chill index was $x_{52} = 8$.
3. β_0, β_1, and β_2 are parameters relating μ_i to x_{i1} and x_{i2}. We will soon interpret the exact meaning of these parameters. For now, note that although we do not know the true values of β_0, β_1, and β_2, we do know that β_1 must be a negative number, because a negative number would express the

fact that the higher x_{i1}, average hourly temperature, is, the lower is μ_i, mean weekly fuel consumption (other factors being equal). We also know that β_2 must be a positive number, because a positive number would express the fact that the higher x_{i2}, the chill index, is, the higher is μ_i, mean weekly fuel consumption (other factors being equal).

In order to clarify some of the subsequent discussion of this example, we assume that (a supernatural power knows that) the true values of β_0, β_1, and β_2 are

$$\beta_0 = 12.930$$

$$\beta_1 = -.087$$

$$\beta_2 = .086$$

Thus, for any observed week and corresponding historical population of potential weekly fuel consumptions in Table 5.9, the true value of μ_i can be calculated (by a supernatural power) by placing the values x_{i1} and x_{i2} into the equation

$$\begin{aligned}\mu_i &= \beta_0 + \beta_1 x_{i1} + \beta_2 x_{i2}\\ &= 12.930 - .087x_{i1} + .086x_{i2}\end{aligned}$$

For example,

$$\begin{aligned}\mu_5 &= \beta_0 + \beta_1 x_{51} + \beta_2 x_{52}\\ &= 12.930 + (-.087)(45.9) + (.086)(8)\\ &= 9.625 \text{ tons of coal}\end{aligned}$$

The true values of all eight mean fuel consumptions are presented in Table 5.9.

4. ϵ_i, which we call *the ith error term*, describes the effects on y_i of all factors that have occurred in OWK_i, the ith observed week, other than x_{i1} and x_{i2}. The average hourly thermostat setting is one such other factor. As another example, recall that the chill index is Weather Forecasts, Inc.'s assessment of the combined effects of all major factors other than average hourly temperature on the chill factor. Although the chill index measures much of the effect of wind velocity on the chill factor, the chill index probably does not completely measure the total effect of wind velocity on the chill factor, since it would be quite possible that two weeks with the same chill index might have somewhat different average hourly wind velocities. Thus, average hourly wind velocity is another factor other than average hourly temperature and the chill index that might influence weekly fuel consumption. Since the equation $y_i = \mu_i + \epsilon_i$ implies that $\epsilon_i = y_i - \mu_i$, we see that ϵ_i measures the distance between y_i and μ_i and thus measures how far y_i deviates from μ_i. It should be noted that although we have observed y_i, we do not

know the true value of ϵ_i, because we do not know the true value of μ_i. However, since (a supernatural power knows that) the true values of μ_1, $\mu_2, \ldots, \mu_8$ are as given in Table 5.9, it follows that the true values of ϵ_1, $\epsilon_2, \ldots, \epsilon_8$ are also as given in Table 5.9, where we have expressed the eight observed fuel consumptions in the form $y_i = \mu_i + \epsilon_i$. For example, since the true value of μ_5 is 9.625, the true value of ϵ_5 is

$$\begin{aligned}\epsilon_5 &= y_5 - \mu_5 \\ &= 9.4 - 9.625 \\ &= -.225\end{aligned}$$

To explain this negative error term, we will suppose that during the fifth observed week the average hourly wind velocity and the average hourly thermostat setting are both *lower than usual* when the average hourly temperature is $x_{51} = 45.9$ and the chill index is $x_{52} = 8$. Thus, these factors cause the fifth observed fuel consumption

$$\begin{aligned}y_5 &= \mu_5 + \epsilon_5 \\ &= 9.625 + (-.225) \\ &= 9.4\end{aligned}$$

to be .225 less than $\mu_5 = 9.625$.

Part 2. Interpreting the Regression Parameters β_0, β_1, and β_2.

The exact interpretations of the parameters β_0, β_1, and β_2 are quite simple. First, suppose that $x_{i1} = 0$ and $x_{i2} = 0$. Then,

$$\begin{aligned}\mu_i &= \beta_0 + \beta_1 x_{i1} + \beta_2 x_{i2} \\ &= \beta_0 + \beta_1(0) + \beta_2(0) = \beta_0\end{aligned}$$

So, β_0 is the mean weekly fuel consumption for all potential weeks having an average hourly temperature of 0°F and a chill index of zero. Often, β_0 is called the **intercept** in the regression model. Although we cannot know the true value of β_0, we assume that (a supernatural power knows that) the true value of β_0 is 12.930. One might wonder if β_0 has any practical interpretation, since it is unlikely that a week having an average hourly temperature of 0°F would also have a chill index of zero. Indeed, sometimes the parameter β_0 and other parameters in a regression analysis do not have practical interpretations, because the situations related to the interpretations would not be likely to occur in practice. In fact, although we will continue to interpret the individual parameters in a regression model, sometimes each parameter does not, by itself, have much practical importance. Rather, the real importance of the parameters in a regression model is that these parameters relate the mean of the dependent variable to the independent variables in an overall sense. For example, in our fuel consumption situation the real importance of the parameters β_0, β_1, and β_2 is that these parameters relate μ_i to x_{i1} and x_{i2} in an overall sense through the equation

$$\mu_i = \beta_0 + \beta_1 x_{i1} + \beta_2 x_{i2}$$

As we will see, this relationship will allow us to find a prediction equation for y_0, the future fuel consumption.

We next interpret the individual meanings of β_1 and β_2. In order to examine the interpretation of β_1, consider two different weeks. Suppose that for the first week average hourly temperature is c and the chill index is *d*. The mean weekly fuel consumption for all such potential weeks, denoted by $\mu_{(c,d)}$, is

$$\mu_{(c,d)} = \beta_0 + \beta_1(c) + \beta_2(d)$$

For the second week, suppose that average hourly temperature is $c + 1$ and the chill index is *d*. The mean weekly fuel consumption for all such potential weeks, denoted by $\mu_{(c+1,d)}$, is

$$\mu_{(c+1,d)} = \beta_0 + \beta_1(c + 1) + \beta_2(d)$$

Now consider the difference between these mean weekly fuel consumptions. This difference is

$$\begin{aligned}\mu_{(c+1,d)} - \mu_{(c,d)} &= [\beta_0 + \beta_1(c + 1) + \beta_2(d)] - [\beta_0 + \beta_1(c) + \beta_2(d)] \\ &= (\beta_0 - \beta_0) + [\beta_1(c + 1) - \beta_1(c)] + [\beta_2(d) - \beta_2(d)] = \beta_1\end{aligned}$$

So, β_1 is equal to the difference between the mean weekly fuel consumptions $\mu_{(c+1,d)}$ and $\mu_{(c,d)}$. But weeks 1 and 2 differ only in that the average hourly temperature during week 2 is 1 degree higher than the average hourly temperature during week 1. Thus, we can interpret the parameter β_1 as the change in mean weekly fuel consumption that is associated with a 1-degree increase in average hourly temperature when the chill index does not change. Since we assume that (a supernatural power knows that) the true value of β_1 is $-.087$, it follows that mean weekly fuel consumption will decrease by .087 tons of coal when average hourly temperature increases by 1 degree and the chill index does not change.

The interpretation of β_2 can be established in a similar fashion. Again consider two different weeks. For the first week, average hourly temperature is c and the chill index is *d*. The mean weekly fuel consumption for all such potential weeks is

$$\mu_{(c,d)} = \beta_0 + \beta_1(c) + \beta_2(d)$$

For the second week, average hourly temperature is c and the chill index is $d + 1$. The mean weekly fuel consumption for all such potential weeks, denoted by $\mu_{(c,d+1)}$, is

$$\mu_{(c,d+1)} = \beta_0 + \beta_1(c) + \beta_2(d + 1)$$

The difference between these mean weekly fuel consumptions is

$$\mu_{(c,d+1)} - \mu_{(c,d)} = [\beta_0 + \beta_1(c) + \beta_2(d + 1)] - [\beta_0 + \beta_1(c) + \beta_2(d)]$$
$$= (\beta_0 - \beta_0) + [\beta_1(c) - \beta_1(c)] + [\beta_2(d + 1) - \beta_2(d)] = \beta_2$$

So β_2 is equal to the difference between the mean weekly fuel consumptions $\mu_{(c,d+1)}$ and $\mu_{(c,d)}$. But weeks 1 and 2 differ only in that week 2 has a chill index one unit higher than the chill index of week 1. Thus, we can interpret the parameter β_2 as the change in mean weekly fuel consumption that is associated with a one-unit increase in the chill index when the average hourly temperature does not change. Since we assume that the true value of β_2 is .086, it follows that mean weekly fuel consumption will increase by .086 tons of coal when the chill index increases by 1 and the average hourly temperature does not change.

Part 3. A Geometric Interpretation of the Regression Model.

We now give a geometrical interpretation of this model. The equation

$$\mu = \beta_0 + \beta_1 x_1 + \beta_2 x_2$$

can be shown to be the equation of a *plane*. Although we will not formally define a plane, intuitively we know that a plane is a flat surface that connects points in three-dimensional space. To see how this fact helps us to interpret our regression model, we define the **experimental region** to be the range of the combinations of the observed values of the independent variables x_1, average hourly temperature, and x_2, the chill index. Referring to Table 5.10, we see that it is reasonable to depict the experimental region as the shaded region in Figure

TABLE 5.10 The Eight Combinations of Observed Values of x_1 and x_2

The *i*th Observed Week (OWK_i)	The *i*th Observed Average Hourly Temperature (°F) (x_{i1})	The *i*th Observed Chill Index (x_{i2})	The *i*th Combination of Observed Values of x_1 and x_2 (x_{i1}, x_{i2})
OWK_1	28.0	18	(28.0, 18)
OWK_2	28.0	14	(28.0, 14)
OWK_3	32.5	24	(32.5, 24)
OWK_4	39.0	22	(39.0, 22)
OWK_5	45.9	8	(45.9, 8)
OWK_6	57.8	16	(57.8, 16)
OWK_7	58.1	1	(58.1, 1)
OWK_8	62.5	0	(62.5, 0)

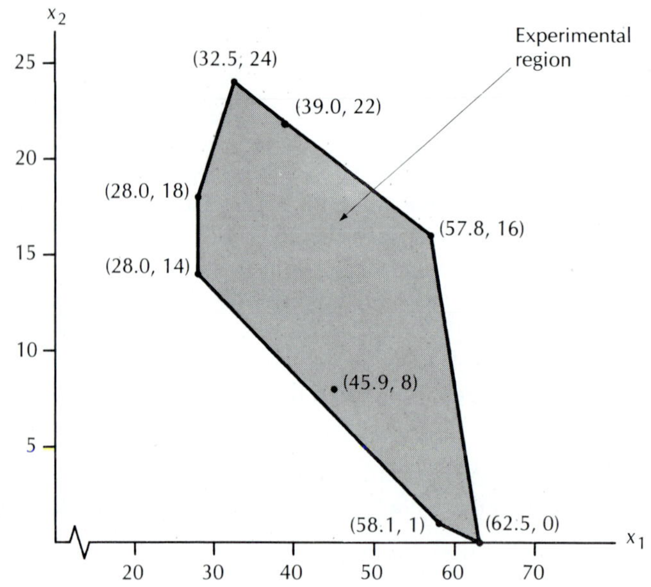

FIGURE 5.7
The Experimental Region

5.7. Next, recalling that the true values of β_0, β_1, and β_2 are respectively 12.930, −.087, and .086, and thus that

$$\begin{aligned}\mu &= \beta_0 + \beta_1 x_1 + \beta_2 x_2 \\ &= 12.930 - .087x_1 + .086x_2\end{aligned}$$

we consider Figure 5.8. In this figure we have plotted the experimental region in the (x_1, x_2) plane. First consider the point representing the combination

$$(x_{51}, x_{52}) = (45.9, 8)$$

of the average hourly temperature and the chill index that have occurred in OWK_5 (the fifth observed week). Above this point we have plotted *as a dot in three-space* the point representing the combination

$$(x_{51}, x_{52}, y_5) = (45.9, 8, 9.4)$$

of the average hourly temperature, the chill index, and the *observed* fuel consumption that have occurred in OWK_5, and we have plotted *as a triangle in three-space* the point representing the combination

$$(x_{51}, x_{52}, \mu_5) = (45.9, 8, 9.625)$$

of the average hourly temperature, the chill index, and the *mean* fuel consumption corresponding to OWK_5, which is

$$\begin{aligned}\mu_5 &= \beta_0 + \beta_1 x_{51} + \beta_2 x_{52} \\ &= 12.930 - .087(45.9) + .086(8) \\ &= 9.625\end{aligned}$$

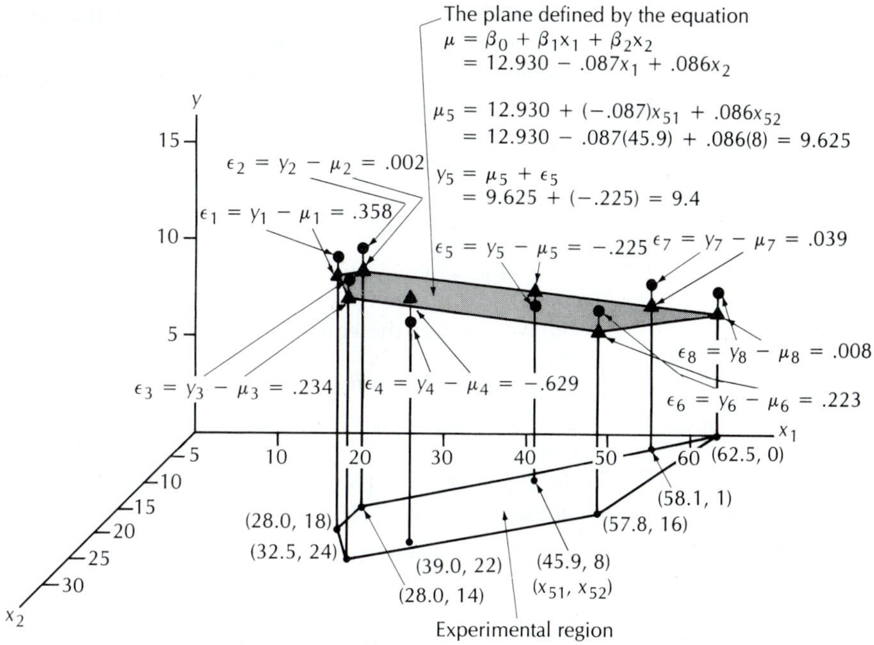

FIGURE 5.8 A Geometrical Interpretation of the Regression Model Relating y to x_1 and x_2

When looking at this figure it is best to pretend that you are sitting high in a football stadium and that you are looking down at the playing field, which is the (x_1, x_2) plane

Also, note from Figure 5.8 that the line segment that equals the distance between $\mu_5 = 9.625$, the fifth mean fuel consumption, and $y_5 = 9.4$, the fifth observed fuel consumption, represents the size of the fifth error term

$$\epsilon_5 = y_5 - \mu_5 = 9.4 - 9.625 = -.225$$

Now, considering all of Figure 5.8, we note the following:

1. We have plotted *as eight dots in three-space* the eight points representing the eight combinations of the average hourly temperatures, the chill indices, and the *observed* fuel consumptions that have occurred in the eight observed weeks.
2. We have plotted *as eight triangles in three-space* the eight points representing the eight combinations of the average hourly temperatures, the chill indices, and the *mean* fuel consumptions corresponding to the eight observed weeks, where the true values of the eight mean fuel consumptions are calculated by the equation

$$\begin{aligned}\mu &= \beta_0 + \beta_1 x_1 + \beta_2 x_2 \\ &= 12.930 - .087x_1 + .086x_2\end{aligned}$$

3. The eight line segments drawn between the triangles and dots are the eight distances between the mean fuel consumptions and the observed fuel consumptions and thus represent the sizes of the eight error terms

$$\begin{aligned}\epsilon_1 &= y_1 - \mu_1\\ \epsilon_2 &= y_2 - \mu_2\\ &\vdots \quad \vdots\\ \epsilon_8 &= y_8 - \mu_8\end{aligned}$$

When we say that the equation

$$\begin{aligned}\mu &= \beta_0 + \beta_1 x_1 + \beta_2 x_2\\ &= 12.930 - .087x_1 + .086x_2\end{aligned}$$

is the equation of a plane, we mean that *the eight points representing the combinations*

$$\begin{aligned}(x_{11}, x_{12}, \mu_1) &= (28.0, 18, 12.042)\\ &\vdots\\ (x_{51}, x_{52}, \mu_5) &= (45.9, 8, 9.625)\\ &\vdots\\ (x_{81}, x_{82}, \mu_8) &= (62.5, 0, 7.492)\end{aligned}$$

and any other points representing combinations of average hourly temperatures, chill indices, and mean fuel consumptions calculated using this equation *lie on, or are connected by, a plane.* The plane connecting these eight points (which are represented by the triangles in Figure 5.8) is the shaded plane. (Note that we illustrate only the portion of this plane corresponding to the experimental region.) We say that this plane is the plane defined by the equation

$$\begin{aligned}\mu &= \beta_0 + \beta_1 x_1 + \beta_2 x_2\\ &= 12.930 - .087x_1 + .086x_2\end{aligned}$$

The regression model

$$\begin{aligned}y &= \mu + \epsilon\\ &= \beta_0 + \beta_1 x_1 + \beta_2 x_2 + \epsilon\end{aligned}$$

can therefore be interpreted geometrically as follows. As illustrated in Figure 5.8, this model says that the eight error terms

$$\epsilon_1, \epsilon_2, \ldots, \epsilon_8$$

cause the eight observed fuel consumptions

$$y_1 = \mu_1 + \epsilon_1$$
$$y_2 = \mu_2 + \epsilon_2$$
$$\vdots \qquad \vdots$$
$$y_8 = \mu_8 + \epsilon_8$$

to deviate from the eight mean fuel consumptions

$$\mu_1, \mu_2, \ldots, \mu_8$$

which exactly lie on the plane defined by the equation

$$\begin{aligned} \mu &= \beta_0 + \beta_1 x_1 + \beta_2 x_2 \\ &= 12.930 - .087x_1 + .086x_2 \end{aligned}$$

For example, we see that the fifth error term, $\epsilon_5 = -.225$, causes the fifth observed fuel consumption

$$\begin{aligned} y_5 &= \mu_5 + \epsilon_5 \\ &= 9.625 + (-.225) \\ &= 9.4 \end{aligned}$$

to be .225 less than the fifth mean fuel consumption $\mu_5 = 9.625$, which exactly lies on the plane defined by the equation

$$\begin{aligned} \mu &= \beta_0 + \beta_1 x_1 + \beta_2 x_2 \\ &= 12.930 - .087x_1 + .086x_2 \end{aligned}$$

Part 4. Another Possible Model Relating y to x_1 and x_2.

To conclude this example, recall that we have obtained the regression model

$$\begin{aligned} y_i &= \mu_i + \epsilon_i \\ &= \beta_0 + \beta_1 x_{i1} + \beta_2 x_{i2} + \epsilon_i \end{aligned}$$

by combining the individual simple linear regression models

$$\begin{aligned} y_i &= \mu_i + \epsilon_i \\ &= \beta_0 + \beta_1 x_{i1} + \epsilon_i \end{aligned} \qquad \text{and} \qquad \begin{aligned} y_i &= \mu_i + \epsilon_i \\ &= \beta_0 + \beta_1 x_{i2} + \epsilon_i \end{aligned}$$

which were motivated by the data plots in Figures 5.1 and 5.5. Although this procedure is reasonable, we could also add the cross-product term $x_{i1}x_{i2}$ to our model to form the model

$$\begin{aligned} y_i &= \mu_i + \epsilon_i \\ &= \beta_0 + \beta_1 x_{i1} + \beta_2 x_{i2} + \beta_3 x_{i1} x_{i2} + \epsilon_i \end{aligned}$$

which also relates y_i to x_{i1} and x_{i2}. It can be shown that the cross-product term $x_{i1}x_{i2}$ should be included in the model if interaction exists between x_1, average hourly temperature, and x_2, the chill index, as these independent variables affect y (weekly fuel consumption). In Section 5.7 we discuss the meaning of interaction, and in Chapter 7 we find (by using several model-building techniques) that the cross-product term $x_{i1}x_{i2}$ should not be included in the model relating y_i to x_{i1} and x_{i2}. In fact, we find in Chapter 7 that

$$\begin{aligned} y_i &= \mu_i + \epsilon_i \\ &= \beta_0 + \beta_1 x_{i1} + \beta_2 x_{i2} + \epsilon_i \end{aligned}$$

is probably the most appropriate model relating y_i to x_{i1} and x_{i2}.

We now reconsider Table 5.3 and define the linear regression model relating a dependent variable y to the independent variables $x_1, x_2, \ldots, x_p$.

The *linear regression model* relating y_i to $x_{i1}, x_{i2}, \ldots, x_{ip}$ is

$$\begin{aligned} y_i &= \mu_i + \epsilon_i \\ &= \beta_0 + \beta_1 x_{i1} + \beta_2 x_{i2} + \cdots + \beta_p x_{ip} + \epsilon_i \end{aligned}$$

1. y_i, *the ith observed value of the dependent variable,* is the value of the dependent variable that we have actually observed when a *random selection* has been made from *the ith historical population of potential values of the dependent variable,* which is the infinite population of potential values of the dependent variable that we could have observed when the values of the independent variables $x_1, x_2, \ldots, x_p$ were $x_{i1}, x_{i2}, \ldots, x_{ip}$.
2. $\mu_i = \beta_0 + \beta_1 x_{i1} + \beta_2 x_{i2} + \cdots + \beta_p x_{ip}$ is the *mean* of the ith historical population of potential values of the dependent variable. Intuitively, μ_i, which we call *the ith mean value of the dependent variable,* is the average of all the potential values of the dependent variable that we could have observed when the values of the independent variables were $x_{i1}, x_{i2}, \ldots, x_{ip}$. This mean describes the effects on y_i of the values $x_{i1}, x_{i2}, \ldots, x_{ip}$ of the independent variables $x_1, x_2, \ldots, x_p$.
3. $\beta_0, \beta_1, \beta_2, \ldots, \beta_p$ are (unknown) *parameters* relating μ_i to $x_{i1}, x_{i2}, \ldots, x_{ip}$.
4. ϵ_i, which we call *the ith error term,* describes the effects on y_i of all factors that have occurred in OST_i, the ith observed situation, other than the values $x_{i1}, x_{i2}, \ldots, x_{ip}$ of the independent variables $x_1, x_2, \ldots, x_p$.

The reason that we call this model a *linear* regression model is that the equation

$$\mu_i = \beta_0 + \beta_1 x_{i1} + \beta_2 x_{i2} + \cdots + \beta_p x_{ip}$$

expresses μ_i *as a linear function of the parameters* $\beta_0, \beta_1, \beta_2, \ldots, \beta_p$. For example, the equation

$$\mu_i = \beta_0 + \beta_1 x_{i1} + \beta_2 x_{i2}$$

is a linear equation because it expresses μ_i as a linear function of the parameters β_0, β_1, and β_2. Furthermore, although the equation

$$\mu_i = \beta_0 + \beta_1 x_{i1} + \beta_2 x_{i2} + \beta_3 x_{i1}^2 + \beta_4 x_{i1} x_{i2}$$

utilizes x_{i1}^2 and $x_{i1}x_{i2}$, this equation is also linear because it expresses μ_i as a linear function of the parameters β_0, β_1, β_2, β_3, and β_4. However, the equation

$$\mu_i = \beta_0 + \beta_1 x_{i1}^{\beta_2}$$

is not a linear equation, because it does not express μ_i as a linear function of the parameters β_0, β_1, and β_2. The reason we emphasize the concept of a linear equation relating μ_i to $x_{i1}, x_{i2}, \ldots, x_{ip}$ is that the techniques of regression analysis are easiest to use and best developed when we assume that a linear equation relates μ_i to x_{i1}, $x_{i2}, \ldots, x_{ip}$.

In this book we study various techniques that can be used to determine an appropriate equation relating μ_i to $x_{i1}, x_{i2}, \ldots, x_{ip}$. It can be shown that if it is reasonable to believe that certain assumptions hold (see Chapter 6), then it is reasonable to believe that for $i = 1, 2, \ldots, n$, y_i has been *randomly selected* from the ith historical population of potential values of the dependent variable. Therefore, calling

$$SPL = \{y_1, y_2, \ldots, y_n\}$$

the *observed sample*, we say that the observed sample has been randomly selected from the n historical populations of potential values of the dependent variable.

5.4 THE LEAST SQUARES POINT ESTIMATES

Since we cannot know the true values of the parameters $\beta_0, \beta_1, \beta_2, \ldots, \beta_p$ relating μ to $x_1, x_2, \ldots, x_p$ in the regression model

$$\begin{aligned} y &= \mu + \epsilon \\ &= \beta_0 + \beta_1 x_1 + \beta_2 x_2 + \cdots + \beta_p x_p + \epsilon \end{aligned}$$

we must estimate these parameters by using the observed data summarized in Table 5.3. For $j = 0, 1, 2, \ldots, p$, we let b_j denote the point estimate of β_j. In general, we can calculate the least squares point estimates $b_0, b_1, b_2, \ldots, b_p$ of the parameters $\beta_0, \beta_1, \beta_2, \ldots, \beta_p$ in this model by using the following matrix algebra formula.

The least squares point estimates $b_0, b_1, b_2, \ldots, b_p$ are calculated by using the formula

$$\begin{bmatrix} b_0 \\ b_1 \\ b_2 \\ \vdots \\ b_p \end{bmatrix} = \mathbf{b} = (\mathbf{X}'\mathbf{X})^{-1}\mathbf{X}'\mathbf{y}$$

where $\mathbf{y}$ and $\mathbf{X}$ are the following column vector and matrix:

$$\mathbf{y} = \begin{bmatrix} y_1 \\ y_2 \\ \vdots \\ y_n \end{bmatrix} \quad \text{and} \quad \mathbf{X} = \begin{array}{c} \begin{array}{cccccc} 0 & 1 & 2 & \cdots & p \\ & x_1 & x_2 & \cdots & x_p \end{array} \\ \begin{bmatrix} 1 & x_{11} & x_{12} & \cdots & x_{1p} \\ 1 & x_{21} & x_{22} & \cdots & x_{2p} \\ \vdots & \vdots & \vdots & \vdots & \vdots \\ 1 & x_{n1} & x_{n2} & \cdots & x_{np} \end{bmatrix} \end{array}$$

Thus, $\mathbf{y}$ is a column vector of the n observed values of the dependent variable, $y_1, y_2, \ldots, y_n$. To define the matrix $\mathbf{X}$, consider the regression model

$$y = \beta_0 + \beta_1 x_1 + \beta_2 x_2 + \cdots + \beta_p x_p + \epsilon$$

If k is the number of parameters ($\beta_0, \beta_1, \beta_2, \ldots, \beta_p$) in this model, then the matrix $\mathbf{X}$ will consist of k columns. The columns in the matrix $\mathbf{X}$ contain the observed values of the independent variables corresponding to (that is, multiplied by) the k parameters $\beta_0, \beta_1, \beta_2, \ldots, \beta_p$, and the columns of this matrix are numbered in the same manner as the parameters $\beta_0, \beta_1, \beta_2, \ldots, \beta_p$ are numbered (see the preceding $\mathbf{X}$ matrix). In the following example we demonstrate how to calculate the least squares point estimates, and after this example we discuss the meaning of the term *least squares*.

EXAMPLE 5.6

The least squares point estimates b_0, b_1, and b_2 of the parameters β_0, β_1, and β_2 in the fuel consumption model

$$y = \beta_0 + \beta_1 x_1 + \beta_2 x_2 + \epsilon$$

are calculated by using the formula

$$\begin{bmatrix} b_0 \\ b_1 \\ b_2 \end{bmatrix} = \mathbf{b} = (\mathbf{X}'\mathbf{X})^{-1}\mathbf{X}'\mathbf{y}$$

where

$$\mathbf{y} = \begin{bmatrix} y_1 \\ y_2 \\ y_3 \\ y_4 \\ y_5 \\ y_6 \\ y_7 \\ y_8 \end{bmatrix} = \begin{bmatrix} 12.4 \\ 11.7 \\ 12.4 \\ 10.8 \\ 9.4 \\ 9.5 \\ 8.0 \\ 7.5 \end{bmatrix} \quad \text{and} \quad \mathbf{X} = \begin{array}{c} \begin{array}{ccc} 0 & 1 & 2 \\ & x_1 & x_2 \end{array} \\ \begin{bmatrix} 1 & x_{11} & x_{12} \\ 1 & x_{21} & x_{22} \\ 1 & x_{31} & x_{32} \\ 1 & x_{41} & x_{42} \\ 1 & x_{51} & x_{52} \\ 1 & x_{61} & x_{62} \\ 1 & x_{71} & x_{72} \\ 1 & x_{81} & x_{82} \end{bmatrix} \end{array} = \begin{array}{c} \begin{array}{ccc} 0 & 1 & 2 \\ & x_1 & x_2 \end{array} \\ \begin{bmatrix} 1 & 28.0 & 18 \\ 1 & 28.0 & 14 \\ 1 & 32.5 & 24 \\ 1 & 39.0 & 22 \\ 1 & 45.9 & 8 \\ 1 & 57.8 & 16 \\ 1 & 58.1 & 1 \\ 1 & 62.5 & 0 \end{bmatrix} \end{array}$$

Here, the column vector $\mathbf{y}$ is simply a vector of the observed weekly fuel consumptions, and the three columns of the $\mathbf{X}$ matrix contain the observed values of the independent variables corresponding to (that is, multiplied by) the three parameters in the model. Therefore, since the number 1 is multiplied by β_0, the column of the $\mathbf{X}$ matrix corresponding to β_0 is a column of 1's. Since the independent variable x_1 (the average hourly temperature) is multiplied by β_1, the column of the $\mathbf{X}$ matrix corresponding to β_1 is a column containing the observed average hourly temperatures (see Table 5.2). The independent variable x_2 (the chill index) is multiplied by β_2, and thus the column of the $\mathbf{X}$ matrix corresponding to β_2 is a column containing the observed chill indices (see Table 5.2).

In order to calculate $\mathbf{b} = (\mathbf{X}'\mathbf{X})^{-1}\mathbf{X}'\mathbf{y}$ we first find

$$\mathbf{X}'\mathbf{X} = \begin{bmatrix} 1 & 1 & 1 & 1 & 1 & 1 & 1 & 1 \\ 28.0 & 28.0 & 32.5 & 39.0 & 45.9 & 57.8 & 58.1 & 62.5 \\ 18 & 14 & 24 & 22 & 8 & 16 & 1 & 0 \end{bmatrix} \begin{bmatrix} 1 & 28.0 & 18 \\ 1 & 28.0 & 14 \\ 1 & 32.5 & 24 \\ 1 & 39.0 & 22 \\ 1 & 45.9 & 8 \\ 1 & 57.8 & 16 \\ 1 & 58.1 & 1 \\ 1 & 62.5 & 0 \end{bmatrix}$$

$$= \begin{bmatrix} 8.0 & 351.8 & 103.0 \\ 351.8 & 16874.76 & 3884.1 \\ 103.0 & 3884.1 & 1901.0 \end{bmatrix}$$

Since the columns of the matrix $\mathbf{X}'\mathbf{X}$ can be verified to be linearly independent of each other, the matrix $\mathbf{X}'\mathbf{X}$ possesses an inverse matrix $(\mathbf{X}'\mathbf{X})^{-1}$, which can be calculated using a standard computer program to be

$$(\mathbf{X}'\mathbf{X})^{-1} = \begin{bmatrix} 8.0 & 351.8 & 103.0 \\ 351.8 & 16874.76 & 3884.1 \\ 103.0 & 3884.1 & 1901.0 \end{bmatrix}^{-1}$$

$$= \begin{bmatrix} 5.43405 & -.085930 & -.118856 \\ -.085930 & .00147070 & .00165094 \\ -.118856 & .00165094 & .00359276 \end{bmatrix}$$

We next calculate

$$\mathbf{X}'\mathbf{y} = \begin{bmatrix} 1 & 1 & 1 & 1 & 1 & 1 & 1 & 1 \\ 28.0 & 28.0 & 32.5 & 39.0 & 45.9 & 57.8 & 58.1 & 62.5 \\ 18 & 14 & 24 & 22 & 8 & 16 & 1 & 0 \end{bmatrix} \begin{bmatrix} 12.4 \\ 11.7 \\ 12.4 \\ 10.8 \\ 9.4 \\ 9.5 \\ 8.0 \\ 7.5 \end{bmatrix}$$

$$= \begin{bmatrix} 81.7 \\ 3413.11 \\ 1157.4 \end{bmatrix}$$

Finally, we compute

$$\begin{bmatrix} b_0 \\ b_1 \\ b_2 \end{bmatrix} = \mathbf{b} = (\mathbf{X}'\mathbf{X})^{-1}\mathbf{X}'\mathbf{y}$$

$$= \begin{bmatrix} 5.43405 & -.085930 & -.118856 \\ -.085930 & .00147070 & .00165094 \\ -.118856 & .00165094 & .00359276 \end{bmatrix} \begin{bmatrix} 81.7 \\ 3413.11 \\ 1157.4 \end{bmatrix}$$

$$= \begin{bmatrix} 13.109 \\ -.0900 \\ .0825 \end{bmatrix}$$

What is the meaning of these point estimates? Recall that β_0 is the mean weekly fuel consumption for all potential weeks having an average hourly temperature of 0°F and a chill index of zero. We have found that the point estimate of β_0

is $b_0 = 13.109$. So we estimate that the mean weekly fuel consumption for all potential weeks having an average hourly temperature of 0°F and a chill index of zero is 13.109 tons of coal. Also, recall that β_1 is the change in mean weekly fuel consumption that is associated with a 1-degree increase in average hourly temperature when the chill index does not change, while β_2 is the change in mean weekly fuel consumption that is associated with a one-unit increase in the chill index when average hourly temperature does not change. Using the point estimates $b_1 = -.0900$ and $b_2 = .0825$ of β_1 and β_2, we estimate that mean weekly fuel consumption decreases (since b_1 is negative) by .0900 tons of coal when average hourly temperature increases by 1 degree and the chill index does not change, and we estimate that mean weekly fuel consumption increases (since b_2 is positive) by .0825 tons of coal when there is a one-unit increase in the chill index and average hourly temperature does not change. Thus, we can use the point estimates b_0, b_1, and b_2 to describe the relationship between the independent variables x_1 (average hourly temperature) and x_2 (the chill index) and the mean weekly fuel consumption.

We now explain why we use the matrix algebra formula

$$\mathbf{b} = (\mathbf{X}'\mathbf{X})^{-1}\mathbf{X}'\mathbf{y}$$

to calculate the point estimates $b_0, b_1, b_2, \ldots, b_p$ of the parameters $\beta_0, \beta_1, \beta_2, \ldots, \beta_p$. First note that

$$\hat{y}_i = b_0 + b_1x_{i1} + b_2x_{i2} + \cdots + b_px_{ip}$$

is the point estimate of the ith mean value of the dependent variable

$$\mu_i = \beta_0 + \beta_1x_{i1} + \beta_2x_{i2} + \cdots + \beta_px_{ip}$$

and since we will predict the ith error term, ϵ_i, to be zero, $\hat{y}_i$ is also the point prediction of the ith observed value of the dependent variable

$$\begin{aligned} y_i &= \mu_i + \epsilon_i \\ &= \beta_0 + \beta_1x_{i1} + \beta_2x_{i2} + \cdots + \beta_px_{ip} + \epsilon_i \end{aligned}$$

Next we define, for $i = 1, 2, \ldots, n$, *the* i*th residual* to be $e_i = y_i - \hat{y}_i$, which is the difference between y_i, the ith observed value of the dependent variable, and $\hat{y}_i$, *the* i*th predicted value of the dependent variable*. Also, we define, for $i = 1, 2, \ldots, n$, the *sum of squared residuals* to be

$$\begin{aligned} SSE = \sum_{i=1}^{n} e_i^2 &= \sum_{i=1}^{n} (y_i - \hat{y}_i)^2 \\ &= \sum_{i=1}^{n} (y_i - (b_0 + b_1x_{i1} + b_2x_{i2} + \cdots + b_px_{ip}))^2 \end{aligned}$$

Intuitively, if any particular values of $b_0, b_1, b_2, \ldots, b_p$ are good point estimates of $\beta_0, \beta_1, \beta_2, \ldots, \beta_p$, they will make the ith predicted value of the dependent variable

$$\hat{y}_i = b_0 + b_1x_{i1} + b_2x_{i2} + \cdots + b_px_{ip}$$

fairly close to

$$y_i = \beta_0 + \beta_1 x_{i1} + \beta_2 x_{i2} + \cdots + \beta_p x_{ip} + \epsilon_i$$

which is the ith observed value of the dependent variable, and thus they will make the ith residual, $e_i = y_i - \hat{y}_i$, and *SSE*, the sum of squared residuals, fairly small.

We use the formula

$$\mathbf{b} = (\mathbf{X}'\mathbf{X})^{-1}\mathbf{X}'\mathbf{y}$$

to calculate the values of the point estimates $b_0, b_1, b_2, \ldots, b_p$ of the parameters $\beta_0, \beta_1, \beta_2, \ldots, \beta_p$ because it can be proved (see Appendix B) that the point estimates $b_0, b_1, b_2, \ldots, b_p$ given by this formula give a value of *SSE* that is smaller than would be given by any other values of $b_0, b_1, b_2, \ldots, b_p$. Since the point estimates $b_0, b_1, b_2, \ldots, b_p$ given by this formula minimize *SSE*, we call these point estimates the *least squares point estimates*. It should be noted that instead of choosing the values of $b_0, b_1, b_2, \ldots, b_p$ that minimize the sum of squared residuals, we could choose the values of $b_0, b_1, b_2, \ldots, b_p$ that minimize some other criterion, such as the sum of the absolute values of the residuals

$$\sum_{i=1}^{n} |e_i| = \sum_{i=1}^{n} |y_i - \hat{y}_i|$$

$$= \sum_{i=1}^{n} |y_i - (b_0 + b_1 x_{i1} + b_2 x_{i2} + \cdots + b_p x_{ip})|$$

There are theoretical reasons, however, for choosing the values of the point estimates that minimize the sum of squared residuals. We study some of these reasons in Chapter 6.

EXAMPLE 5.7

In this example we illustrate the fact that the least squares point estimates of the parameters β_0, β_1, and β_2 in the fuel consumption model

$$y_i = \beta_0 + \beta_1 x_{i1} + \beta_2 x_{i2} + \epsilon_i$$

which we calculated to be

$$\begin{bmatrix} b_0 \\ b_1 \\ b_2 \end{bmatrix} = \mathbf{b} = (\mathbf{X}'\mathbf{X})^{-1}\mathbf{X}'\mathbf{y} = \begin{bmatrix} 13.109 \\ -.0900 \\ .0825 \end{bmatrix}$$

minimize

$$SSE = \sum_{i=1}^{n} e_i^2 = \sum_{i=1}^{n} (y_i - \hat{y}_i)^2$$

$$= \sum_{i=1}^{8} (y_i - (b_0 + b_1 x_{i1} + b_2 x_{i2}))^2$$

TABLE 5.11 The Values of *SSE* Given by the Least Squares Point Estimates $b_0 = 13.109$, $b_1 = -.0900$, and $b_2 = .0825$ and by the Point Estimates $b_0 = 12.54$, $b_1 = -.095$, and $b_2 = .077$

Predictions Using the Least Squares Point Estimates $b_0 = 13.109$, $b_1 = -.0900$, and $b_2 = .0825$					
The *i*th Observed Week (OWK_i)	The *i*th Observed Average Hourly Temperature (°F) (x_{i1})	The *i*th Observed Chill Index (x_{i2})	The *i*th Observed Fuel Consumption (Tons) (y_i)	The *i*th Predicted Fuel Consumption ($\hat{y}_i = b_0 + b_1x_{i1} + b_2x_{i2}$ $= 13.109 - .0900x_{i1} + .0825x_{i2}$)	The *i*th Residual ($e_i = y_i - \hat{y}_i$)
OWK_1	28.0	18	12.4	12.0733	.3267
OWK_2	28.0	14	11.7	11.7433	−.0433
OWK_3	32.5	24	12.4	12.1632	.2368
OWK_4	39.0	22	10.8	11.4131	−.6131
OWK_5	45.9	8	9.4	9.6371	−.2371
OWK_6	57.8	16	9.5	9.2259	.2741
OWK_7	58.1	1	8.0	7.9614	.0386
OWK_8	62.5	0	7.5	7.4829	.0171
					$SSE = \sum_{i=1}^{8} e_i^2 = .674$

Predictions Using the Point Estimates $b_0 = 12.54$, $b_1 = -.095$, and $b_2 = .077$					
The *i*th Observed Week (OWK_i)	The *i*th Observed Average Hourly Temperature (°F) (x_{i1})	The *i*th Observed Chill Index (x_{i2})	The *i*th Observed Fuel Consumption (Tons) (y_i)	The *i*th Predicted Fuel Consumption ($\hat{y}_i = b_0 + b_1x_{i1} + b_2x_{i2}$ $= 12.54 - .095x_{i1} + .077x_{i2}$)	The *i*th Residual ($e_i = y_i - \hat{y}_i$)
OWK_1	28.0	18	12.4	11.2660	1.1340
OWK_2	28.0	14	11.7	10.9580	.7420
OWK_3	32.5	24	12.4	11.3005	1.0995
OWK_4	39.0	22	10.8	10.5290	.2710
OWK_5	45.9	8	9.4	8.7955	.6045
OWK_6	57.8	16	9.5	8.2810	1.2190
OWK_7	58.1	1	8.0	7.0975	.9025
OWK_8	62.5	0	7.5	6.6025	.8975
					$SSE = \sum_{i=1}^{8} e_i^2 = 6.590$

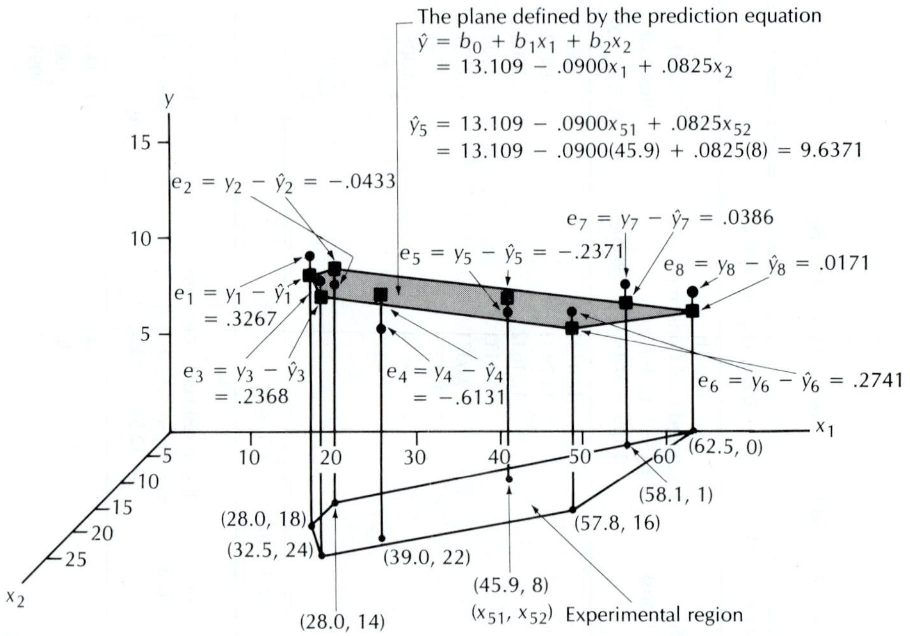

FIGURE 5.9
A Geometrical Interpretation of the Prediction Equation Relating $\hat{y}$ to x_1 and x_2

Compare the upper and lower parts of Table 5.11. We see that $SSE = .674$, the value given by the least squares point estimates $b_0 = 13.109$, $b_1 = -.0900$, and $b_2 = .0825$, is less than $SSE = 6.590$, given by $b_0 = 12.54$, $b_1 = -.095$, and $b_2 = .077$, which are not the least squares point estimates. Hence, using one set of values of b_0, b_1, and b_2 different from the least squares point estimates, we have illustrated that the least squares point estimates give a value of SSE smaller than would be given by any other values of b_0, b_1, and b_2. (There is nothing special about the values 12.54, −.095, and .077. We could choose *any* values of b_0, b_1, and b_2 different from the least squares point estimates and obtain a similar result. See Appendix B for a proof of this.)

Since

$$\begin{aligned}\hat{y}_i &= b_0 + b_1x_{i1} + b_2x_{i2}\\ &= 13.109 - .0900x_{i1} + .0825x_{i2}\end{aligned}$$

is the point prediction of y_i, the ith observed fuel consumption, we call the equation

$$\begin{aligned}\hat{y} &= b_0 + b_1x_1 + b_2x_2\\ &= 13.109 - .0900x_1 + .0825x_2\end{aligned}$$

a prediction equation. This equation defines a plane. To better understand this prediction equation, consider Figure 5.9, in which we have plotted the experimental region in the (x_1, x_2) plane. First consider the point representing the combination

$$(x_{51}, x_{52}) = (45.9, 8)$$

of the average hourly temperature and the chill index that have occurred in OWK_5 (the fifth observed week). Above this point we have plotted *as a dot in three-space* the point representing the combination

$$(x_{51}, x_{52}, y_5) = (45.9, 8, 9.4)$$

of the average hourly temperature, the chill index, and the *observed* fuel consumption that have occurred in OWK_5, and we have plotted *as a square in three-space* the point representing the combination

$$(x_{51}, x_{52}, \hat{y}_5) = (45.9, 8, 9.6371)$$

of the average hourly temperature, the chill index, and the *predicted* fuel consumption corresponding to OWK_5, where $\hat{y}_5$ has been calculated as follows:

$$\begin{aligned}\hat{y}_5 &= 13.109 - .0900x_{51} + .0825x_{52} \\ &= 13.109 - .0900(45.9) + .0825(8) \\ &= 9.6371\end{aligned}$$

Now, considering the entire figure, we note the following:

1. We have plotted *as eight dots in three-space* the eight points representing the eight combinations of the average hourly temperatures, the chill indices, and the *observed* fuel consumptions that have occurred in the eight observed weeks.

2. We have plotted *as eight squares in three-space* the eight points representing the eight combinations of the average hourly temperatures, the chill indices, and the *predicted* fuel consumptions corresponding to the eight observed weeks, where the eight predicted fuel consumptions are calculated by the prediction equation

$$\begin{aligned}\hat{y} &= b_0 + b_1x_1 + b_2x_2 \\ &= 13.109 - .0900x_1 + .0825x_2\end{aligned}$$

This prediction equation is the equation of a plane. When we say this, we mean that *the eight points representing the combinations*

$$\begin{aligned}(x_{11}, x_{12}, \hat{y}_1) &= (28.0, 18, 12.0733) \\ &\vdots \\ (x_{51}, x_{52}, \hat{y}_5) &= (45.9, 8, 9.6371) \\ &\vdots \\ (x_{81}, x_{82}, \hat{y}_8) &= (62.5, 0, 7.4829)\end{aligned}$$

and any other points representing combinations of average hourly temperatures, chill indices, and predicted fuel consumptions calculated by using this prediction equation lie on a plane. The plane connecting these eight points (which are represented by the squares in Figure 5.9) is the shaded plane. (Note that we illustrate only the portion of this plane corresponding to the experimental region.) We say that this plane is the plane defined by the prediction equation

$$\begin{aligned}\hat{y} &= b_0 + b_1x_1 + b_2x_2 \\ &= 13.109 - .0900x_1 + .0825x_2\end{aligned}$$

Examining Figure 5.9, we see that whereas the eight predicted fuel consumptions

$$\begin{aligned}\hat{y}_1 &= 12.0733 \\ \hat{y}_2 &= 11.7433 \\ \vdots \quad & \quad \vdots \\ \hat{y}_8 &= 7.4829\end{aligned}$$

which are represented by the squares, lie exactly on the plane defined by the prediction equation, the eight observed fuel consumptions

$$\begin{aligned}y_1 &= 12.4 \\ y_2 &= 11.7 \\ \vdots \quad & \quad \vdots \\ y_8 &= 7.5\end{aligned}$$

which are represented by the dots, fluctuate around the plane defined by this prediction equation. Moreover, the eight line segments drawn between the squares and dots are the eight distances between the predicted fuel consumptions and the observed fuel consumptions and thus represent the sizes of the eight residuals

$$\begin{aligned}e_1 &= y_1 - \hat{y}_1 = .3267 \\ \vdots \quad & \qquad\qquad \vdots \\ e_5 &= y_5 - \hat{y}_5 = -.2371 \\ \vdots \quad & \qquad\qquad \vdots \\ e_8 &= y_8 - \hat{y}_8 = .0171\end{aligned}$$

It follows that we may interpret

$$\begin{aligned}SSE &= \sum_{i=1}^{8} e_i^2 = \sum_{i=1}^{8} (y_i - \hat{y}_i)^2 \\ &= \sum_{i=1}^{8} (y_i - (b_0 + b_1x_{i1} + b_2x_{i2}))^2 \\ &= \sum_{i=1}^{8} (y_i - (13.109 - .0900x_{i1} + .0825x_{i2}))^2 \\ &= (.3267)^2 + (-.0433)^2 + \cdots + (.0171)^2 = .674\end{aligned}$$

which is the sum of squared residuals given by the least squares point estimates, as the sum of squared distances between the observed fuel consumptions and the predicted fuel consumptions calculated by using the prediction equation

$$\begin{aligned}\hat{y} &= b_0 + b_1x_1 + b_2x_2 \\ &= 13.109 - .0900x_1 + .0825x_2\end{aligned}$$

Since the least squares point estimates minimize *SSE*, we may interpret the least squares point estimates as "positioning" the plane defined by the prediction equation

$$\begin{aligned}\hat{y} &= b_0 + b_1x_1 + b_2x_2 \\ &= 13.109 - .0900x_1 + .0825x_2\end{aligned}$$

so that the sum of squared distances between the observed fuel consumptions and the predicted fuel consumptions calculated by using this prediction equation is smaller than the sum of squared distances between the observed fuel consumptions and the predicted fuel consumptions that would be given by any other plane defined by any other prediction equation. In this sense, then, we might say that the plane defined by the prediction equation employing the least squares point estimates is the best plane that can be positioned between the points representing the eight combinations of the observed values of average hourly temperature, the chill index, and weekly fuel consumption.

We conclude this example by comparing the shaded planes illustrated in Figures 5.8 and 5.9. Geometrically, the shaded plane of Figure 5.9, which connects the eight predicted fuel consumptions $\hat{y}_1, \hat{y}_2, \ldots, \hat{y}_8$ and which is defined by the prediction equation

$$\begin{aligned}\hat{y} &= b_0 + b_1x_1 + b_2x_2 \\ &= 13.109 - .0900x_1 + .0825x_2\end{aligned}$$

is the estimate of the shaded plane of Figure 5.8, which connects the eight mean fuel consumptions $\mu_1, \mu_2, \ldots, \mu_8$ and which is defined by the equation

$$\begin{aligned}\mu &= \beta_0 + \beta_1x_1 + \beta_2x_2 \\ &= 12.930 - .087x_1 + .086x_2\end{aligned}$$

In concluding this section, we note that the column vector **y** and the matrix **X** used to calculate the least squares point estimates b_0 and b_1 of the parameters β_0 and β_1 in the simple linear regression model

$$\begin{aligned}y_i &= \mu_i + \epsilon_i \\ &= \beta_0 + \beta_1x_i + \epsilon_i\end{aligned}$$

are

$$\mathbf{y} = \begin{bmatrix} y_1 \\ y_2 \\ \vdots \\ y_n \end{bmatrix} \quad \text{and} \quad \mathbf{X} = \begin{bmatrix} 1 & x_1 \\ 1 & x_2 \\ \vdots & \vdots \\ 1 & x_n \end{bmatrix}$$

Using this **y** vector and **X** matrix it can be shown that

$$\begin{bmatrix} b_0 \\ b_1 \end{bmatrix} = \mathbf{b} = (\mathbf{X}'\mathbf{X})^{-1}\mathbf{X}'\mathbf{y} = \begin{bmatrix} \bar{y} - b_1\bar{x} \\ \dfrac{n\sum_{i=1}^{n} x_i y_i - \left(\sum_{i=1}^{n} x_i\right)\left(\sum_{i=1}^{n} y_i\right)}{n\sum_{i=1}^{n} x_i^2 - \left(\sum_{i=1}^{n} x_i\right)^2} \end{bmatrix}$$

Therefore, for the simple linear regression model, the formula

$$\mathbf{b} = (\mathbf{X}'\mathbf{X})^{-1}\mathbf{X}'\mathbf{y}$$

yields the same least squares point estimates b_0 and b_1 yielded by the formulas of Section 5.2.

5.5 POINT ESTIMATES AND POINT PREDICTIONS

We will now see how we can use the regression model

$$y = \beta_0 + \beta_1 x_1 + \beta_2 x_2 + \cdots + \beta_p x_p + \epsilon$$

and the least squares point estimates of the parameters $\beta_0, \beta_1, \beta_2, \ldots, \beta_p$ to find a point prediction of y_0, the future (individual) value of the dependent variable (which will occur in the future situation when the values of the p independent variables will be $x_{01}, x_{02}, \ldots, x_{0p}$). We can express y_0 by the equation

$$y_0 = \mu_0 + \epsilon_0$$

Here we assume the following:

1. y_0 is the value of the dependent variable that we will actually observe when a random selection is made from the *future population of potential values of the dependent variable,* which is the infinite population of potential values of the dependent variable that we could observe when the future values of the p independent variables will be $x_{01}, x_{02}, \ldots, x_{0p}$.
2. μ_0 is the *mean* of the future population of potential values of the dependent variable. Intuitively, μ_0, which we call the *future mean value of the dependent variable,* is the average of all the potential values of the dependent variable that we could observe when the future values of the p independent variables will be $x_{01}, x_{02}, \ldots, x_{0p}$.
3. ϵ_0, which we call the *future error term,* describes the effects on y_0 of all factors that will occur in the future situation other than $x_{01}, x_{02}, \ldots, x_{0p}$.

Clearly, we do not know what y_0, the future (individual) value of the dependent variable, will be. Moreover, since we do not know and thus cannot average the infinite number of potential future values of the dependent variable, we cannot know the true value of μ_0, the future mean value of the dependent variable. In order to find a point estimate of μ_0 and a point prediction of y_0, we define the experimental region.

> The **experimental region** is the range of the combinations of the observed values of the p independent variables $x_1, x_2, \ldots, x_p$.

We illustrated the concept of the experimental region in Example 5.5 in the context of the fuel consumption problem. Now, in general, assume that we have used the observed data summarized in Table 5.3 to determine that for $i = 1, 2, \ldots, n$, the linear equation

$$\mu_i = \beta_0 + \beta_1 x_{i1} + \beta_2 x_{i2} + \cdots + \beta_p x_{ip}$$

relates μ_i, the ith mean value of the dependent variable, to $x_{i1}, x_{i2}, \ldots, x_{ip}$, the values of the p independent variables that have occurred in OST_i (the ith observed situation). Then, if the point representing the combination of future values of the p independent variables $(x_{01}, x_{02}, \ldots, x_{0p})$ is in the experimental region, it seems reasonable to conclude that the future mean value of the dependent variable, μ_0, is described by the equation

$$\mu_0 = \beta_0 + \beta_1 x_{01} + \beta_2 x_{02} + \cdots + \beta_p x_{0p}$$

and thus the future value of the dependent variable, $y_0 = \mu_0 + \epsilon_0$, is described by the equation

$$y_0 = \beta_0 + \beta_1 x_{01} + \beta_2 x_{02} + \cdots + \beta_p x_{0p} + \epsilon_0$$

Assuming that $b_0, b_1, b_2, \ldots, b_p$ are the least squares point estimates of the parameters $\beta_0, \beta_1, \beta_2, \ldots, \beta_p$, we can obtain the following point estimate and point prediction.

> **A POINT ESTIMATE OF μ_0 AND A POINT PREDICTION OF y_0:**
>
> $$\hat{y}_0 = b_0 + b_1 x_{01} + b_2 x_{02} + \cdots + b_p x_{0p}$$
>
> is *the point estimate of* μ_0, the future mean value of the dependent variable, and is *the point prediction of* $y_0 = \mu_0 + \epsilon_0$, the future (individual) value of the dependent variable.

The logic behind these results is that if we replace the parameters $\beta_0, \beta_1, \beta_2, \ldots, \beta_p$ with their respective least squares point estimates $b_0, b_1, b_2, \ldots, b_p$, then $\hat{y}_0$ is clearly the point estimate of μ_0. Therefore, since we will predict the future error term, ϵ_0, to be zero, $\hat{y}_0$ is also the point prediction of $y_0 = \mu_0 + \epsilon_0$.

EXAMPLE 5.8

Recall from Example 5.1 that in the fuel consumption problem we wish to predict the amount of fuel (in tons of coal) that will be used to heat buildings at State University during the future week, when the average hourly temperature will be $x_{01} = 40.0$ and the chill index will be $x_{02} = 10$. We assume that this future fuel consumption, which can be expressed in the form $y_0 = \mu_0 + \epsilon_0$, is the fuel consumption that we will actually observe when a random selection is made from the *future population of potential weekly fuel consumptions*. Here, μ_0 is the mean of this population, that is, μ_0 is the average of all the potential weekly fuel consumptions that we could observe when the average hourly temperature will be $x_{01} = 40.0$ and the chill index will be $x_{02} = 10$. Now recall that geometrically the shaded plane illustrated in Figures 5.8 and 5.10, which is defined by the equation

$$\mu = \beta_0 + \beta_1 x_1 + \beta_2 x_2$$

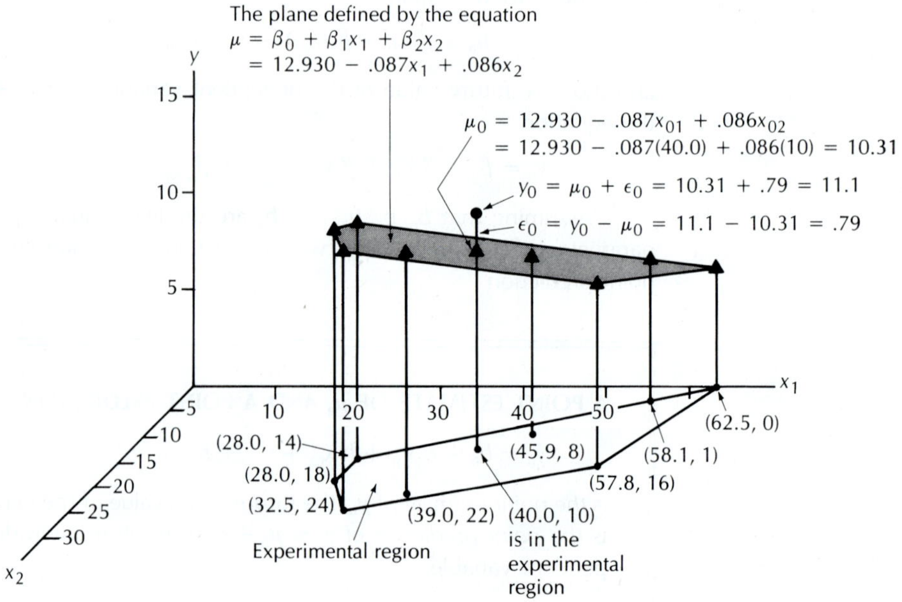

FIGURE 5.10
A Geometrical Interpretation of μ_0, the Future Mean Fuel Consumption, and of $y_0 = \mu_0 + \epsilon_0$, the Future (Individual) Fuel Consumption

is the plane that relates the eight mean fuel consumptions $\mu_1, \mu_2, \ldots, \mu_8$ to the eight points (combinations of observed average hourly temperatures and chill indices)

$$(x_{11}, x_{12}) = (28.0, 18)$$
$$(x_{21}, x_{22}) = (28.0, 14)$$
$$\vdots \qquad \vdots$$
$$(x_{81}, x_{82}) = (62.5, 0)$$

which define the experimental region. By using some elementary geometry, we can show that (as illustrated in Figure 5.10) the point representing the combination of the future average hourly temperature and the future chill index, $(x_{01}, x_{02}) = (40.0, 10)$, is in the experimental region. Therefore, the equation

$$\begin{aligned}\mu_0 &= \beta_0 + \beta_1 x_{01} + \beta_2 x_{02}\\ &= \beta_0 + \beta_1(40.0) + \beta_2(10)\end{aligned}$$

relates μ_0, the future mean fuel consumption, to $x_{01} = 40.0$ and $x_{02} = 10$, and the equation

$$\begin{aligned}y_0 &= \mu_0 + \epsilon_0\\ &= \beta_0 + \beta_1 x_{01} + \beta_2 x_{02} + \epsilon_0\\ &= \beta_0 + \beta_1(40.0) + \beta_2(10) + \epsilon_0\end{aligned}$$

describes the future fuel consumption. Thus, as illustrated in Figure 5.11,

$$\begin{aligned}\hat{y}_0 &= b_0 + b_1 x_{01} + b_2 x_{02}\\ &= b_0 + b_1(40.0) + b_2(10)\\ &= 13.109 + (-.0900)(40.0) + (.0825)(10)\\ &= 10.333 \text{ tons of coal}\end{aligned}$$

is the point estimate of μ_0, which (a supernatural power knows) is equal to

$$\begin{aligned}\mu_0 &= \beta_0 + \beta_1 x_{01} + \beta_2 x_{02}\\ &= 12.930 + (-.087)(40.0) + (.086)(10)\\ &= 10.31 \text{ tons of coal}\end{aligned}$$

and is the point prediction of y_0, which (a supernatural power knows) will be

$$\begin{aligned}y_0 &= \mu_0 + \epsilon_0\\ &= 10.31 + .79\\ &= 11.1 \text{ tons of coal}\end{aligned}$$

In Chapter 6 we explain how to calculate a bound for the farthest $\hat{y}_0$ might be from μ_0, and a bound for the farthest $\hat{y}_0$ might be from y_0.

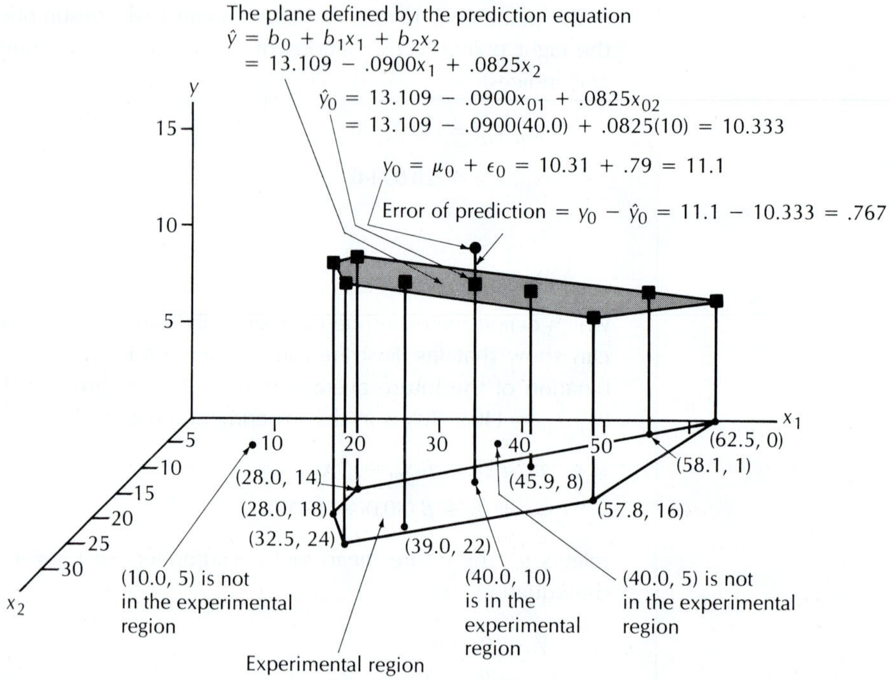

FIGURE 5.11
A Geometrical Interpretation of $\hat{y}_0$

We have said that since the point $(x_{01}, x_{02}) = (40.0, 10)$ is in the experimental region, it is reasonable to use the prediction equation

$$\begin{aligned}\hat{y} &= b_0 + b_1x_1 + b_2x_2 \\ &= 13.109 - .0900x_1 + .0825x_2\end{aligned}$$

to predict y_0. However, it is dangerous (it can lead to unacceptable prediction errors) to extrapolate the plane defined by this prediction equation to predict a future value of y on the basis of a combination of future values of x_1 and x_2 if the point representing this combination is far outside the experimental region. Moreover, the farther outside the experimental region the point representing the combination of future values is, the more likely we are to obtain large prediction errors. For example, since Figure 5.11 reveals that the point representing the combination (10.0, 5) of future values of x_1 and x_2 is far outside the experimental region, extrapolating the plane defined by our prediction equation to predict a future value of y on the basis of this combination might yield an unacceptable prediction error. Intuitively, while it is reasonable to use a prediction equation that is the equation of a plane to predict a future value of y on the basis of future values of x_1 and x_2 that are in the experimental region, it is possibly more appropriate to use a prediction equation that is the equation of a *curved surface* to predict a future value of y on the basis of the combination (10.0, 5). The only way to tell what sort of prediction equation should be used is to observe data so that the point representing the combination (10.0, 5) is in the experimental region. As another example, although a future average hourly temperature of 40.0°F is in the range of the observed average hourly temperatures—28.0°F to 62.5°F—and although a future chill index of 5 is in the range of the observed chill indices—0 to 18—note from Figure 5.11 that the point representing the *combination* (40.0, 5) of future values of x_1 and x_2 is outside the experimental region. Hence, extrapolating the plane defined by the prediction equation

$$\begin{aligned}\hat{y} &= b_0 + b_1x_1 + b_2x_2 \\ &= 13.109 - .0900x_1 + .0825x_2\end{aligned}$$

to predict a future value of y on the basis of the combination (40.0, 5) might also yield an unacceptable prediction error.

5.6 THE QUADRATIC REGRESSION MODEL

In this section we study another useful regression model called the quadratic regression model. Recall that we sometimes wish to consider a regression model relating y, the dependent variable, to a single independent variable, x. Assume that we have observed n situations,

$$OST_1, OST_2, \ldots, OST_n$$

and n values of the independent variable x,

$$x_1, x_2, \ldots, x_n$$

and n values of the dependent variable y,

$$y_1, y_2, \ldots, y_n$$

that have occurred during these n situations. Then

The **quadratic regression model** relating y_i to x_i is expressed as

$$\begin{aligned}y_i &= \mu_i + \epsilon_i \\ &= \beta_0 + \beta_1x_i + \beta_2x_i^2 + \epsilon_i\end{aligned}$$

We call this model a quadratic regression model because the equation

$$\mu_i = \beta_0 + \beta_1x_i + \beta_2x_i^2$$

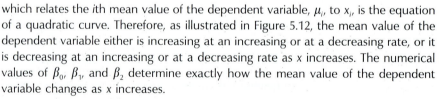

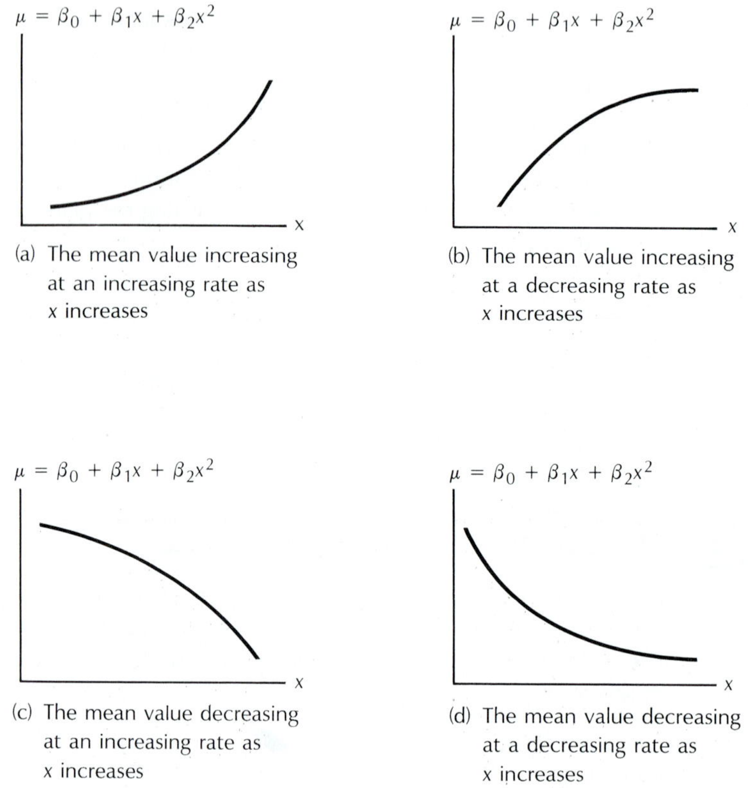

FIGURE 5.12
The Mean Value of the Dependent Variable Changing in a Quadratic Fashion as x Increases

which relates the ith mean value of the dependent variable, μ_i, to x_i, is the equation of a quadratic curve. Therefore, as illustrated in Figure 5.12, the mean value of the dependent variable either is increasing at an increasing or at a decreasing rate, or it is decreasing at an increasing or at a decreasing rate as x increases. The numerical values of β_0, β_1, and β_2 determine exactly how the mean value of the dependent variable changes as x increases.

In Figure 5.13 the quadratic regression model says that the n error terms ϵ_1, $\epsilon_2, \ldots, \epsilon_n$ cause the n observed values of the dependent variable

$$
\begin{aligned}
y_1 &= \mu_1 + \epsilon_1 = \beta_0 + \beta_1 x_1 + \beta_2 x_1^2 + \epsilon_1 \\
y_2 &= \mu_2 + \epsilon_2 = \beta_0 + \beta_1 x_2 + \beta_2 x_2^2 + \epsilon_2 \\
&\vdots \\
y_n &= \mu_n + \epsilon_n = \beta_0 + \beta_1 x_n + \beta_2 x_n^2 + \epsilon_n
\end{aligned}
$$

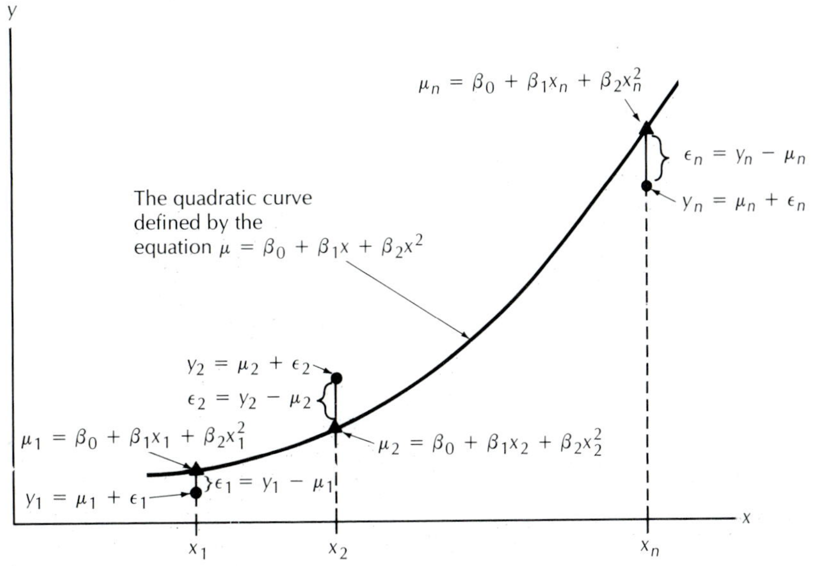

FIGURE 5.13 The Quadratic Regression Model

which are represented by the dots in the figure, to deviate from the n mean values of the dependent variable

$$\mu_1 = \beta_0 + \beta_1 x_1 + \beta_2 x_1^2$$
$$\mu_2 = \beta_0 + \beta_1 x_2 + \beta_2 x_2^2$$
$$\vdots \qquad \vdots$$
$$\mu_n = \beta_0 + \beta_1 x_n + \beta_2 x_n^2$$

which are represented by the triangles in the figure and which exactly lie on the quadratic curve defined by the equation

$$\mu = \beta_0 + \beta_1 x + \beta_2 x^2$$

Since we do not know the true values of the parameters in the quadratic regression model, these parameters must be estimated. If b_0, b_1, and b_2 are the point estimates of the parameters β_0, β_1, and β_2, then, for $i = 1, 2, \ldots, n$,

$$\hat{y}_i = b_0 + b_1 x_i + b_2 x_i^2$$

is the point estimate of the ith mean value of the dependent variable

$$\mu_i = \beta_0 + \beta_1 x_i + \beta_2 x_i^2$$

and the point prediction of the ith observed value of the dependent variable

$$\begin{aligned} y_i &= \mu_i + \epsilon_i \\ &= \beta_0 + \beta_1 x_i + \beta_2 x_i^2 + \epsilon_i \end{aligned}$$

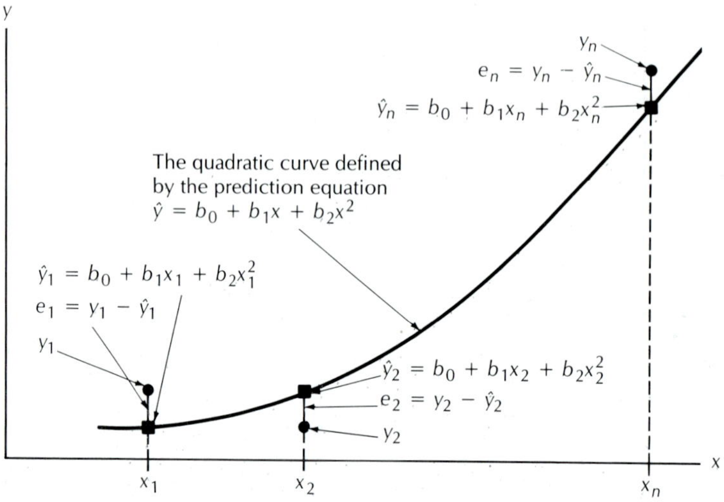

FIGURE 5.14
The Quadratic Prediction Equation and the *n* Residuals

Using

$$\mathbf{y} = \begin{bmatrix} y_1 \\ y_2 \\ \vdots \\ y_n \end{bmatrix} \quad \text{and} \quad \mathbf{X} = \begin{bmatrix} 1 & x_1 & x_1^2 \\ 1 & x_2 & x_2^2 \\ \vdots & \vdots & \vdots \\ 1 & x_n & x_n^2 \end{bmatrix}$$

the least squares point estimates are given by the formula

$$\begin{bmatrix} b_0 \\ b_1 \\ b_2 \end{bmatrix} = \mathbf{b} = (\mathbf{X}'\mathbf{X})^{-1}\mathbf{X}'\mathbf{y}$$

As illustrated in Figure 5.14, the least squares point estimates b_0, b_1, and b_2 position the quadratic prediction equation

$$\hat{y} = b_0 + b_1x + b_2x^2$$

so that we minimize

$$SSE = \sum_{i=1}^{n} (y_i - \hat{y}_i)^2$$

$$= \sum_{i=1}^{n} (y_i - (b_0 + b_1x_i + b_2x_i^2))^2$$

which may be interpreted as the sum of squared distances between the observed values of the dependent variable and the predicted values of the dependent variable given by the preceding prediction equation.

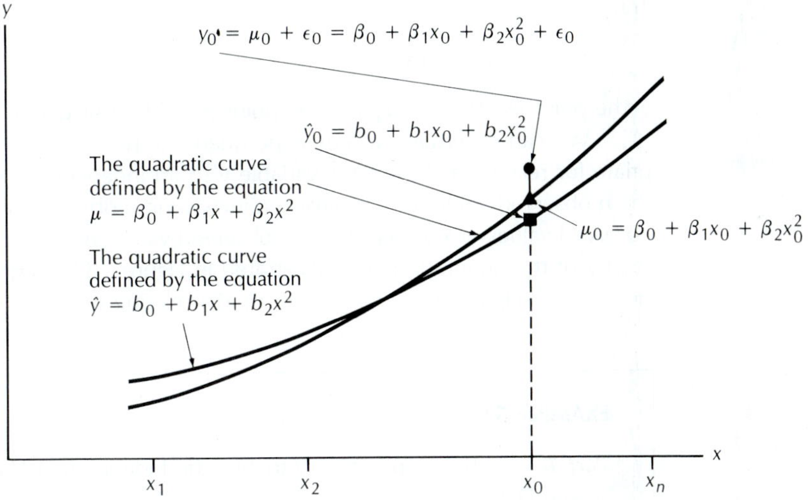

FIGURE 5.15
A Comparison of the Quadratic Curves Defined by $\mu = \beta_0 + \beta_1 x + \beta_2 x^2$, $\hat{y} = b_0 + b_1 x + b_2 x^2$

Next, suppose we wish to predict the future (individual) value of the dependent variable

$$y_0 = \mu_0 + \epsilon_0$$

which is the value of the dependent variable that will occur in the future situation, when the future value of the independent variable will be x_0. If we have concluded that the quadratic curve defined by the equation

$$\mu = \beta_0 + \beta_1 x + \beta_2 x^2$$

relates the n mean values of the dependent variable $\mu_1, \mu_2, \ldots, \mu_n$ to the n observed values of the independent variable $x_1, x_2, \ldots, x_n$, and if (as illustrated in Figure 5.15) x_0, the future value of the independent variable, is in the experimental region, then it seems reasonable to conclude that

1. The equation

$$\mu_0 = \beta_0 + \beta_1 x_0 + \beta_2 x_0^2$$

relates μ_0 to x_0, and thus

2. The future (individual) value of the dependent variable

$$y_0 = \mu_0 + \epsilon_0$$

can be described by the equation

$$\begin{aligned} y_0 &= \mu_0 + \epsilon_0 \\ &= \beta_0 + \beta_1 x_0 + \beta_2 x_0^2 + \epsilon_0 \end{aligned}$$

Thus,

$$\hat{y}_0 = b_0 + b_1x_0 + b_2x_0^2$$

is the point estimate of μ_0 and the point prediction of y_0 (see Figure 5.15).

We can determine whether the quadratic regression model might be appropriate for relating a dependent variable y to an independent variable x by plotting the n observed values of y against the increasing n observed values of x. If, as the n observed values of x increase, the n observed values of y tend to change according to any of the quadratic curves illustrated in Figure 5.12, then the quadratic regression model might be an appropriate model relating y to x.

EXAMPLE 5.9

Part 1. The Dependent Variable, the Independent Variables, and the Observed Data.

Enterprise Industries produces Fresh, a brand of liquid laundry detergent. In order to plan its production schedule more effectively, plan its budget, estimate requirements for producing Fresh, and manage its inventory of Fresh, the company would like to predict the demand for the extra large size bottle of Fresh for future sales periods (where a sales period is defined to be a 4-week period). Thus, "the demand for (the extra large size bottle of) Fresh (in hundreds of thousands of bottles) in a sales period" is the *dependent variable*. We denote this dependent variable by the symbol y. Enterprise Industries believes that demand for Fresh depends upon various independent variables. Some of these independent variables are

1. x_1, the price (in dollars) of Fresh as offered by Enterprise Industries in the sales period;
2. x_2, the average industry price (in dollars) of competitors' similar detergents in the sales period;
3. x_3, Enterprise Industries' advertising expenditure (in hundreds of thousands of dollars) to promote Fresh in the sales period.

Enterprise Industries feels that it can accurately predict demand for Fresh by using regression analysis to develop a prediction equation utilizing only the independent variables x_1, x_2, and x_3. Moreover, the company knows that in order to use the prediction equation it will have to predict future values of these independent variables. Of course, Enterprise Industries knows exactly what x_1, its price for Fresh, and x_3, its advertising expenditure for Fresh, will be in future sales periods. Besides this, it feels that by examining the prices of competitors' similar products immediately prior to a future sales period, it can very accurately predict x_2, the average industry price for competitors' similar detergents for a future sales period. To summarize, therefore, Enterprise Industries can use a

TABLE 5.12 Historical Data, Including Price Differences, Concerning Demand for Fresh Detergent

The *i*th Observed Sales Period (OSP_i)	The *i*th Observed Price for Fresh (dollars) (x_{i1})	The *i*th Observed Average Industry Price (dollars) (x_{i2})	The *i*th Observed Price Difference (dollars) ($x_{i4} = x_{i2} - x_{i1}$)	The *i*th Observed Advertising Expenditure for Fresh (hundreds of thousands of dollars) (x_{i3})	The *i*th Observed Demand for Fresh (hundreds of thousands of bottles) (y_i)
OSP_1	3.85	3.80	−.05	5.50	7.38
OSP_2	3.75	4.00	.25	6.75	8.51
OSP_3	3.70	4.30	.60	7.25	9.52
OSP_4	3.70	3.70	0	5.50	7.50
OSP_5	3.60	3.85	.25	7.00	9.33
OSP_6	3.60	3.80	.20	6.50	8.28
OSP_7	3.60	3.75	.15	6.75	8.75
OSP_8	3.80	3.85	.05	5.25	7.87
OSP_9	3.80	3.65	−.15	5.25	7.10
OSP_{10}	3.85	4.00	.15	6.00	8.00
OSP_{11}	3.90	4.10	.20	6.50	7.89
OSP_{12}	3.90	4.00	.10	6.25	8.15
OSP_{13}	3.70	4.10	.40	7.00	9.10
OSP_{14}	3.75	4.20	.45	6.90	8.86
OSP_{15}	3.75	4.10	.35	6.80	8.90
OSP_{16}	3.80	4.10	.30	6.80	8.87
OSP_{17}	3.70	4.20	.50	7.10	9.26
OSP_{18}	3.80	4.30	.50	7.00	9.00
OSP_{19}	3.70	4.10	.40	6.80	8.75
OSP_{20}	3.80	3.75	−.05	6.50	7.95
OSP_{21}	3.80	3.75	−.05	6.25	7.65
OSP_{22}	3.75	3.65	−.10	6.00	7.27
OSP_{23}	3.70	3.90	.20	6.50	8.00
OSP_{24}	3.55	3.65	.10	7.00	8.50
OSP_{25}	3.60	4.10	.50	6.80	8.75
OSP_{26}	3.65	4.25	.60	6.80	9.21
OSP_{27}	3.70	3.65	−.05	6.50	8.27
OSP_{28}	3.75	3.75	0	5.75	7.67
OSP_{29}	3.80	3.85	.05	5.80	7.93
OSP_{30}	3.70	4.25	.55	6.80	9.26

prediction equation that utilizes x_1, x_2, and x_3, because (for all practical purposes) it knows the future values of these variables.

Although Enterprise Industries believes that it can predict future demand for Fresh accurately enough to meet its needs by utilizing only x_1, x_2, and x_3, the company realizes that other independent variables also affect demand for Fresh. For example, "competitors' average advertising expenditures to promote their similar detergents" in a future sales period also affects demand for Fresh. However, Enterprise Industries will not utilize this independent variable, because its competitors would probably not provide the information needed to determine their average advertising expenditures for future periods.

We will assume that in order to develop a prediction equation relating y to x_1, x_2, and x_3, Enterprise Industries has observed the historical data in Table 5.12, which consist of the values of the independent variables x_1, x_2, and x_3 and the dependent variable y that have occurred during the past $n = 30$ sales periods. In Table 5.12 the subscript i, where $i = 1, 2, \ldots, 30$, denotes the time order of a previously observed sales period, with $i = 1$ referring to the sales period observed first in the time order, or the sales period observed in the most distant past, and with $i = 30$ referring to the sales period observed thirtieth in the time order, or the sales period observed most recently. For $i = 1, 2, \ldots, 30$, we refer to the sales period that has been observed ith in the time order as *the ith observed sales period*, denoted by OSP_i. Moreover, we let x_{i1}, x_{i2}, x_{i3}, and y_i denote, respectively, the price for Fresh, the average industry price, the advertising expenditure for Fresh, and the demand for Fresh that have occurred in the ith observed sales period.

Part 2. Developing a Regression Model.

We now begin to develop a regression model relating y to x_1, x_2, and x_3. Suppose that Enterprise Industries believes on theoretical grounds that the single independent variable

$$x_4 = x_2 - x_1$$

which we call the "price difference" and which is the difference between x_2, the average industry price, and x_1, Enterprise Industries' price for Fresh, adequately describes the effects of x_1 and x_2 on y, demand for Fresh. Therefore, Enterprise Industries has decided to develop a regression model relating y to $x_4 = x_2 - x_1$ and to x_3. In addition to listing the thirty observed values of y, x_1, x_2, and x_3, Table 5.12 lists the thirty observed values of $x_4 = x_2 - x_1$. Here, for $i = 1, 2, \ldots, 30$,

$$x_{i4} = x_{i2} - x_{i1}$$

denotes the price difference that has existed in the ith observed sales period.

Note that the observed values of x_3 (the advertising expenditure for Fresh) are arranged in increasing order, along with the corresponding observed values

TABLE 5.13 Values of x_3, Advertising Expenditure for Fresh Detergent, Arranged in Increasing Order

x_{i3}	y_i	x_{i3}	y_i
5.25	7.87	6.75	8.51
5.25	7.10	6.75	8.75
5.50	7.38	6.80	8.90
5.50	7.50	6.80	8.87
5.75	7.67	6.80	8.75
5.80	7.93	6.80	8.75
6.00	8.00	6.80	9.21
6.00	7.27	6.80	9.26
6.25	8.15	6.90	8.86
6.25	7.65	7.00	9.33
6.50	8.28	7.00	9.10
6.50	7.89	7.00	8.50
6.50	7.95	7.00	9.00
6.50	8.00	7.10	9.26
6.50	8.27	7.25	9.52

of y (the demand for Fresh), in Table 5.13. In Figure 5.16 the observed values of y are plotted against the increasing values of x_3. Inspecting Figure 5.16, we see that as the observed values of x_3 increase, the observed values of y tend to increase in the quadratic fashion illustrated in Figure 5.12a, which shows the mean value of the dependent variable increasing at an increasing rate. This

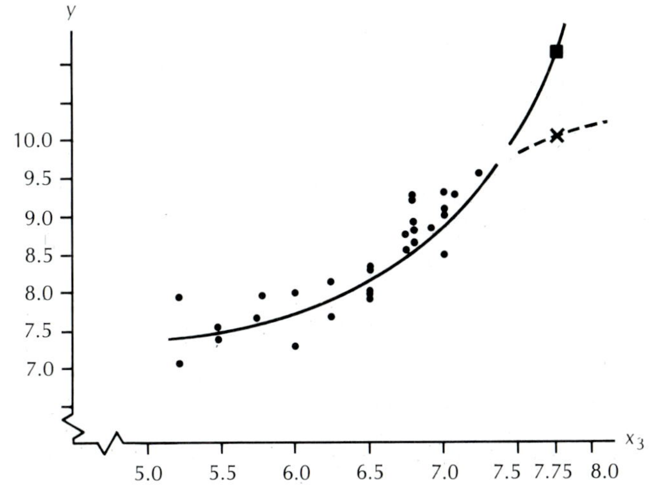

FIGURE 5.16 Plot of the Values of y (Demand for Fresh Detergent) Against Increasing Values of x_3 (Advertising Expenditure for Fresh)

implies that the quadratic regression model illustrated in Figure 5.13,

$$\begin{aligned} y_i &= \mu_i + \epsilon_i \\ &= \beta_0 + \beta_1 x_{i3} + \beta_2 x_{i3}^2 + \epsilon_i \end{aligned}$$

might be an appropriate regression model relating y_i to x_{i3}.

Now, consider predicting a future demand for Fresh using this regression model. As long as the advertising expenditure is in the experimental region (which is the range of advertising expenditures from 5.25 to 7.25—see Table 5.13), this regression model will probably yield a fairly accurate prediction of the future demand for Fresh. However, consider predicting a future demand for Fresh for a sales period in which the advertising expenditure will be 7.75, which, as illustrated in Figure 5.16, is outside the experimental region. If we extrapolate the quadratic prediction equation

$$\hat{y} = b_0 + b_1 x_3 + b_2 x_3^2$$

which is represented as the solid curve in Figure 5.16, we obtain a prediction of demand (represented by the square in Figure 5.16) that is quite large. If we assume that Enterprise Industries' advertising is extremely effective, such a prediction might be reasonable. But, assuming that the law of diminishing returns sets in, it is just as possible that the curve describing demand will change direction, as illustrated by the dashed curve in Figure 5.16. In this case, a much lower prediction of demand (illustrated by the $\times$ in Figure 5.16) is more reasonable. The point we are making here is that since there are no historical data to tell us which prediction is the most reasonable, blindly using the solid curve to predict the demand for Fresh that will occur in a future sales period in which the advertising expenditure for Fresh will be 7.75 might lead to an unacceptably large prediction error. Although it would be unwise to use the prediction equation represented by the solid curve to predict demands for Fresh corresponding to advertising expenditures greater than 7.25 (the end point of the experimental region), the fact that the mean demand for Fresh is increasing at an increasing rate as the advertising expenditure for Fresh increases might mean that Enterprise Industries could significantly increase sales by spending more money on advertising.

Figure 5.16 indicates that the observed values of y tend to increase in a quadratic fashion as the values of x_3 increase, implying that the model

$$\begin{aligned} y_i &= \mu_i + \epsilon_i \\ &= \beta_0 + \beta_1 x_{i3} + \beta_2 x_{i3}^2 + \epsilon_i \end{aligned}$$

relates y_i to x_{i3}. Also, Figure 5.17 and Table 5.14 indicate that the observed values of y tend to increase in a straight line fashion as the values of x_4 (the price difference) increase, implying that the model

$$\begin{aligned} y_i &= \mu_i + \epsilon_i \\ &= \beta_0 + \beta_1 x_{i4} + \epsilon_i \end{aligned}$$

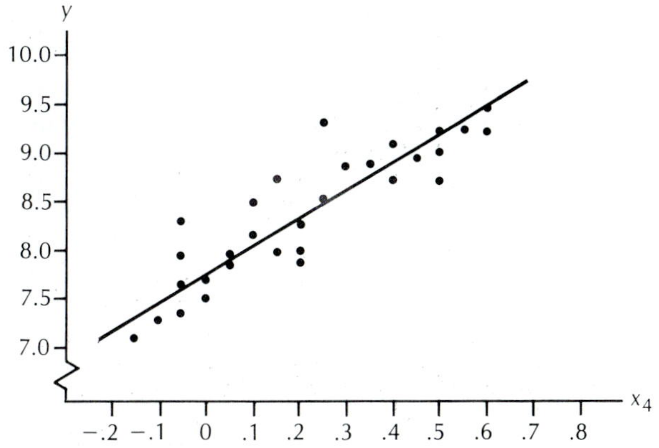

FIGURE 5.17 Plot of the Values of y (Demand for Fresh Detergent) Against Increasing Values of x_4 (Price Difference)

TABLE 5.14 Values of x_4, Price Difference, Arranged in Increasing Order

x_{i4}	y_i	x_{i4}	y_i
−.15	7.10	.20	7.89
−.10	7.27	.20	8.00
−.05	8.27	.25	8.51
−.05	7.38	.25	9.33
−.05	7.95	.30	8.87
−.05	7.65	.35	8.90
0	7.50	.40	9.10
0	7.67	.40	8.75
.05	7.87	.45	8.86
.05	7.93	.50	9.26
.10	8.15	.50	9.00
.10	8.50	.50	8.75
.15	8.75	.55	9.26
.15	8.00	.60	9.52
.20	8.28	.60	9.21

relates y_i to x_{i4}. Thus, it seems reasonable to combine these two models to form the regression model

$$\begin{aligned} y_i &= \mu_i + \epsilon_i \\ &= \beta_0 + \beta_1 x_{i4} + \beta_2 x_{i3} + \beta_3 x_{i3}^2 + \epsilon_i \end{aligned}$$

which relates y_i to x_{i4} and x_{i3}. Note that we could also add the cross-product term, $x_{i4}x_{i3}$, to this model to form the model

$$\begin{aligned} y_i &= \mu_i + \epsilon_i \\ &= \beta_0 + \beta_1 x_{i4} + \beta_2 x_{i3} + \beta_3 x_{i3}^2 + \beta_4 x_{i4} x_{i3} + \epsilon_i \end{aligned}$$

which also relates y_i to x_{i4} and x_{i3}. It can be shown that the cross-product term $x_{i4}x_{i3}$ should be included in the preceding model if interaction exists between x_4, the price difference, and x_3, the advertising expenditure, as these independent variables affect y (the demand for Fresh). We discuss the meaning of interaction in the next section. We find in subsequent discussions that $x_{i4}x_{i3}$ should indeed be included in the preceding model, which implies that

$$\begin{aligned} y_i &= \mu_i + \epsilon_i \\ &= \beta_0 + \beta_1 x_{i4} + \beta_2 x_{i3} + \beta_3 x_{i3}^2 + \beta_4 x_{i4} x_{i3} + \epsilon_i \end{aligned}$$

is a reasonable regression model relating y_i to x_{i4} and x_{i3}. This model assumes that

1. y_i, the ith observed demand for Fresh, is the demand for Fresh that Enterprise Industries has actually observed when a random selection has been made from *the ith historical population of potential demands for Fresh,* which is the infinite population of potential demands for Fresh that Enterprise Industries could have observed when the price difference was x_{i4} and the advertising expenditure was x_{i3}.
2. $\mu_i = \beta_0 + \beta_1 x_{i4} + \beta_2 x_{i3} + \beta_3 x_{i3}^2 + \beta_4 x_{i4} x_{i3}$ is the *mean* of the ith historical population of potential demands for Fresh. Intuitively, μ_i is the average of all of the potential demands for Fresh that Enterprise Industries could have observed when the price difference was x_{i4} and the advertising expenditure was x_{i3}.
3. β_0, β_1, β_2, β_3, and β_4 are (unknown) parameters relating μ_i to x_{i4} and x_{i3}.
4. ϵ_i, the ith error term, describes the effects on y_i of all factors that have occurred in OSP_i other than x_{i4} and x_{i3}. As previously mentioned, "competitors' average advertising expenditures to promote their similar detergents" is one such factor.

Part 3. Computing the Least Squares Point Estimates and Writing the Prediction Equation.

If b_0, b_1, b_2, b_3, and b_4 are the least squares point estimates of the parameters in the model

$$\begin{aligned} y_i &= \mu_i + \epsilon_i \\ &= \beta_0 + \beta_1 x_{i4} + \beta_2 x_{i3} + \beta_3 x_{i3}^2 + \beta_4 x_{i4} x_{i3} + \epsilon_i \end{aligned}$$

then, for $i = 1, 2, \ldots, 30$,

$$\hat{y}_i = b_0 + b_1 x_{i4} + b_2 x_{i3} + b_3 x_{i3}^2 + b_4 x_{i4} x_{i3}$$

is the point estimate of the ith mean demand for Fresh,

$$\mu_i = \beta_0 + \beta_1 x_{i4} + \beta_2 x_{i3} + \beta_3 x_{i3}^2 + \beta_4 x_{i4} x_{i3}$$

and the point prediction of the ith observed demand for Fresh,

$$\begin{aligned} y_i &= \mu_i + \epsilon_i \\ &= \beta_0 + \beta_1 x_{i4} + \beta_2 x_{i3} + \beta_3 x_{i3}^2 + \beta_4 x_{i4} x_{i3} + \epsilon_i \end{aligned}$$

By referring to Table 5.12, and by using the column vector

$$\mathbf{y} = \begin{bmatrix} y_1 \\ y_2 \\ y_3 \\ \vdots \\ y_{30} \end{bmatrix} = \begin{bmatrix} 7.38 \\ 8.51 \\ 9.52 \\ \vdots \\ 9.26 \end{bmatrix}$$

and the matrix

$$\mathbf{X} = \begin{array}{c} \\ \\ \end{array} \begin{array}{ccccc} 1 & x_4 & x_3 & x_3^2 & x_4 x_3 \end{array}$$

$$\mathbf{X} = \begin{bmatrix} 1 & -.05 & 5.50 & (5.50)^2 & (-.05)(5.50) \\ 1 & .25 & 6.75 & (6.75)^2 & (.25)(6.75) \\ 1 & .60 & 7.25 & (7.25)^2 & (.60)(7.25) \\ \vdots & \vdots & \vdots & \vdots & \vdots \\ 1 & .55 & 6.80 & (6.80)^2 & (.55)(6.80) \end{bmatrix}$$

$$\begin{array}{ccccc} 1 & x_4 & x_3 & x_3^2 & x_4 x_3 \end{array}$$

$$= \begin{bmatrix} 1 & -.05 & 5.50 & 30.25 & -.275 \\ 1 & .25 & 6.75 & 45.5625 & 1.6875 \\ 1 & .60 & 7.25 & 52.5625 & 4.35 \\ \vdots & \vdots & \vdots & \vdots & \vdots \\ 1 & .55 & 6.80 & 46.24 & 3.74 \end{bmatrix}$$

we can calculate $(\mathbf{X}'\mathbf{X})^{-1}$ and $\mathbf{X}'\mathbf{y}$ to be

$$(\mathbf{X}'\mathbf{X})^{-1} = \begin{bmatrix} 1315.261 & 543.4463 & -433.586 & 35.50156 & -83.4036 \\ 543.4463 & 464.2447 & -179.952 & 14.80313 & -69.5252 \\ -433.586 & -179.952 & 143.1914 & -11.7449 & 27.67939 \\ 35.50156 & 14.80313 & -11.7449 & .965045 & -2.28257 \\ -83.4036 & -69.5252 & 27.67939 & -2.28257 & 10.45448 \end{bmatrix}$$

$$\mathbf{X}'\mathbf{y} = \begin{bmatrix} 251.48 \\ 57.646 \\ 1632.781 \\ 10677.4 \\ 397.7442 \end{bmatrix}$$

It follows that the least squares point estimates b_0, b_1, b_2, b_3, and b_4 of the parameters β_0, β_1, β_2, β_3, and β_4 are

$$\begin{bmatrix} b_0 \\ b_1 \\ b_2 \\ b_3 \\ b_4 \end{bmatrix} = \mathbf{b} = (\mathbf{X}'\mathbf{X})^{-1}\mathbf{X}'\mathbf{y} = \begin{bmatrix} 29.1133 \\ 11.1342 \\ -7.6080 \\ .6712 \\ -1.4777 \end{bmatrix}$$

These least squares point estimates yield the prediction equation

$$\begin{aligned} \hat{y} &= b_0 + b_1x_4 + b_2x_3 + b_3x_3^2 + b_4x_4x_3 \\ &= 29.1133 + 11.1342x_4 - 7.6080x_3 + .6712x_3^2 - 1.4777x_4x_3 \end{aligned}$$

and the sum of squared residuals

$$\begin{aligned} SSE &= \sum_{i=1}^{30} e_i^2 = \sum_{i=1}^{30} (y_i - \hat{y}_i)^2 \\ &= 1.0644 \end{aligned}$$

which is a smaller sum of squared residuals than would be given by any other values of the point estimates b_0, b_1, b_2, b_3, and b_4.

Part 4. Predicting Demand for Fresh.

Now, suppose that we wish to predict the number of (extra large size bottles of) Fresh that will be demanded in a future sales period, when the price of Fresh will be $x_{01} = \$3.80$, the average industry price will be $x_{02} = \$3.90$, the price difference will be $x_{04} = x_{02} - x_{01} = \$3.90 - \$3.80 = \$.10$, and the advertising expenditure for Fresh will be $x_{03} = \$6.80$. We assume that this future (individual) demand for Fresh, which we can express in the form

$$y_0 = \mu_0 + \epsilon_0$$

is the demand for Fresh that Enterprise Industries will actually observe when a random selection is made from the *future population of potential demands for Fresh.* μ_0 is the mean of this population—that is, μ_0 is the average of all of the potential demands for Fresh that Enterprise Industries could observe when the future price difference will be $x_{04} = .10$ and the future advertising expenditure for Fresh will be $x_{03} = 6.8$. We have said it is reasonable to assume that the equation

$$\mu = \beta_0 + \beta_1x_4 + \beta_2x_3 + \beta_3x_3^2 + \beta_4x_4x_3$$

relates the thirty mean demands for Fresh $\mu_1, \mu_2, \ldots, \mu_{30}$ to the thirty points

$$\begin{aligned} (x_{14}, x_{13}) &= (-.05, 5.5) \\ (x_{24}, x_{23}) &= (.25, 6.75) \\ &\vdots \\ (x_{30,4}, x_{30,3}) &= (.55, 6.8) \end{aligned}$$

which define the experimental region (since these points represent the thirty combinations of the observed price differences and the observed advertising expenditures for Fresh). Since (one can verify that) the point representing the combination of the future price difference and the future advertising expenditure for Fresh,

$$(x_{04}, x_{03}) = (.10, 6.8)$$

is in the experimental region, it follows that the equation

$$\begin{aligned}\mu_0 &= \beta_0 + \beta_1 x_{04} + \beta_2 x_{03} + \beta_3 x_{03}^2 + \beta_4 x_{04} x_{03} \\ &= \beta_0 + \beta_1(.10) + \beta_2(6.8) + \beta_3(6.8)^2 + \beta_4(.10)(6.8)\end{aligned}$$

relates μ_0, the future mean demand for Fresh, to $x_{04} = .10$ and to $x_{03} = 6.8$; and thus the equation

$$\begin{aligned}y_0 &= \mu_0 + \epsilon_0 \\ &= \beta_0 + \beta_1 x_{04} + \beta_2 x_{03} + \beta_3 x_{03}^2 + \beta_4 x_{04} x_{03} + \epsilon_0 \\ &= \beta_0 + \beta_1(.10) + \beta_2(6.8) + \beta_3(6.8)^2 + \beta_4(.10)(6.8) + \epsilon_0\end{aligned}$$

describes the future (individual) demand for Fresh. Therefore,

$$\begin{aligned}\hat{y}_0 &= b_0 + b_1 x_{04} + b_2 x_{03} + b_3 x_{03}^2 + b_4 x_{04} x_{03} \\ &= 29.1133 + 11.1342(.10) - 7.6080(6.8) + .6712(6.8)^2 - 1.4777(.10)(6.8) \\ &= 8.526 \qquad \text{(852,600 bottles)}\end{aligned}$$

is the point estimate of μ_0, the future mean demand for Fresh, and the point prediction of $y_0 = \mu_0 + \epsilon_0$, the future (individual) demand for Fresh.

In addition to the quadratic regression model, there are other regression models that can be used to model situations in which the n observed values of y tend to change in a curved fashion. One such model is the pth-order polynomial regression model.

The ***p*th-order polynomial regression model** is

$$\begin{aligned}y_i &= \mu_i + \epsilon_i \\ &= \beta_0 + \beta_1 x_i + \beta_2 x_i^2 + \beta_3 x_i^3 + \cdots + \beta_p x_i^p + \epsilon_i\end{aligned}$$

Clearly, the quadratic regression model is a special case of this more general polynomial model, where $p = 2$. Third-order and higher-order polynomial models are used to model situations in which the n observed values of y change according to a curve that displays one or more reversals in curvature. A third-order polynomial model

$$\begin{aligned} y_i &= \mu_i + \epsilon_i \\ &= \beta_0 + \beta_1 x_i + \beta_2 x_i^2 + \beta_3 x_i^3 + \epsilon_i \end{aligned}$$

describes a curve with one reversal in curvature (a curve with one peak and one trough), while a pth-order polynomial model describes a curve that displays a total of $(p - 1)$ peaks and troughs.

Although third- and higher-order polynomial models are occasionally found to be useful, the quadratic regression model seems to appropriately describe most situations in which the observed values of y tend to change in a curved fashion. In addition to polynomial models, however, there are still other models that can be used to describe curved relationships. We study some of these models in Chapter 10.

5.7 INTERACTION

Multiple regression models often contain **interaction variables**. We say that there is **no interaction** between two independent variables if the relationship between the mean value of the dependent variable and each one of the independent variables is independent of the value of the other independent variable. There is said to be **interaction** between two independent variables if the relationship between the mean value of the dependent variable and one of the independent variables is dependent upon the value of the other independent variable. We discuss the concept of interaction in detail in the next example.

EXAMPLE 5.10

In the Fresh Detergent problem, there is no interaction between x_4 (the price difference) and x_3 (the advertising expenditure for Fresh) if the relationship between μ, the mean demand for Fresh, and x_3 is independent of x_4, and if the relationship between μ and x_4 is independent of x_3. There is said to be interaction between x_4 and x_3 if the relationship between μ and x_3 is dependent upon x_4, or if the relationship between μ and x_4 is dependent upon x_3.

To better understand these ideas, consider the equation

$$\mu = \beta_0 + \beta_1 x_4 + \beta_2 x_3 + \beta_3 x_3^2$$

which relates μ, the mean demand for Fresh, to x_4 and x_3. Using this equation, we can calculate

1. The mean demand for Fresh for a sales period in which x_4 is d and x_3 is a to be

$$\begin{aligned} \mu_{[d,a]} &= \beta_0 + \beta_1 x_4 + \beta_2 x_3 + \beta_3 x_3^2 \\ &= \beta_0 + \beta_1 d + \beta_2 a + \beta_3 a^2 \end{aligned}$$

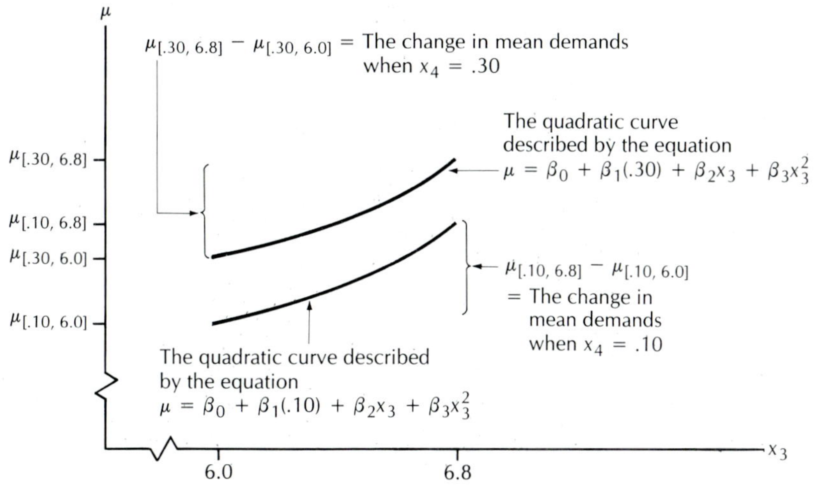

FIGURE 5.18 Two Parallel Quadratic Curves Illustrating No Interaction Between x_3 and x_4

2. The mean demand for Fresh for a sales period in which x_4 is d and x_3 is a higher value, say $a + \Delta$, to be

$$\begin{aligned}\mu_{[d,a+\Delta]} &= \beta_0 + \beta_1 x_4 + \beta_2 x_3 + \beta_3 x_3^2 \\ &= \beta_0 + \beta_1 d + \beta_2(a + \Delta) + \beta_3(a + \Delta)^2\end{aligned}$$

Thus, the equation

$$\mu = \beta_0 + \beta_1 x_4 + \beta_2 x_3 + \beta_3 x_3^2$$

says that (1) the change in the mean demand for Fresh associated with increasing the advertising expenditure for Fresh from a to $a + \Delta$ while holding the price difference constant at d is

$$\begin{aligned}\mu_{[d,a+\Delta]} - \mu_{[d,a]} &= [\beta_0 + \beta_1 d + \beta_2(a + \Delta) + \beta_3(a + \Delta)^2] \\ &\quad -[\beta_0 + \beta_1 d + \beta_2 a + \beta_3 a^2] \\ &= \beta_2\Delta + \beta_3[(a + \Delta)^2 - a^2]\end{aligned}$$

which does not depend on d, the price difference, and (2) the change in the mean demand for Fresh associated with increasing the price difference while holding the advertising expenditure constant does not depend on the advertising expenditure for Fresh. (This can be verified by using techniques similar to the preceding.) It therefore follows that the equation

$$\mu = \beta_0 + \beta_1 x_4 + \beta_2 x_3 + \beta_3 x_3^2$$

assumes that there is no interaction between x_4, the price difference, and x_3, the advertising expenditure for Fresh. This fact is evident from Figure 5.18, which illustrates the following:

1. The quadratic curve defined by the equation

$$\mu = \beta_0 + \beta_1 x_4 + \beta_2 x_3 + \beta_3 x_3^2$$
$$= \beta_0 + \beta_1(.10) + \beta_2 x_3 + \beta_3 x_3^2$$

which describes the relationship between μ and x_3 when x_4 equals .10.

2. The quadratic curve defined by the equation

$$\mu = \beta_0 + \beta_1 x_4 + \beta_2 x_3 + \beta_3 x_3^2$$
$$= \beta_0 + \beta_1(.30) + \beta_2 x_3 + \beta_3 x_3^2$$

which describes the relationship between μ and x_3 when x_4 equals .30.

3. The fact that the two quadratic curves defined by the equations in (1) and (2) increase in a parallel fashion. This results from the fact that the equation

$$\mu = \beta_0 + \beta_1 x_4 + \beta_2 x_3 + \beta_3 x_3^2$$

which assumes no interaction, implies that

$$\mu_{[d,a+\Delta]} - \mu_{[d,a]} = \beta_2 \Delta + \beta_3[(a+\Delta)^2 - a^2]$$

does not depend on the price difference, *d*. Thus, for example,

$$\mu_{[.10,6.8]} - \mu_{[.10,6.0]} = \beta_2(.80) + \beta_3[(6.8)^2 - (6.0)^2]$$
$$= .80\beta_2 + 10.24\beta_3$$

which is the change in the mean demand for Fresh associated with increasing the advertising expenditure from 6.0 to 6.8 while holding the price difference constant at .10, *equals*

$$\mu_{[.30,6.8]} - \mu_{[.30,6.0]} = \beta_2(.80) + \beta_3[(6.8)^2 - (6.0)^2]$$
$$= .80\beta_2 + 10.24\beta_3$$

which is the change in the mean demand for Fresh associated with increasing the advertising expenditure from 6.0 to 6.8 while holding the price difference constant at .30.

We next discuss what is meant when we say there is interaction between x_4 and x_3. Note that by using the equation

$$\mu = \beta_0 + \beta_1 x_4 + \beta_2 x_3 + \beta_3 x_3^2 + \beta_4 x_4 x_3$$

we can calculate

1. The mean demand for Fresh for a sales period in which x_4 is *d* and x_3 is *a* to be

$$\mu_{[d,a]} = \beta_0 + \beta_1 x_4 + \beta_2 x_3 + \beta_3 x_3^2 + \beta_4 x_4 x_3$$
$$= \beta_0 + \beta_1 d + \beta_2 a + \beta_3 a^2 + \beta_4 da$$

2. The mean demand for Fresh for a sales period in which x_4 is d and x_3 is a higher value, say $a + \Delta$, to be

$$\mu_{[d,a+\Delta]} = \beta_0 + \beta_1 x_4 + \beta_2 x_3 + \beta_3 x_3^2 + \beta_4 x_4 x_3$$
$$= \beta_0 + \beta_1 d + \beta_2(a + \Delta) + \beta_3(a + \Delta)^2 + \beta_4 d(a + \Delta)$$

It follows that the equation

$$\mu = \beta_0 + \beta_1 x_4 + \beta_2 x_3 + \beta_3 x_3^2 + \beta_4 x_4 x_3$$

says that the change in the mean demand for Fresh associated with increasing the advertising expenditure for Fresh from a to $a + \Delta$ while holding the price difference constant at d is

$$\mu_{[d,a+\Delta]} - \mu_{[d,a]} = [\beta_0 + \beta_1 d + \beta_2(a + \Delta) + \beta_3(a + \Delta)^2 + \beta_4 d(a + \Delta)]$$
$$- [\beta_0 + \beta_1 d + \beta_2 a + \beta_3 a^2 + \beta_4 da]$$
$$= \beta_2 \Delta + \beta_3[(a + \Delta)^2 - a^2] + \beta_4 d\Delta$$

which does depend on d, the price difference. Therefore, this equation (which includes the cross-product term $x_4 x_3$) assumes that there is interaction between x_4, the price difference, and x_3, the advertising expenditure. To better understand the nature of this interaction, first note that since (from Example 5.9) the least squares point estimates of β_2, β_3, and β_4 are

$$b_2 = -7.6080$$

$$b_3 = .6712$$

$$b_4 = -1.4777$$

the point estimate of

$$\mu_{[d,a+\Delta]} - \mu_{[d,a]} = \beta_2 \Delta + \beta_3[(a + \Delta)^2 - a^2] + \beta_4 d\Delta$$

is

$$b_2 \Delta + b_3[(a + \Delta)^2 - a^2] + b_4 d\Delta = -7.6080\Delta + .6712[(a + \Delta)^2 - a^2]$$
$$- 1.4777 d\Delta$$

Since the point estimate $b_4 = -1.4777$ is negative, the term $-1.4777d\Delta$ in the point estimate of

$$\mu_{[d,a+\Delta]} - \mu_{[d,a]}$$

says that the larger the price difference (d) is, the smaller is the increase in the mean demand for Fresh that is associated with an increase (Δ) in the advertising expenditure for Fresh. This fact seems somewhat logical, because the larger the price difference is (that is, the lower the price for Fresh relative to the average industry price), the higher is the mean demand for Fresh. Thus, when the price

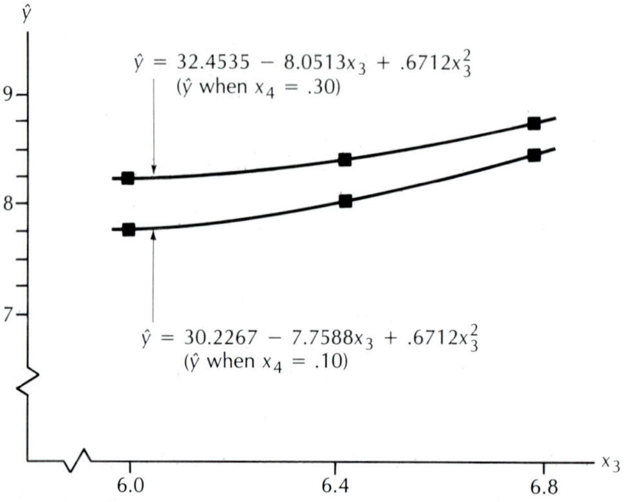

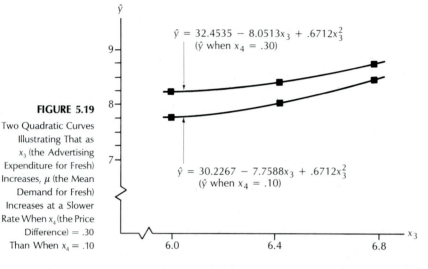

FIGURE 5.19 Two Quadratic Curves Illustrating That as x_3 (the Advertising Expenditure for Fresh) Increases, μ (the Mean Demand for Fresh) Increases at a Slower Rate When x_4 (the Price Difference) = .30 Than When x_4 = .10

difference is large, assuming that there is an upper limit to the mean demand for Fresh (perhaps due to market saturation), there is less opportunity (fewer potential customers who are not buying Fresh) for increased advertising to increase the mean demand for Fresh. And, vice versa, the smaller the price difference is (that is, the higher the price for Fresh relative to the average industry price), the lower is the mean demand for Fresh. Thus, when the price difference is small, there is more opportunity (more potential customers who are not buying Fresh) for increased advertising to increase the mean demand for Fresh.

We now illustrate this type of interaction by plotting the values of the point estimate

$$
\begin{aligned}
\hat{y} &= b_0 + b_1x_4 + b_2x_3 + b_3x_3^2 + b_4x_4x_3 \\
&= 29.1133 + 11.1342x_4 - 7.6080x_3 + .6712x_3^2 - 1.4777x_4x_3
\end{aligned}
$$

of the mean demand for Fresh

$$\mu = \beta_0 + \beta_1x_4 + \beta_2x_3 + \beta_3x_3^2 + \beta_4x_4x_3$$

against increasing values of x_3 when we fix x_4 at two different values—first at $x_4 = .10$ and then at $x_4 = .30$. If we fix x_4 at .10, the point estimate $\hat{y}$ of μ is

$$
\begin{aligned}
\hat{y} &= 29.1133 + 11.1342x_4 - 7.6080x_3 + .6712x_3^2 - 1.4777x_4x_3 \\
&= 29.1133 + 11.1342(.10) - 7.6080x_3 + .6712x_3^2 - 1.4777(.10)x_3 \\
&= 30.2267 - 7.7558x_3 + .6712x_3^2
\end{aligned}
$$

This equation may be interpreted as the point estimate of the equation relating μ (the mean demand for Fresh) to x_3 (the advertising expenditure for Fresh) when x_4 (the price difference) is .10. The quadratic curve defined by this equation is plotted in Figure 5.19 against increasing values of x_3. Note that, for example, when $x_3 = 6.0$,

$$\hat{y} = 30.2267 - 7.7558(6.0) + .6712(6.0)^2 \approx 7.76$$

When $x_3 = 6.4$,

$$\hat{y} = 30.2267 - 7.7558(6.4) + .6712(6.4)^2 \approx 8.08$$

When $x_3 = 6.8$,

$$\hat{y} = 30.2267 - 7.7558(6.8) + .6712(6.8)^2 \approx 8.52$$

Next, if we fix x_4 at .30, the point estimate $\hat{y}$ of μ equals

$$\begin{aligned}\hat{y} &= 29.1133 + 11.1342x_4 - 7.6080x_3 + .6712x_3^2 - 1.4777x_4x_3 \\ &= 29.1133 + 11.1342(.30) - 7.6080x_3 + .6712x_3^2 - 1.4777(.30)x_3 \\ &= 32.4535 - 8.0513x_3 + .6712x_3^2\end{aligned}$$

This equation may be interpreted as the point estimate of the equation relating μ (the mean demand for Fresh) to x_3 (the advertising expenditure for Fresh) when x_4 (the price difference) is .30. The quadratic curve defined by this equation is plotted in Figure 5.19 against increasing values of x_3. Note that, for example, when $x_3 = 6.0$,

$$\hat{y} = 32.4535 - 8.0513(6.0) + .6712(6.0)^2 \approx 8.31$$

When $x_3 = 6.4$,

$$\hat{y} = 32.4535 - 8.0513(6.4) + .6712(6.4)^2 \approx 8.42$$

When $x_3 = 6.8$,

$$\hat{y} = 32.4535 - 8.0513(6.8) + .6712(6.8)^2 \approx 8.74$$

Examining Figure 5.19, we note that, as x_3 increases, the quadratic curve defined by the equation

$$\hat{y} = 32.4535 - 8.0513x_3 + .6712x_3^2$$

which is the point estimate of the equation relating μ to x_3 when x_4 is .30, increases at a slower rate than the quadratic curve defined by the equation

$$\hat{y} = 30.2267 - 7.7558x_3 + .6712x_3^2$$

which is the point estimate of the equation relating μ to x_3 when x_4 is .10. Thus, Figure 5.19 illustrates that because there is less opportunity for increased advertising expenditure to increase the mean demand for Fresh when x_4 equals .30 than when x_4 equals .10, as x_3 increases, μ (the mean demand for Fresh) increases at a slower rate when x_4 equals .30 than when x_4 equals .10.

Recall that the type of interaction that exists between the price difference and the advertising expenditure was estimated by using the observed Fresh

Detergent data in Table 5.12 (we obtained the least squares point estimates using these data). These data (through the preceding analysis) suggest that the larger the price difference is, the smaller is the increase in the mean demand for Fresh that is caused by an increase in the advertising expenditure. However, we can only assume that this type of interaction exists for price differences and advertising expenditures in the experimental region. Examination of Table 5.12 indicates that most of the price differences that have been observed are positive and that the few negative price differences are not large in absolute value. Since this indicates that Enterprise Industries' price for Fresh is usually lower than the average industry price, we can say that Enterprise Industries is usually at a price advantage when selling Fresh. Given this pricing strategy, the type of interaction described is reasonable, (1) because in the "worst case" Enterprise Industries is either at a slight price advantage or at a slight price disadvantage, and in such a situation, increasing the advertising expenditure for Fresh can still be expected to substantially increase the mean demand for Fresh, and (2) the larger the price difference is (that is, the larger Enterprise Industries' price advantage) the smaller is the increase in the mean demand for Fresh associated with increasing the advertising expenditure. If, however, this price situation were to change, the type of interaction we have described might not exist. For example, consider a situation in which (because of frequent changes in pricing strategy) Enterprise Industries is often at a substantial price advantage (that is, the price for Fresh is frequently much lower than the average industry price) and also often at a substantial price disadvantage (that is, the price for Fresh is also frequently much higher than the average industry price). In such a situation, in the "worst case" Enterprise Industries is at a substantial price disadvantage and increasing the advertising expenditure for Fresh would probably not increase the mean demand for Fresh very much. This is because, no matter how much money Enterprise Industries spends on advertising Fresh, customers might not wish to buy a much higher priced product and thus the demand for Fresh might remain

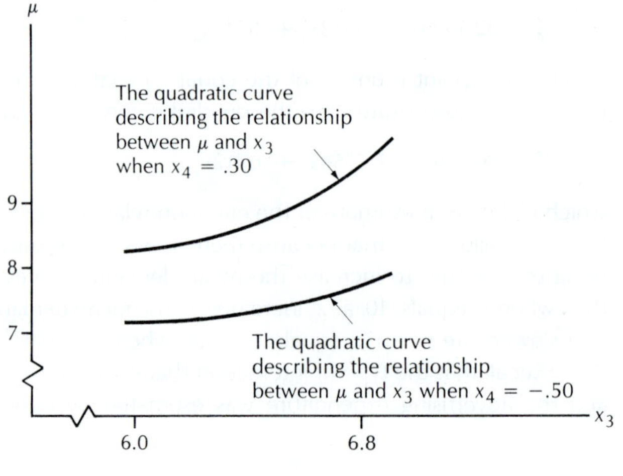

FIGURE 5.20
Two Quadratic Curves Illustrating That as x_3 (the Advertising Expenditure for Fresh) Increases, μ (the Mean Demand for Fresh) Increases at a Faster Rate When x_4 (the Price Difference) $= .30$ Than When $x_4 = -.50$

fairly low. In this situation, the increase in the mean demand for Fresh associated with increasing the advertising expenditure for Fresh would be larger when Enterprise Industries is at a substantial price advantage than when Enterprise Industries is at a substantial price disadvantage. This type of interaction is illustrated in Figure 5.20. In particular, this figure shows that the quadratic curve describing the relationship between μ (the mean demand for Fresh) and x_3 (the advertising expenditure for Fresh) when x_4 (the price difference) equals .30 (that is, when Enterprise Industries is at a substantial price advantage) increases at a faster rate than the quadratic curve describing the relationship between μ and x_3 when x_4 equals $-.50$ (that is, when Enterprise Industries is at a substantial price disadvantage).

In general, interaction between independent variables can be modeled by using cross-product terms (such as x_4x_3 in the Fresh Detergent problem). The need for such terms can sometimes be indicated by the use of data plots. The required plots would be constructed by making separate plots of the observed values of the dependent variable against the increasing values of one of the independent variables for different levels of the other independent variable. If the resulting data patterns appear to be parallel (as in Figure 5.18), then an interaction term is probably not needed. However, if the data patterns do not appear to be parallel (as in Figure 5.19), then an interaction term (the cross-product of the two independent variables) may be needed. Of course, in order to make data plots such as these, the observed data must be "structured." That is, there must be a reasonable number of observations at each level of the appropriate independent variable (for example, in the Fresh Detergent problem, there must be a reasonable number of observations at each of the price differences .10 and .30). This data requirement is most likely to be met when the data have been collected in a designed experiment, where the levels of one of the independent variables can be set by the analyst. (We study designed experiments in detail in Chapter 8.) Since the Fresh Detergent study is not a designed experiment, the required plots cannot be made for this problem. (Examination of the Fresh Detergent data in Table 5.12 shows that there are only a few observations at any one level of the price difference.) In such a case, we must rely on other model-building techniques in order to decide whether or not interaction terms are needed. We examine such techniques in Chapters 6 and 7 and find that these techniques indicate that the interaction term x_4x_3 should be included in the Fresh Detergent model. In Exercises 5.4, 5.7, and 5.8 we give the reader an opportunity to plot data that are structured enough to indicate whether or not interaction exists.

5 EXERCISES

5.1 The chairman of the Accounting Department at State University wishes to emphasize to new accounting majors the important effect of grade-point average in major courses on starting salary after graduation. To do this, records of seven recent accounting graduates are randomly selected:

Accounting Graduate (i)	Grade-Point Average (x_i)	Starting Salary (thousands of dollars) (y_i)
1	3.26	16.6
2	2.60	14.6
3	3.35	16.4
4	2.86	14.9
5	3.82	17.8
6	2.21	13.5
7	3.47	17.3

a. Plot the values of y against the increasing values of x. Note that the plot indicates that the simple linear regression model having a positive slope ($\beta_1 > 0$)

$$\begin{aligned} y_i &= \mu_i + \epsilon_i \\ &= \beta_0 + \beta_1 x_i + \epsilon_i \end{aligned}$$

might be an appropriate regression model relating y_i to x_i.

b. Discuss the meaning of the third historical population of potential starting salaries.

c. Discuss the meaning of μ_3, the third mean starting salary.

d. Discuss the conceptual difference between μ_3 and $y_3 = \mu_3 + \epsilon_3$. What does ϵ_3 measure in this situation?

e. Discuss the meanings of β_0 and β_1 in this model. Why does the interpretation of β_0 fail to make practical sense?

f. Calculate the least squares point estimates b_0 and b_1 of β_0 and β_1.

g. Using the prediction equation $\hat{y}_i = b_0 + b_1 x_i$, calculate a point estimate of μ_3 and a point prediction of $y_3 = \mu_3 + \epsilon_3$.

h. Assume that a future student will obtain a grade-point average of $x_0 = 3.25$. The starting salary of this future student may be expressed in the form $y_0 = \mu_0 + \epsilon_0$.
   (1) Discuss the conceptual difference between μ_0 and $y_0 = \mu_0 + \epsilon_0$.
   (2) Is the grade-point average $x_0 = 3.25$ in the experimental region?
   (3) Using an appropriate prediction equation, calculate a point estimate of μ_0 and a point prediction of y_0.

5.2 Consider Figure 5.5. This plot of the observed values of y (weekly fuel consumption) against the increasing values of x_2 (the chill index) indicates that the simple linear regression model having a positive slope ($\beta_1 > 0$)

$$y = \mu + \epsilon$$
$$= \beta_0 + \beta_1 x_2 + \epsilon$$

might be an appropriate regression model relating y to x_2.

a. Discuss the meanings of β_0 and β_1 in this model.
b. What does the error term measure in the model?
c. Calculate the least squares point estimates b_0 and b_1 of β_0 and β_1.
d. Assume that in a future week the chill index will be $x_0 = 10$. Is $x_0 = 10$ in the experimental region? Use an appropriate prediction equation to calculate a point prediction of $y_0 = \mu_0 + \epsilon_0$, the fuel consumption in the future week.

5.3 Compustat, Inc., sells a certain type of electronic calculator, the CS-22. Compustat has, over the past $n = 25$ sales periods (of four weeks each), spent different amounts of money on advertising and has had three different advertising firms handle the advertising strategy for the CS-22. The three advertising firms have different amounts of experience in the advertising field. One company has had 5 years of experience, the second company has had 10 years of experience, and the third company has had 15 years of experience. Compustat wishes to develop a regression model that relates the dependent variable y_i to the independent variables x_{i1} and x_{i2}.

y_i = demand for the CS-22 during the ith sales period (measured in units of 10,000 calculators sold). Thus, if $y_i = 41.3$, this means that 41,300 calculators were sold in the ith sales period.

x_{i1} = advertising expenditure during the ith sales period (measured in units of \$100,000). Thus, if $x_{i1} = 3$, this means that \$300,000 was spent on advertising in the ith sales period.

x_{i2} = the number of years of experience in the advertising field of the firm handling the advertising strategy for the CS-22 during the ith sales period. Thus, if $x_{i2} = 10$, the advertising firm having 10 years of experience handled the advertising expenditure in the ith sales period.

Compustat wishes to consider using the regression model

$$y_i = \mu_i + \epsilon_i$$
$$= \beta_0 + \beta_1 x_{i1} + \beta_2 x_{i2} + \beta_3 x_{i1}^2 + \beta_4 x_{i1} x_{i2} + \epsilon_i$$

to relate the dependent variable y_i to the independent variables x_{i1} and x_{i2}.

Assume that a sample of $n = 25$ combinations of (y_i, x_{i1}, x_{i2}) values are observed and that the following calculations are made:

$$(\mathbf{X}'\mathbf{X})^{-1} = \begin{bmatrix} .02 & 0 & 0 & 0 & 0 \\ 0 & .01 & 0 & 0 & 0 \\ 0 & 0 & .004 & 0 & 0 \\ 0 & 0 & 0 & .005 & 0 \\ 0 & 0 & 0 & 0 & .002 \end{bmatrix}$$

$$\mathbf{X}'\mathbf{y} = \begin{bmatrix} 1000 \\ 300 \\ 500 \\ 80 \\ 250 \end{bmatrix} \quad \text{and} \quad \sum_{i=1}^{25} y_i^2 = 22{,}102$$

Using the preceding information, answer the following questions:

a. Calculate the least squares point estimates of β_0, β_1, β_2, β_3, and β_4.
b. Assume that in a future sales period Compustat will spend $200,000 on advertising and will have an advertising firm with 5 years of experience handle its advertising strategy for the CS-22. The demand for the CS-22 in this future sales period may be expressed in the form $y_0 = \mu_0 + \epsilon_0$.
   (1) Discuss the meaning of the future population of potential demands for the CS-22.
   (2) Discuss the meaning of μ_0, the future mean demand for the CS-22.
   (3) Discuss the conceptual difference between μ_0 and $y_0 = \mu_0 + \epsilon_0$. What does ϵ_0 measure in this situation?
   (4) Assuming that

$$\mu_0 = \beta_0 + \beta_1 x_{01} + \beta_2 x_{02} + \beta_3 x_{01}^2 + \beta_4 x_{01} x_{02}$$

calculate a point estimate of μ_0 and a point prediction of y_0.

5.4 A study† was undertaken to examine the profit y per sales dollar earned by a construction company and its relationship to the size x_1 of the construction contract (in hundreds of thousands of dollars) and the number x_2 of years of experience of the construction supervisor. Data were obtained from a sample of $n = 18$ construction projects undertaken by the construction company over the past two years:

† The idea and the data for this problem were taken from an example in Mendenhall and Reinmuth (1982). We thank these authors and Duxbury Press for permission to use the idea and the data.

Profit (y_i)	Contract Size (hundreds of thousands of dollars) (x_{i1})	Supervisor Experience (years) (x_{i2})
2.0	5.1	4
3.5	3.5	4
8.5	2.4	2
4.5	4.0	6
7.0	1.7	2
7.0	2.0	2
2.0	5.0	4
5.0	3.2	2
8.0	5.2	6
5.0	4.3	6
6.0	2.9	2
7.5	1.1	2
4.0	2.6	4
4.0	4.0	6
1.0	5.3	4
5.0	4.9	6
6.5	5.0	6
1.5	3.9	4

a. Plot the values of y_i against the increasing values of x_{i1}. Note that the plot indicates that the quadratic regression model

$$\begin{aligned} y_i &= \mu_i + \epsilon_i \\ &= \beta_0 + \beta_1 x_{i1} + \beta_2 x_{i1}^2 + \epsilon_i \end{aligned}$$

might be an appropriate regression model relating y_i to x_{i1}.

b. Plot the values of y_i against the increasing values of x_{i2}. This plot suggests that the relationship between y_i and x_{i2} might be quadratic. However, it is hard to tell, because there are only three levels of x_{i2} in the data (2, 4, and 6).

c. Suppose that the regression model

$$\begin{aligned} y_i &= \mu_i + \epsilon_i \\ &= \beta_0 + \beta_1 x_{i1} + \beta_2 x_{i2} + \beta_3 x_{i1}^2 + \beta_4 x_{i1} x_{i2} + \epsilon_i \end{aligned}$$

is a reasonable regression model relating y_i to x_{i1} and x_{i2}. Note that this model uses the independent variables x_{i1} and x_{i1}^2 suggested by the plot made in part (a) but uses only the linear term x_{i2} and not the quadratic term x_{i2}^2 that is possibly suggested by the plot made in part (b). One reason for not including x_{i2}^2 is that interaction between x_{i1} and x_{i2} is partly causing the plot made in part (b) to suggest that the relationship between y_i and x_{i2} might be quadratic. The preceding regression model takes into account this interaction by utilizing the term $x_{i1}x_{i2}$. To see the need for this interaction term, perform the following graphical analysis:

(1) Plot y_i against x_{i1} when $x_{i2} = 2$.
(2) Plot y_i against x_{i1} when $x_{i2} = 4$.
(3) Plot y_i against x_{i1} when $x_{i2} = 6$.
(4) Combine these plots by graphing them on the same set of axes. Use a different color for each level of x_{i2} (=2, 4, 6).
(5) What do these plots say about whether interaction exists between x_{i1} and x_{i2}? Discuss what interaction between x_{i1} and x_{i2} means in the construction company problem. Intuitively, why does interaction between x_{i1} and x_{i2} make sense?

d. Specify the vector **y** and matrix **X** used to calculate the least squares point estimates of the parameters in the model

$$\begin{aligned} y_i &= \mu_i + \epsilon_i \\ &= \beta_0 + \beta_1 x_{i1} + \beta_2 x_{i2} + \beta_3 x_{i1}^2 + \beta_4 x_{i1} x_{i2} + \epsilon_i \end{aligned}$$

e. If the least squares point estimates of the parameters in the preceding model are calculated, we find that $b_0 = 19.3050$, $b_1 = -1.4866$, $b_2 = -6.3715$, $b_3 = -.7522$, and $b_4 = 1.7171$. By using the least squares point estimates, and entering values of x_{i1} and x_{i2} into the prediction equation,

$$\hat{y}_i = b_0 + b_1 x_{i1} + b_2 x_{i2} + b_3 x_{i1}^2 + b_4 x_{i1} x_{i2}$$

(1) Plot $\hat{y}_i$ against x_{i1} (for $x_{i1} = 3$, 4, and 5) when $x_{i2} = 2$.
(2) Plot $\hat{y}_i$ against x_{i1} (for $x_{i1} = 3$, 4, and 5) when $x_{i2} = 4$.
(3) Plot $\hat{y}_i$ against x_{i1} (for $x_{i1} = 3$, 4, and 5) when $x_{i2} = 6$.
(4) Combine these plots by graphing them on the same set of axes. Use a different color for each level of x_{i2} (=2, 4, 6). What do these plots suggest concerning the policy the company should follow in assigning supervisors to construction projects?

f. Suppose that the construction company has been awarded a "future contract" of $x_{01} = 4.8$ ($480,000) and has decided to assign a supervisor with $x_{02} = 6$ years of experience to this project.

(1) Discuss the meaning of μ_0, the future mean profit for this future contract.
(2) Discuss the conceptual difference between μ_0 and $y_0 = \mu_0 + \epsilon_0$, the future (individual) profit for this future contract.
(3) Plot the eighteen (x_1, x_2) combinations in two dimensions (with x_1 measured on the horizontal axis and x_2 measured on the vertical axis) to form the experimental region.
(4) Is the point $(x_{01} = 4.8, x_{02} = 6)$ in the experimental region?

(5) Using an appropriate prediction equation, calculate a point estimate of μ_0 and a point prediction of y_0.

5.5 Consider Exercise 5.3. The model

$$\begin{aligned} y_i &= \mu_i + \epsilon_i \\ &= \beta_0 + \beta_1 x_{i1} + \beta_2 x_{i2} + \beta_3 x_{i1}^2 + \beta_4 x_{i1} x_{i2} + \epsilon_i \end{aligned}$$

assumes that there is interaction between x_{i1}, advertising expenditure, and x_{i2}, the number of years of experience of the firm handling the advertising strategy, as these independent variables affect y_i, demand for the CS-22.

a. By writing out the appropriate expressions for

$\mu_{[c,d]}$ = the mean demand for all potential sales periods in which the advertising expenditure is c and an advertising firm with *d* years of experience handles the advertising strategy

and

$\mu_{[c+\Delta,d]}$ = the mean demand for all potential sales periods in which the advertising expenditure is $(c + \Delta)$ and an advertising firm with *d* years of experience handles the advertising strategy

and by subtracting these expressions, verify that

$$\mu_{[c+\Delta,d]} - \mu_{[c,d]} = \beta_1 \Delta + \beta_3[(c + \Delta)^2 - c^2] + \beta_4 \Delta d$$

b. Using the facts that

$$\mu_{[c+\Delta,d]} - \mu_{[c,d]} = \beta_1 \Delta + \beta_3[(c + \Delta)^2 - c^2] + \beta_4 \Delta d$$

and that the least squares point estimate of β_4, which is $b_4 = .5$, is a *positive* number, discuss the nature of the interaction between x_{i1}, advertising expenditure, and x_{i2}, the number of years of experience of the firm handling the advertising strategy, as these independent variables affect y_i, demand for the CS-22. Be precise and draw a picture to demonstrate the interaction. To keep your discussion simple, draw your picture by utilizing *two different advertising expenditures* and *two different numbers of years of experience*. Also, discuss why the type of interaction you describe might logically be expected to exist.

c. Using any of the information given previously, find a *point estimate* of

$$\mu_{[3,5]} - \mu_{[2,5]}$$

which is the *change in the mean demand* for the CS-22 associated with increasing the advertising expenditure from \$200,000 to \$300,000 (that is, from 2 to 3) when the experience of the firm handling the advertising strategy is held constant at 5 years.

5.6 Market Planning, Inc., a marketing research firm, has obtained the following prescription sales data for $n = 20$ independent pharmacies.†

Pharmacy	Sales (y)	Floor Space (x_1)	Prescription % (x_2)	Parking (x_3)	Income (x_4)
1	22	4900	9	40	18
2	19	5800	10	50	20
3	24	5000	11	55	17
4	28	4400	12	30	19
5	18	3850	13	42	10
6	21	5300	15	20	22
7	29	4100	20	25	8
8	15	4700	22	60	15
9	12	5600	24	45	16
10	14	4900	27	82	14
11	18	3700	28	56	12
12	19	3800	31	38	8
13	15	2400	36	35	6
14	22	1800	37	28	4
15	13	3100	40	43	6
16	16	2300	41	20	5
17	8	4400	42	46	7
18	6	3300	42	15	4
19	7	2900	45	30	9
20	17	2400	46	16	3

These variables can be described precisely as follows:

y = average weekly prescription sales over the past year (in units of $1,000)

x_1 = floor space (in square feet)

x_2 = percent of floor space allocated to the prescription department

x_3 = number of parking spaces available for the store

x_4 = monthly per capita income for the surrounding community (in units of $100)

a. By plotting y versus x_1, y versus x_2, y versus x_3, and y versus x_4, specify one or more regression models that might be useful in relating y to x_1, x_2, x_3, and x_4.

† The idea for and the data in this problem were taken from an example in Ott (1984). We thank Dr. Ott and Duxbury Press for permission to use the idea and the data.

b. Suppose that $y = \beta_0 + \beta_1 x_1 + \beta_2 x_2 + \epsilon$ is a reasonable regression model describing y. If the least squares point estimates of the parameters in this model are calculated, we find that $b_0 = 48.291$, $b_1 = -.004$, and $b_2 = -.582$. Predict the average weekly prescription sales for a pharmacy having 4,000 sq ft, with 20 percent of this space allocated to the prescription department.

5.7 United Oil Company is attempting to develop a reasonably priced unleaded gasoline that will deliver higher gasoline mileages than can be achieved by its current unleaded gasolines. As part of its development process, United Oil wishes to study the effect of two independent variables—x_1, amount of gasoline additive RST (0, 1, or 2 units), and x_2, amount of gasoline additive XST (0, 1, 2, or 3 units)—on the gasoline mileage y obtained by an automobile called the Empire. For testing purposes, a sample of $n = 22$ Empires is randomly selected. Each Empire is test driven under normal driving conditions. The combinations of x_1 and x_2 used in the experiment, along with the corresponding values of y, are given in the following table.

Gasoline Mileage (mpg) (y_i)	Amount of Gasoline Additive RST (x_{i1})	Amount of Gasoline Additive XST (x_{i2})
17.4	0	0
18.0	0	0
18.6	0	0
19.6	1	0
20.6	1	0
18.6	2	0
19.8	2	0
22.0	0	1
23.0	0	1
23.3	1	1
24.5	1	1
22.3	0	2
23.5	0	2
24.4	1	2
25.0	1	2
25.6	1	2
23.3	2	2
24.0	2	2
24.7	2	2
23.4	1	3
22.0	2	3
23.0	2	3

a. Plot y_i against x_{i1}.
b. Plot y_i against x_{i1} when $x_{i2} = 0$.
c. Plot y_i against x_{i1} when $x_{i2} = 1$.
d. Plot y_i against x_{i1} when $x_{i2} = 2$.
e. Plot y_i against x_{i1} when $x_{i2} = 3$.
f. Combine these plots by graphing them on the same set of axes. Use a different color for each level of x_{i2} ($=0, 1, 2,$ and 3). What do these plots say about whether interaction exists between x_{i1} and x_{i2}?
g. Plot y_i against x_{i2}.
h. Plot y_i against x_{i2} when $x_{i1} = 0$.
i. Plot y_i against x_{i2} when $x_{i1} = 1$.
j. Plot y_i against x_{i2} when $x_{i1} = 2$.
k. Combine these plots by graphing them on the same set of axes. Use a different color for each level of x_{i1} ($=0, 1,$ and 2). What do these plots say about whether interaction exists between x_{i1} and x_{i2}?
l. Suppose that the regression model

$$\begin{aligned} y_i &= \mu_i + \epsilon_i \\ &= \beta_0 + \beta_1 x_{i1} + \beta_2 x_{i1}^2 + \beta_3 x_{i2} + \beta_4 x_{i2}^2 + \epsilon_i \end{aligned}$$

is an appropriate regression model relating y_i to x_{i1} and x_{i2}. Discuss why the graphical analysis in parts (a) through (k) indicates that this model is appropriate.
m. Specify the vector **y** and the matrix **X** used to calculate the least squares point estimates of the parameters in the model in part (l).
n. If the least squares point estimates of the parameters in the model in part (l) are calculated, we find that $b_0 = 18.1589$, $b_1 = 3.3133$, $b_2 = -1.4111$, $b_3 = 5.2752$, and $b_4 = -1.3964$. Noting from the table of values of y_i, x_{i1}, and x_{i2} that the combination of one unit of gasoline additive RST and two units of gasoline additive XST seems to maximize gasoline mileage, assume that in the future United Oil Company will use this combination to make its unleaded gasoline.
   (1) Discuss the meaning of μ_0, the future mean gasoline mileage (corresponding to the future combination $x_{01} = 1$ and $x_{02} = 2$).
   (2) Discuss the conceptual difference between μ_0 and $y_0 = \mu_0 + \epsilon_0$, the future (individual) gasoline mileage corresponding to the above future combination.
   (3) Using an appropriate prediction equation, calculate a point estimate of μ_0 and a point prediction of $y_0 = \mu_0 + \epsilon_0$.

5.8 United Oil Company is attempting to develop a reasonably priced leaded gasoline that will deliver higher gasoline mileages than can be achieved by its current leaded gasolines. As part of its development process, United Oil wishes to study the effect of two independent variables—x_1, amount of gasoline additive WST (0, 1, or 2 units), and x_2, amount of gasoline additive YST (0, 1,

2, or 3 units)—on the gasoline mileage y obtained by an automobile called the Comfort. For testing purposes, a sample of $n = 22$ Comforts is randomly selected. Each Comfort is tested under normal driving conditions. The combinations of x_1 and x_2 used in the experiment, along with the corresponding values of y, are given in the following table.

Gasoline Mileage (mpg) (y_i)	Amount of Gasoline Additive WST (x_{i1})	Amount of Gasoline Additive YST (x_{i2})
18.0	0	0
18.6	0	0
17.4	0	0
23.3	1	0
24.5	1	0
23.0	0	1
22.0	0	1
25.6	1	1
24.4	1	1
25.0	1	1
24.0	2	1
23.3	2	1
24.7	2	1
23.5	0	2
22.3	0	2
23.4	1	2
23.0	2	2
22.0	2	2
19.6	1	3
20.6	1	3
18.6	2	3
19.8	2	3

a. Plot y_i against x_{i1}.
b. Plot y_i against x_{i1} when $x_{i2} = 0$.
c. Plot y_i against x_{i1} when $x_{i2} = 1$.
d. Plot y_i against x_{i1} when $x_{i2} = 2$.
e. Plot y_i against x_{i1} when $x_{i2} = 3$.
f. Combine these plots by graphing them on the same set of axes. Use a different color for each level of x_{i2} ($=0$, 1, 2, and 3). What do these plots say about whether interaction exists between x_{i1} and x_{i2}?
g. Plot y_i against x_{i2}.

h. Plot y_i against x_{i2} when $x_{i1} = 0$.
i. Plot y_i against x_{i2} when $x_{i1} = 1$.
j. Plot y_i against x_{i2} when $x_{i1} = 2$.
k. Combine these plots by graphing them on the same set of axes. Use a different color for each level of x_{i1} $(=0, 1, \text{and } 2)$. What do these plots say about whether interaction exists between x_{i1} and x_{i2}?
l. Suppose that the regression model

$$\begin{aligned} y_i &= \mu_i + \epsilon_i \\ &= \beta_0 + \beta_1 x_{i1} + \beta_2 x_{i1}^2 + \beta_3 x_{i2} + \beta_4 x_{i2}^2 + \beta_5 x_{i1} x_{i2} + \beta_6 x_{i1}^2 x_{i2}^2 + \epsilon_i \end{aligned}$$

is an appropriate regression model relating y_i to x_{i1} and x_{i2}. Discuss why the graphical analysis in parts (a) through (k) indicates that this model is appropriate.
m. Specify the vector **y** and the matrix **X** used to calculate the least squares point estimates of the parameters in the model in part (l).
n. If the least squares point estimates of the parameters in the model in part (l) are calculated, we find that $b_0 = 18.1323$, $b_1 = 7.8004$, $b_2 = -2.2009$, $b_3 = 5.5718$, $b_4 = -1.5557$, $b_5 = -2.9365$, and $b_6 = .2580$. Noting from the table of values of y_i, x_{i1}, and x_{i2} that the combination of one unit of gasoline additive WST and one unit of gasoline additive YST seems to maximize gasoline mileage, assume that in the future United Oil Company will use this combination to make its leaded gasoline.
   (1) Discuss the meaning of μ_0, the future mean gasoline mileage corresponding to the future combination $x_{01} = 1$ and $x_{02} = 1$.
   (2) Discuss the conceptual difference between μ_0 and $y_0 = \mu_0 + \epsilon_0$, the future (individual) gasoline mileage corresponding to the preceding future combination.
   (3) Using an appropriate prediction equation, calculate a point estimate of μ_0 and a point prediction of $y_0 = \mu_0 + \epsilon_0$.

6

STATISTICAL INFERENCE IN REGRESSION ANALYSIS

How can we use probability distributions to make statistical inferences (for example, compute confidence intervals and perform hypothesis tests) in regression analysis? In this chapter we begin to answer this question. In Section 6.1 we study a quantity called the standard error, which will be employed in later sections. Then, in Section 6.2 we present the assumptions that must hold to ensure the validity of the formulas for the confidence intervals and hypothesis tests that we discuss in this and later chapters. Sections 6.3 through 6.6 present a confidence interval and a hypothesis test for the regression parameter β_j and explain how these can be used to help us decide whether an independent variable should be included in a regression model. Then, in Sections 6.7 and 6.8 we explain and present the formulas for a confidence interval for μ_0, the future mean value of the dependent variable, and a prediction interval for y_0, the future (individual) value of the dependent variable. We introduce how to use the Statistical Analysis System (SAS) to perform regression analysis in Section 6.9, while Section 6.10 presents special case formulas for making statistical inferences when employing the simple linear regression model. We conclude this chapter with optional Section 6.11, which explains how to use regression to analyze a randomly selected sample from a single population, and optional Section 6.12, which discusses some advanced hypothesis testing techniques.

6.1 THE STANDARD ERROR

In order to calculate confidence intervals and perform hypothesis tests in regression analysis, we utilize the **standard error**, which is the positive square root of a quantity called the **mean square error**. Denoting the mean square error by the symbol s^2, we define it as follows:

Mean square error $= s^2 = \dfrac{SSE}{n-k} = \dfrac{\sum_{i=1}^{n} e_i^2}{n-k} = \dfrac{\sum_{i=1}^{n} (y_i - \hat{y}_i)^2}{n-k}$

Here

1. n is the number of observed values of the dependent variable $y_1, y_2, \ldots, y_n$.
2. The sum of squared residuals

$$SSE = \sum_{i=1}^{n} e_i^2 = \sum_{i=1}^{n} (y_i - \hat{y}_i)^2$$

$$= \sum_{i=1}^{n} (y_i - (b_0 + b_1 x_{i1} + b_2 x_{i2} + \cdots + b_p x_{ip}))^2$$

is obtained by calculating the n predicted values of the dependent variable $\hat{y}_1, \hat{y}_2, \ldots, \hat{y}_n$ by using the least squares point estimates in the prediction equation

$$\hat{y} = b_0 + b_1x_1 + b_2x_2 + \cdots + b_px_p$$

3. k is the number of least squares point estimates $b_0, b_1, b_2, \ldots, b_p$ used in the preceding prediction equation (or the number of regression parameters $\beta_0, \beta_1, \beta_2, \ldots, \beta_p$).

Since the standard error is the positive square root of the mean square error, s^2, we denote the standard error by the symbol s.

Standard error $= s = \sqrt{s^2} = \sqrt{\dfrac{SSE}{n-k}} = \sqrt{\dfrac{\sum_{i=1}^{n}(y_i - \hat{y}_i)^2}{n-k}}$

EXAMPLE 6.1

Consider the fuel consumption problem of Chapter 5. Since we have collected historical data for eight weeks, there are eight observed fuel consumptions, and thus $n = 8$. Moreover, since there are three least squares point estimates—b_0, b_1, and b_2—in the prediction equation

$$\begin{aligned}\hat{y} &= b_0 + b_1x_1 + b_2x_2 \\ &= 13.109 - .0900x_1 + .0825x_2\end{aligned}$$

it follows that $k = 3$. Table 5.11 showed that

$$\begin{aligned}SSE &= \sum_{i=1}^{8}(y_i - \hat{y}_i)^2 \\ &= (.326745)^2 + (-.0432749)^2 + \cdots + (.017127)^2 \\ &= .674\end{aligned}$$

so the mean square error is

$$\begin{aligned}s^2 = \frac{SSE}{n-k} &= \frac{\sum_{i=1}^{8}(y_i - \hat{y}_i)^2}{8-3} \\ &= \frac{.674}{5} = .1348\end{aligned}$$

and the standard error is

$$s = \sqrt{s^2} = \sqrt{.1348} = .3671$$

We have seen that in order to calculate s, we need to calculate

$$SSE = \sum_{i=1}^{n} (y_i - \hat{y}_i)^2$$

Since the calculation of SSE requires calculating the prediction

$$\hat{y}_i = b_0 + b_1 x_{i1} + b_2 x_{i2} + \cdots + b_p x_{ip}$$

and the residual

$$e_i = y_i - \hat{y}_i$$

for each observed value of the dependent variable, y_i, the calculations using the preceding formula for SSE are quite tedious. Therefore, we wish to present another way to calculate SSE that can be computationally simpler (especially if a prediction equation is constructed using a large amount of historical data). This method says that (as proved in Appendix C)

$$SSE = \sum_{i=1}^{n} (y_i - \hat{y}_i)^2 = \sum_{i=1}^{n} y_i^2 - \mathbf{b}'\mathbf{X}'\mathbf{y}$$

Here,

$$\mathbf{b}' = [b_0 \quad b_1 \quad b_2 \quad \cdots \quad b_p]$$

is a row vector (the transpose of $\mathbf{b}$) containing the least squares point estimates, and $\mathbf{X}'\mathbf{y}$ is the column vector used in calculating the least squares point estimates by the formula

$$\begin{bmatrix} b_0 \\ b_1 \\ b_2 \\ \vdots \\ b_p \end{bmatrix} = \mathbf{b} = (\mathbf{X}'\mathbf{X})^{-1}\mathbf{X}'\mathbf{y}$$

So in terms of computational simplicity, the easiest way to calculate SSE, s^2, and s is to use the following formulas:

Sum of squared residuals $= SSE = \sum_{i=1}^{n} y_i^2 - \mathbf{b}'\mathbf{X}'\mathbf{y}$

Mean square error $= s^2 = \dfrac{SSE}{n-k} = \dfrac{\sum_{i=1}^{n} y_i^2 - \mathbf{b}'\mathbf{X}'\mathbf{y}}{n-k}$

Standard error $= s = \sqrt{\dfrac{SSE}{n-k}} = \sqrt{\dfrac{\sum_{i=1}^{n} y_i^2 - \mathbf{b}'\mathbf{X}'\mathbf{y}}{n-k}}$

EXAMPLE 6.2

In the fuel consumption problem, consider the model

$$y = \beta_0 + \beta_1 x_1 + \beta_2 x_2 + \epsilon$$

In Example 5.6 we used the matrix $(\mathbf{X}'\mathbf{X})^{-1}$ and the column vector

$$\mathbf{X}'\mathbf{y} = \begin{bmatrix} 81.7 \\ 3413.11 \\ 1157.40 \end{bmatrix}$$

to calculate the least squares point estimates b_0, b_1, and b_2 to be

$$\begin{bmatrix} b_0 \\ b_1 \\ b_2 \end{bmatrix} = \mathbf{b} = (\mathbf{X}'\mathbf{X})^{-1}\mathbf{X}'\mathbf{y} = \begin{bmatrix} 13.109 \\ -.0900 \\ .0825 \end{bmatrix}$$

It follows that

$$\begin{aligned} \mathbf{b}'\mathbf{X}'\mathbf{y} &= [13.109 \quad -.0900 \quad .0825] \begin{bmatrix} 81.7 \\ 3413.11 \\ 1157.40 \end{bmatrix} \\ &= 13.109(81.7) + (-.0900)(3413.11) + (.0825)(1157.40) \\ &= 859.236 \end{aligned}$$

Since the eight observed fuel consumptions (see Table 5.2) can be used to calculate

$$\begin{aligned} \sum_{i=1}^{8} y_i^2 &= y_1^2 + y_2^2 + \cdots + y_8^2 \\ &= (12.4)^2 + (11.7)^2 + \cdots + (7.5)^2 = 859.91 \end{aligned}$$

we can calculate SSE, s^2, and s as follows:

$$SSE = \sum_{i=1}^{8} y_i^2 - \mathbf{b'X'y}$$

$$= 859.91 - 859.236 = .674$$

$$s^2 = \frac{SSE}{n-k} = \frac{.674}{8-3} = .1348$$

$$s = \sqrt{s^2} = \sqrt{.1348} = .3671$$

6.2 FOUR ASSUMPTIONS

The validity of most of the formulas for confidence intervals, prediction intervals, and hypothesis tests that we present in this book depends upon whether three assumptions, called the **inference assumptions**, are satisfied. Before presenting the inference assumptions, we should point out that the validity of all the regression results in this book depends upon the underlying assumption that we have determined an *appropriate linear equation*

$$\mu = \beta_0 + \beta_1 x_1 + \beta_2 x_2 + \cdots + \beta_p x_p$$

relating μ, the mean value of the dependent variable, to $x_1, x_2, \ldots, x_p$, the set of the independent variables being utilized. For example, if the true linear equation relating μ to x_1 and x_2 in the fuel consumption problem were

$$\mu = \beta_0 + \beta_1 x_1 + \beta_2 x_2 + \beta_3 x_1^2 + \beta_4 x_1 x_2$$

but we mistakenly used the linear equation

$$\mu = \beta_0 + \beta_1 x_1 + \beta_2 x_2$$

then the results obtained by using the regression model

$$\begin{aligned} y &= \mu + \epsilon \\ &= \beta_0 + \beta_1 x_1 + \beta_2 x_2 + \epsilon \end{aligned}$$

would not be valid. However, we have intuitively reasoned (in Chapter 5) that

$$\mu = \beta_0 + \beta_1 x_1 + \beta_2 x_2$$

is an appropriate linear equation relating μ to x_1 and x_2. Furthermore, later in this chapter and in subsequent chapters we present statistical evidence showing that this equation is appropriate. Therefore, it is reasonable to conclude that the regression model

$$y = \beta_0 + \beta_1 x_1 + \beta_2 x_2 + \epsilon$$

satisfies the preceding underlying assumption. The following are the three inference assumptions:

INFERENCE ASSUMPTION 1 (CONSTANT VARIANCE):

The n historical populations, and any future population, of potential values of the dependent variable have *equal variances.*

INFERENCE ASSUMPTION 2 (INDEPENDENCE):

The randomly selected values of the dependent variable

$$y_1, y_2, \ldots, y_n, \quad \text{and} \quad y_0$$

are all *statistically independent.*

INFERENCE ASSUMPTION 3 (NORMAL POPULATIONS):

The n historical populations, and any future population, of potential values of the dependent variable are *normally distributed.*

To see an illustration of inference assumptions 1 and 3, look at Table 6.2 (in Section 6.4), which illustrates the $n = 8$ historical populations of potential weekly fuel consumptions to be normally distributed with equal variances (that is, the normal curves have the same spread). Denoting the constant variance of the n historical populations, and any future population, of potential values of the dependent variable by the symbol σ^2, we note that it can be proved that the mean square error

$$s^2 = \frac{SSE}{n-k} = \frac{\sum_{i=1}^{n} y_i^2 - \mathbf{b}'\mathbf{X}'\mathbf{y}}{n-k}$$

is the appropriate point estimate of σ^2 (and thus that the standard error, s, is the appropriate point estimate of the constant standard deviation σ).

Intuitively, inference assumption 2 says that there is *no pattern* of greater than average values of the dependent variable being followed (in time) by greater than average values of the dependent variable, and that there is *no pattern* of greater than average values of the dependent variable being followed by less than average values of the dependent variable. A more complete discussion of the meaning of the inference assumptions, along with various techniques that can be used to check their validity, is given in Chapter 9. We assume that the inference assumptions are satisfied for the examples in chapters prior to Chapter 9. Finally, it is important to point out that although most of the formulas for prediction intervals, confidence intervals, and hypothesis tests in this book are strictly valid only when the inference assumptions hold, these formulas are still approximately correct when mild departures from the inference assumptions can be detected. In fact, these assumptions very seldom, if ever, hold exactly in any practical regression problem. Thus, in practice, only pronounced departures from the inference assumptions require attention.

In previous sections we stated that when predicting the n historical values and the future value of the dependent variable

$$\begin{aligned} y_1 &= \mu_1 + \epsilon_1 \\ y_2 &= \mu_2 + \epsilon_2 \\ &\vdots \\ y_n &= \mu_n + \epsilon_n \\ y_0 &= \mu_0 + \epsilon_0 \end{aligned}$$

it is customary to predict the error terms $\epsilon_1, \epsilon_2, \ldots, \epsilon_n$, and ϵ_0 to be zero. The rationale for predicting these error terms to be zero comes from inference assumptions 2 and 3. Since inference assumption 3 says that the n historical populations, and any future population, of potential values of the dependent variable are normally distributed, and since inference assumption 2 says that $y_1, y_2, \ldots, y_n$, and y_0, the values of the dependent variable assumed to be randomly selected from these populations, are statistically independent, it follows that y_i (for $i = 1, 2, \ldots, n$ and 0) has a 50 percent chance of being greater than, and a 50 percent chance of being less than, the corresponding mean μ_i. Hence, each of the error terms

$$\begin{aligned} \epsilon_1 &= y_1 - \mu_1 \\ \epsilon_2 &= y_2 - \mu_2 \\ &\vdots \\ \epsilon_n &= y_n - \mu_n \\ \epsilon_0 &= y_0 - \mu_0 \end{aligned}$$

has a 50 percent chance of being positive and a 50 percent chance of being negative, and thus we predict each of these error terms to be zero.

6.3 A CONFIDENCE INTERVAL FOR A REGRESSION PARAMETER

As illustrated in Chapter 5, the least squares point estimate b_j of the regression parameter β_j will not (unless we are extremely lucky) equal the true value of β_j. In this section we present an intuitive discussion of the concept of a confidence interval for β_j, and in the next section we present a precise discussion of the meaning and derivation of this interval.

We call the difference between b_j and β_j

$$b_j - \beta_j$$

the error of estimation obtained when estimating β_j *by* b_j. Intuitively, this error of estimation expresses how far the point estimate b_j is from β_j. Since we do not know what β_j equals, we cannot know the exact size of the error of estimation (that is, exactly how far b_j is from β_j). However, the science of statistics can provide us with a quantity called the *100(1 − α)% bound on the error of estimation obtained when estimating* β_j *by* b_j. This bound, denoted by $BE_{b_j}[100(1-\alpha)]$, can be computed so that if the inference assumptions are satisfied, we are $100(1-\alpha)$% confident (for example, 95% confident if $\alpha = .05$) that $|b_j - \beta_j|$, the absolute value of the error of estimation obtained when estimating β_j by b_j, is less than or equal to the bound

$$BE_{b_j}[100(1-\alpha)]$$

Mathematically, we are $100(1-\alpha)$% confident that

$$|b_j - \beta_j| \le BE_{b_j}[100(1-\alpha)]$$

This statement is equivalent to the algebraic statement

$$\beta_j - BE_{b_j}[100(1-\alpha)] \le b_j \le \beta_j + BE_{b_j}[100(1-\alpha)]$$

That is, we are $100(1-\alpha)$% confident that the point estimate b_j is within $BE_{b_j}[100(1-\alpha)]$ units of the true value of β_j. Now, the statement

$$\beta_j - BE_{b_j}[100(1-\alpha)] \le b_j \le \beta_j + BE_{b_j}[100(1-\alpha)]$$

can be rearranged to give the algebraic statement

$$b_j - BE_{b_j}[100(1-\alpha)] \le \beta_j \le b_j + BE_{b_j}[100(1-\alpha)]$$

That is, we are $100(1-\alpha)$% confident that β_j is contained in (equals one of the numbers in) the interval

$$[b_j - BE_{b_j}[100(1-\alpha)],\ b_j + BE_{b_j}[100(1-\alpha)]]$$

We call this interval a *100(1 − α)% confidence interval for* β_j. It follows that we are $100(1-\alpha)$% confident that β_j is greater than or equal to the lower bound

$$b_j - BE_{b_j}[100(1-\alpha)]$$

and less than or equal to the upper bound

$$b_j + BE_{b_j}[100(1-\alpha)]$$

of this confidence interval. Since we do not know what β_j equals, we are not absolutely certain (not 100% confident) that β_j is contained in the $100(1-\alpha)\%$ confidence interval for β_j. Exactly what we mean when we say that we are $100(1-\alpha)\%$ confident that β_j is contained in this interval is explained in the next section. For now, suffice it to say that if $100(1-\alpha)\%$ is chosen to be a very high level of confidence (such as 95% confidence), then $100(1-\alpha)\%$ confident means very confident.

A $100(1-\alpha)\%$ CONFIDENCE INTERVAL FOR β_j:

Assume that the linear regression model

$$y = \beta_0 + \beta_1 x_1 + \beta_2 x_2 + \cdots + \beta_p x_p + \epsilon$$

has k parameters $\beta_0, \beta_1, \beta_2, \ldots, \beta_p$, and that, for $j = 0, 1, 2, \ldots, p$, b_j is the least squares point estimate of β_j. If the inference assumptions are satisfied, a $100(1-\alpha)\%$ confidence interval for β_j is

$$[b_j - BE_{b_j}[100(1-\alpha)],\ b_j + BE_{b_j}[100(1-\alpha)]]$$

where $BE_{b_j}[100(1-\alpha)]$ denotes the *$100(1-\alpha)\%$ bound on the error of estimation obtained when estimating β_j by b_j* and is given by the equation

$$BE_{b_j}[100(1-\alpha)] = t_{[\alpha/2]}^{(n-k)} s\sqrt{c_{jj}}$$

Here

$$s = \sqrt{s^2} = \sqrt{\frac{\sum_{i=1}^{n}(y_i - \hat{y}_i)^2}{n-k}} = \sqrt{\frac{\sum_{i=1}^{n} y_i^2 - \mathbf{b}'\mathbf{X}'\mathbf{y}}{n-k}}$$

is the standard error; $t_{[\alpha/2]}^{(n-k)}$ is the point on the scale of the t-distribution having $n-k$ degrees of freedom so that the area in the tail of this t-distribution to the right of $t_{[\alpha/2]}^{(n-k)}$ is $\alpha/2$; and c_{jj} is the diagonal element of the matrix $(\mathbf{X}'\mathbf{X})^{-1}$ corresponding to β_j, when the rows and columns of $(\mathbf{X}'\mathbf{X})^{-1}$ are numbered as the parameters in the regression model are numbered.

Note that, if $n-k$ (the degrees of freedom) is 30 or more, it is sufficient (for reasons discussed in Section 3.3.1) to use the value $z_{[\alpha/2]}$ for $t_{[\alpha/2]}^{(n-k)}$. Here, $z_{[\alpha/2]}$ is (as discussed in Section 3.2.2) the point on the scale of the standard normal curve so that the area under this curve to the right of $z_{[\alpha/2]}$ is $\alpha/2$.

EXAMPLE 6.3

Suppose that we wish to compute confidence intervals for the parameters β_0, β_1, and β_2 in the fuel consumption model

$$\begin{aligned} y_i &= \mu_i + \epsilon_i \\ &= \beta_0 + \beta_1 x_{i1} + \beta_2 x_{i2} + \epsilon_i \end{aligned}$$

Recall that we have previously (in Example 5.6) calculated the least squares point estimates of the parameters β_0, β_1, and β_2 to be

$$\begin{bmatrix} b_0 \\ b_1 \\ b_2 \end{bmatrix} = \mathbf{b} = (\mathbf{X}'\mathbf{X})^{-1}\mathbf{X}'\mathbf{y} = \begin{bmatrix} 13.109 \\ -.0900 \\ .0825 \end{bmatrix}$$

and that we have previously (in Example 6.2) calculated the standard error, s, to be .3671. Now, numbering the rows and columns of $(\mathbf{X}'\mathbf{X})^{-1}$ (which was calculated in Example 5.6) as the parameters β_0, β_1, and β_2 are numbered, we have

$$(\mathbf{X}'\mathbf{X})^{-1} = \begin{array}{c} \text{row} \\ 0 \\ 1 \\ 2 \end{array} \overset{\text{column}}{\begin{array}{ccc} 0 & 1 & 2 \\ \left[\begin{array}{c} 5.43405 \\ -.085930 \\ -.118856 \end{array}\right. & \begin{array}{c} -.085930 \\ .00147070 \\ .00165094 \end{array} & \left.\begin{array}{c} -.118856 \\ .00165094 \\ .00359276 \end{array}\right] \end{array}} = \begin{bmatrix} c_{00} & & \\ & c_{11} & \\ & & c_{22} \end{bmatrix}$$

Thus

1. The diagonal element of $(\mathbf{X}'\mathbf{X})^{-1}$ corresponding to β_0 is $c_{00} = 5.43405 \approx 5.434$.
2. The diagonal element of $(\mathbf{X}'\mathbf{X})^{-1}$ corresponding to β_1 is $c_{11} = .00147070 \approx .00147$.
3. The diagonal element of $(\mathbf{X}'\mathbf{X})^{-1}$ corresponding to β_2 is $c_{22} = .00359276 \approx .0036$.

Part 1. Calculating a Confidence Interval for β_2.

In this example we calculate a confidence interval for β_2, and as an exercise the reader should calculate confidence intervals for β_0 and β_1. The $100(1 - \alpha)\%$ confidence interval for β_2 is

$$\begin{aligned} [b_2 \pm BE_{b_2}[100(1-\alpha)]] &= [b_2 \pm t_{[\alpha/2]}^{(n-k)} s\sqrt{c_{22}}] \\ &= [.0825 \pm t_{[\alpha/2]}^{(8-3)}(.3671)\sqrt{.0036}] \\ &= [.0825 \pm t_{[\alpha/2]}^{(5)}(.022)] \end{aligned}$$

For example, the 95% confidence interval for β_2 is (since $\alpha = .05$)

$$\begin{aligned} [.0825 \pm t_{[.025]}^{(5)}(.022)] &= [.0825 \pm 2.571(.022)] \\ &= [.0825 \pm .0566] \\ &= [.0259, .1391] \end{aligned}$$

Recall from Example 5.5 that

$\beta_2 = \mu_{(c,d+1)} - \mu_{(c,d)} =$ the change in mean weekly fuel consumption associated with a one-unit increase in the chill index when the average hourly temperature does not change

Therefore the 95% confidence interval for β_2, [.0259, .1391], states that we are 95% confident that if the chill index increases by one unit while the average hourly temperature remains constant, the mean weekly fuel consumption will *increase* (because both the lower bound and the upper bound of the confidence interval are positive) by at least .0259 tons of coal and by at most .1391 tons of coal. Since (a supernatural power knows that) the true value of β_2 is .087, the 95% confidence interval for β_2 contains β_2. Since we do not know the true value of β_2, we are not certain (but are 95% confident) that the preceding interval contains β_2.

Part 2. Testing a Hypothesis Concerning β_2.

In Section 6.4 we discuss the exact meaning of 95% confidence. First, however, we discuss how we can use the 95% confidence interval for β_2 in a meaningful way. To do this, consider the null hypothesis

$$H_0: \beta_2 = 0$$

which says that there is *no change* in mean weekly fuel consumption associated with a change in x_2 (the chill index) when x_1 (the average hourly temperature) does not change, and the alternative hypothesis

$$H_1: \beta_2 \neq 0$$

which says that there is a (*positive or negative*) *change* in mean weekly fuel consumption associated with a change in x_2 when x_1 does not change. It would seem reasonable to conclude that x_2 is significantly related to y if we can, at a high level of confidence, reject the null hypothesis $H_0: \beta_2 = 0$ in favor of the alternative hypothesis $H_1: \beta_2 \neq 0$, because this would imply that we can, at a high level of confidence, believe that there is a change in mean weekly fuel consumption associated with a change in x_2 when x_1 does not change. In Section 6.5 we describe in detail testing these hypotheses. For now, we note that since the 95% confidence interval for β_2, [.0259, .1391], does not contain zero, we are (at least) 95% confident that β_2 is not equal to zero. For this reason we can, with at least 95% confidence, reject $H_0: \beta_2 = 0$ in favor of $H_1: \beta_2 \neq 0$. Thus, it would seem reasonable to conclude that x_2 (the chill index) is significantly related to y (weekly fuel consumption) and thus should be included in the regression model

$$y = \beta_0 + \beta_1 x_1 + \beta_2 x_2 + \epsilon$$

To complete this example, recall that we have previously reasoned that if β_2 does not equal zero, then β_2 must be a positive number, because this would express the fact that the larger x_2 (the chill index) is when x_1 does not change, then the larger is the mean weekly fuel consumption. Hence, it would really be more appropriate to decide that x_2 is significantly related to y if we can, at a high level of confidence, reject the null hypothesis $H_0: \beta_2 = 0$ in favor of the *one-sided* alternative hypothesis $H_1: \beta_2 > 0$. Although this is slightly more effective than attempting to reject the null hypothesis $H_0: \beta_2 = 0$ in favor of the *two-sided* alternative hypothesis $H_1: \beta_2 \neq 0$, it makes little practical difference whether we use a one-sided or a two-sided alternative hypothesis in deciding whether an independent variable is significantly related to the dependent variable in a regression model. Consequently, in order to simplify our discussion of hypothesis testing in regression analysis, we will always attempt to reject a null hypothesis in favor of a two-sided alternative hypothesis, even if we can determine an appropriate one-sided alternative hypothesis. Readers who are interested in learning how to test a null hypothesis versus a one-sided alternative hypothesis (for example, $H_0: \beta_2 = 0$ versus $H_1: \beta_2 > 0$) should read the end of Section 6.5 and optional Section 6.12.

6.4 THE STATISTICAL MEANING AND DERIVATION OF THE INTERVAL FOR A REGRESSION PARAMETER

EXAMPLE 6.4

In this example we discuss the meaning of 95% confidence as it pertains to the 95% confidence interval for β_2, [.0259, .1391], which we calculated in Example 6.3 by using the fuel consumptions in the observed sample

$$\begin{aligned} SPL &= \{y_1, y_2, y_3, y_4, y_5, y_6, y_7, y_8\} \\ &= \{12.4, 11.7, 12.4, 10.8, 9.4, 9.5, 8.0, 7.5\} \end{aligned}$$

Recall that we assume that for $i = 1, 2, \ldots, 8$, the ith observed fuel consumption, y_i, has been randomly selected from the ith historical population of potential weekly fuel consumptions. We must now recognize that before this sample was randomly selected, for $i = 1, 2, \ldots, 8$, y_i could have been any of the potential weekly fuel consumptions in the ith historical population, and thus this sample could have been any one of many different sets of eight fuel consumptions. We define the *population of all possible samples of eight fuel consumptions* to be the population of all the different sets of eight fuel consumptions that could potentially have been randomly selected from the eight historical populations. Table 6.1 lists three possible samples in the population of all possible samples. We have arbitrarily denoted these samples SPL_1 (the first possible sample), SPL_2 (the second possible sample), and SPL_3 (the third possible sample). Notice that, given this numbering scheme, SPL_1 is the sample we have actually observed. Also, note that when we number these three possible samples, *we do not mean to imply that the samples in the population of all possible samples*

TABLE 6.1 Three Possible Samples (SPL_1, SPL_2, and SPL_3) of Eight Fuel Consumptions

Week (i)	Average Hourly Temperature (x_{i1})	Chill Index (x_{i2})	Historical Population of Potential Weekly Fuel Consumptions
1	$x_{11} = 28.0$	$x_{12} = 18$	First historical population
2	$x_{21} = 28.0$	$x_{22} = 14$	Second historical population
3	$x_{31} = 32.5$	$x_{32} = 24$	Third historical population
4	$x_{41} = 39.0$	$x_{42} = 22$	Fourth historical population
5	$x_{51} = 45.9$	$x_{52} = 8$	Fifth historical population
6	$x_{61} = 57.8$	$x_{62} = 16$	Sixth historical population
7	$x_{71} = 58.1$	$x_{72} = 1$	Seventh historical population
8	$x_{81} = 62.5$	$x_{82} = 0$	Eighth historical population

SPL_1 $(y_i)_{SPL_1}$	SPL_2 $(y_i)_{SPL_2}$	SPL_3 $(y_i)_{SPL_3}$
$(y_1)_{SPL_1} = 12.4$	$(y_1)_{SPL_2} = 12.0$	$(y_1)_{SPL_3} = 10.7$
$(y_2)_{SPL_1} = 11.7$	$(y_2)_{SPL_2} = 11.8$	$(y_2)_{SPL_3} = 10.2$
$(y_3)_{SPL_1} = 12.4$	$(y_3)_{SPL_2} = 12.3$	$(y_3)_{SPL_3} = 10.5$
$(y_4)_{SPL_1} = 10.8$	$(y_4)_{SPL_2} = 11.5$	$(y_4)_{SPL_3} = 9.8$
$(y_5)_{SPL_1} = 9.4$	$(y_5)_{SPL_2} = 9.1$	$(y_5)_{SPL_3} = 9.5$
$(y_6)_{SPL_1} = 9.5$	$(y_6)_{SPL_2} = 9.2$	$(y_6)_{SPL_3} = 8.9$
$(y_7)_{SPL_1} = 8.0$	$(y_7)_{SPL_2} = 8.5$	$(y_7)_{SPL_3} = 8.5$
$(y_8)_{SPL_1} = 7.5$	$(y_8)_{SPL_2} = 7.2$	$(y_8)_{SPL_3} = 8.0$

are countable. (Under the inference assumptions there are an uncountably infinite number of samples in this population.) When we number SPL_1, SPL_2, and SPL_3, we are simply saying that we can consider three of the possible samples in the population of all possible samples and that *these three samples can be arbitrarily numbered.*

Table 6.2 illustrates eight normal curves, which (in accordance with the inference assumptions) represent the eight historical populations of potential weekly fuel consumptions; and the eight fuel consumptions contained in each of the three samples SPL_1, SPL_2, and SPL_3. Thus, for $i = 1, 2, \ldots, 8$, the ith normal curve depicts the following:

1. The ith historical population of potential weekly fuel consumptions (the infinite population of potential weekly fuel consumptions we could observe when the average hourly temperature is x_{i1} and the chill index is x_{i2}), which has mean

$$\mu_i = \beta_0 + \beta_1 x_{i1} + \beta_2 x_{i2}$$
$$= 12.930 - .087x_{i1} + .086x_{i2}$$

TABLE 6.2 Three Possible Samples (SPL_1, SPL_2, SPL_3) of Eight Fuel Consumptions

First Possible Sample (SPL_1)	Second Possible Sample (SPL_2)	Third Possible Sample (SPL_3)	Historical Populations of Potential Weekly Fuel Consumptions ($\beta_0 = 12.930$, $\beta_1 = -.087$, $\beta_2 = .086$)
First historical population			
$(y_1)_{SPL_1} = \mu_1 + (\epsilon_1)_{SPL_1}$ $= 12.042 + .358 = 12.4$	$(y_1)_{SPL_2} = \mu_1 + (\epsilon_1)_{SPL_2}$ $= 12.042 - .042 = 12.0$	$(y_1)_{SPL_3} = \mu_1 + (\epsilon_1)_{SPL_3}$ $= 12.042 - 1.342 = 10.7$	$\mu_1 = 12.930 - .087(28.0) + .086(18)$ $= 12.042$ $x_{11} = 28.0$, $x_{12} = 18$ $y_1 = \mu_1 + \epsilon_1 = 12.042 + \epsilon_1$
Second historical population			
$(y_2)_{SPL_1} = \mu_2 + (\epsilon_2)_{SPL_1}$ $= 11.698 + .002 = 11.7$	$(y_2)_{SPL_2} = \mu_2 + (\epsilon_2)_{SPL_2}$ $= 11.698 + .102 = 11.8$	$(y_2)_{SPL_3} = \mu_2 + (\epsilon_2)_{SPL_3}$ $= 11.698 - 1.498 = 10.2$	$\mu_2 = 12.930 - .087(28.0) + .086(14)$ $= 11.698$ $x_{21} = 28.0$, $x_{22} = 14$ $y_2 = \mu_2 + \epsilon_2 = 11.698 + \epsilon_2$

First historical population:

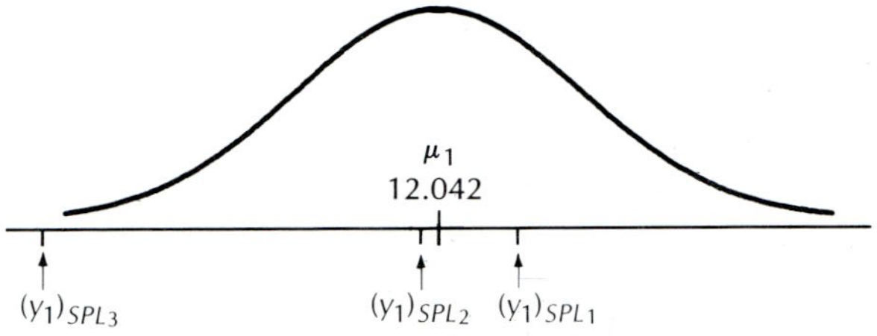

Second historical population:

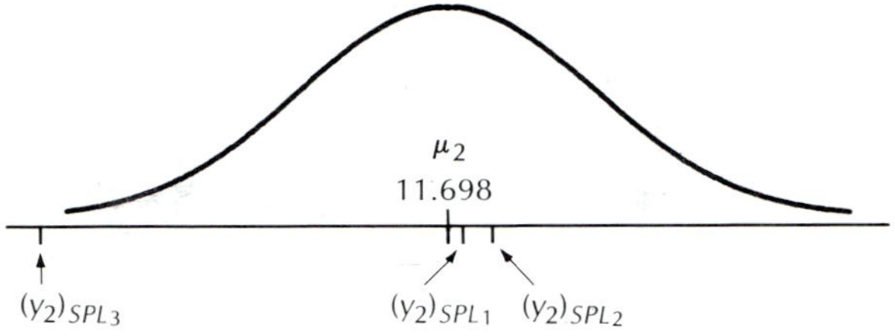

(Continued)

TABLE 6.2 *(Continued)*

First Possible Sample (SPL_1)	Second Possible Sample (SPL_2)	Third Possible Sample (SPL_3)	Historical Populations of Potential Weekly Fuel Consumptions ($\beta_0 = 12.930$, $\beta_1 = -.087$, $\beta_2 = .086$)
Third historical population			
$(y_3)_{SPL_1} = \mu_3 + (\epsilon_3)_{SPL_1}$ $= 12.166 + .234 = 12.4$	$(y_3)_{SPL_2} = \mu_3 + (\epsilon_3)_{SPL_2}$ $= 12.166 + .134 = 12.3$	$(y_3)_{SPL_3} = \mu_3 + (\epsilon_3)_{SPL_3}$ $= 11.698 - 1.198 = 10.5$	$\mu_3 = 12.930 - .087(32.5) + .086(24)$ $= 12.166$ $x_{31} = 32.5$, $x_{32} = 24$ $y_3 = \mu_3 + \epsilon_3 = 12.166 + \epsilon_3$
Fourth historical population			
$(y_4)_{SPL_1} = \mu_4 + (\epsilon_4)_{SPL_1}$ $= 11.429 - .629 = 10.8$	$(y_4)_{SPL_2} = \mu_4 + (\epsilon_4)_{SPL_2}$ $= 11.429 + .071 = 11.5$	$(y_4)_{SPL_3} = \mu_4 + (\epsilon_4)_{SPL_3}$ $= 11.429 - 1.629 = 9.8$	$\mu_4 = 12.930 - .087(39.0) + .086(22)$ $= 11.429$ $x_{41} = 39.0$, $x_{42} = 22$ $y_4 = \mu_4 + \epsilon_4 = 11.429 + \epsilon_4$

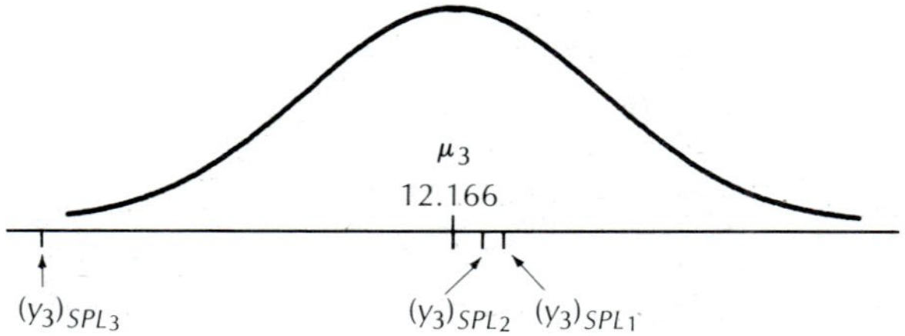

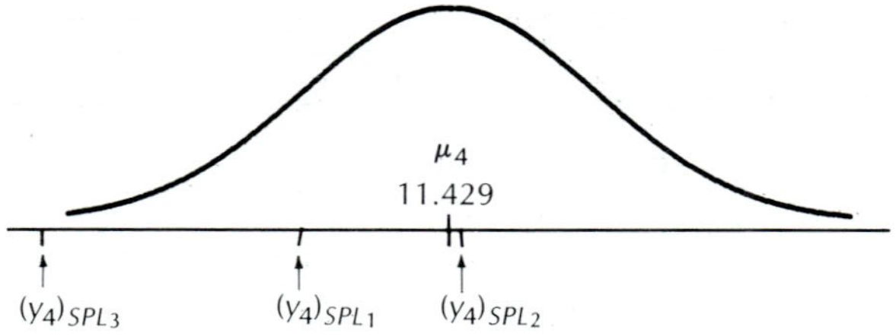

Fifth historical population

$(y_5)_{SPL_1} = \mu_5 + (\epsilon_5)_{SPL_1}$
$= 9.625 - .225 = 9.4$

$(y_5)_{SPL_2} = \mu_5 + (\epsilon_5)_{SPL_2}$
$= 9.625 - .525 = 9.1$

$(y_5)_{SPL_3} = \mu_5 + (\epsilon_5)_{SPL_3}$
$= 9.625 - .125 = 9.5$

$\mu_5 = 12.930 - .087(45.9) + .086(8)$
$= 9.625$

$x_{51} = 45.9,\ x_{52} = 8$

$y_5 = \mu_5 + \epsilon_5 = 9.625 + \epsilon_5$

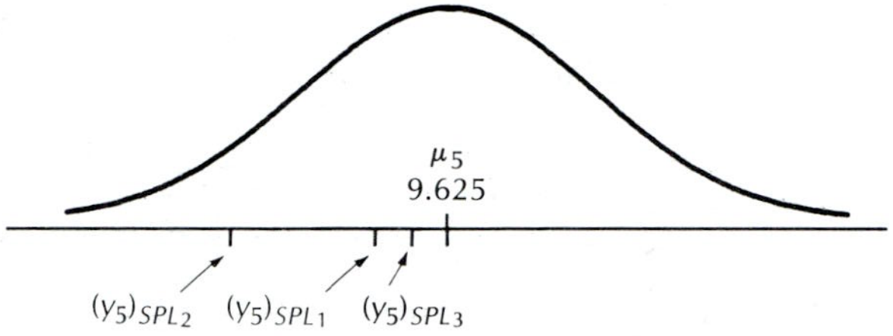

Sixth historical population

$(y_6)_{SPL_1} = \mu_6 + (\epsilon_6)_{SPL_1}$
$= 9.277 + .223 = 9.5$

$(y_6)_{SPL_2} = \mu_6 + (\epsilon_6)_{SPL_2}$
$= 9.277 - .077 = 9.2$

$(y_6)_{SPL_3} = \mu_6 + (\epsilon_6)_{SPL_3}$
$= 9.277 - .377 = 8.9$

$\mu_6 = 12.930 - .087(57.8) + .086(16)$
$= 9.277$

$x_{61} = 57.8,\ x_{62} = 16$

$y_6 = \mu_6 + \epsilon_6 = 9.277 + \epsilon_6$

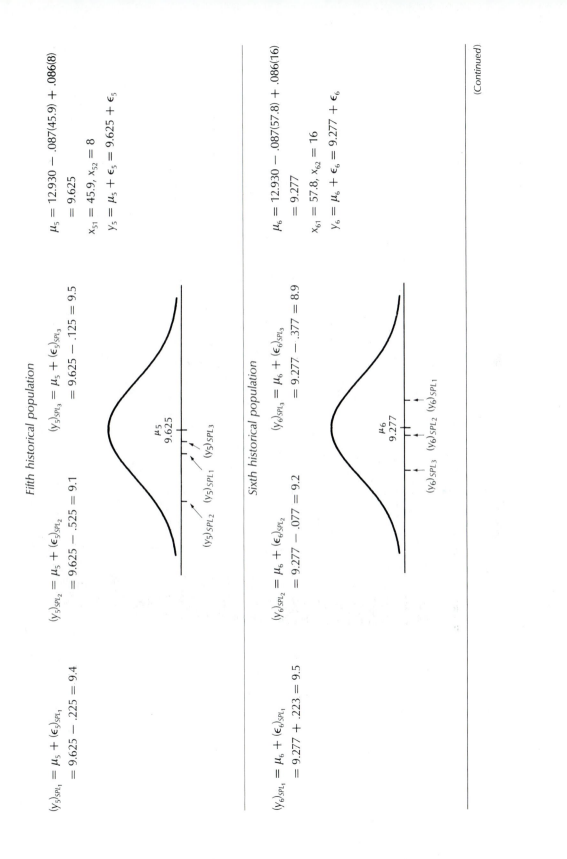

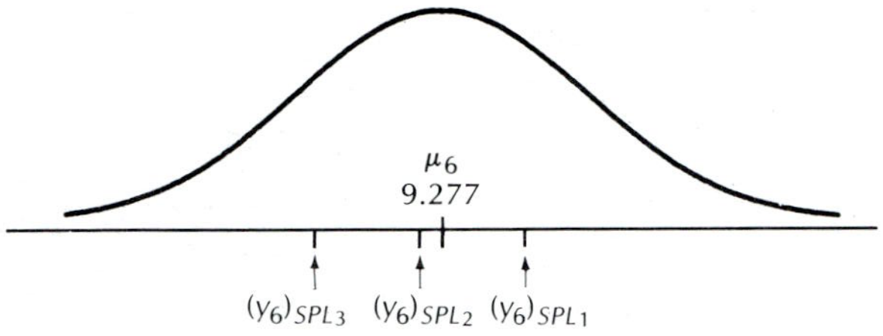

(Continued)

TABLE 6.2 *(Continued)*

First Possible Sample (SPL_1)	Second Possible Sample (SPL_2)	Third Possible Sample (SPL_3)	Historical Populations of Potential Weekly Fuel Consumptions ($\beta_0 = 12.930$, $\beta_1 = -.087$, $\beta_2 = .086$)
	Seventh historical population		
$(y_7)_{SPL_1} = \mu_7 + (\epsilon_7)_{SPL_1}$ $= 7.961 + .039 = 8.0$	$(y_7)_{SPL_2} = \mu_7 + (\epsilon_7)_{SPL_2}$ $= 7.961 + .539 = 8.5$	$(y_7)_{SPL_3} = \mu_7 + (\epsilon_7)_{SPL_3}$ $= 7.961 + .539 = 8.5$	$\mu_7 = 12.930 - .087(58.1) + .086(1)$ $= 7.961$ $x_{71} = 58.1$, $x_{72} = 1$ $y_7 = \mu_7 + \epsilon_7 = 7.961 + \epsilon_7$

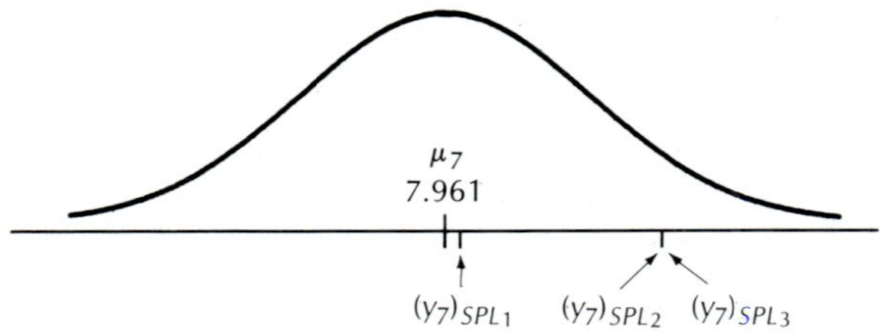

First Possible Sample (SPL_1)	Second Possible Sample (SPL_2)	Third Possible Sample (SPL_3)	Historical Populations of Potential Weekly Fuel Consumptions ($\beta_0 = 12.930$, $\beta_1 = -.087$, $\beta_2 = .086$)
	Eighth historical population		
$(y_8)_{SPL_1} = \mu_8 + (\epsilon_8)_{SPL_1}$ $= 7.492 + .008 = 7.5$	$(y_8)_{SPL_2} = \mu_8 + (\epsilon_8)_{SPL_2}$ $= 7.492 - .292 = 7.2$	$(y_8)_{SPL_3} = \mu_8 + (\epsilon_8)_{SPL_3}$ $= 7.492 + .508 = 8.0$	$\mu_8 = 12.930 - .087(62.5) + .086(0)$ $= 7.492$ $x_{81} = 62.5$, $x_{82} = 0$ $y_8 = \mu_8 + \epsilon_8 = 7.492 + \epsilon_8$

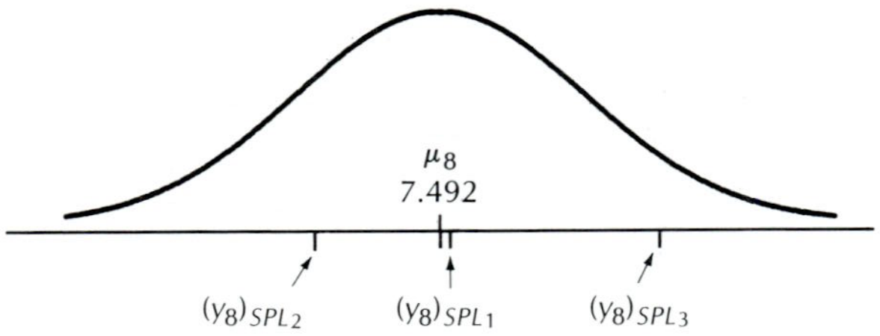

2. The ith fuel consumption in SPL_1, $(y_i)_{SPL_1}$;
3. The ith fuel consumption in SPL_2, $(y_i)_{SPL_2}$;
4. The ith fuel consumption in SPL_3, $(y_i)_{SPL_3}$.

To summarize, before the observed sample, SPL, was randomly selected, SPL could have been SPL_1, SPL_2, SPL_3, or any other sample in the population of all possible samples of eight fuel consumptions. In the present, after SPL has been randomly selected, we see that SPL is actually SPL_1. Therefore, we have actually observed the weekly fuel consumptions in SPL_1.

Now, corresponding to the infinite population of all possible samples of eight fuel consumptions, there are the following regression populations:

1. The infinite population of all possible column vectors of observed fuel consumptions (since each different sample yields a different column vector $\mathbf{y}$);
2. The infinite population of all possible column vectors of least squares point estimates (since each different sample yields a different column vector $\mathbf{b}$ of least squares point estimates); and the infinite population of all possible least squares point estimates of the parameter β_j (since each different sample yields a different least squares point estimate b_j);
3. The infinite population of all possible point estimates of μ_i and point predictions of y_i (since each different sample yields a different point estimate and point prediction $\hat{y}_i$);
4. The infinite population of all possible mean square errors (since each different sample yields a different mean square error s^2);
5. The infinite population of all possible standard errors (since each different sample yields a different standard error s);
6. The infinite population of all possible 95% confidence intervals for β_2 (since each different sample yields a different 95% confidence interval for β_2).

Table 6.3 presents the elements in the preceding populations corresponding to SPL_1, SPL_2, and SPL_3 (as given in Table 6.1). As an example of our notation here, $\mathbf{y}_{SPL_1}$, $\mathbf{b}_{SPL_1}$, $(\hat{y}_i)_{SPL_1}$, $(s^2)_{SPL_1}$, $(s)_{SPL_1}$, and

$$[(b_2)_{SPL_1} \pm t^{(5)}_{[.025]}(s)_{SPL_1}\sqrt{c_{22}}] = [(b_2)_{SPL_1} \pm 2.571(s)_{SPL_1}\sqrt{.0036}]$$

denote, respectively, the $\mathbf{y}$ vector, vector $\mathbf{b}$ of least squares point estimates, point estimate of μ_i (and point prediction of y_i), mean square error, standard error, and 95% confidence interval for β_2 yielded by SPL_1, the first possible sample.

Now consider Figure 6.1. This figure depicts (1) the least squares point estimates of β_2 yielded by SPL_1, SPL_2, and SPL_3 (which are denoted $(b_2)_{SPL_1}$, $(b_2)_{SPL_2}$, and $(b_2)_{SPL_3}$); (2) the 95% confidence intervals for β_2 yielded by SPL_1, SPL_2, and SPL_3 (which are denoted $CI_{\beta_2}[95]_{SPL_1}$, $CI_{\beta_2}[95]_{SPL_2}$, and $CI_{\beta_2}[95]_{SPL_3}$); and (3) the fact that the population of all possible least squares point estimates of β_2 has (if the inference assumptions hold) a normal distribution with mean β_2 (which, a supernatural

TABLE 6.3 The Elements Corresponding to SPL_1, SPL_2, and SPL_3 in Six Regression Populations Related to the Two-Variable Fuel Consumption Model

Three Column Vectors in the Population of All Possible Column Vectors

SPL_1

$$\mathbf{y}_{SPL_1} = \begin{bmatrix} (y_1)_{SPL_1} \\ (y_2)_{SPL_1} \\ (y_3)_{SPL_1} \\ (y_4)_{SPL_1} \\ (y_5)_{SPL_1} \\ (y_6)_{SPL_1} \\ (y_7)_{SPL_1} \\ (y_8)_{SPL_1} \end{bmatrix} = \begin{bmatrix} 12.4 \\ 11.7 \\ 12.4 \\ 10.8 \\ 9.4 \\ 9.5 \\ 8.0 \\ 7.5 \end{bmatrix}$$

SPL_2

$$\mathbf{y}_{SPL_2} = \begin{bmatrix} (y_1)_{SPL_2} \\ (y_2)_{SPL_2} \\ (y_3)_{SPL_2} \\ (y_4)_{SPL_2} \\ (y_5)_{SPL_2} \\ (y_6)_{SPL_2} \\ (y_7)_{SPL_2} \\ (y_8)_{SPL_2} \end{bmatrix} = \begin{bmatrix} 12.0 \\ 11.8 \\ 12.3 \\ 11.5 \\ 9.1 \\ 9.2 \\ 8.5 \\ 7.2 \end{bmatrix}$$

SPL_3

$$\mathbf{y}_{SPL_3} = \begin{bmatrix} (y_1)_{SPL_3} \\ (y_2)_{SPL_3} \\ (y_3)_{SPL_3} \\ (y_4)_{SPL_3} \\ (y_5)_{SPL_3} \\ (y_6)_{SPL_3} \\ (y_7)_{SPL_3} \\ (y_8)_{SPL_3} \end{bmatrix} = \begin{bmatrix} 10.7 \\ 10.2 \\ 10.5 \\ 9.8 \\ 9.5 \\ 8.9 \\ 8.5 \\ 8.0 \end{bmatrix}$$

Three Column Vectors in the Population of All Possible Column Vectors of Least Squares Point Estimates

SPL_1

$$\begin{bmatrix} (b_0)_{SPL_1} \\ (b_1)_{SPL_1} \\ (b_2)_{SPL_1} \end{bmatrix} = \mathbf{b}_{SPL_1} = (\mathbf{X}'\mathbf{X})^{-1}\mathbf{X}'\mathbf{y}_{SPL_1} = \begin{bmatrix} 13.109 \\ -.0900 \\ .0825 \end{bmatrix}$$

SPL_2

$$\begin{bmatrix} (b_0)_{SPL_2} \\ (b_1)_{SPL_2} \\ (b_2)_{SPL_2} \end{bmatrix} = \mathbf{b}_{SPL_2} = (\mathbf{X}'\mathbf{X})^{-1}\mathbf{X}'\mathbf{y}_{SPL_2} = \begin{bmatrix} 12.949 \\ -.0882 \\ .0876 \end{bmatrix}$$

SPL_3

$$\begin{bmatrix} (b_0)_{SPL_3} \\ (b_1)_{SPL_3} \\ (b_2)_{SPL_3} \end{bmatrix} = \mathbf{b}_{SPL_3} = (\mathbf{X}'\mathbf{X})^{-1}\mathbf{X}'\mathbf{y}_{SPL_3} = \begin{bmatrix} 11.593 \\ -.0548 \\ .0256 \end{bmatrix}$$

Three Point Estimates of μ_i (and Point Predictions of y_i) in the Population of All Possible Point Estimates of μ_i (and Point Predictions of y_i)

SPL_1

$$(\hat{y}_i)_{SPL_1} = (b_0)_{SPL_1} + (b_1)_{SPL_1}x_{i1} + (b_2)_{SPL_1}x_{i2} = 13.109 - .0900x_{i1} + .0825x_{i2}$$

SPL_2

$$(\hat{y}_i)_{SPL_2} = (b_0)_{SPL_2} + (b_1)_{SPL_2}x_{i1} + (b_2)_{SPL_2}x_{i2} = 12.949 - .0882x_{i1} + .0876x_{i2}$$

SPL_3

$$(\hat{y}_i)_{SPL_3} = (b_0)_{SPL_3} + (b_1)_{SPL_3}x_{i1} + (b_2)_{SPL_3}x_{i2} = 11.593 - .0548x_{i1} + .0256x_{i2}$$

Three Mean Square Errors in the Population of All Possible Mean Square Errors		
SPL_1	SPL_2	SPL_3
$(s)^2_{SPL_1} = \frac{\sum_{i=1}^{8}((y_i)_{SPL_1} - (\hat{y}_i)_{SPL_1})^2}{8-3}$ $= \frac{\sum_{i=1}^{8}(y_i)^2_{SPL_1} - \mathbf{b}'_{SPL_1}\mathbf{X}'\mathbf{y}_{SPL_1}}{8-3}$ $= \frac{.674}{5} = .1348$	$(s)^2_{SPL_2} = \frac{\sum_{i=1}^{8}((y_i)_{SPL_2} - (\hat{y}_i)_{SPL_2})^2}{8-3}$ $= \frac{\sum_{i=1}^{8}(y_i)^2_{SPL_2} - \mathbf{b}'_{SPL_2}\mathbf{X}'\mathbf{y}_{SPL_2}}{8-3}$ $= \frac{.685}{5} = .137$	$(s)^2_{SPL_3} = \frac{\sum_{i=1}^{8}((y_i)_{SPL_3} - (\hat{y}_i)_{SPL_3})^2}{8-3}$ $= \frac{\sum_{i=1}^{8}(y_i)^2_{SPL_3} - \mathbf{b}'_{SPL_3}\mathbf{X}'\mathbf{y}_{SPL_3}}{8-3}$ $= \frac{.2172}{5} = .0434$

Three Standard Errors in the Population of All Possible Standard Errors		
SPL_1	SPL_2	SPL_3
$(s)_{SPL_1} = \sqrt{(s)^2_{SPL_1}}$ $= \sqrt{.1348} = .3671$	$(s)_{SPL_2} = \sqrt{(s)^2_{SPL_2}}$ $= \sqrt{.137} = .3701$	$(s)_{SPL_3} = \sqrt{(s)^2_{SPL_3}}$ $= \sqrt{.0434} = .2084$

Three 95% Confidence Intervals for β_2 in the Population of All Possible 95% Confidence Intervals for β_2		
SPL_1	SPL_2	SPL_3
$[(b_2)_{SPL_1} \pm 2.571(s)_{SPL_1}\sqrt{.0036}]$ $= [.0825 \pm 2.571(.3671)\sqrt{.0036}]$ $= [.0825 \pm .0566] = [.0259, .1391]$	$[(b_2)_{SPL_2} \pm 2.571(s)_{SPL_2}\sqrt{.0036}]$ $= [.0876 \pm 2.571(.3701)\sqrt{.0036}]$ $= [.0876 \pm .057] = [.0306, .1446]$	$[(b_2)_{SPL_3} \pm 2.571(s)_{SPL_3}\sqrt{.0036}]$ $= [.0256 \pm 2.571(.2084)\sqrt{.0036}]$ $= [.0256 \pm .0321] = [-.0065, .0577]$

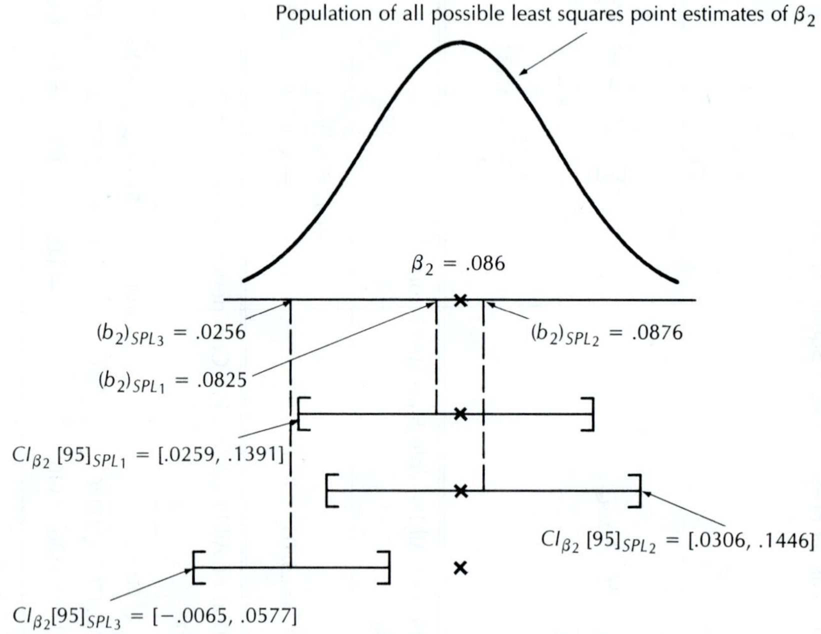

FIGURE 6.1
95% Confidence Intervals (CI) for β_2 Obtained Using the Observations in SPL_1, SPL_2, and SPL_3

power knows, equals .086). Examining Figure 6.1, we see that both of the confidence intervals

$$CI_{\beta_2}[95]_{SPL_1} = [.0259, .1391]$$

$$CI_{\beta_2}[95]_{SPL_2} = [.0306, .1446]$$

contain $\beta_2 = .086$ (denoted by the × in the figure), but the confidence interval

$$CI_{\beta_2}[95]_{SPL_3} = [-.0065, .0577]$$

does not contain $\beta_2 = .086$. Thus, two of the three 95% confidence intervals for β_2 depicted in Figure 6.1 contain β_2. We later prove that the interpretation of 95% confidence here is that 95 percent of the 95% confidence intervals for β_2 in the population of all such intervals contain $\beta_2 = .086$, but 5 percent of the confidence intervals in this population do not contain $\beta_2 = .086$. Thus, when we compute the 95% confidence interval for β_2 by using the observed sample (SPL_1) and obtain the interval [.0259, .1391], we can be 95% confident that this interval contains β_2, because 95 percent of the confidence intervals in the population of all possible 95% confidence intervals for β_2 contain β_2 and because we have obtained one of the confidence intervals in this population.

We now discuss the logic behind deriving the $100(1-\alpha)\%$ confidence interval for β_j. We begin by considering the regression model

$$\begin{aligned} y_i &= \mu_i + \epsilon_i \\ &= \beta_0 + \beta_1 x_{i1} + \cdots + \beta_j x_{ij} + \cdots + \beta_p x_{ip} + \epsilon_i \end{aligned}$$

and the following properties (see Appendix D for a proof of (1), (2), and (3)).

The population of all possible least squares point estimates of β_j

1. Has *mean* $\mu_{b_j} = \beta_j$;
2. Has *variance* $\sigma^2_{b_j} = \sigma^2 c_{jj}$ (if inference assumptions 1 and 2 hold);
3. Has *standard deviation* $\sigma_{b_j} = \sigma\sqrt{c_{jj}}$ (if inference assumptions 1 and 2 hold);
4. Has a *normal distribution* (if inference assumptions 1, 2, and 3 hold).

Here

σ^2 = the constant variance of the historical and future populations of potential values of the dependent variable

c_{jj} = the diagonal element of the matrix $(\mathbf{X}'\mathbf{X})^{-1}$ that corresponds to β_j when the rows and columns of $(\mathbf{X}'\mathbf{X})^{-1}$ are numbered as the parameters in the regression model are numbered

Property (1) says that μ_{b_j}, the mean of the population of all possible least squares point estimates of β_j, equals β_j. For this reason, when we use the observed sample

$$SPL = \{y_1, y_2, \ldots, y_n\}$$

to calculate the least squares point estimate b_j of β_j by the formula

$$\mathbf{b} = (\mathbf{X}'\mathbf{X})^{-1}\mathbf{X}'\mathbf{y}$$

we are using an **unbiased** estimation procedure. This unbiasedness property tells us that although the least squares point estimate b_j that we calculate probably does not equal β_j, the average of all the different possible least squares point estimates of β_j that we could have calculated is equal to β_j.

Property (2) says that $\sigma^2_{b_j}$, the variance of the population of all possible least squares point estimates of β_j, equals $\sigma^2 c_{jj}$. An important theorem, the **Gauss-Markov Theorem**, is related to Property (2). To understand what this theorem says, note that since the least squares point estimate b_j of β_j is calculated by using the equation

$$\mathbf{b} = (\mathbf{X}'\mathbf{X})^{-1}\mathbf{X}'\mathbf{y}$$

it follows that b_j is a linear function of $y_1, y_2, \ldots, y_n$ (the elements contained in $\mathbf{y}$), and thus we are calculating b_j by using a *linear estimation procedure* (which, from the preceding discussion, is also an *unbiased* estimation procedure).

GAUSS-MARKOV THEOREM:

The *variance* of the population of all possible *least squares* point estimates of β_j is *smaller* than the variance of the population of all possible point estimates of β_j that could be obtained by using any other unbiased, linear estimation procedure.

Intuitively, this means that the least squares point estimates of β_j in the population of all such estimates are clustered more closely around β_j than are the point estimates of β_j that could be obtained by using any other unbiased, linear estimation procedure. Thus, when we calculate the least squares point estimate b_j of β_j, we are likely to obtain a point estimate of β_j that is closer to the true value of β_j than would be a point estimate of β_j that we would obtain by using any other unbiased, linear estimation procedure.

Now, the point estimate of σ is s; therefore

$$s_{b_j} = s\sqrt{c_{jj}}$$

which we call the standard error of the estimate b_j, is the point estimate of

$$\sigma_{b_j} = \sigma\sqrt{c_{jj}}$$

the standard deviation of the population of all possible least squares point estimates of β_j. Then, defining the statistic

$$t_{[b_j,\beta_j]} = \frac{b_j - \beta_j}{s_{b_j}} = \frac{b_j - \beta_j}{s\sqrt{c_{jj}}}$$

and considering the population of all possible $t_{[b_j,\beta_j]}$ statistics, we can make the following statement.

If the inference assumptions hold, *the population of all possible* $t_{[b_j,\beta_j]}$ *statistics* has a t-distribution with $n - k$ degrees of freedom, where

n = the number of observed values of the dependent variable in $SPL = \{y_1, y_2, \ldots, y_n\}$

k = the number of parameters in the regression model

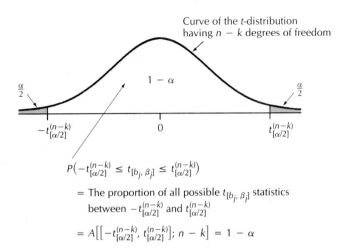

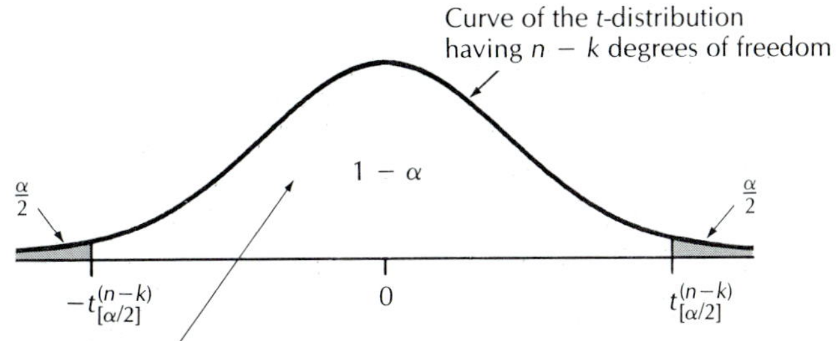

FIGURE 6.2 If the Inference Assumptions Hold, $P(-t_{[\alpha/2]}^{(n-k)} \leq t_{[b_j,\beta_j]} \leq t_{[\alpha/2]}^{(n-k)}) = 1 - \alpha$

$$P(-t_{[\alpha/2]}^{(n-k)} \leq t_{[b_j,\beta_j]} \leq t_{[\alpha/2]}^{(n-k)})$$

= The proportion of all possible $t_{[b_j,\beta_j]}$ statistics between $-t_{[\alpha/2]}^{(n-k)}$ and $t_{[\alpha/2]}^{(n-k)}$

$$= A\left[\left[-t_{[\alpha/2]}^{(n-k)}, t_{[\alpha/2]}^{(n-k)}\right]; n - k\right] = 1 - \alpha$$

Now, consider the t-point $t_{[\alpha/2]}^{(n-k)}$ and Figure 6.2. It follows from the preceding statement that the probability

$$P(-t_{[\alpha/2]}^{(n-k)} \leq t_{[b_j,\beta_j]} \leq t_{[\alpha/2]}^{(n-k)}) = P\left(-t_{[\alpha/2]}^{(n-k)} \leq \frac{b_j - \beta_j}{s_{b_j}} \leq t_{[\alpha/2]}^{(n-k)}\right)$$

$$= A[[-t_{[\alpha/2]}^{(n-k)}, t_{[\alpha/2]}^{(n-k)}]; n - k] = 1 - \alpha$$

which is the area under the curve of the t-distribution having $n - k$ degrees of freedom between $-t_{[\alpha/2]}^{(n-k)}$ and $t_{[\alpha/2]}^{(n-k)}$. Multiplying the inequality in the probability statement

$$P\left(-t_{[\alpha/2]}^{(n-k)} \leq \frac{b_j - \beta_j}{s_{b_j}} \leq t_{[\alpha/2]}^{(n-k)}\right) = 1 - \alpha$$

by s_{b_j} (which is positive), we see that this probability statement is equivalent to

$$P(-t_{[\alpha/2]}^{(n-k)}s_{b_j} \leq b_j - \beta_j \leq t_{[\alpha/2]}^{(n-k)}s_{b_j}) = 1 - \alpha$$

which implies (subtracting b_j through the inequality) that

$$P(-b_j - t_{[\alpha/2]}^{(n-k)}s_{b_j} \leq -\beta_j \leq -b_j + t_{[\alpha/2]}^{(n-k)}s_{b_j}) = 1 - \alpha$$

which in turn implies (multiplying the inequality by -1) that

$$P(b_j + t_{[\alpha/2]}^{(n-k)}s_{b_j} \geq \beta_j \geq b_j - t_{[\alpha/2]}^{(n-k)}s_{b_j}) = 1 - \alpha$$

This probability statement is equivalent to

$$P(b_j - t_{[\alpha/2]}^{(n-k)}s_{b_j} \leq \beta_j \leq b_j + t_{[\alpha/2]}^{(n-k)}s_{b_j}) = 1 - \alpha$$

which, since $s_{b_j} = s\sqrt{c_{jj}}$, says that

$$P(b_j - t_{[\alpha/2]}^{(n-k)}s\sqrt{c_{jj}} \leq \beta_j \leq b_j + t_{[\alpha/2]}^{(n-k)}s\sqrt{c_{jj}}) = 1 - \alpha$$

This probability statement says that the proportion of confidence intervals that contain β_j in the population of all possible $100(1-\alpha)\%$ confidence intervals for β_j *is equal to* $1-\alpha$. That is, if we compute a $100(1-\alpha)\%$ confidence interval for β_j using the formula

$$[b_j \pm BE_{b_j}[100(1-\alpha)]] = [b_j \pm t_{[\alpha/2]}^{(n-k)} s\sqrt{c_{jj}}]$$

then $100(1-\alpha)\%$ (for example, 95%) of the confidence intervals in the population of all possible $100(1-\alpha)\%$ confidence intervals for β_j *contain* β_j, and $100(\alpha)\%$ (for example, 5%) of the confidence intervals in this population do not contain β_j. We illustrated this fact in Example 6.4 (see Figure 6.1).

6.5 TESTING THE IMPORTANCE OF AN INDEPENDENT VARIABLE (TESTING $H_0: \beta_j = 0$)

The independent variable x_j is significantly related to the dependent variable y in the regression model

$$y = \beta_0 + \beta_1 x_1 + \cdots + \beta_j x_j + \cdots + \beta_p x_p + \epsilon$$

if x_j is important, or useful, in describing, predicting, or controlling y when using this regression model. As illustrated in Example 6.3, it would seem reasonable to decide that the independent variable x_j is significantly related to the dependent variable y in this regression model if we can, at a high level of confidence, reject the null hypothesis $H_0: \beta_j = 0$ in favor of the alternative hypothesis $H_1: \beta_j \neq 0$. We now present an example of an intuitive approach to testing $H_0: \beta_j = 0$ versus $H_1: \beta_j \neq 0$.

EXAMPLE 6.5

Consider again the fuel consumption model

$$\begin{aligned} y &= \mu + \epsilon \\ &= \beta_0 + \beta_1 x_1 + \beta_2 x_2 + \epsilon \end{aligned}$$

In Example 6.3 we saw that a $100(1-\alpha)\%$ confidence interval for β_2 is

$$\begin{aligned} [b_2 \pm t_{[\alpha/2]}^{(n-k)} s\sqrt{c_{22}}] &= [.0825 \pm t_{[\alpha/2]}^{(8-3)}(.3671)\sqrt{.0036}] \\ &= [.0825 \pm t_{[\alpha/2]}^{(5)}(.022)] \end{aligned}$$

By using the appropriate $t_{[\alpha/2]}^{(5)}$-points, one can verify that the 95%, 98%, and 99% confidence intervals for β_2 are as illustrated in Figure 6.3. We can determine how confident we can be that we should reject $H_0: \beta_2 = 0$ in favor of $H_1: \beta_2 \neq 0$ by computing the t_{b_2} *statistic* and its related *prob-value*. We define these quan-

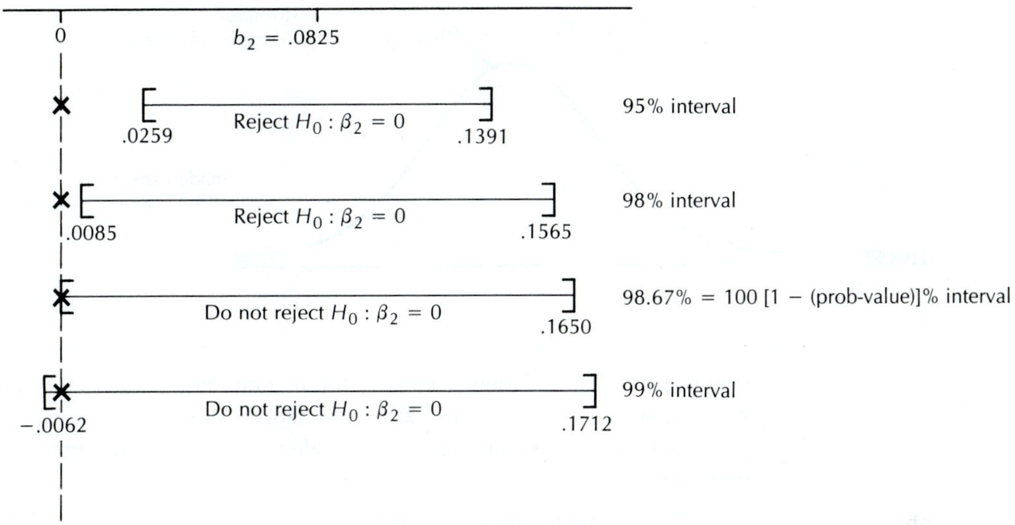

FIGURE 6.3 $H_0: \beta_2 = 0$ Can Be Rejected at Any Level of Confidence Less Than 98.67%

tities as follows:

$$t_{b_2} = \frac{b_2}{s\sqrt{c_{22}}} = \frac{.0825}{.3671\sqrt{.0036}} = 3.75$$

$$\text{prob-value} = 2A[[|t_{b_2}|, \infty); n - k]$$

where $A[[|t_{b_2}|, \infty); n - k]$ is the area under the curve of the t-distribution having $n - k = 8 - 3 = 5$ degrees of freedom to the right of $|t_{b_2}| = |3.75| = 3.75$, the absolute value of the t_{b_2} statistic.† Since this area can be calculated by computer to be .00665 (see Figure 6.4), it follows that

$$\begin{aligned} \text{prob-value} &= 2A[[|t_{b_2}|, \infty); n - k] \\ &= 2A[[|3.75|, \infty); 8 - 3] \\ &= 2A[[3.75, \infty); 5] \\ &= 2(.00665) \\ &= .01330 \end{aligned}$$

We will prove later that (1) for any level of confidence, $100(1 - \alpha)\%$, less than

$$100[1 - (\text{prob-value})]\% = 100[1 - .01330]\% = 98.67\%$$

† Although we have initially defined the prob-value in terms of an *area* under a curve, we will see later that the prob-value is a *probability*.

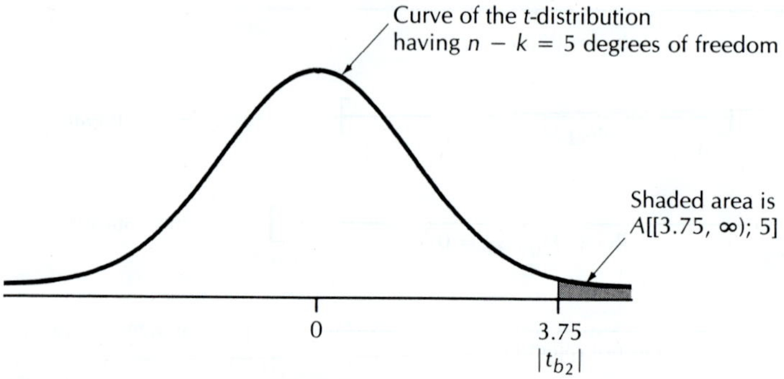

FIGURE 6.4
$A[[|t_{b_2}|, \infty); n - k] = A[[3.75, \infty); 5]$

the $100(1-\alpha)\%$ confidence interval for β_2 does not contain zero and thus we can, with at least $100(1-\alpha)\%$ confidence, reject $H_0: \beta_2 = 0$ in favor of $H_1: \beta_2 \neq 0$; and (2) for any level of confidence, $100(1-\alpha)\%$, greater than or equal to

$$100[1 - (\text{prob-value})]\% = 98.67\%$$

the $100(1-\alpha)\%$ confidence interval for β_2 does contain zero and thus we cannot, with at least $100(1-\alpha)\%$ confidence, reject $H_0: \beta_2 = 0$ in favor of $H_1: \beta_2 \neq 0$. These facts are illustrated in Figure 6.3. This figure depicts the 95%, 98%, 98.67%, and 99% confidence intervals for β_2. Notice that the 95% and 98% confidence intervals ($100(1-\alpha)\%$ less than 98.67%) do not contain zero and hence we can reject $H_0: \beta_2 = 0$ at these levels of confidence. However, the 98.67% confidence interval (which has a lower bound of zero) and the 99% confidence interval ($100(1-\alpha)\%$ greater than 98.67%) do contain zero and hence we cannot reject $H_0: \beta_2 = 0$ at these levels of confidence. To summarize, since the prob-value is .01330, we can be up to

$$100[1 - (\text{prob-value})]\% = 98.67\%$$

confident that the null hypothesis $H_0: \beta_2 = 0$ is false and that the alternative hypothesis $H_1: \beta_2 \neq 0$ is true. This provides substantial evidence that x_2, the chill index, is significantly related to y, weekly fuel consumption.

Example 6.5 has illustrated an intuitive approach to testing $H_0: \beta_j = 0$ versus $H_1: \beta_j \neq 0$ based on confidence intervals, t statistics, and prob-values. Later in this section we discuss this intuitive approach further. First, however, we explain the classical approach to testing $H_0: \beta_j = 0$ versus $H_1: \beta_j \neq 0$, which also makes use of confidence intervals, t statistics, and prob-values but is more formally based on the errors that can be made in hypothesis testing. These errors are summarized in Table 6.4. Two possible states of nature are listed across the top of the table—"the null hypothesis H_0 is true" and "the null hypothesis H_0 is false." We will never know with certainty whether H_0 is true or false. In our attempt to determine which state of nature is the true one, we perform a hypothesis test. Our two possible conclusions,

TABLE 6.4
Errors in Hypothesis Testing

Decisions Made in Hypothesis Test	State of Nature	
	Null hypothesis $H_0: \beta_j = 0$ is true	Null hypothesis $H_0: \beta_j = 0$ is false
Reject null hypothesis $H_0: \beta_j = 0$	Type I error	Correct action
Do not reject null hypothesis $H_0: \beta_j = 0$	Correct action	Type II error

reject or do not reject the null hypothesis, are listed down the side of Table 6.4. If we reject the null hypothesis H_0, there is only one error that can be made—a **Type I error**, which is defined to be the error of *rejecting H_0 when H_0 is true*. The reason that only one error can result from rejecting H_0 is that if we reject H_0 when H_0 is true, we have committed a Type I error, but if we reject H_0 when H_0 is false, we have taken a correct action. If, on the other hand, we do not reject the null hypothesis H_0, again only one error can be made—a **Type II error**, which is defined to be the error of *not rejecting H_0 when H_0 is false*. The reason that only one error can result from not rejecting H_0 is that if we do not reject H_0 when H_0 is true, we have taken a correct action, but if we do not reject H_0 when H_0 is false, we have committed a Type II error.

When performing a hypothesis test, we obviously desire that both the *probability of a Type I error*, denoted by α, and the *probability of a Type II error*, denoted by β, be *small*. One commonly used hypothesis testing procedure is based on taking a sample of a fixed size and setting α equal to a specified value. Here, we sometimes choose α to be as high as .1, but we usually choose α to be between .05 and .01, with .05 being the most frequent choice. The reason we usually do not choose α to be lower than .01 is that setting α extremely small often leads to a probability of a Type II error that is unacceptably large. This is because, generally, for a fixed sample size, the lower α is, the higher is β. How to determine the precise value of α that should be used (say, .05 rather than .01) is discussed in the next section. We now present the following summary about testing $H_0: \beta_j = 0$ versus $H_1: \beta_j \neq 0$.

TESTING $H_0: \beta_j = 0$ VERSUS $H_1: \beta_j \neq 0$:

Define

$$t_{[b_j, \beta_j]} = \frac{b_j - \beta_j}{s\sqrt{c_{jj}}}$$

and assume that the population of all possible $t_{[b_j,\beta_j]}$ statistics has a t-distribution with $n - k$ degrees of freedom. In order to test $H_0: \beta_j = 0$ versus $H_1: \beta_j \neq 0$, define the t_{b_j} *statistic* and its related *prob-value* to be

$$t_{b_j} = \frac{b_j}{s\sqrt{c_{jj}}} \qquad \text{and} \qquad \text{prob-value} = 2A[[|t_{b_j}|, \infty); n - k]$$

where $A[[|t_{b_j}|, \infty); n - k]$ is the area under the curve of the t-distribution having $n - k$ degrees of freedom to the right of $|t_{b_j}|$, the absolute value of the t_{b_j} statistic. Then, we can reject $H_0: \beta_j = 0$ in favor of $H_1: \beta_j \neq 0$ by setting the probability of a Type I error equal to α if and only if any of the following three equivalent conditions hold:

1. $|t_{b_j}| > t_{[\alpha/2]}^{(n-k)}$—that is, if $t_{b_j} > t_{[\alpha/2]}^{(n-k)}$ or $t_{b_j} < -t_{[\alpha/2]}^{(n-k)}$.
2. *Prob-value* $< \alpha$.
3. The $100(1 - \alpha)\%$ confidence interval for β_j, which is $[b_j \pm t_{[\alpha/2]}^{(n-k)} s\sqrt{c_{jj}}]$, does not contain zero.

Moreover,

4. For any level of confidence, $100(1 - \alpha)\%$, less than $100[1 - (\text{prob-value})]\%$, the $100(1 - \alpha)\%$ confidence interval for β_j does not contain zero and thus we can, with at least $100(1 - \alpha)\%$ confidence, reject $H_0: \beta_j = 0$ in favor of $H_1: \beta_j \neq 0$.
5. For any level of confidence, $100(1 - \alpha)\%$, greater than or equal to $100[1 - (\text{prob-value})]\%$, the $100(1 - \alpha)\%$ confidence interval for β_j does contain zero and thus we cannot, with at least $100(1 - \alpha)\%$ confidence, reject $H_0: \beta_j = 0$ in favor of $H_1: \beta_j \neq 0$.

The use of conditions (1), (2), and (3) to determine whether we can reject H_0 in favor of H_1 by setting the probability of a Type I error equal to α is generally considered to be the classical approach to hypothesis testing, whereas the use of (4) and (5)—which say that we can be up to $100[1 - (\text{prob-value})]\%$ confident that H_0 is false and H_1 is true—is the intuitive approach that we illustrated in Example 6.5.

We first consider condition (1). This condition, which says that we can reject $H_0: \beta_j = 0$ in favor of $H_1: \beta_j \neq 0$ by setting the probability of a Type I error equal to α if and only if the t_{b_j} statistic is greater than $t_{[\alpha/2]}^{(n-k)}$ or less than $-t_{[\alpha/2]}^{(n-k)}$, is probably the easiest condition to use from a computational standpoint. It only requires that we calculate the t_{b_j} statistic and look up the value of $t_{[\alpha/2]}^{(n-k)}$ in a t-table. To intuitively understand condition (1), note that the t_{b_j} statistic

$$t_{b_j} = \frac{b_j}{s\sqrt{c_{jj}}} = \frac{b_j - 0}{s\sqrt{c_{jj}}}$$

measures the distance between b_j, the point estimate of β_j, and zero (the value that makes the null hypothesis $H_0: \beta_j = 0$ true). A t_{b_j} statistic nearly equal to (or exactly equal to) zero, which results when b_j, the point estimate of β_j, is nearly equal to (or exactly equal to) zero, supplies little or no evidence to support rejecting H_0: $\beta_j = 0$ in favor of $H_1: \beta_j \neq 0$. However, a positive t_{b_j} statistic substantially greater than zero, which results when b_j is substantially greater than zero, provides evidence to support rejecting $H_0: \beta_j = 0$ in favor of $H_1: \beta_j \neq 0$, since, in this case, the point estimate b_j indicates that β_j is greater than zero. Similarly, a negative t_{b_j} statistic substantially less than zero, which results when b_j is substantially smaller than zero, also provides evidence to support rejecting $H_0: \beta_j = 0$ in favor of $H_1: \beta_j \neq 0$, since, in this case, the point estimate b_j indicates that β_j is smaller than zero. We call the points $-t_{[\alpha/2]}^{(n-k)}$ and $t_{[\alpha/2]}^{(n-k)}$ **rejection points**, because these points tell us how different from zero (positive or negative) the t_{b_j} statistic must be in order for us to be able to reject $H_0: \beta_j = 0$ by setting the probability of a Type I error equal to α. Thus, the rejection point $t_{[\alpha/2]}^{(n-k)}$ is a point that is positive enough so that if the t_{b_j} statistic is greater than $t_{[\alpha/2]}^{(n-k)}$, we consider the evidence that β_j is greater than zero to be so strong that we can reject H_0. Similarly, the rejection point $-t_{[\alpha/2]}^{(n-k)}$ is a point that is negative enough so that if the t_{b_j} statistic is less than (more negative than) $-t_{[\alpha/2]}^{(n-k)}$, we consider the evidence that β_j is smaller than zero to be so strong that we can reject H_0.

We now present an example that demonstrates how to use rejection points—condition (1)—in the context of the fuel consumption problem. In this example we will also see that use of condition (1) guarantees that the probability of a Type I error equals α.

EXAMPLE 6.6

Consider the fuel consumption model

$$\begin{aligned} y &= \mu + \epsilon \\ &= \beta_0 + \beta_1 x_1 + \beta_2 x_2 + \epsilon \end{aligned}$$

and testing $H_0: \beta_1 = 0$ versus $H_1: \beta_1 \neq 0$. Assuming that the inference assumptions hold, and defining the $t_{[b_1, \beta_1]}$ statistic to be

$$t_{[b_1, \beta_1]} = \frac{b_1 - \beta_1}{s\sqrt{c_{11}}}$$

it follows from the discussion of Section 6.4 that the population of all possible $t_{[b_1, \beta_1]}$ statistics has a t-distribution with $n - k$ degrees of freedom. From condition (1), we can reject $H_0: \beta_1 = 0$ in favor of $H_1: \beta_1 \neq 0$ by setting α, the probability of a Type I error, equal to .05 if and only if the absolute value of

$$t_{b_1} = \frac{b_1}{s\sqrt{c_{11}}}$$

is greater than $t_{[\alpha/2]}^{(n-k)} = t_{[.05/2]}^{(8-3)} = t_{[.025]}^{(5)} = 2.571$—that is, if $t_{b_1} > 2.571$ or $t_{b_1} < -2.571$. In order to show that this rejection condition guarantees that the probability of a Type I error equals .05, note that since the population of all possible $t_{[b_1,\beta_1]}$ statistics has a t-distribution with $n - k = 8 - 3 = 5$ degrees of freedom, it follows that if the null hypothesis $H_0: \beta_1 = 0$ is true, then the population of all possible t_{b_1} statistics has a t-distribution with $n - k = 5$ degrees of freedom, because if $H_0: \beta_1 = 0$ is true, then

$$t_{[b_1,\beta_1]} = \frac{b_1 - \beta_1}{s\sqrt{c_{11}}} = \frac{b_1 - 0}{s\sqrt{c_{11}}} = \frac{b_1}{s\sqrt{c_{11}}} = t_{b_1}$$

This fact tells us that if t_{b_1} is the t_{b_1} statistic that will be randomly selected from the population of all such statistics, then

1. $P(t_{b_1}$ will be greater than $t_{[\alpha/2]}^{(n-k)} = t_{[.025]}^{(5)} = 2.571$ when $H_0: \beta_1 = 0$ is true) is given by

   $$A[[t_{[\alpha/2]}^{(n-k)}, \infty); n - k] = A[[t_{[.025]}^{(5)}, \infty); 5] = A[[2.571, \infty); 5]$$

   which is the area under the curve of the t-distribution having $n - k = 5$ degrees of freedom to the right of $t_{[\alpha/2]}^{(n-k)} = t_{[.025]}^{(5)} = 2.571$, and which (by the definition of $t_{[\alpha/2]}^{(n-k)} = t_{[.025]}^{(5)}$) equals $\alpha/2 = .025$ (see Figure 6.5).
2. $P(t_{b_1}$ will be less than $-t_{[\alpha/2]}^{(n-k)} = -t_{[.025]}^{(5)} = -2.571$ when $H_0: \beta_1 = 0$ is true) is given by

   $$A[(-\infty, -t_{[\alpha/2]}^{(n-k)}]; n - k] = A[(-\infty, -t_{[.025]}^{(5)}]; 5] = A[(-\infty, -2.571]; 5]$$

   which is the area under the curve of the t-distribution having $n - k = 5$ degrees of freedom to the left of $-t_{[\alpha/2]}^{(n-k)} = -t_{[.025]}^{(5)} = -2.571$, and which (by the definition of $-t_{[\alpha/2]}^{(n-k)} = -t_{[.025]}^{(5)}$) equals $\alpha/2 = .025$ (see Figure 6.5).

Thus, since the events "$t_{b_1} > t_{[\alpha/2]}^{(n-k)} = 2.571$" and "$t_{b_1} < -t_{[\alpha/2]}^{(n-k)} = -2.571$" are mutually exclusive, it follows that (see Figure 6.5)

$$\begin{aligned}
P\begin{pmatrix}\text{We will commit}\\ \text{a Type I error}\end{pmatrix} &= P\begin{pmatrix}\text{We will reject } H_0: \beta_1 = 0\\ \text{when } H_0: \beta_1 = 0 \text{ is true}\end{pmatrix}\\
&= P(t_{b_1} > t_{[\alpha/2]}^{(n-k)} = 2.571 \quad \text{or} \quad t_{b_1} < -t_{[\alpha/2]}^{(n-k)} = -2.571\\
&\qquad \text{when } H_0: \beta_1 = 0 \text{ is true})\\
&= P\begin{pmatrix}t_{b_1} > t_{[\alpha/2]}^{(n-k)} = 2.571\\ \text{when } H_0: \beta_1 = 0 \text{ is true}\end{pmatrix}\\
&\qquad + P\begin{pmatrix}t_{b_1} < -t_{[\alpha/2]}^{(n-k)} = -2.571\\ \text{when } H_0: \beta_1 = 0 \text{ is true}\end{pmatrix}\\
&= A[[t_{[\alpha/2]}^{(n-k)}, \infty); n - k] + A[(-\infty, -t_{[\alpha/2]}^{(n-k)}]; n - k]\\
&= \alpha/2 + \alpha/2 = .025 + .025\\
&= \alpha = .05
\end{aligned}$$

This probability statement says that if $H_0: \beta_1 = 0$ is true, then $100\alpha\% = 100(.05)\% = 5\%$ of the t_{b_1} statistics in the population of all such statistics are

$P(\text{We will commit a Type I error}) = \frac{\alpha}{2} + \frac{\alpha}{2} = \alpha(= .05) =$ The proportion of t_{b_1} statistics in the population of all possible t_{b_1} statistics which, if H_0 is true, are greater than $t_{[\alpha/2]}^{(n-k)}$ or less than $-t_{[\alpha/2]}^{(n-k)}$ and thus would (if randomly selected) cause us to reject H_0 when H_0 is true

$P(t_{b_1} < -t_{[\alpha/2]}^{(n-k)}$ when H_0 is true)
$= A\left[(-\infty, -t_{[\alpha/2]}^{(n-k)}]; n-k\right]$
$= \frac{\alpha}{2} = .025$

$P(t_{b_1} > t_{[\alpha/2]}^{(n-k)}$ when H_0 is true)
$= A\left[[t_{[\alpha/2]}^{(n-k)}, \infty); n-k\right]$
$= \frac{\alpha}{2} = .025$

Curve of the t-distribution having $n - k = 5$ degrees of freedom

$-t_{[\alpha/2]}^{(n-k)}$
-2.571

0

$t_{[\alpha/2]}^{(n-k)}$
2.571

Five t_{b_1} statistics in the population of all possible t_{b_1} statistics, assuming $H_0: \beta_1 = 0$ is true

$(t_{b_1})_{SPL_1} = -3.186$ $(t_{b_1})_{SPL_2} = -.624$ $(t_{b_1})_{SPL_3} = -.262$ $(t_{b_1})_{SPL_4} = 1.161$ $(t_{b_1})_{SPL_5} = 2.952$

If $t_{b_1} < -t_{[\alpha/2]}^{(n-k)}$, then reject H_0

If $-t_{[\alpha/2]}^{(n-k)} \leq t_{b_1} \leq t_{[\alpha/2]}^{(n-k)}$, then do not reject H_0

If $t_{b_1} > t_{[\alpha/2]}^{(n-k)}$, then reject H_0

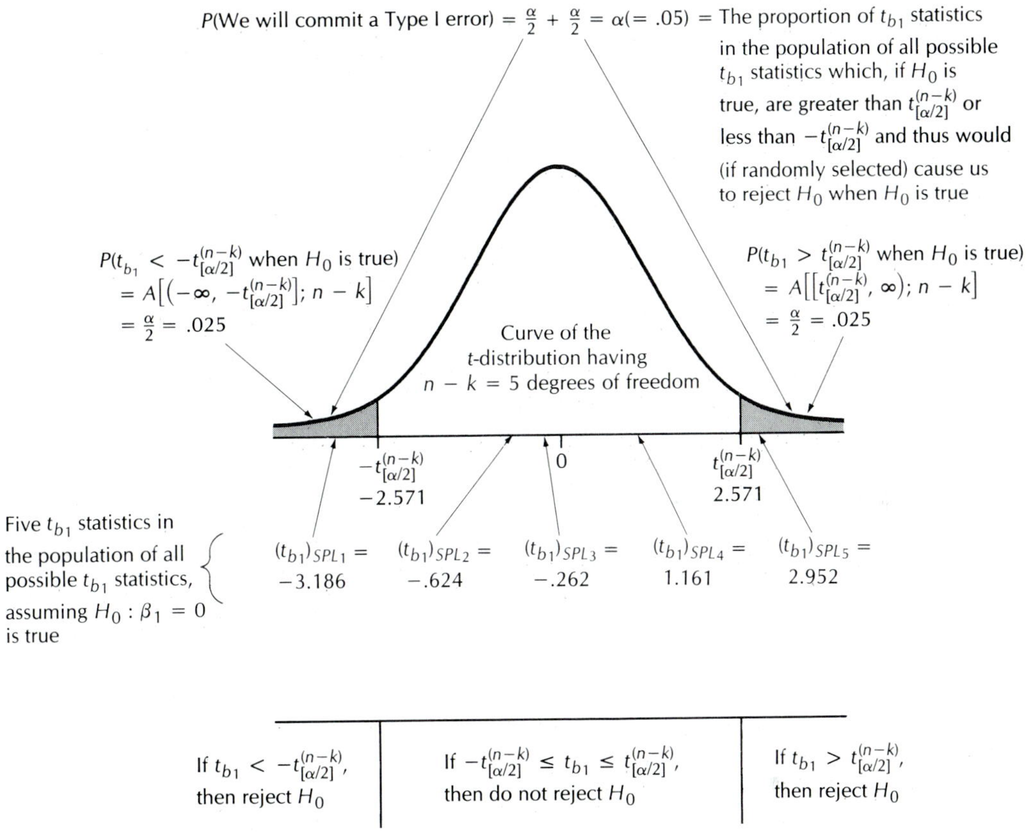

FIGURE 6.5   The Probability of a Type I Error When Testing $H_0: \beta_1 = 0$ Versus $H_1: \beta_1 \neq 0$

greater than the positive rejection point $t_{[\alpha/2]}^{(n-k)} = 2.571$ or less than the negative rejection point $-t_{[\alpha/2]}^{(n-k)} = -2.571$ and thus would (if randomly selected) cause us to reject $H_0: \beta_1 = 0$ when $H_0: \beta_1 = 0$ is true (that is, cause us to commit a Type I error).[†]

Now, using the information in Examples 5.4 and 6.2, we can calculate the t_{b_1} statistic to be

$$t_{b_1} = \frac{b_1}{s\sqrt{c_{11}}} = \frac{-.0900}{.3671\sqrt{.00147}} = -6.39$$

Since $t_{b_1} = -6.39 < -2.571 = -t_{[.025]}^{(5)}$ (that   is, $|t_{b_1}| = |-6.39| = 6.39 >$ $2.571 = t_{[.025]}^{(5)}$), we can reject $H_0: \beta_1 = 0$ by setting α equal to .05.

[†] Although we show here that use of condition (1) implies that the probability of a Type I error equals α when testing $H_0: \beta_1 = 0$ versus $H_1: \beta_1 \neq 0$, our discussion applies in total generality to testing $H_0: \beta_j = 0$ versus $H_1: \beta_j \neq 0$.

We next discuss condition (2), which says that we can reject $H_0: \beta_j = 0$ in favor of $H_1: \beta_j \neq 0$ by setting the probability of a Type I error equal to α if and only if

$$\text{prob-value} = 2A[[|t_{b_j}|, \infty); n - k]$$

is less than α. Since condition (2) requires that we calculate the area

$$A[[|t_{b_j}|, \infty); n - k]$$

this condition is more complicated than condition (1) from a computational standpoint. However, condition (2) has the following advantage over condition (1). If there were several hypothesis testers, all of whom wished to use different values of α, the probability of a Type I error, then if condition (1) were used, each hypothesis tester would have to look up a different rejection point $t_{[\alpha/2]}^{(n-k)}$ to decide whether to reject H_0 at the hypothesis tester's particular chosen value of α. However, if condition (2) is used, only the prob-value needs to be calculated, and each hypothesis tester knows that if the prob-value is less than his or her particular chosen value of α, then H_0 should be rejected. With respect to calculating $A[[|t_{b_j}|, \infty); n - k]$, recall from Section 3.3.1 that as the number of degrees of freedom approaches infinity, the probability curve of the t-distribution approaches (becomes shaped more and more like) the probability curve of a standard normal distribution. Therefore, if $n - k$ (the number of degrees of freedom) is 30 or more, it is sufficient (when calculating the prob-value) to approximate

$$A[[|t_{b_j}|, \infty); n - k]$$

by

$$A[[|t_{b_j}|, \infty); N(0, 1)]$$

which is the area under the standard normal curve to the right of $|t_{b_j}|$.

We now present an example illustrating the use of condition (2) and then show that using this condition does indeed guarantee that the probability of a Type I error equals α.

EXAMPLE 6.7

In Table 6.5 we summarize the calculation of the t_{b_j} statistics and related prob-values for testing the importance of the independent variables (and the intercept) in the fuel consumption model

$$y = \beta_0 + \beta_1 x_1 + \beta_2 x_2 + \epsilon$$

Since the prob-value for testing $H_0: \beta_2 = 0$ versus $H_1: \beta_2 \neq 0$ is .01330, and since this prob-value is less than .05, less than .02, and not less than .01, it follows by condition (2) that we can reject $H_0: \beta_2 = 0$ in favor of $H_1: \beta_2 \neq 0$ by setting α, the probability of a Type I error, equal to .05 or .02, but we cannot reject H_0 in favor of H_1 if we set α equal to .01. Moreover, since the prob-values

TABLE 6.5 The Calculations and the SAS Printout of the t_{b_j} Statistics and Prob-Values for Testing $H_0: \beta_0 = 0$, $H_0: \beta_1 = 0$, and $H_0: \beta_2 = 0$ in the Fuel Consumption Model $y = \beta_0 + \beta_1 x_1 + \beta_2 x_2 + \epsilon$

Independent Variable	b_j	$s\sqrt{c_{jj}}$	$t_{b_j} = \frac{b_j}{s\sqrt{c_{jj}}}$	Prob-Value $= 2A[[\lvert t_{b_j}\rvert, \infty); n - k]$
Intercept	$b_0 = 13.11$	$s\sqrt{c_{00}} = .8557$	$t_{b_0} = \frac{b_0}{s\sqrt{c_{00}}} = 15.32$	$2A[[\lvert 15.32\rvert, \infty); 5] = 2(.00001) = .00002$
x_1	$b_1 = -.0900$	$s\sqrt{c_{11}} = .0141$	$t_{b_1} = \frac{b_1}{s\sqrt{c_{11}}} = -6.39$	$2A[[\lvert -6.39\rvert, \infty); 5] = 2(.000695) = .00139$
x_2	$b_2 = .0825$	$s\sqrt{c_{22}} = .0220$	$t_{b_2} = \frac{b_2}{s\sqrt{c_{22}}} = 3.75$	$2A[[\lvert 3.75\rvert, \infty); 5] = 2(.00665) = .01330$

VARIABLE	DF	PARAMETER ESTIMATE[a]	STANDARD ERROR[b]	T FOR H0: PARAMETER=0[c]	PROB > |T|[d]
INTERCEP	1	13.108737	0.855698	15.319	0.0001
X1	1	-0.090014	0.014077	-6.394	0.0014
X2	1	0.082495	0.022003	3.749	0.0133

[a] $b_j: b_0, b_1, b_2$ [b] $s\sqrt{c_{jj}}$ [c] t_{b_j} [d] Prob-value

for testing $H_0: \beta_0 = 0$ and $H_0: \beta_1 = 0$ are less than .01 (see Table 6.5), we can reject each of these hypotheses by setting α equal to .01. Thus, we have substantial evidence that the intercept β_0 and the independent variables x_1 and x_2 are important in describing the dependent variable y in the preceding model. To conclude this example, note that Table 6.5 also presents a computer printout of the quantities calculated in the table. This printout results from carrying out a regression analysis of the fuel consumption data by using the Statistical Analysis System (SAS). Briefly, SAS is a computer system that can be used to analyze data. Among other things (such as data storage and retrieval, report writing, and file handling), SAS can be used to carry out a regression analysis. In Section 6.9 we present a brief description of how to perform regression in SAS. In Table 6.5 note that when SAS output indicates that a prob-value is equal to .0001, this means that the prob-value is less than or equal to .0001.

Now, having shown in Example 6.6 that use of condition (1) implies that the probability of a Type I error equals α, we can show that use of condition (2) also implies that the probability of a Type I error equals α by showing that conditions (1) and (2) are equivalent. To do this, we show that if condition (2) holds, then condition (1) holds, and if condition (2) does not hold, then condition (1) does not hold. If condition (2) holds, then

$$\text{prob-value} = 2A[[|t_{b_j}|, \infty); n - k]$$

is less than α, which implies that

$$A[[|t_{b_j}|, \infty); n - k] < \frac{\alpha}{2}$$

Looking at Figures 6.6a and 6.6b, we see that this fact implies that condition (1),

$$|t_{b_j}| > t_{[\alpha/2]}^{(n-k)}$$

holds. If condition (2) does not hold, then

$$\text{prob-value} = 2A[[|t_{b_j}|, \infty); n - k]$$

is greater than or equal to α, which implies that

$$A[[|t_{b_j}|, \infty); n - k] \geq \frac{\alpha}{2}$$

Looking at Figures 6.6a and 6.6c, we see that this fact implies that

$$|t_{b_j}| \leq t_{[\alpha/2]}^{(n-k)}$$

which says that condition (1) does not hold. Thus, we have shown that conditions (1) and (2) are equivalent.

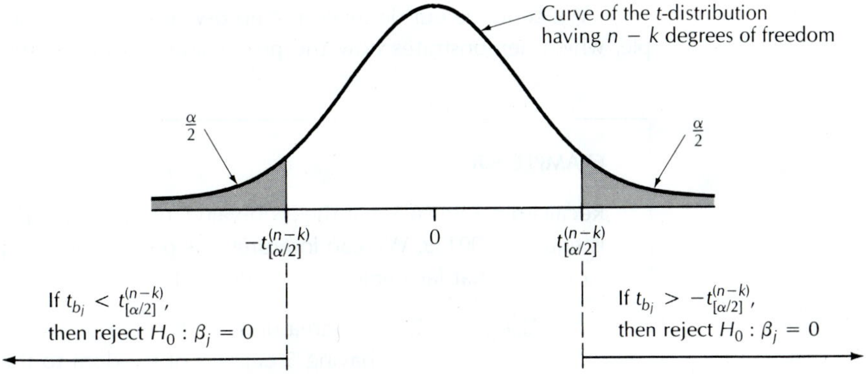

(a) The rejection points for testing $H_0: \beta_j = 0$ versus $H_1: \beta_j \neq 0$ based on setting α

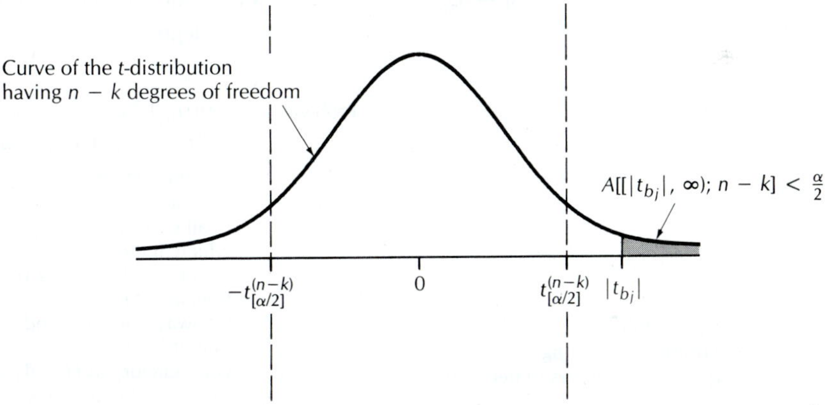

(b) If prob-value $= 2A[[|t_{b_j}|, \infty); n-k] < \alpha$, then $A[[|t_{b_j}|, \infty); n-k] < \alpha/2$, and $|t_{b_j}| > t_{[\alpha/2]}^{(n-k)}$. Reject $H_0: \beta_j = 0$

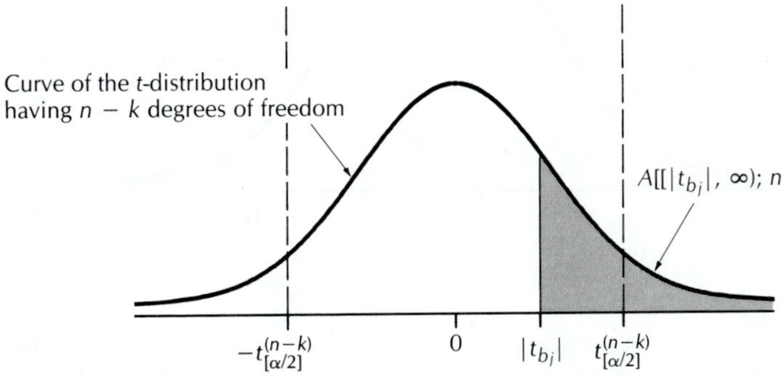

FIGURE 6.6
Testing $H_0: \beta_j = 0$ Versus $H_1: \beta_j \neq 0$ Using Prob-Values

(c) If prob-value $= 2A[[|t_{b_j}|, \infty); n-k] \geq \alpha$, then $A[[|t_{b_j}|, \infty); n-k] \geq \alpha/2$, and $|t_{b_j}| \leq t_{[\alpha/2]}^{(n-k)}$. Do not reject $H_0: \beta_j = 0$

We continue our discussion of prob-values by presenting the following example, which demonstrates how the prob-value can be interpreted as a probability.

EXAMPLE 6.8

Recall from Table 6.5 that the prob-value for testing $H_0: \beta_1 = 0$ versus $H_1: \beta_1 \neq 0$ is equal to .00139. We can interpret this prob-value as a probability as follows. First, note that (as depicted in Figure 6.7)

$$A[[|t_{b_1}|, \infty); 5] = \text{the area under the curve of the } t\text{-distribution}$$
$$\text{having 5 degrees of freedom to the right of } |t_{b_1}|$$

is equal to

$$A[(-\infty, -|t_{b_1}|]; 5] = \text{the area under the curve of the } t\text{-distribution}$$
$$\text{having 5 degrees of freedom to the left of } -|t_{b_1}|$$

$$\begin{aligned}
\text{Prob-value} &= 2A[[|t_{b_1}|, \infty); 5] = 2(.000695) = .00139 \\
&= A[[|t_{b_1}|, \infty); 5] + A[(-\infty, -|t_{b_1}|]; 5] \\
&= \text{The proportion of } t_{b_1} \\
&\quad \text{statistics in the population} \\
&\quad \text{of all such statistics} \\
&\quad \text{which, if the null} \\
&\quad \text{hypothesis } H_0 : \beta_1 = 0 \text{ is} \\
&\quad \text{true, are at least as} \\
&\quad \text{far away from zero and} \\
&\quad \text{thus at least as} \\
&\quad \text{contradictory to } H_0 : \beta_1 = 0 \\
&\quad \text{as } -6.39, \text{ the observed} \\
&\quad t_{b_1} \text{ statistic}
\end{aligned}$$

Curve of the
t-distribution having
$n - k = 5$ degrees of freedom

$A[(-\infty, -|t_{b_1}|]; 5]$
  $= A[(-\infty, -6.39]; 5]$
  $= .000695$

$A[[|t_{b_1}|, \infty); 5]$
  $= A[[6.39, \infty); 5]$
  $= .000695$

$-|t_{b_1}|$
-6.39

0

$|t_{b_1}|$

$(t_{b_1})_{SPL_1} = -6.39$ $(t_{b_1})_{SPL_2} = -.929$ $(t_{b_1})_{SPL_3} = .614$ $\Big\}$ Three t_{b_1} statistics in the population of all possible t_{b_1} statistics, assuming $H_0 : \beta_1 = 0$ is true

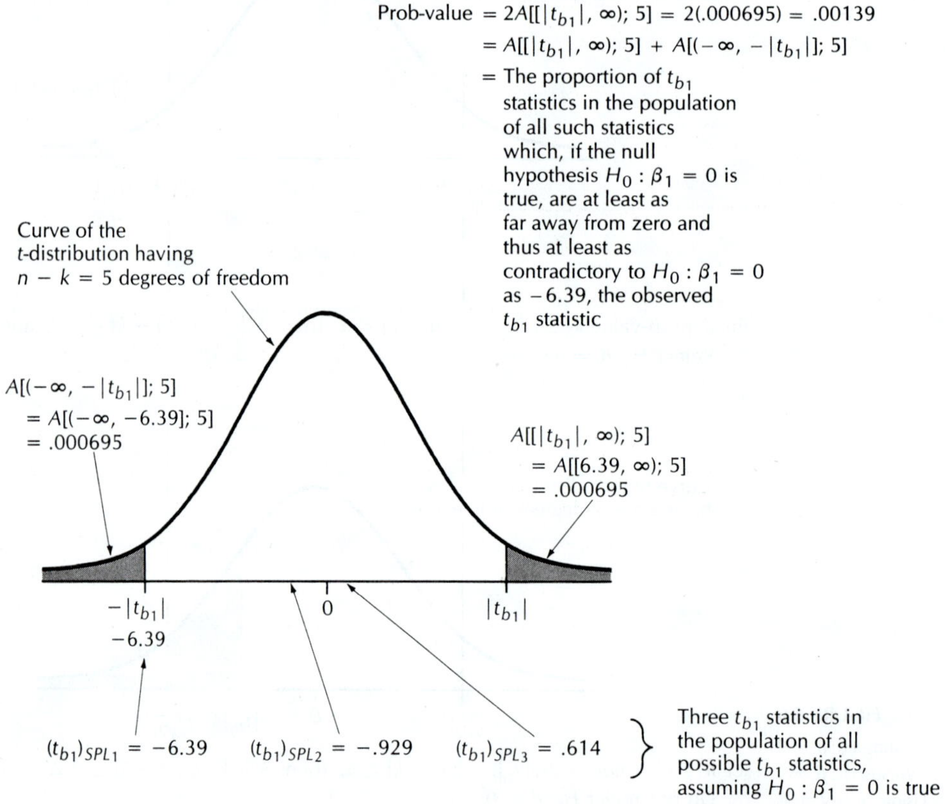

FIGURE 6.7 The Prob-Value for Testing $H_0: \beta_1 = 0$ Versus $H_0: \beta_j \neq 0$ in the Two-Variable Fuel Consumption Model

It follows that

$$\begin{aligned}\text{prob-value} &= 2A[[|t_{b_1}|, \infty); 5] \\ &= A[[|t_{b_1}|, \infty); 5] + A[(-\infty, -|t_{b_1}|]; 5] \\ &= A[[|-6.39|, \infty); 5] + A[(-\infty, -|-6.39|]; 5] \\ &= A[[6.39, \infty); 5] + A[(-\infty, -6.39]; 5]\end{aligned}$$

Now, as stated in Example 6.6, assuming that the inference assumptions hold, we know that if $H_0: \beta_1 = 0$ is true, then the population of all possible t_{b_1} statistics has a t-distribution with $n - k = 5$ degrees of freedom. Therefore, the preceding prob-value is equal to the proportion of t_{b_1} statistics in the population of all such statistics which, if $H_0: \beta_1 = 0$ is true, are greater than or equal to $|t_{b_1}| = 6.39$ or less than or equal to $-|t_{b_1}| = -6.39$, where $t_{b_1} = -6.39$ is the observed t_{b_1} statistic (that is, the t_{b_1} statistic that we calculated using the observed sample, *SPL*). Noting that the farther the t_{b_1} statistic is from zero, the more this statistic contradicts the null hypothesis $H_0: \beta_1 = 0$, we can say that this prob-value is equal to the proportion of t_{b_1} statistics in the population of all such statistics which, if $H_0: \beta_1 = 0$ is true, are at least as far from zero and thus at least as contradictory to $H_0: \beta_1 = 0$ as -6.39, the observed t_{b_1} statistic.

Since the prob-value equals .00139 $\approx$ .0014 (see Table 6.5), we know that if the null hypothesis $H_0: \beta_1 = 0$ is true, then only .14 percent (or 14 in 10,000) of the t_{b_1} statistics in the population of all such statistics are at least as far away from zero and thus at least as contradictory to $H_0: \beta_1 = 0$ as -6.39, the observed t_{b_1} statistic. Therefore, the prob-value of .0014 says that if $H_0: \beta_1 = 0$ is true, then we have observed a t_{b_1} statistic (-6.39) which is so rare that its occurrence can be described as a 14 in 10,000 chance. This interpretation helps us to reach one of two possible conclusions: (1) The null hypothesis $H_0: \beta_1 = 0$ is true, and we have observed a t_{b_1} statistic (-6.39) so rare that its occurrence can be described as a 14 in 10,000 chance; or (2) The null hypothesis $H_0: \beta_1 = 0$ is not true. Since any reasonable person would find it very difficult to believe that a 14 in 10,000 chance has actually occurred, and since we must believe that this has happened if we are to believe that $H_0: \beta_1 = 0$ is true, we would have substantial evidence to conclude that $H_0: \beta_1 = 0$ is not true. The fact that the prob-value of .0014 is so small casts great doubt on the validity of the null hypothesis $H_0: \beta_1 = 0$ and thus lends great support to the validity of the alternative hypothesis $H_1: \beta_1 \neq 0$.

In general, if we assume that when the null hypothesis $H_0: \beta_j = 0$ is true, the population of all possible t_{b_j} statistics has a t-distribution with $n - k$ degrees of freedom, it follows that the prob-value for testing $H_0: \beta_j = 0$ versus $H_1: \beta_j \neq 0$

$$\begin{aligned}\text{prob-value} &= 2A[[|t_{b_j}|, \infty); n - k] \\ &= A[[|t_{b_j}|, \infty); n - k] + A[(-\infty, -|t_{b_j}|]; n - k]\end{aligned}$$

is equal to the proportion of t_{b_j} statistics in the population of all such statistics which, if the null hypothesis $H_0: \beta_j = 0$ is true, are at least as far away from zero and thus at least as contradictory to $H_0: \beta_j = 0$ as the observed t_{b_j} statistic. From the reasoning demonstrated in the preceding example, it is clear that a prob-value for testing $H_0: \beta_j = 0$ versus $H_1: \beta_j \neq 0$ that is smaller than .0014 would cast even more doubt on the validity of $H_0: \beta_j = 0$ than does the prob-value of .0014. For example, if the prob-value for testing $H_0: \beta_j = 0$ were .0001, then the occurrence of the t_{b_j} statistic would be so rare that it could be described as a 1 in 10,000 chance, and a conclusion that $H_0: \beta_j = 0$ is true would require that we believe this 1 in 10,000 chance has actually occurred. On the other hand, a prob-value for testing $H_0: \beta_j = 0$ versus $H_1: \beta_j \neq 0$ that is larger than .0014 would cast less doubt on the validity of the null hypothesis $H_0: \beta_j = 0$. For example, if the prob-value for testing $H_0: \beta_j = 0$ were .25, then the occurrence of the t_{b_j} statistic could be described as a 25 percent, or a 1 in 4, chance. Therefore, a conclusion that $H_0: \beta_j = 0$ is true would require that we believe a 1 in 4 chance has occurred. Since it is not particularly difficult to believe a 1 in 4 chance has occurred, the prob-value of .25 would not cast a great deal of doubt on the null hypothesis $H_0: \beta_j = 0$. Using this reasoning, it is clear that interpreting the prob-value as a probability tells us that *the smaller the prob-value is, the more we doubt the validity of the null hypothesis $H_0: \beta_j = 0$ and the more confident we are that we should reject the null hypothesis $H_0: \beta_j = 0$ in favor of the alternative hypothesis $H_1: \beta_j \neq 0$.*

Next, in order to prove that using condition (3) implies that the probability of a Type I error equals α, we show that conditions (1) and (3) are equivalent. To do this, we show that if condition (3) holds, then condition (1) holds, and if condition (3) does not hold, then condition (1) does not hold. If condition (3) holds, then the $100(1 - \alpha)\%$ confidence interval for β_j,

$$[b_j - t_{[\alpha/2]}^{(n-k)} s\sqrt{c_{jj}},\ b_j + t_{[\alpha/2]}^{(n-k)} s\sqrt{c_{jj}}]$$

does not contain zero. This implies that

$$0 < b_j - t_{[\alpha/2]}^{(n-k)} s\sqrt{c_{jj}} \qquad \text{or} \qquad 0 > b_j + t_{[\alpha/2]}^{(n-k)} s\sqrt{c_{jj}}$$

which implies (by algebraic manipulation) that

$$\frac{b_j}{s\sqrt{c_{jj}}} > t_{[\alpha/2]}^{(n-k)} \qquad \text{or} \qquad \frac{b_j}{s\sqrt{c_{jj}}} < -t_{[\alpha/2]}^{(n-k)}$$

which implies that

$$\left|\frac{b_j}{s\sqrt{c_{jj}}}\right| > t_{[\alpha/2]}^{(n-k)}$$

which implies that condition (1),

$$|t_{b_j}| > t_{[\alpha/2]}^{(n-k)}$$

holds. Next, if condition (3) does not hold, then the $100(1 - \alpha)\%$ confidence interval for β_j,

$$[b_j - t_{[\alpha/2]}^{(n-k)} s\sqrt{c_{jj}},\ b_j + t_{[\alpha/2]}^{(n-k)} s\sqrt{c_{jj}}]$$

does contain zero. This implies that

$$b_j - t_{[\alpha/2]}^{(n-k)} s\sqrt{c_{jj}} \leq 0 \leq b_j + t_{[\alpha/2]}^{(n-k)} s\sqrt{c_{jj}}$$

which implies (by algebraic manipulation) that

$$-t_{[\alpha/2]}^{(n-k)} \leq \frac{b_j}{s\sqrt{c_{jj}}} \leq t_{[\alpha/2]}^{(n-k)}$$

which implies that

$$\left|\frac{b_j}{s\sqrt{c_{jj}}}\right| \leq t_{[\alpha/2]}^{(n-k)}$$

which implies that

$$|t_{b_j}| \leq t_{[\alpha/2]}^{(n-k)}$$

which says that condition (1) does not hold. Thus, we have shown that conditions (1) and (3) are equivalent. The equivalency of conditions (1) and (3), along with the previously proved equivalency of conditions (1) and (2), imply, in summary, that conditions (1), (2), and (3) are equivalent. Moreover, since condition (3)—rejecting $H_0: \beta_j = 0$ if and only if zero is not contained in the $100(1 - \alpha)\%$ confidence interval for β_j—is the condition illustrated in Example 6.5 for rejecting $H_0: \beta_j = 0$ with at least $100(1 - \alpha)\%$ confidence, and since we have just shown that use of condition (3) implies that the probability of a Type I error equals α, we can say that *deciding whether we can reject $H_0: \beta_j = 0$ by setting the probability of a Type I error equal to α* (for example, .05) *is equivalent to deciding whether we can reject $H_0: \beta_j = 0$ with at least $100(1 - \alpha)\%$* (for example, 95%) *confidence*.

Next, we will prove (4) and (5) in the previous box (entitled "Testing $H_0: \beta_j = 0$ versus $H_1: \beta_j \neq 0$"). To prove (4), note that if we assume that the level of confidence, $100(1 - \alpha)\%$, is less than 100[1 − (prob-value)]%, this implies that the prob-value is less than α, which implies (since condition (2) implies condition (3)) that the $100(1 - \alpha)\%$ confidence interval for β_j does not contain zero, and thus we can, with at least $100(1 - \alpha)\%$ confidence, reject $H_0: \beta_j = 0$. To prove (5), note that if we assume that the level of confidence, $100(1 - \alpha)\%$, is greater than or equal to 100[1 − (prob-value)]%, this implies that the prob-value is greater than or equal to α, which implies (since condition (2) not holding implies condition (3) not holding) that the $100(1 - \alpha)\%$ confidence interval for β_j does contain zero, and thus we cannot, with at least $100(1 - \alpha)\%$ confidence, reject $H_0: \beta_j = 0$.

To conclude, we summarize the discussion of this section. As previously stated, the use of conditions (1), (2), and (3) to determine whether we can reject $H_0: \beta_j = 0$

in favor of $H_1: \beta_j \neq 0$ by setting the probability of a Type I error equal to α is generally considered to be the classical approach to hypothesis testing, whereas the use of (4) and (5)—which say that we can be up to 100[1 − (prob-value)]% confident that $H_0: \beta_j = 0$ is false and that $H_1: \beta_j \neq 0$ is true—is the intuitive approach illustrated in Example 6.5.

Since the smaller the prob-value is, the larger is 100[1 − (prob-value)]%, we conclude from the intuitive approach that the smaller the prob-value is, the more confident we are that we should reject the null hypothesis $H_0: \beta_j = 0$ in favor of the alternative hypothesis $H_1: \beta_j \neq 0$. This conclusion is further supported by our interpretation of the prob-value as a probability—the smaller the prob-value is, the more we doubt the validity of the null hypothesis $H_0: \beta_j = 0$. Reporting the prob-value as the result of a statistical test (rather than merely whether we "reject H_0" or "do not reject H_0" based on setting the probability of a Type I error equal to α) is becoming increasingly popular, because the result of the statistical test may be used by several people to make different decisions at different times. Thus, by reporting the prob-value (the smallness of which represents the weight of evidence against the null hypothesis H_0), the individual performing the statistical test permits any other potential decision makers to reach their own conclusions.

EXAMPLE 6.9

Table 6.6 summarizes the calculations and SAS printout of the t_{b_j} statistics and the prob-values for testing the importance of the independent variables (and the intercept) in the Fresh detergent model

$$y = \beta_0 + \beta_1 x_4 + \beta_2 x_3 + \beta_3 x_3^2 + \beta_4 x_4 x_3 + \epsilon$$

Note that in making the calculations in Table 6.6, we have used from Example 5.9:

1. The least squares point estimates b_0, b_1, b_2, b_3, and b_4 of the parameters β_0, β_1, β_2, β_3, and β_4;
2. The sum of squared residuals $SSE = 1.0644$, which we utilize to calculate the standard error

$$s = \sqrt{\frac{SSE}{n-k}} = \sqrt{\frac{1.0644}{30-5}} = .2063$$

3. The diagonal elements c_{00}, c_{11}, c_{22}, c_{33}, and c_{44} of $(\mathbf{X}'\mathbf{X})^{-1}$.

Since the prob-values for testing the hypotheses in Table 6.6 are all less than .05, we can reject each of these hypotheses by setting the probability of a Type I error equal to $\alpha = .05$. Thus, we have substantial evidence that the intercept β_0 and the independent variables x_4, x_3, x_3^2, and x_4x_3 are important in describing the dependent variable y in the Fresh detergent model. Note that although in Chapter 5 we utilized data plots to justify the importance of the independent

TABLE 6.6 The Calculations and the SAS Printout of the t_{b_j} Statistics and Prob-Values for Testing $H_0: \beta_0 = 0$, $H_0: \beta_1 = 0$, $H_0: \beta_2 = 0$, $H_0: \beta_3 = 0$, and $H_0: \beta_4 = 0$ in the Fresh Detergent Model $y = \beta_0 + \beta_1 x_4 + \beta_2 x_3 + \beta_3 x_3^2 + \beta_4 x_4 x_3 + \epsilon$

Independent Variable	b_j	$s\sqrt{c_{jj}}$	$t_{b_j} = \frac{b_j}{s\sqrt{c_{jj}}}$	Prob-Value $= 2A[[\lvert t_{b_j}\rvert, \infty); 25]$
Intercept	$b_0 = 29.1133$	$s\sqrt{c_{00}} = 7.4832$	$t_{b_0} = \frac{b_0}{s\sqrt{c_{00}}} = \frac{29.1133}{7.4832} = 3.8905$	$2A[[\lvert 3.8905\rvert, \infty); 25] = .00066$
x_4	$b_1 = 11.1342$	$s\sqrt{c_{11}} = 4.4459$	$t_{b_1} = \frac{b_1}{s\sqrt{c_{11}}} = \frac{11.1342}{4.4459} = 2.5044$	$2A[[\lvert 2.5044\rvert, \infty); 25] = .01915$
x_3	$b_2 = -7.6080$	$s\sqrt{c_{22}} = 2.4691$	$t_{b_2} = \frac{b_2}{s\sqrt{c_{22}}} = \frac{-7.6080}{2.4691} = -3.0813$	$2A[[\lvert -3.0813\rvert, \infty); 25] = .00496$
x_3^2	$b_3 = .6712$	$s\sqrt{c_{33}} = .2027$	$t_{b_3} = \frac{b_3}{s\sqrt{c_{33}}} = \frac{.6712}{.2027} = 3.3115$	$2A[[\lvert 3.3115\rvert, \infty); 25] = .00282$
$x_4 x_3$	$b_4 = -1.4777$	$s\sqrt{c_{44}} = .6672$	$t_{b_4} = \frac{b_4}{s\sqrt{c_{44}}} = \frac{-1.4777}{.6672} = -2.2149$	$2A[[\lvert -2.2149\rvert, \infty); 25] = .03610$

VARIABLE	DF	PARAMETER ESTIMATE[a]	STANDARD ERROR[b]	T FOR HO: PARAMETER=0[c]	PROB > |T|[d]
INTERCEP	1	29.113287	7.483206	3.890	0.0007
X4	1	11.134226	4.445854	2.504	0.0192
X3	1	-7.608007	2.469109	-3.081	0.0050
X3SQ	1	0.671247	0.202701	3.312	0.0028
X43	1	-1.477717	0.667165	-2.215	0.0361

[a] $b_j: b_0, b_1, b_2, b_3, b_4$ [b] $s\sqrt{c_{jj}}$ [c] t_{b_j} [d] Prob-value

variables x_4, x_3, and x_3^2, we were not able to utilize data plots to justify the importance of the interaction term x_4x_3. Therefore, the smallness of the prob-value for testing $H_0: \beta_4 = 0$ versus $H_1: \beta_4 \neq 0$ is our first convincing justification of the importance of x_4x_3.

Before leaving this section, recall that we sometimes wish to test a null hypothesis versus a one-sided alternative hypothesis. For example, considering the fuel consumption model

$$y = \beta_0 + \beta_1 x_1 + \beta_2 x_2 + \epsilon$$

we have previously reasoned that if β_2 does not equal zero, then β_2 must be positive. Thus, we would be interested in testing $H_0: \beta_2 = 0$ versus $H_1: \beta_2 > 0$.

TESTING $H_0: \beta_j = 0$ VERSUS $H_1: \beta_j > 0$; AND
TESTING $H_0: \beta_j = 0$ VERSUS $H_1: \beta_j < 0$:

Define

$$t_{[b_j, \beta_j]} = \frac{b_j - \beta_j}{s\sqrt{c_{jj}}}$$

and assume that the population of all possible $t_{[b_j, \beta_j]}$ statistics has a t-distribution with $n - k$ degrees of freedom. In order to test $H_0: \beta_j = 0$, define the t_{b_j} statistic to be

$$t_{b_j} = \frac{b_j}{s\sqrt{c_{jj}}}$$

Then

1. We can reject $H_0: \beta_j = 0$ in favor of $H_1: \beta_j > 0$ by setting the probability of a Type I error equal to α if and only if $t_{b_j} > t_{[\alpha]}^{(n-k)}$, where $t_{[\alpha]}^{(n-k)}$ is the point on the scale of the t-distribution having $n - k$ degrees of freedom so that the area under the curve of this t-distribution to the right of $t_{[\alpha]}^{(n-k)}$ is α.
2. We can reject $H_0: \beta_j = 0$ in favor of $H_1: \beta_j < 0$ by setting the probability of a Type I error equal to α if and only if $t_{b_j} < -t_{[\alpha]}^{(n-k)}$, where $-t_{[\alpha]}^{(n-k)}$ is the point on the scale of the t-distribution having $n - k$ degrees of freedom so that the area under the curve of this t-distribution to the left of $-t_{[\alpha]}^{(n-k)}$ is α.

We give a more complete discussion of testing one-sided alternative hypotheses—including the rationale for the above procedure—in optional Section 6.12.

6.6 DECIDING WHETHER TO INCLUDE AN INDEPENDENT VARIABLE IN A REGRESSION MODEL

In general, the more confident we can be that we should reject the null hypothesis $H_0: \beta_j = 0$ in favor of the alternative hypothesis $H_1: \beta_j \neq 0$, the more confident we can be that the independent variable x_j is significantly related to the dependent variable y in the regression model

$$y = \beta_0 + \beta_1 x_1 + \cdots + \beta_j x_j + \cdots + \beta_p x_p + \epsilon$$

and thus the more confident we can be that we should seriously consider including x_j in a final regression model. In this context, we use the expression "seriously consider" because in addition to how confident we can be that we should reject H_0: $\beta_j = 0$ in favor of $H_1: \beta_j \neq 0$, there are other considerations involved in determining whether to include x_j in a final regression model.

Table 6.7 summarizes the implications (with respect to deciding whether to include an independent variable in a regression model) of making Type I and Type II errors when testing $H_0: \beta_j = 0$ versus $H_1: \beta_j \neq 0$. Since we usually set the probability of a Type I error, α, between .05 and .01, we usually conclude that x_j is significantly related to y if we can, at a level of confidence set between 95% and 99%, reject $H_0: \beta_j = 0$ in favor of $H_1: \beta_j \neq 0$. Here, however, it is important also to use *prior belief* about the significance of the independent variable x_j, along with how confident we can be that we should reject $H_0: \beta_j = 0$ in favor of $H_1: \beta_j \neq 0$, in order to decide whether x_j should be included in a final regression model. In particular,

TABLE 6.7
The Implications of Making Type I and Type II Errors When Testing $H_0: \beta_j = 0$

Decisions Made in Hypothesis Test	State of Nature	
	Null hypothesis $H_0: \beta_j = 0$ is true: x_j is not significantly related to y	Null hypothesis $H_0: \beta_j = 0$ is false: x_j is significantly related to y
Reject null hypothesis $H_0: \beta_j = 0$. Include x_j in final regression model	Type I error: Include x_j in final regression model when x_j is not significantly related to y	Correct action: Include x_j in final regression model when x_j is significantly related to y
Do not reject null hypothesis $H_0: \beta_j = 0$. Do not include x_j in final regression model	Correct action: Do not include x_j in final regression model when x_j is not significantly related to y	Type II error: Do not include x_j in final regression model when x_j is significantly related to y

suppose we believe on theoretical grounds that the independent variable x_j is significantly related to the dependent variable y, which says that we believe on theoretical grounds that $H_0: \beta_j = 0$ is false and $H_1: \beta_j \neq 0$ is true. In the fuel consumption problem, for example, we would believe that the average hourly temperature during the week is significantly related to the weekly fuel consumption. If we believe that $H_0: \beta_j = 0$ is false, we do not believe we will make a Type I error (rejecting $H_0: \beta_j = 0$ when $H_0: \beta_j = 0$ is true). It follows that we might allow a higher probability of a Type I error (say, an α of .05, or even higher, instead of an α of .01), because this would mean that there would be a lower probability of a Type II error (not including x_j in a final regression model, when x_j is significantly related to y), which we wish to guard against, since we believe that x_j is significantly related to y. To summarize, if we believe on theoretical grounds that x_j is significantly related to y, we might allow α to be .05, or even higher, which would imply that we should conclude that x_j is significantly related to y if we can, at a 95% (or possibly even lower) level of confidence, reject $H_0: \beta_j = 0$ in favor of $H_1: \beta_j \neq 0$. For example, suppose that we are firmly convinced on theoretical grounds that the independent variable x_j is significantly related to y. Suppose, moreover, that we can reject the null hypothesis H_0: $\beta_j = 0$ in favor of the alternative hypothesis $H_1: \beta_j \neq 0$ with up to 88% confidence, but we cannot reject $H_0: \beta_j = 0$ in favor of $H_1: \beta_j \neq 0$ with at least 95% confidence. Then, to decide that x_j is not significantly related to y and thus not to include x_j in a final regression model would contradict both our prior belief and the statistical evidence (which at least mildly supports our prior belief, since we are up to 88% confident that we should reject $H_0: \beta_j = 0$ in favor of $H_1: \beta_j \neq 0$). Thus, we should seriously consider including the independent variable x_j in a final regression model. On the other hand, if there are no strong theoretical grounds for believing that x_j is significantly related to y, and if x_j is just being tried out to see whether it might be significantly related to y, then the fact that we are up to 88% confident that we should reject $H_0: \beta_j = 0$ in favor of $H_1: \beta_j \neq 0$ probably would not be enough statistical evidence to lead us to conclude that x_j is significantly related to y and that x_j should be included in a final regression model. In general, if prior belief does not indicate that the independent variable x_j is significantly related to the dependent variable y, which implies that we believe that $H_0: \beta_j = 0$ may be true, then we wish to guard against making a Type I error (rejecting $H_0: \beta_j = 0$ when $H_0: \beta_j = 0$ is true), because we do not wish to include x_j in a final regression model when x_j is not significantly related to y (see Table 6.7). Therefore, if we do not believe that x_j is significantly related to y, we might require α to be .01, or even lower, which would imply that we should conclude that x_j is significantly related to y only if we can, at a 99% (or possibly even higher) level of confidence, reject $H_0: \beta_j = 0$ in favor of $H_1: \beta_j \neq 0$.

Before concluding this section, we should make two comments. First, if we can, at a high level of confidence, reject the null hypothesis $H_0: \beta_0 = 0$ in favor of the alternative hypothesis $H_1: \beta_0 \neq 0$, we should include the intercept β_0 in the regression model

$$\begin{aligned} y &= \mu + \epsilon \\ &= \beta_0 + \beta_1 x_1 + \cdots + \beta_p x_p + \epsilon \end{aligned}$$

If we cannot, at a high level of confidence, reject the null hypothesis $H_0: \beta_0 = 0$, this might indicate that the intercept β_0 should not be included in the regression model. However, since

$$\mu = \beta_0 + \beta_1 x_1 + \cdots + \beta_p x_p$$

we see that β_0 equals the mean value of the dependent variable when all the independent variables equal zero. If, logically speaking, μ did not equal zero when all the independent variables equaled zero (for example, in the fuel consumption problem, the mean fuel consumption would not equal zero when the average hourly temperature equals zero), it follows that β_0 would not equal zero. In such a situation, it is common practice to include the intercept β_0 in the regression model even if we cannot, at a high level of confidence, reject $H_0: \beta_0 = 0$ in favor of $H_1: \beta_0 \neq 0$.

Second, we show in Section 7.2 that in most regression analyses the size of the t_{b_j} statistic measures the *additional importance* of the independent variable x_j *over and above* the combined importance of the other independent variables in the regression model. Therefore, the t_{b_j} statistic must be used with caution in assessing the importance of an independent variable.

6.7 CONFIDENCE INTERVALS FOR MEANS AND PREDICTION INTERVALS FOR INDIVIDUAL VALUES

As illustrated in Chapter 5,

$$\hat{y}_0 = b_0 + b_1 x_{01} + b_2 x_{02} + \cdots + b_p x_{0p}$$

which is the point estimate of the future mean value of the dependent variable

$$\mu_0 = \beta_0 + \beta_1 x_{01} + \beta_2 x_{02} + \cdots + \beta_p x_{0p}$$

and the point prediction of the future (individual) value of the dependent variable

$$\begin{aligned} y_0 &= \mu_0 + \epsilon_0 \\ &= \beta_0 + \beta_1 x_{01} + \beta_2 x_{02} + \cdots + \beta_p x_{0p} + \epsilon_0 \end{aligned}$$

will (unless we are extremely lucky) differ from both μ_0 and y_0. The following are the formulas for a $100(1-\alpha)\%$ confidence interval for μ_0 and a $100(1-\alpha)\%$ prediction interval for y_0.

A $100(1-\alpha)\%$ CONFIDENCE INTERVAL FOR μ_0, AND A $100(1-\alpha)\%$ PREDICTION INTERVAL FOR y_0:

1. *A $100(1-\alpha)\%$ confidence interval for* μ_0 is

$$[\hat{y}_0 - BE_{\hat{y}_0}[100(1-\alpha)],\ \hat{y}_0 + BE_{\hat{y}_0}[100(1-\alpha)]]$$

where $BE_{\hat{y}_0}[100(1-\alpha)]$ denotes the *$100(1-\alpha)$% bound on the error of estimation obtained when estimating μ_0 by $\hat{y}_0$* and is given by the equation

$$BE_{\hat{y}_0}[100(1-\alpha)] = t_{[\alpha/2]}^{(n-k)} s\sqrt{\mathbf{x}_0'(\mathbf{X}'\mathbf{X})^{-1}\mathbf{x}_0}$$

2. *A $100(1-\alpha)$% prediction interval for y_0* is

$$[\hat{y}_0 - BP_{\hat{y}_0}[100(1-\alpha)],\ \hat{y}_0 + BP_{\hat{y}_0}[100(1-\alpha)]]$$

where $BP_{\hat{y}_0}[100(1-\alpha)]$ denotes the *$100(1-\alpha)$% bound on the error of prediction obtained when predicting y_0 by $\hat{y}_0$* and is given by the equation

$$BP_{\hat{y}_0}[100(1-\alpha)] = t_{[\alpha/2]}^{(n-k)} s\sqrt{1 + \mathbf{x}_0'(\mathbf{X}'\mathbf{X})^{-1}\mathbf{x}_0}$$

Here

$$s = \sqrt{\frac{\sum_{i=1}^{n}(y_i - \hat{y}_i)^2}{n-k}} = \sqrt{\frac{\sum_{i=1}^{n} y_i^2 - \mathbf{b}'\mathbf{X}'\mathbf{y}}{n-k}}$$

is the standard error, $t_{[\alpha/2]}^{(n-k)}$ is the point on the scale of the t-distribution having $(n-k)$ degrees of freedom so that the area in the tail of this t-distribution to the right of $t_{[\alpha/2]}^{(n-k)}$ is $\alpha/2$, and $\mathbf{x}_0'$ is a row vector containing the future values of the independent variables. More specifically,

$$\mathbf{x}_0' = [1 \quad x_{01} \quad x_{02} \quad \cdots \quad x_{0p}]$$

which says that $\mathbf{x}_0'$ is a row vector containing the numbers multiplied by the least squares point estimates $b_0, b_1, b_2, \ldots, b_p$ in the point estimate (and prediction)

$$\hat{y}_0 = b_0 + b_1x_{01} + b_2x_{02} + \cdots + b_px_{0p}$$

EXAMPLE 6.10

In the fuel consumption problem, recall that Weather Forecasts, Inc., has told us that the average hourly temperature in the future week will be $x_{01} = 40.0$ and the chill index in the future week will be $x_{02} = 10$. We saw in Example 5.8 that

$$\begin{aligned}\hat{y}_0 &= b_0 + b_1x_{01} + b_2x_{02} \\ &= 13.109 - .0900(40.0) + .0825(10) \\ &= 10.333 \text{ tons of coal}\end{aligned}$$

is the point estimate of the future mean fuel consumption

$$\begin{aligned}\mu_0 &= \beta_0 + \beta_1x_{01} + \beta_2x_{02} \\ &= \beta_0 + \beta_1(40.0) + \beta_2(10)\end{aligned}$$

and is the point prediction of the future (individual) fuel consumption

$$\begin{aligned} y_0 &= \mu_0 + \epsilon_0 \\ &= \beta_0 + \beta_1 x_{01} + \beta_2 x_{02} + \epsilon_0 \\ &= \beta_0 + \beta_1(40.0) + \beta_2(10) + \epsilon_0 \end{aligned}$$

In order to calculate the 100(1 − α)% bound on the error of estimation obtained when estimating μ_0 by $\hat{y}_0$,

$$BE_{\hat{y}_0}[100(1-\alpha)] = t_{[\alpha/2]}^{(n-k)} s\sqrt{\mathbf{x}_0'(\mathbf{X}'\mathbf{X})^{-1}\mathbf{x}_0}$$

and the 100(1 − α)% bound on the error of prediction obtained when predicting y_0 by $\hat{y}_0$,

$$BP_{\hat{y}_0}[100(1-\alpha)] = t_{[\alpha/2]}^{(n-k)} s\sqrt{1 + \mathbf{x}_0'(\mathbf{X}'\mathbf{X})^{-1}\mathbf{x}_0}$$

we need to calculate the matrix algebra expression $\mathbf{x}_0'(\mathbf{X}'\mathbf{X})^{-1}\mathbf{x}_0$, where $\mathbf{x}_0'$ is a row vector containing the numbers multiplied by the least squares point estimates b_0, b_1, and b_2 in the point estimate (and prediction)

$$\begin{aligned} \hat{y}_0 &= b_0 + b_1 x_{01} + b_2 x_{02} \\ &= 13.109 - .0900(40.0) + .0825(10) \\ &= 10.333 \end{aligned}$$

Since 1 is multiplied by b_0, and $x_{01} = 40.0$, the future value of the first independent variable, average hourly temperature, is multiplied by b_1, and $x_{02} = 10$, the future value of the second independent variable, the chill index, is multiplied by b_2, it follows that

$$\begin{aligned} \mathbf{x}_0' &= [1 \quad x_{01} \quad x_{02}] \\ &= [1 \quad 40 \quad 10] \end{aligned} \qquad \text{and} \qquad \mathbf{x}_0 = \begin{bmatrix} 1 \\ x_{01} \\ x_{02} \end{bmatrix} = \begin{bmatrix} 1 \\ 40 \\ 10 \end{bmatrix}$$

Hence, since we have previously calculated $(\mathbf{X}'\mathbf{X})^{-1}$ (see Example 5.6), it follows that

$$\mathbf{x}_0'(\mathbf{X}'\mathbf{X})^{-1}\mathbf{x}_0 = [1 \quad 40 \quad 10] \begin{bmatrix} 5.43405 & -.085930 & -.118856 \\ -.085930 & .00147070 & .00165094 \\ -.118856 & .00165094 & .00359276 \end{bmatrix} \begin{bmatrix} 1 \\ 40 \\ 10 \end{bmatrix}$$

$$= [.80828 \quad -.0105926 \quad -.0168908] \begin{bmatrix} 1 \\ 40 \\ 10 \end{bmatrix} = .2157$$

We previously found

$$t_{[.025]}^{(n-k)} = t_{[.025]}^{(8-3)} = t_{[.025]}^{(5)} = 2.571$$

$$s = .3671 \qquad \text{(see Example 6.1)}$$

so the 95% bound on the error of estimation obtained when estimating μ_0 by $\hat{y}_0$ is

$$\begin{aligned} BE_{\hat{y}_0}[95] &= t^{(5)}_{[.025]} s\sqrt{\mathbf{x}_0'(\mathbf{X}'\mathbf{X})^{-1}\mathbf{x}_0} \\ &= 2.571(.3671)\sqrt{.2157} \\ &= .438 \end{aligned}$$

and the 95% confidence interval for μ_0 is

$$\begin{aligned} [\hat{y}_0 - BE_{\hat{y}_0}[95], \hat{y}_0 + BE_{\hat{y}_0}[95]] &= [10.333 - .438, 10.333 + .438] \\ &= [9.895, 10.771] \end{aligned}$$

Thus, we are 95% confident that μ_0, the future mean fuel consumption, is greater than or equal to 9.895 tons of coal and is less than or equal to 10.771 tons of coal. Since (a supernatural power knows that) the true value of μ_0 is

$$\begin{aligned} \mu_0 &= \beta_0 + \beta_1 x_{01} + \beta_2 x_{02} \\ &= 12.930 + (-.087)(40.0) + (.086)(10) \\ &= 10.31 \end{aligned}$$

it follows that our 95% confidence interval contains μ_0. Of course, since we do not know the true value of μ_0, we are not certain (but are 95% confident) that this interval contains μ_0.

Next, since the 95% bound on the error of prediction obtained when predicting y_0 by $\hat{y}_0$ is

$$\begin{aligned} BP_{\hat{y}_0}[95] &= t^{(5)}_{[.025]} s\sqrt{1 + \mathbf{x}_0'(\mathbf{X}'\mathbf{X})^{-1}\mathbf{x}_0} \\ &= 2.571(.3671)\sqrt{1.2157} \\ &= 1.04 \end{aligned}$$

it follows that the 95% prediction interval for y_0 is

$$\begin{aligned} [\hat{y}_0 - BP_{\hat{y}_0}[95], \hat{y}_0 + BP_{\hat{y}_0}[95]] &= [10.333 - 1.04, 10.333 + 1.04] \\ &= [9.293, 11.373] \end{aligned}$$

This interval says that we are 95% confident that y_0, the future (individual) fuel consumption, will be greater than or equal to 9.293 tons of coal and less than or equal to 11.373 tons of coal. The upper bound of this interval is extremely important, because it says that if we plan to acquire 11.373 tons of coal in the future week, then we can be very confident that we will have enough fuel to adequately heat buildings at State University (since we are very confident that y_0 will be no greater than 11.373 tons of coal). As will be illustrated in Example 6.11, there are many regression problems in which both the upper bound and the lower bound of a $100(1 - \alpha)\%$ prediction interval for y_0 are important. However, in this problem, although the upper bound of the 95% prediction interval for y_0 is important, the lower bound is probably not very important.

Since (a supernatural power knows that) the future (individual) fuel consumption, y_0, will be

$$\begin{aligned} y_0 &= \mu_0 + \epsilon_0 \\ &= 10.31 + .79 \\ &= 11.1 \text{ tons of coal} \end{aligned}$$

it follows that y_0 will be contained in the 95% prediction interval for y_0, [9.293, 11.373]. So if we acquire 11.373 tons of coal, we will have enough coal to adequately heat the buildings at State University next week (since $y_0 = 11.1$ will be less than 11.373). Since we do not know the value of y_0, we do not know with certainty that y_0 will be contained in the preceding interval and hence are not absolutely sure that a supply of 11.373 tons of coal will be adequate. However, since we are 95% confident that y_0 will be contained in the interval, we are very confident that a supply of 11.373 tons of coal will be adequate.

We now compare the confidence interval for the *future mean value* of the dependent variable, μ_0, with the prediction interval for the *future (individual) value* of the dependent variable, y_0. Note that we have used the term *confidence interval* to refer to the interval for μ_0, and we have used the term *prediction interval* to refer to the interval for $y_0 = \mu_0 + \epsilon_0$. The reason we have used different terms to refer to these intervals is that μ_0 and $y_0 = \mu_0 + \epsilon_0$ are fundamentally different quantities. First, μ_0, the future mean value of the dependent variable, is a population parameter, since it is the mean, and thus a descriptive measure, of the infinite population of potential values of the dependent variable that we could observe when the future values of the p independent variables will be $x_{01}, x_{02}, \ldots, x_{0p}$. In general, it is standard practice to use the term *point estimate* to refer to a single numerical guess of a population parameter and to use the term *confidence interval* to refer to an interval for a population parameter. However, the future (individual) value of the dependent variable

$$y_0 = \mu_0 + \epsilon_0$$

is not a population parameter. Rather, y_0 is simply a single element that will be randomly selected from a population of potential values of the dependent variable. In general, it is standard practice to use the term *point prediction* to refer to a single numerical guess and the term *prediction interval* to refer to an interval for a single (or individual) value of the dependent variable that will be randomly selected from a future population (in the fuel consumption problem the actual amount of fuel that will be consumed during the future week). Also note that, since we do not know and cannot average the infinite number of potential future values of the dependent variable, we will never know the true value of the population mean μ_0, and thus

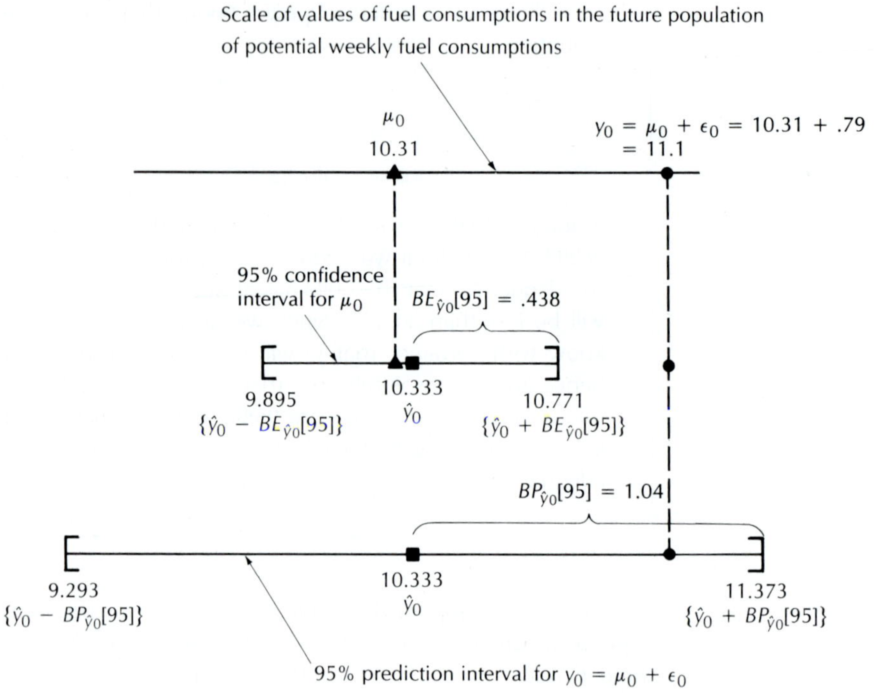

FIGURE 6.8
A Comparison of a Confidence Interval for μ_0 and a Prediction Interval for $y_0 = \mu_0 + \epsilon_0$

we will never know whether μ_0 is contained in the confidence interval. However, we will eventually learn whether or not the future value of the dependent variable, y_0, is in the prediction interval. In the fuel consumption problem, for example, in the current week we do not know the value of the future fuel consumption y_0, and hence we have calculated a point prediction and 95% prediction interval for y_0. But, at the end of the future week, we will find that 11.1 tons of coal have been consumed, and so we will know that the future fuel consumption y_0 is in our 95% prediction interval, [9.293, 11.373]. Of course, this knowledge comes too late for planning purposes, so we must use our point prediction and 95% prediction interval to help plan our fuel acquisitions.

Next, as shown in Figure 6.8 (which illustrates the 95% confidence interval and the 95% prediction interval calculated in the fuel consumption problem), the confidence interval for μ_0 and the prediction interval for y_0 are both centered at

$$\hat{y}_0 = b_0 + b_1x_{01} + b_2x_{02} + \cdots + b_px_{0p}$$

which is both the point estimate of μ_0 and the point prediction of y_0. However, as illustrated in this figure, the bound on the error of prediction obtained when predicting y_0 by $\hat{y}_0$

$$BP_{\hat{y}_0}[100(1-\alpha)] = t_{[\alpha/2]}^{(n-k)} s\sqrt{1 + \mathbf{x}_0'(\mathbf{X}'\mathbf{X})^{-1}\mathbf{x}_0}$$

is larger than the bound on the error of estimation obtained when estimating μ_0 by $\hat{y}_0$

$$BE_{\hat{y}_0}[100(1-\alpha)] = t_{[\alpha/2]}^{(n-k)} s\sqrt{\mathbf{x}_0'(\mathbf{X}'\mathbf{X})^{-1}\mathbf{x}_0}$$

(The only difference between the two formulas is that the formula for $BP_{\hat{y}_0}[100(1-\alpha)]$ has an additional 1 under the radical, making this bound larger.) Intuitively, the reason for this is that the bound on the error of estimation,

$$BE_{\hat{y}_0}[100(1-\alpha)]$$

obtained when estimating

$$\mu_0 = \beta_0 + \beta_1 x_{01} + \beta_2 x_{02} + \cdots + \beta_p x_{0p}$$

accounts for the uncertainty caused by the fact that the least squares point estimates $b_0, b_1, \ldots, b_p$ used to calculate the point estimate

$$\hat{y}_0 = b_0 + b_1 x_{01} + b_2 x_{02} + \cdots + b_p x_{0p}$$

differ from the unknown parameters $\beta_0, \beta_1, \ldots, \beta_p$, whereas the bound on the error of prediction,

$$BP_{\hat{y}_0}[100(1-\alpha)]$$

obtained when predicting

$$\begin{aligned} y_0 &= \mu_0 + \epsilon_0 \\ &= \beta_0 + \beta_1 x_{01} + \beta_2 x_{02} + \cdots + \beta_p x_{0p} + \epsilon_0 \end{aligned}$$

accounts for *both* the uncertainty caused by the fact that the least squares point estimates $b_0, b_1, \ldots, b_p$ differ from the unknown parameters $\beta_0, \beta_1, \ldots, \beta_p$, *and* the uncertainty caused by the fact that the future error term ϵ_0 (which is predicted to be equal to zero, although it probably will not be zero) is unknown. Thus, since the bound on the error of prediction takes into account more uncertainty than does the bound on the error of estimation, $BP_{\hat{y}_0}[100(1-\alpha)]$ is larger than $BE_{\hat{y}_0}[100(1-\alpha)]$, and hence the $100(1-\alpha)\%$ prediction interval for y_0

$$[\hat{y}_0 \pm BP_{\hat{y}_0}[100(1-\alpha)]] = [\hat{y}_0 \pm t_{[\alpha/2]}^{(n-k)} s\sqrt{1 + \mathbf{x}_0'(\mathbf{X}'\mathbf{X})^{-1}\mathbf{x}_0}]$$

is longer than the $100(1-\alpha)\%$ confidence interval for μ_0

$$[\hat{y}_0 \pm BE_{\hat{y}_0}[100(1-\alpha)]] = [\hat{y}_0 \pm t_{[\alpha/2]}^{(n-k)} s\sqrt{\mathbf{x}_0'(\mathbf{X}'\mathbf{X})^{-1}\mathbf{x}_0}]$$

Referring to Figure 6.8, we see that in the fuel consumption problem $\hat{y}_0 = 10.333$ (the point estimate of μ_0 and the point prediction of y_0) differs from both

$$\mu_0 = 10.31 \qquad \text{and} \qquad y_0 = \mu_0 + \epsilon_0 = 10.31 + .79 = 11.1$$

We also can see that the 95% confidence interval for the future mean fuel consumption contains μ_0. However, the future (individual) fuel consumption y_0 is not contained in the 95% confidence interval for μ_0, because this interval is not "long enough." But, the 95% prediction interval for y_0 is long enough, and y_0 is contained in the 95% prediction interval for y_0. Of course, we should point out that the situation illustrated in Figure 6.8 is only typical. The relative positions of μ_0, y_0, and $\hat{y}_0$ will be different in different situations. However, Figure 6.8 does illustrate the very important fact that we must make certain not to omit the extra 1 under the radical when computing the $100(1-\alpha)\%$ prediction interval. If the extra 1 is mistakenly omitted, we will be computing the $100(1-\alpha)\%$ confidence interval for μ_0, which very possibly is not long enough to contain $y_0 = \mu_0 + \epsilon_0$, the future (individual) value of the dependent variable.

EXAMPLE 6.11

In Example 5.9 we saw that the equation

$$\begin{aligned}\mu_0 &= \beta_0 + \beta_1 x_{04} + \beta_2 x_{03} + \beta_3 x_{03}^2 + \beta_4 x_{04} x_{03} \\ &= \beta_0 + \beta_1(.10) + \beta_2(6.80) + \beta_3(6.80)^2 + \beta_4(.10)(6.80)\end{aligned}$$

relates μ_0, the future mean demand for Fresh, to $x_{04} = .10$ and $x_{03} = 6.80$, the values of the price difference and the advertising expenditure for Fresh in the future sales period, and thus that

$$\begin{aligned}\hat{y}_0 &= b_0 + b_1 x_{04} + b_2 x_{03} + b_3 x_{03}^2 + b_4 x_{04} x_{03} \\ &= 29.1133 + 11.1342(.10) - 7.6080(6.80) + .6712(6.80)^2 \\ &\quad - 1.4777(.10)(6.80) \\ &= 8.526 \qquad (852{,}600 \text{ bottles})\end{aligned}$$

is the point estimate of μ_0 and the point prediction of $y_0 = \mu_0 + \epsilon_0$, the future (individual) demand for Fresh. Moreover,

$$\begin{aligned}\mathbf{x}_0' &= [1 \quad .10 \quad 6.80 \quad (6.80)^2 \quad (.10)(6.80)] \\ &= [1 \quad .10 \quad 6.80 \quad 46.24 \quad .68]\end{aligned}$$

is the row vector containing the numbers multiplied by b_0, b_1, b_2, b_3, and b_4 in the preceding prediction equation and hence (it can be shown that) $\mathbf{x}_0'(\mathbf{X}'\mathbf{X})^{-1}\mathbf{x}_0 = .1609$. Also,

$$t_{[.025]}^{(n-k)} = t_{[.025]}^{(30-5)} = t_{[.025]}^{(25)} = 2.06$$

$$s = .2063 \qquad \text{(see Example 6.9)}$$

Therefore,

$$\begin{aligned}[\hat{y}_0 \pm t_{[.025]}^{(25)} s\sqrt{\mathbf{x}_0'(\mathbf{X}'\mathbf{X})^{-1}\mathbf{x}_0}] &= [8.526 \pm 2.06(.2063)\sqrt{.1609}] \\ &= [8.526 \pm .171] \\ &= [8.355, 8.697] \\ &= [835{,}500 \text{ bottles}, 869{,}700 \text{ bottles}]\end{aligned}$$

is a *95% confidence interval for* μ_0, *the future mean demand* for Fresh, and

$$
\begin{aligned}
[\hat{y}_0 \pm t^{(25)}_{[.025]} s\sqrt{1 + \mathbf{x}_0'(\mathbf{X}'\mathbf{X})^{-1}\mathbf{x}_0}] &= [8.526 \pm 2.06(.2063)\sqrt{1.1609}] \\
&= [8.526 \pm .458] \\
&= [8.068, 8.984] \\
&= [806{,}800 \text{ bottles}, 898{,}400 \text{ bottles}]
\end{aligned}
$$

is a *95% prediction interval for* y_0, *the future (individual) demand* for Fresh. Since Enterprise Industries needs to predict the number of extra large size bottles of Fresh that will be demanded in the future sales period, the *95% prediction interval* for y_0 (the future demand for Fresh), [8.068, 8.984], is very useful. Enterprise Industries is 95% confident that the future demand for Fresh will be greater than or equal to 806,800 bottles and less than or equal to 898,400 bottles. The upper limit of this prediction interval is very important, because if Enterprise Industries plans to have a supply of 898,400 extra large size bottles of Fresh on hand during the future sales period, it can be very confident that this supply will adequately meet the demand for Fresh. The lower limit of this prediction interval is also very important, because Enterprise Industries can be very confident that it will sell at least 806,800 bottles in the future sales period. Since each bottle will sell for $x_{01} = 3.80$ in the future sales period, Enterprise Industries is very confident that sales revenue from the extra large size bottles of Fresh will be at least 806,800 × \$3.80 = \$3,065,840 in the future sales period. This minimum revenue figure will help Enterprise Industries to better understand its cash flow situation.

6.8 THE STATISTICAL MEANINGS AND DERIVATIONS OF THE INTERVALS FOR MEANS AND INDIVIDUAL VALUES

EXAMPLE 6.12

In this example we return to the fuel consumption problem. Here we discuss the meaning of 95% confidence as it pertains to the 95% confidence interval for

$$\mu_0 = \beta_0 + \beta_1 x_{01} + \beta_2 x_{02} = \beta_0 + \beta_1(40.0) + \beta_2(10)$$

and the 95% prediction interval for

$$y_0 = \mu_0 + \epsilon_0 = \beta_0 + \beta_1(40.0) + \beta_2(10) + \epsilon_0$$

both of which were calculated in Example 6.10. We begin by noting that, corresponding to the infinite population of all possible samples of eight fuel consumptions, there are

1. The infinite population of all possible point estimates of μ_0 and point predictions of y_0 (since each different sample yields a different point estimate and point prediction $\hat{y}_0$);
2. The infinite population of all possible 95% confidence intervals for μ_0 (since each different sample yields a different 95% confidence interval for μ_0);
3. The infinite population of all possible 95% prediction intervals for y_0 (since each different sample yields a different 95% prediction interval for y_0).

Table 6.8 presents the elements in these populations corresponding to SPL_1, SPL_2, and SPL_3 (as given in Table 6.1). Now, consider Figure 6.9. The normal curve at the top of this figure shows that if the inference assumptions hold, the population of all possible point estimates of μ_0 and point predictions of y_0 has a normal distribution with mean μ_0, which (a supernatural power knows) is equal to 10.31. Figure 6.9 also depicts (1) the point estimates of μ_0 and point predictions of y_0 yielded by SPL_1, SPL_2, and SPL_3 (which are denoted $(\hat{y}_0)_{SPL_1}$, $(\hat{y}_0)_{SPL_2}$, and $(\hat{y}_0)_{SPL_3}$); (2) the 95% confidence intervals for μ_0 yielded by SPL_1, SPL_2, and SPL_3 (which are denoted $CI_{\mu_0}[95]_{SPL_1}$, $CI_{\mu_0}[95]_{SPL_2}$, and $CI_{\mu_0}[95]_{SPL_3}$); and (3) the 95% prediction intervals for y_0 yielded by SPL_1, SPL_2, and SPL_3 (which are denoted $PI_{y_0}[95]_{SPL_1}$, $PI_{y_0}[95]_{SPL_2}$, and $PI_{y_0}[95]_{SPL_3}$).

Examining the confidence intervals in the figure, we see that the confidence intervals

$$CI_{\mu_0}[95]_{SPL_1} = [9.895, 10.771]$$

$$CI_{\mu_0}[95]_{SPL_2} = [9.857, 10.739]$$

contain $\mu_0 = 10.31$ (denoted by ▲ in the figure), while the confidence interval

$$CI_{\mu_0}[95]_{SPL_3} = [9.066, 10.248]$$

does not contain $\mu_0 = 10.31$. Thus, two of the three 95% confidence intervals for μ_0 depicted in Figure 6.9 contain μ_0. We will prove (after we conclude this example) that the interpretation of 95% confidence here is that 95 percent of the 95% confidence intervals for μ_0 in the population of all such intervals contain $\mu_0 = 10.31$, and 5 percent of the confidence intervals in this population do not contain $\mu_0 = 10.31$. Thus, when we compute the 95% confidence interval for μ_0 by using the observed sample (which, we have seen, is SPL_1) and obtain the interval [9.895, 10.771], we can be 95% confident that this interval contains μ_0, because 95 percent of the confidence intervals in the population of all possible 95% confidence intervals for μ_0 contain μ_0, and because we have obtained one of the confidence intervals in this population.

The normal curve at the bottom of Figure 6.9 depicts the future population of all potential weekly fuel consumptions that could be observed when $x_{01} = 40.0$ and $x_{02} = 10$. This population is depicted as having a normal distribution in accordance with inference assumption 3. The mean of this normal

TABLE 6.8 The Elements Corresponding to SPL_1, SPL_2, and SPL_3 in Three Regression Populations Related to the Two-Variable Fuel Consumption Model

Three Point Estimates of μ_0 (and Point Predictions of y_0) in the Population of All Possible Point Estimates of μ_0 (and Point Predictions of y_0)		
SPL_1	SPL_2	SPL_3
$(\hat{y}_0)_{SPL_1} = (b_0)_{SPL_1} + (b_1)_{SPL_1}x_{01} + (b_2)_{SPL_1}x_{02}$ $= 13.109 - .0900(40) + .0825(10)$ $= 10.333$	$(\hat{y}_0)_{SPL_2} = (b_0)_{SPL_2} + (b_1)_{SPL_2}x_{01} + (b_2)_{SPL_2}x_{02}$ $= 12.949 - .0882(40) + .0876(10)$ $= 10.298$	$(\hat{y}_0)_{SPL_3} = (b_0)_{SPL_3} + (b_1)_{SPL_3}x_{01} + (b_2)_{SPL_3}x_{02}$ $= 11.593 - .0548(40) + .0256(10)$ $= 9.657$
Three 95% Confidence Intervals for μ_0 in the Population of All Possible 95% Confidence Intervals for μ_0		
SPL_1	SPL_2	SPL_3
$[(\hat{y}_0)_{SPL_1} \pm 2.571(s)_{SPL_1}\sqrt{.2157}]$ $= [10.333 \pm 2.571(.3671)\sqrt{.2157}]$ $= [10.333 \pm .438] = [9.895, 10.771]$	$[(\hat{y}_0)_{SPL_2} \pm 2.571(s)_{SPL_2}\sqrt{.2157}]$ $= [10.298 \pm 2.571(.3701)\sqrt{.2157}]$ $= [10.298 \pm .441] = [9.857, 10.739]$	$[(\hat{y}_0)_{SPL_3} \pm 2.571(s)_{SPL_3}\sqrt{.2157}]$ $= [9.657 \pm 2.571(.2084)\sqrt{.2157}]$ $= [9.657 \pm .249] = [9.408, 9.906]$
Three 95% Prediction Intervals for y_0 in the Population of All Possible 95% Prediction Intervals for y_0		
SPL_1	SPL_2	SPL_3
$[(\hat{y}_0)_{SPL_1} \pm 2.571(s)_{SPL_1}\sqrt{1.2157}]$ $= [10.333 \pm 2.571(.3671)\sqrt{1.2157}]$ $= [10.333 \pm 1.04] = [9.293, 11.373]$	$[(\hat{y}_0)_{SPL_2} \pm 2.571(s)_{SPL_2}\sqrt{1.2157}]$ $= [10.298 \pm 2.571(.3701)\sqrt{1.2157}]$ $= [10.298 \pm 1.049] = [9.249, 11.347]$	$[(\hat{y}_0)_{SPL_3} \pm 2.571(s)_{SPL_3}\sqrt{1.2157}]$ $= [9.657 \pm 2.571(.2084)\sqrt{1.2157}]$ $= [9.657 \pm .591] = [9.066, 10.248]$

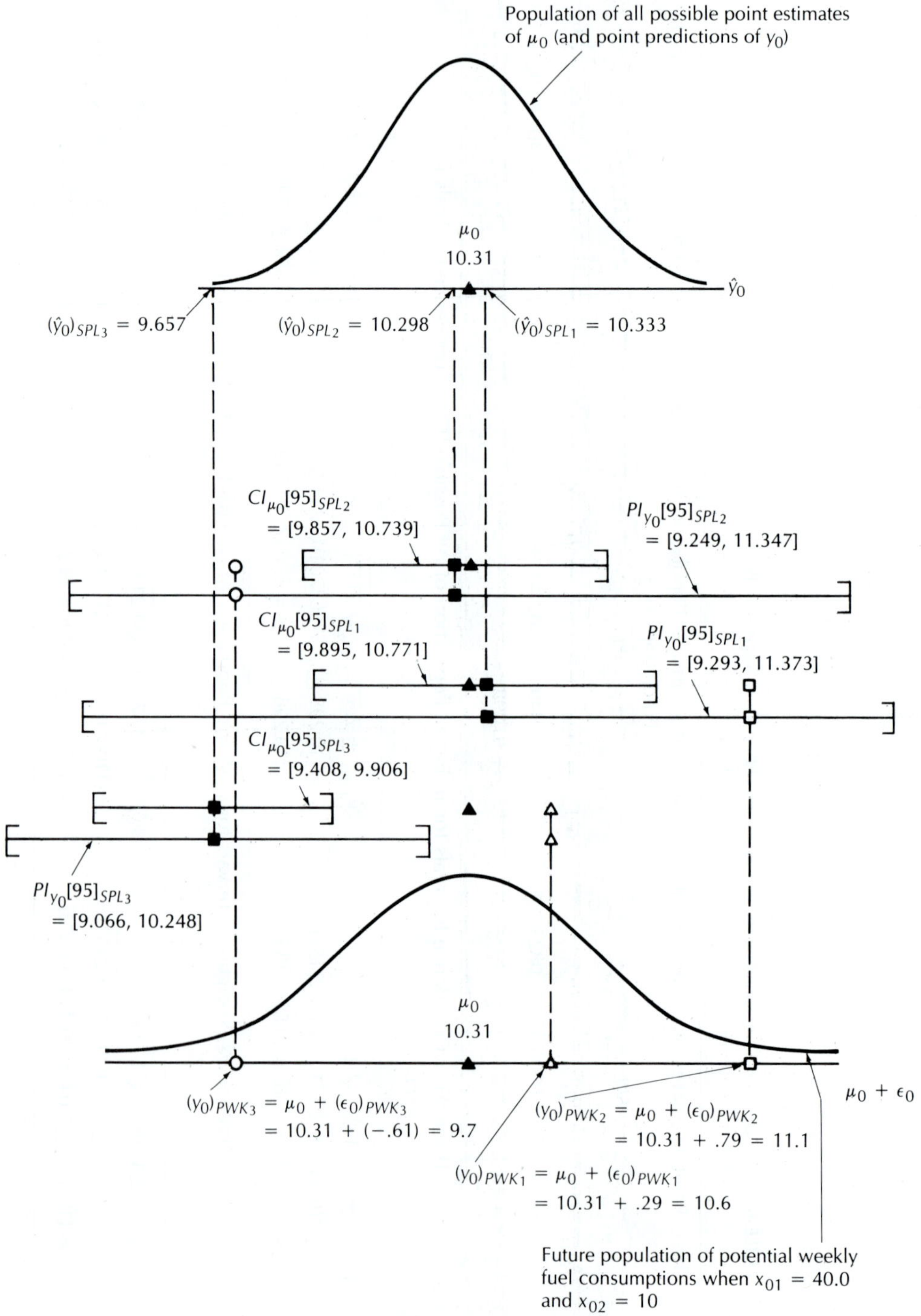

FIGURE 6.9 An Illustration of the Meaning of Confidence Intervals (*CI*) and Prediction Intervals (*PI*) for the Two-Variable Fuel Consumption Model When $x_{01} = 40.0$ and $x_{02} = 10$

distribution is the future mean fuel consumption, $\mu_0 = 10.31$. Shown along the scale of this normal curve (as △, □, and ○) are three different future (individual) fuel consumptions in the future population of all potential weekly fuel consumptions that could be observed when $x_{01} = 40.0$ and $x_{02} = 10$. These fuel consumptions are denoted $(y_0)_{PWK_1}$, $(y_0)_{PWK_2}$, and $(y_0)_{PWK_3}$. Here the subscripts PWK_1, PWK_2, and PWK_3 are used to represent potential week 1, potential week 2, and potential week 3, which are three different potential future weeks having an average hourly temperature of $x_{01} = 40.0$ and a chill index of $x_{02} = 10$. The fact that these potential weeks are different implies that the fuel consumptions $(y_0)_{PWK_1}$, $(y_0)_{PWK_2}$, and $(y_0)_{PWK_3}$ are different. Figure 6.9 assumes that

1. After using the observations in SPL_1 to compute

   $$PI_{y_0}[95]_{SPL_1} = [9.293, 11.373]$$

   the future fuel consumption $(y_0)_{PWK_2} = 11.1$ is subsequently observed.
2. After using the observations in SPL_2 to compute

   $$PI_{y_0}[95]_{SPL_2} = [9.249, 11.347]$$

   the future fuel consumption $(y_0)_{PWK_3} = 9.7$ is subsequently observed.
3. After using the observations in SPL_3 to compute

   $$PI_{y_0}[95]_{SPL_3} = [9.066, 10.248]$$

   the future fuel consumption $(y_0)_{PWK_1} = 10.6$ is subsequently observed.

That is, the figure depicts the following prediction interval–future fuel consumption combinations:

$$[PI_{y_0}[95]_{SPL_1} = [9.293, 11.373]; (y_0)_{PWK_2} = 11.1]$$

$$[PI_{y_0}[95]_{SPL_2} = [9.249, 11.347]; (y_0)_{PWK_3} = 9.7]$$

$$[PI_{y_0}[95]_{SPL_3} = [9.066, 10.248]; (y_0)_{PWK_1} = 10.6]$$

We now consider the infinite *population of all possible prediction interval–future fuel consumption combinations*. This population consists of an infinite number of subpopulations, each subpopulation containing combinations of one particular 95% prediction interval paired with all the potential future fuel consumptions that could be observed when $x_{01} = 40.0$ and $x_{02} = 10$ (see Table 6.9). In general, we call a prediction interval–future fuel consumption combination successful if the future fuel consumption that we observe (after having observed our sample) falls in the prediction interval that we have calculated by using our sample. We call this combination unsuccessful if the future fuel consumption does not fall in the prediction interval. Looking at Figure 6.9, we can see that the prediction interval–future fuel consumption combinations

$$[PI_{y_0}[95]_{SPL_1} = [9.293, 11.373]; (y_0)_{PWK_2} = 11.1]$$

$$[PI_{y_0}[95]_{SPL_2} = [9.249, 11.347]; (y_0)_{PWK_3} = 9.7]$$

TABLE 6.9 Nine Elements in the Population of All Possible Prediction Interval–Future Fuel Consumption Combinations for the Two-Variable Fuel Consumption Model When $x_{01} = 40.0$ and $x_{02} = 10$

Three elements in the subpopulation corresponding to $PI_{y_0}[95]_{SPL_1}$	$[PI_{y_0}[95]_{SPL_1} = [9.293, 11.373]; (y_0)_{PWK_1} = 10.6]$
	$[PI_{y_0}[95]_{SPL_1} = [9.293, 11.373]; (y_0)_{PWK_2} = 11.1]$
	$[PI_{y_0}[95]_{SPL_1} = [9.293, 11.373]; (y_0)_{PWK_3} = 9.7]$
Three elements in the subpopulation corresponding to $PI_{y_0}[95]_{SPL_2}$	$[PI_{y_0}[95]_{SPL_2} = [9.249, 11.347]; (y_0)_{PWK_1} = 10.6]$
	$[PI_{y_0}[95]_{SPL_2} = [9.249, 11.347]; (y_0)_{PWK_2} = 11.1]$
	$[PI_{y_0}[95]_{SPL_2} = [9.249, 11.347]; (y_0)_{PWK_3} = 9.7]$
Three elements in the subpopulation corresponding to $PI_{y_0}[95]_{SPL_3}$	$[PI_{y_0}[95]_{SPL_3} = [9.066, 10.248]; (y_0)_{PWK_1} = 10.6]$
	$[PI_{y_0}[95]_{SPL_3} = [9.066, 10.248]; (y_0)_{PWK_2} = 11.1]$
	$[PI_{y_0}[95]_{SPL_3} = [9.066, 10.248]; (y_0)_{PWK_3} = 9.7]$

are successful, while the prediction interval–future fuel consumption combination

$$[PI_{y_0}[95]_{SPL_3} = [9.066, 10.248]; (y_0)_{PWK_1} = 10.6]$$

is unsuccessful. Thus, two of the three prediction interval–future fuel consumption combinations depicted in Figure 6.9 are successful. We will prove (after we conclude this example) that the interpretation of 95% confidence here is that 95 percent of the prediction interval–future fuel consumption combinations in the population of all such combinations (where the prediction intervals are based on 95% confidence) are successful, while 5 percent are unsuccessful. Therefore, before we compute the 95% prediction interval for y_0 by using the observed sample (SPL_1) and obtain the interval [9.293, 11.373], and before we observe y_0, we can be 95% confident that we will be successful (that is, obtain a 95% prediction interval such that y_0 falls in our interval), since 95 percent of the prediction interval–future fuel consumption combinations in the population of all such combinations are successful, and since we know that we will obtain one prediction interval–future fuel consumption combination in this population.

We now discuss the logic behind the derivation of the $100(1 - \alpha)\%$ confidence interval for the future mean value of the dependent variable

$$\mu_0 = \beta_0 + \beta_1 x_{01} + \beta_2 x_{02} + \cdots + \beta_p x_{0p}$$

We begin by considering the following properties (see Appendix D for a proof of (1), (2), and (3)):

The population of all possible point estimates of μ_0

1. Has *mean* $\mu_{\hat{y}_0} = \mu_0$;
2. Has *variance* $\sigma^2_{\hat{y}_0} = \sigma^2 \mathbf{x}_0'(\mathbf{X}'\mathbf{X})^{-1}\mathbf{x}_0$ (if inference assumptions 1 and 2 hold);
3. Has *standard deviation* $\sigma_{\hat{y}_0} = \sigma\sqrt{\mathbf{x}_0'(\mathbf{X}'\mathbf{X})^{-1}\mathbf{x}_0}$ (if inference assumptions 1 and 2 hold);
4. Has a *normal distribution* (if inference assumptions 1, 2, and 3 hold).

Property (1) says that $\mu_{\hat{y}_0}$, the mean of the population of all possible point estimates of μ_0, is equal to μ_0. For this reason, we say that when we use the observed sample $SPL = \{y_1, y_2, \ldots, y_n\}$ to calculate the least squares point estimates $b_0, b_1, b_2, \ldots, b_p$ and the point estimate $\hat{y}_0$ of μ_0

$$\hat{y}_0 = b_0 + b_1x_{01} + b_2x_{02} + \cdots + b_px_{0p}$$

we are using an *unbiased* estimation procedure. This unbiasedness property tells us that although the point estimate $\hat{y}_0$ that we calculate (by using the observed sample) probably does not equal μ_0, the average of all the different possible point estimates of μ_0 that we could have calculated is equal to μ_0.

Now, since the point estimate of σ is s, the quantity

$$s_{\hat{y}_0} = s\sqrt{\mathbf{x}_0'(\mathbf{X}'\mathbf{X})^{-1}\mathbf{x}_0}$$

which we call the *standard error of the estimate* $\hat{y}_0$, is the point estimate of

$$\sigma_{\hat{y}_0} = \sigma\sqrt{\mathbf{x}_0'(\mathbf{X}'\mathbf{X})^{-1}\mathbf{x}_0}$$

the standard deviation of the population of all possible point estimates of μ_0. Then, defining the statistic

$$t_{[\hat{y}_0, \mu_0]} = \frac{\hat{y}_0 - \mu_0}{s_{\hat{y}_0}} = \frac{\hat{y}_0 - \mu_0}{s\sqrt{\mathbf{x}_0'(\mathbf{X}'\mathbf{X})^{-1}\mathbf{x}_0}}$$

we can make the following statement.

If the inference assumptions hold, *the population of all possible* $t_{[\hat{y}_0, \mu_0]}$ *statistics* has a t-distribution with $n - k$ degrees of freedom.

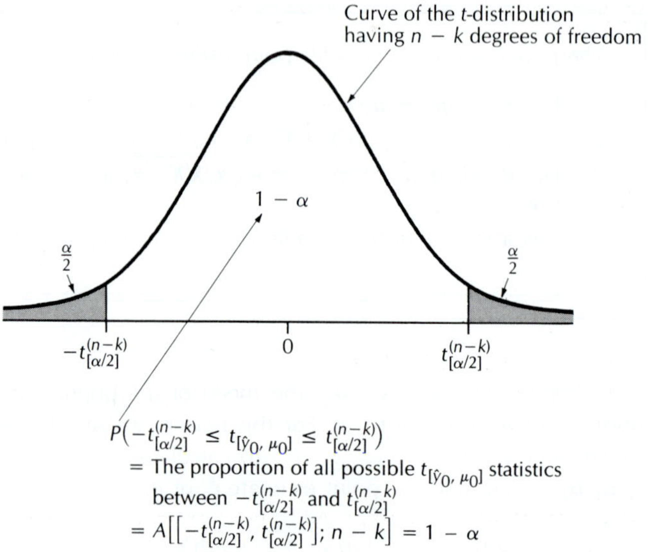

FIGURE 6.10

If the Inference
Assumptions Hold,
$P(-t_{[\alpha/2]}^{(n-k)}) \leq t_{[\hat{y}_0,\mu_0]} \leq t_{[\alpha/2]}^{(n-k)}$
$= 1 - \alpha$

$$P\left(-t_{[\alpha/2]}^{(n-k)} \leq t_{[\hat{y}_0,\, \mu_0]} \leq t_{[\alpha/2]}^{(n-k)}\right)$$
$$= \text{The proportion of all possible } t_{[\hat{y}_0,\, \mu_0]} \text{ statistics}$$
$$\text{between } -t_{[\alpha/2]}^{(n-k)} \text{ and } t_{[\alpha/2]}^{(n-k)}$$
$$= A\left[\left[-t_{[\alpha/2]}^{(n-k)},\, t_{[\alpha/2]}^{(n-k)}\right]; n - k\right] = 1 - \alpha$$

Thus, if the inference assumptions hold, then (as illustrated in Figure 6.10) the probability

$$P\left(-t_{[\alpha/2]}^{(n-k)} \leq t_{[\hat{y}_0,\mu_0]} \leq t_{[\alpha/2]}^{(n-k)}\right) = P\left(-t_{[\alpha/2]}^{(n-k)} \leq \frac{\hat{y}_0 - \mu_0}{s_{\hat{y}_0}} \leq t_{[\alpha/2]}^{(n-k)}\right)$$

$$= A[[-t_{[\alpha/2]}^{(n-k)},\, t_{[\alpha/2]}^{(n-k)}]; n - k] = 1 - \alpha$$

which is the area under the curve of the t-distribution having $n - k$ degrees of freedom between $-t_{[\alpha/2]}^{(n-k)}$ and $t_{[\alpha/2]}^{(n-k)}$. By using algebraic manipulations analogous to those carried out in Section 6.4 (where we derived a $100(1 - \alpha)\%$ confidence interval for β_j), we can show the probability statement

$$P\left(-t_{[\alpha/2]}^{(n-k)} \leq \frac{\hat{y}_0 - \mu_0}{s_{\hat{y}_0}} \leq t_{[\alpha/2]}^{(n-k)}\right) = 1 - \alpha$$

to be equivalent to the expression

$$P(\hat{y}_0 - t_{[\alpha/2]}^{(n-k)} s\sqrt{\mathbf{x}_0'(\mathbf{X}'\mathbf{X})^{-1}\mathbf{x}_0} \leq \mu_0 \leq \hat{y}_0 + t_{[\alpha/2]}^{(n-k)} s\sqrt{\mathbf{x}_0'(\mathbf{X}'\mathbf{X})^{-1}\mathbf{x}_0}) = 1 - \alpha$$

This probability statement says that the proportion of confidence intervals that contain μ_0 in the population of all possible $100(1 - \alpha)\%$ confidence intervals for μ_0 is equal to $1 - \alpha$. That is, if we compute a $100(1 - \alpha)\%$ confidence interval for μ_0 using the formula

$$[\hat{y}_0 \pm BE_{\hat{y}_0}[100(1 - \alpha)]] = [\hat{y}_0 \pm t_{[\alpha/2]}^{(n-k)} s\sqrt{\mathbf{x}_0'(\mathbf{X}'\mathbf{X})^{-1}\mathbf{x}_0}]$$

then $100(1 - \alpha)\%$ (for example, 95%) of the confidence intervals in the population of all possible $100(1 - \alpha)\%$ confidence intervals for μ_0 contain μ_0, while $100(\alpha)\%$

(for example, 5%) of the confidence intervals in this population do not contain μ_0. We illustrated this fact in Example 6.12 (see Figure 6.9).

We next discuss the logic behind the derivation of the $100(1 - \alpha)\%$ prediction interval for the future (individual) value of the dependent variable

$$\begin{aligned} y_0 &= \mu_0 + \epsilon_0 \\ &= \beta_0 + \beta_1 x_{01} + \beta_2 x_{02} + \cdots + \beta_p x_{0p} + \epsilon_0 \end{aligned}$$

Recall that the point prediction of y_0 is

$$\hat{y}_0 = b_0 + b_1 x_{01} + b_2 x_{02} + \cdots + b_p x_{0p}$$

and that the prediction error obtained when predicting y_0 by $\hat{y}_0$ is $y_0 - \hat{y}_0$. Then, since there are an infinite population of possible point predictions of y_0, and an infinite population of potential future (individual) values of the dependent variable, there is an infinite population of prediction errors, which we will call the *population of all possible prediction errors.*

The population of all possible prediction errors

1. Has *mean* $\mu_{(y_0 - \hat{y}_0)} = 0$;
2. Has *variance* $\sigma^2_{(y_0 - \hat{y}_0)} = \sigma^2(1 + \mathbf{x}_0'(\mathbf{X}'\mathbf{X})^{-1}\mathbf{x}_0)$ (if inference assumptions 1 and 2 hold);
3. Has *standard deviation* $\sigma_{(y_0 - \hat{y}_0)} = \sigma\sqrt{1 + \mathbf{x}_0'(\mathbf{X}'\mathbf{X})^{-1}\mathbf{x}_0}$ (if inference assumptions 1 and 2 hold);
4. Has a *normal distribution* (if inference assumptions 1, 2, and 3 hold).

(See Appendix D for a proof of (1), (2), and (3).)

Property (1) says that $\mu_{(y_0 - \hat{y}_0)}$, the mean of the population of all possible prediction errors, equals zero. For this reason, we say that we are using an *unbiased* prediction procedure. This unbiasedness property tells us that when we use the observed sample to compute the point prediction $\hat{y}_0$, and when we subsequently observe the future value y_0, although the prediction error we obtain will probably not equal zero (the point prediction $\hat{y}_0$ will probably not equal y_0), the average of all the prediction errors that we could obtain is equal to zero.

The point estimate of σ is s; therefore

$$s_{(y_0 - \hat{y}_0)} = s\sqrt{1 + \mathbf{x}_0'(\mathbf{X}'\mathbf{X})^{-1}\mathbf{x}_0}$$

which we call the *standard error of the prediction error* $(y_0 - \hat{y}_0)$, is the point estimate of

$$\sigma_{(y_0 - \hat{y}_0)} = \sigma\sqrt{1 + \mathbf{x}_0'(\mathbf{X}'\mathbf{X})^{-1}\mathbf{x}_0}$$

the standard deviation of the population of all possible prediction errors. Then, defining the statistic

$$t_{[\hat{y}_0, y_0]} = \frac{y_0 - \hat{y}_0}{s_{(y_0 - \hat{y}_0)}} = \frac{y_0 - \hat{y}_0}{s\sqrt{1 + \mathbf{x}_0'(\mathbf{X}'\mathbf{X})^{-1}\mathbf{x}_0}}$$

we can make the following statement:

> If the inference assumptions hold, *the population of all possible* $t_{[\hat{y}_0, y_0]}$ *statistics* has a t-distribution with $n - k$ degrees of freedom.

Then, if the inference assumptions hold, the probability

$$P(-t_{[\alpha/2]}^{(n-k)} \leq t_{[\hat{y}_0, y_0]} \leq t_{[\alpha/2]}^{(n-k)}) = P\left(-t_{[\alpha/2]}^{(n-k)} \leq \frac{y_0 - \hat{y}_0}{s_{(y_0 - \hat{y}_0)}} \leq t_{[\alpha/2]}^{(n-k)}\right)$$

$$= A[[-t_{[\alpha/2]}^{(n-k)}, t_{[\alpha/2]}^{(n-k)}]; n - k] = 1 - \alpha$$

which is the area under the curve of the t-distribution having $n - k$ degrees of freedom between $-t_{[\alpha/2]}^{(n-k)}$ and $t_{[\alpha/2]}^{(n-k)}$ (see Figure 6.11). By using algebraic manipulations analogous to those carried out in Section 6.4, we can show the probability statement

$$P\left(-t_{[\alpha/2]}^{(n-k)} \leq \frac{y_0 - \hat{y}_0}{s_{(y_0 - \hat{y}_0)}} \leq t_{[\alpha/2]}^{(n-k)}\right) = 1 - \alpha$$

to be equivalent to the expression

$$P(\hat{y}_0 - t_{[\alpha/2]}^{(n-k)} s\sqrt{1 + \mathbf{x}_0'(\mathbf{X}'\mathbf{X})^{-1}\mathbf{x}_0} \leq y_0 \leq \hat{y}_0 + t_{[\alpha/2]}^{(n-k)} s\sqrt{1 + \mathbf{x}_0'(\mathbf{X}'\mathbf{X})^{-1}\mathbf{x}_0}) = 1 - \alpha$$

Now, consider the *population of all possible prediction interval–future (individual) value combinations*. (This population was discussed in detail for the fuel consumption problem in Example 6.12.) We will call a prediction interval–future value combination successful if the future value of the dependent variable that we observe after having observed our sample falls in the prediction interval that we have calculated by using our sample, and we will call this combination unsuccessful if the future value does not fall in the prediction interval. Then, the probability statement

$$P(\hat{y}_0 - t_{[\alpha/2]}^{(n-k)} s\sqrt{1 + \mathbf{x}_0'(\mathbf{X}'\mathbf{X})^{-1}\mathbf{x}_0} \leq y_0 \leq \hat{y}_0 + t_{[\alpha/2]}^{(n-k)} s\sqrt{1 + \mathbf{x}_0'(\mathbf{X}'\mathbf{X})^{-1}\mathbf{x}_0}) = 1 - \alpha$$

says that $100(1 - \alpha)\%$ of the prediction interval–future value combinations in the population of all such combinations are successful, but $100(\alpha)\%$ of the prediction interval–future value combinations in this population are unsuccessful. We illustrated this fact in Example 6.12 (see Figure 6.9).

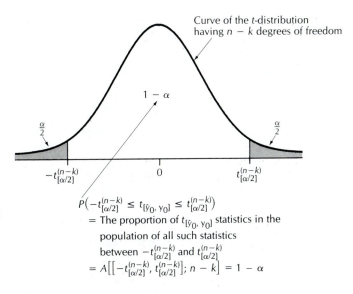

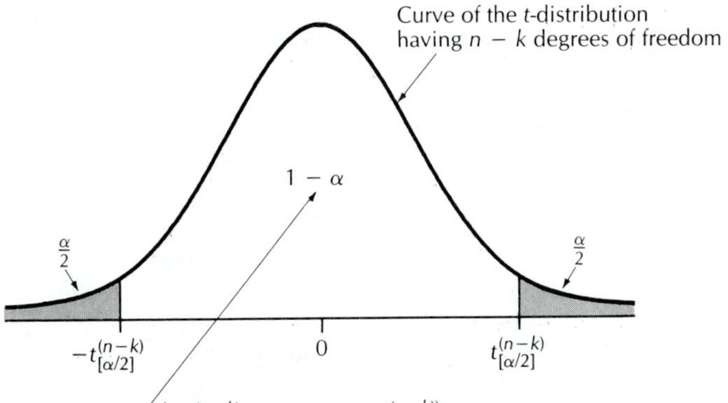

$P(-t_{[\alpha/2]}^{(n-k)} \leq t_{[\hat{y}_0, y_0]} \leq t_{[\alpha/2]}^{(n-k)})$
= The proportion of $t_{[\hat{y}_0, y_0]}$ statistics in the population of all such statistics between $-t_{[\alpha/2]}^{(n-k)}$ and $t_{[\alpha/2]}^{(n-k)}$
$= A[[-t_{[\alpha/2]}^{(n-k)}, t_{[\alpha/2]}^{(n-k)}]; n - k] = 1 - \alpha$

FIGURE 6.11 If the Inference Assumptions Hold, $P(-t_{[\alpha/2]}^{(n-k)} \leq t_{[\hat{y}_0, y_0]} \leq t_{[\alpha/2]}^{(n-k)}) = 1 - \alpha$

6.9 USING THE STATISTICAL ANALYSIS SYSTEM (SAS) TO PERFORM REGRESSION ANALYSIS

In this section we demonstrate how to carry out a regression analysis of the Fresh detergent demand data by using the Statistical Analysis System (SAS). (We do not give detailed instructions for using SAS. Readers interested in learning more about SAS should consult the various SAS user's guides.)

In Table 6.10 we show the SAS statements used to run a regression analysis of the Fresh demand problem. Notes to the right of these statements describe the purpose of each statement. Job control statements are omitted. In particular, running this problem requires the use of transformations to create the squared term x_3^2 and the interaction term x_4x_3.

Figure 6.12 shows the SAS output for the Fresh detergent demand problem that is generated by the statements in Table 6.10. Important quantities on the output are labeled. Notice that the SAS output does not include the $100(1 - \alpha)\%$ confidence interval for β_j, but it does give b_j and $s\sqrt{c_{jj}}$, so we can use these quantities to calculate this confidence interval by the formula

$$[b_j \pm t_{[\alpha/2]}^{(n-k)} s\sqrt{c_{jj}}]$$

Some of the quantities on the SAS output (explained variation, total variation, analysis of variance table, R^2, and $\bar{R}^2$) will be explained in later chapters.

6.10 INFERENCES FOR SIMPLE LINEAR REGRESSION

We now give special case formulas for simple linear regression (see p. 236).

TABLE 6.10
SAS Statements for a Regression Analysis of the Fresh Detergent Demand Data

Statement	Explanation
`DATA DETR;`	Names the data file (here the name assigned is DETR)
`INPUT Y X4 X3;`	Lists variable names (here Y = demand for Fresh, X4 = price difference, X3 = advertising expenditure)
`X3SQ = X3*X3;`	Transformation defines the squared term x_3^2 and assigns the name X3SQ to this variable
`X43 = X4*X3;`	Transformation defines the interaction term x_4x_3 and assigns the name X43 to this variable
`CARDS;`	Identifies data cards (or lines)
`7.38 -0.05 5.50`	30 data cards (or lines). No semicolon follows.

`8.51 0.25 6.75`
`9.52 0.60 7.25`
`.`
`.`
`.`
	Data for each observed sales period are on a separate line listed in the same order as given on the INPUT statement (see Table 5.12 for the Fresh Detergent demand data)

`9.26 0.55 6.80`	
`. 0.10 6.80`	Future advertising expenditure $x_{03} = 6.80$; future price difference $x_{04} = .10$ Decimal point in first column indicates a missing value. Used to obtain point prediction, confidence interval, and prediction interval when $x_{03} = 6.80$ and $x_{04} = .10$
`PROC PRINT;`	Prints the data
`PROC REG DATA = DETR;`	Calls regression procedure. Data file to be used is DETR

```
MODEL Y = X4 X3 X3SQ X43 / P XPX I CLM CLI;
```

Specifies model
$y = \beta_0 + \beta_1x_4 + \beta_2x_3 + \beta_3x_3^2 + \beta_4x_4x_3 + \epsilon$
(Intercept β_0 is assumed)

P = Residuals desired
XPX = **X′X** and **X′y** desired
I = $(\mathbf{X'X})^{-1}$ desired
CLM = 95% confidence interval desired
CLI = 95% prediction interval desired

FIGURE 6.12 SAS Output for Fresh Detergent Demand Problem

S T A T I S T I C A L   A N A L Y S I S   S Y S T E M

OBS	Y	X4	X3	X3SQ	X43
1	7.38	-0.05	5.50	30.2500	-0.2750
2	8.51	0.25	6.75	45.5625	1.6875
3	9.52	0.60	7.25	52.5625	4.3500
4	7.50	0.00	5.50	30.2500	0.0000
5	9.33	0.25	7.00	49.0000	1.7500
6	8.28	0.20	6.50	42.2500	1.3000
7	8.75	0.15	6.75	45.5625	1.0125
8	7.87	0.05	5.25	27.5625	0.2625
9	7.10	-0.15	5.25	27.5625	-0.7875
10	8.00	0.15	6.00	36.0000	0.9000
11	7.89	0.20	6.50	42.2500	1.3000
12	8.15	0.10	6.25	39.0625	0.6250
13	9.10	0.40	7.00	49.0000	2.8000
14	8.86	0.45	6.90	47.6100	3.1050
15	8.90	0.35	6.80	46.2400	2.3800
16	8.87	0.30	6.80	46.2400	2.0400
17	9.26	0.50	7.10	50.4100	3.5500
18	9.00	0.50	7.00	49.0000	3.5000
19	8.75	0.40	6.80	46.2400	2.7200
20	7.95	-0.05	6.50	42.2500	-0.3250
21	7.65	-0.05	6.25	39.0625	-0.3125
22	7.27	-0.10	6.00	36.0000	-0.6000
23	8.00	0.20	6.50	42.2500	1.3000
24	8.50	0.10	7.00	49.0000	0.7000
25	8.75	0.50	6.80	46.2400	3.4000
26	9.21	0.60	6.80	46.2400	4.0800
27	8.27	-0.05	6.50	42.2500	-0.3250
28	7.67	0.00	5.75	33.0625	0.0000
29	7.93	0.05	5.80	33.6400	0.2900
30	9.26	0.55	6.80	46.2400	3.7400
31	.	0.10	6.80	46.2400	0.6800

} Data printout

(Continued)

FIGURE 6.12 (Continued)

MODEL CROSSPRODUCTS X' X X' Y Y' Y

X' X		INTERCEP	X4	X3	X3SQ	X43		Y
INTERCEP		30	6.4	193.6	1258.85	44.1675		251.48
X4		6.4	2.865	44.1675	304.7814	19.71662		57.646
X3	$\mathbf{X'X} =$	193.6	44.1675	1258.85	8241.968	304.7814	$\mathbf{X'y} =$	1632.781
X3SQ		1258.85	304.7814	8241.968	54298.76	2103.453		10677.4
X43		44.1675	19.71662	304.7814	2103.453	135.8881		397.7442
Y		251.48	57.646	1632.781	10677.4	397.7442		2121.532[a]

X' X INVERSE, B, SSE

INVERSE		INTERCEP	X4	X3	X3SQ	X43		Y
INTERCEP		1315.261	543.4463	-433.586	35.50156	-83.4036		29.11329
X4		543.4463	464.2447	-179.952	14.80313	-69.5252		11.13423
X3	$(\mathbf{X'X})^{-1} =$	433.586	-179.952	143.1914	-11.7449	27.67939	$\mathbf{b} =$	-7.60801
X3SQ		35.50156	14.80313	-11.7449	0.965045	-2.28257		0.6712472
X43		-83.4036	-69.5252	27.67939	-2.28257	10.45448		-1.47772
Y		29.11329	11.13423	-7.60801	0.6712472	-1.47772		1.064397[b]

[a] $\sum_{i=1}^{n} y_i^2$ [b] SSE

DEP VARIABLE: Y

SOURCE	DF	SUM OF SQUARES	MEAN SQUARE	F VALUE	PROB>F	
MODEL	4	12.394190[a]	3.098548	72.777	0.0001	Analysis of variance table
ERROR	25	1.064397[b]	0.042576[f]			
C TOTAL	29	13.458587[c]				

ROOT MSE	0.206339[d]	R-SQUARE	0.9209[h]
DEP MEAN	8.382667	ADJ R-SQ	0.9083[i]
C.V.	2.461498		

VARIABLE	DF	PARAMETER ESTIMATE[e]	STANDARD ERROR[g]	T FOR H0: PARAMETER=0[j]	PROB > !T![k]
INTERCEP	1	29.113287	7.483206	3.890	0.0007
X4	1	11.134226	4.445854	2.504	0.0192
X3	1	-7.608007	2.469109	-3.081	0.0050
X3SQ	1	0.671247	0.202701	3.312	0.0028
X43	1	-1.477717	0.667165	-2.215	0.0361

OBS	ACTUAL	PREDICT VALUE	STD ERR PREDICT	LOWER95% MEAN	UPPER95% MEAN	LOWER95% PREDICT	UPPER95% PREDICT	RESIDUAL
1	7.380	7.424	0.081140	7.257	7.591	6.968	7.881	-.044139
2	8.510	8.833	0.049414	8.531	8.735	8.196	9.070	-.122850
3	9.520	9.490	0.127991	9.227	9.754	8.990	9.990	0.029866
4	7.500	7.574	0.079322	7.411	7.738	7.119	8.030	-.074478
5	9.330	8.946	0.086038	8.769	9.123	8.485	9.406	0.384097
6	8.280	8.327	0.056341	8.211	8.443	7.887	8.768	-.047250
7	8.750	8.517	0.063947	8.385	8.649	8.072	8.962	0.233113
8	7.870	7.841	0.163020	7.506	8.177	7.300	8.383	0.028687
9	7.100	7.166	0.144177	6.869	7.463	6.648	7.684	-.066071
10	8.000	7.970	0.080969	7.804	8.137	7.514	8.427	0.029666
11	7.890	8.327	0.056341	8.211	8.443	7.887	8.768	-.437250
12	8.150	7.974	0.062179	7.846	8.102	7.530	8.418	0.176312
13	9.100	9.064	0.060524	8.940	9.189	8.622	9.507	0.035566
14	8.860	8.998	0.055135	8.885	9.112	8.558	9.438	-.138209
15	8.900	8.797	0.047450	8.700	8.895	8.361	9.233	0.102677
16	8.870	8.743	0.047649	8.645	8.841	8.307	9.179	0.126964
17	9.260	9.255	0.080749	9.089	9.422	8.799	9.712	0.004773
18	9.000	9.143	0.067340	9.005	9.282	8.696	9.590	-.143455
19	8.750	8.652	0.051799	8.745	8.958	8.413	9.290	-.101611
20	7.950	7.945	0.082413	7.775	8.115	7.487	8.403	0.005016
21	7.650	7.689	0.082596	7.519	7.859	7.231	8.147	-.038914
22	7.270	7.403	0.103346	7.191	7.616	6.928	7.879	-.133353
23	8.000	8.327	0.056341	8.211	8.443	7.887	8.768	-.327250
24	8.500	8.827	0.131141	8.557	9.097	8.324	9.331	-.327373
25	8.750	8.960	0.069988	8.816	9.104	8.511	9.409	-.210185
26	9.210	9.069	0.094423	8.874	9.263	8.601	9.536	0.141240
27	8.270	7.945	0.082413	7.775	8.115	7.487	8.403	0.325016
28	7.670	7.560	0.065802	7.425	7.696	7.114	8.006	0.109641
29	7.930	7.696	0.065635	7.561	7.831	7.250	8.142	0.234223
30	9.260	9.014	0.081742	8.846	9.183	8.557	9.472	0.245527
31		8.526[l]	0.082777[m]	8.355	8.696	8.068	8.984	

SUM OF RESIDUALS 1.00586E-13
SUM OF SQUARED RESIDUALS 1.064397[n]

95% confidence interval (LOWER95% MEAN – UPPER95% MEAN)

95% prediction interval (LOWER95% PREDICT – UPPER95% PREDICT)

[a] Explained variation [b] SSE [c] Total variation [d] s

[e] $b_j: b_0, b_1, b_2, b_3, b_4$ [f] s^2 [g] $s\sqrt{c_{jj}}$ [h] R^2

[i] $\bar{R}^2$ [j] t_{b_j} [k] Prob-value [l] Point prediction $\hat{y}_0$

[m] $s\sqrt{\mathbf{x}_0'(\mathbf{X}'\mathbf{X})^{-1}\mathbf{x}_0}$ [n] SSE

For the simple linear regression model

$$y = \mu + \epsilon$$
$$= \beta_0 + \beta_1 x + \epsilon$$

1. The *standard error* is

$$s = \sqrt{\frac{SSE}{n-2}} = \sqrt{\frac{\sum_{i=1}^{n} y_i^2 - [b_0 \sum_{i=1}^{n} y_i + b_1 \sum_{i=1}^{n} x_i y_i]}{n-2}}$$

Moreover, if the inference assumptions are satisfied,

2. The *100(1 − α)% confidence interval for* β_0 is

$$[b_0 - BE_{b_0}[100(1-\alpha)],\ b_0 + BE_{b_0}[100(1-\alpha)]]$$

where

$$BE_{b_0}[100(1-\alpha)] = t_{[\alpha/2]}^{(n-2)} s \sqrt{\frac{1}{n} + \frac{\bar{x}^2}{\sum_{i=1}^{n} x_i^2 - n\bar{x}^2}}$$

3. The *100(1 − α)% confidence interval for* β_1 is

$$[b_1 - BE_{b_1}[100(1-\alpha)],\ b_1 + BE_{b_1}[100(1-\alpha)]]$$

where

$$BE_{b_1}[100(1-\alpha)] = t_{[\alpha/2]}^{(n-2)} s \sqrt{\frac{1}{\sum_{i=1}^{n} x_i^2 - n\bar{x}^2}}$$

4. $\hat{y}_0 = b_0 + b_1 x_0$ is the *point estimate* of the future mean value of the dependent variable, $\mu_0 = \beta_0 + \beta_1 x_0$, and is the *point prediction* of the future (individual) value of the dependent variable, $y_0 = \mu_0 + \epsilon_0$.
5. The *100(1 − α)% confidence interval for* μ_0 is

$$[\hat{y}_0 - BE_{\hat{y}_0}[100(1-\alpha)],\ \hat{y}_0 + BE_{\hat{y}_0}[100(1-\alpha)]]$$

where

$$BE_{\hat{y}_0}[100(1-\alpha)] = t_{[\alpha/2]}^{(n-2)} s \sqrt{\frac{1}{n} + \frac{(x_0 - \bar{x})^2}{\sum_{i=1}^{n} x_i^2 - n\bar{x}^2}}$$

6. The *100(1 − α)% prediction interval for* y_0 is

$$[\hat{y}_0 - BP_{\hat{y}_0}[100(1-\alpha)],\ \hat{y}_0 + BP_{\hat{y}_0}[100(1-\alpha)]]$$

where

$$BP_{\hat{y}_0}[100(1-\alpha)] = t_{[\alpha/2]}^{(n-2)} s \sqrt{1 + \frac{1}{n} + \frac{(x_0 - \bar{x})^2}{\sum_{i=1}^{n} x_i^2 - n\bar{x}^2}}$$

The preceding formula for s follows because

$$\mathbf{b} = \begin{bmatrix} b_0 \\ b_1 \end{bmatrix} \qquad \mathbf{X} = \begin{bmatrix} 1 & x_1 \\ 1 & x_2 \\ \vdots & \vdots \\ 1 & x_n \end{bmatrix} \qquad \mathbf{y} = \begin{bmatrix} y_1 \\ y_2 \\ \vdots \\ y_n \end{bmatrix}$$

Thus, since

$$\mathbf{X}'\mathbf{y} = \begin{bmatrix} 1 & 1 & \cdots & 1 \\ x_1 & x_2 & \cdots & x_n \end{bmatrix} \begin{bmatrix} y_1 \\ y_2 \\ \vdots \\ y_n \end{bmatrix} = \begin{bmatrix} \sum_{i=1}^{n} y_i \\ \sum_{i=1}^{n} x_i y_i \end{bmatrix}$$

and

$$\mathbf{b}'\mathbf{X}'\mathbf{y} = [b_0 \quad b_1] \begin{bmatrix} \sum_{i=1}^{n} y_i \\ \sum_{i=1}^{n} x_i y_i \end{bmatrix} = b_0 \sum_{i=1}^{n} y_i + b_1 \sum_{i=1}^{n} x_i y_i$$

it follows that

$$s = \sqrt{\frac{SSE}{n-k}} = \sqrt{\frac{\sum_{i=1}^{n} y_i^2 - \mathbf{b}'\mathbf{X}'\mathbf{y}}{n-k}} = \sqrt{\frac{\sum_{i=1}^{n} y_i^2 - [b_0 \sum_{i=1}^{n} y_i + b_1 \sum_{i=1}^{n} x_i y_i]}{n-2}}$$

The formulas for $BE_{b_0}[100(1-\alpha)]$ and $BE_{b_1}[100(1-\alpha)]$ follow because, using elementary matrix algebra, we can show that the diagonal elements of $(\mathbf{X}'\mathbf{X})^{-1}$ corresponding to β_0 and β_1 are

$$c_{00} = \frac{1}{n} + \frac{\bar{x}^2}{\sum_{i=1}^{n} x_i^2 - n\bar{x}^2} \qquad \text{and} \qquad c_{11} = \frac{1}{\sum_{i=1}^{n} x_i^2 - n\bar{x}^2}$$

Hence,

$$BE_{b_0}[100(1-\alpha)] = t_{[\alpha/2]}^{(n-k)} s\sqrt{c_{00}}$$

$$= t_{[\alpha/2]}^{(n-2)} s\sqrt{\frac{1}{n} + \frac{\bar{x}^2}{\sum_{i=1}^{n} x_i^2 - n\bar{x}^2}}$$

$$BE_{b_1}[100(1-\alpha)] = t_{[\alpha/2]}^{(n-k)} s\sqrt{c_{11}}$$

$$= t_{[\alpha/2]}^{(n-2)} s\sqrt{\frac{1}{\sum_{i=1}^{n} x_i^2 - n\bar{x}^2}}$$

Finally, the formulas for $BE_{\hat{y}_0}[100(1-\alpha)]$ and $BP_{\hat{y}_0}[100(1-\alpha)]$ follow. Since

$$\mathbf{x}_0' = [1 \quad x_0]$$

for the simple linear regression model, it can be shown by using elementary matrix algebra that

$$\mathbf{x}_0'(\mathbf{X}'\mathbf{X})^{-1}\mathbf{x}_0 = \frac{1}{n} + \frac{(x_0 - \bar{x})^2}{\sum_{i=1}^{n} x_i^2 - n\bar{x}^2}$$

Hence,

$$BE_{\hat{y}_0}[100(1-\alpha)] = t_{[\alpha/2]}^{(n-k)} s\sqrt{\mathbf{x}_0'(\mathbf{X}'\mathbf{X})^{-1}\mathbf{x}_0}$$

$$= t_{[\alpha/2]}^{(n-k)} s\sqrt{\frac{1}{n} + \frac{(x_0 - \bar{x})^2}{\sum_{i=1}^{n} x_i^2 - n\bar{x}^2}}$$

$$BP_{\hat{y}_0}[100(1-\alpha)] = t_{[\alpha/2]}^{(n-k)} s\sqrt{1 + \mathbf{x}_0'(\mathbf{X}'\mathbf{X})^{-1}\mathbf{x}_0}$$

$$= t_{[\alpha/2]}^{(n-k)} s\sqrt{1 + \frac{1}{n} + \frac{(x_0 - \bar{x})^2}{\sum_{i=1}^{n} x_i^2 - n\bar{x}^2}}$$

Note that the box in Section 6.5 entitled "Testing $H_0: \beta_j = 0$ versus $H_1: \beta_j \neq 0$" applies to testing $H_0: \beta_0 = 0$ and $H_0: \beta_1 = 0$, where s, c_{00}, and c_{11} are as given here.

EXAMPLE 6.13

Reconsider the Comp-U-Systems problem in Example 5.4. In Figure 6.13 we present the SAS output resulting from using the model

$$y = \mu + \epsilon$$
$$= \beta_0 + \beta_1 x + \epsilon$$

FIGURE 6.13 SAS Output for a Regression Analysis of the Comp-U-Systems Data

DEP VARIABLE: Y

SOURCE	DF	SUM OF SQUARES	MEAN SQUARE	F VALUE	PROB>F	
MODEL	1	19918.844[a]	19918.844	935.149	0.0001	Analysis of variance table
ERROR	9	191.702[b]	21.300184[f]			
C TOTAL	10	20110.545[c]				

ROOT MSE	4.615212[d]	R-SQUARE	0.9905[h]
DEP MEAN	107.636	ADJ R-SQ	0.9894[i]
C.V.	4.287782		

VARIABLE	DF	PARAMETER ESTIMATE[e]	STANDARD ERROR[g]	T FOR HO: PARAMETER=0[j]	PROB > !T![k]
INTERCEP	1	11.464088	3.439026	3.334	0.0087
X	1	24.602210	0.804514	30.580	0.0001

x	OBS	ACTUAL[l]	PREDICT VALUE	STD ERR PREDICT	LOWER95% MEAN	UPPER95% MEAN	LOWER95% PREDICT	UPPER95% PREDICT	RESIDUAL
1	1	37.000	36.066	2.723	29.907	42.226	23.944	48.188	0.933702
2	2	58.000	60.669	2.073	55.980	65.357	49.224	72.113	-2.669
2	3	68.000	60.669	2.073	55.980	65.357	49.224	72.113	7.331
3	4	82.000	85.271	1.572	81.715	88.827	74.241	96.300	-3.271
4	5	103.000	109.873	1.393	106.721	113.025	98.967	120.779	-6.873
4	6	109.000	109.873	1.393	106.721	113.025	98.967	120.779	-.872928
4	7	112.000	109.873	1.393	106.721	113.025	98.967	120.779	2.127
5	8	134.000	134.475	1.645	130.753	138.197	123.391	145.559	-.475138
5	9	138.000	134.475	1.645	130.753	138.197	123.391	145.559	3.525
6	10	154.000	159.077	2.183	154.139	164.016	147.528	170.627	-5.077
7	11	189.000	183.680[m]	2.850[n]	177.233	190.126	171.409	195.950	5.320
					95% confidence interval		95% prediction interval		

SUM OF RESIDUALS -1.06581E-14
SUM OF SQUARED RESIDUALS 191.7017[p]

[a] Explained variation [b] SSE [c] Total variation [d] s
[e] $b_j: b_0, b_1$ [f] s^2 [g] $s\sqrt{c_{jj}}$ [h] R^2
[i] $\bar{R}^2$ [j] t_{b_j} [k] Prob-value [l] Observed values $= y$
[m] Point prediction $\hat{y}_0$ [n] $s\sqrt{\mathbf{x}_0'(\mathbf{X}'\mathbf{X})^{-1}\mathbf{x}_0}$ [p] SSE

to perform a regression analysis of the $n = 11$ Comp-U-Systems service times in Table 5.8. In Example 5.4 we found that the slope β_1 in this model is the change in the mean service time (in minutes) associated with each additional microcomputer serviced on a call; and Figure 6.13 shows that $b_1 = 24.6022$ and $s\sqrt{c_{11}} = .8045$. It follows that a 95% confidence interval for β_1 is

$$\begin{aligned}[b_1 \pm t_{[.025]}^{(n-k)} s\sqrt{c_{11}}] &= [24.6022 \pm t_{[.025]}^{(11-2)}(.8045)] \\ &= [24.6022 \pm 2.262(.8045)] \\ &= [22.7824, 26.422]\end{aligned}$$

This interval says that Comp-U-Systems is 95% confident that the mean service time increases by between 22.7824 and 26.422 minutes for each additional microcomputer serviced on a call. (This interval can be hand-calculated by using the data in Table 5.8 and the formulas in the preceding box.)

Next, note that the lower portion of Figure 6.13 gives the point estimates (and point predictions), the 95% confidence intervals, and the 95% prediction intervals pertaining to the mean and the future (individual) service times for x microcomputers ($x = 1, 2, 3, 4, 5, 6$, and 7). The estimates and predictions (which can be hand-calculated by using the formulas in the preceding box) are arranged according to increasing observed service times (since the regression analysis was performed by rearranging the data in Table 5.8 according to increasing observed service times). We have indicated a column of x values (number of microcomputers serviced) corresponding to the observed service times.

Then, for example, considering the infinite population of all potential times to service four microcomputers, it follows that

$$\begin{aligned}\hat{y}_0 &= b_0 + b_1 x_0 \\ &= 11.4641 + 24.6022(4) \\ &= 109.873 \text{ minutes}\end{aligned}$$

is the point estimate of

$$\begin{aligned}\mu_0 &= \beta_0 + \beta_1 x_0 \\ &= \beta_0 + \beta_1(4)\end{aligned}$$

the mean time required to service four microcomputers, and is the point prediction of

$$\begin{aligned}y_0 &= \mu_0 + \epsilon_0 \\ &= \beta_0 + \beta_1(4) + \epsilon_0\end{aligned}$$

the future (individual) time required to service four microcomputers. Moreover, the 95% confidence interval for μ_0, [106.721, 113.025], says that Comp-U-Systems is 95% confident that the mean service time for four microcomputers is between 106.721 and 113.025 minutes, and the 95% prediction interval for $y_0 = \mu_0 + \epsilon_0$, [98.967, 120.779], says that Comp-U-Systems is 95% confident that an individual service time for four microcomputers will be between 98.967 and 120.779 minutes.

We conclude this section with an example that illustrates the usefulness of a confidence interval for the slope (β_1) in the simple linear regression model.

EXAMPLE 6.14

In analysis of the stock market, we sometimes use the model

$$\begin{aligned} y &= \mu + \epsilon \\ &= \beta_0 + \beta_1 x + \epsilon \end{aligned}$$

to relate

y = the rate of return on a particular stock

to

x = the rate of return on the overall stock market

When using the preceding model, we can interpret β_1 to be the percentage point change in the mean (or expected) rate of return on the particular stock that is associated with an increase of one percentage point in the rate of return on the overall stock market.

1. If regression analysis can be used to conclude (at a high level of confidence) that β_1 is greater than 1 (for example, if the 95% confidence interval for β_1 were [1.1826, 1.4723]), this would indicate that the mean rate of return on the particular stock changes more quickly than the rate of return on the overall stock market. Such a stock is called an *aggressive stock,* since gains for such a stock tend to be greater than overall market gains (which occur when the market is bullish), whereas losses for such a stock tend to be greater than overall market losses (which occur when the market is bearish). Such a stock should be purchased if you expect the market to rise and avoided if you expect the market to fall.
2. If regression analysis can be used to conclude (at a high level of confidence) that β_1 is less than 1 (for example, if the 95% confidence interval for β_1 were [.4729, .7861]), this would indicate that the mean rate of return on the particular stock changes more slowly than the rate of return on the overall stock market. Such a stock is called a *defensive stock,* since losses for such a stock tend to be less than overall market losses, whereas gains for such

a stock tend to be less than overall market gains. Such a stock should be held if you expect the market to fall and sold off if you expect the market to rise.

3. If the least squares point estimate b_1 of β_1 is nearly equal to 1, and if the 95% confidence interval for β_1 contains 1, this might indicate that the mean rate of return on the particular stock changes at roughly the same rate as the rate of return on the overall stock market.

For most stocks, positive values of β_1 describe the relationships between the mean rates of return on these stocks and the rate of return on the overall stock market. This is because general economic conditions that affect the overall market tend to affect most individual stocks in the same direction. Although stocks having zero or negative β_1 values are rare, they do exist. For example, gold stocks have negative values of β_1, because when the overall market falls, the value of gold rises, and when the overall market rises, the value of gold falls.

Since the parameter β_0 equals the mean (or "expected") rate of return on the particular stock when the rate of return on the overall stock market is zero, the value of β_0 is determined by factors specific to the firm associated with the stock. For instance, if regression analysis can be used to conclude (at a high level of confidence) that β_0 is greater than zero (say, if the 95% confidence interval for β_0 were [.1381, .3742]), this would indicate that there is a positive mean rate of return on the particular stock when the rate of return on the overall market is zero.

*6.11 INFERENCES FOR A SINGLE POPULATION

If we have randomly selected a sample of n element values

$$SPL = \{y_1, y_2, \ldots, y_n\}$$

from a single population having mean μ, variance σ^2, and standard deviation σ, then, for $i = 1, 2, \ldots, n$, we can express y_i by the regression model

$$\begin{aligned} y_i &= \mu + \epsilon_i \\ &= \beta_0 + \epsilon_i \end{aligned}$$

The column vector $\mathbf{y}$ and the matrix $\mathbf{X}$ used to calculate the least squares point estimate b_0 of $\beta_0 = \mu$ are

$$\mathbf{y} = \begin{bmatrix} y_1 \\ y_2 \\ \vdots \\ y_n \end{bmatrix} \quad \text{and} \quad \mathbf{X} = \begin{bmatrix} 1 \\ 1 \\ \vdots \\ 1 \end{bmatrix}$$

Thus,

$$\mathbf{X'X} = [1 \quad 1 \quad \cdots \quad 1]\begin{bmatrix} 1 \\ 1 \\ \vdots \\ 1 \end{bmatrix} = n$$

$$(\mathbf{X'X})^{-1} = \frac{1}{n}$$

$$\mathbf{X'y} = [1 \quad 1 \quad \cdots \quad 1]\begin{bmatrix} y_1 \\ y_2 \\ \vdots \\ y_n \end{bmatrix} = \sum_{i=1}^{n} y_i$$

It follows that the least squares point estimate of $\beta_0 = \mu$ is

$$b_0 = (\mathbf{X'X})^{-1}\mathbf{X'y} = \left(\frac{1}{n}\right)\left(\sum_{i=1}^{n} y_i\right) = \bar{y}$$

This value of b_0 minimizes the sum of squared residuals

$$SSE = \sum_{i=1}^{n} e_i^2 = \sum_{i=1}^{n} (y_i - \hat{y}_i)^2$$
$$= \sum_{i=1}^{n} (y_i - b_0)^2$$

where $\hat{y}_i = b_0$ is the point prediction of $y_i = \mu + \epsilon_i = \beta_0 + \epsilon_i$. Hence, we have shown that the least squares point estimate, b_0, of the population mean, μ, is the sample mean, $\bar{y}$. This implies that

$$SSE = \sum_{i=1}^{n} (y_i - \bar{y})^2$$

which in turn implies that the standard error is

$$s = \sqrt{\frac{SSE}{n-k}} = \sqrt{\frac{\sum_{i=1}^{n} (y_i - \bar{y})^2}{n-1}}$$

which is the sample standard deviation.

We can obtain the results shown in the following box by using the formulas presented in Sections 6.7 and 6.8, because

1. The point prediction of a future element value that will be randomly selected from the single population

$$y_0 = \mu + \epsilon_0$$
$$= \beta_0 + \epsilon_0$$

is

$$\hat{y}_0 = b_0 = \bar{y}$$

2. The number 1 is multiplied by b_0 in the preceding point prediction, which implies that $\mathbf{x}_0' = [1]$ and thus that

$$\mathbf{x}_0'(\mathbf{X}'\mathbf{X})^{-1}\mathbf{x}_0 = [1]\frac{1}{n}[1] = \frac{1}{n}$$

3. When we randomly select a sample from a single population, μ_0 (the future mean value of the dependent variable) equals μ (the mean of the single population).

STATISTICAL INFERENCES CONCERNING A SINGLE POPULATION:

Assume that $\bar{y}$ is the *mean* and s is the *standard deviation* of a sample randomly selected from a single population having mean μ, variance σ^2, and standard deviation σ. Then,

1. $\bar{y}$ is the *point estimate* of μ and is the *point prediction* of the future (individual) element value $y_0 = \mu + \epsilon_0$.

2. The *population of all possible sample means* (that is, point estimates of μ)
   a. Has *mean* $\mu_{\bar{y}} = \mu_{\hat{y}_0} = \mu_0 = \mu$—that is, has *mean* $\mu_{\bar{y}} = \mu$;
   b. Has *variance* $\sigma^2_{\bar{y}} = \sigma^2_{\hat{y}_0} = \sigma^2\mathbf{x}_0'(\mathbf{X}'\mathbf{X})^{-1}\mathbf{x}_0 = \sigma^2(1/n) = \sigma^2/n$—that is, has variance $\sigma^2_{\bar{y}} = \sigma^2/n$ (if inference assumption 2 holds);
   c. Has *standard deviation* $\sigma_{\bar{y}} = \sigma/\sqrt{n}$ (if inference assumption 2 holds);
   d. Has a *normal distribution* (if inference assumptions 2 and 3 hold).

3. If we define the $t_{[\hat{y}_0,\mu_0]}$ *statistic* to be

$$t_{[\hat{y}_0,\mu_0]} = \frac{\hat{y}_0 - \mu_0}{s_{\hat{y}_0}} = \frac{\hat{y}_0 - \mu_0}{s\sqrt{\mathbf{x}_0'(\mathbf{X}'\mathbf{X})^{-1}\mathbf{x}_0}}$$

$$= \frac{\bar{y} - \mu}{s\sqrt{1/n}} = \frac{\bar{y} - \mu}{s/\sqrt{n}} = \frac{\bar{y} - \mu}{s_{\bar{y}}} = t_{[\bar{y},\mu]}$$

then the *population of all possible* $t_{[\hat{y}_0,\mu_0]}$ *statistics* (that is, $t_{[\bar{y},\mu]}$ statistics) has a *t-distribution with $n - k = n - 1$ degrees of freedom* (if inference assumptions 2 and 3 hold).

4. A *100(1 − α)% confidence interval for* $\mu_0 = \mu$ is

$$[\hat{y}_0 \pm t^{(n-k)}_{[\alpha/2]}s\sqrt{\mathbf{x}_0'(\mathbf{X}'\mathbf{X})^{-1}\mathbf{x}_0}] = \left[\bar{y} \pm t^{(n-1)}_{[\alpha/2]}s\sqrt{\frac{1}{n}}\right] = \left[\bar{y} \pm t^{(n-1)}_{[\alpha/2]}\left(\frac{s}{\sqrt{n}}\right)\right]$$

5. A $100(1-\alpha)\%$ prediction interval for $y_0 = \mu + \epsilon_0$ is

$$[\hat{y}_0 \pm t_{[\alpha/2]}^{(n-k)} s\sqrt{1 + \mathbf{x}_0'(\mathbf{X}'\mathbf{X})^{-1}\mathbf{x}_0}] = \left[\bar{y} \pm t_{[\alpha/2]}^{(n-1)} s\sqrt{1 + \frac{1}{n}}\right]$$

Results (2), (3), and (4) in this box are discussed in more detail in optional Section 3.4. Inference assumption 1 (Constant Variance) is not mentioned here because this assumption automatically holds when we are sampling from a single population. Inference assumption 2, which says that $y_1, y_2, \ldots, y_n$, and y_0 are statistically independent, will probably hold if the population sampled is infinite (so that the result of one random selection will not affect the result of another random selection). Finally, it can be shown that inference assumption 3, which states that the population sampled is normally distributed, is more crucial to the validity of the prediction interval for $y_0 = \mu + \epsilon_0$ than to the validity of the confidence interval for μ, which approximately holds if the population sampled is described by a probability curve that is mound-shaped (even if this curve is somewhat skewed to the right or left).

EXAMPLE 6.15

In Example 2.6 we saw that $\bar{y} = 31.2$ mpg is the mean and $s = .7517$ mpg is the standard deviation of

$$\begin{aligned} SPL &= \{y_1, y_2, y_3, y_4, y_5\} \\ &= \{30.7, 31.8, 30.2, 32.0, 31.3\} \end{aligned}$$

which National Motors randomly selected from the infinite population of all Gas-Mizer mileages (which we assume is normally distributed). It follows that $\bar{y} = 31.2$ is the point estimate of μ, the mean of all Gas-Mizer mileages, and that a 95% confidence interval for μ is

$$\begin{aligned} \left[\bar{y} \pm t_{[.025]}^{(n-1)}\left(\frac{s}{\sqrt{n}}\right)\right] &= \left[\bar{y} \pm t_{[.025]}^{(5-1)}\left(\frac{s}{\sqrt{5}}\right)\right] \\ &= \left[31.2 \pm 2.776\left(\frac{.7517}{\sqrt{5}}\right)\right] \\ &= [31.2 \pm .9333] \\ &= [30.3, 32.1] \qquad \text{(rounded to one decimal place)} \end{aligned}$$

Suppose that new federal gasoline mileage standards state that in order to avoid paying a heavy fine, National Motors must convince the government that μ is at least 30 mpg. Since the 95% confidence interval for μ makes the company 95% confident that μ is greater than or equal to the lower bound 30.3, and since this lower bound is itself greater than 30 mpg, National Motors can be at least 95% confident that μ is greater than 30 mpg. Thus, it can report to the federal government that there is strong evidence that the Gas-Mizer not only meets but exceeds current gasoline mileage standards.

Next, note that we can assume that purchasing a Gas-Mizer is equivalent to randomly selecting a Gas-Mizer from the infinite population of all Gas-Mizers. Thus, the gasoline mileage that will be obtained by this randomly selected (purchased) Gas-Mizer, which we denote as

$$y_0 = \mu + \epsilon_0$$

is a future Gas-Mizer mileage that will be randomly selected from the infinite population of all Gas-Mizer mileages. Since $\bar{y} = 31.2$ is the point prediction of $y_0 = \mu + \epsilon_0$, it follows that a 95% prediction interval for y_0 is

$$\left[\bar{y} \pm t_{[.025]}^{(n-1)} s \sqrt{1 + \frac{1}{n}}\right] = \left[\bar{y} \pm t_{[.025]}^{(5-1)} s \sqrt{1 + \frac{1}{5}}\right]$$

$$= [31.2 \pm 2.776(.7517)\sqrt{1.2}]$$

$$= [31.2 \pm 2.29]$$

$$= [28.9, 33.5] \quad \text{(rounded to one decimal place)}$$

This 95% prediction interval says that the purchaser of a (randomly selected) Gas-Mizer can be 95% confident that his or her Gas-Mizer will obtain (in 50,000 miles of driving) a gasoline mileage somewhere between a minimum of 28.9 mpg and a maximum of 33.5 mpg.

EXAMPLE 6.16

If National Motors desires to utilize the sample of $n = 49$ Gas-Mizer mileages listed in Table 3.1, which has mean $\bar{y} = 31.5531$ and standard deviation $s = .799$, then, assuming that the population of all Gas-Mizer mileages is normally distributed and replacing $t_{[\alpha/2]}^{(n-1)}$ by $z_{[\alpha/2]}$ (since $n = 49$ is large), it follows that a 95% confidence interval for μ is

$$\left[\bar{y} \pm z_{[.025]}\left(\frac{s}{\sqrt{n}}\right)\right] = \left[31.5531 \pm 1.96\left(\frac{.799}{\sqrt{49}}\right)\right]$$

$$= [31.5531 \pm .2237]$$

$$= [31.3, 31.8]$$

and that a 95% prediction interval for $y_0 = \mu + \epsilon_0$ is

$$\left[\bar{y} \pm z_{[.025]} s \sqrt{1 + \frac{1}{n}}\right] = \left[31.5531 \pm 1.96(.799)\sqrt{1 + \frac{1}{49}}\right]$$

$$= [31.5531 \pm 1.5819]$$

$$= [30.0, 33.1]$$

The Central Limit Theorem (discussed in detail in optional Section 3.4) tells us that since $n = 49$ is a large sample size, this confidence interval for μ is approximately correct no matter what probability distribution describes the population of all Gas-Mizer mileages. However, even though $n = 49$ is large, the validity of the preceding prediction interval for $y_0 = \mu + \epsilon_0$ still depends on the assumption that the population of all Gas-Mizer mileages is normally distributed.

*6.12 ADVANCED HYPOTHESIS TESTING TECHNIQUES

In this section we extend our discussion of hypothesis testing.

CONFIDENCE INTERVAL RESULT I:

Assume that θ (for example, β_j) is a *parameter*, $\hat{\theta}$ (for example, b_j) is a *point estimate* of θ, and $s_{\hat{\theta}}$ (for example, $s_{b_j} = s\sqrt{c_{jj}}$) is a quantity (called the *standard error of the estimate* $\hat{\theta}$) which is such that if we define the $t_{[\hat{\theta},\theta]}$ *statistic* to be

$$t_{[\hat{\theta},\theta]} = \frac{\hat{\theta} - \theta}{s_{\hat{\theta}}}$$

then the *population of all possible* $t_{[\hat{\theta},\theta]}$ *statistics* has a *t-distribution with df degrees of freedom*. Then

1. A *two-sided* $100(1 - \alpha)\%$ *confidence interval for* θ is

   $$[\hat{\theta} - t_{[\alpha/2]}^{(df)} s_{\hat{\theta}}, \hat{\theta} + t_{[\alpha/2]}^{(df)} s_{\hat{\theta}}]$$

   Here, $t_{[\alpha/2]}^{(df)}$ is the point on the scale of the t-distribution having *df* degrees of freedom so that the area under the curve of this t-distribution to the right of $t_{[\alpha/2]}^{(df)}$ is $\alpha/2$.
2. Each of

   $$[\hat{\theta} - t_{[\alpha]}^{(df)} s_{\hat{\theta}}, \infty) \qquad \text{and} \qquad (-\infty, \hat{\theta} + t_{[\alpha]}^{(df)} s_{\hat{\theta}}]$$

   is a *one-sided* $100(1 - \alpha)\%$ *confidence interval for* θ. Here, $t_{[\alpha]}^{(df)}$ and $-t_{[\alpha]}^{(df)}$ are the points on the scale of the t-distribution having *df* degrees of freedom so that the area under the curve of this t-distribution to the right of $t_{[\alpha]}^{(df)}$ is α and to the left of $-t_{[\alpha]}^{(df)}$ is α.

In Table 6.11 we summarize two-sided and one-sided confidence intervals based on the t-distribution for several parameters we have thus far considered, along with the $t_{[\hat{\theta},c]}$ statistics used to test hypotheses about these parameters. The one-sided

TABLE 6.11 Confidence Intervals and $t_{[\hat{\theta},c]}$ Statistics Related to Testing $H_0\colon \theta = c$

Parameter	Point Estimate	Standard Error of the Estimate	$t_{[\hat{\theta},\theta]}$ Statistic	Degrees of Freedom	$100(1-\alpha)\%$ Confidence Intervals	$t_{[\hat{\theta},c]}$ Statistic
θ	$\hat{\theta}$	$s_{\hat{\theta}}$	$t_{[\hat{\theta},\theta]} = \dfrac{\hat{\theta}-\theta}{s_{\hat{\theta}}}$	df	$[\hat{\theta} \pm t_{[\alpha/2]}^{(df)} s_{\hat{\theta}}]$ $[\hat{\theta} - t_{[\alpha]}^{(df)} s_{\hat{\theta}}, \infty)$ $(-\infty, \hat{\theta} + t_{[\alpha]}^{(df)} s_{\hat{\theta}}]$	$t_{[\hat{\theta},c]} = \dfrac{\hat{\theta}-c}{s_{\hat{\theta}}}$
μ	$\bar{y}$	$s_{\bar{y}} = \dfrac{s}{\sqrt{n}}$	$t_{[\bar{y},\mu]} = \dfrac{\bar{y}-\mu}{s/\sqrt{n}}$	$n-1$	$\left[\bar{y} \pm t_{[\alpha/2]}^{(n-1)}\left(\dfrac{s}{\sqrt{n}}\right)\right]$ $\left[\bar{y} - t_{[\alpha]}^{(n-1)}\left(\dfrac{s}{\sqrt{n}}\right), \infty\right)$ $\left(-\infty, \bar{y} + t_{[\alpha]}^{(n-1)}\left(\dfrac{s}{\sqrt{n}}\right)\right]$	$t_{[\bar{y},c]} = \dfrac{\bar{y}-c}{s/\sqrt{n}}$
β_j	b_j	$s_{b_j} = s\sqrt{c_{jj}}$	$t_{[b_j,\beta_j]} = \dfrac{b_j-\beta_j}{s\sqrt{c_{jj}}}$	$n-k$	$[b_j \pm t_{[\alpha/2]}^{(n-k)} s\sqrt{c_{jj}}]$ $[b_j - t_{[\alpha]}^{(n-k)} s\sqrt{c_{jj}}, \infty)$ $(-\infty, b_j + t_{[\alpha]}^{(n-k)} s\sqrt{c_{jj}}]$	$t_{[b_j,c]} = \dfrac{b_j-c}{s\sqrt{c_{jj}}}$
μ_0	$\hat{y}_0$	$s_{\hat{y}_0} = s\sqrt{\mathbf{x}_0'(\mathbf{X}'\mathbf{X})^{-1}\mathbf{x}_0}$	$t_{[\hat{y}_0,\mu_0]} = \dfrac{\hat{y}_0-\mu_0}{s\sqrt{\mathbf{x}_0'(\mathbf{X}'\mathbf{X})^{-1}\mathbf{x}_0}}$	$n-k$	$[\hat{y}_0 \pm t_{[\alpha/2]}^{(n-k)} s\sqrt{\mathbf{x}_0'(\mathbf{X}'\mathbf{X})^{-1}\mathbf{x}_0}]$ $[\hat{y}_0 - t_{[\alpha]}^{(n-k)} s\sqrt{\mathbf{x}_0'(\mathbf{X}'\mathbf{X})^{-1}\mathbf{x}_0}, \infty)$ $(-\infty, \hat{y}_0 + t_{[\alpha]}^{(n-k)} s\sqrt{\mathbf{x}_0'(\mathbf{X}'\mathbf{X})^{-1}\mathbf{x}_0}]$	$t_{[\hat{y}_0,c]} = \dfrac{\hat{y}_0-c}{s\sqrt{\mathbf{x}_0'(\mathbf{X}'\mathbf{X})^{-1}\mathbf{x}_0}}$

confidence intervals are useful in testing one-sided alternative hypotheses, and the preceding two-sided confidence interval is useful in testing $H_0: \theta = c$ versus $H_1: \theta \neq c$, where c is a constant.

HYPOTHESIS TESTING RESULT I:

Define

$$t_{[\hat{\theta},\theta]} = \frac{\hat{\theta} - \theta}{s_{\hat{\theta}}}$$

and assume that the population of all possible $t_{[\hat{\theta},\theta]}$ statistics has a t-distribution with df degrees of freedom. In order to test $H_0: \theta = c$ versus $H_1: \theta \neq c$, define the $t_{[\hat{\theta},c]}$ statistic and its related prob-value to be

$$t_{[\hat{\theta},c]} = \frac{\hat{\theta} - c}{s_{\hat{\theta}}} \qquad \text{and} \qquad \text{prob-value} = 2A[[|t_{[\hat{\theta},c]}|, \infty); df]$$

where $A[[|t_{[\hat{\theta},c]}|, \infty); df]$ is the area under the curve of the t-distribution having df degrees of freedom to the right of $|t_{[\hat{\theta},c]}|$, the absolute value of the $t_{[\hat{\theta},c]}$ statistic. Then, we can reject $H_0: \theta = c$ in favor of $H_1: \theta \neq c$ *by setting the probability of a Type I error equal to* α if and only if any of the following three equivalent conditions hold:

1. $|t_{[\hat{\theta},c]}| > t_{[\alpha/2]}^{(df)}$—that is, if $t_{[\hat{\theta},c]} > t_{[\alpha/2]}^{(df)}$ or $t_{[\hat{\theta},c]} < -t_{[\alpha/2]}^{(df)}$.
2. Prob-value $< \alpha$.
3. The $100(1 - \alpha)\%$ confidence interval for θ, $[\hat{\theta} \pm t_{[\alpha/2]}^{(df)} s_{\hat{\theta}}]$, does not contain c.

Moreover,

4. For any level of confidence, $100(1 - \alpha)\%$, less than $100[1 - (\text{prob-value})]\%$, the $100(1 - \alpha)\%$ confidence interval for θ *does not contain* c, and thus we *can*, with at least $100(1 - \alpha)\%$ confidence, *reject* $H_0: \theta = c$ in favor of $H_1: \theta \neq c$.
5. For any level of confidence, $100(1 - \alpha)\%$, greater than or equal to $100[1 - (\text{prob-value})]\%$, the $100(1 - \alpha)\%$ confidence interval for θ *does contain* c, and thus we *cannot*, with at least $100(1 - \alpha)\%$ confidence, *reject* $H_0: \theta = c$ in favor of $H_1: \theta \neq c$.

Hypothesis Testing Result I says that in order to test $H_0: \theta = c$ versus $H_1: \theta \neq c$, we first select a $t_{[\hat{\theta},\theta]}$ statistic that involves the parameter θ and so that the population of all possible $t_{[\hat{\theta},\theta]}$ statistics has a t-distribution with df degrees of freedom. We now illustrate the use of condition (1) in Hypothesis Testing Result I.

EXAMPLE 6.17

The G & P Corporation produces a 16-ounce bottle of Frizzy Shampoo, filled by an automated bottle-filling process. If, for a particular adjustment of the bottle-filling process, this process is substantially overfilling bottles of Frizzy (which results in lost profits for G & P) or underfilling bottles of Frizzy (which is unfair to customers), then this process must be shut down and readjusted. For a given adjustment of the bottle-filling process, we consider the infinite population of all bottles of Frizzy that could potentially be produced by G & P Corporation. For each bottle of Frizzy, there is a corresponding bottle fill, which we define to be the amount of Frizzy Shampoo (in ounces) contained in the bottle. We let μ denote the mean of the infinite population of all the bottle fills that could potentially be produced by (the particular adjustment of) the bottle-filling process. G & P has decided that it will shut down and readjust the process if it can, by setting α, the probability of a Type I error, equal to .05, reject the null hypothesis $H_0: \mu = 16$ (the bottle-filling process is working properly) in favor of the alternative hypothesis $H_1: \mu \neq 16$ (the mean fill is above or below the desired 16 ounces). To interpret the meaning of α here, consider Table 6.12, which summarizes the Type I and Type II errors in this problem. It is clear that setting α equal to .05 implies that G & P's hypothesis testing procedure will cause the process to be readjusted 5 percent of the time when the process is working properly. We assume that G & P has decided to set α at .05 rather than .01 because the company feels that making a Type II error (not readjusting the process when the process is not working properly) is very serious, and setting α at .05 makes the probability of a Type II error smaller than setting α at .01.

In order to test $H_0: \mu = 16$ versus $H_1: \mu \neq 16$ (that is, $H_0: \theta = c$ versus $H_1: \theta \neq c$), we assume that G & P will randomly select a sample of $n = 6$ bottle fills from (the particular adjustment of) the Frizzy bottle-filling process. Then, assuming that the population of all possible bottle fills is normally distributed, and defining the $t_{[\hat{\theta},\theta]}$ statistic to be

$$t_{[\hat{\theta},\theta]} = \frac{\hat{\theta} - \theta}{s_{\hat{\theta}}} = \frac{\bar{y} - \mu}{s/\sqrt{n}} = t_{[\bar{y},\mu]}$$

we see from the discussion of Section 6.11 that the population of all possible $t_{[\hat{\theta},\theta]}$ statistics (that is, $t_{[\bar{y},\mu]}$ statistics) has a t-distribution with $df = n - 1 = 6 - 1 = 5$ degrees of freedom. Therefore, by condition (1) in Hypothesis Testing Result I, we can reject $H_0: \mu = 16$ $(\theta = c)$ in favor of $H_1: \mu \neq 16$ $(\theta \neq c)$ by setting α, the probability of a Type I error, equal to .05 if and only if the absolute value of

$$t_{[\hat{\theta},c]} = \frac{\hat{\theta} - c}{s_{\hat{\theta}}} = \frac{\bar{y} - 16}{s/\sqrt{6}} = t_{[\bar{y},16]}$$

is greater than $t_{[\alpha/2]}^{(df)} = t_{[\alpha/2]}^{(n-1)} = t_{[.05/2]}^{(6-1)} = t_{[.025]}^{(5)} = 2.571$—that is, if and only if $t_{[\hat{\theta},c]} = t_{[\bar{y},16]}$ is greater than $t_{[\alpha/2]}^{(df)} = 2.571$ or less than $-t_{[\alpha/2]}^{(df)} = -2.571$. To understand this

TABLE 6.12 Errors in Frizzy Bottle Fill Hypothesis Testing

Decisions Made in Hypothesis Test	State of Nature	
	Null hypothesis $H_0: \mu = 16$ is true: Process is king properly	Null Hypothesis $H_0: \mu = 16$ is false: Process is not working properly
Reject null hypothesis $H_0: \mu = 16$. Readjust the process	Type I error: Readjust the process when the process is working properly	Correct action: Readjust the process when the process is not working properly
Do not reject null hypothesis $H_0: \mu = 16$. Do not readjust the process	Correct action: Do not readjust the process when the process is working properly	Type II error: Do not readjust the process when the process is not working properly

condition, note that the $t_{[\bar{y},16]}$ statistic

$$t_{[\bar{y},16]} = \frac{\bar{y} - 16}{s/\sqrt{6}}$$

measures the distance between $\bar{y}$, the point estimate of μ, and 16 (the value that makes the null hypothesis $H_0: \mu = 16$ true). It follows that condition (1)—reject $H_0: \mu = 16$ in favor of $H_1: \mu \neq 16$ if the $t_{[\bar{y},16]}$ statistic is substantially greater than zero (that is, greater than $t^{(5)}_{[.025]} = 2.571$) or substantially less than zero (that is, less than $-t^{(5)}_{[.025]} = -2.571$)—is intuitively reasonable:

1. A $t_{[\bar{y},16]}$ statistic nearly or exactly equal to zero, which results when $\bar{y}$ (the point estimate of μ) is nearly or exactly equal to 16, supplies little or no evidence to support rejecting $H_0: \mu = 16$ in favor of $H_1: \mu \neq 16$, since, in this case, $\bar{y}$ indicates that μ is nearly or exactly equal to 16.
2. A $t_{[\bar{y},16]}$ statistic substantially greater than zero, which results when $\bar{y}$ is substantially greater than 16, provides evidence to support rejecting $H_0: \mu = 16$ in favor of $H_1: \mu \neq 16$, since, in this case, $\bar{y}$ indicates that μ is greater than 16.
3. A $t_{[\bar{y},16]}$ statistic substantially less than zero, which results when $\bar{y}$ is substantially smaller than 16, provides evidence to support rejecting $H_0: \mu = 16$ in favor of $H_1: \mu \neq 16$, since, in this case, $\bar{y}$ indicates that μ is smaller than 16.

Now, assume that when G & P randomly selects the sample of $n = 6$ bottle fills, it obtains the sample

$$SPL = \{y_1, y_2, y_3, y_4, y_5, y_6\}$$
$$= \{15.68, 16.00, 15.61, 15.93, 15.86, 15.72\}$$

which (it can be verified) has mean $\bar{y} = 15.8$ and standard deviation $s = .1532$. It follows that

$$t_{[\bar{y},16]} = \frac{\bar{y} - 16}{s/\sqrt{n}} = \frac{15.8 - 16}{.1532/\sqrt{6}} = -3.2$$

Since $t_{[\bar{y},16]} = -3.2 < -2.571 = -t^{(5)}_{[.025]}$ (that is, since $|t_{[\bar{y},16]}| = |-3.2| = 3.2 > 2.571 = t^{(5)}_{[.025]}$), we can reject $H_0: \mu = 16$ by setting α equal to .05.

We now prove that, in general, using condition (1)—reject $H_0: \theta = c$ in favor of $H_1: \theta \neq c$ if

$$t_{[\hat{\theta},c]} = \frac{\hat{\theta} - c}{s_{\hat{\theta}}}$$

is greater than $t^{(df)}_{[\alpha/2]}$ or less than $-t^{(df)}_{[\alpha/2]}$—guarantees that the probability of a Type I error equals α. Since we assume that the population of all possible $t_{[\hat{\theta},\theta]}$ statistics has a t-distribution with df degrees of freedom, it follows that if the null hypothesis $H_0: \theta = c$ is true, the population of all possible $t_{[\hat{\theta},c]}$ statistics has a t-distribution with df degrees of freedom, because, if $H_0: \theta = c$ is true, then

$$t_{[\hat{\theta},\theta]} = \frac{\hat{\theta} - \theta}{s_{\hat{\theta}}} = \frac{\hat{\theta} - c}{s_{\hat{\theta}}} = t_{[\hat{\theta},c]}$$

This fact tells us that if $t_{[\hat{\theta},c]}$ is the $t_{[\hat{\theta},c]}$ statistic that will be randomly selected from the population of all such statistics, then

1. $P(t_{[\hat{\theta},c]}$ will be greater than $t^{(df)}_{[\alpha/2]}$ when $H_0: \theta = c$ is true) is given by

   $$A[[t^{(df)}_{[\alpha/2]}, \infty); df]$$

   which is the area under the curve of the t-distribution having df degrees of freedom to the right of $t^{(df)}_{[\alpha/2]}$, and which (by the definition of $t^{(df)}_{[\alpha/2]}$) equals $\alpha/2$ (see Figure 6.14a).
2. $P(t_{[\hat{\theta},c]}$ will be less than $-t^{(df)}_{[\alpha/2]}$ when $H_0: \theta = c$ is true) is given by

   $$A[(-\infty, -t^{(df)}_{[\alpha/2]}]; df]$$

   which is the area under the curve of the t-distribution having df degrees of freedom to the left of $-t^{(df)}_{[\alpha/2]}$, and which (by the definition of $-t^{(df)}_{[\alpha/2]}$) equals $\alpha/2$ (see Figure 6.14a).

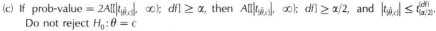

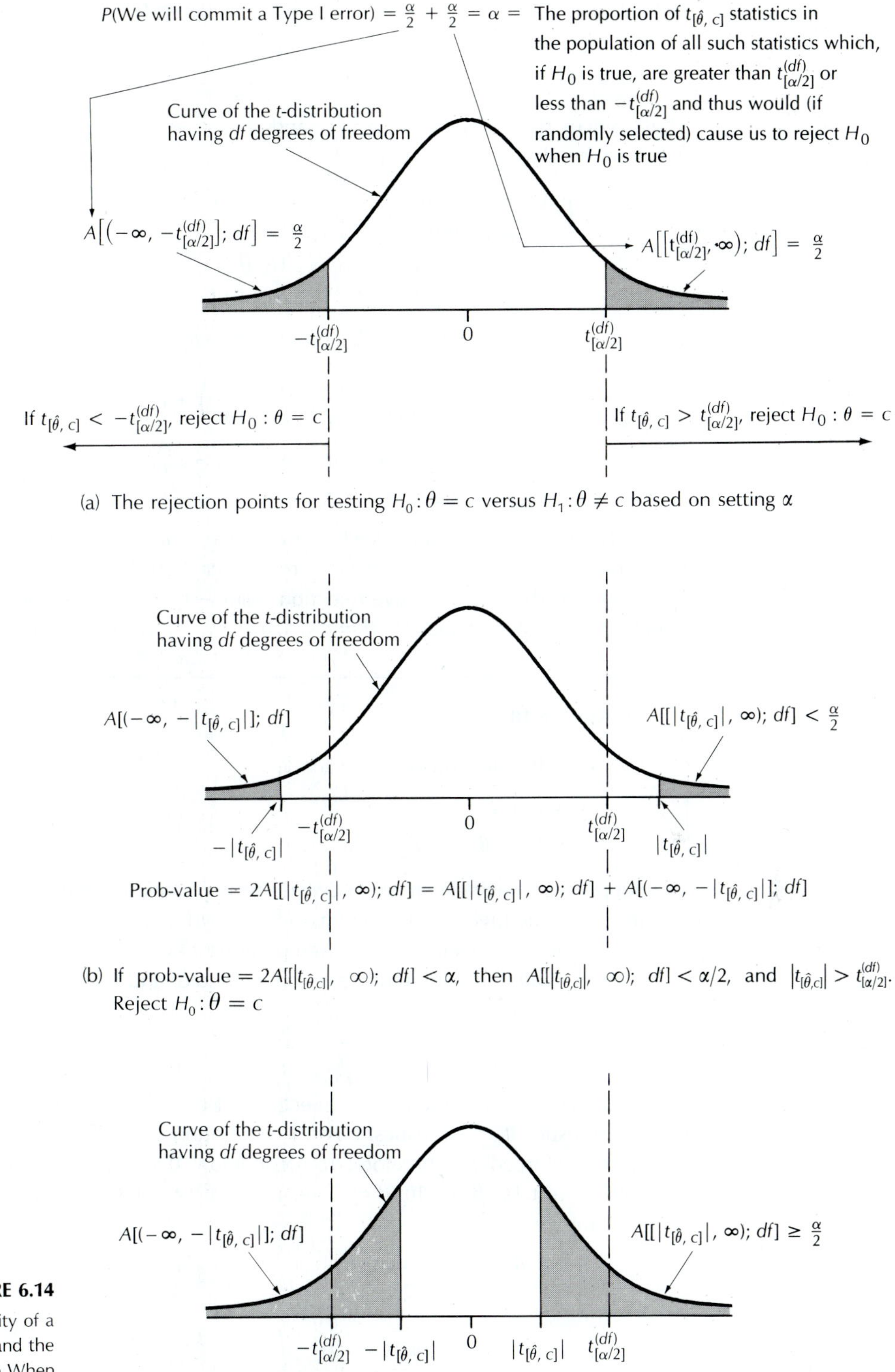

P(We will commit a Type I error) $= \frac{\alpha}{2} + \frac{\alpha}{2} = \alpha =$ The proportion of $t_{[\hat{\theta}, c]}$ statistics in the population of all such statistics which, if H_0 is true, are greater than $t^{(df)}_{[\alpha/2]}$ or less than $-t^{(df)}_{[\alpha/2]}$ and thus would (if randomly selected) cause us to reject H_0 when H_0 is true
Curve of the t-distribution having df degrees of freedom
$A[(-\infty, -t^{(df)}_{[\alpha/2]}]; df] = \frac{\alpha}{2}$
$A[[t^{(df)}_{[\alpha/2]}, \infty); df] = \frac{\alpha}{2}$
$-t^{(df)}_{[\alpha/2]}$
0
$t^{(df)}_{[\alpha/2]}$
If $t_{[\hat{\theta}, c]} < -t^{(df)}_{[\alpha/2]}$, reject $H_0 : \theta = c$
If $t_{[\hat{\theta}, c]} > t^{(df)}_{[\alpha/2]}$, reject $H_0 : \theta = c$
Curve of the t-distribution having df degrees of freedom
$A[(-\infty, -|t_{[\hat{\theta}, c]}|]; df]$
$A[[|t_{[\hat{\theta}, c]}|, \infty); df] < \frac{\alpha}{2}$
$-|t_{[\hat{\theta}, c]}|$
$-t^{(df)}_{[\alpha/2]}$
0
$t^{(df)}_{[\alpha/2]}$
$|t_{[\hat{\theta}, c]}|$
Prob-value $= 2A[[|t_{[\hat{\theta}, c]}|, \infty); df] = A[[|t_{[\hat{\theta}, c]}|, \infty); df] + A[(-\infty, -|t_{[\hat{\theta}, c]}|]; df]$
Curve of the t-distribution having df degrees of freedom
$A[(-\infty, -|t_{[\hat{\theta}, c]}|]; df]$
$A[[|t_{[\hat{\theta}, c]}|, \infty); df] \geq \frac{\alpha}{2}$
$-t^{(df)}_{[\alpha/2]}$
$-|t_{[\hat{\theta}, c]}|$
0
$|t_{[\hat{\theta}, c]}|$
$t^{(df)}_{[\alpha/2]}$

(a) The rejection points for testing $H_0 : \theta = c$ versus $H_1 : \theta \neq c$ based on setting α

(b) If prob-value $= 2A[[|t_{[\hat{\theta},c]}|, \infty); df] < \alpha$, then $A[[|t_{[\hat{\theta},c]}|, \infty); df] < \alpha/2$, and $|t_{[\hat{\theta},c]}| > t^{(df)}_{[\alpha/2]}$. Reject $H_0 : \theta = c$

(c) If prob-value $= 2A[[|t_{[\hat{\theta},c]}|, \infty); df] \geq \alpha$, then $A[[|t_{[\hat{\theta},c]}|, \infty); df] \geq \alpha/2$, and $|t_{[\hat{\theta},c]}| \leq t^{(df)}_{[\alpha/2]}$. Do not reject $H_0 : \theta = c$

FIGURE 6.14

The Probability of a Type I Error and the Prob-Value When Testing $H_0 : \theta = c$ Versus $H_1 : \theta \neq c$

Thus, since the events $t_{[\hat{\theta},c]} > t^{(df)}_{[\alpha/2]}$ and $t_{[\hat{\theta},c]} < -t^{(df)}_{[\alpha/2]}$ are mutually exclusive, it follows (see Figure 6.14a) that

$$\begin{aligned}
P&\begin{pmatrix}\text{We will commit}\\ \text{a Type I error}\end{pmatrix}\\
&= P\begin{pmatrix}\text{We will reject } H_0:\theta = c\\ \text{when } H_0:\theta = c \text{ is true}\end{pmatrix}\\
&= P(t_{[\hat{\theta},c]} > t^{(df)}_{[\alpha/2]} \quad \text{or} \quad t_{[\hat{\theta},c]} < -t^{(df)}_{[\alpha/2]} \quad \text{when } H_0:\theta = c \text{ is true})\\
&= P\begin{pmatrix}t_{[\hat{\theta},c]} > t^{(df)}_{[\alpha/2]}\\ \text{when } H_0:\theta = c \text{ is true}\end{pmatrix} + P\begin{pmatrix}t_{[\hat{\theta},c]} < -t^{(df)}_{[\alpha/2]}\\ \text{when } H_0:\theta = c \text{ is true}\end{pmatrix}\\
&= A[[t^{(df)}_{[\alpha/2]}, \infty); df] + A[(-\infty, -t^{(df)}_{[\alpha/2]}]; df]\\
&= \alpha/2 + \alpha/2 = \alpha
\end{aligned}$$

Thus, if $H_0:\theta = c$ is true, then $100\alpha\%$ (for example, 5% if $\alpha = .05$) of the $t_{[\hat{\theta},c]}$ statistics in the population of all such statistics are greater than the positive rejection point $t^{(df)}_{[\alpha/2]}$ or are less than the negative rejection point $-t^{(df)}_{[\alpha/2]}$ and thus would (if randomly selected) cause us to reject $H_0:\theta = c$ when $H_0:\theta = c$ is true (a Type I error).

EXAMPLE 6.18

Consider the fuel consumption model

$$\begin{aligned}
y &= \mu + \epsilon\\
&= \beta_0 + \beta_1 x_1 + \beta_2 x_2 + \epsilon
\end{aligned}$$

Suppose an engineer at State University hypothesizes that β_2, the change in mean weekly fuel consumption associated with a one-unit increase in the chill index when the average hourly temperature does not change, is equal to .10 (tons of coal). If the inference assumptions hold, and the $t_{[\hat{\theta},\theta]}$ statistic is

$$t_{[\hat{\theta},\theta]} = \frac{\hat{\theta} - \theta}{s_{\hat{\theta}}} = \frac{b_2 - \beta_2}{s\sqrt{c_{22}}} = t_{[b_2,\beta_2]}$$

it follows from the discussion of Section 6.4 that the population of all possible $t_{[\hat{\theta},\theta]}$ statistics ($t_{[b_2,\beta_2]}$ statistics) has a t-distribution with $df = n - k = 8 - 3 = 5$ degrees of freedom. Therefore, from condition (1) in Hypothesis Testing Result I, we can test $H_0:\beta_2 = .10$ ($\theta = c$) versus $H_1:\beta_2 \neq .10$ ($\theta \neq c$) by setting α equal to .05 if we calculate

$$\begin{aligned}
t_{[\hat{\theta},c]} &= \frac{\hat{\theta} - c}{s_{\hat{\theta}}} = \frac{b_2 - .10}{s\sqrt{c_{22}}} = t_{[b_2,.10]}\\
&= \frac{.0825 - .10}{.3671\sqrt{.0036}}\\
&= -.7945
\end{aligned}$$

and use the rejection points

$$t_{[\alpha/2]}^{(df)} = t_{[.05/2]}^{(n-k)} = t_{[.025]}^{(8-3)} = t_{[.025]}^{(5)} = 2.571$$
$$-t_{[\alpha/2]}^{(df)} = -2.571$$

Since $t_{[b_2,.10]} = -.7945$ is between the rejection points -2.571 and 2.571, we cannot reject $H_0: \beta_2 = .10$ in favor of $H_1: \beta_2 \neq .10$ by setting α equal to .05. Note, however, that our failure to reject $H_0: \beta_2 = .10$ does not mean that we believe this null hypothesis is literally true. Rather, it means that there is not sufficient statistical evidence (based on setting α equal to .05) to reject $H_0: \beta_2 = .10$. In fact, recall that (a supernatural power knows that) the true value of β_2 is .086 (tons of coal).

We next present an example of using condition (2) in Hypothesis Testing Result I.

EXAMPLE 6.19

Recall from Example 6.17 that the $t_{[\hat{\theta},c]}$ statistic for testing $H_0: \mu = 16$ versus $H_1: \mu \neq 16$ is $t_{[\hat{\theta},c]} = t_{[\bar{y},16]} = -3.2$. Therefore, condition (2) in Hypothesis Testing Result I says that we can reject $H_0: \mu = 16$ $(\theta = c)$ in favor of $H_1: \mu \neq 16$ $(\theta \neq c)$ by setting the probability of a Type I error equal to α if and only if

$$\begin{aligned} \text{prob-value} &= 2A[[|t_{[\hat{\theta},c]}|, \infty); df] \\ &= 2A[[|t_{[\bar{y},16]}|, \infty); n-1] \\ &= 2A[[|-3.2|, \infty); 6-1] \\ &= 2A[[3.2, \infty); 5] \\ &= 2(.01312) \\ &= .02624 \end{aligned}$$

is less than α. For example, since prob-value $= .02624$ is less than .05, not less than .02, and not less than .01, it follows by condition (2) that we can reject $H_0: \mu = 16$ in favor of $H_1: \mu \neq 16$ if we set α, the probability of a Type I error, equal to .05, but we cannot reject H_0 in favor of H_1 if we set α equal to .02 or .01.

Since we have previously shown that use of condition (1) implies that the probability of a Type I error equals α, we can prove that use of condition (2) also implies that the probability of a Type I error equals α by showing that conditions (1) and (2) are equivalent. To do this, we show that if condition (2) holds, then condition (1) holds, and if condition (2) does not hold, then condition (1) does not hold. If condition (2) holds, then

$$\text{prob-value} = 2A[[|t_{[\hat{\theta},c]}|, \infty); df]$$

is less than α, which implies that

$$A[[|t_{[\hat{\theta},c]}|, \infty); df] < \alpha/2$$

Looking at Figures 6.14a and 6.14b, we see that this fact implies that condition (1),

$$|t_{[\hat{\theta},c]}| > t_{[\alpha/2]}^{(df)}$$

holds. If condition (2) does not hold, then

$$\text{prob-value} = 2A[[|t_{[\hat{\theta},c]}|, \infty); df]$$

is greater than or equal to α, which implies that

$$A[[|t_{[\hat{\theta},c]}|, \infty); df] \geq \alpha/2$$

Looking at Figures 6.14a and 6.14c, we see that this fact implies that

$$|t_{[\hat{\theta},c]}| \leq t_{[\alpha/2]}^{(df)}$$

which means that condition (1) does not hold. Thus, we have shown that conditions (1) and (2) are equivalent.

We next present an example of interpreting the prob-value as a probability and of using (3), (4), and (5) in Hypothesis Testing Result I. (The proof of the validity of (3), (4), and (5) is analogous to the proof of the validity of these results in the box entitled "Testing $H_0: \beta_j = 0$ versus $H_1: \beta_j \neq 0$.")

EXAMPLE 6.20

Recall from Example 6.17 that the $t_{[\hat{\theta},c]}$ statistic for testing $H_0: \mu = 16$ versus $H_1: \mu \neq 16$ is $t_{[\hat{\theta},c]} = t_{[\bar{y},16]} = -3.2$. Now, as depicted in Figure 6.14b,

$A[[|t_{[\hat{\theta},c]}|, \infty); df]$ = the area under the curve of the t-distribution having df degrees of freedom to the right of $|t_{[\hat{\theta},c]}|$

is equal to

$A[(-\infty, -|t_{[\hat{\theta},c]}|]; df]$ = the area under the curve of the t-distribution having df degrees of freedom to the left of $-|t_{[\hat{\theta},c]}|$

It follows that

$$\begin{aligned}
\text{prob-value} &= 2A[[|t_{[\hat{\theta},c]}|, \infty); df] \\
&= A[[|t_{[\hat{\theta},c]}|, \infty); df] + A[(-\infty, -|t_{[\hat{\theta},c]}|]; df] \\
&= A[[|t_{[\bar{y},16]}|, \infty); n-1] + A[(-\infty, -|t_{[\bar{y},16]}|]; n-1] \\
&= A[[|-3.2|, \infty); 6-1] + A[(-\infty, -|-3.2|]; 6-1] \\
&= A[[3.2, \infty); 5] + A[(-\infty, -3.2]; 5]
\end{aligned}$$

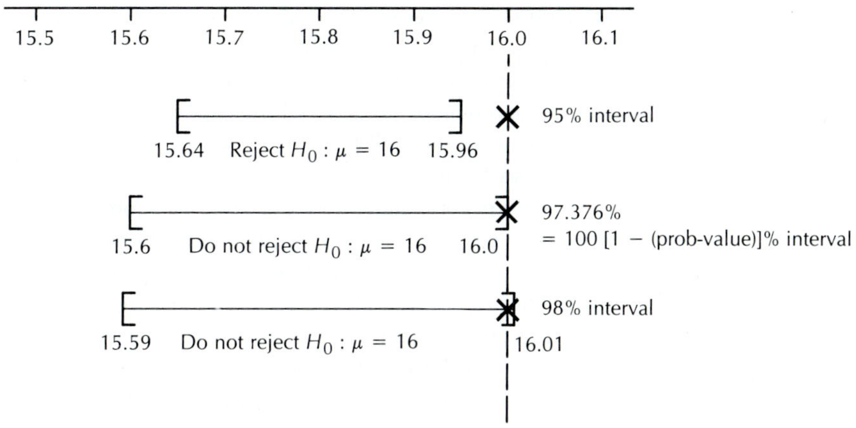

FIGURE 6.15
$H_0: \mu = 16$ Can Be Rejected at Any Level of Confidence Less Than 97.376%

Since this population of all possible $t_{[\bar{y},\mu]}$ statistics has a t-distribution with $n - 1 = 6 - 1 = 5$ degrees of freedom, it follows that if $H_0: \mu = 16$ is true, the population of all possible $t_{[\bar{y},16]}$ statistics has a t-distribution with $n - 1 = 5$ degrees of freedom. Therefore, this prob-value equals the proportion of $t_{[\bar{y},16]}$ statistics in the population of all such statistics which, if $H_0: \mu = 16$ is true, are at least as far away from zero and thus at least as contradictory to $H_0: \mu = 16$ as -3.2, the observed $t_{[\bar{y},16]}$ statistic. Since this prob-value, which was calculated in Example 6.19 to be .02624, says that G & P Corporation must believe that a 2.624 percent chance has occurred if it is to believe that $H_0: \mu = 16$ is true, the company has substantial evidence that $H_0: \mu = 16$ is false and thus that $H_1: \mu \neq 16$ is true.

Finally, we note that it follows from (4) in Hypothesis Testing Result I that for any level of confidence, $100(1 - \alpha)\%$, less than

$$100[1 - (\text{prob-value})]\% = 100[1 - .02624]\% = 97.376\%$$

the $100(1 - \alpha)\%$ confidence interval for μ

$$\left[\bar{y} \pm t_{[\alpha/2]}^{(n-1)}\left(\frac{s}{\sqrt{n}}\right)\right] = \left[15.8 \pm t_{[\alpha/2]}^{(6-1)}\left(\frac{.1532}{\sqrt{6}}\right)\right]$$
$$= [15.8 \pm t_{[\alpha/2]}^{(5)}(.0625)]$$

does not contain 16. (For example, the 95% interval

$$[15.8 \pm t_{[.025]}^{(5)}(.0625)] = [15.8 \pm 2.571(.0625)]$$
$$= [15.64, 15.96]$$

does not contain 16. See Figure 6.15.) Thus, for such a level of confidence G & P *can*, with at least $100(1 - \alpha)\%$ confidence, *reject* $H_0: \mu = 16$ in favor of $H_1: \mu \neq 16$. It follows from (5) in Hypothesis Testing Result I that for any level of confidence, $100(1 - \alpha)\%$, greater than or equal to

$$100[1 - (\text{prob-value})]\% = 97.376\%$$

the $100(1 - \alpha)\%$ confidence interval for μ *does contain 16*. (For example, the 97.376% and 98% intervals contain 16. See Figure 6.15.) Thus, for such a level of confidence G & P *cannot*, with at least $100(1 - \alpha)\%$ confidence, *reject* H_0: $\mu = 16$ in favor of $H_1 : \mu \neq 16$. To summarize, G & P can be up to 97.376% confident that the null hypothesis $H_0 : \mu = 16$ is false and that the alternative hypothesis $H_1 : \mu \neq 16$ is true.

Before concluding this example, we comment briefly about the difference between statistical significance and practical significance. Since we have seen that G & P can reject $H_0 : \mu = 16$ in favor of $H_1 : \mu \neq 16$ with at least 95% confidence, we say that G & P's hypothesis test has statistical significance at the 95% confidence level. Whether this result has practical significance, however, is somewhat questionable. Looking at the 95% confidence interval for μ, [15.64, 15.96], we see that G & P is 95% confident that μ, the mean bottle fill, is between 15.64 ounces and 15.96 ounces. So, in this case, while the mean bottle fill is probably somewhat below the desired 16 ounces, the customers are not being seriously cheated. However, since the company wishes to maintain good customer relations, although the practical significance of G & P's result here is somewhat questionable, the company will nevertheless readjust the bottling process in accordance with a policy of readjusting the process if it can reject $H_0 : \mu = 16$ in favor of $H_1 : \mu \neq 16$ with at least 95% confidence (that is, by setting α, the probability of a Type I error, equal to .05).

In the following box we consider testing $H_0 : \theta = c$ versus $H_1 : \theta > c$ (which, for the situations studied in this book, is equivalent to testing $H_0 : \theta \leq c$ versus $H_1 : \theta > c$).

HYPOTHESIS TESTING RESULT II:

Define

$$t_{[\hat{\theta},\theta]} = \frac{\hat{\theta} - \theta}{s_{\hat{\theta}}}$$

and assume that the population of all possible $t_{[\hat{\theta},\theta]}$ statistics has a *t*-distribution with *df* degrees of freedom. In order to test $H_0 : \theta = c$ versus $H_1 : \theta > c$, define the $t_{[\hat{\theta},c]}$ statistic and its related prob-value to be

$$t_{[\hat{\theta},c]} = \frac{\hat{\theta} - c}{s_{\hat{\theta}}} \quad \text{and} \quad \text{prob-value} = A[[t_{[\hat{\theta},c]}, \infty); df]$$

where $A[[t_{[\hat{\theta},c]}, \infty); df]$ is the area under the curve of the t-distribution having df degrees of freedom to the right of $t_{[\hat{\theta},c]}$. Then, we can reject $H_0 : \theta = c$ in favor of $H_1 : \theta > c$ *by setting the probability of a Type I error equal to* α if and only if any of the following three equivalent conditions hold:

1. $t_{[\hat{\theta},c]} > t_{[\alpha]}^{(df)}$.
2. Prob-value $< \alpha$.
3. The one-sided $100(1 - \alpha)\%$ confidence interval for θ, $[\hat{\theta} - t_{[\alpha]}^{(df)} s_{\hat{\theta}}, \infty)$, does not contain c.

Moreover,

4. For any level of confidence, $100(1 - \alpha)\%$, less than $100[1 - (\text{prob-value})]\%$, the one-sided $100(1 - \alpha)\%$ confidence interval for θ *does not contain* c, and thus we *can*, with at least $100(1 - \alpha)\%$ confidence, *reject* $H_0 : \theta = c$ in favor of $H_1 : \theta > c$.
5. For any level of confidence, $100(1 - \alpha)\%$, greater than or equal to $100[1 - (\text{prob-value})]\%$, the one-sided $100(1 - \alpha)\%$ confidence interval for θ *does contain* c, and thus we *cannot*, with at least $100(1 - \alpha)\%$ confidence, *reject* $H_0 : \theta = c$ in favor of $H_1 : \theta > c$.

We now illustrate the use of condition (1) in Hypothesis Testing Result II.

EXAMPLE 6.21

In the Gas-Mizer problem recall that federal gasoline mileage standards state that μ, the mean of all Gas-Mizer mileages, must be at least 30 mpg if National Motors wishes to avoid paying a fine. Here, we might be tempted to say that National Motors can "prove" that $\mu \geq 30$ if it can accept the null hypothesis $H_0 : \mu \geq 30$ instead of the alternative hypothesis $H_1 : \mu < 30$. However, *since hypothesis testing seeks to find how confident we can be that the null hypothesis should be rejected in favor of the alternative hypothesis*, and not how confident we can be that the null hypothesis should be accepted, we cannot use hypothesis testing to "prove" that a null hypothesis is true. Thus, we cannot make the statement that National Motors wishes to prove—that $\mu \geq 30$—the null hypothesis. It might, therefore, be tempting to make $\mu \geq 30$ the alternative hypothesis. But we cannot do this either, because in hypothesis testing the alternative hypothesis always involves a strict inequality ($>$ or $<$), and thus the null hypothesis involves an equality ($=$) or a nonstrict inequality ($\geq$ or $\leq$). The only way out of this predicament is to use the following procedure:

1. State what you wish to prove, so that this statement is written as a *strict inequality* ($<$, $>$, or $\neq$), and make it the *alternative hypothesis* H_1.
2. State what is reasonably possible if the alternative hypothesis is false, and make this statement the *null hypothesis* H_0.

National Motors can use this procedure by rewriting what it wishes to prove ($\mu \geq 30$) as the inequality $\mu > 30$ and making it the alternative hypothesis, and by making $\mu \leq 30$ (which is reasonably possible if $\mu > 30$ is false) the null hypothesis.

Suppose, then, that the government has decided to conclude that the Gas-Mizer meets the federal mileage standard if National Motors can, by setting α, the probability of a Type I error, equal to .05, reject $H_0: \mu \leq 30$ in favor of $H_1: \mu > 30$, which says that the Gas-Mizer exceeds the federal mileage standard. In order to test H_0 versus H_1, we assume that National Motors will randomly select a sample of $n = 5$ Gas-Mizer mileages from the population of all such mileages and that the population of all Gas-Mizer mileages is normally distributed. We define the $t_{[\hat{\theta},\theta]}$ statistic to be

$$t_{[\hat{\theta},\theta]} = \frac{\hat{\theta} - \theta}{s_{\hat{\theta}}} = \frac{\bar{y} - \mu}{s/\sqrt{n}} = t_{[\bar{y},\mu]}$$

Then, from the discussion of Section 6.11, the population of all possible $t_{[\hat{\theta},\theta]}$ statistics (that is, $t_{[\bar{y},\mu]}$ statistics) has a t-distribution with $df = n - 1 = 5 - 1 = 4$ degrees of freedom. Testing $H_0: \mu \leq 30$ versus $H_1: \mu > 30$ is equivalent to testing $H_0: \mu = 30$ versus $H_1: \mu > 30$. It follows from condition (1) in Hypothesis Testing Result II that we can reject $H_0: \mu = 30$ ($\theta = c$) in favor of $H_1: \mu > 30$ ($\theta > c$) by setting α equal to .05 if and only if

$$t_{[\hat{\theta},c]} = \frac{\hat{\theta} - c}{s_{\hat{\theta}}} = \frac{\bar{y} - 30}{s/\sqrt{n}} = t_{[\bar{y},30]}$$

is greater than

$$t_{[\alpha]}^{(df)} = t_{[.05]}^{(n-1)} = t_{[.05]}^{(5-1)} = t_{[.05]}^{(4)} = 2.132$$

Use of condition (1) is intuitively reasonable, because it says that a $t_{[\bar{y},30]}$ statistic substantially greater than zero, which results when $\bar{y}$, the point estimate of μ, is substantially greater than 30, provides substantial evidence to support rejecting $H_0: \mu = 30$ in favor of $H_1: \mu > 30$.

Now, from Example 6.15, $\bar{y} = 31.2$ is the mean and $s = .7517$ is the standard deviation of the sample of $n = 5$ Gas-Mizer mileages that National Motors has randomly selected from the population of all such mileages. It follows that

$$t_{[\bar{y},30]} = \frac{\bar{y} - 30}{s/\sqrt{n}} = \frac{31.2 - 30}{.7517/\sqrt{5}} = 3.569$$

Since $t_{[\bar{y},30]} = 3.569 > 2.132 = t^{(4)}_{[.05]}$, we can reject $H_0: \mu = 30$ in favor of $H_1: \mu > 30$ by setting α equal to .05.

We now prove that, in general, using condition (1)—reject $H_0: \theta = c$ in favor of $H_1: \theta > c$ if

$$t_{[\hat{\theta},c]} = \frac{\hat{\theta} - c}{s_{\hat{\theta}}}$$

is greater than $t^{(df)}_{[\alpha]}$—guarantees that the probability of a Type I error equals α. Since we assume that the population of all possible $t_{[\hat{\theta},\theta]}$ statistics has a t-distribution with df degrees of freedom, it follows that, if the null hypothesis $H_0: \theta = c$ is true, the population of all possible $t_{[\hat{\theta},c]}$ statistics has a t-distribution with df degrees of freedom. This fact tells us (see Figure 6.16a) that

$$\begin{aligned} P\begin{pmatrix}\text{We will commit}\\ \text{a Type I error}\end{pmatrix} &= P\begin{pmatrix}\text{We will reject } H_0: \theta = c\\ \text{when } H_0: \theta = c \text{ is true}\end{pmatrix} \\ &= P(t_{[\hat{\theta},c]} > t^{(df)}_{[\alpha]} \quad \text{when } H_0: \theta = c \text{ is true}) \\ &= A[[t^{(df)}_{[\alpha]}, \infty); df] \\ &= \alpha \end{aligned}$$

This probability statement says that if $H_0: \theta = c$ is true, then 100α% of the $t_{[\hat{\theta},c]}$ statistics in the population of all such statistics are greater than the rejection point $t^{(df)}_{[\alpha]}$ and thus would (if randomly selected) cause us to reject $H_0: \theta = c$ when $H_0: \theta = c$ is true (a Type I error).

We next present an example of using condition (2) in Hypothesis Testing Result II.

EXAMPLE 6.22

Recall from Example 6.21 that the $t_{[\hat{\theta},c]}$ statistic for testing $H_0: \mu = 30$ versus $H_1: \mu > 30$ is $t_{[\hat{\theta},c]} = t_{[\bar{y},30]} = 3.569$. Therefore, condition (2) in Hypothesis Testing Result II says that we can reject $H_0: \mu = 30$ ($\theta = c$) in favor of $H_1: \mu > 30$ ($\theta > c$) by setting the probability of a Type I error equal to α if and only if

$$\begin{aligned} \text{prob-value} &= A[[t_{[\hat{\theta},c]}, \infty); df] \\ &= A[[t_{[\bar{y},30]}, \infty); n - 1] \\ &= A[[3.569, \infty); 5 - 1 = 4] \\ &= .01275 \end{aligned}$$

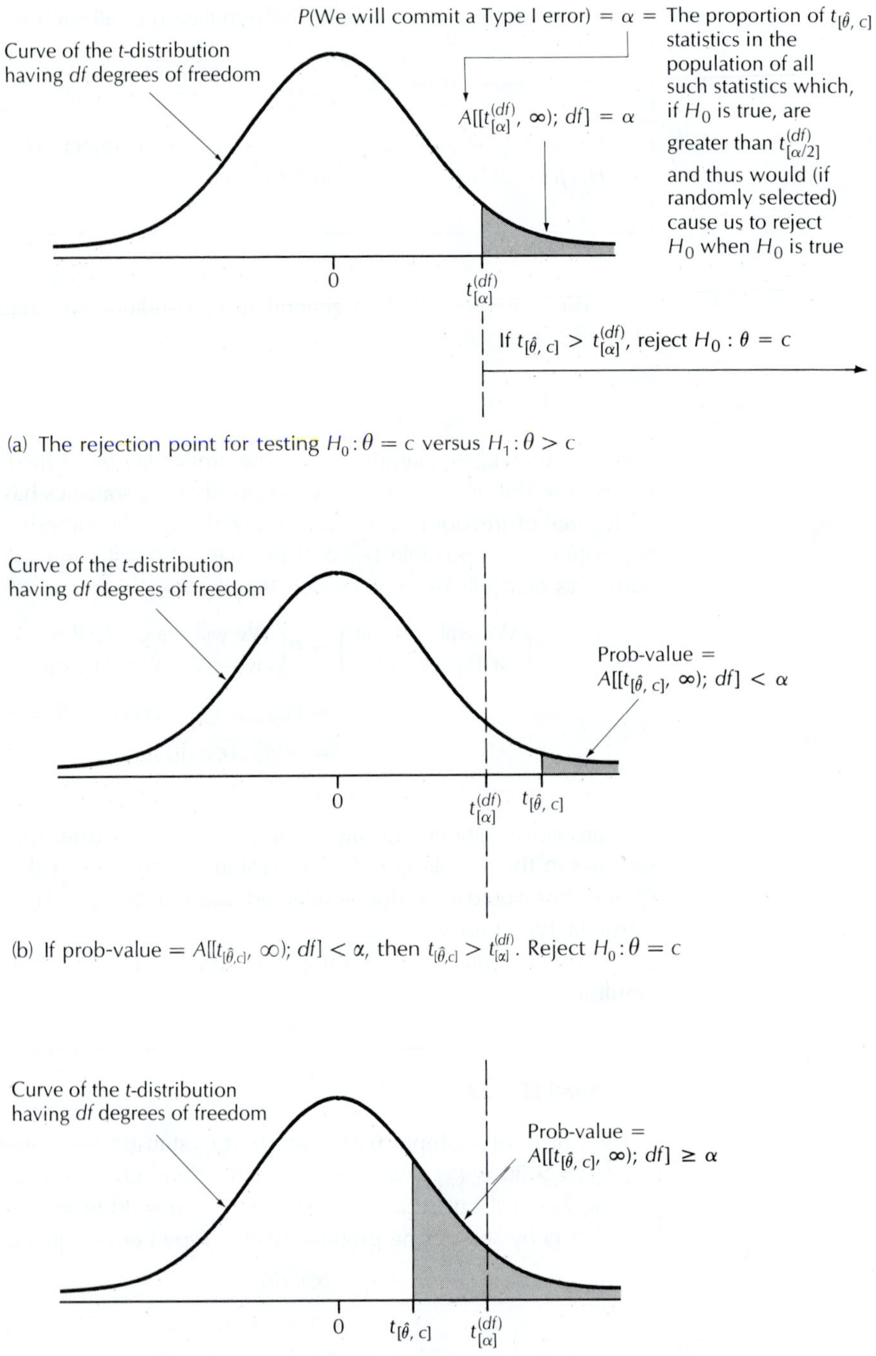
P(We will commit a Type I error) = α = The proportion of $t_{[\hat{\theta}, c]}$ statistics in the population of all such statistics which, if H_0 is true, are greater than $t_{[\alpha/2]}^{(df)}$ and thus would (if randomly selected) cause us to reject H_0 when H_0 is true
Curve of the t-distribution having df degrees of freedom
$A[[t_{[\alpha]}^{(df)}, \infty); df] = \alpha$
0
$t_{[\alpha]}^{(df)}$
If $t_{[\hat{\theta}, c]} > t_{[\alpha]}^{(df)}$, reject $H_0 : \theta = c$
Curve of the t-distribution having df degrees of freedom
Prob-value = $A[[t_{[\hat{\theta}, c]}, \infty); df] < \alpha$
0
$t_{[\alpha]}^{(df)}$
$t_{[\hat{\theta}, c]}$
Curve of the t-distribution having df degrees of freedom
Prob-value = $A[[t_{[\hat{\theta}, c]}, \infty); df] \geq \alpha$
0
$t_{[\hat{\theta}, c]}$
$t_{[\alpha]}^{(df)}$

(a) The rejection point for testing $H_0 : \theta = c$ versus $H_1 : \theta > c$

(b) If prob-value = $A[[t_{[\hat{\theta},c]}, \infty); df] < \alpha$, then $t_{[\hat{\theta},c]} > t_{[\alpha]}^{(df)}$. Reject $H_0 : \theta = c$

(c) If prob-value = $A[[t_{[\hat{\theta},c]}, \infty); df] \geq \alpha$, then $t_{[\hat{\theta},c]} \leq t_{[\alpha]}^{(df)}$. Do not reject $H_0 : \theta = c$

FIGURE 6.16
The Probability of a Type I Error and the Prob-Value When Testing $H_0 : \theta = c$ Versus $H_1 : \theta > c$

is less than α. Since this prob-value is less than .05, less than .02, and not less than .01, it follows by condition (2) that we can reject $H_0: \mu = 30$ in favor of $H_1: \mu > 30$ if we set α equal to .05 or .02, but we cannot reject H_0 in favor of H_1 if we set α equal to .01.

Since we have previously shown that use of condition (1) implies that the probability of a Type I error equals α, we can prove that the use of condition (2) also implies that the probability of a Type I error equals α by showing that conditions (1) and (2) are equivalent. If condition (2) holds, then

$$\text{prob-value} = A[[t_{[\hat{\theta},c]}, \infty); df]$$

is less than α. Looking at Figures 6.16a and 6.16b, we see that this fact implies that condition (1),

$$t_{[\hat{\theta},c]} > t_{[\alpha]}^{(df)}$$

holds. If condition (2) does not hold, then

$$\text{prob-value} = A[[t_{[\hat{\theta},c]}, \infty); df]$$

is greater than or equal to α. Looking at Figures 6.16a and 6.16c, we see that this fact implies that

$$t_{[\hat{\theta},c]} \leq t_{[\alpha]}^{(df)}$$

which means that condition (1) does not hold. Thus, we have shown that conditions (1) and (2) are equivalent.

We next present an example of interpreting the prob-value as a probability and of using (3), (4), and (5) in Hypothesis Testing Result II.

EXAMPLE 6.23

Consider the Gas-Mizer problem and the prob-value calculated in Example 6.22. Since the population of all possible $t_{[\bar{y},\mu]}$ statistics has a t-distribution with $n - 1 = 5 - 1 = 4$ degrees of freedom, it follows that, if $H_0: \mu = 30$ is true, the population of all possible $t_{[\bar{y},30]}$ statistics has a t-distribution with $n - 1 = 4$ degrees of freedom. Therefore,

$$\begin{aligned} \text{prob-value} &= A[[t_{[\bar{y},30]}, \infty); n - 1] \\ &= A[[3.569, \infty); 4] \\ &= .01275 \\ &\approx .013 \end{aligned}$$

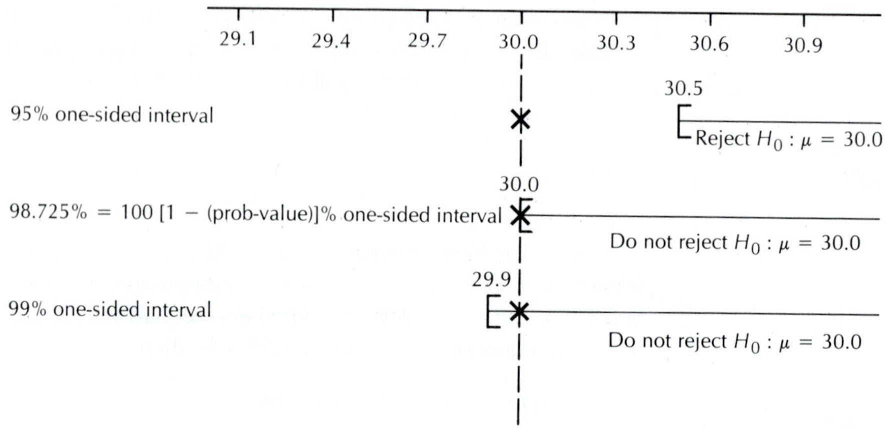

FIGURE 6.17
$H_0: \mu = 30$ Can Be Rejected in Favor of $H_1: \mu > 30$ at Any Level of Confidence Less Than 98.725%

equals the proportion of $t_{[\bar{y},30]}$ statistics in the population of all such statistics which, if the null hypothesis $H_0: \mu = 30$ is true, are at least as large as and thus at least as contradictory to $H_0: \mu = 30$ as 3.569, the observed $t_{[\bar{y},30]}$ statistic. Since this prob-value, which equals .013, says that National Motors must believe that a 13 in 1,000 chance has occurred if it is to believe that $H_0: \mu = 30$ is true, the company has substantial evidence to conclude that $H_0: \mu = 30$ is false and thus that $H_1: \mu > 30$ is true.

Finally, we note that it follows from (4) in Hypothesis Testing Result II that for any level of confidence, $100(1 - \alpha)\%$, less than

$$100[1 - (\text{prob-value})]\% = 100(1 - .01275)\% = 98.725\%$$

the one-sided $100(1 - \alpha)\%$ confidence interval for μ,

$$\left[\bar{y} - t_{[\alpha]}^{(n-1)}\left(\frac{s}{\sqrt{n}}\right), \infty\right) = \left[31.2 - t_{[\alpha]}^{(5-1)}\left(\frac{.7517}{\sqrt{5}}\right), \infty\right)$$

$$= [31.2 - t_{[\alpha]}^{(4)}(.3362), \infty)$$

does not contain 30. (For example, the 95% interval

$$[31.2 - t_{[.05]}^{(4)}(.3362), \infty) = [31.2 - 2.132(.3362), \infty)$$
$$= [31.2 - .72, \infty)$$
$$= [30.5, \infty)$$

does not contain 30. See Figure 6.17.) Thus, for such a level of confidence National Motors *can*, with at least $100(1 - \alpha)\%$ confidence, *reject* $H_0: \mu = 30$ in favor of $H_1: \mu > 30$. It follows from (5) in Hypothesis Testing Result II that for any level of confidence, $100(1 - \alpha)\%$, greater than or equal to

$$100[1 - (\text{prob-value})]\% = 98.725\%$$

the one-sided $100(1-\alpha)\%$ confidence interval for μ *does contain 30.* (For example, the 98.725% interval, which can be calculated to be $[30, \infty)$, does contain 30. See Figure 6.17.) Thus, for such a level of confidence National Motors *cannot,* with at least $100(1-\alpha)\%$ confidence, *reject* $H_0: \mu = 30$ in favor of $H_1: \mu > 30$. To summarize, National Motors can be up to 98.725% confident that $H_0: \mu = 30$ is false and that $H_1: \mu > 30$ is true. Specifically, the one-sided 95% confidence interval for μ, $[30.5, \infty)$, says that National Motors can be 95% confident that μ, the mean Gas-Mizer mileage, is at least 30.5 mpg.

We next consider testing $H_0: \theta = c$ versus $H_1: \theta < c$ (which, for the situations studied in this book, is equivalent to testing $H_0: \theta \geq c$ versus $H_1: \theta < c$).

HYPOTHESIS TESTING RESULT III:

Define

$$t_{[\hat{\theta},\theta]} = \frac{\hat{\theta} - \theta}{s_{\hat{\theta}}}$$

and assume that the population of all possible $t_{[\hat{\theta},\theta]}$ statistics has a *t*-distribution with *df* degrees of freedom. In order to test $H_0: \theta = c$ versus $H_1: \theta < c$, define the $t_{[\hat{\theta},c]}$ statistic and its related prob-value to be

$$t_{[\hat{\theta},c]} = \frac{\hat{\theta} - c}{s_{\hat{\theta}}} \qquad \text{and} \qquad \text{prob-value} = A[(-\infty, t_{[\hat{\theta},c]}];\, df]$$

where $A[(-\infty, t_{[\hat{\theta},c]}];\, df]$ is the area under the curve of the *t*-distribution having *df* degrees of freedom to the left of $t_{[\hat{\theta},c]}$. Then, we can reject $H_0: \theta = c$ in favor of $H_1: \theta < c$ *by setting the probability of a Type I error equal to* α if and only if any of the following three equivalent conditions hold:

1. $t_{[\hat{\theta},c]} < -t_{[\alpha]}^{(df)}$.
2. Prob-value $< \alpha$.
3. The one-sided $100(1-\alpha)\%$ confidence interval for θ, $(-\infty, \hat{\theta} + t_{[\alpha]}^{(df)} s_{\hat{\theta}}]$, does not contain c.

Moreover,

4. For any level of confidence, $100(1-\alpha)\%$, less than $100[1 - (\text{prob-value})]\%$, the one-sided $100(1-\alpha)\%$ confidence interval for θ *does not contain c,* and thus we *can,* with at least $100(1-\alpha)\%$ confidence, *reject* $H_0: \theta = c$ in favor of $H_1: \theta < c$.

5. For any level of confidence, $100(1-\alpha)\%$, greater than or equal to $100[1-(\text{prob-value})]\%$, the one-sided $100(1-\alpha)\%$ confidence interval for θ *does contain* c, and thus we *cannot*, with at least $100(1-\alpha)\%$ confidence, *reject* $H_0: \theta = c$ in favor of $H_1: \theta < c$.

We now illustrate using condition (1) in Hypothesis Testing Result III.

EXAMPLE 6.24

Consider the fuel consumption regression model

$$\begin{aligned} y &= \mu + \epsilon \\ &= \beta_0 + \beta_1 x_1 + \beta_2 x_2 + \epsilon \end{aligned}$$

We have previously reasoned that if β_1 does not equal zero, then β_1 must be a negative number, because a negative number would express the fact that the larger x_1, the average hourly temperature, is, the smaller is the mean weekly fuel consumption. Suppose we have decided to conclude that x_1 is significantly related to y if we can, by setting α, the probability of a Type I error, equal to .05, reject $H_0: \beta_1 = 0$ in favor of $H_1: \beta_1 < 0$. Assuming that the inference assumptions hold, and defining the $t_{[\hat{\theta},\theta]}$ statistic to be

$$t_{[\hat{\theta},\theta]} = \frac{\hat{\theta} - \theta}{s_{\hat{\theta}}} = \frac{b_1 - \beta_1}{s\sqrt{c_{11}}} = t_{[b_1,\beta_1]}$$

we find, from the discussion of Section 6.4, that the population of all possible $t_{[\hat{\theta},\theta]}$ statistics (that is, $t_{[b_1,\beta_1]}$ statistics) has a t-distribution with $df = n - k = 8 - 3 = 5$ degrees of freedom. Therefore, it follows from condition (1) in Hypothesis Testing Result III that we can reject $H_0: \beta_1 = 0$ $(\theta = c)$ in favor of $H_1: \beta_1 < 0$ $(\theta < c)$ by setting α equal to .05 if and only if

$$t_{[\hat{\theta},c]} = \frac{\hat{\theta} - c}{s_{\hat{\theta}}} = \frac{b_1 - 0}{s\sqrt{c_{11}}} = \frac{b_1}{s\sqrt{c_{11}}} = t_{b_1}$$

is less than $-t_{[\alpha]}^{(df)} = -t_{[.05]}^{(n-k)} = -t_{[.05]}^{(8-3)} = -t_{[.05]}^{(5)} = -2.015$. Using condition (1) is intuitively reasonable, because it says that a t_{b_1} statistic substantially less than zero, which results when b_1, the point estimate of β_1, is substantially smaller than zero, provides evidence to support rejecting $H_0: \beta_1 = 0$ in favor of $H_1: \beta_1 < 0$.

Now, recall that in Table 6.5 we calculated the t_{b_1} statistic to be

$$t_{b_1} = \frac{b_1}{s\sqrt{c_{11}}} = \frac{-.0900}{.3671\sqrt{.00147}} = -6.39$$

Since $t_{b_1} = -6.39 < -2.015 = -t_{[.05]}^{(5)}$, we can reject $H_0: \beta_1 = 0$ in favor of $H_1: \beta_1 < 0$ by setting α equal to .05.

We now prove that, in general, using condition (1)—reject $H_0: \theta = c$ in favor of $H_1: \theta < c$ if

$$t_{[\hat{\theta},c]} = \frac{\hat{\theta} - c}{s_{\hat{\theta}}}$$

is less than $-t_{[\alpha]}^{(df)}$—guarantees that the probability of a Type I error equals α. Since we assume that the population of all possible $t_{[\hat{\theta},\theta]}$ statistics has a t-distribution with df degrees of freedom, it follows that, if the null hypothesis $H_0: \theta = c$ is true, the population of all possible $t_{[\hat{\theta},c]}$ statistics has a t-distribution with df degrees of freedom. This fact tells us (see Figure 6.18a) that

$$\begin{aligned} P\begin{pmatrix}\text{We will commit}\\ \text{a Type I error}\end{pmatrix} &= P\begin{pmatrix}\text{We will reject } H_0: \theta = c\\ \text{when } H_0: \theta = c \text{ is true}\end{pmatrix} \\ &= P(t_{[\hat{\theta},c]} < -t_{[\alpha]}^{(df)} \quad \text{when } H_0: \theta = c \text{ is true}) \\ &= A[(-\infty, -t_{[\alpha]}^{(df)}]; df] \\ &= \alpha \end{aligned}$$

This probability statement says that if $H_0: \theta = c$ is true, then $100\alpha\%$ of the $t_{[\hat{\theta},c]}$ statistics in the population of all such statistics are less than the rejection point $-t_{[\alpha]}^{(df)}$ and thus would (if randomly selected) cause us to reject $H_0: \theta = c$ when $H_0: \theta = c$ is true (a Type I error).

We next present an example of using condition (2) in Hypothesis Testing Result III.

EXAMPLE 6.25

Recall from Example 6.24 that the $t_{[\hat{\theta},c]}$ statistic for testing $H_0: \beta_1 = 0$ versus $H_1: \beta_1 < 0$ is $t_{[\hat{\theta},c]} = t_{b_1} = -6.39$. Therefore, condition (2) in Hypothesis Testing Result III says that we can reject $H_0: \beta_1 = 0$ ($\theta = c$) in favor of $H_1: \beta_1 < 0$ ($\theta < c$) by setting the probability of a Type I error equal to α if and only if

$$\begin{aligned} \text{prob-value} &= A[(-\infty, t_{[\hat{\theta},c]}]; df] \\ &= A[(-\infty, t_{b_1}]; n - k] \\ &= A[(-\infty, -6.39]; 8 - 3 = 5] \\ &= .000695 \end{aligned}$$

is less than α. Since this prob-value is less than .05 and .01, it follows that we can reject H_0 in favor of H_1 if we set α equal to .05 or .01.

Since we have previously shown that use of condition (1) implies that the probability of a Type I error equals α, we can prove that the use of condition (2) also implies that the probability of a Type I error equals α by showing that conditions (1)

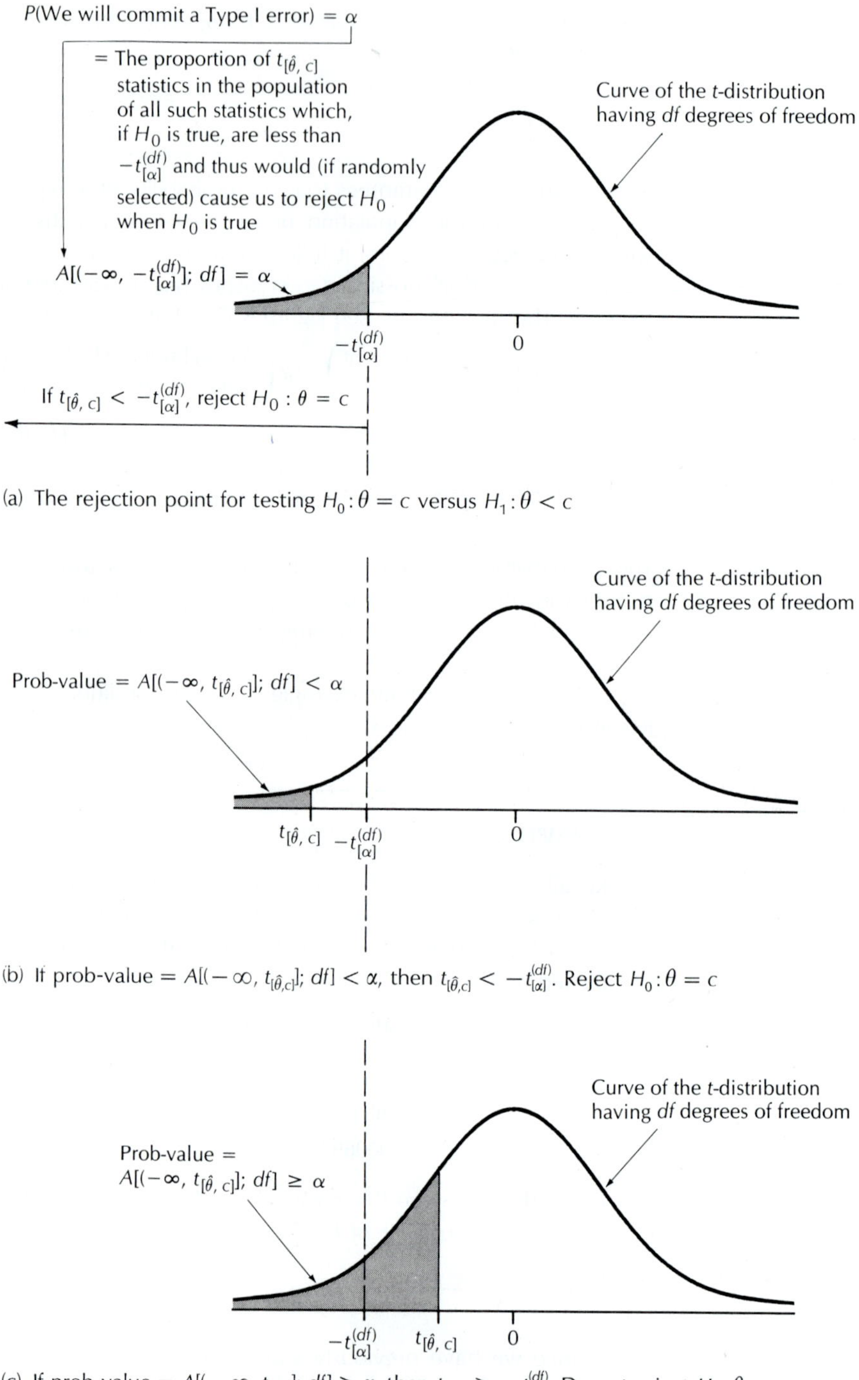
P(We will commit a Type I error) = α
= The proportion of $t_{[\hat{\theta}, c]}$ statistics in the population of all such statistics which, if H_0 is true, are less than $-t_{[\alpha]}^{(df)}$ and thus would (if randomly selected) cause us to reject H_0 when H_0 is true
Curve of the t-distribution having df degrees of freedom
$A[(-\infty, -t_{[\alpha]}^{(df)}]; df] = \alpha$
$-t_{[\alpha]}^{(df)}$
0
If $t_{[\hat{\theta}, c]} < -t_{[\alpha]}^{(df)}$, reject $H_0 : \theta = c$
Curve of the t-distribution having df degrees of freedom
Prob-value = $A[(-\infty, t_{[\hat{\theta}, c]}]; df] < \alpha$
$t_{[\hat{\theta}, c]}$
$-t_{[\alpha]}^{(df)}$
0
Curve of the t-distribution having df degrees of freedom
Prob-value = $A[(-\infty, t_{[\hat{\theta}, c]}]; df] \geq \alpha$
$-t_{[\alpha]}^{(df)}$
$t_{[\hat{\theta}, c]}$
0

(a) The rejection point for testing $H_0 : \theta = c$ versus $H_1 : \theta < c$

(b) If prob-value $= A[(-\infty, t_{[\hat{\theta},c]}]; df] < \alpha$, then $t_{[\hat{\theta},c]} < -t_{[\alpha]}^{(df)}$. Reject $H_0 : \theta = c$

(c) If prob-value $= A[(-\infty, t_{[\hat{\theta},c]}]; df] \geq \alpha$, then $t_{[\hat{\theta},c]} \geq -t_{[\alpha]}^{(df)}$. Do not reject $H_0 : \theta = c$

FIGURE 6.18
The Probability of a Type I Error and the Prob-Value When Testing $H_0 : \theta = c$ Versus $H_1 : \theta < c$

and (2) are equivalent. If condition (2) holds, then

$$\text{prob-value} = A[(-\infty, t_{[\hat{\theta},c]}]; df]$$

is less than α. Looking at Figures 6.18a and 6.18b, we see that this fact implies that condition (1),

$$t_{[\hat{\theta},c]} < -t_{[\alpha]}^{(df)}$$

holds. If condition (2) does not hold, then

$$\text{prob-value} = A[(-\infty, t_{[\hat{\theta},c]}]; df]$$

is greater than or equal to α. Looking at Figures 6.18a and 6.18c, we see that this fact implies that

$$t_{[\hat{\theta},c]} \geq -t_{[\alpha]}^{(df)}$$

which means that condition (1) does not hold. Thus, we have shown that conditions (1) and (2) are equivalent.

We next present an example of interpreting the prob-value as a probability and of using (3), (4), and (5) in Hypothesis Testing Result III.

EXAMPLE 6.26

Consider the fuel consumption problem and the prob-value calculated in Example 6.25. In Section 6.5 we saw that if $H_0 : \beta_1 = 0$ is true, then the population of all possible t_{b_1} statistics has a t-distribution with $n - k = 5$ degrees of freedom. Therefore,

$$\begin{aligned}\text{prob-value} &= A[(-\infty, t_{[\hat{\theta},c]}]; df]\\ &= A[(-\infty, t_{b_1}]; n - k]\\ &= A[(-\infty, -6.39]; 5]\\ &= .000695\end{aligned}$$

equals the proportion of t_{b_1} statistics in the population of all such statistics which, if the null hypothesis $H_0 : \beta_1 = 0$ is true, are at least as small as and thus at least as contradictory to $H_0 : \beta_1 = 0$ as -6.39, the observed t_{b_1} statistic. Since this prob-value, which equals .000695, says that we must believe that a 695 in 1,000,000 chance has occurred if we are to believe that $H_0 : \beta_1 = 0$ is true, we would have substantial evidence to conclude that $H_0 : \beta_1 = 0$ is false and thus that $H_1 : \beta_1 < 0$ is true.

Finally, we note that it follows from (4) in Hypothesis Testing Result III that for any level of confidence, $100(1 - \alpha)\%$, less than

$$100[1 - (\text{prob-value})]\% = 100[1 - .000695]\% = 99.9305\%$$

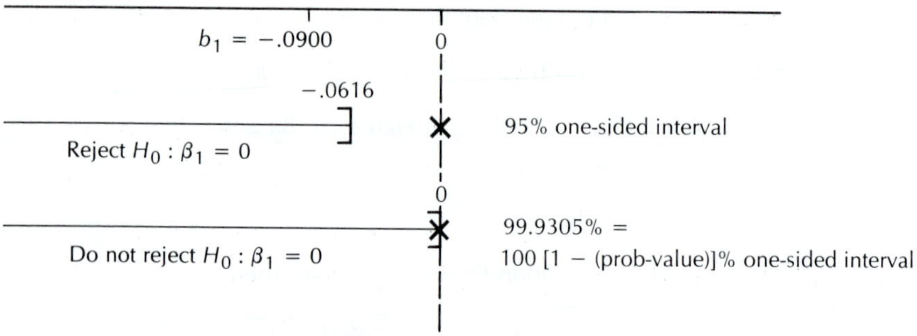

FIGURE 6.19
$H_0 : \beta_1 = 0$ Can Be Rejected in Favor of $H_1 : \beta_1 < 0$ at Any Level of Confidence Less Than 99.9305%

the one-sided $100(1-\alpha)\%$ confidence interval for β_1,

$$\begin{aligned}(-\infty, b_1 + t_{[\alpha]}^{(n-k)} s\sqrt{c_{11}}] &= (-\infty, -.0900 + t_{[\alpha]}^{(8-3)}(.3671)\sqrt{.00147}] \\ &= (-\infty, -.0900 + t_{[\alpha]}^{(5)}(.0141)]\end{aligned}$$

does not contain zero. (For example, the 95% interval

$$\begin{aligned}(-\infty, -.0900 + t_{[.05]}^{(5)}(.0141)] &= (-\infty, -.0900 + 2.015(.0141)] \\ &= (-\infty, -.0616]\end{aligned}$$

does not contain zero. See Figure 6.19.) Thus, for such a level of confidence we *can*, with at least $100(1-\alpha)\%$ confidence, *reject* $H_0 : \beta_1 = 0$ in favor of $H_1 : \beta_1 < 0$. It follows from (5) in Hypothesis Testing Result III that for any level of confidence, $100(1-\alpha)\%$, greater than or equal to

$$100[1 - (\text{prob-value})]\% = 99.9305\%$$

the one-sided $100(1-\alpha)\%$ confidence interval for β_1 *does contain zero.* (For example, the 99.9305% interval, which can be calculated to be $(-\infty, 0]$, does contain zero. See Figure 6.19.) Thus, for such a level of confidence we *cannot*, with at least $100(1-\alpha)\%$ confidence, *reject* $H_0 : \beta_1 = 0$ in favor of $H_1 : \beta_1 < 0$. To summarize, we can be up to 99.9305% confident that $H_0 : \beta_1 = 0$ is false and that $H_1 : \beta_1 < 0$ is true. Specifically, this one-sided 95% confidence interval for β_1, $(-\infty, -.0616]$, says that we can be 95% confident that β_1 is less than or equal to $-.0616$ (tons of coal).

In order to understand the rest of this section, which explains testing hypotheses by using the normal distribution, the reader should have studied optional Section 3.4. We define the $z_{[\hat{\theta},c]}$ statistic to be

$$z_{[\hat{\theta},c]} = \frac{\hat{\theta} - c}{\sigma_{\hat{\theta}}} \quad \text{or} \quad z_{[\hat{\theta},c]} = \frac{\hat{\theta} - c}{s_{\hat{\theta}}}$$

TABLE 6.13 Confidence Intervals and $z_{[\hat{\theta},c]}$ Statistics Related to Testing $H_0: \theta = c$

Parameter	Point Estimate	Standard Deviation (Actual or Estimated)	$z_{[\hat{\theta},\theta]}$ Statistic	$100(1-\alpha)\%$ Confidence Intervals	$z_{[\hat{\theta},c]}$ Statistic
θ	$\hat{\theta}$	$\sigma_{\hat{\theta}}$ or $s_{\hat{\theta}}$	$z_{[\hat{\theta},\theta]} = \frac{\hat{\theta}-\theta}{\sigma_{\hat{\theta}}}$ or $z_{[\hat{\theta},\theta]} = \frac{\hat{\theta}-\theta}{s_{\hat{\theta}}}$	$[\hat{\theta} \pm z_{[\alpha/2]}\sigma_{\hat{\theta}}]$ or $[\hat{\theta} \pm z_{[\alpha/2]}s_{\hat{\theta}}]$ $[\hat{\theta} - z_{[\alpha]}\sigma_{\hat{\theta}}, \infty)$ or $[\hat{\theta} - z_{[\alpha]}s_{\hat{\theta}}, \infty)$ $(-\infty, \hat{\theta} + z_{[\alpha]}\sigma_{\hat{\theta}}]$ or $(-\infty, \hat{\theta} + z_{[\alpha]}s_{\hat{\theta}}]$	$z_{[\hat{\theta},c]} = \frac{\hat{\theta}-c}{\sigma_{\hat{\theta}}}$ or $z_{[\hat{\theta},c]} = \frac{\hat{\theta}-c}{s_{\hat{\theta}}}$
μ	$\bar{y}$	$\sigma_{\bar{y}} = \frac{\sigma}{\sqrt{n}}$ or $s_{\bar{y}} = \frac{s}{\sqrt{n}}$	$z_{[\bar{y},\mu]} = \frac{\bar{y}-\mu}{\sigma/\sqrt{n}}$ or $z_{[\bar{y},\mu]} = \frac{\bar{y}-\mu}{s/\sqrt{n}}$	$\left[\bar{y} \pm z_{[\alpha/2]}\left(\frac{\sigma}{\sqrt{n}}\right)\right]$ or $\left[\bar{y} \pm z_{[\alpha/2]}\left(\frac{s}{\sqrt{n}}\right)\right]$ $\left[\bar{y} - z_{[\alpha]}\left(\frac{\sigma}{\sqrt{n}}\right), \infty\right)$ or $\left[\bar{y} - z_{[\alpha]}\left(\frac{s}{\sqrt{n}}\right), \infty\right)$ $\left(-\infty, \bar{y} + z_{[\alpha]}\left(\frac{\sigma}{\sqrt{n}}\right)\right]$ or $\left(-\infty, \bar{y} + z_{[\alpha]}\left(\frac{s}{\sqrt{n}}\right)\right]$	$z_{[\bar{y},c]} = \frac{\bar{y}-c}{\sigma/\sqrt{n}}$ or $z_{[\bar{y},c]} = \frac{\bar{y}-c}{s/\sqrt{n}}$
p	$\hat{p}$	$\sigma_{\hat{p}} = \sqrt{\frac{p(1-p)}{n}}$ or $s_{\hat{p}} = \sqrt{\frac{\hat{p}(1-\hat{p})}{n-1}}$	$z_{[\hat{p},p]} = \frac{\hat{p}-p}{\sqrt{p(1-p)/n}}$ or $z_{[\hat{p},p]} = \frac{\hat{p}-p}{\sqrt{\hat{p}(1-\hat{p})/(n-1)}}$	$\left[\hat{p} \pm z_{[\alpha/2]}\sqrt{\frac{p(1-p)}{n}}\right]$ or $\left[\hat{p} \pm z_{[\alpha/2]}\sqrt{\frac{\hat{p}(1-\hat{p})}{n-1}}\right]$ $\left[\hat{p} - z_{[\alpha]}\sqrt{\frac{p(1-p)}{n}}, \infty\right)$ or $\left[\hat{p} - z_{[\alpha]}\sqrt{\frac{\hat{p}(1-\hat{p})}{n-1}}, \infty\right)$ $\left(-\infty, \hat{p} + z_{[\alpha]}\sqrt{\frac{p(1-p)}{n}}\right]$ or $\left(-\infty, \hat{p} + z_{[\alpha]}\sqrt{\frac{\hat{p}(1-\hat{p})}{n-1}}\right]$	$z_{[\hat{p},c]} = \frac{\hat{p}-c}{\sqrt{p(1-p)/n}}$ or $z_{[\hat{p},c]} = \frac{\hat{p}-c}{\sqrt{\hat{p}(1-\hat{p})/(n-1)}}$

where $s_{\hat{\theta}}$ is the point estimate of $\sigma_{\hat{\theta}}$ (see Table 6.13). In the following box we summarize how to use the $z_{[\hat{\theta},c]}$ statistic (examples of which are given in Table 6.13) and the appropriate rejection point conditions, prob-values, and confidence intervals to test $H_0: \theta = c$ versus $H_1: \theta \neq c$; $H_0: \theta = c$ versus $H_1: \theta > c$; and $H_0: \theta = c$ versus $H_1: \theta < c$.

HYPOTHESIS TESTING RESULT IV:

Define

$$z_{[\hat{\theta},\theta]} = \frac{\hat{\theta} - \theta}{\sigma_{\hat{\theta}}} \quad \text{or} \quad z_{[\hat{\theta},\theta]} = \frac{\hat{\theta} - \theta}{s_{\hat{\theta}}}$$

and assume that the population of all possible $z_{[\hat{\theta},\theta]}$ statistics has a standard normal distribution. In order to test $H_0: \theta = c$, define the $z_{[\hat{\theta},c]}$ statistic to be

$$z_{[\hat{\theta},c]} = \frac{\hat{\theta} - c}{\sigma_{\hat{\theta}}} \quad \text{or} \quad z_{[\hat{\theta},c]} = \frac{\hat{\theta} - c}{s_{\hat{\theta}}}$$

and consider the following alternative hypotheses, rejection point conditions, prob-values, and $100(1 - \alpha)\%$ confidence intervals, where we replace $\sigma_{\hat{\theta}}$ by $s_{\hat{\theta}}$ in the confidence intervals if we use the $z_{[\hat{\theta},c]}$ statistic utilizing $s_{\hat{\theta}}$.

Alternative Hypothesis	Rejection Point Condition	Prob-Value	Confidence Interval
$H_1: \theta \neq c$	$\lvert z_{[\hat{\theta},c]}\rvert > z_{[\alpha/2]}$ that is, $z_{[\hat{\theta},c]} > z_{[\alpha/2]}$ or $z_{[\hat{\theta},c]} < -z_{[\alpha/2]}$	$2A[[\lvert z_{[\hat{\theta},c]}\rvert, \infty); N(0, 1)]$	$[\hat{\theta} \pm z_{[\alpha/2]}\sigma_{\hat{\theta}}]$
$H_1: \theta > c$	$z_{[\hat{\theta},c]} > z_{[\alpha]}$	$A[[z_{[\hat{\theta},c]}, \infty); N(0, 1)]$	$[\hat{\theta} - z_{[\alpha]}\sigma_{\hat{\theta}}, \infty)$
$H_1: \theta < c$	$z_{[\hat{\theta},c]} < -z_{[\alpha]}$	$A[(-\infty, z_{[\hat{\theta},c]}]; N(0, 1)]$	$(-\infty, \hat{\theta} + z_{[\alpha]}\sigma_{\hat{\theta}}]$

Then, we can reject $H_0: \theta = c$ in favor of an appropriate alternative hypothesis *by setting the probability of a Type I error equal to* α if and only if any of the following three equivalent conditions hold:

1. The appropriate rejection point condition holds.
2. The appropriate prob-value is less than α.
3. The appropriate $100(1 - \alpha)\%$ confidence interval for θ does not contain c.

Moreover,

4. For any level of confidence, $100(1-\alpha)\%$, less than $100[1-(\text{prob-value})]\%$, the appropriate $100(1-\alpha)\%$ confidence interval for θ *does not contain* c, and thus we *can*, with at least $100(1-\alpha)\%$ confidence, *reject* $H_0: \theta = c$ in favor of the alternative hypothesis.
5. For any level of confidence, $100(1-\alpha)\%$, greater than or equal to $100[1-(\text{prob-value})]\%$, the appropriate $100(1-\alpha)\%$ confidence interval for θ *does contain* c, and thus we *cannot*, with at least $100(1-\alpha)\%$ confidence, *reject* $H_0: \theta = c$ in favor of the alternative hypothesis.

Examining the results in Hypothesis Testing Result IV, we note that $A[[|z_{[\hat{\theta},c]}|, \infty); N(0, 1)]$, $A[[z_{[\hat{\theta},c]}, \infty); N(0, 1)]$, and $A[(-\infty, z_{[\hat{\theta},c]}]; N(0, 1)]$ are the areas under the standard normal curve to the right of $|z_{[\hat{\theta},c]}|$, to the right of $z_{[\hat{\theta},c]}$, and to the left of $z_{[\hat{\theta},c]}$, respectively. In the following example we illustrate the use of Hypothesis Testing Result IV.

EXAMPLE 6.27

Recent medical research has sought to develop drugs to lessen the severity and duration of viral infections. Virol, a popular drug, is known to do this for 70 percent of all patients. Pharmco, Inc., has developed a new drug, Phantol, and wishes to prove that Phantol is more effective than Virol (that is, lessens the severity and duration of viral infections for more than 70 percent of all patients). Pharmco will use a sample of $n = 300$ patients to attempt to reject the null hypothesis $H_0: p \leq .70$ in favor of the alternative hypothesis $H_1: p > .70$, where p is the true proportion of all patients taking Phantol for whom the severity and duration of viral infections would be lessened. Define the $z_{[\hat{\theta},\theta]}$ statistic to be

$$z_{[\hat{\theta},\theta]} = \frac{\hat{\theta}-\theta}{s_{\hat{\theta}}} = \frac{\hat{p}-p}{\sqrt{\hat{p}(1-\hat{p})/(n-1)}} = z_{[\hat{p},p]}$$

It follows from the discussion of Section 3.4 that since the sample size $n = 300$ is large and Pharmco believes that the true value of p is between .70 and .85 (implying that the interval $[p \pm 3\sqrt{p(1-p)/n}]$ does not contain zero or 1), the population of all possible $z_{[\hat{\theta},\theta]}$ statistics (that is, $z_{[\hat{p},p]}$ statistics) approximately has a standard normal distribution. Testing $H_0: p \leq .70$ versus $H_1: p > .70$ is equivalent to testing $H_0: p = .70$ versus $H_1: p > .70$. Therefore, from condition (1) in Hypothesis Testing Result IV, we can reject $H_0: p = .70$ $(\theta = c)$ in favor of $H_1: p > .70$ $(\theta > c)$ by setting α, the probability of a Type I error, equal to .05

if and only if

$$z_{[\hat{\theta},c]} = \frac{\hat{\theta} - c}{s_{\hat{\theta}}} = \frac{\hat{p} - .70}{\sqrt{\hat{p}(1 - \hat{p})/(n - 1)}} = z_{[\hat{p},.70]}$$

is greater than $z_{[\alpha]} = z_{[.05]} = 1.645$. Now, suppose that 231 of the 300 patients taking Phantol experience less severe and shorter viral infections. Since this implies that $\hat{p} = 231/300 = .77$, we can calculate the $z_{[\hat{p},.70]}$ statistic to be

$$z_{[\hat{p},.70]} = \frac{\hat{p} - .70}{\sqrt{\hat{p}(1 - \hat{p})/(n - 1)}} = \frac{.77 - .70}{\sqrt{.77(.23)/(300 - 1)}} = 2.88$$

Since $z_{[\hat{p},.70]} = 2.88 > 1.645 = z_{[.05]}$, we can reject $H_0: p = .70$ in favor of $H_1: p > .70$ by setting α equal to .05.

Next, we note that

$$\begin{aligned} \text{prob-value} &= A[[z_{[\hat{\theta},c]}, \infty); N(0, 1)] \\ &= A[[z_{[\hat{p},.70]}, \infty); N(0, 1)] \\ &= A[[2.88, \infty); N(0, 1)] \\ &= .002 \end{aligned}$$

Since this prob-value is less than .05 and less than .01, it follows by condition (2) that we can reject $H_0: p = .70$ in favor of $H_1: p > .70$ if we set α equal to .05 or .01.

In order to interpret the prob-value as a probability, note that since the population of all possible $z_{[\hat{p},p]}$ statistics has a standard normal distribution, it follows that if $H_0: p = .70$ is true, then the population of all possible $z_{[\hat{p},.70]}$ statistics has a standard normal distribution. Therefore,

$$\begin{aligned} \text{prob-value} &= A[[z_{[\hat{p},.70]}, \infty); N(0, 1)] \\ &= A[[2.88, \infty); N(0, 1)] \\ &= .002 \end{aligned}$$

equals the proportion of $z_{[\hat{p},.70]}$ statistics in the population of all such statistics which, if the null hypothesis $H_0: p = .70$ is true, are at least as large as and thus at least as contradictory to $H_0: p = .70$ as 2.88, the observed $z_{[\hat{p},.70]}$ statistic. Since this prob-value means that Pharmco must believe that a 2 in 1,000 chance has occurred if it is to believe that $H_0: p = .70$ is true, the company would have substantial evidence to conclude that $H_0: p = .70$ is false and thus that $H_1: p > .70$ is true.

Finally, it follows from (4) in Hypothesis Testing Result IV that for any level of confidence, $100(1 - \alpha)\%$, less than

$$100[1 - (\text{prob-value})]\% = 100[1 - .002]\% = 99.8\%$$

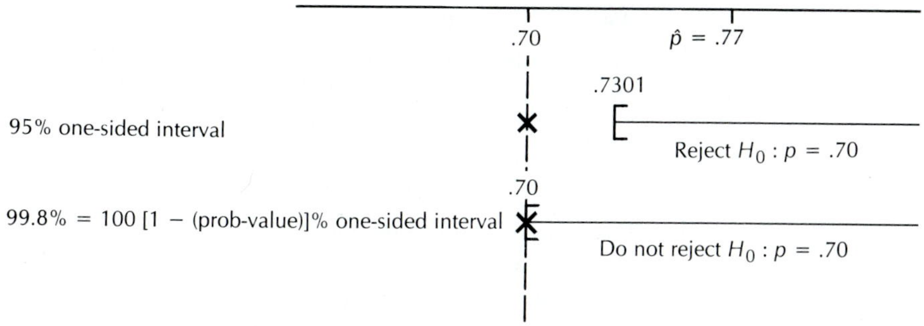

FIGURE 6.20
$H_0: p = .70$ Can Be Rejected in Favor of $H_1: p > .70$ at Any Level of Confidence Less Than 99.8%

the one-sided confidence interval for p,

$$\left[\hat{p} - z_{[\alpha]}\sqrt{\frac{\hat{p}(1-\hat{p})}{n-1}}, \infty\right) = \left[.77 - z_{[\alpha]}\sqrt{\frac{(.77)(.23)}{300-1}}, \infty\right)$$

$$= [.77 - z_{[\alpha]}(.0243), \infty)$$

does not contain .70. (For example, the 95% interval

$$\begin{aligned}[.77 - z_{[.05]}(.0243), \infty) &= [.77 - 1.64(.0243), \infty) \\ &= [.77 - .0399, \infty) \\ &= [.7301, \infty)\end{aligned}$$

does not contain .70. See Figure 6.20.) Thus, for such levels of confidence Pharmco *can*, with at least $100(1-\alpha)\%$ confidence, *reject* $H_0: p = .70$ in favor of $H_1: p > .70$. It follows from (5) in Hypothesis Testing Result IV that for any level of confidence, $100(1-\alpha)\%$, greater than or equal to

$$100[1 - (\text{prob-value})]\% = 99.8\%$$

the one-sided $100(1-\alpha)\%$ confidence interval for p *does contain .70.* (For example, the 99.8% interval does contain .70. See Figure 6.20.) Thus, for such levels of confidence Pharmco *cannot*, with at least $100(1-\alpha)\%$ confidence, *reject* $H_0: p = .70$ in favor of $H_1: p > .70$. To summarize, Pharmco can be up to 99.8% confident that $H_0: p = .70$ is false and that $H_1: p > .70$ is true. Specifically, the 95% confidence interval for p, $[.7301, \infty)$, says that Pharmco can be 95% confident that at least 73.01 percent of all patients taking Phantol will experience less severe and shorter viral infections. To complete this example, we note that we have based the test of $H_0: p = .70$ versus $H_1: p > .70$ on the statistic

$$z_{[\hat{p},p]} = \frac{\hat{p} - p}{\sqrt{\hat{p}(1-\hat{p})/(n-1)}}$$

rather than on the statistic

$$z_{[\hat{p},p]} = \frac{\hat{p} - p}{\sqrt{p(1-p)/n}}$$

We have done this because the former $z_{[\hat{p},p]}$ statistic employs the quantity $\sqrt{\hat{p}(1-\hat{p})/(n-1)}$, which is also utilized in the $100(1-\alpha)\%$ confidence interval for p,

$$\left[\hat{p} - z_{[\alpha]}\sqrt{\frac{\hat{p}(1-\hat{p})}{n-1}}, \infty\right)$$

This allows us to use (4) and (5) in Hypothesis Testing Result IV to tell us how confident we can be that $H_0: p = .70$ is false and $H_1: p > .70$ is true.

If we do use the latter $z_{[\hat{p},p]}$ statistic (which is presented in most basic statistics texts), then

$$z_{[\hat{p},.70]} = \frac{\hat{p} - .70}{\sqrt{(.70)(1-.70)/300}} = \frac{.77 - .70}{\sqrt{(.70)(1-.70)/300}} = 2.65$$

and

$$\begin{aligned} \text{prob-value} &= A[[z_{[\hat{p},.70]}, \infty); N(0, 1)] \\ &= A[[2.65, \infty); N(0, 1)] \\ &= .004 \end{aligned}$$

which differs slightly from the prob-value we have previously computed.

6 EXERCISES

6.1 Consider Exercise 5.3 and the model

$$\begin{aligned} y_i &= \mu_i + \epsilon_i \\ &= \beta_0 + \beta_1 x_{i1} + \beta_2 x_{i2} + \beta_3 x_{i1}^2 + \beta_4 x_{i1} x_{i2} + \epsilon_i \end{aligned}$$

Using the information in Exercise 5.3, do the following:

a. Verify that $\mathbf{b}'\mathbf{X}'\mathbf{y} = 22{,}057$.
b. Verify that $s^2 = 2.25$ and thus that $s = 1.5$.
c. Calculate 95% confidence intervals for β_0, β_1, β_2, β_3, and β_4. Based on these intervals, which of the hypotheses $H_0: \beta_0 = 0$, $H_0: \beta_1 = 0$, $H_0: \beta_2 = 0$, $H_0: \beta_3 = 0$, and $H_0: \beta_4 = 0$ can we reject with at least 95% confidence?
d. Calculate the t_{b_j} statistics used to test $H_0: \beta_0 = 0$, $H_0: \beta_1 = 0$, $H_0: \beta_2 = 0$, $H_0: \beta_3 = 0$, and $H_0: \beta_4 = 0$. By using the appropriate rejection points, determine which of these hypotheses we can reject by setting α, the probability of a Type I error, equal to .05. By using the appropriate rejection points, determine which of these hypotheses we can reject by setting α equal to .01.

6.2 Consider Exercise 5.4 and the model

$$\begin{aligned} y_i &= \mu_i + \epsilon_i \\ &= \beta_0 + \beta_1 x_{i1} + \beta_2 x_{i2} + \beta_3 x_{i1}^2 + \beta_4 x_{i1} x_{i2} + \epsilon_i \end{aligned}$$

a. If we use this model to perform a regression analysis of the data in Exercise 5.4, we obtain an unexplained variation of $SSE = 12.2480$. Using this quantity, calculate s.

b. The SAS output of the t_{b_j} statistics and prob-values that can be used to test $H_0: \beta_0 = 0$, $H_0: \beta_1 = 0$, $H_0: \beta_2 = 0$, $H_0: \beta_3 = 0$, and $H_0: \beta_4 = 0$ follows (STANDARD ERROR $= s\sqrt{c_{jj}}$, T $= t_{b_j}$, and PROB = prob-value).

VARIABLE	DF	PARAMETER ESTIMATE	STANDARD ERROR	T FOR H0: PARAMETER=0	PROB > |T|
INTERCEP	1	19.304957	2.052057	9.408	0.0001
X1	1	-1.486602	1.177734	-1.262	0.2290
X2	1	-6.371452	1.042308	-6.113	0.0001
X1SQ	1	-0.752248	0.225237	-3.340	0.0053
X12	1	1.717053	0.253779	6.766	0.0001

(1) Using the fact that $s\sqrt{c_{33}} = .225237$, find c_{33}, the diagonal element of $(\mathbf{X}'\mathbf{X})^{-1}$ corresponding to β_3.

(2) Demonstrate how $t_{b_3} = -3.340$ has been calculated. By using the appropriate rejection point, determine whether we can reject $H_0: \beta_3 = 0$ in favor of $H_1: \beta_3 \neq 0$ by setting α equal to .05.

(3) Explain how the prob-value for testing $H_0: \beta_3 = 0$ versus $H_1: \beta_3 \neq 0$ has been calculated. By using this prob-value, determine whether we can reject $H_0: \beta_3 = 0$ versus $H_1: \beta_3 \neq 0$ by setting α equal to .05; to .01; to .001.

(4) Write a short essay describing what the prob-value for testing $H_0: \beta_3 = 0$ versus $H_1: \beta_3 \neq 0$ says about a confidence interval for β_3 and about how confident we can be that $H_0: \beta_3 = 0$ should be rejected in favor of $H_1: \beta_3 \neq 0$. To aid your discussion, calculate and depict the 95%, 99%, and 99.47% confidence intervals for β_3.

(5) Interpret the prob-value for testing $H_0: \beta_3 = 0$ versus $H_1: \beta_3 \neq 0$ as a probability.

(6) What do the prob-values for testing $H_0: \beta_0 = 0$, $H_0: \beta_1 = 0$, $H_0: \beta_2 = 0$, $H_0: \beta_3 = 0$, and $H_0: \beta_4 = 0$ tell us about the importance of the independent variables in the preceding regression model? Notice here that the prob-value for testing $H_0: \beta_1 = 0$ does not cast substantial doubt on this hypothesis. This might indicate that x_{i1} should not be included in the preceding model. However, the prob-values for testing $H_0: \beta_3 = 0$ and $H_0: \beta_4 = 0$ indicate that x_{i1}^2 and $x_{i1}x_{i2}$ are important. In such a situation, we often include the linear term x_{i1} even though the prob-value for testing $H_0: \beta_1 = 0$ indicates that this term may not be important. This is because of a phenomenon called *multicollinearity*, which exists when independent variables are related to each other (here x_{i1}, x_{i1}^2, and $x_{i1}x_{i2}$ are related). Intuitively, this causes the importance of x_{i1} to be spread between x_{i1}, x_{i1}^2, and $x_{i1}x_{i2}$, making x_{i1} appear to have less importance than it might actually have. Therefore, it may

make sense to include x_{i1} in the regression model. We discuss multicollinearity in detail in Chapter 7.

c. Two other regression models that might be used to relate y_i to x_{i1} and x_{i2} are

$$y_i = \mu_i + \epsilon_i$$
$$= \underset{(.0001)}{\beta_0} + \underset{(.2355)}{\beta_1 x_{i1}} + \underset{(.0002)}{\beta_2 x_{i2}} + \underset{(.0903)}{\beta_3 x_{i1}^2} + \underset{(.6865)}{\beta_4 x_{i2}^2} + \underset{(.0135)}{\beta_5 x_{i1} x_{i2}} + \epsilon_i$$

$$y_i = \mu_i + \epsilon_i$$
$$= \underset{(.2696)}{\beta_0} + \underset{(.1306)}{\beta_1 x_{i1}} + \underset{(.7845)}{\beta_2 x_{i2}} + \underset{(.0032)}{\beta_3 x_{i1}^2} + \underset{(.4056)}{\beta_4 x_{i1} x_{i2}} + \underset{(.0471)}{\beta_5 x_{i1}^2 x_{i2}} + \epsilon_i$$

The prob-values related to the importance of the independent variables in these models are placed under the variables. Discuss what these prob-values say about adding x_{i2}^2 or $x_{i1}^2 x_{i2}$ to the model

$$y_i = \mu_i + \epsilon_i$$
$$= \beta_0 + \beta_1 x_{i1} + \beta_2 x_{i2} + \beta_3 x_{i1}^2 + \beta_4 x_{i1} x_{i2} + \epsilon_i$$

Note that when x_{i2}^2 or $x_{i1}^2 x_{i2}$ is added to the model, the prob-values related to some of the other independent variables in this model make these variables look less important than do the previously presented prob-values in the SAS output. Again, this is because of *multicollinearity* (discussed in detail in Chapter 7).

6.3 Consider Exercise 5.1 and the model

$$y_i = \mu_i + \epsilon_i$$
$$= \beta_0 + \beta_1 x_i + \epsilon_i$$

Using the data in Exercise 5.1, calculate a 95% confidence interval for β_1. Interpret the results.

6.4 Consider Exercise 5.7 and the following models:

$$y_i = \mu_i + \epsilon_i$$
$$= \underset{(.0001)}{\beta_0} + \underset{(.0001)}{\beta_1 x_{i1}} + \underset{(.0001)}{\beta_2 x_{i1}^2} + \underset{(.0001)}{\beta_3 x_{i2}} + \underset{(.0001)}{\beta_4 x_{i2}^2} + \epsilon_i$$

$$y_i = \mu_i + \epsilon_i$$
$$= \underset{(.0001)}{\beta_0} + \underset{(.0001)}{\beta_1 x_{i1}} + \underset{(.0002)}{\beta_2 x_{i1}^2} + \underset{(.0001)}{\beta_3 x_{i2}} + \underset{(.0001)}{\beta_4 x_{i2}^2} + \underset{(.9777)}{\beta_5 x_{i1} x_{i2}} + \epsilon_i$$

If we use these models to perform regression analyses of the data in Exercise 5.7, the prob-values related to the importance of the independent variables are as placed under these variables. Based on these prob-values, which of the preceding two models seems to describe best the data in Exercise 5.7?

6.5 Consider Exercise 5.8 and the following models:

$$y_i = \mu_i + \epsilon_i$$

$$= \underset{(.0001)}{\beta_0} + \underset{(.0001)}{\beta_1 x_{i1}} + \underset{(.0056)}{\beta_2 x_{i1}^2} + \underset{(.0001)}{\beta_3 x_{i2}} + \underset{(.0005)}{\beta_4 x_{i2}^2} + \underset{(.0198)}{\beta_5 x_{i1} x_{i2}} + \epsilon_i$$

$$y_i = \mu_i + \epsilon_i$$

$$= \underset{(.0001)}{\beta_0} + \underset{(.0001)}{\beta_1 x_{i1}} + \underset{(.0001)}{\beta_2 x_{i1}^2} + \underset{(.0001)}{\beta_3 x_{i2}} + \underset{(.0001)}{\beta_4 x_{i2}^2} + \underset{(.0001)}{\beta_5 x_{i1} x_{i2}}$$

$$+ \underset{(.0005)}{\beta_6 x_{i1}^2 x_{i2}^2} + \epsilon_i$$

$$y_i = \mu_i + \epsilon_i$$

$$= \underset{(.0001)}{\beta_0} + \underset{(.0001)}{\beta_1 x_{i1}} + \underset{(.0150)}{\beta_2 x_{i1}^2} + \underset{(.0001)}{\beta_3 x_{i2}} + \underset{(.0036)}{\beta_4 x_{i2}^2} + \underset{(.0350)}{\beta_5 x_{i1} x_{i2}} + \underset{(.4931)}{\beta_6 x_{i1}^2 x_{i2}^2}$$

$$+ \underset{(.2509)}{\beta_7 x_{i1} x_{i2}^2} + \underset{(.2680)}{\beta_8 x_{i1}^2 x_{i2}} + \epsilon_i$$

If we use these models to perform regression analyses of the data in Exercise 5.8, the prob-values related to the importance of the independent variables are as placed under these variables. Based on these prob-values, which of the three models seems to describe best the data in Exercise 5.8?

6.6 Using SPL_1, SPL_2, and SPL_3 in Table 6.1, calculate three 95% confidence intervals for β_1 in the fuel consumption model $y = \beta_0 + \beta_1 x_1 + \beta_2 x_2 + \epsilon$. What percentage of these three confidence intervals contain β_1 $(= -.087)$? Discuss the meaning of 95% confidence here.

6.7 In the fuel consumption problem, suppose that in a future week the average hourly temperature will be $x_{01} = 37.0$ and the chill index will be $x_{02} = 16$. Using the fuel consumption model

$$y = \mu + \epsilon$$
$$= \beta_0 + \beta_1 x_1 + \beta_2 x_2 + \epsilon$$

a. Calculate three 95% confidence intervals for μ_0, the future mean fuel consumption, by utilizing SPL_1, SPL_2, and SPL_3 in Table 6.1. By using the true values of β_0, β_1, and β_2, calculate the true value of μ_0. What percentage of the three confidence intervals contain μ_0? Discuss the meaning of 95% confidence here.

b. Calculate three 95% prediction intervals for $y_0 = \mu_0 + \epsilon_0$, the future individual fuel consumption, by utilizing SPL_1, SPL_2, and SPL_3 in Table 6.1. If

   (1) after using SPL_1 to calculate a prediction interval, we observe the future fuel consumption $y_0 = 11.9$,

(2) after using SPL_2 to calculate a prediction interval, we observe the future fuel consumption $y_0 = 10.5$,

(3) after using SPL_3 to calculate a prediction interval, we observe the future fuel consumption $y_0 = 11.4$,

what percentage of the three prediction interval–future fuel consumption combinations are successful? Discuss the meaning of 95% confidence here.

In Exercises 6.8 to 6.12 use the information you have obtained in preceding exercises, the indicated model, and the future value(s) of the independent variable(s) to calculate (a) a 95% confidence interval for μ_0 (the future mean value of the dependent variable) and (b) a 95% prediction interval for y_0 (the future individual value of the dependent variable). Also, interpret the practical meaning of the calculated intervals.

6.8 Exercises: 5.3, 6.1
Model: $y = \beta_0 + \beta_1 x_1 + \beta_2 x_2 + \beta_3 x_1^2 + \beta_4 x_1 x_2 + \epsilon$
Future values: $x_{01} = 2.0$, $x_{02} = 5$

6.9 Exercises: 5.1, 6.3
Model: $y = \beta_0 + \beta_1 x + \epsilon$
Future value: $x_0 = 3.25$
Hint: Use the formulas in Section 6.10 to hand-calculate the intervals.

6.10 Exercises: 5.4, 6.2
Model: $y = \beta_0 + \beta_1 x_1 + \beta_2 x_2 + \beta_3 x_1^2 + \beta_4 x_1 x_2 + \epsilon$
Future values: $x_{01} = 4.8$, $x_{02} = 6$
Hint: When we use the appropriate $\mathbf{x}_0'$ vector, we find that $\mathbf{x}_0'(\mathbf{X}'\mathbf{X})^{-1}\mathbf{x}_0 = .2052$. What is $\mathbf{x}_0'$?

6.11 Exercises: 5.7, 6.4
Model: $y = \beta_0 + \beta_1 x_1 + \beta_2 x_1^2 + \beta_3 x_2 + \beta_4 x_2^2 + \epsilon$
Future values: $x_{01} = 1.0$, $x_{02} = 2$
Hint: If we use the model to perform a regression analysis of the data in Exercise 5.7, we obtain $SSE = 6.7590$. Moreover, when we use the appropriate $\mathbf{x}_0'$ vector, we find that $\mathbf{x}_0'(\mathbf{X}'\mathbf{X})^{-1}\mathbf{x}_0 = .1565$. What is $\mathbf{x}_0'$?

6.12 Exercises: 5.8, 6.5
Model: $y = \beta_0 + \beta_1 x_1 + \beta_2 x_1^2 + \beta_3 x_2 + \beta_4 x_2^2 + \beta_5 x_1 x_2 + \beta_6 x_1^2 x_2^2 + \epsilon$
Future values: $x_{01} = 1.0$, $x_{02} = 1$
Hint: If we use the model to perform a regression analysis of the data in Exercise 5.8, we obtain $SSE = 6.9087$. Moreover, when we use the appropriate $\mathbf{x}_0'$ vector, we obtain $\mathbf{x}_0'(\mathbf{X}'\mathbf{X})^{-1}\mathbf{x}_0 = .2280$. What is $\mathbf{x}_0'$?

6.13 Zenex Radio Corporation has developed a new way to assemble an electrical component used in the manufacture of radios. The company wishes to determine whether μ, the mean assembly time of this component using the new method, is less than 20 minutes, which is known to be the mean assembly time for the component using the current method of assembly. Suppose that Zenex Radio randomly selects a sample of $n = 6$ employees, thoroughly trains each employee to use the new assembly method, has each employee assemble one component using the new method, records the assembly times, and calculates the mean and standard deviation of the sample of $n = 6$ assembly times to be $\bar{y} = 14.29$ (minutes) and $s = 2.19$ (minutes). Assuming that the population of all assembly times has a normal distribution, calculate a 95% confidence interval for μ. Using this confidence interval, can Zenex Radio be at least 95% confident that μ is less than 20 minutes? Justify your answer. Also, calculate a 95% prediction interval for y_0, a future (individual) assembly time.

The remaining problems in this chapter pertain to testing $H_0: \theta = c$ versus one of the alternative hypotheses $H_1: \theta \neq c$, $H_1: \theta > c$, or $H_1: \theta < c$. For each problem,

1. Define an appropriate $t_{[\hat{\theta},c]}$ statistic or $z_{[\hat{\theta},c]}$ statistic, and use an appropriate rejection point(s) to determine whether we would reject $H_0: \theta = c$ by setting α, the probability of a Type I error, equal to .05; and to .01.
2. Calculate the appropriate prob-value for testing $H_0: \theta = c$.
3. Using the prob-value calculated in part (2), determine whether we would reject $H_0: \theta = c$ by setting α equal to .05; to .03; and to .01.
4. Interpret the prob-value calculated in part (2) as a probability, and discuss how much doubt this prob-value casts upon $H_0: \theta = c$.
5. Discuss what the prob-value calculated in part (2) says about a confidence interval for θ and about how confident we can be that we should reject $H_0: \theta = c$. Illustrate by calculating and depicting the appropriate 95% and 100[1 − (prob-value)]% confidence intervals for θ.

6.14 Consider the Frizzy Shampoo problem, and test $H_0: \mu = 16$ versus $H_1: \mu \neq 16$ by using the fact that a sample of $n = 6$ bottle fills randomly selected (from the particular adjustment) of the Frizzy bottle-filling process has mean $\bar{y} = 15.7665$ and standard deviation $s = .1524$. Use the fact that $A[[3.75, \infty); 5] = .00665$.

6.15 Consider the fuel consumption model

$$y = \beta_0 + \beta_1 x_1 + \beta_2 x_2 + \epsilon$$

and use the information in Table 6.5 to test $H_0: \beta_2 = 0$ versus $H_1: \beta_2 > 0$.

6.16 Use the sample information in Exercise 6.13 to test $H_0: \mu = 20$ versus $H_1: \mu < 20$. Use the fact that $A[(-\infty, -6.39]; 5] = .000695$.

6.17 Use the sample information in Exercise 3.9 to test $H_0: \mu = 60$ versus $H_1: \mu < 60$.

6.18 Use the sample information in Example 3.7 to test $H_0: p = .10$ versus $H_1: p > .10$.

7

MODEL BUILDING

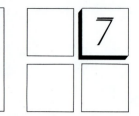

Having studied several commonly used regression models in Chapter 5, we now address the following important question: How can we construct a multiple regression model that will accurately describe and predict a dependent variable?

Recall that the hypothesis test of $H_0: \beta_j = 0$ versus $H_1: \beta_j \neq 0$ where β_j is the regression parameter multiplied by x_j in the regression model

$$y = \beta_0 + \beta_1 x_1 + \cdots + \beta_j x_j + \cdots + \beta_p x_p + \epsilon$$

is useful in assessing the importance of the single independent variable x_j and therefore in deciding whether to include x_j in a final regression model. However, we see in this chapter that a problem called *multicollinearity*, which exists when the independent variables in a regression model are related to each other, can hinder the ability of the hypothesis test of $H_0: \beta_j = 0$ versus $H_1: \beta_j \neq 0$ to tell whether or not an independent variable is important. Consequently, we need to develop measures of the combined importance of all the independent variables in a regression model. Section 7.1 presents several such measures—the *multiple coefficient of determination, R^2*; the *multiple correlation coefficient, R*; the *simple coefficient of determination, r^2*; and the *simple correlation coefficient, r*. Section 7.1 also discusses two hypothesis tests (an *F-test* and a *t*-test) related to the utility of the overall regression model. After studying multicollinearity in Section 7.2, we discuss in Sections 7.3 and 7.4 how to go about comparing the utility of overall regression models on the basis of various criteria—the multiple coefficient of determination, R^2; the mean square error, s^2; the lengths of prediction intervals; *corrected R^2*; and the *C statistic*. Then in Section 7.5 we explain several automatic screening procedures (one of which is *stepwise regression*) that can be used to identify a set or sets of independent variables that might be important in describing and predicting the dependent variable when there are a large number of potentially important independent variables in a regression problem. In Sections 7.6 and 7.7 we study two further *F*-tests: a *partial F-test*, which is used to assess the importance of a portion of a regression model (that is, a set of one or more of the independent variables), and an *F-test of lack of fit*, which can be used to help decide whether the functional form of a regression model is adequate.

Recall that in Chapter 1 we defined a *quantitative* independent variable to be an independent variable that assumes *numerical* values corresponding to the points on a line and a *qualitative* independent variable to be an independent variable that is not quantitative. Whereas the levels of a quantitative independent variable are numerical, the levels of a qualitative independent variable can be defined by describing them. In Section 7.8 we present an example of using *dummy variables*, which are quantitative independent variables used to measure the effects of qualitative independent variables.

7.1 MEASURING THE UTILITY OF THE OVERALL REGRESSION MODEL

7.1.1 Multiple Coefficients of Determination and Correlation: R^2 and R

In this section we begin to study measures of the combined importance of all the independent variables $x_1, x_2, \ldots, x_p$ in describing the dependent variable y in a regression model. We say that such measures are measures of the importance, or utility, of the *overall regression model* in describing the dependent variable. Recall that the least squares point estimates $b_0, b_1, \ldots, b_p$ of the parameters $\beta_0, \beta_1, \ldots, \beta_p$ give the prediction

$$\hat{y}_i = b_0 + b_1 x_{i1} + \cdots + b_p x_{ip}$$

of

$$y_i = \beta_0 + \beta_1 x_{i1} + \cdots + \beta_p x_{ip} + \epsilon_i$$

the ith observed value of the dependent variable. Moreover, the least squares point estimates $b_0, b_1, \ldots, b_p$ give point predictions $\hat{y}_1, \hat{y}_2, \ldots, \hat{y}_n$ of the n observed values of the dependent variable $y_1, y_2, \ldots, y_n$ so that the value of the sum of squared residuals (which we now also call the *unexplained variation*),

$$SSE = \sum_{i=1}^{n} (y_i - \hat{y}_i)^2$$

is smaller than the value that would be given by any other point estimates. The unexplained variation is one measure of the combined importance of all the independent variables $x_1, x_2, \ldots, x_p$ in describing the dependent variable y. The smaller SSE is, the smaller are the differences between the predicted and observed values of the dependent variable, and hence the greater is the utility of the overall regression model in describing the dependent variable. However, since the value of

$$\sum_{i=1}^{n} (y_i - \hat{y}_i)^2$$

depends on the units in which the observations $y_1, y_2, \ldots, y_n$ are measured, it can be somewhat difficult to tell whether a given value of SSE is a small number. Hence, we need a relative measure of the utility of the overall regression model. That is, we need a measure that is independent of the size of the units in which the observations $y_1, y_2, \ldots, y_n$ are measured. One such measure is called the *multiple coefficient of determination*, denoted by R^2.

To introduce R^2, assume we have the n observed values of the dependent variable y, but we do not have the observed values of the independent variables $x_1, x_2, \ldots, x_p$ with which to predict y_i, the ith observed value of the dependent variable. In such a case, the only reasonable prediction of y_i would be

$$\bar{y} = \frac{\sum_{i=1}^{n} y_i}{n}$$

which is the average of the n observed values $y_1, y_2, \ldots, y_n$. The error of prediction would then be $y_i - \bar{y}$. However, in reality, we do have the observed values of the independent variables $x_1, x_2, \ldots, x_p$ to use in predicting y_i. The prediction of y_i is

$$\hat{y}_i = b_0 + b_1 x_{i1} + \cdots + b_p x_{ip}$$

which means that the error of prediction is $y_i - \hat{y}_i$. Now consider the following question: How much has the historical information concerning the independent variables improved the error of prediction? We can see that the error of prediction equals $y_i - \bar{y}$ without using this historical information and equals $y_i - \hat{y}_i$ using this information. Therefore, using the independent variables has decreased the error of prediction from $y_i - \bar{y}$ to $y_i - \hat{y}_i$, or by an amount equal to

$$(y_i - \bar{y}) - (y_i - \hat{y}_i) = (\hat{y}_i - \bar{y})$$

It can be shown (Searle (1971)) that in general

$$\sum_{i=1}^{n} (y_i - \bar{y})^2 - \sum_{i=1}^{n} (y_i - \hat{y}_i)^2 = \sum_{i=1}^{n} (\hat{y}_i - \bar{y})^2$$

The quantity

$$\sum_{i=1}^{n} (y_i - \bar{y})^2$$

is called the **total variation** and equals the sum of squared prediction errors that would be obtained if we did not use the observed values of the independent variables to predict the n observed values of the dependent variable. The quantity

$$\sum_{i=1}^{n} (y_i - \hat{y}_i)^2$$

is called the **unexplained variation** and equals the sum of squared prediction errors that is obtained when we use the observed values of the independent variables to predict the n observed values of the dependent variable. The quantity

$$\sum_{i=1}^{n} (\hat{y}_i - \bar{y})^2$$

is called the **explained variation**. Since

$$\sum_{i=1}^{n} (\hat{y}_i - \bar{y})^2 = \sum_{i=1}^{n} (y_i - \bar{y})^2 - \sum_{i=1}^{n} (y_i - \hat{y}_i)^2$$

this explained variation represents the reduction in the sum of squared prediction errors that has been accomplished by using the historical information concerning the independent variables in predicting the n observed values of the dependent variable. It follows that

$$\sum_{i=1}^{n} (y_i - \bar{y})^2 = \sum_{i=1}^{n} (\hat{y}_i - \bar{y})^2 + \sum_{i=1}^{n} (y_i - \hat{y}_i)^2$$

$$\text{Total variation} = \text{Explained variation} + \text{Unexplained variation}$$

We now define the **multiple coefficient of determination**, R^2, by the equation

$$R^2 = \frac{\sum_{i=1}^{n} (\hat{y}_i - \bar{y})^2}{\sum_{i=1}^{n} (y_i - \bar{y})^2} = \frac{\text{Explained variation}}{\text{Total variation}}$$

That is, R^2 is the proportion of the total variation in the n observed values of the dependent variable that is explained by the overall regression model. Since neither the explained variation nor the total variation can be a negative number (both of these quantities are sums of squares), R^2 cannot be a negative number. Since the explained variation cannot be more than the total variation, it follows that

$$R^2 = \frac{\text{Explained variation}}{\text{Total variation}}$$

cannot be more than 1.

Another way to express R^2 is

$$\begin{aligned} R^2 = \frac{\text{Explained variation}}{\text{Total variation}} &= \frac{\text{Total variation} - \text{Unexplained variation}}{\text{Total variation}} \\ &= \frac{\text{Total variation}}{\text{Total variation}} - \frac{\text{Unexplained variation}}{\text{Total variation}} \\ &= 1 - \frac{\text{Unexplained variation}}{\text{Total variation}} \end{aligned}$$

Thus, we see that the nearer R^2 is to 1, the smaller is the proportion of the total variation that is the unexplained variation. Since the unexplained variation equals the sum of squared prediction errors, the nearer R^2 is to 1, the smaller are the prediction errors (as compared to the total variation) and the greater is the utility of the overall regression model in describing the dependent variable. Most standard multiple regression computer packages calculate both R^2, the multiple coefficient of determination, and the positive square root of the multiple coefficient of determination, $R = \sqrt{R^2}$, which is called the **multiple correlation coefficient**.

Recall that a formula which often provides a simpler way to calculate the unexplained variation is

$$SSE = \sum_{i=1}^{n} (y_i - \hat{y}_i)^2 = \sum_{i=1}^{n} y_i^2 - \mathbf{b}'\mathbf{X}'\mathbf{y}$$

It can also be shown that there are formulas which often provide simpler ways to calculate the total variation and explained variation.

For the regression model

$$y = \beta_0 + \beta_1 x_1 + \cdots + \beta_p x_p + \epsilon$$

1. Total variation $= \sum_{i=1}^{n} (y_i - \bar{y})^2 = \sum_{i=1}^{n} y_i^2 - n\bar{y}^2$.
2. Explained variation $= \sum_{i=1}^{n} (\hat{y}_i - \bar{y})^2 = \mathbf{b}'\mathbf{X}'\mathbf{y} - n\bar{y}^2$.
3. Unexplained variation $= \sum_{i=1}^{n} (y_i - \hat{y}_i)^2 = \sum_{i=1}^{n} y_i^2 - \mathbf{b}'\mathbf{X}'\mathbf{y}$.
4. Total variation = Explained variation + Unexplained variation.
5. Multiple coefficient of determination $= R^2 = \dfrac{\text{Explained variation}}{\text{Total variation}}$

$$= 1 - \frac{\text{Unexplained variation}}{\text{Total variation}}.$$

6. R^2 is the proportion of the total variation in the n observed values of the dependent variable that is explained by the overall regression model.
7. Multiple correlation coefficient $= R = \sqrt{R^2}$.

EXAMPLE 7.1

Consider the two-variable fuel consumption model

$$y = \beta_0 + \beta_1 x_1 + \beta_2 x_2 + \epsilon$$

Using the observed fuel consumption data, we previously made the following calculations:

$$\sum_{i=1}^{8} y_i^2 = 859.91 \qquad \mathbf{b}'\mathbf{X}'\mathbf{y} = 859.236 \qquad \bar{y} = \frac{\sum_{i=1}^{8} y_i}{8} = 10.21$$

$$\text{Unexplained variation} = SSE = \sum_{i=1}^{8} (y_i - \hat{y}_i)^2 = \sum_{i=1}^{8} y_i^2 - \mathbf{b}'\mathbf{X}'\mathbf{y}$$

$$= 859.91 - 859.236 = .674$$

DEP VARIABLE: Y

SOURCE	DF	SUM OF SQUARES	MEAN SQUARE	F VALUE	PROB>F
MODEL	2	24.875018[a]	12.437509[e]	92.303[g]	0.0001[i]
ERROR	5	0.673732[b]	0.134746[f]		
C TOTAL	7	25.548750[c]			

ROOT MSE	0.367078[d]	R-SQUARE	0.9736[h]
DEP MEAN	10.212500	ADJ R-SQ	0.9631
C.V.	3.594401		

[a] Explained variation = SS_{model} [b] SSE [c] Total variation
[d] s [e] MS_{model} [f] $MSE = s^2$
[g] F(model) [h] R^2 [i] Prob-value

FIGURE 7.1 SAS Output of the Analysis of Variance Table for the Two-Variable Fuel Consumption Model

We can calculate the total variation to be

$$\text{Total variation} = \sum_{i=1}^{8} (y_i - \bar{y})^2 = \sum_{i=1}^{8} y_i^2 - 8\bar{y}^2$$

$$= 859.91 - 8(10.21)^2 = 25.55$$

Moreover, we can calculate the explained variation by either of the following two methods:

$$\text{Explained variation} = \text{Total variation} - \text{Unexplained variation}$$
$$= 25.55 - .674 = 24.876$$

or

$$\text{Explained variation} = \sum_{i=1}^{8} (\hat{y}_i - \bar{y})^2$$
$$= \mathbf{b}'\mathbf{X}'\mathbf{y} - 8\bar{y}^2$$
$$= 859.236 - 8(10.21)^2 = 24.876$$

Next, we can calculate the multiple coefficient of determination by either of the following two methods:

$$R^2 = \frac{\text{Explained variation}}{\text{Total variation}} = \frac{24.876}{25.55} = .974$$

or

$$R^2 = 1 - \frac{\text{Unexplained variation}}{\text{Total variation}} = 1 - \frac{.674}{25.55} = .974$$

This value of R^2, the multiple coefficient of determination, says that the regression model

$$y = \beta_0 + \beta_1 x_1 + \beta_2 x_2 + \epsilon$$

explains 97.4% of the total variation in the eight observed fuel consumptions. Also, an R^2 of .974 implies that the multiple correlation coefficient is

$$R = \sqrt{R^2} = \sqrt{.974} = .9869$$

In Figure 7.1 we present the SAS printout of the **analysis of variance table** summarizing the total variation, the explained variation, the unexplained variation (SSE), and R^2 for this model. Some of the other quantities in Figure 7.1, such as MS_{model} and F(model), will be explained later.

7.1.2 An *F*-Test for the Overall Model

In this section we present an F-test related to the utility of the overall regression model

$$y = \beta_0 + \beta_1 x_1 + \cdots + \beta_p x_p + \epsilon$$

Specifically, we test the null hypothesis

$$H_0: \beta_1 = \beta_2 = \cdots = \beta_p = 0$$

which says that none of the independent variables $x_1, x_2, \ldots, x_p$ affect y, versus the alternative hypothesis

$$H_1: \text{At least one of } \beta_1, \beta_2, \ldots, \beta_p \text{ does not equal zero}$$

which says that at least one of the independent variables $x_1, x_2, \ldots, x_p$ affects y. If we can reject H_0 in favor of H_1 by specifying a *small* probability of a Type I error, then it is reasonable to conclude that at least one of $x_1, x_2, \ldots, x_p$ *significantly* affects y. In this case, we should use t_{b_j} statistics (see Section 6.5) and other techniques to determine *which* of these variables significantly affect y. Recalling that k denotes the number of parameters in the overall regression model, and denoting the *explained variation* (also called the **sum of squares due to the overall model**) by SS_{model}, we summarize the F-test of H_0 versus H_1.

TESTING $H_0: \beta_1 = \beta_2 = \cdots = \beta_p = 0$ VERSUS
H_1: AT LEAST ONE OF $\beta_1, \beta_2, \ldots, \beta_p$ DOES NOT EQUAL ZERO:

Assume the inference assumptions are satisfied, and define the *overall F-statistic* to be

$$F(\text{model}) = \frac{MS_{\text{model}}}{MSE}$$

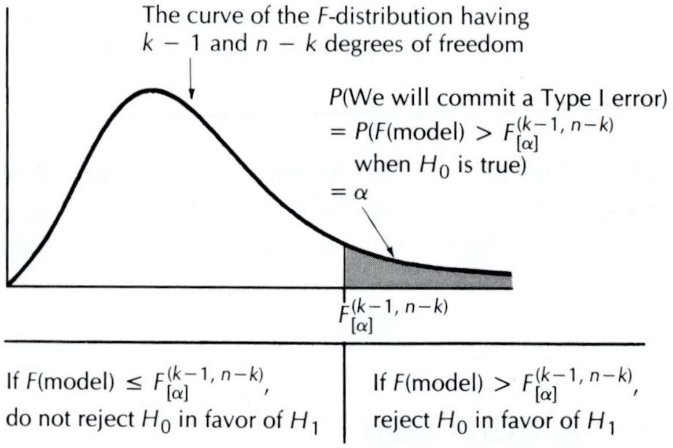

(a) The rejection point $F_{[\alpha]}^{(k-1,n-k)}$ based on setting the probability of a Type I error equal to α

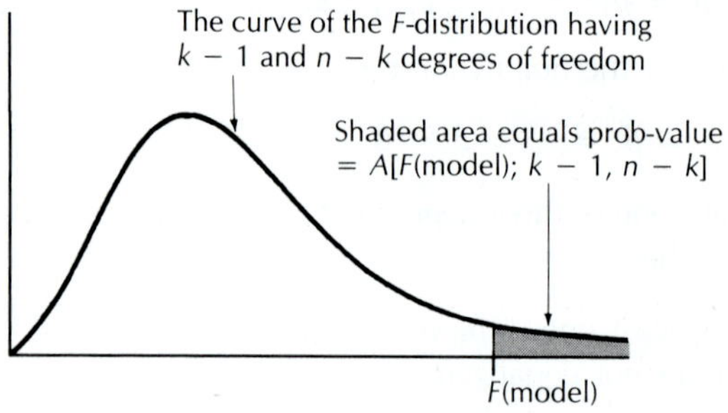

(b) If prob-value is smaller than α, then $F(\text{model}) > F_{[\alpha]}^{(k-1,n-k)}$. Reject H_0

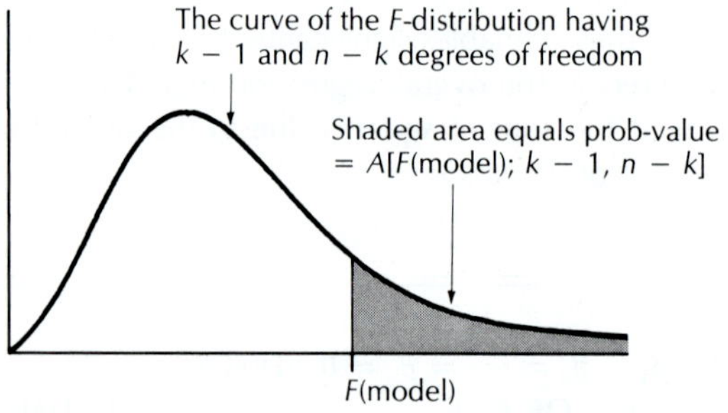

(c) If prob-value is greater than or equal to α, then $F(\text{model}) \leq F_{[\alpha]}^{(k-1,n-k)}$. Do not reject H_0

FIGURE 7.2
Testing
$H_0: \beta_1 = \cdots = \beta_p = 0$

where

$$MS_{\text{model}} = \frac{SS_{\text{model}}}{k-1}$$

$$= \frac{\text{Explained variation}}{k-1}$$

$$MSE = \frac{SSE}{n-k}$$

$$= s^2$$

(MS denotes mean square, and E denotes error) and define

$$\text{prob-value} = A[F(\text{model}); k-1, n-k]$$

which is the area to the right of F(model) under the curve of the F-distribution having $k-1$ and $n-k$ degrees of freedom (see Figure 7.2b and Figure 7.2c). Then, we can reject H_0 in favor of H_1 by setting the probability of a Type I error equal to α if and only if either of the following equivalent conditions hold:

1. $F(\text{model}) > F_{[\alpha]}^{(k-1,n-k)}$, where $F_{[\alpha]}^{(k-1,n-k)}$ is the point on the scale of the F-distribution having $k-1$ and $n-k$ degrees of freedom so that the area under this curve to the right of $F_{[\alpha]}^{(k-1,n-k)}$ is α (see Figure 7.2a).
2. Prob-value $< \alpha$ (see Figure 7.2b and Figure 7.2c).

First consider condition (1). This condition, which says we should reject H_0 in favor of H_1 if

$$F(\text{model}) = \frac{MS_{\text{model}}}{MSE} = \frac{\text{Explained variation}/(k-1)}{SSE/(n-k)}$$

is large, is intuitively reasonable, because a large value of F(model) would be caused by an explained variation that is large relative to the unexplained variation (SSE). This would occur if at least one of the independent variables in the regression model significantly affects the dependent variable, which would imply that H_0 is false and that H_1 is true. To show that use of condition (1) guarantees that the probability of a Type I error equals α, consider the population of all possible samples. Since different samples yield different overall F-statistics, it is clear that an infinite population of overall F-statistics exists. We can make the following statement about this population.

If the inference assumptions hold, and if the null hypothesis $H_0: \beta_1 = \beta_2 = \cdots = \beta_p = 0$ is true, the *population of all possible overall F statistics* has an *F-distribution* with $k-1$ and $n-k$ degrees of freedom.

This result tells us that if F(model) is the overall F statistic that will be randomly selected from the population of all such statistics, then

$$P\begin{pmatrix}\text{We will commit} \\ \text{a Type I error}\end{pmatrix} = P\begin{pmatrix}\text{We will reject } H_0: \beta_1 = \beta_2 = \cdots = \beta_p = 0 \\ \text{when } H_0 \text{ is true}\end{pmatrix}$$

$$= P(F(\text{model}) > F_{[\alpha]}^{(k-1,n-k)} \text{ when } H_0 \text{ is true})$$

$$= A[F_{[\alpha]}^{(k-1,n-k)}; k-1, n-k]$$

$$= \alpha$$

(see Figure 7.2a). Therefore, since the population of all possible overall F statistics has an F-distribution with $k - 1$ and $n - k$ degrees of freedom when H_0 is true, testing H_0 by setting the probability of a Type I error equal to α means that, if H_0 is true, then $100\alpha\%$ of the overall F statistics in the population of all such statistics are greater than the rejection point $F_{[\alpha]}^{(k-1,n-k)}$ and thus would (if randomly selected) cause us to reject H_0 when H_0 is true (a Type I error).

Next, we note that the validity of condition (2) and the equivalence of conditions (1) and (2) follow by examining Figures 7.2a, 7.2b, and 7.2c. Moreover, since the population of all possible overall F statistics has an F-distribution with $k - 1$ and $n - k$ degrees of freedom when H_0 is true,

$$\text{prob-value} = A[F(\text{model}); k-1, n-k]$$

equals the proportion of overall F statistics in the population of all such statistics which, if the null hypothesis H_0 is true, are at least as large as and thus at least as contradictory to H_0 as F(model), the observed overall F statistic.

EXAMPLE 7.2

When we use the fuel consumption model

$$y_i = \beta_0 + \beta_1 x_{i1} + \beta_2 x_{i2} + \epsilon_i$$

(which has $k = 3$ parameters) to carry out a regression analysis of the data in Table 5.2, we obtain the following quantities (see Figure 7.1 for the SAS printout of the analysis of variance table summarizing these quantities):

$$\text{Total variation} = \sum_{i=1}^{8} (y_i - \bar{y})^2 = 25.5488$$

$$SS_{\text{model}} = \text{Explained variation} = \sum_{i=1}^{8} (\hat{y}_i - \bar{y})^2 = 24.8750$$

$$MS_{\text{model}} = \frac{SS_{\text{model}}}{k-1} = \frac{24.8750}{3-1} = 12.4375$$

$$SSE = \text{Unexplained variation} = \sum_{i=1}^{8} (y_i - \hat{y}_i)^2 = .6737$$

$$s^2 = MSE = \frac{SSE}{n-k} = \frac{.6737}{8-3} = .1347$$

$$F(\text{model}) = \frac{MS_{\text{model}}}{MSE} = \frac{12.4375}{.1347} = 92.303$$

$$\begin{aligned} \text{prob-value} &= A[F(\text{model}); k-1, n-k] \\ &= A[92.303; 3-1=2, 8-3=5] \end{aligned}$$

which is the area to the right of 92.303 under the curve of the F-distribution having 2 and 5 degrees of freedom and which can be shown to be less than .0001 (for the purposes of subsequent discussion, we suppose that the prob-value is equal to .0001). If we wish to use condition (1) to determine whether we can reject

$$H_0: \beta_1 = \beta_2 = 0$$

in favor of

H_1: At least one of β_1 or β_2
does not equal zero

by setting α, the probability of a Type I error, equal to .05, then we would use the rejection point

$$F_{[\alpha]}^{(k-1,n-k)} = F_{[.05]}^{(2,5)} = 5.79$$

Since $F(\text{model}) = 92.303 > 5.79 = F_{[.05]}^{(2,5)}$, we can reject H_0 in favor of H_1 by setting α equal to .05. Alternatively, since prob-value = .0001 is less than .05 and .01, it follows by condition (2) that we can reject H_0 in favor of H_1 by setting α equal to .05 or .01. Therefore, we conclude that at least one of the independent variables x_1 (average hourly temperature) or x_2 (the chill index) significantly affects y (weekly fuel consumption). Thus, we should use t_{b_j} statistics and other techniques to determine which of the independent variables significantly affects y (see Example 6.7).

In order to interpret the prob-value as a probability, recall that if the null hypothesis $H_0: \beta_1 = \beta_2 = 0$ is true, then the population of all possible overall F statistics has an F-distribution with $k-1=2$ and $n-k=5$ degrees of freedom. Therefore,

$$\begin{aligned} \text{prob-value} &= A[F(\text{model}); k-1, n-k] \\ &= A[92.303; 2, 5] \\ &= .0001 \end{aligned}$$

says that if the null hypothesis H_0 is true, then only .01% (or, equivalently, 1 in 10,000) of the overall F statistics in the population of all such statistics are at least as large as and thus at least as contradictory to H_0 as 92.303, the observed overall F statistic. Thus, the prob-value .0001 says that if the null hypothesis H_0 is true, then we have observed an overall F statistic so rare that its occurrence can be interpreted as a 1 in 10,000 chance. This interpretation helps us to reach one of two possible conclusions:

1. The null hypothesis $H_0: \beta_1 = \beta_2 = 0$ is true, and we have observed an overall F statistic so rare that its occurrence can be described as a 1 in 10,000 chance; or
2. The null hypothesis $H_0: \beta_1 = \beta_2 = 0$ is not true.

Since any reasonable person would find it very difficult to believe that a 1 in 10,000 chance has actually occurred, we would have substantial evidence to conclude that $H_0: \beta_1 = \beta_2 = 0$ is not true. The fact that the prob-value .0001 is so small casts great doubt on the validity of the null hypothesis $H_0: \beta_1 = \beta_2 = 0$ and lends great support to the validity of the alternative hypothesis H_1: At least one of β_1 or β_2 does not equal zero.

7.1.3 Simple Coefficients of Determination and Correlation: r^2 and r

If we are considering the simple linear regression model

$$y = \beta_0 + \beta_1 x + \epsilon$$

we call R^2 the **simple coefficient of determination** and denote this quantity by r^2. It follows that r^2 is the proportion of the total variation in the n observed values of the dependent variable that is explained by the simple linear regression model.

We next define the **simple correlation coefficient**, denoted by r, to be $r = \pm\sqrt{r^2}$, where we define $r = +\sqrt{r^2}$ if b_1, the least squares point estimate of the slope β_1, is a positive number, and $r = -\sqrt{r^2}$ if b_1 is a negative number.

Since for the simple linear regression model the prediction of y_i is

$$\hat{y}_i = b_0 + b_1 x_i$$

it can be shown (by utilizing the algebraic formulas for the least squares point estimates b_0 and b_1) that

$$r^2 = \frac{\sum_{i=1}^{n} (\hat{y}_i - \bar{y})^2}{\sum_{i=1}^{n} (y_i - \bar{y})^2}$$

is equal to

$$\frac{\left[\sum_{i=1}^{n} (x_i - \bar{x})(y_i - \bar{y})\right]^2}{\sum_{i=1}^{n} (x_i - \bar{x})^2 \sum_{i=1}^{n} (y_i - \bar{y})^2}$$

where

$$\bar{x} = \frac{\sum_{i=1}^{n} x_i}{n}$$

is the average of the n observed values of the independent variable x, and

$$\bar{y} = \frac{\sum_{i=1}^{n} y_i}{n}$$

is the average of the n observed values of the dependent variable y. Furthermore, taking square roots implies that

The **simple correlation coefficient** is

$$r = \frac{\sum_{i=1}^{n} (x_i - \bar{x})(y_i - \bar{y})}{\left[\sum_{i=1}^{n} (x_i - \bar{x})^2 \sum_{i=1}^{n} (y_i - \bar{y})^2\right]^{1/2}}$$

It can be shown that this formula for r is equivalent to the formula

$$r = b_1 \frac{s_x}{s_y}$$

where

$$s_x = \sqrt{\frac{\sum_{i=1}^{n} (x_i - \bar{x})^2}{n-1}} \quad \text{and} \quad s_y = \sqrt{\frac{\sum_{i=1}^{n} (y_i - \bar{y})^2}{n-1}}$$

and b_1 is the least squares point estimate of the slope β_1 in the simple linear regression model.

It follows that this formula for r automatically yields a value of the simple correlation coefficient which is positive if b_1 is positive or negative if b_1 is negative. Since r^2 is always between zero and 1, it follows that r is always between -1 and 1. A value of r close to 1 indicates that the independent variable x and the dependent variable y have a strong tendency to move together in a linear fashion with a positive slope and therefore that x and y are highly related and **positively correlated**. A value of r close to -1 indicates that x and y have a strong tendency to move together in a linear fashion with a negative slope and that the independent variables x and y are highly related and **negatively correlated**.

7.1.4 A *t*-Test for the Population Correlation Coefficient

We have seen that the simple correlation coefficient, r, measures the linear relationship, or correlation, between the n observed values of the independent variable, x, and the n observed values of the dependent variable, y, which constitute the sample. A similar coefficient of linear correlation can be defined for the population of all possible combinations of observed values of x and y. We call this coefficient the **population correlation coefficient** and denote it by ρ. We use r as the point estimate of ρ. Moreover, if we assume that the population of all possible combinations of observed values of x and y has a bivariate normal probability distribution (see Wonnacott and Wonnacott (1981) for a discussion of this distribution), we can use r to test the null hypothesis $H_0: \rho = 0$, which says that there is no linear relationship between x and y, versus the alternative hypothesis $H_1: \rho \neq 0$, which says that there is a (positive or negative) linear relationship between x and y. However, it can be shown that using r to test $H_0: \rho = 0$ versus $H_1: \rho \neq 0$ is equivalent to using the t_{b_1} statistic (and the related prob-value) to test $H_0: \beta_1 = 0$ versus $H_1: \beta_1 \neq 0$, where β_1 is the slope parameter—that is, the parameter describing the *linear relationship* between x and y—in the simple linear regression model.

TESTING A HYPOTHESIS CONCERNING β_1 IN THE SIMPLE LINEAR REGRESSION MODEL, OR TESTING A HYPOTHESIS CONCERNING ρ:

If

$$t_{b_1} = \frac{b_1}{s \Big/ \sqrt{\sum_{i=1}^{n} (x_i - \bar{x})^2}} \quad \text{and} \quad \text{prob-value} = 2A[[|t_{b_1}|, \infty); n - 2]$$

we can reject $H_0: \beta_1 = 0$ (or $H_0: \rho = 0$) in favor of $H_1: \beta_1 \neq 0$ (or $H_1: \rho \neq 0$) by setting the probability of a Type I error equal to α if and only if either of the following equivalent conditions hold:

1. $|t_{b_1}| > t_{[\alpha/2]}^{(n-2)}$, where $t_{[\alpha/2]}^{(n-2)}$ is the point on the scale of the t-distribution having $n - 2$ degrees of freedom so that the area under this curve to the right of $t_{[\alpha/2]}^{(n-2)}$ is $\alpha/2$.
2. Prob-value $< \alpha$.

This result follows from the fact that it can be shown by using matrix algebra that c_{11}, the diagonal element of $(\mathbf{X}'\mathbf{X})^{-1}$ corresponding to the parameter β_1 in the model

$$y = \beta_0 + \beta_1 x + \epsilon$$

is

$$\frac{1}{\sum_{i=1}^{n} (x_i - \bar{x})^2}$$

Thus, it follows that

$$t_{b_1} = \frac{b_1}{s\sqrt{c_{11}}} = \frac{b_1}{s \bigg/ \sqrt{\sum_{i=1}^{n} (x_i - \bar{x})^2}}$$

Notice that although the mechanics involved in testing $H_0: \rho = 0$ and $H_0: \beta_1 = 0$ are the same, the assumptions on which these tests are based are different. Testing $H_0: \rho = 0$ is based on the bivariate normality assumption, and testing $H_0: \beta_1 = 0$ is based on the inference assumptions.

EXAMPLE 7.3

Consider the fuel consumption problem and the simple linear regression model

$$y = \beta_0 + \beta_1 x_1 + \epsilon$$

which relates y, weekly fuel consumption, to x_1, average hourly temperature. The eight observed average hourly temperatures and the eight observed weekly fuel consumptions are listed and plotted in Figure 7.3. Using the historical data in Figure 7.3, we make the following calculations:

$$\bar{x}_1 = \frac{\sum_{i=1}^{8} x_{i1}}{8} = \frac{351.8}{8} = 43.98$$

$$\bar{y} = \frac{\sum_{i=1}^{8} y_i}{8} = \frac{81.7}{8} = 10.21$$

$$r = \frac{\sum_{i=1}^{8} (x_{i1} - \bar{x}_1)(y_i - \bar{y})}{\left[\sum_{i=1}^{8} (x_{i1} - \bar{x}_1)^2 \sum_{i=1}^{8} (y_i - \bar{y})^2\right]^{1/2}}$$

$$= \frac{(x_{11} - \bar{x}_1)(y_1 - \bar{y}) + \cdots + (x_{81} - \bar{x}_1)(y_8 - \bar{y})}{[[(x_{11} - \bar{x}_1)^2 + \cdots + (x_{81} - \bar{x}_1)^2][(y_1 - \bar{y})^2 + \cdots + (y_8 - \bar{y})^2]]^{1/2}}$$

$$= \frac{(28.0 - 43.98)(12.4 - 10.21) + \cdots + (62.5 - 43.98)(7.5 - 10.21)}{[[(28.0 - 43.98)^2 + \cdots + (62.5 - 43.98)^2][(12.4 - 10.21)^2 + \cdots + (7.5 - 10.21)^2]]^{1/2}}$$

$$= -.9487$$

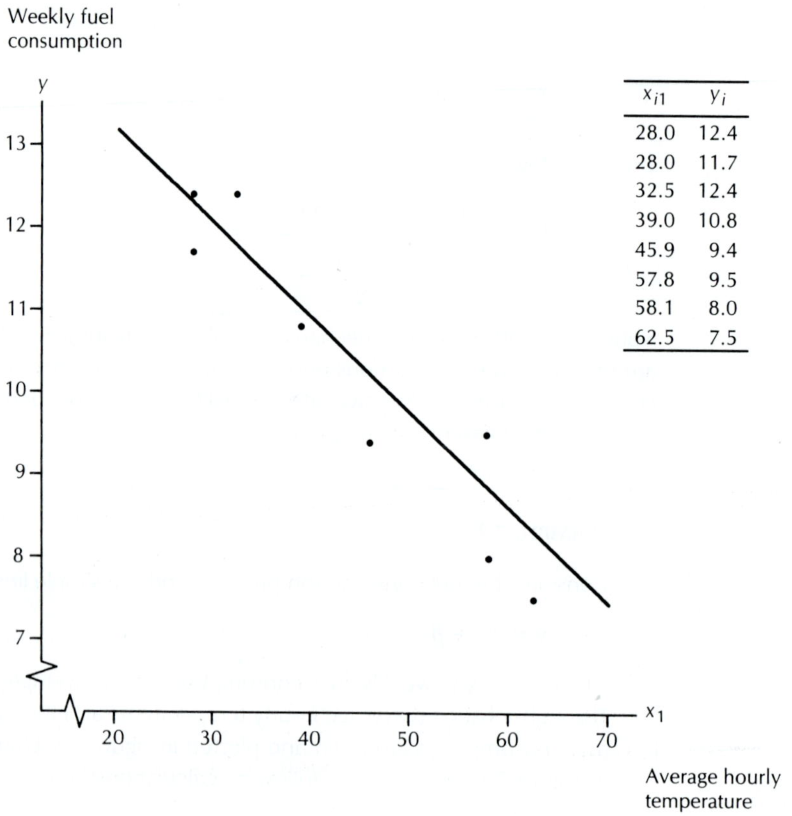

x_{i1}	y_i
28.0	12.4
28.0	11.7
32.5	12.4
39.0	10.8
45.9	9.4
57.8	9.5
58.1	8.0
62.5	7.5

FIGURE 7.3 Plot of the Observed Values of y (Weekly Fuel Consumption) Against Increasing Values of x_1 (Average Hourly Temperature)

The simple correlation coefficient of $r = -.9487$ indicates that (as illustrated in Figure 7.3) x_1 and y have a strong tendency to move together in a linear fashion with a negative slope. An r of $-.9487$ also means that the simple coefficient of determination equals

$$r^2 = (-.9487)^2 = .90$$

which says that the simple linear regression model

$$y = \beta_0 + \beta_1 x_1 + \epsilon$$

explains 90% of the total variation in the eight observed fuel consumptions.

Next, since we can use the data in Figure 7.3 to calculate the least squares point estimate of β_1 in this simple linear regression model to be $-.1279$ and the standard error to be .654, and since we can use the fuel consumption data to compute

$$\sum_{i=1}^{8} (x_{i1} - \bar{x}_1)^2 = 1404.371$$

it follows that

$$t_{b_1} = \frac{b_1}{s\Big/\sqrt{\sum_{i=1}^{8}(x_{i1}-\bar{x}_1)^2}} = \frac{-.1279}{(.654)/\sqrt{1404.371}} = -7.33$$

and

$$\begin{aligned}\text{prob-value} &= 2A[[|-7.33|, \infty); 8-2] \\ &= 2A[[7.33, \infty); 6] \\ &= .0003\end{aligned}$$

Thus, under the appropriate assumptions, since the prob-value is very small, we have very strong evidence that $H_0: \beta_1 = 0$ (or $H_0: \rho = 0$) is false and $H_1: \beta_1 \neq 0$ (or $H_1: \rho \neq 0$) is true, and thus that there is a negative linear relationship, or negative correlation, between x_1 and y.

7.1.5 A Comparison of Correlation Analysis and Regression Analysis

In this section we briefly compare correlation analysis and regression analysis. The key question answered by correlation analysis is whether any relationship exists between two variables—that is, how confident we can be that we should reject $H_0: \rho = 0$ in favor of $H_1: \rho \neq 0$. Regression analysis not only answers this question (by determining how confident we can be that we should reject $H_0: \beta_1 = 0$ in favor of $H_1: \beta_1 \neq 0$) but also answers more interesting questions—for example, how x and y are related and how x can be used to predict y.

For example, if we were to use the regression model

$$\begin{aligned} y &= \mu + \epsilon \\ &= \beta_0 + \beta_1 x + \epsilon \end{aligned}$$

to relate

y = the rate of return on a particular stock

to

x = the rate of return on the overall stock market

then, while $r = .74$ might indicate a moderate positive linear relationship, or correlation, between x and y, we could use regression analysis to learn about the nature of this relationship.

As another example, if we were to use the regression model

$$y = \beta_0 + \beta_1 x + \epsilon$$

to relate

y = the productivity of a word processing specialist

to

x = the score on a word processing aptitude test

then, while $r = .84$ might indicate a fairly strong positive linear relationship, or correlation, between x and y, regression analysis could be used to predict the productivity of prospective word processing specialists on the basis of their aptitude test scores.

In conclusion, regression analysis is more informative than (and thus preferred to) correlation analysis. Correlation analysis is usually used to understand better certain aspects of regression analysis (for example, multicollinearity, which we discuss in the next section).

7.2 THE PROBLEM OF MULTICOLLINEARITY

When the independent variables in a regression model are interrelated or are dependent on each other, **multicollinearity** is said to exist among the independent variables. We illustrate multicollinearity in the following example.

EXAMPLE 7.4

In the fuel consumption model

$$y = \beta_0 + \beta_1 x_1 + \beta_2 x_2 + \epsilon$$

it is logical that the independent variables x_1 (average hourly temperature) and x_2 (the chill index) might be related to each other, because the passage of weather fronts during a week might bring to State University both a low average hourly temperature and high wind velocities and other factors causing a high chill index. Alternatively, the absence of weather fronts during a week might bring to State University both a higher average hourly temperature and lower wind velocities and other factors causing a low chill index. Thus we might expect low values of x_1 (average hourly temperature) to be associated with high values of x_2 (the chill index) and higher values of x_1 to be associated with lower values of x_2. There are various ways to use the observed values of x_1 and x_2 to determine the existence and extent of multicollinearity between x_1 and x_2. One simple method is to plot the values of x_2 against the increasing values of x_1. The eight combinations of observed values of x_1 and x_2 are listed in Table 7.1, and the plot of the values of x_2 against the increasing values of x_1 is given in Figure 7.4. Figure 7.4 indicates that as x_1 increases, x_2 tends to decrease in a linear fashion with a negative slope. Thus, our expectations are confirmed—low values of x_1 are associated with high values of x_2 and higher values of x_1

TABLE 7.1 The Eight Combinations of Observed Values of x_1 and x_2

The *i*th Observed Week (OWK_i)	The *i*th Observed Average Hourly Temperature (x_{i1})	The *i*th Observed Chill Index (x_{i2})	The *i*th Combination of Observed Values of x_1 and x_2 (x_{i1}, x_{i2})
OWK_1	28.0	18	(28.0, 18)
OWK_2	28.0	14	(28.0, 14)
OWK_3	32.5	24	(32.5, 24)
OWK_4	39.0	22	(39.0, 22)
OWK_5	45.9	8	(45.9, 8)
OWK_6	57.8	16	(57.8, 16)
OWK_7	58.1	1	(58.1, 1)
OWK_8	62.5	0	(62.5, 0)

$$\sum_{i=1}^{8} x_{i1} = 351.8 \qquad \sum_{i=1}^{8} x_{i2} = 103$$

$$\bar{x}_1 = \frac{351.8}{8} = 43.98 \qquad \bar{x}_2 = \frac{103}{8} = 12.875$$

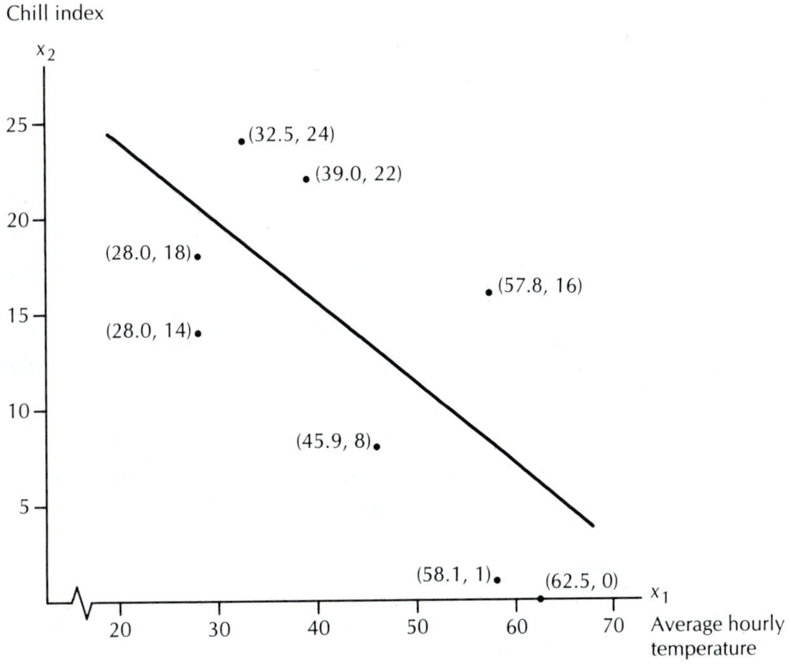

FIGURE 7.4 Multicollinearity Between x_1 (Average Hourly Temperature) and x_2 (Chill Index)

are associated with lower values of x_2. One way to measure the *extent* of the multicollinearity between x_1 and x_2 is to compute the *simple correlation coefficient* between x_1 and x_2. This quantity, denoted by r_{x_1,x_2}, is computed using the following formula:

$$r_{x_1,x_2} = \frac{\sum_{i=1}^{n} (x_{i1} - \bar{x}_1)(x_{i2} - \bar{x}_2)}{\left[\sum_{i=1}^{n} (x_{i1} - \bar{x}_1)^2 \sum_{i=1}^{n} (x_{i2} - \bar{x}_2)^2\right]^{1/2}}$$

where

$$\bar{x}_1 = \frac{\sum_{i=1}^{n} x_{i1}}{n} \quad \text{and} \quad \bar{x}_2 = \frac{\sum_{i=1}^{n} x_{i2}}{n}$$

Using the data in Table 7.1, we make the following calculations:

$$\bar{x}_1 = \frac{\sum_{i=1}^{8} x_{i1}}{8} = \frac{351.8}{8} = 43.98$$

$$\bar{x}_2 = \frac{\sum_{i=1}^{8} x_{i2}}{8} = \frac{103}{8} = 12.875$$

$$r_{x_1,x_2} = \frac{\sum_{i=1}^{8} (x_{i1} - \bar{x}_1)(x_{i2} - \bar{x}_2)}{\left[\sum_{i=1}^{8} (x_{i1} - \bar{x}_1)^2 \sum_{i=1}^{8} (x_{i2} - \bar{x}_2)^2\right]^{1/2}}$$

$$= \frac{(x_{11} - \bar{x}_1)(x_{12} - \bar{x}_2) + \cdots + (x_{81} - \bar{x}_1)(x_{82} - \bar{x}_2)}{[[(x_{11} - \bar{x}_1)^2 + \cdots + (x_{81} - \bar{x}_1)^2][(x_{12} - \bar{x}_2)^2 + \cdots + (x_{82} - \bar{x}_2)^2]]^{1/2}}$$

$$= \frac{(28.0 - 43.98)(18 - 12.875) + \cdots + (62.5 - 43.98)(0 - 12.875)}{[[(28.0 - 43.98)^2 + \cdots + (62.5 - 43.98)^2][(18 - 12.875)^2 + \cdots + (0 - 12.875)^2]]^{1/2}}$$

$$= -.7182$$

We next see how to interpret what $r_{x_1,x_2} = -.7182$ says about the extent of the multicollinearity between x_1 and x_2.

In general, consider the linear regression model

$$y = \beta_0 + \beta_1 x_1 + \cdots + \beta_j x_j + \cdots + \beta_k x_k + \cdots + \beta_p x_p + \epsilon$$

As illustrated in the preceding example, we can use the n observed values of the independent variables x_j and x_k to plot the values of x_k against the increasing values of x_j or to compute the *simple correlation coefficient* between x_j and x_k.

The simple correlation coefficient between x_j and x_k is

$$r_{x_j,x_k} = \frac{\sum_{i=1}^{n} (x_{ij} - \bar{x}_j)(x_{ik} - \bar{x}_k)}{\left[\sum_{i=1}^{n} (x_{ij} - \bar{x}_j)^2 \sum_{i=1}^{n} (x_{ik} - \bar{x}_k)^2\right]^{1/2}}$$

where

$$\bar{x}_j = \frac{\sum_{i=1}^{n} x_{ij}}{n} \quad \text{and} \quad \bar{x}_k = \frac{\sum_{i=1}^{n} x_{ik}}{n}$$

On comparing the formula for r_{x_j,x_k} with the formula for the simple correlation coefficient between x and y (see Section 7.1.3) r_{x_j,x_k} may be seen as a special case of r if we consider x_j to be the independent variable x and x_k to be the dependent variable y. From our interpretation of r in Section 7.1.3 it follows that r_{x_j,x_k} is a measure of the linear relationship, or correlation, between x_j and x_k. For example, the simple correlation coefficient between x_1 and x_2 in the fuel consumption problem, which is $r_{x_1,x_2} = -.7182$, indicates (as illustrated in Figure 7.4) that the independent variables x_1 and x_2 have some tendency to move together in a straight line fashion with a negative slope and therefore that the independent variables x_1 and x_2 are somewhat negatively correlated. Most standard multiple regression computer packages calculate the simple correlation coefficient for each pair of independent variables in the multiple regression model. However, these simple correlation coefficients are not the only measures of the multicollinearity between the independent variables. We discuss one other measure of multicollinearity later in this section.

In general, when two or more independent variables in a regression model are related to each other (or *correlated*), they contribute redundant information. That is, even though each independent variable contributes information for describing and predicting the dependent variable, some of the information overlaps. The consequences of multicollinearity in a regression analysis vary from slight to quite substantial, depending on the extent of the multicollinearity and the aspect of the regression analysis one considers. Since multicollinearity is often encountered in a regression analysis, we now summarize its effects.

7.2.1 The First Effect of Multicollinearity

If multicollinearity exists among the independent variables in a regression analysis, the values of the least squares point estimates of the parameters in the model depend on the particular independent variables that are included in the model. That is, when an independent variable is added to a regression model, and when the independent variable being added is related to the independent variable(s) already in the model, the least squares point estimates of the regression parameters will change. As an example, consider the one- and two-variable fuel consumption models

$$y = \beta_0 + \beta_1 x_1 + \epsilon \qquad \text{and} \qquad y = \beta_0 + \beta_1 x_1 + \beta_2 x_2 + \epsilon$$

which relate the dependent variable y, weekly fuel consumption, to the independent variables x_1, average hourly temperature, and x_2, the chill index. The least squares point estimates of the parameters β_0 and β_1 in the model

$$y = \beta_0 + \beta_1 x_1 + \epsilon$$

have been calculated to be $b_0 = 15.84$ and $b_1 = -.1279$. However, when the independent variable x_2, which is related to x_1, is added to the fuel consumption model, the least squares point estimates become $b_0 = 13.109$, $b_1 = -.0900$, and $b_2 = .0825$. Thus, we see that adding a correlated independent variable results in altered least squares point estimates b_0 and b_1. Thus, when multicollinearity exists between the independent variables in a regression analysis, the least squares point estimates are conditional—they depend upon the correlated independent variables included in the regression model. Because of this, when multicollinearity exists, a particular least squares point estimate b_j does not really measure the influence of the independent variable x_j upon the mean value of the dependent variable. Rather, b_j measures a partial influence of the independent variable x_j upon the mean value of the dependent variable, and this estimate b_j depends upon which of the correlated independent variables are included in the regression model. Extreme cases of multicollinearity can cause the least squares point estimates $b_0, b_1, b_2, \ldots, b_p$ of the parameters $\beta_0, \beta_1, \beta_2, \ldots, \beta_p$ to be far from the true values of the parameters. In fact, some of the least squares point estimates may have the wrong sign—that is, the sign (positive or negative) of a least squares point estimate may be different from the sign of the true value of the parameter.

It should also be mentioned that when multicollinearity exists, the practical meaning of the true regression parameters in a regression model becomes clouded. For example, in the fuel consumption model

$$\begin{aligned} y &= \mu + \epsilon \\ &= \beta_0 + \beta_1 x_1 + \beta_2 x_2 + \epsilon \end{aligned}$$

recall that

$$\beta_1 = \mu_{(c+1,d)} - \mu_{(c,d)} = \text{the change in mean weekly fuel consumption associated with a 1-degree increase in average hourly temperature when the chill index does not change}$$

Thus, the parameter β_1 measures the influence on mean weekly fuel consumption of a change in the average hourly temperature *when the chill index does not change*.

However, since the multicollinearity between x_1 and x_2 tells us that a change in average hourly temperature is often accompanied by a change in the chill index, the practical meaning of β_1 becomes uncertain.

7.2.2 The Second Effect of Multicollinearity

Perhaps the most serious effect of multicollinearity is that it can hinder our ability to use the $100(1-\alpha)\%$ confidence interval for β_j, or the t_{b_j} statistic, or the prob-value to determine whether an independent variable x_j is significantly related to the dependent variable. Since most standard multiple regression computer packages present the t_{b_j} statistic and prob-value for testing $H_0: \beta_j = 0$, the following discussion concerns these quantities. Recall that we can, by setting the probability of a Type I error equal to α, reject $H_0: \beta_j = 0$ in favor of $H_1: \beta_j \neq 0$ if and only if

$$|t_{b_j}| > t_{[\alpha/2]}^{(n-k)}$$

or, equivalently, if and only if

$$\text{prob-value} = 2A[[|t_{b_j}|, \infty); n-k] < \alpha$$

Thus, the larger (in absolute value) the t_{b_j} statistic is and the smaller the prob-value is, the stronger is the evidence that we should reject $H_0: \beta_j = 0$ in favor of $H_1: \beta_j \neq 0$. Therefore, intuitively, the size of the t_{b_j} statistic (how large it is) and the size of the prob-value (how small it is) measure the importance of the independent variable x_j in describing and predicting the dependent variable y. When multicollinearity exists, the sizes of the t_{b_j} statistic and of the related prob-value *measure the additional importance of the independent variable x_j over the combined importance of the other independent variables in the regression model.* Since two or more correlated independent variables contribute redundant information, multicollinearity often causes the t_{b_j} statistics obtained by running a regression analysis relating a dependent variable to a set of correlated independent variables to be smaller (in absolute value) than the t_{b_j} statistics that would be obtained if separate regression analyses were run, where each separate regression analysis relates the dependent variable to a smaller set (for example, only one) of the correlated independent variables. Thus, multicollinearity can cause some of the correlated independent variables to appear to be less important than they really are. In extreme cases, the t_{b_j} statistics and the prob-values for a set of correlated independent variables can make each and every variable look unimportant (that is, for each β_j we would not reject $H_0: \beta_j = 0$ with a small probability of a Type I error) when in fact the correlated independent variables taken together accurately describe and predict the dependent variable.

EXAMPLE 7.5

Consider the fuel consumption problem and the regression model

$$\begin{aligned} y_i &= \mu_i + \epsilon_i \\ &= \beta_0 + \beta_1 x_{i1} + \beta_2 x_{i2} + \epsilon_i \end{aligned}$$

TABLE 7.2 The Effect of Multicollinearity on the t_{b_j} Statistics and the Prob-Values in the Fuel Consumption Problem

The t_{b_j} Statistics and Prob-Values for Testing $H_0: \beta_0 = 0$, $H_0: \beta_1 = 0$, and $H_0: \beta_2 = 0$ in the Model $y = \beta_0 + \beta_1 x_1 + \beta_2 x_2 + \epsilon$

Independent Variable	b_j	$s\sqrt{c_{jj}}$
Intercept	$b_0 = 13.11$	$s\sqrt{c_{00}} = .8557$
x_1	$b_1 = -.0900$	$s\sqrt{c_{11}} = .0141$
x_2	$b_2 = .0825$	$s\sqrt{c_{22}} = .0220$

The t_{b_j} Statistics and Prob-Values for Testing $H_0: \beta_0 = 0$, $H_0: \beta_1 = 0$, $H_0: \beta_2 = 0$, and $H_0: \beta_3 = 0$ in the Model $y = \beta_0 + \beta_1 x_1 + \beta_2 x_2 + \beta_3 x_3 + \epsilon$

Independent Variable	b_j	$s\sqrt{c_{jj}}$
Intercept	$b_0 = 10.79$	$.365\sqrt{c_{00}} = .365\sqrt{43.4032} = 2.404$
x_1	$b_1 = -.07601$	$.365\sqrt{c_{11}} = .365\sqrt{.00285994} = .0195$
x_2	$b_2 = .06852$	$.365\sqrt{c_{22}} = .365\sqrt{.00497613} = .0257$
x_3	$b_3 = .1967$	$.365\sqrt{c_{33}} = .365\sqrt{.274305} = .1912$

$t_{b_j} = \dfrac{b_j}{s\sqrt{c_{jj}}}$	Prob-Value $= 2A[[\lvert t_{b_j}\rvert, \infty); n - k = 8 - 3 = 5]$
$t_{b_0} = \dfrac{b_0}{s\sqrt{c_{00}}} = 15.32$	$2A[[\lvert 15.32\rvert, \infty); 5] = .00002$
$t_{b_1} = \dfrac{b_1}{s\sqrt{c_{11}}} = -6.39$	$2A[[\lvert -6.39\rvert, \infty); 5] = .00139$
$t_{b_2} = \dfrac{b_2}{s\sqrt{c_{22}}} = 3.75$	$2A[[\lvert 3.75\rvert, \infty); 5] = .01330$

$t_{b_j} = \dfrac{b_j}{s\sqrt{c_{jj}}}$	Prob-Value $= 2A[[\lvert t_{b_j}\rvert, \infty); n - k = 8 - 4 = 4]$
$t_{b_0} = \dfrac{b_0}{s\sqrt{c_{00}}} = \dfrac{10.79}{2.404} = 4.49$	$2A[[4.49, \infty); 4] = 2(.00545) = .0109$
$t_{b_1} = \dfrac{b_1}{s\sqrt{c_{11}}} = \dfrac{-.07601}{.0195} = -3.90$	$2A[[\lvert -3.90\rvert, \infty); 4] = 2(.0088) = .0176$
$t_{b_2} = \dfrac{b_2}{s\sqrt{c_{22}}} = \dfrac{.06852}{.0257} = 2.67$	$2A[[\lvert 2.67\rvert, \infty); 4] = 2(.02815) = .0563$
$t_{b_3} = \dfrac{b_3}{s\sqrt{c_{33}}} = \dfrac{.1967}{.1912} = 1.03$	$2A[[\lvert 1.03\rvert, \infty); 4] = 2(.18075) = .3615$

The calculation of the t_{b_j} statistics and the prob-values for testing the importance of the independent variables in this model were summarized in Table 6.5, which is repeated here in Table 7.2. Remember that although the independent variable x_2, the chill index, measures much of the effect of wind velocity on the "chill factor," the chill index probably does not measure the total effect of the wind velocity on the chill factor, since, for instance, it is quite possible that two weeks with the same chill index might have somewhat different average hourly wind velocities. For this reason we might wish to consider a linear regression model that explicitly describes the effect of the average hourly wind velocity on fuel consumption. Letting x_3 denote the independent variable average hourly wind velocity and letting x_{i3} denote the average hourly wind velocity that has occurred in the ith observed week, we might consider the linear regression model

$$y_i = \mu_i + \epsilon_i$$
$$= \beta_0 + \beta_1 x_{i1} + \beta_2 x_{i2} + \beta_3 x_{i3} + \epsilon_i$$

In addition to the values of y, x_1, and x_2 that have occurred during the past $n = 8$ weeks, assume we have observed the values of average hourly wind velocity that have occurred during these weeks. The eight combinations of observed values of y, x_1, x_2, and x_3 are listed in Table 7.3. Using

$$
\mathbf{y} = \begin{bmatrix} 12.4 \\ 11.7 \\ 12.4 \\ 10.8 \\ 9.4 \\ 9.5 \\ 8.0 \\ 7.5 \end{bmatrix}
\quad \text{and} \quad
\mathbf{X} = \begin{bmatrix}
& x_1 & x_2 & x_3 \\
1 & 28.0 & 18.0 & 10.2 \\
1 & 28.0 & 14.0 & 11.5 \\
1 & 32.5 & 24.0 & 12.1 \\
1 & 39.0 & 22.0 & 9.5 \\
1 & 45.9 & 8.0 & 8.9 \\
1 & 57.8 & 16.0 & 9.2 \\
1 & 58.1 & 1.0 & 8.0 \\
1 & 62.5 & 0.0 & 7.0
\end{bmatrix}
$$

we can calculate the least squares point estimates of the parameters β_0, β_1, β_2, and β_3 in the preceding model to be

$$
\begin{bmatrix} b_0 \\ b_1 \\ b_2 \\ b_3 \end{bmatrix} = \mathbf{b} = (\mathbf{X'X})^{-1}\mathbf{X'y} = \begin{bmatrix} 10.79 \\ -.07601 \\ .06852 \\ .1967 \end{bmatrix}
$$

Table 7.3 shows the SAS output of these point estimates and other important quantities resulting from a regression analysis using this model. The calculation of the t_{b_j} statistics and the prob-values for testing the importance of x_1, x_2, x_3 and the intercept is summarized in Table 7.2. Note that in these calculations we

TABLE 7.3 The Observed Fuel Consumption Data, Including Observed Average Hourly Wind Velocities, and the SAS Output Resulting From a Regression Analysis Using the Model $y = \beta_0 + \beta_1 x_1 + \beta_2 x_2 + \beta_3 x_3 + \epsilon$

The *i*th Observed Week (OWK_i)	The *i*th Observed Average Hourly Temperature (x_{i1})	The *i*th Observed Chill Index (x_{i2})	The *i*th Observed Average Hourly Wind Velocity (x_{i3})	The *i*th Observed Fuel Consumption (y_i)
OWK_1	$x_{11} = 28.0$	$x_{12} = 18$	$x_{13} = 10.2$	$y_1 = 12.4$
OWK_2	$x_{21} = 28.0$	$x_{22} = 14$	$x_{23} = 11.5$	$y_2 = 11.7$
OWK_3	$x_{31} = 32.5$	$x_{32} = 24$	$x_{33} = 12.1$	$y_3 = 12.4$
OWK_4	$x_{41} = 39.0$	$x_{42} = 22$	$x_{43} = 9.5$	$y_4 = 10.8$
OWK_5	$x_{51} = 45.9$	$x_{52} = 8$	$x_{53} = 8.9$	$y_5 = 9.4$
OWK_6	$x_{61} = 57.8$	$x_{62} = 16$	$x_{63} = 9.2$	$y_6 = 9.5$
OWK_7	$x_{71} = 58.1$	$x_{72} = 1$	$x_{73} = 8.0$	$y_7 = 8.0$
OWK_8	$x_{81} = 62.5$	$x_{82} = 0$	$x_{83} = 7.0$	$y_8 = 7.5$

DEP VARIABLE: Y

SOURCE	DF	SUM OF SQUARES	MEAN SQUARE	F VALUE	PROB>F
MODEL	3	25.016129[a]	8.338710	62.624	0.0008
ERROR	4	0.532621[b]	0.133155		
C TOTAL	7	25.548750[c]			

ROOT MSE	0.364904[d]	R-SQUARE	0.9792[g]
DEP MEAN	10.212500	ADJ R-SQ	0.9635
C.V.	3.573116		

VARIABLE	DF	PARAMETER ESTIMATE[e]	STANDARD ERROR[f]	T FOR H0: PARAMETER=0[h]	PROB > |T|[i]
INTERCEP	1	10.794031	2.404032	4.490	0.0109
X1	1	-0.076013	0.019515	-3.895	0.0176
X2	1	0.068523	0.025741	2.662	0.0563
X3	1	0.196742	0.191115	1.029	0.3614

[a] Explained variation [b] *SSE* [c] Total variation [d] s
[e] $b_j: b_0, b_1, b_2, b_3$ [f] $s\sqrt{c_{jj}}$ [g] R^2 [h] t_{b_j} [i] Prob-value

have used the standard error, s, which is

$$s = \sqrt{\frac{\sum_{i=1}^{n} (y_i - \hat{y}_i)^2}{n-k}} = \sqrt{\frac{\sum_{i=1}^{n} y_i^2 - \mathbf{b}'\mathbf{X}'\mathbf{y}}{n-k}}$$

$$= \sqrt{\frac{.5326}{8-4}}$$

$$= .365$$

and the diagonal elements of $(\mathbf{X}'\mathbf{X})^{-1}$, where it can be shown that

$$(\mathbf{X}'\mathbf{X})^{-1} = \begin{array}{c} \\ 0 \\ 1 \\ 2 \\ 3 \end{array} \begin{array}{c} \begin{array}{cccc} 0 & 1 & 2 & 3 \end{array} \\ \begin{bmatrix} 43.4032 & -.31560 & .110327 & -3.22725 \\ -.31560 & .00285994 & .000264639 & .0195211 \\ .110327 & .000264639 & .00497613 & -.0194798 \\ -3.22725 & .0195211 & -.0194798 & .274305 \end{bmatrix} \end{array}$$

$$= \begin{bmatrix} c_{00} & & & \\ & c_{11} & & \\ & & c_{22} & \\ & & & c_{33} \end{bmatrix}$$

In order to interpret the results in Table 7.2, first note that we would expect multicollinearity to exist between x_1, x_2, and x_3. We have already discussed why we would expect multicollinearity between x_1 and x_2. In addition, we would expect x_1 (average hourly temperature) and x_3 (average hourly wind velocity) to be related, since a high average hourly wind velocity might have been due to the passage of weather fronts, which might also result in a low average hourly temperature. If we use the data in Table 7.3 to calculate r_{x_1,x_3}, the simple correlation coefficient between x_1 and x_3, we find that

$$r_{x_1,x_3} = \frac{\sum_{i=1}^{8} (x_{i1} - \bar{x}_1)(x_{i3} - \bar{x}_3)}{\left[\sum_{i=1}^{8} (x_{i1} - \bar{x}_1)^2 \sum_{i=1}^{8} (x_{i3} - \bar{x}_3)^2\right]^{1/2}}$$

$$= -.8659$$

which indicates that the independent variables x_1 and x_3 have a tendency to move together in a straight line fashion with a negative slope and therefore that x_1 and x_3 are somewhat negatively correlated. Next, recall that x_2, the chill index, is defined to measure much of the effect of wind velocity on the chill factor. Thus, the independent variables x_2 and x_3 are certainly related. In particular, a high average hourly wind velocity should result in a high chill index. If we use the data in Table 7.3 to calculate r_{x_2,x_3}, the simple correlation coefficient between x_2 and x_3, we find that

$$r_{x_2,x_3} = \frac{\sum_{i=1}^{8} (x_{i2} - \bar{x}_2)(x_{i3} - \bar{x}_3)}{\left[\sum_{i=1}^{8} (x_{i2} - \bar{x}_2)^2 \sum_{i=1}^{8} (x_{i3} - \bar{x}_3)^2\right]^{1/2}}$$

$$= .8054$$

which indicates that the independent variables x_2 and x_3 have a tendency to move together in a straight line fashion with a positive slope and therefore that x_2 and x_3 are somewhat positively correlated. The fact that there is multicollinearity between x_3 and x_1 and between x_3 and x_2 means that x_3 contributes information (for the description and prediction of weekly fuel consumption) that is already partially contributed by x_1 and by x_2. In fact, since we have defined the chill index as expressing, for a given average hourly temperature, the combined effects of all the major factors other than average hourly temperature on the general chill factor, we would hope that the chill index measures most of the effect of wind velocity on the general chill factor. Since the t_{b_j} statistic and prob-value for testing the hypothesis $H_0: \beta_3 = 0$ in the model

$$y = \beta_0 + \beta_1 x_1 + \beta_2 x_2 + \beta_3 x_3 + \epsilon$$

are (see Table 7.2)

$$t_{b_3} = 1.03 \qquad \text{and} \qquad \text{prob-value} = 2A[[|1.03|, \infty); 4] = .3615$$

we cannot reject $H_0: \beta_3 = 0$ with a small probability of a Type I error (say $\alpha = .05$ or $\alpha = .01$). Thus, we would conclude that the independent variable x_3 does not have substantial additional importance over the combined importance of the independent variables x_1 and x_2. This at least partly confirms our belief that we have defined the chill index so that it measures most of the effect of wind velocity and also implies that we probably should not include the average hourly wind velocity in a final regression model if we have included the chill index.

Next, using Table 7.2 to compare the t_{b_j} statistics and the corresponding prob-values for the hypotheses $H_0: \beta_0 = 0$, $H_0: \beta_1 = 0$, and $H_0: \beta_2 = 0$ as related to the two- and three-variable fuel consumption models, we find that for the two-variable model (which does not include x_3) the t_{b_j} statistics are larger than the corresponding t_{b_j} statistics for the three-variable model (which does include x_3), and we find that the corresponding prob-values are smaller for the two-variable model than for the three-variable model. This says that the t_{b_j} statistics and prob-values indicate that the independent variables x_1 and x_2 and the intercept β_0 have more additional importance in the two-variable model, which does not include the correlated independent variable x_3, than they have in the three-variable model, which does include x_3. Since we have concluded that x_3 does not have substantial additional importance over the combined importance of x_1 and x_2 in the regression model

$$y = \beta_0 + \beta_1 x_1 + \beta_2 x_2 + \beta_3 x_3 + \epsilon$$

and since removing x_3 from the model would eliminate the multicollinearity between x_3 and x_1 and x_2, we should use the model

$$y = \beta_0 + \beta_1 x_1 + \beta_2 x_2 + \epsilon$$

rather than the three-variable model to judge the importance of x_1 and x_2. By examining Table 7.2 and recalling our previous discussions concerning the t_{b_j} statistics and prob-values in this table, we conclude that we are confident that x_1 and x_2 are significantly related to y and that it is reasonable to use the regression model

$$y = \beta_0 + \beta_1 x_1 + \beta_2 x_2 + \epsilon$$

to describe and predict weekly fuel consumption.

The reason that multicollinearity can cause the t_{b_j} statistic

$$t_{b_j} = \frac{b_j}{s\sqrt{c_{jj}}}$$

and the related prob-value to give a misleading impression of the importance of the independent variable x_j is that c_{jj}, the diagonal element of $(\mathbf{X}'\mathbf{X})^{-1}$ corresponding to the parameter β_j, is used to calculate the t_{b_j} statistic. It can be shown that if we consider the regression model

$$y_i = \beta_0 + \beta_1 x_{i1} + \cdots + \beta_j x_{ij} + \cdots + \beta_p x_{ip} + \epsilon_i$$

and if we let

$$\bar{x}_j = \frac{\sum_{i=1}^{n} x_{ij}}{n}$$

then

$$c_{jj} = \frac{1}{\sum_{i=1}^{n} (x_{ij} - \bar{x}_j)^2 (1 - R_j^2)}$$

where R_j^2 is the multiple coefficient of determination that would be calculated by running a regression analysis using the model

$$x_{ij} = \beta_0 + \beta_1 x_{i1} + \cdots + \beta_{j-1} x_{i,j-1} + \beta_{j+1} x_{i,j+1} + \cdots + \beta_p x_{ip}$$

which expresses the independent variable x_j as a function of the remaining independent variables $x_1, \ldots, x_{j-1}, x_{j+1}, \ldots, x_p$. Since it follows from the discussion of Section 7.1 that R_j^2 is the proportion of the total variation in the n observed values of the independent variable x_j that is explained by the overall regression model, it follows that R_j^2 is a measure of the multicollinearity between the n observed values of x_j and the n combinations of observed values of the independent variables $x_1, \ldots, x_{j-1}, x_{j+1}, \ldots, x_p$. The greater this multicollinearity is, the greater is R_j^2, the larger is

$$c_{jj} = \frac{1}{\sum_{i=1}^{n} (x_{ij} - \bar{x}_j)^2(1 - R_j^2)}$$

and thus the larger is the denominator of the t_{b_j} statistic

$$t_{b_j} = \frac{b_j}{s\sqrt{c_{jj}}}$$

If the t_{b_j} statistic has a large denominator, then the t_{b_j} statistic might, depending on the size of b_j, be small. Thus, strong multicollinearity between the independent variable x_j and the remaining independent variables $x_1, \ldots, x_{j-1}, x_{j+1}, \ldots, x_p$ can cause the t_{b_j} statistic to be small (and thus the related prob-value to be large), which would give the impression that x_j is not important (even if it really is important) in describing and predicting the dependent variable y.

Recalling from Section 6.4 that the variance of the population of all possible least squares point estimates of the parameter β_j is

$$\sigma_{b_j}^2 = \sigma^2 c_{jj}$$

it follows that

$$\sigma_{b_j}^2 = \sigma^2 \frac{1}{\sum_{i=1}^{n} (x_{ij} - \bar{x}_j)^2(1 - R_j^2)}$$

$$= \frac{\sigma^2}{\sum_{i=1}^{n} (x_{ij} - \bar{x}_j)^2} \left(\frac{1}{1 - R_j^2}\right)$$

$$= \frac{\sigma^2}{\sum_{i=1}^{n} (x_{ij} - \bar{x}_j)^2} VIF_j$$

where

$$VIF_j = \frac{1}{1 - R_j^2}$$

is the *variance inflation factor for* b_j. If $R_j^2 = 0$, that is, if x_j is not related to the other independent variables through the regression model

$$x_j = \beta_0 + \beta_1 x_1 + \cdots + \beta_{j-1} x_{j-1} + \beta_{j+1} x_{j+1} + \cdots + \beta_p x_p + \epsilon$$

then the variance inflation factor VIF_j is equal to 1, which implies that

$$\sigma_{b_j}^2 = \frac{\sigma^2}{\sum_{i=1}^{n} (x_{ij} - \bar{x}_j)^2}$$

If $R_j^2 > 0$, that is, if x_j is related to the other independent variables through the model, then $1 - R_j^2$ is less than 1, and thus

$$VIF_j = \frac{1}{1 - R_j^2}$$

is greater than 1, which implies that

$$\sigma_{b_j}^2 = \frac{\sigma^2}{\sum_{i=1}^{n} (x_{ij} - \bar{x}_j)^2} VIF_j$$

is inflated beyond the value of $\sigma_{b_j}^2$ when $R_j^2 = 0$. Both the largest variance inflation factor among the independent variables and the mean of the variance inflation factors for the independent variables

$$\overline{VIF} = \frac{\sum_{j=1}^{p} VIF_j}{p}$$

are used as indicators of the severity of multicollinearity. If the largest variance inflation factor is greater than 10, or if the mean of the variance inflation factors is substantially greater than 1, then multicollinearity may be seriously influencing the least squares point estimates.

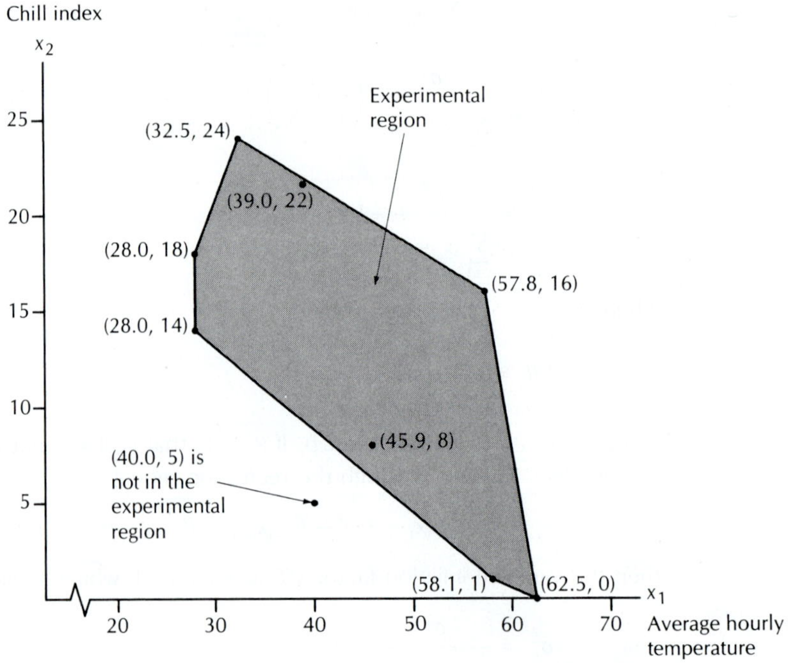

FIGURE 7.5 The Effect of Multicollinearity on the Experimental Region

7.2.3 The Third Effect of Multicollinearity

Generally speaking, multicollinearity between the independent variables in a regression model does not hinder the model's ability to predict a future value of the dependent variable based on a combination of future values of the independent variables if the combination of future values of the independent variables is in the experimental region. However, as illustrated in the following example, multicollinearity can cause certain combinations of future values of the independent variables that one might expect to be in the experimental region to actually not be in the experimental region.

EXAMPLE 7.6

Consider the fuel consumption model

$$y = \beta_0 + \beta_1 x_1 + \beta_2 x_2 + \epsilon$$

In Figure 7.5 we illustrate the experimental region, which is the shaded region surrounded by the points representing the eight combinations of the observed values of x_1 (average hourly temperature) and x_2 (the chill index). Consider the combination (40.0, 5) of a future average hourly temperature equal to 40.0 and a future chill index equal to 5. Since the future average hourly temperature of 40.0 is in the range of observed average hourly temperatures (which is 28.0 to 62.5) and since the future chill index of 5 is in the range of observed chill indices (which is 0 to 18) we might expect the combination (40.0, 5) to be in the experimental region. However, examination of Figure 7.5 indicates that the multicollinearity between x_1 and x_2 causes the experimental region to be a diagonal (or slanted) region that does not include the combination (40.0, 5).

7.2.4 The Fourth Effect of Multicollinearity

The last effect of multicollinearity that we will discuss is that the presence of strong multicollinearity among the independent variables in a regression model can lead to serious rounding errors in the calculation of the least squares point estimates of the regression parameters, even when a computer is being used to perform the calculations.[†] Many standard multiple regression computer programs are set up to minimize such errors, but serious rounding errors are more likely when strong multicollinearity exists. In order to lessen the chance of serious rounding errors, the analyst can "scale" the observed values of the independent and dependent variables (see Draper and Smith (1981)).

[†] The reason for this is that strong multicollinearity causes the columns of $\mathbf{X}$ to be nearly linearly dependent. This can cause substantial rounding errors in calculating $(\mathbf{X}'\mathbf{X})^{-1}$ and hence $\mathbf{b} = (\mathbf{X}'\mathbf{X})^{-1}\mathbf{X}'\mathbf{y}$.

The best solution to the multicollinearity problem is to try to avoid it by not including redundant independent variables in the regression model (variables that provide the same information). If we can identify redundancy among independent variables already in the regression model, several remedies can be used to attempt to lessen the influence of the multicollinearity. One possible solution is to remove one or more of the highly correlated independent variables in order to lessen the multicollinearity. However, one must be careful not to remove an independent variable that is important in describing and predicting the dependent variable. Such an independent variable should not be removed unless it can be replaced by another independent variable that is equally important in describing and predicting the dependent variable and that does not cause the same strong multicollinearity. This approach, however, has two disadvantages. The first disadvantage is that when the independent variable is removed from the regression model, the possibility of gathering any further information about the importance of that variable is eliminated. The second disadvantage is that the values of the least squares point estimates of the regression parameters corresponding to the variables remaining in the model are influenced by the fact that there are correlated independent variables that are not included in the regression model. A second possible solution to the multicollinearity problem is to add more observations to the data used in building the regression model, so that the multicollinearity is (possibly) lessened.

Finally, there are estimation procedures which are modifications of the least squares estimation procedure and which, when multicollinearity exists, are capable of producing point estimates of the parameters that are "better" than the least squares point estimates in the sense that they are closer to the true values of the parameters. One such procedure is called *ridge regression* and is discussed in Draper and Smith (1981).

7.3 COMPARING REGRESSION MODELS USING VARIOUS CRITERIA

7.3.1 Comparing Regression Models Using R^2, s^2, and the Length of a Prediction Interval

When building a regression model, we wish to obtain a final model that is *reasonably easy and inexpensive to use* (in terms for example, of being able to accurately and inexpensively determine future values of the independent variables) and that *accurately describes, predicts, and controls* the dependent variable. In previous sections we saw that prior belief about the significance of the independent variables, data plots, t_{b_j} statistics, and prob-values can help determine an appropriate set of independent variables to use in a final regression model. However, since multicollinearity can cause the t_{b_j} statistics and prob-values to give misleading impressions of the importance of the independent variables, we need other procedures for deciding which independent variables should be included in a final regression model. In this section we study several procedures of this kind.

One procedure to help determine an appropriate set of independent variables is to consider all reasonable regression models and compare them on the basis of various criteria. One such criterion is R^2, the multiple coefficient of determination. We have seen that the larger R^2 is for a particular regression model, the larger is the proportion of the total variation that is explained by the regression model. However, we must balance the magnitude of R^2, or in general, the "goodness" of any criterion, against the difficulty and expense of using the regression model to describe, predict, and control the dependent variable. Generally speaking, the size of R^2 does not necessarily indicate whether a regression model will accurately describe, predict, and control the dependent variable. For example, although a value of R^2 close to 1 indicates that the regression model explains a high proportion of the total variation in the n observed values of the dependent variable, a value of R^2 close to 1 does not necessarily imply that if we are interested in predicting future values of the dependent variable, the regression model will produce predictions accurate enough for the application at hand. Moreover, when using R^2 to *compare* overall regression models, we must be aware that (it can be proved that) *adding any independent variable to a regression model,* even an independent variable totally unrelated to the dependent variable,

1. Will decrease the unexplained variation

$$SSE = \sum_{i=1}^{n} (y_i - \hat{y}_i)^2$$

2. Will leave the total variation unchanged, since

$$\text{Total variation} = \sum_{i=1}^{n} (y_i - \bar{y})^2$$

   is a function of only the n observed values of the dependent variable $(y_1, y_2, \ldots, y_n)$ and thus is independent of the regression model under consideration.
3. Will increase

$$R^2 = 1 - \frac{\text{Unexplained variation}}{\text{Total variation}}$$

Since adding any independent variable to a regression model will increase R^2, it would be absurd to continue to add independent variables until R^2 decreases, because R^2 will never decrease!

One important consideration in predicting y_0, the future (individual) value of the dependent variable, is the *length of the prediction interval*

$$[\hat{y}_0 - BP_{\hat{y}_0}[100(1-\alpha)],\ \hat{y}_0 + BP_{\hat{y}_0}[100(1-\alpha)]]$$

which is

$$[\hat{y}_0 + BP_{\hat{y}_0}[100(1-\alpha)]] - [\hat{y}_0 - BP_{\hat{y}_0}[100(1-\alpha)]] = 2BP_{\hat{y}_0}[100(1-\alpha)]$$
$$= 2t_{[\alpha/2]}^{(n-k)} s\sqrt{1 + \mathbf{x}_0'(\mathbf{X}'\mathbf{X})^{-1}\mathbf{x}_0}$$

Since a shorter prediction interval provides a more precise guess of the future value y_0 than does a longer prediction interval, a shorter prediction interval is more desirable. A criterion related to the length of the prediction interval is the mean square error

$$s^2 = \frac{SSE}{n-k} = \frac{\sum_{i=1}^{n}(y_i - \hat{y}_i)^2}{n-k} = \frac{\sum_{i=1}^{n} y_i^2 - \mathbf{b}'\mathbf{X}'\mathbf{y}}{n-k}$$

Since we want the length of the prediction interval to be small, we want s, and thus s^2, to be small. Consequently, we want R^2 to be large and s^2 to be small. The mean square error, s^2, is a better criterion than R^2 for comparing regression models, because adding *any* independent variable to a regression model will increase R^2 but adding an "unimportant" independent variable (that is, an independent variable that is not significantly related to y) to a regression model can increase s^2. As we will demonstrate in Example 7.7, this can happen when the decrease in the unexplained variation caused by the addition of the independent variable is not enough to offset the decrease in the denominator of s^2, $n - k$, caused by the addition of the independent variable. If addition of an extra independent variable does increase s^2 (and thus s), this indicates that the length of the $100(1-\alpha)\%$ prediction interval for y_0 will almost surely increase and thus that the independent variable should not be included in a final regression model. Thus, when comparing models, s^2 is superior to R^2 because an increase in s^2 caused by the addition of an independent variable tells us that we (almost surely) should not include the independent variable in a final regression model, whereas R^2 cannot tell us that an independent variable added to a model should not be included in a final regression model (since R^2 always increases).

Since the inclusion of an additional independent variable in a regression model can actually increase the mean square error, it might be tempting to keep adding independent variables until the mean square error is not decreased. However, it would not be wise to use such a procedure. To see why, look more carefully at the relationship between s and the length of the $100(1-\alpha)\%$ prediction interval for y_0, which is

$$2t_{[\alpha/2]}^{(n-k)} s\sqrt{1 + \mathbf{x}_0'(\mathbf{X}'\mathbf{X})^{-1}\mathbf{x}_0}$$

Adding an independent variable

1. Increases k, the number of parameters in the regression model.
2. Decreases $n - k$, the number of degrees of freedom upon which the $t_{[\alpha/2]}^{(n-k)}$-point in the prediction interval is based.
3. Increases the $t_{[\alpha/2]}^{(n-k)}$-point (as will be demonstrated in Example 7.7).

The fact that adding an independent variable to a regression model increases $t_{[\alpha/2]}^{(n-k)}$, plus the fact that adding an independent variable usually does not decrease (but often increases) $\sqrt{1 + \mathbf{x}_0'(\mathbf{X}'\mathbf{X})^{-1}\mathbf{x}_0}$, means that if adding an independent variable to a regression model increases s^2 (and thus s), then the length of the 100(1 − α)% prediction interval for y_0 will almost surely increase, and if adding an independent variable to a regression model decreases s^2 (and thus s), then the length of the 100(1 − α)% prediction interval for y_0 will decrease if and only if the decrease in s is enough to offset the increase in $t_{[\alpha/2]}^{(n-k)}$ and the possible increase in $\sqrt{1 + \mathbf{x}_0'(\mathbf{X}'\mathbf{X})^{-1}\mathbf{x}_0}$. To summarize,

1. The independent variable x_j should not be included in the final regression model unless x_j reduces s^2 (and thus s) enough to reduce the length of the 100(1 − α)% prediction interval for y_0.
2. Prediction of a future value of the dependent variable would require knowledge of the corresponding future value of the independent variable x_j. Thus, in order to decide whether to include x_j in the final regression model, we must decide whether the inclusion of x_j reduces s enough so that the length of the 100(1 − α)% prediction interval for y_0 is reduced enough to offset the potential errors caused by possible inaccurate determination of future values of x_j, or the possible expense of accurately (or inaccurately) determining future values of x_j.

Since the key factor in determining a final regression model for use in prediction is the length of the 100(1 − α)% prediction interval for y_0, one might wonder why we do not use the length of the prediction interval directly in comparing regression models. While it is extremely useful to compare the lengths of prediction intervals, the length of the 100(1 − α)% prediction interval for y_0 depends upon $\mathbf{x}_0'$, the row vector containing the future values of the independent variables. Since we often wish to compute prediction intervals for different future situations, and since these future situations often have different combinations of future values of the independent variables and thus different $\mathbf{x}_0'$ vectors, we would compute prediction intervals having slightly different lengths. But s does not change for different future situations; s is a constant factor with respect to the length of prediction intervals for different future situations (as long as we are considering the same regression model). Thus, it is common practice to compare regression models on the basis of s^2 (or s).

EXAMPLE 7.7

Consider the fuel consumption problem. Recall that y = weekly fuel consumption, x_1 = average hourly temperature during the week, x_2 = the weekly chill index, and x_3 = average hourly wind velocity during the week. In Table 7.4 we summarize the total variation, the unexplained variation, the explained variation,

TABLE 7.4 R^2, s^2, and s for Eight Regression Models

Model	Total Variation	Unexplained Variation	Explained Variation
$y = \beta_0 + \beta_1 x_1 + \epsilon$	25.55	2.57	22.98
$y = \beta_0 + \beta_1 x_2 + \epsilon$	25.55	6.18	19.37
$y = \beta_0 + \beta_1 x_3 + \epsilon$	25.55	3.71	21.84
$y = \beta_0 + \beta_1 x_1 + \beta_2 x_2 + \epsilon$	25.55	.674	24.876
$y = \beta_0 + \beta_1 x_1 + \beta_2 x_3 + \epsilon$	25.55	1.48	24.07
$y = \beta_0 + \beta_1 x_2 + \beta_2 x_3 + \epsilon$	25.55	2.55	23.0
$y = \beta_0 + \beta_1 x_1 + \beta_2 x_2 + \beta_3 x_1 x_2 + \epsilon$	25.55	.660	24.89
$y = \beta_0 + \beta_1 x_1 + \beta_2 x_2 + \beta_3 x_3 + \epsilon$	25.55	.533	25.017

TABLE 7.5 The Lengths of 95% Prediction Intervals for Four Regression Models

Model	Point Prediction When $x_{01} = 40.0$, $x_{02} = 10$, $x_{03} = 11$
(1) $y = \beta_0 + \beta_1 x_1 + \epsilon$	$\hat{y}_0 = b_0 + b_1(40) = 10.72$
(2) $y = \beta_0 + \beta_1 x_1 + \beta_2 x_2 + \epsilon$	$\hat{y}_0 = b_0 + b_1(40) + b_2(10) = 10.33$
(3) $y = \beta_0 + \beta_1 x_1 + \beta_2 x_2 + \beta_3 x_1 x_2 + \epsilon$	$\hat{y}_0 = b_0 + b_1(40) + b_2(10) + b_3(40)(10) = 10.27$
(4) $y = \beta_0 + \beta_1 x_1 + \beta_2 x_2 + \beta_3 x_3 + \epsilon$	$\hat{y}_0 = b_0 + b_1(40) + b_2(10) + b_3(11) = 10.60$

Model	$\sqrt{1 + \mathbf{x}_0'(\mathbf{X}'\mathbf{X})^{-1}\mathbf{x}_0}$	$BP_{\hat{y}_0}[95]$	95% Prediction Interval $[\hat{y}_0 \pm BP_{\hat{y}_0}[95]]$
(1)	$\sqrt{1.14}$	1.71	[9.01, 12.43]
(2)	$\sqrt{1.21}$	1.04	[9.29, 11.37]
(3)	$\sqrt{1.5}$	1.38	[8.89, 11.65]
(4)	$\sqrt{1.75}$	1.34	[9.26, 11.94]

R^2, s^2, and s for eight regression models—three possible models with one independent variable each, three possible models with two independent variables each, and two possible models with three independent variables each. Note that the model

$$y = \beta_0 + \beta_1 x_1 + \epsilon$$

$R^2 = \frac{\text{Explained Variation}}{\text{Total Variation}}$	k	$n - k = 8 - k$	$s^2 = \frac{\text{Unexplained Variation}}{n - k}$	$s = \sqrt{s^2}$
.90	2	6	.428	.654
.758	2	6	1.03	1.015
.855	2	6	.618	.786
.974	3	5	.135	.367
.942	3	5	.295	.543
.90	3	5	.511	.715
.9742	4	4	.165	.406
.979	4	4	.133	.365

k	$n - k = 8 - k$	$t_{[.025]}^{(8-k)}$	s	Length of 95% Prediction Interval $2t_{[\alpha/2]}^{(8-k)} s\sqrt{1 + \mathbf{x}_0'(\mathbf{X}'\mathbf{X})^{-1}\mathbf{x}_0}$
2	6	2.447	.654	3.42
3	5	2.571	.367	2.08
4	4	2.776	.406	2.76
4	4	2.776	.365	2.68

has the largest R^2 and smallest s of the three one-variable models, and the model

$$y = \beta_0 + \beta_1 x_1 + \beta_2 x_2 + \epsilon$$

has the largest R^2 and smallest s of the three two-variable models. Table 7.5 summarizes information concerning 95% prediction intervals given by the four models

$$y = \beta_0 + \beta_1 x_1 + \epsilon$$
$$y = \beta_0 + \beta_1 x_1 + \beta_2 x_2 + \epsilon$$
$$y = \beta_0 + \beta_1 x_1 + \beta_2 x_2 + \beta_3 x_1 x_2 + \epsilon$$
$$y = \beta_0 + \beta_1 x_1 + \beta_2 x_2 + \beta_3 x_3 + \epsilon$$

This table contains, for each of these models, the 95% prediction interval for y_0, a future (individual) fuel consumption that will occur in a future week having an average hourly temperature of $x_{01} = 40.0$, a chill index of $x_{02} = 10$, and an average hourly wind velocity of $x_{03} = 11$; and the length of this 95% prediction interval.

First, note from Tables 7.4 and 7.5 that adding the independent variable x_2 to the regression model

$$y = \beta_0 + \beta_1 x_1 + \epsilon$$

to form the regression model

$$y = \beta_0 + \beta_1 x_1 + \beta_2 x_2 + \epsilon$$

1. Increases R^2 from .90 to .974.
2. Decreases s from .654 to .367.
3. Increases $t_{[.025]}^{(8-k)}$ from $t_{[.025]}^{(6)} = 2.447$ to $t_{[.025]}^{(5)} = 2.571$.
4. Increases $\sqrt{1 + \mathbf{x}_0'(\mathbf{X}'\mathbf{X})^{-1}\mathbf{x}_0}$ from $\sqrt{1.14}$ to $\sqrt{1.21}$.
5. Reduces the length of the 95% prediction interval for y_0 from 3.42 to 2.08, because the decrease in s (from .654 to .367) caused by the addition of x_2 has been enough to offset the increases in $t_{[.025]}^{(8-k)}$ and $\sqrt{1 + \mathbf{x}_0'(\mathbf{X}'\mathbf{X})^{-1}\mathbf{x}_0}$.

Thus, since we have strong evidence that $H_0: \beta_2 = 0$ is false (because the prob-value related to this hypothesis is .01330—see Table 7.2), and since the addition of the independent variable x_2 decreases s^2 and the length of the 95% prediction interval for y_0, we conclude that the independent variable x_2 does have substantial additional importance over the importance of the independent variable x_1 in the regression model

$$y = \beta_0 + \beta_1 x_1 + \beta_2 x_2 + \epsilon$$

Hence, we conclude that the independent variable x_2 should be included in a final regression model.

Next, note that the regression model

$$y = \beta_0 + \beta_1 x_1 + \beta_2 x_2 + \beta_3 x_1 x_2 + \epsilon$$

utilizes the cross-product term x_1x_2, which should be included in the regression model if interaction exists between x_1 and x_2 as these independent variables affect y. Using

$$\mathbf{y} = \begin{bmatrix} 12.4 \\ 11.7 \\ 12.4 \\ 10.8 \\ 9.4 \\ 9.5 \\ 8.0 \\ 7.5 \end{bmatrix} \quad \text{and} \quad \mathbf{X} = \begin{array}{c} \begin{matrix} 1 & x_1 & x_2 & x_1x_2 \end{matrix} \\ \begin{bmatrix} 1 & 28.0 & 18 & (28.0)(18) \\ 1 & 28.0 & 14 & (28.0)(14) \\ 1 & 32.5 & 24 & (32.5)(24) \\ 1 & 39.0 & 22 & (39.0)(22) \\ 1 & 45.9 & 8 & (45.9)(8) \\ 1 & 57.8 & 16 & (57.8)(16) \\ 1 & 58.1 & 1 & (58.1)(1) \\ 1 & 62.5 & 0 & (62.5)(0) \end{bmatrix} \end{array}$$

$$= \begin{array}{c} \begin{matrix} 1 & x_1 & x_2 & x_1x_2 \end{matrix} \\ \begin{bmatrix} 1 & 28.0 & 18 & 504 \\ 1 & 28.0 & 14 & 392 \\ 1 & 32.5 & 24 & 780 \\ 1 & 39.0 & 22 & 858 \\ 1 & 45.9 & 8 & 367.2 \\ 1 & 57.8 & 16 & 924.8 \\ 1 & 58.1 & 1 & 58.1 \\ 1 & 62.5 & 0 & 0 \end{bmatrix} \end{array}$$

we can calculate the least squares point estimates b_0, b_1, b_2, and b_3 of the parameters β_0, β_1, β_2, and β_3, the unexplained variation, s^2, and s as follows:

$$\begin{bmatrix} b_0 \\ b_1 \\ b_2 \\ b_3 \end{bmatrix} = \mathbf{b} = (\mathbf{X}'\mathbf{X})^{-1}\mathbf{X}'\mathbf{y} = \begin{bmatrix} 12.6926 \\ -.0825 \\ .1102 \\ -.0006 \end{bmatrix}$$

$$SSE = \sum_{i=1}^{8} y_i^2 - \mathbf{b}'\mathbf{X}'\mathbf{y} = .66$$

$$s^2 = \frac{SSE}{n-k} = \frac{.66}{8-4} = \frac{.66}{4} = .165$$

$$s = \sqrt{s^2} = \sqrt{.165} = .406$$

Since the mean square error for the model

$$y = \beta_0 + \beta_1x_1 + \beta_2x_2 + \beta_3x_1x_2 + \epsilon$$

is $s^2 = .165$, and since the mean square error for the model

$$y = \beta_0 + \beta_1 x_1 + \beta_2 x_2 + \epsilon$$

is $s^2 = .135$ (see Table 7.4), we see that adding the cross-product term x_1x_2 has increased s^2 from .135 to .165. In this case, although the addition of x_1x_2 has decreased the unexplained variation, this decrease (from .674 to .66) has not been enough to offset the change in the denominator of s^2, which decreases from 5 to 4. Examining Tables 7.4 and 7.5, we see that adding the cross-product term x_1x_2 to the model

$$y = \beta_0 + \beta_1 x_1 + \beta_2 x_2 + \epsilon$$

to form the model

$$y = \beta_0 + \beta_1 x_1 + \beta_2 x_2 + \beta_3 x_1 x_2 + \epsilon$$

1. Increases R^2 from .974 to .9742.
2. Increases s^2 from .135 to .165, and increases s from .367 to .406.
3. Increases $t_{[.025]}^{(8-k)}$ from $t_{[.025]}^{(5)} = 2.571$ to $t_{[.025]}^{(4)} = 2.776$.
4. Increases $\sqrt{1 + \mathbf{x}_0'(\mathbf{X}'\mathbf{X})^{-1}\mathbf{x}_0}$ from $\sqrt{1.21}$ to $\sqrt{1.5}$.
5. Increases the length of the 95% prediction interval for y_0 from 2.08 to 2.76, because the addition of x_1x_2 increases $t_{[.025]}^{(8-k)}$, s, and $\sqrt{1 + \mathbf{x}_0'(\mathbf{X}'\mathbf{X})^{-1}\mathbf{x}_0}$.

Moreover, since the t_{b_j} statistic and the prob-value for the hypothesis $H_0: \beta_3 = 0$ in the model

$$y = \beta_0 + \beta_1 x_1 + \beta_2 x_2 + \beta_3 x_1 x_2 + \epsilon$$

are

$$t_{b_3} = \frac{b_3}{s\sqrt{c_{33}}} = \frac{-.0006}{.406\sqrt{.000022}} = -.2894$$

and

$$\text{prob-value} = 2A[[|-.2894|, \infty); 4] = .7866$$

we cannot reject $H_0: \beta_3 = 0$ with a small probability of a Type I error (say $\alpha = .05$). Because of this, and because the addition of the cross-product term x_1x_2 increases s^2 and the length of the 95% prediction interval for y_0, we conclude that the independent variable x_1x_2 does not have substantial additional importance over the combined importance of the independent variables x_1 and x_2 in the regression model

$$y = \beta_0 + \beta_1 x_1 + \beta_2 x_2 + \beta_3 x_1 x_2 + \epsilon$$

and we should not include the independent variable x_1x_2 in a final regression model.

Last, note from Tables 7.4 and 7.5 that adding the independent variable x_3 to the regression model

$$y = \beta_0 + \beta_1 x_1 + \beta_2 x_2 + \epsilon$$

to form the regression model

$$y = \beta_0 + \beta_1 x_1 + \beta_2 x_2 + \beta_3 x_3 + \epsilon$$

1. Increases R^2 from .974 to .979.
2. Decreases s from .367 to .365.
3. Increases $t_{[.025]}^{(8-k)}$ from $t_{[.025]}^{(5)} = 2.571$ to $t_{[.025]}^{(4)} = 2.776$.
4. Increases $\sqrt{1 + \mathbf{x}_0'(\mathbf{X}'\mathbf{X})^{-1}\mathbf{x}_0}$ from $\sqrt{1.21}$ to $\sqrt{1.75}$.
5. Increases the length of the 95% prediction interval for y_0 from 2.08 to 2.68, because the decrease in s (from .367 to .365) caused by the addition of x_3 has not been enough to offset the increases in $t_{[.025]}^{(8-k)}$ and $\sqrt{1 + \mathbf{x}_0'(\mathbf{X}'\mathbf{X})^{-1}\mathbf{x}_0}$.

It should be noted that the large increase in $\sqrt{1 + \mathbf{x}_0'(\mathbf{X}'\mathbf{X})^{-1}\mathbf{x}_0}$ from $\sqrt{1.21}$ to $\sqrt{1.75}$ is partly caused by the fact that $x_{03} = 11$ causes the point $(x_{01}, x_{02}, x_{03}) = (40.0, 10, 11)$ to be quite far outside the experimental region, which is the range of the eight points representing the eight combinations of the observed values of x_1, x_2, and x_3 (see Table 7.3). However, even if there had been no increase in $\sqrt{1 + \mathbf{x}_0'(\mathbf{X}'\mathbf{X})^{-1}\mathbf{x}_0}$, the decrease in s (from .367 to .365) caused by the addition of x_3 would not have been enough to offset the increase in $t_{[.025]}^{(8-k)}$ (from 2.571 to 2.776). Thus, since we cannot reject $H_0\colon \beta_3 = 0$ with a small probability of a Type I error (from Table 7.2 the prob-value for this hypothesis test is .3615), and since the addition of the independent variable x_3 does not decrease s enough to decrease the length of the 95% prediction interval for y_0, we conclude that the independent variable x_3 does not have substantial additional importance over the combined importance of the independent variables x_1 and x_2 in the regression model

$$y = \beta_0 + \beta_1 x_1 + \beta_2 x_2 + \beta_3 x_3 + \epsilon$$

In addition, inclusion of x_3 in a final regression model would make the model more difficult and possibly more expensive to use (since future values of x_3 would have to be predicted). Therefore, the independent variable x_3 should not be included in a final regression model.

To summarize, we have decided that it is worthwhile to include the independent variables x_1, average hourly temperature, and x_2, the chill index, in a final regression model but that it is not worthwhile to include the independent variables x_1x_2 or x_3, average hourly wind velocity, in a final regression model. Thus, a reasonable final regression model to use in predicting weekly fuel consumption is

$$y = \beta_0 + \beta_1 x_1 + \beta_2 x_2 + \epsilon$$

7.3.2 Corrected R^2

Since the inclusion of an independent variable that is unrelated to the dependent variable will increase R^2 to some extent, it is sometimes useful to correct for this by reducing R^2 appropriately. We do this by considering a corrected R^2, denoted by $\bar{R}^2$ and defined as follows.

Assume that the regression model under consideration

$$y = \beta_0 + \beta_1 x_1 + \cdots + \beta_{k-1} x_{k-1} + \epsilon$$

includes $k - 1$ independent variables (and thus, because of the intercept β_0, utilizes k parameters). Then, the **corrected multiple coefficient of determination** (*corrected* R^2) is

$$\bar{R}^2 = \left(R^2 - \frac{k-1}{n-1}\right)\left(\frac{n-1}{n-k}\right)$$

where R^2 is the multiple coefficient of determination, and n is the number of observed values of the dependent variable.

To understand the reasoning behind this definition of $\bar{R}^2$, first note that it can be shown that if the values of the $k - 1$ independent variables are completely random (that is, randomly chosen from a population of numbers), they will still explain enough of the total variation in the observed values of the dependent variable to make R^2 equal to, on the average, $(k - 1)/(n - 1)$. Therefore, our first step in correcting R^2 is to subtract this random explanation and form the quantity

$$R^2 - \frac{k-1}{n-1}$$

If the values of the independent variables are completely random, then this corrected version of R^2 is equal to zero. However, if the values of the independent variables are not completely random, then the quantity

$$R^2 - \frac{k-1}{n-1}$$

reduces R^2 too much. To see why, note that if R^2 is equal to 1, then the preceding corrected version of R^2 is not equal to 1 but is equal to

$$1 - \frac{k-1}{n-1} = \frac{n-k}{n-1}$$

which is less than 1, since $n - k < n - 1$. To define a corrected R^2 that is equal to 1 if R^2 is equal to 1, we multiply

$$R^2 - \frac{k-1}{n-1}$$

by the factor

$$\frac{n-1}{n-k}$$

to form the quantity

$$\left(R^2 - \frac{k-1}{n-1}\right)\left(\frac{n-1}{n-k}\right)$$

which is the previously defined $\bar{R}^2$. Before presenting an example illustrating $\bar{R}^2$, note that it can be shown that the mean square error, s^2, and $\bar{R}^2$ are related by the equation

$$s^2 = (1 - \bar{R}^2)s_y^2$$

where

$$s_y^2 = \frac{\sum_{i=1}^{n} (y_i - \bar{y})^2}{n-1}$$

Since s_y^2 is independent of the model under consideration (because s_y^2 depends only on the observed values of the dependent variable $y_1, y_2, \ldots, y_n$), the equation

$$s^2 = (1 - \bar{R}^2)s_y^2$$

implies that s^2 *decreases if and only if* $\bar{R}^2$ *increases.*

EXAMPLE 7.8

In Table 7.6 we summarize R^2, $\bar{R}^2$, and s^2 for three regression models in the fuel consumption problem and present the SAS output of R^2, $\bar{R}^2$, and s^2 for the model

$$y = \beta_0 + \beta_1 x_1 + \beta_2 x_2 + \epsilon$$

Examining Table 7.6, we see that adding the independent variable x_2 to the regression model

$$y = \beta_0 + \beta_1 x_1 + \epsilon$$

TABLE 7.6 R^2, $\bar{R}^2$, and s^2 for Three Regression Models, and the SAS Output of R^2, $\bar{R}^2$, and s^2 for the Model $y = \beta_0 + \beta_1 x_1 + \beta_2 x_2 + \epsilon$

Model	k	R^2	$\bar{R}^2 = \left(R^2 - \frac{k-1}{n-1}\right)\left(\frac{n-1}{n-k}\right) = \left(R^2 - \frac{k-1}{7}\right)\left(\frac{7}{8-k}\right)$	s^2
$y = \beta_0 + \beta_1 x_1 + \epsilon$	2	.90	$\bar{R}^2 = \left(.90 - \frac{1}{7}\right)\left(\frac{7}{6}\right) = .883$	.428
$y = \beta_0 + \beta_1 x_1 + \beta_2 x_2 + \epsilon$	3	.974	$\bar{R}^2 = \left(.974 - \frac{2}{7}\right)\left(\frac{7}{5}\right) = .963$	.135
$y = \beta_0 + \beta_1 x_1 + \beta_2 x_2 + \beta_3 x_1 x_2 + \epsilon$	4	.9742	$\bar{R}^2 = \left(.9742 - \frac{3}{7}\right)\left(\frac{7}{4}\right) = .955$	.165

```
DEP VARIABLE: Y

                      SUM OF            MEAN
SOURCE       DF      SQUARES           SQUARE      F VALUE     PROB>F

MODEL         2    24.875018a       12.437509       92.303     0.0001
ERROR         5     0.673732b        0.134746e
C TOTAL       7    25.548750c

    ROOT MSE        0.367078d       R-SQUARE      0.9736f
    DEP MEAN       10.212500        ADJ R-SQ      0.9631g
    C.V.            3.594401
```

[a] Explained variation [b] SSE [c] Total variation [d] s
[e] s^2 [f] R^2 [g] $\bar{R}^2$

to form the model

$$y = \beta_0 + \beta_1 x_1 + \beta_2 x_2 + \epsilon$$

increases R^2, increases $\bar{R}^2$, and decreases s^2; and adding the cross-product term $x_1 x_2$ to the regression model

$$y = \beta_0 + \beta_1 x_1 + \beta_2 x_2 + \epsilon$$

to form the model

$$y = \beta_0 + \beta_1 x_1 + \beta_2 x_2 + \beta_3 x_1 x_2 + \epsilon$$

increases R^2, decreases $\bar{R}^2$, and increases s^2.

7.3.3 The *C* Statistic

Another criterion that can be used to compare overall regression models is called the C statistic. The C statistic is often called the C_k statistic in the statistical literature, but we omit the k for notational simplicity. The C statistic is defined as follows.

> Suppose we are attempting to choose an appropriate set of independent variables (to form a final regression model) from p potential independent variables. Then, if the regression model
>
> $$y = \beta_0 + \beta_1 x_1 + \cdots + \beta_{k-1} x_{k-1} + \epsilon$$
>
> includes $k - 1$ independent variables (and, because of the intercept β_0, utilizes k parameters), we define the ***C* statistic** to be
>
> $$C = \frac{SSE}{s_p^2} - (n - 2k)$$
>
> Here, n is the number of observed values of the dependent variable, SSE is the unexplained variation calculated from the model
>
> $$y = \beta_0 + \beta_1 x_1 + \cdots + \beta_{k-1} x_{k-1} + \epsilon$$
>
> and s_p^2 is the mean square error calculated from the model
>
> $$y = \beta_0 + \beta_1 x_1 + \cdots + \beta_p x_p + \epsilon$$
>
> which includes all p independent variables (and thus utilizes $p + 1$ parameters).

Since

$$C = \frac{SSE}{s_p^2} - (n - 2k)$$

is a function of SSE, the unexplained variation calculated from the model

$$y = \beta_0 + \beta_1 x_1 + \cdots + \beta_{k-1} x_{k-1} + \epsilon$$

and since we want SSE to be small, *we want C to be small.* Although adding even an unimportant independent variable to a regression model will decrease the unexplained variation, adding an unimportant independent variable can increase C. As we show in Example 7.9, this can happen when the decrease in the unexplained variation caused by the addition of the extra independent variable is not enough to offset the decrease in $n - 2k$ caused by the addition of the extra independent variable (which increases k by 1). It should be noted that although adding an unimportant independent variable to a regression model can increase both s^2 and C, there is no exact relationship between s^2 and C. For example, as we show in Example 7.9, adding an independent variable to a regression model can decrease s^2 and increase C. We have seen that we want to find a model for which C is small. In addition, it can be shown from the theory behind the C statistic that *we also wish to find a model for which the C statistic roughly equals k* (k equals the number of parameters in the model). If a model has a C statistic substantially greater than k, it can be shown that this model has substantial bias and is undesirable. Thus, although we want to find a model for which C is as small as possible, if C for such a model is substantially greater than k, we may prefer to choose a different model for which C is slightly larger and more nearly equal to the number of parameters in that (different) model.

EXAMPLE 7.9

For the fuel consumption problem, assume that we are attempting to choose an appropriate set of independent variables (to form a final regression model) from the $p = 3$ potential independent variables x_1, average hourly temperature, x_2, the chill index, and x_3, average hourly wind velocity. The mean square error calculated from the model

$$y = \beta_0 + \beta_1 x_1 + \beta_2 x_2 + \beta_3 x_3 + \epsilon$$

which includes all $p = 3$ independent variables is

$$s_p^2 = \frac{SSE}{n - (p + 1)} = \frac{.532}{8 - (3 + 1)} = .133$$

Thus, the C statistic for the three regression models

$$y = \beta_0 + \beta_1 x_1 + \epsilon$$

$$y = \beta_0 + \beta_1 x_1 + \beta_2 x_2 + \epsilon$$

$$y = \beta_0 + \beta_1 x_1 + \beta_2 x_2 + \beta_3 x_3 + \epsilon$$

are as summarized in Table 7.7.

Since $s^2 = .428$ and $C = 15.323$ for the model

$$y = \beta_0 + \beta_1 x_1 + \epsilon$$

and since $s^2 = .135$ and $C = 3.068$ for the model

$$y = \beta_0 + \beta_1 x_1 + \beta_2 x_2 + \epsilon$$

we see that adding the independent variable x_2 to the model

$$y = \beta_0 + \beta_1 x_1 + \epsilon$$

TABLE 7.7 The C Statistics for Three Regression Models

Model	SSE	k	s^2	$C = \frac{SSE}{s_p^2} - (n - 2k) = \frac{SSE}{.133} - (8 - 2k)$
$y = \beta_0 + \beta_1 x_1 + \epsilon$	2.57	2	.428	$C = \frac{2.57}{.133} - (8 - 2(2)) = 19.323 - 4 = 15.323$
$y = \beta_0 + \beta_1 x_1 + \beta_2 x_2 + \epsilon$	.674	3	.135	$C = \frac{.674}{.133} - (8 - 2(3)) = 5.068 - 2 = 3.068$
$y = \beta_0 + \beta_1 x_1 + \beta_2 x_2 + \beta_3 x_3 + \epsilon$	.532	4	.133	$C = \frac{.532}{.133} - (8 - 2(4)) = 4.0 - 0 = 4.0$

has decreased both s^2 and C. Since $s^2 = .133$ and $C = 4.0$ for the model

$$y = \beta_0 + \beta_1 x_1 + \beta_2 x_2 + \beta_3 x_3 + \epsilon$$

we see that adding the independent variable x_3 to the model

$$y = \beta_0 + \beta_1 x_1 + \beta_2 x_2 + \epsilon$$

has decreased s^2 (from .135 to .133) but has increased C (from 3.068 to 4.0). In this case, note from Table 7.7 that although the addition of x_3 has decreased the unexplained variation, this decrease (from .674 to .532) has not been enough to offset the change in $n - 2k$, which decreases from $n - 2k = 8 - 2(3) = 2$ for the model

$$y = \beta_0 + \beta_1 x_1 + \beta_2 x_2 + \epsilon$$

to $n - 2k = 8 - 2(4) = 0$ for the model

$$y = \beta_0 + \beta_1 x_1 + \beta_2 x_2 + \beta_3 x_3 + \epsilon$$

Thus, since the model

$$y = \beta_0 + \beta_1 x_1 + \beta_2 x_2 + \epsilon$$

has the smallest C statistic, $C = 3.068$, which roughly equals the number of parameters in this model, $k = 3$, the model

$$y = \beta_0 + \beta_1 x_1 + \beta_2 x_2 + \epsilon$$

is the best of the three models in Table 7.7 if we use the C statistic as a criterion.

7.4 MODEL BUILDING FOR THE FRESH DETERGENT PROBLEM

We now consider finding a final model to use to predict demand for Fresh. Table 7.8 summarizes R^2 and s^2 for several possible regression models in the Fresh Detergent problem. Consider, for example, the model

$$y = \beta_0 + \beta_1 x_4 + \beta_2 x_3 + \beta_3 x_3^2 + \beta_4 x_4 x_3 + \epsilon$$

Since the total variation of the 30 observed demands for Fresh in Table 5.12 is

$$\sum_{i=1}^{30} (y_i - \bar{y})^2 = \sum_{i=1}^{30} y_i^2 - 30\bar{y}^2 = 13.4586$$

and since the explained variation for this model can be calculated to be

$$\sum_{i=1}^{30} (\hat{y}_i - \bar{y})^2 = \mathbf{b}'\mathbf{X}'\mathbf{y} - 30\bar{y}^2 = 12.3942$$

it follows that the multiple coefficient of determination for this model is

$$R^2 = \frac{\text{Explained variation}}{\text{Total variation}} = \frac{12.3942}{13.4586} = .9209$$

An R^2 of .9209 says that this regression model explains 92.09% of the total variation in the 30 observed demands for Fresh.

Most of the models in Table 7.8 involve x_1, x_2, x_3, or x_4 (as defined in Table 5.12). However, we also include a model involving x_5, competitors' average advertising expenditures. In Example 5.9 we said that when forecasting demand for Fresh, Enterprise Industries would not use the independent variable "competitors' average advertising expenditures," because the competitors probably would not reveal their advertising budgets for future sales periods. However, in order to illustrate some of the model-building concepts that we have presented in this chapter, we will suppose that in addition to having observed the Fresh Detergent data presented in Table 5.12, Enterprise Industries has analyzed the advertising campaigns of its two main competitors, Dynamic Industries and G & P Corporation, over the past 30 sales periods and has estimated for these sales periods the values of the independent variable

x_5 = competitors' average advertising expenditures measured in hundreds of thousands of dollars

The 30 estimated values of x_5 are listed in Table 7.9, where, for $i = 1, 2, \ldots, 30$, x_{i5} denotes the estimated competitors' average advertising expenditures in the ith observed sales period.

If we consider adding the independent variable x_5 to the model

$$y = \beta_0 + \beta_1 x_4 + \beta_2 x_3 + \epsilon$$

TABLE 7.8

R^2 and s^2 for Several Regression Models in the Fresh Detergent Problem

Model	R^2	s^2
$y = \beta_0 + \beta_1 x_1 + \epsilon$	.2202	.3784
$y = \beta_0 + \beta_1 x_1 + \beta_2 x_1^2 + \epsilon$	.2286	.3845
$y = \beta_0 + \beta_1 x_2 + \epsilon$	.5490	.2168
$y = \beta_0 + \beta_1 x_2 + \beta_2 x_2^2 + \epsilon$	.5590	.2198
$y = \beta_0 + \beta_1 x_3 + \epsilon$	.7673	.1119
$y = \beta_0 + \beta_1 x_3 + \beta_2 x_3^2 + \epsilon$	.8380	.0808
$y = \beta_0 + \beta_1 x_4 + \epsilon$	.7915	.1002
$y = \beta_0 + \beta_1 x_4 + \beta_2 x_4^2 + \epsilon$	.8043	.0975
$y = \beta_0 + \beta_1 x_1 + \beta_2 x_2 + \epsilon$	.8288	.0854
$y = \beta_0 + \beta_1 x_1 + \beta_2 x_3 + \epsilon$	.7717	.1138
$y = \beta_0 + \beta_1 x_2 + \beta_2 x_3 + \epsilon$	.8377	.0809
$y = \beta_0 + \beta_1 x_1 + \beta_2 x_2 + \beta_3 x_3 + \epsilon$	.8936	.0551
$y = \beta_0 + \beta_1 x_1 + \beta_2 x_2 + \beta_3 x_3 + \beta_4 x_3^2 + \epsilon$	.9084	.0493
$y = \beta_0 + \beta_1 x_1 + \beta_2 x_1^2 + \beta_3 x_2 + \beta_4 x_2^2 + \beta_5 x_3 + \beta_6 x_3^2 + \epsilon$	.9161	.0491
$y = \beta_0 + \beta_1 x_4 + \beta_2 x_3 + \epsilon$	.8860	.0568
$y = \beta_0 + \beta_1 x_4 + \beta_2 x_3 + \beta_3 x_5 + \epsilon$	.8860	.0590
$y = \beta_0 + \beta_1 x_4 + \beta_2 x_3 + \beta_3 x_3^2 + \epsilon$	.9054	.0490
$y = \beta_0 + \beta_1 x_4 + \beta_2 x_4^2 + \beta_3 x_3 + \beta_4 x_3^2 + \epsilon$	.9106	.0481
$y = \beta_0 + \beta_1 x_4 + \beta_2 x_3 + \beta_3 x_3^2 + \beta_4 x_4 x_3 + \epsilon$	.9209	.0426
$y = \beta_0 + \beta_1 x_4 + \beta_2 x_3 + \beta_3 x_3^2 + \beta_4 x_4 x_3 + \beta_5 x_4 x_3^2 + \epsilon$	.9225	.0434

and thus forming the model

$$y = \beta_0 + \beta_1 x_4 + \beta_2 x_3 + \beta_3 x_5 + \epsilon$$

we see that s^2 increases from .0568 to .0590. Thus, the addition of x_5 is not warranted. One reason for the lack of significance of x_5 may be that the values of x_5 do not vary a great deal (see Table 7.9). This lack of variability may be due to the fact that in a given sales period, "above average advertising expenditures" by one competitor and "below average advertising expenditures" by the other competitor tend to average out. Of course, the lack of variability in the x_5 values may also be due to the fact that neither competitor changed advertising expenditures much from period to period. Enterprise Industries would have to study this situation thoroughly before concluding that variations in "competitors' average advertising expenditures" do not significantly affect demand for Fresh.

Next, consider using the term $x_4 x_3^2$, in addition to the term $x_4 x_3$, to model the interaction between x_4 and x_3. Examination of Table 7.8 reveals that adding $x_4 x_3^2$ to the model

$$y = \beta_0 + \beta_1 x_4 + \beta_2 x_3 + \beta_3 x_3^2 + \beta_4 x_4 x_3 + \epsilon$$

TABLE 7.9
The 30 (Estimated) Competitors' Average Advertising Expenditures

The ith Observed Sales Period (OSP_i)	The ith Estimated Competitors' Average Advertising Expenditures (x_{i5})	The ith Observed Sales Period (OSP_i)	The ith Estimated Competitors' Average Advertising Expenditures (x_{i5})
OSP_1	6.40	OSP_{16}	6.30
OSP_2	6.30	OSP_{17}	6.50
OSP_3	6.20	OSP_{18}	6.30
OSP_4	6.40	OSP_{19}	6.25
OSP_5	6.30	OSP_{20}	6.30
OSP_6	6.30	OSP_{21}	6.45
OSP_7	6.25	OSP_{22}	6.50
OSP_8	6.50	OSP_{23}	6.45
OSP_9	6.40	OSP_{24}	6.35
OSP_{10}	6.25	OSP_{25}	6.25
OSP_{11}	6.25	OSP_{26}	6.40
OSP_{12}	6.35	OSP_{27}	6.35
OSP_{13}	6.20	OSP_{28}	6.45
OSP_{14}	6.30	OSP_{29}	6.35
OSP_{15}	6.35	OSP_{30}	6.45

to form the model

$$y = \beta_0 + \beta_1 x_4 + \beta_2 x_3 + \beta_3 x_3^2 + \beta_4 x_4 x_3 + \beta_5 x_4 x_3^2 + \epsilon$$

increases s^2 from .0426 to .0434. Thus, the addition of $x_4 x_3^2$ is not warranted. It should also be noted that since the independent variables $x_4 x_3$ and $x_4 x_3^2$ are related to each other and to the independent variables x_4, x_3, and x_3^2, substantial multicollinearity certainly exists between the independent variables in this model. In fact, so much multicollinearity exists that the prob-values related to the hypotheses $H_0 : \beta_4 = 0$ and $H_0 : \beta_5 = 0$ in this model are both larger than .10 and thus make the interaction terms $x_4 x_3$ and $x_4 x_3^2$ appear to be unimportant. If we were to conclude from this that little or no interaction exists between x_4 and x_3, we would be making a mistake, because we recall (see Table 6.6) that the prob-value related to the hypothesis $H_0 : \beta_4 = 0$ in the model

$$y = \beta_0 + \beta_1 x_4 + \beta_2 x_3 + \beta_3 x_3^2 + \beta_4 x_4 x_3 + \epsilon$$

is less than .05. It follows that the interaction term $x_4 x_3$ really is important. Thus, we should include the interaction term $x_4 x_3$ in the Fresh Detergent demand model. This illustrates that when we are considering using squared terms and interaction terms in a regression model, multicollinearity can hinder the ability of t_{b_j} statistics and prob-values to indicate accurately which of these terms are important.

TABLE 7.10 The Lengths of 95% Prediction Intervals for Three Regression Models in the Fresh Detergent Problem

Model	Point Prediction When $x_{01} = 3.80$, $x_{02} = 3.90$, $x_{04} = .10$, $x_{03} = 6.80$
(1) $y = \beta_0 + \beta_1 x_1 + \beta_2 x_2 + \beta_3 x_3 + \epsilon$	$\hat{y}_0 = b_0 + b_1(3.80) + b_2(3.90) + b_3(6.80) = 8.325$
(2) $y = \beta_0 + \beta_1 x_4 + \beta_2 x_3 + \beta_3 x_3^2 + \epsilon$	$\hat{y}_0 = b_0 + b_1(.10) + b_2(6.80) + b_3(6.80)^2 = 8.445$
(3) $y = \beta_0 + \beta_1 x_4 + \beta_2 x_3 + \beta_3 x_3^2 + \beta_4 x_4 x_3 + \epsilon$	$\hat{y}_0 = b_0 + b_1(.10) + b_2(6.80) + b_3(6.80)^2 + b_4(.10)(6.80) = 8.526$

Model	$BP_{\hat{y}_0}[95]$	95% Prediction Interval $[\hat{y}_0 \pm BP_{\hat{y}_0}[95]]$	Length of 95% Prediction Interval $2BP_{\hat{y}_0}[95]$
(1)	.523	[7.80, 8.85]	1.05
(2)	.483	[7.96, 8.93]	.97
(3)	.458	[8.07, 8.98]	.91

In order to decide upon a final model, note that the model

$$y = \beta_0 + \beta_1 x_4 + \beta_2 x_3 + \beta_3 x_3^2 + \beta_4 x_4 x_3 + \epsilon$$

has the smallest s^2 of any model in Table 7.8. It also yields the shortest prediction interval of any model in Table 7.10, and the prob-values in Table 6.6 indicate that the intercept β_0 and the independent variables x_4, x_3, x_3^2, and x_4x_3 are important in describing demand for Fresh. All these considerations imply that this model is a reasonable final model to use to predict demand for Fresh Detergent.

7.5 STEPWISE REGRESSION, FORWARD SELECTION, BACKWARD ELIMINATION, AND MAXIMUM R^2 IMPROVEMENT

In the fuel consumption problem we have attempted to choose an appropriate set of independent variables to form a final regression model from the three potential independent variables: x_1, the average hourly temperature, x_2, the chill index, and x_3, the average hourly wind velocity. In a regression problem where the number of potential independent variables is not large, we can fairly easily compare all reasonable regression models with respect to various criteria (such as R^2, s^2, and C). However, if we are attempting to choose an appropriate set of independent variables from a large number of potential independent variables, comparing all reasonable regression models can be quite unwieldy. In this case, it is useful to employ a screening procedure that can be used to identify one set (or several sets) of the "most important" independent variables. We now present four such screening procedures.

7.5.1 Stepwise Regression

Stepwise regression is generally carried out on a computer and is available in most standard regression computer packages. There are slight variations in the way that different computer packages carry out stepwise regression. Assuming that y is the dependent variable and $x_1, x_2, \ldots, x_p$ are the p potential independent variables (where p will generally be large), we explain how most of the computer packages carry out stepwise regression. To make our description as concise as possible, we need to introduce some new terminology. Stepwise regression uses t_{b_j} statistics (and related prob-values) to determine the importance (or significance) of the independent variables in various regression models. In this context, *the t_{b_j} statistic indicates that the independent variable x_j is significant at the α level if and only if the related prob-value is less than α* (which implies that we can reject $H_0: \beta_j = 0$ in favor of $H_1: \beta_j \neq 0$ by allowing the probability of a Type I error to be equal to α). Then stepwise regression is performed as follows.

Choice of α_{entry} and α_{stay}: Before beginning the stepwise procedure, we choose a value of α_{entry}, which we call "the probability of a Type I error related to entering an independent variable into the regression model," and a value of α_{stay}, which we call "the probability of a Type I error related to retaining an independent variable that was previously entered into the model." We discuss the considerations involved in choosing these values after our description of the stepwise procedure. For now, suffice it to say that it is common practice to set both α_{entry} and α_{stay} equal to .05.

Step 1: The stepwise procedure considers the p possible one-independent-variable regression models of the form

$$y = \beta_0 + \beta_1 x_j + \epsilon$$

Each different model includes a different potential independent variable. For each model, the t_{b_1} statistic (and prob-value) related to testing $H_0: \beta_1 = 0$ versus $H_1: \beta_1 \neq 0$ is calculated. Denoting the independent variable giving the largest absolute value of the t_{b_1} statistic (and the smallest prob-value) by the symbol $x_{[1]}$, we consider the model

$$y = \beta_0 + \beta_1 x_{[1]} + \epsilon$$

If the t_{b_1} statistic does not indicate that the independent variable $x_{[1]}$ is significant at the α_{entry} level in the model, then the stepwise procedure terminates by choosing the model

$$y = \beta_0 + \epsilon$$

If the t_{b_1} statistic indicates that the independent variable $x_{[1]}$ is significant at the α_{entry} level in the model, then $x_{[1]}$ is retained for use in Step 2.

Step 2: The stepwise procedure considers the $p - 1$ possible two-independent-variable regression models of the form

$$y = \beta_0 + \beta_1 x_{[1]} + \beta_2 x_j + \epsilon$$

Each different model includes $x_{[1]}$, the independent variable chosen in Step 1, and a different potential independent variable chosen from the remaining $p - 1$ independent variables that were not chosen in Step 1. For each model, the t_{b_2} statistic (and prob-value) related to testing $H_0: \beta_2 = 0$ versus $H_1: \beta_2 \neq 0$ is calculated. Denoting the independent variable giving the largest absolute value of the t_{b_2} statistic (and the smallest prob-value) by the symbol $x_{[2]}$, we consider the model

$$y = \beta_0 + \beta_1 x_{[1]} + \beta_2 x_{[2]} + \epsilon$$

If the t_{b_2} statistic indicates that the independent variable $x_{[2]}$ is significant at the α_{entry} level in the model, then $x_{[2]}$ is retained in this model, and the stepwise procedure checks to see if the independent variable $x_{[1]}$ should be allowed to stay in the model.

This check should be made, because multicollinearity will probably cause the t_{b_1} statistic calculated for the model

$$y = \beta_0 + \beta_1 x_{[1]} + \beta_2 x_{[2]} + \epsilon$$

to be different from the t_{b_1} statistic calculated for the model

$$y = \beta_0 + \beta_1 x_{[1]} + \epsilon$$

If the t_{b_1} statistic does not indicate that the independent variable $x_{[1]}$ is significant at the α_{stay} level in the model

$$y = \beta_0 + \beta_1 x_{[1]} + \beta_2 x_{[2]} + \epsilon$$

then the stepwise procedure returns to the beginning of Step 2, and starting with a new one-independent-variable model that uses the new significant independent variable $x_{[2]}$, the stepwise procedure attempts to find a new two-independent-variable model

$$y = \beta_0 + \beta_1 x_{[2]} + \beta_2 x_j + \epsilon$$

If the t_{b_1} statistic indicates that the independent variable $x_{[1]}$ is significant at the α_{stay} level in the model

$$y = \beta_0 + \beta_1 x_{[1]} + \beta_2 x_{[2]} + \epsilon$$

then both the independent variables $x_{[1]}$ and $x_{[2]}$ are retained for use in Step 3.

Further steps: The stepwise procedure continues by adding independent variables one at a time to the model. At each step an independent variable is added to the model if and only if it has the largest (in absolute value) t_{b_j} statistic of the independent variables not in the model, and if its t_{b_j} statistic indicates that it is significant at the α_{entry} level. After adding an independent variable to the model, the stepwise procedure checks all the independent variables already included in the model and removes any independent variable that does not produce a t_{b_j} statistic indicating that the variable is significant at the α_{stay} level. Only after the necessary removals are made does the stepwise procedure attempt to add another independent variable to the model. The stepwise procedure terminates when all the independent variables not in the model have t_{b_j} statistics indicating that these variables are insignificant at the α_{entry} level or when the variable to be added to the model is the one just removed from it.

Regarding the choice of α_{entry} and α_{stay}, Draper and Smith (1981) state that it is usually best to choose α_{entry} equal to α_{stay}. It is not recommended that α_{stay} be chosen less than α_{entry}, because this makes it too likely that an independent variable that has just been added to the model will (in subsequent steps) be removed from the model. Sometimes, however, it is reasonable to choose α_{stay} to be greater than α_{entry}, because this makes it more likely that an independent variable whose significance decreases as new independent variables are added to the model will be allowed to stay in the model. Draper and Smith go on to suggest that α_{entry} and α_{stay} be set equal to .05 or .10.

However, we should point out that setting α_{entry} and α_{stay} higher than .10 is also reasonable, because this will cause more independent variables to be included in the model and thus will give the analyst an opportunity to consider additional independent variables. Indeed, though the model obtained by the stepwise procedure may be reasonable, it should not necessarily be regarded as the best final regression model. First, since the choices of α_{entry} and α_{stay} are arbitrary, and since the many hypothesis tests performed by the stepwise procedure imply that Type I and Type II errors might be committed, the stepwise procedure might include some unimportant independent variables in the model and exclude some important independent variables from the model. Second, it is sometimes appropriate to include powers (such as squared values) of the independent variables and interaction terms in a final regression model. While such terms can be included in the set of p potential independent variables to be considered by stepwise regression, we often omit them so that the (probably) already large list of potential independent variables is not unduly increased. Thus, if we do omit powers and interaction terms from consideration, and if some of these terms are important, the stepwise procedure will omit some important independent variables. In general, then, stepwise regression should be regarded as a screening procedure that can be used to find at least some of the most important independent variables. Once stepwise regression identifies these independent variables, we should then carefully use the other model-building techniques discussed in this book to arrive at an appropriate final regression model.

EXAMPLE 7.10

In the fuel consumption problem, recall that y, weekly fuel consumption, is the dependent variable. We let x_1, the average hourly temperature, x_2, the chill index, and x_3, the average hourly wind velocity, be the $p = 3$ potential independent variables to be considered in stepwise regression, and we set both α_{entry} and α_{stay} equal to .05.

Step 1: The $p = 3$ possible one-independent-variable regression models of the form

$$y = \beta_0 + \beta_1 x_j + \epsilon$$

are

$$y = \beta_0 + \underset{(.00033)}{\beta_1 x_1} + \epsilon$$

$$y = \beta_0 + \underset{(.00490)}{\beta_1 x_2} + \epsilon$$

$$y = \beta_0 + \underset{(.00101)}{\beta_1 x_3} + \epsilon$$

The prob-value related to testing $H_0: \beta_1 = 0$ versus $H_1: \beta_1 \neq 0$ is given below the $\beta_1 x_j$ term for each model. Since the independent variable giving the smallest prob-value is x_1, we consider the model

$$y = \beta_0 + \underset{(.00033)}{\beta_1 x_1} + \epsilon$$

Since .00033, the prob-value related to $H_0: \beta_1 = 0$, is less than $\alpha_{\text{entry}} = .05$, the independent variable x_1 is significant at the α_{entry} level and is retained for use in Step 2.

Step 2: The $p - 1 = 3 - 1 = 2$ possible two-independent-variable regression models of the form

$$y = \beta_0 + \beta_1 x_{[1]} + \beta_2 x_j + \epsilon$$

are

$$y = \beta_0 + \underset{(.00139)}{\beta_1 x_1} + \underset{(.01330)}{\beta_2 x_2} + \epsilon$$

$$y = \beta_0 + \underset{(.04040)}{\beta_1 x_1} + \underset{(.11250)}{\beta_2 x_3} + \epsilon$$

The prob-values related to testing $H_0: \beta_1 = 0$ and $H_0: \beta_2 = 0$ are given below the $\beta_1 x_1$ and $\beta_2 x_j$ terms for each model. Since the independent variable giving the smallest prob-value related to $H_0: \beta_2 = 0$ is x_2, we consider the model

$$y = \beta_0 + \underset{(.00139)}{\beta_1 x_1} + \underset{(.01330)}{\beta_2 x_2} + \epsilon$$

Since .01330, the prob-value related to $H_0: \beta_2 = 0$, is less than $\alpha_{\text{entry}} = .05$, the independent variable x_2 is significant at the α_{entry} level. Since .00139, the prob-value related to $H_0: \beta_1 = 0$, is less than $\alpha_{\text{stay}} = .05$, the independent variable x_1 is significant at the α_{stay} level. Thus, both x_1 and x_2 are retained for use in Step 3.

Step 3: The $p - 2 = 3 - 2 = 1$ three-independent-variable regression model of the form

$$y = \beta_0 + \beta_1 x_{[1]} + \beta_2 x_{[2]} + \beta_3 x_j + \epsilon$$

is

$$y = \beta_0 + \underset{(.01761)}{\beta_1 x_1} + \underset{(.05627)}{\beta_2 x_2} + \underset{(.36145)}{\beta_3 x_3} + \epsilon$$

The prob-values related to testing $H_0: \beta_1 = 0$, $H_0: \beta_2 = 0$, and $H_0: \beta_3 = 0$ are summarized below the $\beta_1 x_1$, $\beta_2 x_2$, and $\beta_3 x_3$ terms in this model. Since .36145, the prob-value related to $H_0: \beta_3 = 0$, is not less than $\alpha_{\text{entry}} = .05$, the independent variable x_3 is not significant at the α_{entry} level. Thus, the stepwise procedure terminates by choosing the regression model

$$y = \beta_0 + \beta_1 x_1 + \beta_2 x_2 + \epsilon$$

Note that this model is the same model chosen by the other model-building techniques presented in this chapter.

Although we have used only $p = 3$ potential independent variables in the fuel consumption problem, remember that stepwise regression is most profitably used as a screening procedure in a regression problem having a large number of potential independent variables. We now illustrate such a use of stepwise regression.

EXAMPLE 7.11

In the Fresh Detergent problem, recall that y, demand for Fresh, is the dependent variable. We let x_1, the price for Fresh; x_2, the average industry price; x_3, the advertising expenditure for Fresh; $x_4 = x_2 - x_1$, the price difference; x_5, competitors' average advertising expenditures; and $x_6 = x_3 - x_5$, the advertising expenditure difference, be the $p = 6$ potential independent variables to be considered in stepwise regression. When both α_{entry} and α_{stay} are set equal to .05, the stepwise regression procedure that we have described (1) adds x_4 on the first step, (2) adds x_3 (and retains x_4) on the second step, and (3) terminates after Step 2 when no more independent variables can be added. Thus, the stepwise regression procedure arrives at the model

$$y = \beta_0 + \beta_1 x_4 + \beta_2 x_3 + \epsilon$$

Three comments should be made here. First, although we can calculate the least squares point estimates of the parameters in the models

$$y = \beta_0 + \beta_1 x_1 + \beta_2 x_4 + \epsilon$$

$$y = \beta_0 + \beta_1 x_2 + \beta_2 x_4 + \epsilon$$

we cannot calculate the least squares point estimates of the parameters in the model

$$y = \beta_0 + \beta_1 x_1 + \beta_2 x_2 + \beta_3 x_4 + \epsilon$$

because in this model the independent variable $x_4 = x_2 - x_1$ is a linear combination of x_1 and x_2, and thus the columns of the **X** matrix are linearly dependent, which implies that $(\mathbf{X}'\mathbf{X})^{-1}$ does not exist. However, the stepwise procedure adds x_4 on the first step, and since x_4, x_1, and x_2 contribute redundant information, we do not reach a point where we consider a model containing x_4, x_1, and x_2 (similar comments can be made concerning x_3, x_5, and x_6). Second, the stepwise procedure has added x_4 instead of x_1 or x_2. This confirms Enterprise Industries' belief (mentioned in Example 5.1) that the single independent variable $x_4 = x_2 - x_1$ adequately describes the effects of x_1 and x_2 on y. Third, since the stepwise procedure has not considered powers and interaction terms, the model arrived at by this procedure,

$$y = \beta_0 + \beta_1 x_4 + \beta_2 x_3 + \epsilon$$

should not be considered the best final regression model. In fact, previous analysis (see, for example, Table 7.9) has indicated that

$$y = \beta_0 + \beta_1 x_4 + \beta_2 x_3 + \beta_3 x_3^2 + \beta_4 x_4 x_3 + \epsilon$$

is a more appropriate model.

7.5.2 Forward Selection

This procedure works in the same way as stepwise regression, *except that once an independent variable is entered into the model, it is never removed.* Forward selection is generally considered to be less effective than stepwise regression but to be useful in some problems.

7.5.3 Backward Elimination

A regression analysis is performed by using a regression model containing all the p potential independent variables. Then, the independent variable having the smallest (in absolute value) t_{b_j} statistic is chosen. If the t_{b_j} statistic indicates that this independent variable is significant at the α_{stay} level (α_{stay} is chosen prior to the beginning of the procedure), then the procedure terminates by choosing the regression model containing all p independent variables. If the t_{b_j} statistic does not indicate that this independent variable is significant at the α_{stay} level, then this independent variable is removed from the model and a regression analysis is performed by using a regression model containing all the remaining independent variables. The procedure continues by removing independent variables one at a time from the model. At each step, an independent variable is removed from the model if it has the smallest (in absolute value) t_{b_j} statistic of the independent variables remaining in the model and if its t_{b_j} statistic does not indicate that it is significant at the α_{stay} level. The procedure terminates when no independent variable remaining in the model can be removed. Backward elimination is generally considered to be a reasonable procedure, especially for analysts who like to start with all possible independent variables in the model so that they will not "miss any important variables."

7.5.4 Maximum R^2 Improvement

The following description of this procedure is quoted from the *SAS User's Guide, 1982 Edition*, which calls this procedure the MAXR method.

> Unlike the three techniques above, this method does not settle on a single model. Instead, it looks for the "best" one-variable model, the "best" two-variable model, and so forth.
>
> The MAXR method begins by finding the one-variable model producing the highest R^2. Then another variable, the one that would yield the greatest increase in R^2, is added.
>
> Once the two-variable model is obtained, each of the variables in the model is compared to each variable not in the model. For each comparison, MAXR determines if removing one variable and replacing it with the other variable would increase R^2. After comparing all possible switches, the one that produces the largest increase in R^2 is made.
>
> Comparisons begin again, and the process continues until MAXR finds that no switch could increase R^2. The two-variable model thus achieved is considered the "best" two-variable model the technique can find.
>
> Another variable is then added to the model, and the comparing and switching process is repeated to find the "best" three-variable model, and so forth.
>
> The difference between the stepwise technique and the maximum R^2 improvement method is that all switches are evaluated before any switch is made in the MAXR method. In the stepwise method, the "worst" variable may be removed without considering what adding the "best" remaining variable might accomplish.

The MAXR method is becoming increasingly popular and is generally considered to be superior to stepwise regression, forward selection, and backward elimination.

7.6 AN *F*-TEST FOR PORTIONS OF A MODEL

In this section we present an *F*-test related to the utility of a portion of a regression model. For example, if we consider the Fresh Detergent model

$$y = \beta_0 + \beta_1 x_4 + \beta_2 x_3 + \beta_3 x_3^2 + \beta_4 x_4 x_3 + \epsilon$$

it might be useful to test the null hypothesis

$$H_0: \beta_3 = \beta_4 = 0$$

which says that neither of the higher-order terms x_3^2 and x_4x_3 affect y, versus the alternative hypothesis

H_1: At least one of β_3 and β_4 does not equal zero

which says that at least one of the higher-order terms x_3^2 and x_4x_3 affects y. In general, consider the regression model

$$y = \beta_0 + \beta_1 x_1 + \cdots + \beta_g x_g + \beta_{g+1} x_{g+1} + \cdots + \beta_p x_p + \epsilon$$

and consider testing the null hypothesis

$$H_0: \beta_{g+1} = \beta_{g+2} = \cdots = \beta_p = 0$$

which says that none of the independent variables $x_{g+1}, x_{g+2}, \ldots, x_p$ affect y, versus the alternative hypothesis

H_1: At least one of $\beta_{g+1}, \beta_{g+2}, \ldots, \beta_p$ does not equal zero

which says that at least one of the independent variables $x_{g+1}, x_{g+2}, \ldots, x_p$ affects y. If we can reject H_0 in favor of H_1 by specifying a *small* probability of a Type I error, then it is reasonable to conclude that at least one of $x_{g+1}, x_{g+2}, \ldots, x_p$ *significantly* affects y. In this case, we should use t_{b_j} statistics and other techniques to determine which of $x_{g+1}, x_{g+2}, \ldots, x_p$ significantly affect y. In order to test H_0 versus H_1, consider the following two models:

Complete model: $y = \beta_0 + \beta_1 x_1 + \cdots + \beta_g x_g + \beta_{g+1} x_{g+1} + \cdots + \beta_p x_p + \epsilon$

Reduced model: $y = \beta_0 + \beta_1 x_1 + \cdots + \beta_g x_g + \epsilon$

Here, the complete model is assumed to have k parameters, the reduced model is the complete model under the assumption that H_0 is true, and $(p - g)$ denotes the number of regression parameters we have set equal to zero in the statement of the null hypothesis H_0. Using regression analysis, we calculate SSE_C, the unexplained variation for the complete model, and SSE_R, the unexplained variation for the reduced model. Then, we consider the difference

$$SS_{\text{drop}} = SSE_R - SSE_C$$

which we call the *drop in the unexplained variation attributable to the independent variables* $x_{g+1}, x_{g+2}, \ldots, x_p$. Now, it can be shown that the "extra" independent variables $x_{g+1}, x_{g+2}, \ldots, x_p$ will always make SSE_C somewhat smaller than SSE_R and hence will always make SS_{drop} positive. The question is whether this difference is large enough to conclude that at least one of the independent variables $x_{g+1}, x_{g+2}, \ldots, x_p$ significantly affects y. Since the value of SS_{drop} depends on the units in which the observed values of the dependent variable are measured, we need to modify SS_{drop} and formulate a unitless measure of the additional importance of the set $x_{g+1}, x_{g+2}, \ldots, x_p$. Such a statistic is called the *partial F statistic* and is denoted by

$$F(x_{g+1}, \ldots, x_p | x_1, x_2, \ldots, x_g)$$

In the following box we define this statistic and show how it is used to test H_0 versus H_1.

TESTING $H_0: \beta_{g+1} = \beta_{g+2} = \cdots = \beta_p = 0$ VERSUS H_1: AT LEAST ONE OF $\beta_{g+1}, \beta_{g+2}, \ldots, \beta_p$ DOES NOT EQUAL ZERO:

Assume the inference assumptions are satisfied, and define the partial F statistic to be

$$F(x_{g+1}, \ldots, x_p | x_1, \ldots, x_g) = \frac{MS_{\text{drop}}}{MSE_C}$$

where

$$MS_{\text{drop}} = \frac{SS_{\text{drop}}}{p-g} = \frac{\{SSE_R - SSE_C\}}{p-g}$$

$$MSE_C = \frac{SSE_C}{n-k}$$

(Note that MSE_C equals s^2 for the complete model.)

Define

$$\text{prob-value} = A[F(x_{g+1}, \ldots, x_p | x_1, \ldots, x_g); p-g, n-k]$$

which is the area to the right of $F(x_{g+1}, \ldots, x_p | x_1, \ldots, x_g)$ under the curve of the F-distribution having $p-g$ and $n-k$ degrees of freedom.

Then we can reject H_0 in favor of H_1 by setting the probability of a Type I error equal to α if and only if either of the following equivalent conditions hold:

1. $F(x_{g+1}, \ldots, x_p | x_1, \ldots, x_g) > F_{[\alpha]}^{(p-g, n-k)}$, where $F_{[\alpha]}^{(p-g, n-k)}$ is the point on the scale of the F-distribution having $p-g$ and $n-k$ degrees of freedom so that the area under this curve to the right of $F_{[\alpha]}^{(p-g, n-k)}$ is α.
2. Prob-value $< \alpha$.

Condition 1, which says we should reject H_0 in favor of H_1 if

$$F(x_{g+1}, \ldots, x_p | x_1, \ldots, x_g) = \frac{\{SSE_R - SSE_C\}/(p-g)}{SSE_C/(n-k)}$$

is large, is reasonable, because a large value of

$$F(x_{g+1}, \ldots, x_p | x_1, \ldots, x_g)$$

would be caused by a large value of $\{SSE_R - SSE_C\}$, which would occur if at least one of the independent variables $x_{g+1}, x_{g+2}, \ldots, x_p$ makes SSE_C substantially smaller than SSE_R, which would indicate that $H_0: \beta_{g+1} = \beta_{g+2} = \cdots = \beta_p = 0$ is false and that H_1: At least one of $\beta_{g+1}, \beta_{g+2}, \ldots, \beta_p$ does not equal zero is true.

Also, note that there is a relationship between the t_{b_j} statistic and the partial F statistic. It can be proved that

$$(t_{b_j})^2 = F(x_j | x_1, \ldots, x_{j-1}, x_{j+1}, \ldots, x_p)$$

and that

$$(t_{[\alpha/2]}^{(n-k)})^2 = F_{[\alpha]}^{(1,n-k)}$$

Hence, the conclusion that

$$|t_{b_j}| > t_{[\alpha/2]}^{(n-k)}$$

which leads us to reject $H_0: \beta_j = 0$ in favor of $H_1: \beta_j \neq 0$, will be made if and only if

$$F(x_j | x_1, \ldots, x_{j-1}, x_{j+1}, \ldots, x_p) > F_{[\alpha]}^{(1,n-k)}$$

which again leads us to reject $H_0: \beta_j = 0$ in favor of $H_1: \beta_j \neq 0$. Thus, the conditions for rejecting $H_0: \beta_j = 0$ in favor of $H_1: \beta_j \neq 0$ using the t_{b_j} statistic and partial F statistic are equivalent. Furthermore, it can be shown that the prob-value for testing $H_0: \beta_j = 0$ versus $H_1: \beta_j \neq 0$ can be computed in either of the following two ways:

$$\text{prob-value} = A[[|t_{b_j}|, \infty); n-k]$$

or

$$\text{prob-value} = A[F(x_j | x_1, \ldots, x_{j-1}, x_{j+1}, \ldots, x_p); 1, n-k]$$

EXAMPLE 7.12

When we use the Fresh Detergent model

$$\text{Complete model: } y_i = \beta_0 + \beta_1 x_{i4} + \beta_2 x_{i3} + \beta_3 x_{i3}^2 + \beta_4 x_{i4} x_{i3} + \epsilon_i$$

which has $k = 5$ parameters, to carry out a regression analysis of the data in Table 5.12, we obtain an unexplained variation of $SSE_C = 1.0644$. In order to test

$$H_0: \beta_3 = \beta_4 = 0$$

versus

$$H_1: \text{At least one of } \beta_3 \text{ and } \beta_4 \text{ does not equal zero}$$

note that $(p - g) = 2$, since two parameters (β_3 and β_4) are set equal to zero in the statement of the null hypothesis H_0. Also, under the assumption that H_0 is true, the complete model becomes the following reduced model:

$$\text{Reduced model: } y_i = \beta_0 + \beta_1 x_{i4} + \beta_2 x_{i3} + \epsilon_i$$

for which the unexplained variation is $SSE_R = 1.5337$. Thus, in order to test H_0 versus H_1 we use the following partial F statistic and prob-value:

$$F(x_{i3}^2, x_{i4}x_{i3} | x_{i4}, x_{i3}) = \frac{MS_{\text{drop}}}{MSE_C} = \frac{.2347}{.0426} = 5.5094$$

since

$$MS_{\text{drop}} = \frac{SS_{\text{drop}}}{p - g} = \frac{\{SSE_R - SSE_C\}}{p - g} = \frac{\{1.5337 - 1.0644\}}{2} = \frac{.4693}{2} = .2347$$

$$MSE_C = \frac{SSE_C}{n - k} = \frac{1.0644}{30 - 5} = \frac{1.0644}{25} = .0426$$

and

$$\begin{aligned} \text{prob-value} &= A[F(x_{i3}^2, x_{i4}x_{i3} | x_{i4}, x_{i3}); p - g, n - k] \\ &= A[5.5094; 2, 25] \\ &= .0111 \end{aligned}$$

which is the area to the right of 5.5094 under the curve of the F-distribution having 2 and 25 degrees of freedom. If we wish to use condition (1) to determine whether we can reject H_0 in favor of H_1 by setting α, the probability of a Type I error, equal to .05, then we would use the rejection point

$$F_{[\alpha]}^{(p-g, n-k)} = F_{[.05]}^{(2,25)} = 3.39 \qquad \text{(see Appendix Table E-3)}$$

Since

$$F(x_{i3}^2, x_{i4}x_{i3} | x_{i4}, x_{i3}) = 5.5094 > 3.39 = F_{[.05]}^{(2,25)}$$

we can reject $H_0: \beta_3 = \beta_4 = 0$ in favor of H_1: At least one of β_3 and β_4 does not equal zero by setting α equal to .05. Alternatively, since prob-value = .0111 is less than .05 and .02, we can reject H_0 in favor of H_1 by setting α equal to .05 or .02. However, since prob-value = .0111 is not less than .01, we cannot reject H_0 in favor of H_1 by setting α equal to .01. In summary, the smallness of the prob-value provides substantial evidence that $H_0: \beta_3 = \beta_4 = 0$ is false and that

H_1: At least one of β_3 and β_4 does not equal zero is true. Therefore, we have substantial evidence that at least one of the higher-order terms x_3^2 and x_4x_3 significantly affects y (demand for Fresh). Thus, we should use t_{b_j} statistics and other techniques to determine which of these higher-order terms significantly affect demand for Fresh (see Example 6.9 and Section 7.4).

Note that multicollinearity can hinder the ability of the partial F statistic to indicate the significance of at least one of the independent variables $x_{g+1}, x_{g+2}, \ldots, x_p$ (although, in many cases, multicollinearity affects this partial F statistic less than it affects the individual t_{b_j} statistics related to $x_{g+1}, x_{g+2}, \ldots, x_p$). In Section 7.8 and Chapter 8 we explain how the partial F statistic can be used to help compare population means.

7.7 AN *F*-TEST OF LACK OF FIT

We have seen that building a multiple regression model involves both specifying an appropriate set of independent variables and determining the functional form of the regression relationship between the dependent variable and the *quantitative* independent variables. Determining the functional form involves making such decisions as whether the dependent variable is related to a particular quantitative independent variable in a linear fashion or in a quadratic fashion, and whether two or more independent variables should be multiplied together to form interaction variables. We have seen that graphical analysis, the t_{b_j} statistic, the partial F statistic, and the overall F statistic are tools useful for determining the proper functional form. When more than one value of the dependent variable is observed at some of the combinations of values of the independent variables, we can perform what is called a **lack-of-fit test** of the functional form of the overall model. Performing this test requires calculating a quantity called the **sum of squares due to pure error**, denoted by SS_{PE} and given by the equation

$$SS_{PE} = \sum_{l=1}^{m} \left[\sum_{k=1}^{n_l} (y_{l,k} - \bar{y}_l)^2 \right]$$

In this equation m is the total number of combinations of values of the independent variables at which at least one corresponding value of the dependent variable has been observed, and

$$\bar{y}_l = \frac{\sum_{k=1}^{n_l} y_{l,k}}{n_l}$$

is the average of the n_l values $y_{l,1}, y_{l,2}, \ldots, y_{l,n_l}$ of the dependent variable that have been observed at the lth combination of values of the independent variables. The calculation of this quantity is illustrated in Example 7.13. In the following box we show how SS_{PE} can be used to perform a lack-of-fit test.

AN *F*-TEST OF LACK OF FIT:

Assume the inference assumptions are satisfied, and consider a regression model having k parameters and testing

H_0: The functional form of the regression model is correct

versus

H_1: The functional form of the regression model is not correct

Define the *lack-of-fit F statistic* to be

$$F(LF) = \frac{MS_{LF}}{MS_{PE}}$$

where

$$MS_{LF} = \frac{SS_{LF}}{m - k}$$

$$SS_{LF} = SSE - SS_{PE}$$

SSE = the unexplained variation for the regression model having k parameters

$$SS_{PE} = \sum_{l=1}^{m}\left[\sum_{k=1}^{n_l}(y_{l,k} - \bar{y}_l)^2\right]$$

$$MS_{PE} = \frac{SS_{PE}}{n - m}$$

and define

$$\text{prob-value} = A[F(LF); m - k, n - m]$$

Then, we can reject H_0 in favor of H_1 by setting the probability of a Type I error equal to α if and only if either of following two equivalent conditions hold:

1. $F(LF) > F_{[\alpha]}^{(m-k,n-m)}$.
2. Prob-value $< \alpha$.

Condition (1), which says that we should reject H_0 in favor of H_1 if $F(LF)$ is large, is reasonable, because a large value of $F(LF)$ would be caused by a large value of MS_{LF}. This would be caused by a large value of

$$SS_{LF} = SSE - SS_{PE}$$

which would occur if the regression model yields an unexplained variation (SSE) that is much larger than SS_{PE}, the sum of squares due to pure error. This would imply that the regression model under consideration *lacks the appropriate terms needed to fit the observed data.*

EXAMPLE 7.13

In Example 5.4 Comp-U-Systems has used the simple linear regression model

$$\begin{aligned} y_i &= \mu_i + \epsilon_i \\ &= \beta_0 + \beta_1 x_i + \epsilon_i \end{aligned}$$

to relate

x_i = the number of microcomputers serviced on the ith service call

to

y_i = the number of minutes required to perform service on the ith service call

Examining the Comp-U-Systems service call data in Table 5.8, we see that more than one value of the dependent variable was observed at some of the values of the independent variable. Therefore, letting x_l denote the lth smallest number of microcomputers serviced, and letting $y_{l,k}$ denote the number of minutes required to perform service on the kth service call involving x_l microcomputers, we can rearrange the data in Table 5.8 as illustrated in Table 7.11. Then, since there are $n = 11$ values of the dependent variable in Table 7.11, and since there are $m = 7$ values of the independent variable at which at least one corresponding value of the dependent variable has been observed, the sum of squares due to pure error is

$$\begin{aligned} SS_{PE} &= \sum_{l=1}^{7} \left[\sum_{k=1}^{n_l} (y_{l,k} - \bar{y}_l)^2 \right] \\ &= [(y_{2,1} - \bar{y}_2)^2 + (y_{2,2} - \bar{y}_2)^2] \\ &\quad + [(y_{4,1} - \bar{y}_4)^2 + (y_{4,2} - \bar{y}_4)^2 + (y_{4,3} - \bar{y}_4)^2] \\ &\quad + [(y_{5,1} - \bar{y}_5)^2 + (y_{5,2} - \bar{y}_5)^2] \\ &= [(58 - 63)^2 + (68 - 63)^2] \\ &\quad + [(103 - 108)^2 + (109 - 108)^2 + (112 - 108)^2] \\ &\quad + [(134 - 136)^2 + (138 - 136)^2] \\ &= 100 \end{aligned}$$

TABLE 7.11 Comp-U-Systems Service Call Data

The *l*th Smallest Number of Microcomputers Serviced (x_l)	Number of Minutes Required to Perform Service ($y_{l,k}$)			Mean of $y_{l,k}$'s ($\bar{y}_l$)
$x_1 = 1$	$y_{1,1} = 37$			$\bar{y}_1 = 37$
$x_2 = 2$	$y_{2,1} = 58$	$y_{2,2} = 68$		$\bar{y}_2 = 63$
$x_3 = 3$	$x_{3,1} = 82$			$\bar{y}_3 = 82$
$x_4 = 4$	$y_{4,1} = 103$	$y_{4,2} = 109$	$y_{4,3} = 112$	$\bar{y}_4 = 108$
$x_5 = 5$	$y_{5,1} = 134$	$y_{5,2} = 138$		$\bar{y}_5 = 136$
$x_6 = 6$	$y_{6,1} = 154$			$\bar{y}_6 = 154$
$x_7 = 7$	$y_{7,1} = 189$			$\bar{y}_7 = 189$

and thus

$$\begin{aligned} MS_{PE} &= \frac{SS_{PE}}{n - m} \\ &= \frac{100}{11 - 7} \\ &= \frac{100}{4} \\ &= 25 \end{aligned}$$

If we consider the simple linear regression model

$$y_{l,k} = \beta_0 + \beta_1 x_l + \epsilon_{l,k}$$

and note from Figure 6.14 that the unexplained variation resulting from using this model to perform a regression analysis of the data in Table 7.11 is $SSE = 191.702$, we can make the following calculations

$$SS_{LF} = SSE - SS_{PE} = 191.702 - 100 = 91.702$$

$$MS_{LF} = \frac{SS_{LF}}{m - k} = \frac{91.702}{7 - 2} = \frac{91.702}{5} = 18.3404$$

$$F(LF) = \frac{MS_{LF}}{MS_{PE}} = \frac{18.3404}{25} = .7336$$

If we wish to use condition (1) to determine whether we can reject

H_0: The functional form of the regression model $y_{l,k} = \beta_0 + \beta_1 x_l + \epsilon_{l,k}$ is correct

in favor of

H_1: The functional form of the regression model $y_{l,k} = \beta_0 + \beta_1 x_l + \epsilon_{l,k}$ is not correct

by setting α, the probability of a Type I error, equal to .05, then we would use the rejection point

$$F_{[\alpha]}^{(m-k,n-m)} = F_{[.05]}^{(5,4)} = 6.26$$

Since $F(LF) = .7336 < 6.26 = F_{[.05]}^{(5,4)}$, we cannot reject H_0 in favor of H_1 by setting α equal to .05. Thus, there is not much evidence that the functional form of the model

$$y_{l,k} = \beta_0 + \beta_1 x_l + \epsilon_{l,k}$$

is incorrect. Intuitively, this conclusion should be expected, because the plot of the Comp-U-Systems data in Figure 5.8 indicates that as the values of x increase, the values of y tend to increase in a straight line fashion. If the plot of the Comp-U-Systems data had indicated that as the values of x increase, the values of y tend to increase in a quadratic fashion, which suggests that the quadratic regression model

$$y_{l,k} = \beta_0 + \beta_1 x_l + \beta_2 x_l^2 + \epsilon_{l,k}$$

is appropriate, then the F-test of lack of fit would probably have indicated that the functional form of the simple linear model is not correct.

7.8 AN EXAMPLE OF USING DUMMY VARIABLES

EXAMPLE 7.14

In this example we reconsider the Fresh Detergent demand example. Recall that Enterprise Industries has employed the historical data of Table 5.12 to develop the regression model

$$\begin{aligned} y_i &= \mu_i + \epsilon_i \\ &= \beta_0 + \beta_1 x_{i4} + \beta_2 x_{i3} + \beta_3 x_{i3}^2 + \beta_4 x_{i4} x_{i3} + \epsilon_i \end{aligned}$$

which relates the dependent variable y (demand for Fresh) to the independent variables x_3 (Enterprise Industries' advertising expenditure) and x_4 (the price difference).

In order to ultimately increase the demand for Fresh, Enterprise Industries' marketing department is conducting a study in which it wishes to compare the effectiveness of three different advertising campaigns. These campaigns are denoted as campaigns A, B, and C. Here, campaign A consists entirely of television commercials, campaign B consists of a balanced mixture of television

TABLE 7.12 Advertising Campaigns Used by Enterprise Industries

The *i*th Observed Sales Period (OSP_i)	Advertising Campaign Used in OSP_i	The *i*th Observed Sales Period (OSP_i)	Advertising Campaign Used in OSP_i
OSP_1	B	OSP_{16}	B
OSP_2	B	OSP_{17}	B
OSP_3	B	OSP_{18}	A
OSP_4	A	OSP_{19}	B
OSP_5	C	OSP_{20}	B
OSP_6	A	OSP_{21}	C
OSP_7	C	OSP_{22}	A
OSP_8	C	OSP_{23}	A
OSP_9	B	OSP_{24}	A
OSP_{10}	C	OSP_{25}	A
OSP_{11}	A	OSP_{26}	B
OSP_{12}	C	OSP_{27}	C
OSP_{13}	C	OSP_{28}	B
OSP_{14}	A	OSP_{29}	C
OSP_{15}	B	OSP_{30}	C

and radio commercials, and campaign C consists of a balanced mixture of television, radio, newspaper, and magazine ads. In order to conduct this study, Enterprise Industries has randomly selected an advertising campaign to be used during each of the past 30 sales periods. Table 7.12 lists the campaigns (A, B, or C) used during these sales periods. After a campaign has been selected for a (four-week) sales period, Enterprise Industries has employed the campaign for the first three weeks of the sales period (no campaign is used during the last week, so that the effect of the advertising campaign used in the current sales period will not carry over to the next sales period).

In this example, we explain how Enterprise Industries can use **dummy variables** to study and compare the effectiveness of advertising campaigns A, B, and C. In order to do this, we define two dummy variables:

$$D_{i,B} = \begin{cases} 1 & \text{if campaign B is used in sales period } i \\ 0 & \text{otherwise} \end{cases}$$

$$D_{i,C} = \begin{cases} 1 & \text{if campaign C is used in sales period } i \\ 0 & \text{otherwise} \end{cases}$$

Using these dummy variables, we now consider a regression model relating y_i to x_{i3}, x_{i4}, $D_{i,B}$, and $D_{i,C}$. This model is

$$\begin{aligned} y_i &= \mu_i + \epsilon_i \\ &= \beta_0 + \beta_1 x_{i4} + \beta_2 x_{i3} + \beta_3 x_{i3}^2 + \beta_4 x_{i4} x_{i3} + \beta_5 D_{i,B} + \beta_6 D_{i,C} + \epsilon_i \end{aligned}$$

In order to see what the dummy variables do, we wish to interpret the meanings of the parameters β_5 and β_6 in this model (and also the meaning of the difference $\beta_6 - \beta_5$). Consider a sales period in which the price difference (x_{i4}) is d; Enterprise Industries' advertising expenditure (x_{i3}) is a; and advertising campaign A is used (which implies that both $D_{i,B}$ and $D_{i,C}$ are equal to zero). The mean demand for Fresh for all such sales periods, denoted as $\mu_{[d,a,A]}$, is (employing the preceding model)

$$\begin{aligned}\mu_{[d,a,A]} &= \beta_0 + \beta_1 d + \beta_2 a + \beta_3 a^2 + \beta_4 da + \beta_5(0) + \beta_6(0) \\ &= \beta_0 + \beta_1 d + \beta_2 a + \beta_3 a^2 + \beta_4 da\end{aligned}$$

Next, consider a sales period in which the price difference (x_{i4}) is d; Enterprise Industries' advertising expenditure (x_{i3}) is a; and advertising campaign B is used (which implies that $D_{i,B} = 1$ and $D_{i,C} = 0$). The mean demand for all such sales periods, denoted $\mu_{[d,a,B]}$, is

$$\begin{aligned}\mu_{[d,a,B]} &= \beta_0 + \beta_1 d + \beta_2 a + \beta_3 a^2 + \beta_4 da + \beta_5(1) + \beta_6(0) \\ &= \beta_0 + \beta_1 d + \beta_2 a + \beta_3 a^2 + \beta_4 da + \beta_5\end{aligned}$$

Last, consider a sales period in which the price difference (x_{i4}) is d; Enterprise Industries' advertising expenditure (x_{i3}) is a; and advertising campaign C is used (which implies that $D_{i,B} = 0$ and $D_{i,C} = 1$). The mean demand for all such sales periods, denoted $\mu_{[d,a,C]}$, is

$$\begin{aligned}\mu_{[d,a,C]} &= \beta_0 + \beta_1 d + \beta_2 a + \beta_3 a^2 + \beta_4 da + \beta_5(0) + \beta_6(1) \\ &= \beta_0 + \beta_1 d + \beta_2 a + \beta_3 a^2 + \beta_4 da + \beta_6\end{aligned}$$

In order to interpret the meaning of β_5, consider the difference between $\mu_{[d,a,B]}$ and $\mu_{[d,a,A]}$, which is

$$\begin{aligned}\mu_{[d,a,B]} - \mu_{[d,a,A]} &= (\beta_0 + \beta_1 d + \beta_2 a + \beta_3 a^2 + \beta_4 da + \beta_5) \\ &\quad - (\beta_0 + \beta_1 d + \beta_2 a + \beta_3 a^2 + \beta_4 da) \\ &= \beta_5\end{aligned}$$

Since this difference equals β_5, the parameter β_5 is the change in mean demand for Fresh associated with changing from advertising campaign A to advertising campaign B while the price difference and advertising expenditure remain constant. Thus, intuitively, β_5 measures the effect on mean demand for Fresh of changing from advertising campaign A to advertising campaign B.

We can interpret the meaning of β_6 by considering the difference between $\mu_{[d,a,C]}$ and $\mu_{[d,a,A]}$, which is

$$\begin{aligned}\mu_{[d,a,C]} - \mu_{[d,a,A]} &= (\beta_0 + \beta_1 d + \beta_2 a + \beta_3 a^2 + \beta_4 da + \beta_6) \\ &\quad - (\beta_0 + \beta_1 d + \beta_2 a + \beta_3 a^2 + \beta_4 da) \\ &= \beta_6\end{aligned}$$

Since this difference equals β_6, the parameter β_6 can be interpreted as the change in mean demand for Fresh associated with changing from advertising campaign A to advertising campaign C while the price difference and advertising expenditure remain constant. So, intuitively, β_6 measures the effect on mean demand for Fresh of changing from advertising campaign A to advertising campaign C.

Finally, we can interpret the meaning of the difference $\beta_6 - \beta_5$ by considering the difference between $\mu_{[d,a,C]}$ and $\mu_{[d,a,B]}$, which is

$$\begin{aligned}\mu_{[d,a,C]} - \mu_{[d,a,B]} &= (\beta_0 + \beta_1 d + \beta_2 a + \beta_3 a^2 + \beta_4 da + \beta_6)\\ &\quad -(\beta_0 + \beta_1 d + \beta_2 a + \beta_3 a^2 + \beta_4 da + \beta_5)\\ &= \beta_6 - \beta_5\end{aligned}$$

This says that $\beta_6 - \beta_5$ can be interpreted as the change in mean demand for Fresh associated with changing from advertising campaign B to advertising campaign C while the price difference and advertising expenditure remain constant. Thus, intuitively, $\beta_6 - \beta_5$ measures the effect on mean demand for Fresh of changing from advertising campaign B to advertising campaign C.

In summary, then, when we employ the dummy variables $D_{i,B}$ and $D_{i,C}$, the parameters

$$\begin{aligned}\beta_5 &= \mu_{[d,a,B]} - \mu_{[d,a,A]}\\ \beta_6 &= \mu_{[d,a,C]} - \mu_{[d,a,A]}\\ \beta_6 - \beta_5 &= \mu_{[d,a,C]} - \mu_{[d,a,B]}\end{aligned}$$

measure, respectively, the effects on mean demand for Fresh of (1) changing from advertising campaign A to advertising campaign B; (2) changing from advertising campaign A to advertising campaign C; and (3) changing from advertising campaign B to advertising campaign C.

We are able to compare the effects of the three advertising campaigns (A, B, and C) by using two dummy variables ($D_{i,B}$ and $D_{i,C}$), because β_5 and β_6 (the parameters multiplied by $D_{i,B}$ and $D_{i,C}$) express the effects of campaigns B and C with respect to the effect of campaign A. We do not employ three dummy variables ($D_{i,A}$, $D_{i,B}$, and $D_{i,C}$, where $D_{i,A} = 1$ if campaign A is used in sales period i, and $D_{i,A} = 0$ otherwise), because if we do, the columns of the **X** matrix can be shown to be linearly dependent and the least squares point estimates of the model parameters cannot be computed using the methods presented in this book. Instead of using $D_{i,B}$ and $D_{i,C}$, we could use any two of the dummy variables $D_{i,A}$, $D_{i,B}$, and $D_{i,C}$. If we used $D_{i,A}$ and $D_{i,B}$, the parameters β_5 and β_6 would express the effects of campaigns A and B with respect to the effect of campaign C. If we used $D_{i,A}$ and $D_{i,C}$, the parameters β_5 and β_6 would express the effects of campaigns A and C with respect to the effect of campaign B.

The least squares point estimates b_0, b_1, b_2, b_3, b_4, b_5, and b_6 of the parameters in the regression model

$$y_i = \beta_0 + \beta_1 x_{i4} + \beta_2 x_{i3} + \beta_3 x_{i3}^2 + \beta_4 x_{i4} x_{i3} + \beta_5 D_{i,B} + \beta_6 D_{i,C} + \epsilon_i$$

can be obtained by employing the formula

$$\mathbf{b} = (\mathbf{X}'\mathbf{X})^{-1}\mathbf{X}'\mathbf{y}$$

where the column vector $\mathbf{y}$ is

$$\mathbf{y} = \begin{bmatrix} y_1 \\ y_2 \\ y_3 \\ \vdots \\ y_{30} \end{bmatrix} = \begin{bmatrix} 7.38 \\ 8.51 \\ 9.52 \\ \vdots \\ 9.26 \end{bmatrix}$$

and the matrix $\mathbf{X}$ is (see Tables 5.12 and 7.12)

$$\mathbf{X} = \begin{array}{c} \\ \end{array}\begin{array}{ccccccc} 1 & x_4 & x_3 & x_3^2 & x_4x_3 & D_{i,B} & D_{i,C} \end{array}$$

$$\mathbf{X} = \begin{bmatrix} 1 & -.05 & 5.50 & (5.50)^2 & (-.05)(5.50) & 1 & 0 \\ 1 & .25 & 6.75 & (6.75)^2 & (.25)(6.75) & 1 & 0 \\ 1 & .60 & 7.25 & (7.25)^2 & (.60)(7.25) & 1 & 0 \\ 1 & 0 & 5.50 & (5.50)^2 & (0)(5.50) & 0 & 0 \\ 1 & .25 & 7.00 & (7.00)^2 & (.25)(7.00) & 0 & 1 \\ 1 & .20 & 6.50 & (6.50)^2 & (.20)(6.50) & 0 & 0 \\ \vdots & \vdots & \vdots & \vdots & \vdots & \vdots & \vdots \\ 1 & .55 & 6.80 & (6.80)^2 & (.55)(6.80) & 0 & 1 \end{bmatrix}$$

When the appropriate calculations are done, we find that

$$\begin{bmatrix} b_0 \\ b_1 \\ b_2 \\ b_3 \\ b_4 \\ b_5 \\ b_6 \end{bmatrix} = \mathbf{b} = (\mathbf{X}'\mathbf{X})^{-1}\mathbf{X}'\mathbf{y} = \begin{bmatrix} 25.6127 \\ 9.0587 \\ -6.5377 \\ .5844 \\ -1.1565 \\ .2137 \\ .3818 \end{bmatrix}$$

These least squares point estimates yield the prediction equation

$$\begin{aligned} \hat{y}_i &= b_0 + b_1x_{i4} + b_2x_{i3} + b_3x_{i3}^2 + b_4x_{i4}x_{i3} + b_5D_{i,B} + b_6D_{i,C} \\ &= 25.6127 + 9.0587x_{i4} - 6.5377x_{i3} + .5844x_{i3}^2 - 1.1565x_{i4}x_{i3} + .2137D_{i,B} \\ &\quad + .3818D_{i,C} \end{aligned}$$

In addition, the unexplained variation is found to be $SSE = .3936$, the standard error is found to be $s = .1308$, and the multiple coefficient of determination is found to be $R^2 = .9708$. This value of R^2 says that the model we have specified explains 97.08% of the total variation in the 30 observed demands for Fresh. Figure 7.6 presents the SAS output of the t_{b_j} statistics and the prob-values for

FIGURE 7.6 SAS Output of the t_{b_j} Statistic and Prob-Value for Testing the Hypotheses $H_0: \beta_j = 0$ (for $j = 0, 1, 2, 3, 4, 5, 6$), and the 95% Confidence Intervals for β_5 and β_6

VARIABLE	PARAMETER ESTIMATE[a]	T FOR HO: PARAMETER=0[b]	PROB > |T|[c]
INTERCEP	25.61269602	5.34	0.0001
X4	9.05868432	2.99	0.0066
X3	-6.53767133	-4.13	0.0004
X3SQ	0.58444394	4.50	0.0002
X43	-1.15648054	-2.54	0.0184
DB	0.21368626	3.44	0.0023
DC	0.38177617	6.23	0.0001

[a] $b_j: b_0, b_1, b_2, b_3, b_4, b_5, b_6$ [b] t_{b_j} [c] Prob-value [d] $s\sqrt{c_{jj}}$

testing $H_0: \beta_j = 0$ (for $j = 0, 1, 2, 3, 4, 5, 6$). Since the prob-value for testing each of these hypotheses is less than .05, we can reject each of these hypotheses by setting α, the probability of a Type I error, equal to .05. This implies that we are confident that the intercept β_0 and each of the independent variables x_{i4}, x_{i3}, x_{i3}^2, $x_{i4}x_{i3}$, $D_{i,B}$, and $D_{i,C}$ has substantial importance over the combined importance of the other independent variables in the regression model and that each of these independent variables should be included in a final regression model. Finally, the 95% confidence intervals for the parameters β_5 and β_6 in the regression model can be computed to be, respectively, [.0851, .3423] and [.2550, .5085]. Note (see Figure 7.6) that these intervals are computed using the formula

$$[b_j \pm t_{[.025]}^{(n-k)} s\sqrt{c_{jj}}] = [b_j \pm t_{[.025]}^{(30-7)} s\sqrt{c_{jj}}]$$
$$= [b_j \pm 2.069 s\sqrt{c_{jj}}]$$

We now use the preceding information to study the effects of advertising compaigns A, B, and C. Since $b_5 = .2137$ is the least squares point estimate of

$$\beta_5 = \mu_{[d,a,B]} - \mu_{[d,a,A]}$$

which is the change in mean demand for Fresh associated with changing from advertising campaign A to advertising campaign B while the price difference and advertising expenditure remain constant, Enterprise Industries estimates that the effect of changing from advertising campaign A to advertising campaign B is to increase (since b_5 is positive) mean demand for Fresh by .2137 (or 21,370 bottles). Moreover, the 95% confidence interval for β_5, [.0851, .3423], makes Enterprise Industries 95% confident that the effect of changing from advertising campaign A to advertising campaign B is to increase mean demand for Fresh by between .0851 (8,510 bottles) and .3423 (34,230 bottles). In addition, the prob-value of .0023 associated with the null hypothesis

$$H_0: \beta_5 = \mu_{[d,a,B]} - \mu_{[d,a,A]} = 0$$

STD ERROR OF ESTIMATE[d]	$[b_j \pm 2.069 s\sqrt{c_{jj}}] =$ 95% Confidence Interval for β_j
4.79378249	
3.03170457	
1.58136655	
0.12987222[f]	
0.45573648	
0.06215362	[.0851, .3423]
0.06125253	[.2550, .5085]

says that we have very strong evidence that

$$\beta_5 = \mu_{[d,a,B]} - \mu_{[d,a,A]} > 0$$

or that $\mu_{[d,a,B]} > \mu_{[d,a,A]}$. That is, Enterprise Industries has very strong evidence that advertising campaign B is more effective than advertising campaign A.

Since $b_6 = .3818$ is the least squares point estimate of

$$\beta_6 = \mu_{[d,a,C]} - \mu_{[d,a,A]}$$

which is the change in mean demand for Fresh associated with changing from advertising campaign A to advertising campaign C while the price difference and advertising expenditure remain constant, Enterprise Industries estimates that the effect of changing from advertising campaign A to advertising campaign C is to increase (since b_6 is positive) mean demand for Fresh by .3818 (or 38,180 bottles). Moreover, the 95% confidence interval for β_6, [.2550, .5085], makes Enterprise Industries 95% confident that the effect of changing from advertising campaign A to advertising campaign C is to increase mean demand for Fresh by between .2550 (25,500 bottles) and .5085 (50,850 bottles). In addition, since the prob-value associated with the null hypothesis

$$H_0: \beta_6 = \mu_{[d,a,C]} - \mu_{[d,a,A]} = 0$$

is less than .0001, we have overwhelming evidence that

$$\beta_6 = \mu_{[d,a,C]} - \mu_{[d,a,A]} > 0$$

or that $\mu_{[d,a,C]} > \mu_{[d,a,A]}$. That is, Enterprise Industries has overwhelming evidence that advertising campaign C is more effective than advertising campaign A.

Since $b_6 - b_5 = .3818 - .2137 = .1681$ is the point estimate of

$$\beta_6 - \beta_5 = \mu_{[d,a,C]} - \mu_{[d,a,B]}$$

which is the change in mean demand for Fresh associated with changing from advertising campaign B to advertising campaign C while the price difference and advertising expenditure remain constant, Enterprise Industries estimates that the effect of changing from advertising campaign B to advertising campaign C is to increase (since $b_6 - b_5$ is positive) mean demand for Fresh by .1681 (or 16,810 bottles). We later explain how to find a 95% confidence interval for $\beta_6 - \beta_5$ and how to test the null hypothesis $H_0: \beta_6 - \beta_5 = 0$. We will find that a 95% confidence interval for $\beta_6 - \beta_5$ $(= \mu_{[d,a,C]} - \mu_{[d,a,B]})$ is [.0363, .2999]. This interval makes Enterprise Industries 95% confident that the effect of changing from advertising campaign B to advertising campaign C is to increase mean demand for Fresh by between .0363 (3,630 bottles) and .2999 (29,990 bottles). In addition, we will find that the prob-value for testing the null hypothesis

$$H_0: \beta_6 - \beta_5 = \mu_{[d,a,C]} - \mu_{[d,a,B]} = 0$$

is .0147. Since this prob-value is fairly small, we have strong evidence that

$$\beta_6 - \beta_5 = \mu_{[d,a,C]} - \mu_{[d,a,B]} > 0$$

or that $\mu_{[d,a,C]} > \mu_{[d,a,B]}$. That is, Enterprise Industries has strong evidence that advertising campaign C is more effective than advertising campaign B.

In Section 8.2 of Chapter 8 we explain how to find a confidence interval for $\beta_6 - \beta_5$ $(= \mu_{[d,a,C]} - \mu_{[d,a,B]})$ and how to test $H_0: \beta_6 - \beta_5 = 0$. However, it is important to note that if we consider the following dummy variable model

$$\begin{aligned} y_i &= \mu_i + \epsilon_i \\ &= \beta_0 + \beta_1 x_{i4} + \beta_2 x_{i3} + \beta_3 x_{i3}^2 + \beta_4 x_{i4} x_{i3} + \beta_5 D_{i,A} + \beta_6 D_{i,C} + \epsilon_i \end{aligned}$$

it can be verified that

$$\mu_{[d,a,C]} - \mu_{[d,a,B]} = \beta_6$$

Thus, by using the formula

$$[b_6 \pm t_{[\alpha/2]}^{(n-k)} s\sqrt{c_{66}}]$$

we can calculate a $100(1 - \alpha)\%$ confidence interval for

$$\beta_6 = \mu_{[d,a,C]} - \mu_{[d,a,B]}$$

The reader will have an opportunity to do this in Exercise 7.9.

Suppose that on the basis of the preceding analyses Enterprise Industries concludes that campaign C is the most effective advertising campaign and decides to employ advertising campaign C in future sales periods. Now, recall that Enterprise Industries wishes to predict the future (individual) demand for Fresh, y_0, when the price of Fresh will be $x_{01} = \$3.80$, the average industry price will be $x_{02} = \$3.90$, the price difference will be $x_{04} = x_{02} - x_{01} = \$.10$, and the advertising expenditure for Fresh will be $x_{03} = 6.80$ (or \$680,000). Since we can express the future demand in the form

$$\begin{aligned} y_0 &= \mu_0 + \epsilon_0 \\ &= \beta_0 + \beta_1 x_{04} + \beta_2 x_{03} + \beta_3 x_{03}^2 + \beta_4 x_{04} x_{03} + \beta_5 D_{0,B} + \beta_6 D_{0,C} + \epsilon_0 \end{aligned}$$

a point prediction of the future demand for Fresh is

$$\begin{aligned} \hat{y}_0 &= b_0 + b_1 x_{04} + b_2 x_{03} + b_3 x_{03}^2 + b_4 x_{04} x_{03} + b_5 D_{0,B} + b_6 D_{0,C} \\ &= 25.6127 + 9.0587 x_{04} - 6.5377 x_{03} + .5844 x_{03}^2 \\ &\quad - 1.1565 x_{04} x_{03} + .2137 D_{0,B} + .3818 D_{0,C} \\ &= 25.6127 + 9.0587(.10) - 6.5377(6.80) + .5844(6.80)^2 \\ &\quad - 1.1565(.10)(6.80) + .2137(0) + .3818(1) \\ &= 8.6825 \text{ (or 868,250 bottles)} \end{aligned}$$

Here, $D_{0,B} = 0$ and $D_{0,C} = 1$, since Enterprise Industries will use advertising campaign C in the future sales periods (see the initial definitions of the dummy variables $D_{i,B}$ and $D_{i,C}$). A 95% prediction interval for the future demand, y_0, is [8.385, 8.980], or [838,500 bottles, 898,000 bottles]. This prediction interval is computed using the formula

$$\left[\hat{y}_0 \pm t_{[.025]}^{(n-k)} s \sqrt{1 + \mathbf{x}_0'(\mathbf{X}'\mathbf{X})^{-1}\mathbf{x}_0}\right]$$

where

$$\hat{y}_0 = 8.6825$$

$$t_{[.025]}^{(n-k)} = t_{[.025]}^{(30-7)} = t_{[.025]}^{(23)} = 2.069$$

and

$$\begin{aligned} \mathbf{x}_0' &= [1 \quad .10 \quad 6.80 \quad (6.80)^2 \quad (.10)(6.80) \quad 0 \quad 1] \\ &= [1 \quad .10 \quad 6.80 \quad 46.24 \quad .68 \quad 0 \quad 1] \end{aligned}$$

is the row vector containing the numbers multiplied by b_0, b_1, b_2, b_3, b_4, b_5, and b_6 in the preceding prediction equation. This prediction interval says that if the price difference will be $x_{04} = \$.10$, if the advertising expenditure for Fresh will be $x_{03} = \$680{,}000$, and if Enterprise Industries will use advertising campaign C, then the company can be 95% confident that the future demand for Fresh will be no less than 838,500 bottles and no more than 898,000 bottles.

Next, notice that the regression model we have used,

$$\begin{aligned} y_i &= \mu_i + \epsilon_i \\ &= \beta_0 + \beta_1 x_{i4} + \beta_2 x_{i3} + \beta_3 x_{i3}^2 + \beta_4 x_{i4} x_{i3} + \beta_5 D_{i,B} + \beta_6 D_{i,C} + \epsilon_i \end{aligned}$$

does not contain any interaction terms involving the dummy variables $D_{i,B}$ and $D_{i,C}$. If such interaction terms are included, however, these terms do not have much importance over the combined importance of the other independent

FIGURE 7.7 SAS Output of the Least Squares Point Estimates, t_{b_j} Statistics, and Prob-Values for the Model $y_i = \beta_0 + \beta_1 x_{i4} + \beta_2 x_{i3} + \beta_3 x_{i3}^2 + \beta_4 x_{i4} x_{i3} + \beta_5 D_{i,B} + \beta_6 D_{i,C} + \beta_7 x_{i4} D_{i,B} + \beta_8 x_{i4} D_{i,C} + \beta_9 x_{i3} D_{i,B} + \beta_{10} x_{i3} D_{i,C} + \beta_{11} x_{i4} x_{i3} D_{i,B} + \beta_{12} x_{i4} x_{i3} D_{i,C} + \epsilon_i$

VARIABLE	DF	PARAMETER ESTIMATE	STANDARD ERROR	T FOR H0: PARAMETER=0	PROB > |T|
INTERCEP	1	32.590707	5.740052	5.678	0.0001
X4	1	5.887626	5.658572	1.040	0.3127
X3	1	-8.603161	1.838238	-4.680	0.0002
X3SQ	1	0.735327	0.147080	5.000	0.0001
X43	1	-0.665556	0.831766	-0.800	0.4347
DB	1	-0.761000	1.169902	-0.650	0.5241
DC	1	-2.415693	1.398314	-1.728	0.1022
X4DB	1	8.277369	6.093695	1.358	0.1921
X4DC	1	11.961760	9.746570	1.227	0.2364
X3DB	1	0.156574	0.192191	0.815	0.4265
X3DC	1	0.447779	0.224349	1.996	0.0622
X43DB	1	-1.231316	0.887758	-1.387	0.1834
X43DC	1	-1.836136	1.445220	-1.270	0.2210

variables. For example, consider the model

$$\begin{aligned} y_i &= \mu_i + \epsilon_i \\ &= \beta_0 + \beta_1 x_{i4} + \beta_2 x_{i3} + \beta_3 x_{i3}^2 + \beta_4 x_{i4} x_{i3} + \beta_5 D_{i,B} + \beta_6 D_{i,C} + \beta_7 x_{i4} D_{i,B} \\ &\quad + \beta_8 x_{i4} D_{i,C} + \beta_9 x_{i3} D_{i,B} + \beta_{10} x_{i3} D_{i,C} + \beta_{11} x_{i4} x_{i3} D_{i,B} + \beta_{12} x_{i4} x_{i3} D_{i,C} + \epsilon_i \end{aligned}$$

This model contains the two-factor interaction terms $x_{i4}D_{i,B}$, $x_{i4}D_{i,C}$, $x_{i3}D_{i,B}$, and $x_{i3}D_{i,C}$, as well as the three-factor interaction terms $x_{i4}x_{i3}D_{i,B}$ and $x_{i4}x_{i3}D_{i,C}$, in addition to the terms in the model we have previously discussed. Although three-factor interaction terms can be very difficult to interpret, intuitively, the term $x_{i4}x_{i3}D_{i,B}$ for example, measures the amount of interaction between the interaction variable $x_{i4}x_{i3}$ and the dummy variable $D_{i,B}$, which represents the effect of advertising campaign B. Figure 7.7 gives the SAS output of the least squares point estimates of the parameters in the expanded model (these estimates are obtained using the data in Tables 5.12 and 7.12), along with the t_{b_j} statistics and associated prob-values for testing $H_0: \beta_j = 0$ ($j = 0, 1, \ldots, 12$). Looking at Figure 7.7, we see that the prob-values for all the interaction terms in the expanded model are quite large (most are substantially greater than .05). In addition, the standard error for this expanded model is $s = .1311$, which is larger than the standard error (.1308) for our previous model. This larger standard error, and the large prob-values associated with the interaction terms in the expanded model, indicate that the inclusion of the additional interaction terms is not warranted. In order to reduce the degree of multicollinearity, we might wish to try some other models with fewer interaction terms. For example, we might consider the model

$$\begin{aligned} y_i &= \beta_0 + \beta_1 x_{i4} + \beta_2 x_{i3} + \beta_3 x_{i3}^2 + \beta_4 x_{i4} x_{i3} + \beta_5 D_{i,B} + \beta_6 D_{i,C} + \beta_7 x_{i4} D_{i,B} \\ &\quad + \beta_8 x_{i4} D_{i,C} + \beta_9 x_{i3} D_{i,B} + \beta_{10} x_{i3} D_{i,C} + \epsilon_i \end{aligned}$$

FIGURE 7.8 SAS Output of the Least Squares Point Estimates, t_{b_j} Statistics, and Prob-Values for the Model $y_i = \beta_0 + \beta_1 x_{i4} + \beta_2 x_{i3} + \beta_3 x_{i3}^2 + \beta_4 x_{i4} x_{i3} + \beta_5 D_{i,B} + \beta_6 D_{i,C} + \beta_7 x_{i4} D_{i,B} + \beta_8 x_{i4} D_{i,C} + \beta_9 x_{i3} D_{i,B} + \beta_{10} x_{i3} D_{i,C} + \epsilon_i$

VARIABLE	DF	PARAMETER ESTIMATE	STANDARD ERROR	T FOR H0: PARAMETER=0	PROB > |T|
INTERCEP	1	30.727132	5.611330	5.476	0.0001
X4	1	12.100134	3.600116	3.361	0.0033
X3	1	-8.016542	1.797726	-4.459	0.0003
X3SQ	1	0.689466	0.143943	4.790	0.0001
X43	1	-1.575823	0.531068	-2.967	0.0079
DB	1	-0.722749	1.176255	-0.614	0.5462
DC	1	-1.680605	1.139724	-1.475	0.1567
X4DB	1	-0.132349	0.454120	-0.291	0.7739
X4DC	1	-0.443757	0.442450	-1.003	0.3285
X3DB	1	0.149000	0.193337	0.771	0.4504
X3DC	1	0.331765	0.184911	1.794	0.0887

However, examination of Figure 7.8, which presents the SAS output of key quantities related to this model, indicates that inclusion of the additional interaction terms in this model is not warranted. Moreover, when other possible models are analyzed, we find that the inclusion of interaction terms other than $x_{i4}x_{i3}$ is not warranted.

To complete this example, we perform a partial *F*-test that is useful in studying the effects of advertising campaigns A, B, and C. Consider the following complete model:

$$\text{Complete model: } y_i = \beta_0 + \beta_1 x_{i4} + \beta_2 x_{i3} + \beta_3 x_{i3}^2 + \beta_4 x_{i4} x_{i3} + \beta_5 D_{i,B} + \beta_6 D_{i,C} + \epsilon_i$$

which has $k = 7$ parameters and for which the unexplained variation is $SSE_C = .3936$. Since we have previously seen that

$$\beta_5 = \mu_{[d,a,B]} - \mu_{[d,a,A]}$$

$$\beta_6 = \mu_{[d,a,C]} - \mu_{[d,a,A]}$$

the null hypothesis

$$H_0: \beta_5 = \beta_6 = 0 \quad \text{or} \quad H_0: \beta_5 = 0 \text{ and } \beta_6 = 0$$

is equivalent to

$$H_0: \mu_{[d,a,B]} - \mu_{[d,a,A]} = 0 \text{ and } \mu_{[d,a,C]} - \mu_{[d,a,A]} = 0$$

which is equivalent to

$$H_0: \mu_{[d,a,A]} = \mu_{[d,a,B]} = \mu_{[d,a,C]}$$

In order to test

$$H_0: \beta_5 = \beta_6 = 0$$

or equivalently,

$$\mu_{[d,a,A]} = \mu_{[d,a,B]} = \mu_{[d,a,C]}$$

which says that advertising campaigns A, B, and C have the same effects on the mean demand for Fresh, versus

H_1: At least one of β_5 and β_6 does not equal zero

or equivalently,

H_1: At least two of $\mu_{[d,a,A]}$, $\mu_{[d,a,B]}$, and $\mu_{[d,a,C]}$ differ from each other

which says that at least two of advertising campaigns A, B, and C have different effects on the mean demand for Fresh, first note that $p - g = 2$, since two parameters (β_5 and β_6) are set equal to zero in the statement of the null hypothesis H_0. Also note that under the assumption that H_0 is true, the complete model becomes the following reduced model:

$$\text{Reduced model: } y_i = \beta_0 + \beta_1 x_{i4} + \beta_2 x_{i3} + \beta_3 x_{i3}^2 + \beta_4 x_{i4} x_{i3} + \epsilon_i$$

for which the unexplained variation is $SSE_R = 1.0644$. Thus, in order to test H_0 versus H_1 we use the following partial F statistic and prob-value:

$$F(D_{i,B}, D_{i,C} \mid x_{i4}, x_{i3}, x_{i3}^2, x_{i4}x_{i3}) = \frac{MS_{\text{drop}}}{MSE_C} = \frac{.3354}{.0171} = 19.614$$

since

$$MS_{\text{drop}} = \frac{SS_{\text{drop}}}{p-g} = \frac{\{SSE_R - SSE_C\}}{p-g} = \frac{\{1.0644 - .3936\}}{2} = \frac{.6708}{2} = .3354$$

$$MSE_C = \frac{SSE_C}{n - k}$$

$$= \frac{.3936}{30 - 7}$$

$$= \frac{.3936}{23}$$

$$= .0171$$

and

$$\text{prob-value} = A[F(D_{i,B}, D_{i,C} | x_{i4}, x_{i3}, x_{i3}^2, x_{i4}x_{i3}); p - g, n - k]$$
$$= A[19.614; 2, 23]$$

which is the area to the right of 19.614 under the curve of the F-distribution having 2 and 23 degrees of freedom, and which can be computer calculated to be less than .0001. (For the purposes of subsequent discussion, we assume that the prob-value equals .0001.) If we wish to use condition (1)—see the partial F-test in Section 7.6—to determine whether we can reject H_0 in favor of H_1 by setting α equal to .05, then we would use the rejection point

$$F_{[\alpha]}^{(p-g,n-k)} = F_{[.05]}^{(2,23)} = 3.42$$

Since

$$F(D_{i,B}, D_{i,C} | x_{i4}, x_{i3}, x_{i3}^2, x_{i4}x_{i3}) = 19.614 > 3.42 = F_{[.05]}^{(2,23)}$$

we can reject H_0 in favor of H_1 by setting α equal to .05. Alternatively, since prob-value = .0001 is less than .05 and .01, it follows by condition (2)—again see Section 7.6—that we can reject H_0 in favor of H_1 by setting α equal to .05 or .01. These facts provide substantial evidence that at least two of advertising campaigns A, B, and C have different effects on the mean demand for Fresh.

Using the partial F-test has an advantage over determining whether there are differences in the effects of advertising campaigns A, B, and C by using t statistics (and related prob-values) to perform individual tests of

$$H_0: \beta_5 = \mu_{[d,a,B]} - \mu_{[d,a,A]} = 0 \quad \text{versus} \quad H_1: \beta_5 = \mu_{[d,a,B]} - \mu_{[d,a,A]} \neq 0$$
$$H_0: \beta_6 = \mu_{[d,a,C]} - \mu_{[d,a,A]} = 0 \quad \text{versus} \quad H_1: \beta_6 = \mu_{[d,a,C]} - \mu_{[d,a,A]} \neq 0$$
$$H_0: \beta_6 - \beta_5 = \mu_{[d,a,C]} - \mu_{[d,a,B]} = 0 \quad \text{versus} \quad H_1: \beta_6 - \beta_5 = \mu_{[d,a,C]} - \mu_{[d,a,B]} \neq 0$$

This is because, although we can set the probability of a Type I error equal to .05 for each individual test, the probability of falsely rejecting H_0 in at least one of these tests can be shown to be greater than .05. On the other hand, we can use the partial F statistic to determine whether there are differences in the effects of advertising campaigns A, B, and C by testing the single hypothesis

$$H_0: \mu_{[d,a,A]} = \mu_{[d,a,B]} = \mu_{[d,a,C]}$$

and we can do this by setting the (overall) probability of a Type I error equal to .05. If the partial F-test indicates that such differences exist, then we should use confidence intervals and t statistics to investigate the exact nature of these differences.

7 EXERCISES

7.1 Consider Exercise 5.3 and the model

$$\begin{aligned} y_i &= \mu_i + \epsilon_i \\ &= \beta_0 + \beta_1 x_{i1} + \beta_2 x_{i2} + \beta_3 x_{i1}^2 + \beta_4 x_{i1} x_{i2} + \epsilon_i \end{aligned}$$

Using the fact that

$$\sum_{i=1}^{25} y_i^2 = 22{,}102 \qquad 25\bar{y}^2 = 21{,}802 \qquad \mathbf{b'X'y} = 22{,}057$$

a. Calculate the total variation $= \sum_{i=1}^{25} (y_i - \bar{y})^2$.

b. Calculate the unexplained variation $= \sum_{i=1}^{25} (y_i - \hat{y}_i)^2$.

c. Calculate the explained variation $= \sum_{i=1}^{25} (\hat{y}_i - \bar{y})^2$. Show how this can be calculated two ways.

d. Calculate the multiple coefficient of determination, R^2. Show how this can be calculated two ways.

e. Calculate the overall F statistic. Using this statistic and the appropriate rejection point, determine whether we can, by setting α equal to .05, reject $H_0: \beta_1 = \beta_2 = \beta_3 = \beta_4 = 0$ in favor of H_1: At least one of β_1, β_2, β_3, and β_4 does not equal zero.

7.2 Consider Exercise 5.1 and the model

$$\begin{aligned} y_i &= \mu_i + \epsilon_i \\ &= \beta_0 + \beta_1 x_i + \epsilon_i \end{aligned}$$

Using the data in Exercise 5.1,

a. Calculate the simple coefficient of determination, r^2.

b. Calculate the t_{b_1} statistic and use the appropriate rejection points to determine whether we can, by setting α equal to .05, reject $H_0: \beta_1 = 0$ (or $H_0: \rho = 0$) in favor of $H_1: \beta_1 \neq 0$ (or $H_1: \rho \neq 0$).

7.3 Consider Exercise 5.4. Following is the SAS output of the analysis of variance table that results when the model

$$y = \beta_0 + \beta_1 x_1 + \beta_2 x_2 + \beta_3 x_1^2 + \beta_4 x_1 x_2 + \beta_5 x_1^2 x_2 + \epsilon$$

is used to perform a regression analysis of the data in Exercise 5.4.

```
DEP VARIABLE: Y

                     SUM OF          MEAN
SOURCE      DF      SQUARES        SQUARE     F VALUE    PROB>F

MODEL        5    80.576926     16.115385      22.226    0.0001
ERROR       12     8.700852      0.725071
C TOTAL     17    89.277778
```

Discuss the meaning of the quantities on this output. Be sure to explain how the overall F statistic and the related prob-value have been calculated. Based on this prob-value, determine whether we can reject $H_0: \beta_1 = \beta_2 = \beta_3 = \beta_4 = \beta_5 = 0$ in favor of H_1: At least one of β_1, β_2, β_3, β_4, and β_5 does not equal zero by setting α equal to .05 or .01.

7.4 Consider Exercise 5.4. The following table summarizes the values of R^2, $\bar{R}^2$, and s^2, and the prediction intervals (for the future combination $x_{01} = 4.8$ and $x_{02} = 6$) that result when various models are used to perform regression analyses of the data in Exercise 5.4.

Model	R^2	$\bar{R}^2$	s^2	Prediction Interval
$y = \beta_0 + \beta_1 x_1 + \beta_2 x_2 + \epsilon$	.3284	.2389	3.9971	[−.2618, 8.834]
$y = \beta_0 + \beta_1 x_1 + \beta_2 x_2 + \beta_3 x_1^2 + \epsilon$	.3797	.2468	3.9556	[−.1358, 8.986]
$y = \beta_0 + \beta_1 x_1 + \beta_2 x_2 + \beta_3 x_1^2 + \beta_4 x_1 x_2 + \epsilon$	.8628	.8206	.9422	[3.758, 8.362]
$y = \beta_0 + \beta_1 x_1 + \beta_2 x_2 + \beta_3 x_1^2 + \beta_4 x_1 x_2 + \beta_5 x_1^2 x_2 + \epsilon$	.9025	.8619	.7251	[3.968, 8.043]
$y = \beta_0 + \beta_1 x_1 + \beta_2 x_2 + \beta_3 x_1^2 + \beta_4 x_1 x_2 + \beta_5 x_2^2 + \epsilon$	.8647	.8084	1.0063	[3.620, 8.432]

a. Consider the model

$$y = \beta_0 + \beta_1 x_1 + \beta_2 x_2 + \beta_3 x_1^2 + \beta_4 x_1 x_2 + \epsilon$$

Using the fact that $R^2 = .8628$, demonstrate how $\bar{R}^2 = .8206$ has been calculated.

b. Based on the preceding table, which model seems best? Justify your answer. Is the model you have chosen the same model that you chose in part (c) of Exercise 6.2? Discuss.

7.5 Consider Exercise 5.6. Following is the SAS output of the values of R^2 and the C statistic that result when regression analyses of the data in Exercise 5.6 are performed by using regression models with all combinations of the independent variables defined in Exercise 5.6. Based on the C statistic, which model seems best? Why?

REGRESSION ANALYSES
PROC RSQUARE-ALL POSSIBLE SUBSETS ANALYSIS

N=20 REGRESSION MODELS FOR DEPENDENT VARIABLE VOLUME

NUMBER IN MODEL	R-SQUARE	C(P)	VARIABLES IN MODEL
1	0.00480421	30.45388047	PARKING
1	0.03353172	29.11293360	FLOOR_SP
1	0.14798995	23.77023759	INCOME
1	0.43933184	10.17094219	PRESC_PCT
2	0.06855667	29.47803470	FLOOR_SP PARKING
2	0.20543099	23.08899693	PARKING INCOME
2	0.23487329	21.71468547	FLOOR_SP INCOME
2	0.53142435	7.87223587	PRESC_PCT PARKING
2	0.54748785	7.12242198	PRESC_PCT INCOME
2	0.66566267	1.60624219	FLOOR_SP PRESC_PCT
3	0.25569607	22.74271718	FLOOR_SP PARKING INCOME
3	0.60243233	6.55771633	PRESC_PCT PARKING INCOME
3	0.66641145	3.57129027	FLOOR_SP PRESC_PCT INCOME
3	0.67943313	2.96346249	FLOOR_SP PRESC_PCT PARKING
4	0.68058567	4.90966443	FLOOR_SP PRESC_PCT PARKING INCOME

7.6 Consider Exercise 5.6. Following is the SAS output of a stepwise regression of the data in Exercise 5.6. Here, the potential independent variables are all the independent variables defined in Exercise 5.6, and both α_{entry} and α_{stay} have been set equal to .15.

```
STEPWISE REGRESSION PROCEDURE FOR DEPENDENT VARIABLE VOLUME

STEP 1    VARIABLE PRESC_PCT
          ENTERED                R SQUARE=0.43933184        C(P)=10.17094219

                                 DF    SUM OF SQUARES    MEAN SQUARE      F     PROB>F

          REGRESSION              1      329.74051403   329.74051403   14.10   0.0014
          ERROR                  18      420.80948597    23.37830478
          TOTAL                  19      750.55000000

                              B VALUE       STD ERROR     TYPE II SS      F     PROB>F

          INTERCEPT        25.98133346
          PRESC_PCT        -0.32055657     0.08535423   329.74051403   14.10   0.0014
----------------------------------------------------------------------------------------
STEP 2    VARIABLE FLOOR_SP
          ENTERED                R SQUARE=0.66566267         C(P)=1.60624219

                                 DF    SUM OF SQUARES    MEAN SQUARE      F     PROB>F

          REGRESSION              2      499.61311336   249.80655668   16.92   0.0001
          ERROR                  17      250.93688664    14.76099333
          TOTAL                  19      750.55000000

                              B VALUE       STD ERROR     TYPE II SS      F     PROB>F

          INTERCEPT        48.29085530
          FLOOR_SP         -0.00384228     0.00113262   169.87259933   11.51   0.0035
          PRESC_PCT        -0.58189034     0.10263739   474.44587802   32.14   0.0001
----------------------------------------------------------------------------------------
NO OTHER VARIABLES MET THE 0.1500 SIGNIFICANCE LEVEL
FOR ENTRY INTO THE MODEL
```

a. What is the first independent variable entered?
b. What is the second independent variable entered? Has the first independent entered been retained? Why?
c. What is the final model arrived at by stepwise regression? Is this model the same model that you chose in Exercise 7.5?

7.7 Consider Exercise 5.8 and the following models and corresponding unexplained variations and mean square errors (which are obtained by using the models to perform regression analyses of the data in Exercise 5.8).

Model	*SSE*	s^2
$y_i = \beta_0 + \beta_1 x_{i1} + \beta_2 x_{i1}^2 + \beta_3 x_{i2} + \beta_4 x_{i2}^2 + \epsilon_i$	22.4124	1.3184
$y_i = \beta_0 + \beta_1 x_{i1} + \beta_2 x_{i1}^2 + \beta_3 x_{i2} + \beta_4 x_{i2}^2 + \beta_5 x_{i1} x_{i2} + \epsilon$	15.7971	.9873
$y_i = \beta_0 + \beta_1 x_{i1} + \beta_2 x_{i1}^2 + \beta_3 x_{i2} + \beta_4 x_{i2}^2 + \beta_5 x_{i1} x_{i2} + \beta_6 x_{i1}^2 x_{i2}^2 + \epsilon_i$	6.9087	.4606
$y_i = \beta_0 + \beta_1 x_{i1} + \beta_2 x_{i1}^2 + \beta_3 x_{i2} + \beta_4 x_{i2}^2 + \beta_5 x_{i1} x_{i2} + \beta_6 x_{i1}^2 x_{i2}^2 + \beta_7 x_{i1} x_{i2}^2 + \beta_8 x_{i1}^2 x_{i2} + \epsilon_i$	6.1569	.4736

a. Consider the model

$$y_i = \beta_0 + \beta_1 x_{i1} + \beta_2 x_{i1}^2 + \beta_3 x_{i2} + \beta_4 x_{i2}^2 + \beta_5 x_{i1}x_{i2} + \beta_6 x_{i1}^2 x_{i2}^2 + \beta_7 x_{i1} x_{i2}^2 + \beta_8 x_{i1}^2 x_{i2} + \epsilon_i$$

and calculate the partial F statistic used to test

$H_0: \beta_5 = \beta_6 = \beta_7 = \beta_8 = 0$

versus

H_1: At least one of β_5, β_6, β_7, and β_8 does not equal zero

Using this statistic and the appropriate rejection point, determine whether we can reject H_0 in favor of H_1 by setting α equal to .05. Also, calculate the partial F statistic used to test

$H_0: \beta_7 = \beta_8 = 0$

versus

H_1: At least one of β_7 and β_8 does not equal zero

Using this statistic and the appropriate rejection point, determine whether we can reject H_0 in favor of H_1 by setting α equal to .05.

b. Consider the model

$$y_i = \beta_0 + \beta_1 x_{i1} + \beta_2 x_{i1}^2 + \beta_3 x_{i2} + \beta_4 x_{i2}^2 + \beta_5 x_{i1}x_{i2} + \beta_6 x_{i1}^2 x_{i2}^2 + \epsilon_i$$

and calculate the partial F statistic used to test

$H_0: \beta_5 = \beta_6 = 0$

versus

H_1: At least one of β_5 and β_6 does not equal zero

Using this statistic and the appropriate rejection point, determine whether we can reject H_0 in favor of H_1 by setting α equal to .05. The prob-value for testing H_0 versus H_1 is calculated to be .0002. Discuss how this prob-value has been calculated, and using this prob-value, determine whether we can reject H_0 in favor of H_1 by setting α equal to .05 or .01.

c. Based upon the results obtained in parts (a) and (b) and upon the mean square errors given, which of the preceding four models seems best? Justify your answer. Is the model you have chosen the same model that you chose in Exercise 6.5. Discuss.

7.8 Examining the data in Exercise 5.8, we see that more than one value of the dependent variable has been observed at some of the combinations of values of the independent variables. Therefore, letting $y_{x_1 x_2,k}$ denote the kth gasoline mileage obtained by using x_1 units of gasoline additive WST and x_2 units of gasoline additive YST, we can rearrange the data in Exercise 5.8 as follows.

Combination	x_1	x_2	$y_{x_1x_2,k}$			$\bar{y}_{x_1x_2}$
1	0	0	$y_{00,1} = 18.0$	$y_{00,2} = 18.6$	$y_{00,3} = 17.4$	$\bar{y}_{00} = 18.0$
2	1	0	$y_{10,1} = 23.3$	$y_{10,2} = 24.5$		$\bar{y}_{10} = 23.9$
3	0	1	$y_{01,1} = 23.0$	$y_{01,2} = 22.0$		$\bar{y}_{01} = 22.5$
4	1	1	$y_{11,1} = 25.6$	$y_{11,2} = 24.4$	$y_{11,3} = 25.0$	$\bar{y}_{11} = 25.0$
5	2	1	$y_{21,1} = 24.0$	$y_{21,2} = 23.3$	$y_{21,3} = 24.7$	$\bar{y}_{21} = 24.0$
6	0	2	$y_{02,1} = 23.5$	$y_{02,2} = 22.3$		$\bar{y}_{02} = 22.9$
7	1	2	$y_{12,1} = 23.4$			$\bar{y}_{12} = 23.4$
8	2	2	$y_{22,1} = 23.0$	$y_{22,2} = 22.0$		$\bar{y}_{22} = 22.5$
9	1	3	$y_{13,1} = 19.6$	$y_{13,2} = 20.6$		$\bar{y}_{13} = 20.1$
10	2	3	$y_{23,1} = 18.6$	$y_{23,2} = 19.8$		$\bar{y}_{23} = 19.2$

a. Verify that SS_{PE}, the sum of squares due to pure error, is 6.08.

b. Noting that there are $n = 22$ values of the dependent variable and that there are $m = 10$ combinations of values of the independent variables at which at least one corresponding value of the dependent variable has been observed, calculate MS_{PE}.

c. Using the fact that $SSE = 15.7971$ is the unexplained variation resulting from using the model

$$y = \beta_0 + \beta_1 x_1 + \beta_2 x_1^2 + \beta_3 x_2 + \beta_4 x_2^2 + \beta_5 x_1 x_2 + \epsilon$$

to perform a regression analysis of the data in Exercise 5.8, calculate the lack-of-fit F statistic related to the preceding model. By using this statistic and the appropriate rejection point, determine whether we can, by setting α equal to .05, reject the null hypothesis that the functional form of this model is correct. The prob-value related to this hypothesis can be calculated to be .0153. Discuss how this prob-value has been calculated, and, using this prob-value, determine whether we can reject this hypothesis by setting α equal to .05 or .01.

d. Using the fact that $SSE = 6.9087$ is the unexplained variation resulting from using the model

$$y = \beta_0 + \beta_1 x_1 + \beta_2 x_1^2 + \beta_3 x_2 + \beta_4 x_2^2 + \beta_5 x_1 x_2 + \beta_6 x_1^2 x_2^2 + \epsilon$$

to perform a regression analysis of the data in Exercise 5.8, calculate the lack-of-fit F statistic related to this model. By using this statistic and the appropriate rejection point, determine whether we can, by setting α equal to .05, reject the null hypothesis that the functional form of this model is correct.

e. Based upon the results obtained in parts (c) and (d), which of the models in parts (c) and (d) seems best?

7.9 Consider the Fresh Detergent problem in Section 7.8 and the model

$$\begin{aligned} y_i &= \mu_i + \epsilon_i \\ &= \beta_0 + \beta_1 x_{i4} + \beta_2 x_{i3} + \beta_3 x_{i3}^2 + \beta_4 x_{i4} x_{i3} + \beta_5 D_{i,A} + \beta_6 D_{i,C} + \epsilon_i \end{aligned}$$

where $D_{i,A} = 1$ if campaign A is used in sales period i, and $D_{i,A} = 0$ otherwise, and where $D_{i,C} = 1$ if campaign C is used in sales period i, and $D_{i,C} = 0$ otherwise. When this model is used to perform a regression analysis of the data in Tables 5.12 and 7.12, we find that the least squares point estimates of the parameters β_5 and β_6 are

$$b_5 = -.2137 \qquad \text{and} \qquad b_6 = .1681$$

the diagonal elements of $(\mathbf{X}'\mathbf{X})^{-1}$ corresponding to β_5 and β_6 are

$$c_{55} = .2258 \qquad \text{and} \qquad c_{66} = .2372$$

and the standard error is $s = .1308$.

a. Using the fact that

$$\mu_i = \beta_0 + \beta_1 x_{i4} + \beta_2 x_{i3} + \beta_3 x_{i3}^2 + \beta_4 x_{i4} x_{i3} + \beta_5 D_{i,A} + \beta_6 D_{i,C}$$

show that

$$\mu_{[d,a,A]} - \mu_{[d,a,B]} = \beta_5$$

$$\mu_{[d,a,C]} - \mu_{[d,a,B]} = \beta_6$$

$$\mu_{[d,a,C]} - \mu_{[d,a,A]} = \beta_6 - \beta_5$$

b. Calculate 95% confidence intervals for $\mu_{[d,a,A]} - \mu_{[d,a,B]}$ and $\mu_{[d,a,C]} - \mu_{[d,a,B]}$. Interpret these intervals. What is the relationship between the interval for $\mu_{[d,a,A]} - \mu_{[d,a,B]}$ you have just calculated and the interval for $\mu_{[d,a,B]} - \mu_{[d,a,A]}$ calculated in Example 7.14? Note that the interval for $\mu_{[d,a,C]} - \mu_{[d,a,B]}$ you have calculated is the same interval discussed in Example 7.14.

c. Calculate the t_{b_6} statistic used to test

$$H_0: \beta_6 = \mu_{[d,a,C]} - \mu_{[d,a,B]} = 0$$

versus

$$H_1: \beta_6 = \mu_{[d,a,C]} - \mu_{[d,a,B]} \neq 0$$

If

$$A[[|t_{b_6}|, \infty); 23] = .00735$$

calculate the prob-value for testing H_0 versus H_1.

7.10 International Oil, Inc., is attempting to develop a reasonably priced leaded gasoline that will deliver higher gasoline mileages than can be achieved by its current leaded gasolines. As part of its development process, International Oil wishes to study the effect of one qualitative independent variable—x_1, leaded gasoline type (A, B, or C)—and of one quantitative independent variable—x_2, amount of gasoline additive VST (0, 1, 2, or 3 units)—on the gasoline mileage y obtained by an automobile called the Encore. For testing purposes a sample of $n = 22$ Encores is randomly selected and driven under normal driving conditions.

The combinations of x_1 and x_2 used in the experiment, along with the corresponding values of y, are given in the following table.

Gasoline Mileage (in mpg) (y_i)	Leaded Gasoline Type (x_{i1})	Amount of Gasoline Additive VST (x_{i2})
18.0	A	0
18.6	A	0
17.4	A	0
23.3	B	0
24.5	B	0
23.0	A	1
22.0	A	1
25.6	B	1
24.4	B	1
25.0	B	1
24.0	C	1
23.3	C	1
24.7	C	1
23.5	A	2
22.3	A	2
23.4	B	2
23.0	A	2
22.0	A	2
19.6	B	3
20.6	B	3
18.6	C	3
19.8	C	3

a. Plot y_i against x_{i1} (= A, B, and C).
b. Plot y_i against x_{i1} when $x_{i2} = 0$.
c. Plot y_i against x_{i1} when $x_{i2} = 1$.

d. Plot y_i against x_{i1} when $x_{i2} = 2$.
e. Plot y_i against x_{i1} when $x_{i2} = 3$.
f. Combine the plots you made in parts (b), (c), (d), and (e) by drawing these plots on the same set of axes. Use a different color for each level of x_{i2} (=0, 1, 2, and 3). What do these plots say about whether interaction exists between x_{i1} and x_{i2}?
g. Plot y_i against x_{i2}.
h. Plot y_i against x_{i2} when $x_{i1} = A$.
i. Plot y_i against x_{i2} when $x_{i1} = B$.
j. Plot y_i against x_{i2} when $x_{i1} = C$.
k. Combine the plots you made in parts (h), (i), and (j) by drawing these plots on the same set of axes. Use a different color for each level of x_{i1} (=A, B, and C). What do these plots say about whether interaction exists between x_{i1} and x_{i2}?
l. Further analysis indicates that the regression model

$$\begin{aligned} y_i &= \mu_i + \epsilon_i \\ &= \beta_0 + \beta_1 D_{i,B} + \beta_2 D_{i,C} + \beta_3 x_{i2} + \beta_4 x_{i2}^2 + \beta_5 D_{i,B} x_{i2} + \beta_6 D_{i,C} x_{i2} \\ &\quad + \beta_7 D_{i,B} x_{i2}^2 + \beta_8 D_{i,C} x_{i2}^2 + \epsilon_i \end{aligned}$$

is a reasonable regression model relating y_i to x_{i1} and x_{i2}. Here, $D_{i,B} = 1$ if gasoline type B was used to obtain y_i, and $D_{i,B} = 0$ otherwise, and $D_{i,C} = 1$ if gasoline type C was used to obtain y_i, and $D_{i,C} = 0$ otherwise. Discuss why the graphical analysis in parts (a) through (k) indicates that this model is reasonable.
m. Specify the vector **y** and the matrix **X** used to calculate the least squares point estimates of the parameters in the model in part (l).
n. If the least squares point estimates of the parameters in the model in part (l) are calculated, we find that

$$b_0 = 18.0 \quad b_1 = 5.9385 \quad b_2 = 5.7 \quad b_3 = 6.55 \quad b_4 = -2.05$$

$$b_5 = -4.427 \quad b_6 = -5.35 \quad b_7 = .9115 \quad b_8 = 1.15$$

and that the unexplained variation, *SSE*, is 6.1569. Noting from parts (a) through (k) that gasoline type B seems to maximize gasoline mileage, suppose that International Oil decides to produce gasoline type B. Using the model in part (l), find a point estimate of $\mu_{[B,x_2]}$, the mean Encore mileage that would be obtained by using gasoline type B and additive amount x_2. This point estimate is denoted by $\hat{y}_{[B,x_2]}$. It can be shown (for example, by using differential calculus) that the value of x_2 maximizing $\hat{y}_{[B,x_2]}$ is .93. Calculate a point estimate and a 95% confidence interval for $\mu_{[B,x_2]}$, and calculate a point prediction and a 95% prediction interval for

$$y_{[B,x_2]} = \mu_{[B,x_2]} + \epsilon_{[B,x_2]}$$

a future (individual) Encore mileage that will be observed when using gasoline type B and additive amount x_2. In your calculations, use the fact that

$$\mathbf{x}'_{[B,.93]}(\mathbf{X}'\mathbf{X})^{-1}\mathbf{x}_{[B,.93]} = .2430$$

What is $\mathbf{x}'_{[B,.93]}$?

o. Express the following differences in terms of $\beta_1, \beta_2, \ldots, \beta_8$ and x_2:

$$\mu_{[B,x_2]} - \mu_{[A,x_2]}$$

$$\mu_{[C,x_2]} - \mu_{[A,x_2]}$$

$$\mu_{[C,x_2]} - \mu_{[B,x_2]}$$

$$\frac{\mu_{[C,x_2]} + \mu_{[B,x_2]}}{2} - \mu_{[A,x_2]}$$

Then, assuming that $x_2 = .93$, find point estimates of these differences.

p. Consider the following models (prob-values shown under variables):

$$\underset{(.0001)}{y_i = \beta_0} + \underset{(.0001)}{\beta_1 D_{i,B}} + \underset{(.0293)}{\beta_2 D_{i,C}} + \underset{(.0001)}{\beta_3 x_{i2}} + \underset{(.0001)}{\beta_4 x_{i2}^2} + \epsilon_i \qquad SSE = 22.4124$$

$$\underset{(.0001)}{y_i = \beta_0} + \underset{(.0001)}{\beta_1 D_{i,B}} + \underset{(.0004)}{\beta_2 D_{i,C}} + \underset{(.0001)}{\beta_3 x_{i2}} + \underset{(.0001)}{\beta_4 x_{i2}^2}$$
$$+ \underset{(.0001)}{\beta_5 D_{i,B} x_{i2}} + \underset{(.0012)}{\beta_6 D_{i,C} x_{i2}} + \epsilon_i \qquad SSE = 7.3260$$

$$\underset{(.0001)}{y_i = \beta_0} + \underset{(.0001)}{\beta_1 D_{i,B}} + \underset{(.0132)}{\beta_2 D_{i,C}} + \underset{(.0001)}{\beta_3 x_{i2}} + \underset{(.0036)}{\beta_4 x_{i2}^2} + \underset{(.0078)}{\beta_5 D_{i,B} x_{i2}}$$
$$+ \underset{(.0588)}{\beta_6 D_{i,C} x_{i2}} + \underset{(.1709)}{\beta_7 D_{i,B} x_{i2}^2} + \underset{(.1838)}{\beta_8 D_{i,C} x_{i2}^2} + \epsilon_i \qquad SSE = 6.1569$$

Based on the prob-values, which of the models seems to best describe the preceding data? By using the indicated unexplained variations to calculate the standard errors for the models, determine which model has the smallest standard error.

7.11 Consider Exercise 7.10 and the model

$$y_i = \beta_0 + \beta_1 D_{i,B} + \beta_2 D_{i,C} + \beta_3 x_{i2} + \beta_4 x_{i2}^2 + \beta_5 D_{i,B} x_{i2}$$
$$+ \beta_6 D_{i,C} x_{i2} + \beta_7 D_{i,B} x_{i2}^2 + \beta_8 D_{i,C} x_{i2}^2 + \epsilon_i$$

a. Using the fact that the total variation and the unexplained variation for this model are, respectively, 127.4727 and 6.1569, calculate the overall F statistic. Using this statistic and the appropriate rejection point, determine whether we can, by setting α equal to .05, reject

$$H_0: \beta_1 = \beta_2 = \beta_3 = \beta_4 = \beta_5 = \beta_6 = \beta_7 = \beta_8 = 0$$

in favor of

H_1: At least one of β_1, β_2, β_3, β_4, β_5, β_6, β_7, and β_8 does not equal zero

The prob-value for testing H_0 versus H_1 is calculated to be .0001. Demonstrate how this prob-value has been calculated, and use it to determine whether we can reject H_0 in favor of H_1 by setting α equal to .05 or .01.

b. Using the unexplained variations presented in part (p) of Exercise 7.10, calculate the partial F statistics used to test

$$H_0: \beta_5 = \beta_6 = \beta_7 = \beta_8 = 0$$

versus

H_1: At least one of β_5, β_6, β_7, and β_8 does not equal zero

and to test

$$H_0: \beta_7 = \beta_8 = 0$$

versus

H_1: At least one of β_7 and β_8 does not equal zero

Then, use these partial F statistics and the appropriate rejection points to determine whether we can reject the preceding null hypotheses by setting α equal to .05. The prob-values related to these null hypotheses are calculated to be, respectively, .0013 and .3230. Demonstrate how these prob-values have been calculated, and use them to determine whether we can reject the null hypotheses by setting α equal to .05 or .01.

7.12 An analyst at National Motors is interested in developing a regression model for predicting automobile sales in Ohio for the standard and luxury models of the Lance.† To do this, monthly data have been observed over the past 18 months on the following variables:

y_i = number of Lances sold in month i (measured in thousands of cars)

x_{i1} = (average) price per gallon of gasoline in month i

x_{i2} = (average) interest rate in month i

$$x_{i3} = \begin{cases} 1 & \text{if we are considering standard model sales} \\ 0 & \text{if we are considering luxury model sales} \end{cases}$$

† The idea and data for this problem were taken from an example in Ott (1984). We thank Dr. Ott and Duxbury Press for permission to use the idea and data.

The data are as follows:

i	y_i	x_{i1}	x_{i2}	x_{i3}
1	22.1	1.39	12.1	1
1	7.2	1.39	12.1	0
2	15.4	1.44	12.2	1
2	5.4	1.44	12.2	0
3	11.7	1.45	12.3	1
3	7.6	1.45	12.3	0
4	10.3	1.32	14.2	1
4	2.5	1.32	14.2	0
5	11.4	1.35	15.8	1
5	2.4	1.35	15.8	0
6	7.5	1.28	16.3	1
6	1.7	1.28	16.3	0
7	13.0	1.26	16.5	1
7	4.3	1.26	16.5	0
8	12.8	1.26	14.7	1
8	3.7	1.26	14.7	0
9	14.6	1.25	13.4	1
9	3.9	1.25	13.4	0
10	18.9	1.24	12.9	1
10	7.0	1.24	12.9	0
11	19.3	1.20	11.2	1
11	6.8	1.20	11.2	0
12	30.1	1.20	10.9	1
12	10.1	1.20	10.9	0
13	28.2	1.18	10.3	1
13	9.4	1.18	10.3	0
14	25.6	1.10	9.7	1
14	7.9	1.10	9.7	0
15	37.5	1.11	9.6	1
15	14.1	1.11	9.6	0
16	36.1	1.14	9.1	1
16	14.5	1.14	9.1	0
17	39.8	1.17	7.8	1
17	14.9	1.17	7.8	0
18	44.3	1.18	8.3	1
18	15.6	1.18	8.3	0

a. Using all the model-building techniques discussed in this book, develop a regression model relating y_i to x_{i1} and x_{i2} for the standard model (that is, when $x_{i3} = 1$).

b. Using all the model-building techniques discussed in this book, develop a regression model relating y_i to x_{i1} and x_{i2} for the luxury model (that is, when $x_{i3} = 0$).
c. Compare the regression models you have developed in parts (a) and (b).
d. Using all the model-building techniques discussed in this book, develop *one* regression model relating y_i to x_{i1} and x_{i2} for both the standard and luxury models. *Hint:* Use a dummy variable.

8

THE ANALYSIS OF DESIGNED EXPERIMENTS

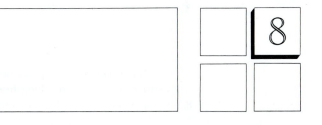

In this chapter we explain how to use regression analysis and other techniques to analyze data resulting from designed experiments. We begin in Section 8.1 by taking another look at the Fresh Detergent problem. In Sections 8.2 through 8.5 we show how to compare two or more population means using regression analysis. Specifically, Section 8.2 discusses confidence intervals and hypothesis tests for a linear combination of regression parameters; Section 8.3 explains a new kind of confidence interval—simultaneous confidence intervals; and Sections 8.4 and 8.5 present several regression models that are useful in comparing population means.

In Section 8.6 we review and extend our discussion of experimental design. In particular, we review the completely randomized design and introduce the *randomized block design*. We conclude this chapter with optional Section 8.7, which discusses some special methods for comparing two population means. Note that in some experimental situations, in addition to (or instead of) regression analysis we can use *analysis of variance* to analyze experimental data. For this reason, we present both the regression approach and the analysis of variance approach to analyzing experimental situations. However, since the analysis of variance approach is less general, we make the study of this topic optional.

8.1 ANOTHER LOOK AT THE FRESH DETERGENT PROBLEM

EXAMPLE 8.1

Table 8.1 summarizes the Fresh Detergent demand data previously presented in Tables 5.12 and 7.12. Figure 8.1 presents the SAS output of the results obtained by using the model

$$\begin{aligned} y_i &= \mu_i + \epsilon_i \\ &= \beta_0 + \beta_1 x_{i4} + \beta_2 x_{i3} + \beta_3 x_{i3}^2 + \beta_4 x_{i4} x_{i3} + \beta_5 D_{i,B} + \beta_6 D_{i,C} + \epsilon_i \end{aligned}$$

to perform a regression analysis of the data in Table 8.1. Table 8.3 summarizes the least squares point estimates of various differences in mean demands for Fresh obtained by using advertising campaigns A, B, and C. To obtain point estimates of these differences, instead of using regression analysis, it might be tempting to arrange the Fresh demands according to the advertising campaign used, as shown in Table 8.2, and then to calculate the sample means and sample mean differences presented in Tables 8.2 and 8.3. Using the least squares point estimates is more appropriate, however, because the least squares point estimates take into account the values of x_4 and x_3, while the sample means do not. For example, note from Table 8.3 that using the inappropriate sample mean difference $\bar{y}_C - \bar{y}_B = -.04$ would cause us to estimate that $\mu_{[d,a,C]}$ is 4,000 bottles less than $\mu_{[d,a,B]}$, but using the appropriate least squares point

TABLE 8.1 Historical Data for Fresh Detergent Demand

The *i*th Observed Sales Period (OSP_i)	The *i*th Observed Price Difference (dollars) ($x_{i4} = x_{i2} - x_{i1}$)	The *i*th Observed Advertising Expenditure for Fresh (hundreds of thousands of dollars) (x_{i3})	Advertising Campaign Used in OSP_i	The *i*th Observed Demand for Fresh (hundreds of thousands of bottles) (y_i)
OSP_1	−.05	5.50	B	7.38
OSP_2	.25	6.75	B	8.51
OSP_3	.60	7.25	B	9.52
OSP_4	0	5.50	A	7.50
OSP_5	.25	7.00	C	9.33
OSP_6	.20	6.50	A	8.28
OSP_7	.15	6.75	C	8.75
OSP_8	.05	5.25	C	7.87
OSP_9	−.15	5.25	B	7.10
OSP_{10}	.15	6.00	C	8.00
OSP_{11}	.20	6.50	A	7.89
OSP_{12}	.10	6.25	C	8.15
OSP_{13}	.40	7.00	C	9.10
OSP_{14}	.45	6.90	A	8.86
OSP_{15}	.35	6.80	B	8.90
OSP_{16}	.30	6.80	B	8.87
OSP_{17}	.50	7.10	B	9.26
OSP_{18}	.50	7.00	A	9.00
OSP_{19}	.40	6.80	B	8.75
OSP_{20}	−.05	6.50	B	7.95
OSP_{21}	−.05	6.25	C	7.65
OSP_{22}	−.10	6.00	A	7.27
OSP_{23}	.20	6.50	A	8.00
OSP_{24}	.10	7.00	A	8.50
OSP_{25}	.50	6.80	A	8.75
OSP_{26}	.60	6.80	B	9.21
OSP_{27}	−.05	6.50	C	8.27
OSP_{28}	0	5.75	B	7.67
OSP_{29}	.05	5.80	C	7.93
OSP_{30}	.55	6.80	C	9.26

TABLE 8.2 Fresh Detergent Demands Arranged According to Advertising Campaign Used

Demands Observed When Using Campaign A	Demands Observed When Using Campaign B	Demands Observed When Using Campaign C
7.50	7.38	9.33
8.28	8.51	8.75
7.89	9.52	7.87
8.86	7.10	8.00
9.00	8.90	8.15
7.27	8.87	9.10
8.00	9.26	7.65
8.50	8.75	8.27
8.75	7.95	7.93
	9.21	9.26
	7.67	
$\bar{y}_A = \frac{74.05}{9} = 8.23$	$\bar{y}_B = \frac{93.12}{11} = 8.47$	$\bar{y}_C = \frac{84.31}{10} = 8.43$

estimate $b_6 - b_5 = .1681$ causes us to estimate that $\mu_{[d,a,C]}$ is 16,810 bottles greater than $\mu_{[d,a,B]}$. In this situation, sample mean differences are not appropriate point estimates, but in some situations they would be. We will consider situations in which sample mean differences are appropriate later in this chapter.

8.2 STATISTICAL INFERENCES FOR A LINEAR COMBINATION OF REGRESSION PARAMETERS

8.2.1 A Confidence Interval

In Example 7.14 we considered the Fresh Detergent model

$$y_i = \mu_i + \epsilon_i$$
$$= \beta_0 + \beta_1 x_{i4} + \beta_2 x_{i3} + \beta_3 x_{i3}^2 + \beta_4 x_{i4} x_{i3} + \beta_5 D_{i,B} + \beta_6 D_{i,C} + \epsilon_i$$

and stated that a 95% confidence interval for

$$\mu_{[d,a,C]} - \mu_{[d,a,B]} = \beta_6 - \beta_5$$

is [.0363, .2999]. To see how we calculate this interval, note that $\beta_6 - \beta_5$ can be expressed as a linear combination of the parameters in this model. That is, we can

FIGURE 8.1 SAS Output of the t_{b_j} Statistics and Prob-Values for Testing the Hypotheses $H_0: \beta_j = 0$ (for $j = 0, 1, 2, 3, 4, 5, 6$) and the 95% Confidence Intervals for β_5 and β_6

VARIABLE	PARAMETER ESTIMATE[a]	T FOR HO: PARAMETER=0[b]	PROB > |T|[c]	STD ERROR OF ESTIMATE[d]	95% Confidence Interval for β_j $[b_j \pm 2.069 s\sqrt{c_{jj}}]$
INTERCEPT	25.61269602	5.34	0.0001	4.79378249	
X4	9.05868432	2.99	0.0066	3.03170457	
X3	-6.53767133	-4.13	0.0004	1.58136655	
X3SQ	0.58444394	4.5	0.0002	0.12987222	
X43	-1.15648054	-2.54	0.0184	0.45573648	
DB	0.21368626	3.44	0.0023	0.06215362	[.0851, .3423]
DC	0.38177617	6.23	0.0001	0.06125253	[.2550, .5085]

[a] $b_j: b_0, b_1, b_2, b_3, b_4, b_5, b_6$ [b] t_{b_j} [c] Prob-value [d] $s\sqrt{c_{jj}}$

TABLE 8.3 Point Estimates of Differences in Mean Demands for Fresh Obtained by Least Squares Regression Analysis and Sample Mean Differences

Difference in Mean Demands	Least Squares Point Estimate	Difference in Sample Means
$\mu_{[d,a,B]} - \mu_{[d,a,A]} = \beta_5$	$b_5 = .2137$	$\bar{y}_B - \bar{y}_A = 8.47 - 8.23 = .24$
$\mu_{[d,a,C]} - \mu_{[d,a,A]} = \beta_6$	$b_6 = .3818$	$\bar{y}_C - \bar{y}_A = 8.43 - 8.23 = .20$
$\mu_{[d,a,C]} - \mu_{[d,a,B]} = \beta_6 - \beta_5$	$b_6 - b_5 = .3818 - .2137 = .1681$	$\bar{y}_C - \bar{y}_B = 8.43 - 8.47 = -.04$

express $\beta_6 - \beta_5$ as

$$\begin{aligned}
\beta_6 - \beta_5 &= 0 \cdot \beta_0 + 0 \cdot \beta_1 + 0 \cdot \beta_2 + 0 \cdot \beta_3 + 0 \cdot \beta_4 + (-1) \cdot \beta_5 + 1 \cdot \beta_6 \\
&= [0 \quad 0 \quad 0 \quad 0 \quad 0 \quad -1 \quad 1] \begin{bmatrix} \beta_0 \\ \beta_1 \\ \beta_2 \\ \beta_3 \\ \beta_4 \\ \beta_5 \\ \beta_6 \end{bmatrix} \\
&= \boldsymbol{\lambda}'\boldsymbol{\beta}
\end{aligned}$$

where

$$\boldsymbol{\lambda}' = [0 \quad 0 \quad 0 \quad 0 \quad 0 \quad -1 \quad 1] \quad \text{and} \quad \boldsymbol{\beta} = \begin{bmatrix} \beta_0 \\ \beta_1 \\ \beta_2 \\ \beta_3 \\ \beta_4 \\ \beta_5 \\ \beta_6 \end{bmatrix}$$

In this section we present the formula for a confidence interval for a linear combination $\boldsymbol{\lambda}'\boldsymbol{\beta}$ of the parameters in a regression model. Then, since the difference $\mu_{[d,a,C]} - \mu_{[d,a,B]} = \beta_6 - \beta_5$ can be expressed as such a linear combination, we can use this formula to compute a confidence interval for the difference $\beta_6 - \beta_5$.

To compute a $100(1 - \alpha)\%$ confidence interval for a linear combination of the parameters $\beta_0, \beta_1, \beta_2, \ldots, \beta_p$ in the regression model

$$\begin{aligned}
y &= \mu + \epsilon \\
&= \beta_0 + \beta_1 x_1 + \beta_2 x_2 + \cdots + \beta_p x_p + \epsilon
\end{aligned}$$

consider the linear combination of regression parameters

$$\boldsymbol{\lambda}'\boldsymbol{\beta} = \sum_{j=0}^{p} \lambda_j \beta_j = [\lambda_0 \quad \lambda_1 \quad \lambda_2 \quad \cdots \quad \lambda_p] \begin{bmatrix} \beta_0 \\ \beta_1 \\ \beta_2 \\ \vdots \\ \beta_p \end{bmatrix}$$

where

$$\boldsymbol{\lambda}' = [\lambda_0 \quad \lambda_1 \quad \lambda_2 \quad \cdots \quad \lambda_p] \quad \text{and} \quad \boldsymbol{\beta} = \begin{bmatrix} \beta_0 \\ \beta_1 \\ \beta_2 \\ \vdots \\ \beta_p \end{bmatrix}$$

Assume that the linear regression model

$$y = \beta_0 + \beta_1 x_1 + \beta_2 x_2 + \cdots + \beta_p x_p + \epsilon$$

has k parameters $\beta_0, \beta_1, \beta_2, \ldots, \beta_p$ and that $\mathbf{b}$ is a column vector containing the least squares point estimates $b_0, b_1, b_2, \ldots, b_p$. Then:

1. A *point estimate* of the linear combination

$$\boldsymbol{\lambda}'\boldsymbol{\beta} = \sum_{j=0}^{p} \lambda_j \beta_j$$

is

$$\boldsymbol{\lambda}'\mathbf{b} = \sum_{j=0}^{p} \lambda_j b_j$$

2. If the inference assumptions are satisfied, a *100(1 − α)% confidence interval for* $\boldsymbol{\lambda}'\boldsymbol{\beta}$ is

$$[\boldsymbol{\lambda}'\mathbf{b} - BE_{\boldsymbol{\lambda}'\mathbf{b}}[100(1-\alpha)], \boldsymbol{\lambda}'\mathbf{b} + BE_{\boldsymbol{\lambda}'\mathbf{b}}[100(1-\alpha)]]$$

where $BE_{\boldsymbol{\lambda}'\mathbf{b}}[100(1-\alpha)]$ denotes the *100(1 − α)% bound on the error of estimation obtained when estimating* $\boldsymbol{\lambda}'\boldsymbol{\beta}$ *by* $\boldsymbol{\lambda}'\mathbf{b}$ and is given by the equation

$$BE_{\boldsymbol{\lambda}'\mathbf{b}}[100(1-\alpha)] = t_{[\alpha/2]}^{(n-k)} s \sqrt{\boldsymbol{\lambda}'(\mathbf{X}'\mathbf{X})^{-1}\boldsymbol{\lambda}}$$

Here s is the standard error, and $t_{[\alpha/2]}^{(n-k)}$ is the point on the scale of the t-distribution having $n - k$ degrees of freedom so that the area in the tail of this t-distribution to the right of $t_{[\alpha/2]}^{(n-k)}$ is $\alpha/2$.

EXAMPLE 8.2

In this example we consider the Fresh detergent model

$$\begin{aligned} y_i &= \mu_i + \epsilon_i \\ &= \beta_0 + \beta_1 x_{i4} + \beta_2 x_{i3} + \beta_3 x_{i3}^2 + \beta_4 x_{i4} x_{i3} + \beta_5 D_{i,B} + \beta_6 D_{i,C} + \epsilon_i \end{aligned}$$

and calculate a 95% confidence interval for

$$\begin{aligned}\mu_{[d,a,C]} - \mu_{[d,a,B]} &= \beta_6 - \beta_5 \\ &= 0 \cdot \beta_0 + 0 \cdot \beta_1 + 0 \cdot \beta_2 + 0 \cdot \beta_3 \\ &\quad + 0 \cdot \beta_4 + (-1) \cdot \beta_5 + 1 \cdot \beta_6 \\ &= [0 \quad 0 \quad 0 \quad 0 \quad 0 \quad -1 \quad 1] \begin{bmatrix} \beta_0 \\ \beta_1 \\ \beta_2 \\ \beta_3 \\ \beta_4 \\ \beta_5 \\ \beta_6 \end{bmatrix} \\ &= \boldsymbol{\lambda}'\boldsymbol{\beta} \qquad \text{where } \boldsymbol{\lambda}' = [0 \quad 0 \quad 0 \quad 0 \quad 0 \quad -1 \quad 1]\end{aligned}$$

Since we saw in Example 7.14 that the least squares point estimates of β_5 and β_6 are $b_5 = .2137$ and $b_6 = .3818$ and the standard error is $s = .1308$, it follows that a point estimate of $\mu_{[d,a,C]} - \mu_{[d,a,B]} = \beta_6 - \beta_5 = \boldsymbol{\lambda}'\boldsymbol{\beta}$ is

$$\begin{aligned}\boldsymbol{\lambda}'\mathbf{b} &= [0 \quad 0 \quad 0 \quad 0 \quad 0 \quad -1 \quad 1] \begin{bmatrix} b_0 \\ b_1 \\ b_2 \\ b_3 \\ b_4 \\ b_5 \\ b_6 \end{bmatrix} \\ &= b_6 - b_5 \\ &= .3818 - .2137 \\ &= .1681\end{aligned}$$

and a 95% confidence interval for $\mu_{[d,a,C]} - \mu_{[d,a,B]} = \beta_6 - \beta_5 = \boldsymbol{\lambda}'\boldsymbol{\beta}$ is

$$\begin{aligned}[\boldsymbol{\lambda}'\mathbf{b} \pm t_{[.025]}^{(n-k)} s \sqrt{\boldsymbol{\lambda}'(\mathbf{X}'\mathbf{X})^{-1}\boldsymbol{\lambda}}] &= [\boldsymbol{\lambda}'\mathbf{b} \pm t_{[.025]}^{(30-7)} s \sqrt{\boldsymbol{\lambda}'(\mathbf{X}'\mathbf{X})^{-1}\boldsymbol{\lambda}}] \\ &= [.1681 \pm 2.069(.1308)\sqrt{.2372}] \\ &= [.1681 \pm .1318] \\ &= [.0363, .2999]\end{aligned}$$

As discussed in Example 7.14, this interval makes Enterprise Industries 95% confident that the effect of changing from advertising campaign B to advertising campaign C is to increase mean demand for Fresh by between .0363 (3,630 bottles) and .2999 (29,990 bottles).

The justification for the formula for $BE_{\boldsymbol{\lambda}'\mathbf{b}}[100(1-\alpha)]$ comes from the following.

The population of all possible point estimates of $\boldsymbol{\lambda}'\boldsymbol{\beta}$ (that is, values of $\boldsymbol{\lambda}'\mathbf{b}$)

1. Has *mean* $\mu_{\boldsymbol{\lambda}'\mathbf{b}} = \boldsymbol{\lambda}'\boldsymbol{\beta}$.
2. Has *variance* $\sigma^2_{\boldsymbol{\lambda}'\mathbf{b}} = \sigma^2\boldsymbol{\lambda}'(\mathbf{X}'\mathbf{X})^{-1}\boldsymbol{\lambda}$ (if inference assumptions 1 and 2 hold).
3. Has *standard deviation* $\sigma_{\boldsymbol{\lambda}'\mathbf{b}} = \sigma\sqrt{\boldsymbol{\lambda}'(\mathbf{X}'\mathbf{X})^{-1}\boldsymbol{\lambda}}$ (if inference assumptions 1 and 2 hold).
4. Has a *normal distribution* (if inference assumptions 1, 2, and 3 hold).

Moreover, if we define the $t_{[\boldsymbol{\lambda}'\mathbf{b},\boldsymbol{\lambda}'\boldsymbol{\beta}]}$ statistic to be

$$t_{[\boldsymbol{\lambda}'\mathbf{b},\boldsymbol{\lambda}'\boldsymbol{\beta}]} = \frac{\boldsymbol{\lambda}'\mathbf{b} - \boldsymbol{\lambda}'\boldsymbol{\beta}}{s\sqrt{\boldsymbol{\lambda}'(\mathbf{X}'\mathbf{X})^{-1}\boldsymbol{\lambda}}}$$

then the *population of all possible* $t_{[\boldsymbol{\lambda}'\mathbf{b},\boldsymbol{\lambda}'\boldsymbol{\beta}]}$ *statistics* has a *t-distribution* with $n-k$ degrees of freedom (if inference assumptions 1, 2, and 3 hold).

As an exercise, the reader should use these results to prove that

$$[\boldsymbol{\lambda}'\mathbf{b} \pm t_{[\alpha/2]}^{(n-k)} s\sqrt{\boldsymbol{\lambda}'(\mathbf{X}'\mathbf{X})^{-1}\boldsymbol{\lambda}}]$$

is a $100(1-\alpha)\%$ confidence interval for $\boldsymbol{\lambda}'\boldsymbol{\beta}$. As another exercise, the reader who has studied Appendix D should prove (1), (2), and (3).

Finally, we wish to show that the previously discussed confidence intervals for β_j and μ_0 are special cases of the confidence interval for $\boldsymbol{\lambda}'\boldsymbol{\beta}$. Note that the single parameter β_j in the regression model

$$y = \beta_0 + \beta_1 x_1 + \cdots + \beta_{j-1}x_{j-1} + \beta_j x_j + \beta_{j+1}x_{j+1} + \cdots + \beta_p x_p + \epsilon$$

can be written as

$$\beta_j = 0\cdot\beta_0 + 0\cdot\beta_1 + \cdots + 0\cdot\beta_{j-1} + 1\cdot\beta_j + 0\cdot\beta_{j+1} + \cdots + 0\cdot\beta_p$$

$$= [0 \quad 0 \quad \cdots \quad 0 \quad 1 \quad 0 \quad \cdots \quad 0]\begin{bmatrix}\beta_0\\ \beta_1\\ \vdots\\ \beta_{j-1}\\ \beta_j\\ \beta_{j+1}\\ \vdots\\ \beta_p\end{bmatrix}$$

$$= \boldsymbol{\lambda}'\boldsymbol{\beta}, \qquad \text{where } \boldsymbol{\lambda}' = [0 \quad 0 \quad \cdots \quad 0 \quad 1 \quad 0 \quad \cdots \quad 0]$$

Thus, the single parameter β_j can be considered a special case of the linear combination $\boldsymbol{\lambda}'\boldsymbol{\beta}$. This implies that the formula for a 100(1 − α)% confidence interval for β_j,

$$[b_j \pm t_{[\alpha/2]}^{(n-k)} s\sqrt{c_{jj}}]$$

is a special case of the formula for a 100(1 − α)% confidence interval for $\boldsymbol{\lambda}'\boldsymbol{\beta}$,

$$[\boldsymbol{\lambda}'\mathbf{b} \pm t_{[\alpha/2]}^{(n-k)} s\sqrt{\boldsymbol{\lambda}'(\mathbf{X}'\mathbf{X})^{-1}\boldsymbol{\lambda}}]$$

In this case $\boldsymbol{\lambda}'$ is a row vector containing the number 1 in position $(j + 1)$ and containing zeroes elsewhere, and

$$\boldsymbol{\lambda}'\mathbf{b} = b_j \qquad \text{and} \qquad \boldsymbol{\lambda}'(\mathbf{X}'\mathbf{X})^{-1}\boldsymbol{\lambda} = c_{jj}$$

Next, since

$$\begin{aligned}
\mu_0 &= \beta_0 + \beta_1 x_{01} + \beta_2 x_{02} + \cdots + \beta_p x_{0p} \\
&= [1 \quad x_{01} \quad x_{02} \quad \cdots \quad x_{0p}] \begin{bmatrix} \beta_0 \\ \beta_1 \\ \vdots \\ \beta_p \end{bmatrix} \\
&= \boldsymbol{\lambda}'\boldsymbol{\beta}, \qquad \text{where } \boldsymbol{\lambda}' = [1 \quad x_{01} \quad x_{02} \quad \cdots \quad x_{0p}]
\end{aligned}$$

it follows that μ_0, the future mean value of the dependent variable, is also a special case of the linear combination $\boldsymbol{\lambda}'\boldsymbol{\beta}$. This implies that the formula for a 100(1 − α)% confidence interval for μ_0

$$[\hat{y}_0 \pm t_{[\alpha/2]}^{(n-k)} s\sqrt{\mathbf{x}_0'(\mathbf{X}'\mathbf{X})^{-1}\mathbf{x}_0}]$$

where

$$\mathbf{x}_0' = [1 \quad x_{01} \quad x_{02} \quad \cdots \quad x_{0p}]$$

is a special case of the formula for a 100(1 − α)% confidence interval for $\boldsymbol{\lambda}'\boldsymbol{\beta}$, because

$$\begin{aligned}
\boldsymbol{\lambda}'\mathbf{b} &= [1 \quad x_{01} \quad x_{02} \quad \cdots \quad x_{0p}] \begin{bmatrix} b_0 \\ b_1 \\ b_2 \\ \vdots \\ b_p \end{bmatrix} \\
&= b_0 + b_1 x_{01} + b_2 x_{02} + \cdots + b_p x_{0p} \\
&= \hat{y}_0
\end{aligned}$$

and

$$\boldsymbol{\lambda}'(\mathbf{X}'\mathbf{X})^{-1}\boldsymbol{\lambda} = \mathbf{x}_0'(\mathbf{X}'\mathbf{X})^{-1}\mathbf{x}_0$$

8.2.2 Hypothesis Testing

Next we discuss hypothesis tests concerning linear combinations of regression parameters. If $\boldsymbol{\lambda}'\boldsymbol{\beta}$ is a linear combination of the parameters in the regression model

$$y = \beta_0 + \beta_1 x_1 + \beta_2 x_2 + \cdots + \beta_p x_p + \epsilon$$

it is often useful to test the null hypothesis $H_0: \boldsymbol{\lambda}'\boldsymbol{\beta} = 0$ versus the alternative hypothesis $H_1: \boldsymbol{\lambda}'\boldsymbol{\beta} \neq 0$. For example, in the Fresh Detergent problem it would be useful to test

$$H_0: \beta_6 - \beta_5 = \mu_{[d,a,C]} - \mu_{[d,a,B]} = 0$$

versus

$$H_1: \beta_6 - \beta_5 = \mu_{[d,a,C]} - \mu_{[d,a,B]} \neq 0$$

which says that there is a difference between the mean demands for Fresh obtained by advertising campaigns B and C.

TESTING $H_0: \boldsymbol{\lambda}'\boldsymbol{\beta} = 0$ VERSUS $H_1: \boldsymbol{\lambda}'\boldsymbol{\beta} \neq 0$:

Define

$$t_{[\boldsymbol{\lambda}'\mathbf{b}, \boldsymbol{\lambda}'\boldsymbol{\beta}]} = \frac{\boldsymbol{\lambda}'\mathbf{b} - \boldsymbol{\lambda}'\boldsymbol{\beta}}{s\sqrt{\boldsymbol{\lambda}'(\mathbf{X}'\mathbf{X})^{-1}\boldsymbol{\lambda}}}$$

and assume that the population of all possible $t_{[\boldsymbol{\lambda}'\mathbf{b}, \boldsymbol{\lambda}'\boldsymbol{\beta}]}$ statistics has a t-distribution with $n - k$ degrees of freedom. In order to test $H_0: \boldsymbol{\lambda}'\boldsymbol{\beta} = 0$ versus $H_1: \boldsymbol{\lambda}'\boldsymbol{\beta} \neq 0$, define the $t_{\boldsymbol{\lambda}'\mathbf{b}}$ statistic and its related prob-value to be

$$t_{\boldsymbol{\lambda}'\mathbf{b}} = \frac{\boldsymbol{\lambda}'\mathbf{b}}{s\sqrt{\boldsymbol{\lambda}'(\mathbf{X}'\mathbf{X})^{-1}\boldsymbol{\lambda}}} \quad \text{and} \quad \text{prob-value} = 2A[[|t_{\boldsymbol{\lambda}'\mathbf{b}}|, \infty); n - k]$$

where $A[[|t_{\boldsymbol{\lambda}'\mathbf{b}}|, \infty); n - k]$ is the area under the curve of the t-distribution having $n - k$ degrees of freedom to the right of $|t_{\boldsymbol{\lambda}'\mathbf{b}}|$, the absolute value of the $t_{\boldsymbol{\lambda}'\mathbf{b}}$ statistic. Then, we can reject $H_0: \boldsymbol{\lambda}'\boldsymbol{\beta} = 0$ in favor of $H_1: \boldsymbol{\lambda}'\boldsymbol{\beta} \neq 0$ by setting the probability of a Type I error equal to α if and only if any of the following three equivalent conditions hold.

1. $|t_{\boldsymbol{\lambda}'\mathbf{b}}| > t_{[\alpha/2]}^{(n-k)}$—that is, if $t_{\boldsymbol{\lambda}'\mathbf{b}} > t_{[\alpha/2]}^{(n-k)}$ or $t_{\boldsymbol{\lambda}'\mathbf{b}} < -t_{[\alpha/2]}^{(n-k)}$.
2. Prob-value $< \alpha$.
3. The $100(1 - \alpha)\%$ confidence interval for $\boldsymbol{\lambda}'\boldsymbol{\beta}$—$[\boldsymbol{\lambda}'\mathbf{b} \pm t_{[\alpha/2]}^{(n-k)} s\sqrt{\boldsymbol{\lambda}'(\mathbf{X}'\mathbf{X})^{-1}\boldsymbol{\lambda}}]$—does not contain zero.

Moreover

4. For any level of confidence, $100(1-\alpha)\%$, less than $100[1-(\text{prob-value})]\%$, the $100(1-\alpha)\%$ confidence interval for $\boldsymbol{\lambda}'\boldsymbol{\beta}$ does not contain zero, and thus we can, with at least $100(1-\alpha)\%$ confidence, reject $H_0: \boldsymbol{\lambda}'\boldsymbol{\beta} = 0$ in favor of $H_1: \boldsymbol{\lambda}'\boldsymbol{\beta} \neq 0$.
5. For any level of confidence, $100(1-\alpha)\%$, greater than or equal to $100[1-(\text{prob-value})]\%$, the $100(1-\alpha)\%$ confidence interval for $\boldsymbol{\lambda}'\boldsymbol{\beta}$ does contain zero, and thus we cannot, with at least $100(1-\alpha)\%$ confidence, reject $H_0: \boldsymbol{\lambda}'\boldsymbol{\beta} = 0$ in favor of $H_1: \boldsymbol{\lambda}'\boldsymbol{\beta} \neq 0$.

Note that the proof of the results in the preceding box is exactly the same as the proof given in Section 6.5 of the results in the box entitled "Testing $H_0: \beta_j = 0$ versus $H_1: \beta_j \neq 0$" (replacing β_j, b_j, $s\sqrt{c_{jj}}$, $t_{[b_j, \beta_j]}$, and t_{b_j} in Section 6.5 by $\boldsymbol{\lambda}'\boldsymbol{\beta}$, $\boldsymbol{\lambda}'\mathbf{b}$, $s\sqrt{\boldsymbol{\lambda}'(\mathbf{X}'\mathbf{X})^{-1}\boldsymbol{\lambda}}$, $t_{[\boldsymbol{\lambda}'\mathbf{b}, \boldsymbol{\lambda}'\boldsymbol{\beta}]}$, and $t_{\boldsymbol{\lambda}'\mathbf{b}}$, respectively).

EXAMPLE 8.3

We consider the Fresh Detergent model

$$\begin{aligned} y_i &= \mu_i + \epsilon_i \\ &= \beta_0 + \beta_1 x_{i4} + \beta_2 x_{i3} + \beta_3 x_{i3}^2 + \beta_4 x_{i4} x_{i3} + \beta_5 D_{i,B} + \beta_6 D_{i,C} + \epsilon_i \end{aligned}$$

and test

$$H_0: \beta_6 - \beta_5 = \mu_{[d,a,C]} - \mu_{[d,a,B]} = 0$$

versus

$$H_1: \beta_6 - \beta_5 = \mu_{[d,a,C]} - \mu_{[d,a,B]} \neq 0$$

Assuming that the inference assumptions hold, and defining the $t_{[\boldsymbol{\lambda}'\mathbf{b}, \boldsymbol{\lambda}'\boldsymbol{\beta}]}$ statistic to be

$$t_{[\boldsymbol{\lambda}'\mathbf{b}, \boldsymbol{\lambda}'\boldsymbol{\beta}]} = \frac{\boldsymbol{\lambda}'\mathbf{b} - \boldsymbol{\lambda}'\boldsymbol{\beta}}{s\sqrt{\boldsymbol{\lambda}'(\mathbf{X}'\mathbf{X})^{-1}\boldsymbol{\lambda}}} = \frac{b_6 - b_5 - (\beta_6 - \beta_5)}{s\sqrt{\boldsymbol{\lambda}'(\mathbf{X}'\mathbf{X})^{-1}\boldsymbol{\lambda}}}$$

it follows from the discussion of Section 8.2.1 that the population of all possible $t_{[\boldsymbol{\lambda}'\mathbf{b}, \boldsymbol{\lambda}'\boldsymbol{\beta}]}$ statistics has a t-distribution with $n - k = 30 - 7 = 23$ degrees of freedom. Therefore, from condition (1) in the preceding box, we can test H_0 versus

H_1 by setting α, the probability of a Type I error, equal to .05 if we calculate

$$\begin{aligned} t_{\boldsymbol{\lambda}'\mathbf{b}} &= \frac{\boldsymbol{\lambda}'\mathbf{b}}{s\sqrt{\boldsymbol{\lambda}'(\mathbf{X}'\mathbf{X})^{-1}\boldsymbol{\lambda}}} \\ &= \frac{b_6 - b_5}{s\sqrt{\boldsymbol{\lambda}'(\mathbf{X}'\mathbf{X})^{-1}\boldsymbol{\lambda}}} \\ &= \frac{.1681}{.1308\sqrt{.2372}} \\ &= 2.6389 \end{aligned}$$

and use the rejection points

$$\begin{aligned} t_{[\alpha/2]}^{(n-k)} &= t_{[.05/2]}^{(30-7)} = t_{[.025]}^{(23)} = 2.069 \\ -t_{[\alpha/2]}^{(n-k)} &= -2.069 \end{aligned}$$

Since $t_{\boldsymbol{\lambda}'\mathbf{b}} = 2.6389 > 2.069 = t_{[.025]}^{(23)}$, we can reject H_0 in favor of H_1 by setting α equal to .05.

Next, we note that

$$\begin{aligned} \text{prob-value} &= 2A[[|t_{\boldsymbol{\lambda}'\mathbf{b}}|, \infty); n-k] \\ &= 2A[[|2.6389|, \infty); 30-7] \\ &= 2A[[2.6389, \infty); 23] \\ &= 2(.00735) \\ &= .0147 \end{aligned}$$

where $A[[2.6389, \infty); 23]$ is the area under the curve of the t-distribution having $n - k = 23$ degrees of freedom to the right of $|t_{\boldsymbol{\lambda}'\mathbf{b}}| = 2.6389$, the absolute value of the $t_{\boldsymbol{\lambda}'\mathbf{b}}$ statistic. Since this prob-value is less than .05 and .02, but not less than .01, we can reject H_0 in favor of H_1 if we set α equal to .05 or .02, but we cannot reject H_0 in favor of H_1 if we set α equal to .01.

In order to interpret the prob-value as a probability, note that since the population of all possible $t_{[\boldsymbol{\lambda}'\mathbf{b},\boldsymbol{\lambda}'\boldsymbol{\beta}]}$ statistics has a t-distribution with 23 degrees of freedom, it follows that if

$$H_0: \boldsymbol{\lambda}'\boldsymbol{\beta} = \beta_6 - \beta_5 = \mu_{[d,a,C]} - \mu_{[d,a,B]} = 0$$

is true, then the population of all possible $t_{\boldsymbol{\lambda}'\mathbf{b}}$ statistics has a t-distribution with 23 degrees of freedom, since if $H_0: \boldsymbol{\lambda}'\boldsymbol{\beta} = 0$ is true, then

$$t_{[\boldsymbol{\lambda}'\mathbf{b},\boldsymbol{\lambda}'\boldsymbol{\beta}]} = \frac{\boldsymbol{\lambda}'\mathbf{b} - \boldsymbol{\lambda}'\boldsymbol{\beta}}{s\sqrt{\boldsymbol{\lambda}'(\mathbf{X}'\mathbf{X})^{-1}\boldsymbol{\lambda}}} = \frac{\boldsymbol{\lambda}'\mathbf{b} - 0}{s\sqrt{\boldsymbol{\lambda}'(\mathbf{X}'\mathbf{X})^{-1}\boldsymbol{\lambda}}} = \frac{\boldsymbol{\lambda}'\mathbf{b}}{s\sqrt{\boldsymbol{\lambda}'(\mathbf{X}'\mathbf{X})^{-1}\boldsymbol{\lambda}}} = t_{\boldsymbol{\lambda}'\mathbf{b}}$$

Therefore,

$$\begin{aligned}\text{prob-value} &= 2A[[|t_{\lambda'b}|, \infty); n-k] \\ &= 2A[[2.6389, \infty); 23] \\ &= .0147\end{aligned}$$

equals the proportion of $t_{\lambda'b}$ statistics in the population of all such statistics which, if the null hypothesis

$$H_0: \mu_{[d,a,C]} - \mu_{[d,a,B]} = 0$$

is true, are at least as far away from zero and thus at least as contradictory to

$$H_0: \mu_{[d,a,C]} - \mu_{[d,a,B]} = 0$$

as 2.6389, the observed $t_{\lambda'b}$ statistic. Since this prob-value, which equals .0147, says that Enterprise Industries must believe that a 147 in 10,000 chance has occurred if it is to believe that H_0 is true, the company has substantial evidence to conclude that

$$H_0: \mu_{[d,a,C]} - \mu_{[d,a,B]} = 0$$

is false and thus that

$$H_1: \mu_{[d,a,C]} - \mu_{[d,a,B]} \neq 0$$

is true.

Finally, we note that it follows from (4) in the preceding box that for any level of confidence, $100(1-\alpha)\%$, less than

$$100[1 - (\text{prob-value})]\% = 100[1 - .0147]\% = 98.53\%$$

the $100(1-\alpha)\%$ confidence interval for $\mu_{[d,a,C]} - \mu_{[d,a,B]}$ does not contain zero, and thus Enterprise Industries can, with at least $100(1-\alpha)\%$ confidence, reject

$$H_0: \mu_{[d,a,C]} - \mu_{[d,a,B]} = 0$$

in favor of

$$H_1: \mu_{[d,a,C]} - \mu_{[d,a,B]} \neq 0$$

It follows from (5) that for any level of confidence, $100(1-\alpha)\%$, greater than or equal to

$$100[1 - (\text{prob-value})]\% = 98.53\%$$

the $100(1-\alpha)\%$ confidence interval for $\mu_{[d,a,C]} - \mu_{[d,a,B]}$ does contain zero, and thus Enterprise Industries cannot, with at least $100(1-\alpha)\%$ confidence, reject

$$H_0: \mu_{[d,a,C]} - \mu_{[d,a,B]} = 0$$

FIGURE 8.2 $H_0: \mu_{[d,a,C]} - \mu_{[d,a,B]} = 0$ Can Be Rejected at Any Level of Confidence Less Than 98.53%

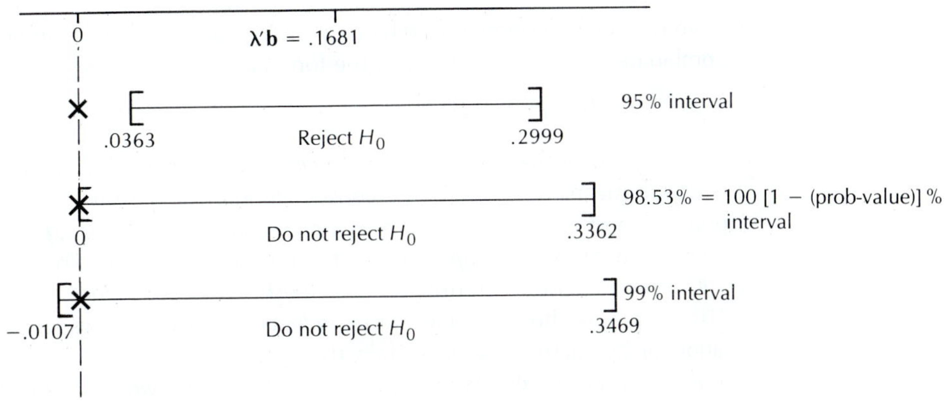

in favor of

$$H_1: \mu_{[d,a,C]} - \mu_{[d,a,B]} \neq 0$$

These facts are illustrated in Figure 8.2, which depicts the 95%, 98.53%, and 99% confidence intervals for $\mu_{[d,a,C]} - \mu_{[d,a,B]}$. Notice that the 95% confidence interval (100(1 − α)% less than 98.53%) does not contain zero, and hence we can reject

$$H_0: \mu_{[d,a,C]} - \mu_{[d,a,B]} = 0$$

at this level of confidence. However, the 98.53% confidence interval (which has a lower bound of zero) and the 99% confidence interval (100(1 − α)% greater than 98.53%) do contain zero. Thus, we cannot reject H_0 at these levels of confidence. To summarize, since the prob-value is .0147, we can be up to

$$100[1 - (\text{prob-value})]\% = 98.53\%$$

confident that the null hypothesis

$$H_0: \mu_{[d,a,C]} - \mu_{[d,a,B]} = 0$$

is false and that the alternative hypothesis

$$H_1: \mu_{[d,a,C]} - \mu_{[d,a,B]} \neq 0$$

is true.

8.3 SIMULTANEOUS CONFIDENCE INTERVALS

From now on, we sometimes refer to a 100(1 − α)% confidence interval for a linear combination $\boldsymbol{\lambda}'\boldsymbol{\beta}$ computed using the formula

$$[\boldsymbol{\lambda}'\mathbf{b} \pm t_{[\alpha/2]}^{(n-k)} s\sqrt{\boldsymbol{\lambda}'(\mathbf{X}'\mathbf{X})^{-1}\boldsymbol{\lambda}}]$$

as an *individual 100(1 − α)% confidence interval* for $\boldsymbol{\lambda}'\boldsymbol{\beta}$. In order to make clear what we mean by an individual confidence interval, we say that a sample in the population of all possible samples is *individually successful* for $\boldsymbol{\lambda}'\boldsymbol{\beta}$ if a 100(1 − α)% confidence interval for $\boldsymbol{\lambda}'\boldsymbol{\beta}$ computed using the sample contains the true value of $\boldsymbol{\lambda}'\boldsymbol{\beta}$. When we use the term *individual 100(1 − α)% confidence*, we are saying that 100(1 − α)% of the individual 100(1 − α)% confidence intervals for $\boldsymbol{\lambda}'\boldsymbol{\beta}$ in the population of all such intervals contain the true value of $\boldsymbol{\lambda}'\boldsymbol{\beta}$. Equivalently, the term *individual 100(1 − α)% confidence* means that, if we were to compute individual 100(1 − α)% confidence intervals for $\boldsymbol{\lambda}'\boldsymbol{\beta}$ by using each of the samples in the population of all possible samples, then 100(1 − α)% of these samples would be individually successful for $\boldsymbol{\lambda}'\boldsymbol{\beta}$.

In contrast to the concept of *an individual 100(1 − α)% confidence interval*, there is the concept of a *simultaneous 100(1 − α)% confidence interval*. There are several formulas for simultaneous 100(1 − α)% confidence intervals, including one developed by Scheffé (1959). We say that a sample in the population of all possible samples is *simultaneously successful* for all the linear combinations of regression parameters in a set of linear combinations if all the 100(1 − α)% confidence intervals for the linear combinations in the set computed using the sample contain their respective linear combinations. Scheffé (1959) has shown that if we were to compute *Scheffé simultaneous 100(1 − α)% confidence intervals* for all the linear combinations in a set of linear combinations by applying the Scheffé simultaneous 100(1 − α)% confidence interval formula to each of the samples in the population of all possible samples, then 100(1 − α)% of these samples would be simultaneously successful for all the linear combinations in the set.

To explain more fully the difference between individual and simultaneous success, we look at a very simple example. Suppose we wish to compute $\frac{2}{3} \times 100\% = 66.67\%$ confidence intervals for two parameters β_1 and β_2. Assume here that (a supernatural power knows that) $\beta_1 = 2$ and $\beta_2 = 5$, and consider three possible samples (denoted SPL_1, SPL_2, and SPL_3). The upper part of Table 8.4 lists hypothetical *individual* 66.67% confidence intervals for β_1 and β_2 computed using these three samples. Here we see that

1. 66.67% (two out of three) of these samples are individually successful for β_1.
2. 66.67% (two out of three) of these samples are individually successful for β_2.
3. Only 33.33% (one out of three) of these samples are simultaneously successful for β_1 and β_2.

This illustrates the general fact that if we were to compute *individual* 100(1 − α)% confidence intervals for each parameter in a set of parameters by applying the

TABLE 8.4 Individual 66.67% Confidence Intervals and Simultaneous 66.67% Confidence Intervals for β_1 and β_2

Sample Used to Calculate Intervals	Individual 66.67% Confidence Intervals for		Result
	$\beta_1 = 2$	$\beta_2 = 5$	
SPL_1	*[1.1, 3.1]	*[3.9, 5.9]	Simultaneously successful for β_1 and β_2
SPL_2	*[1.5, 3.5]	[5.4, 7.4]	Not simultaneously successful for β_1 and β_2
SPL_3	[2.1, 4.1]	*[4.2, 6.2]	Not simultaneously successful for β_1 and β_2
	$\frac{2}{3} \times 100\% = 66.67\%$ of the samples are individually successful for β_1	$\frac{2}{3} \times 100\% = 66.67\%$ of the samples are individually successful for β_2	$\frac{1}{3} \times 100\% = 33.33\%$ of the samples are simultaneously successful for β_1 and β_2

Sample Used to Calculate Intervals	Simultaneous 66.67% Confidence Intervals for		Result
	$\beta_1 = 2$	$\beta_2 = 5$	
SPL_1	*[.8, 3.4]	*[3.6, 6.2]	Simultaneously successful for β_1 and β_2
SPL_2	*[1.2, 3.8]	[5.1, 7.7]	Not simultaneously successful for β_1 and β_2
SPL_3	*[1.8, 4.4]	*[3.9, 6.5]	Simultaneously successful for β_1 and β_2
	$\frac{3}{3} \times 100\% = 100\%$ of the samples are individually successful for β_1	$\frac{2}{3} \times 100\% = 66.67\%$ of the samples are individually successful for β_2	$\frac{2}{3} \times 100\% = 66.67\%$ of the samples are simultaneously successful for β_1 and β_2

Note: An asterisk beside an interval indicates that the interval contains the parameter of interest.

individual $100(1 - \alpha)\%$ confidence interval formula to each of the samples in the population of all possible samples, $100(1 - \alpha)\%$ of these samples would be individually successful for each parameter in the set, but less than $100(1 - \alpha)\%$ of these samples would be simultaneously successful for all of the parameters in the set. The lower part of Table 8.4 lists hypothetical *simultaneous* 66.67% confidence intervals for β_1 and β_2 computed using three possible samples (SPL_1, SPL_2, and SPL_3). Here we see that

1. 66.67% (two out of three) of these samples are simultaneously successful for β_1 and β_2.
2. 100% (all three) of these samples are individually successful for β_1.
3. 66.67% (two out of three) of these samples are individually successful for β_2.

This illustrates the general fact that if we were to compute Scheffé simultaneous $100(1-\alpha)\%$ confidence intervals for all the parameters in a set of parameters by applying the Scheffé simultaneous $100(1-\alpha)\%$ confidence interval formula to each of the samples in the population of all possible samples, $100(1-\alpha)\%$ of these samples would be simultaneously successful for all the parameters in the set, and at least $100(1-\alpha)\%$ of these samples would be individually successful for each parameter in the set.

Comparing the individual and simultaneous intervals in Table 8.4, we see that the Scheffé simultaneous 66.67% confidence intervals are longer than the individual 66.67% confidence intervals. In general, Scheffé simultaneous $100(1-\alpha)\%$ confidence intervals are longer than individual $100(1-\alpha)\%$ confidence intervals computed for the same linear combinations of regression parameters. Thus, intuitively, we are "paying for" simultaneous confidence by obtaining longer (and thus less precise) confidence intervals. Some analysts argue that longer intervals are too high a price to pay for simultaneous confidence, particularly when the simultaneous confidence intervals are so imprecise that they provide little or no meaningful information. Others are willing to pay this price for simultaneous confidence. We suggest calculating both individual $100(1-\alpha)\%$ confidence intervals and Scheffé simultaneous $100(1-\alpha)\%$ confidence intervals. Conclusions (about, for example, differences in population means) can be drawn by examining both the individual and simultaneous intervals that have been computed.

We now explain how to calculate Scheffé simultaneous $100(1-\alpha)\%$ confidence intervals. Consider the linear regression model

$$y = \beta_0 + \beta_1 x_1 + \beta_2 x_2 + \cdots + \beta_p x_p + \epsilon$$

and assume that we wish to calculate Scheffé simultaneous $100(1-\alpha)\%$ confidence intervals for all the linear combinations of regression parameters in a set of linear combinations, where q denotes the number of the regression parameters $\beta_0, \beta_1, \beta_2, \ldots, \beta_p$ that are *involved nontrivially* in at least one of the linear combinations in the set. Here, any parameter β_j is said to be involved nontrivially in a linear combination.

$$\boldsymbol{\lambda}'\boldsymbol{\beta} = \sum_{j=0}^{p} \lambda_j \beta_j$$

if $\lambda_j \neq 0$. Then, the Scheffé simultaneous $100(1-\alpha)\%$ confidence interval for any linear combination in the set is as follows.

SCHEFFÉ SIMULTANEOUS CONFIDENCE INTERVALS:

If the inference assumptions are satisfied, the Scheffé simultaneous $100(1-\alpha)\%$ confidence interval for $\boldsymbol{\lambda}'\boldsymbol{\beta}$ is

$$\left[\boldsymbol{\lambda}'\mathbf{b} - \sqrt{qF_{[\alpha]}^{(q,n-k)}}\,s\sqrt{\boldsymbol{\lambda}'(\mathbf{X}'\mathbf{X})^{-1}\boldsymbol{\lambda}},\ \boldsymbol{\lambda}'\mathbf{b} + \sqrt{qF_{[\alpha]}^{(q,n-k)}}\,s\sqrt{\boldsymbol{\lambda}'(\mathbf{X}'\mathbf{X})^{-1}\boldsymbol{\lambda}}\right]$$

Here

1. $F_{[\alpha]}^{(q,n-k)}$ is the point on the scale of the F-distribution having q and $(n-k)$ degrees of freedom so that the area under this curve to the right of $F_{[\alpha]}^{(q,n-k)}$ is α.
2. $\mathbf{b}$ is a column vector containing the least squares estimates $b_0, b_1, b_2, \ldots, b_p$.
3. s is the standard error.
4. n is the number of observed values of the dependent variable.
5. k is the number of parameters in the regression model.
6. q is the number of the regression parameters $\beta_0, \beta_1, \beta_2, \ldots, \beta_p$ that are *involved nontrivially* in at least one of the linear combinations for which simultaneous intervals are being computed.

EXAMPLE 8.4

We now demonstrate how Enterprise Industries can calculate Scheffé simultaneous 95% confidence intervals for all the linear combinations of regression parameters in the following set (Set I).

Set I

$$\mu_{[d,a,B]} - \mu_{[d,a,A]} = \beta_5$$

$$\mu_{[d,a,C]} - \mu_{[d,a,A]} = \beta_6$$

$$\mu_{[d,a,C]} - \mu_{[d,a,B]} = \beta_6 - \beta_5$$

Since there are two regression parameters, β_5 and β_6, from the regression model

$$y_i = \beta_0 + \beta_1 x_{i4} + \beta_2 x_{i3} + \beta_3 x_{i3}^2 + \beta_4 x_{i4} x_{i3} + \beta_5 D_{i,B} + \beta_6 D_{i,C} + \epsilon_i$$

involved nontrivially in at least one of the linear combinations in Set I, it follows that $q = 2$ and that

$$\begin{aligned}\sqrt{qF_{[\alpha]}^{(q,n-k)}} &= \sqrt{2F_{[.05]}^{(2,30-7)}}\\ &= \sqrt{2F_{[.05]}^{(2,23)}}\\ &= \sqrt{2(3.42)}\\ &= 2.6153\end{aligned}$$

TABLE 8.5 Scheffé Simultaneous 95% Confidence Intervals

Linear Combination ($\boldsymbol{\lambda}'\boldsymbol{\beta}$)	Point Estimate ($\boldsymbol{\lambda}'\mathbf{b}$)	$s\sqrt{\boldsymbol{\lambda}'(\mathbf{X}'\mathbf{X})^{-1}\boldsymbol{\lambda}}$	Individual 95% Confidence Interval for $\boldsymbol{\lambda}'\boldsymbol{\beta}$ ($[\boldsymbol{\lambda}'\mathbf{b} \pm (2.069)s\sqrt{\boldsymbol{\lambda}'(\mathbf{X}'\mathbf{X})^{-1}\boldsymbol{\lambda}}]$)
$\mu_{[d,a,B]} - \mu_{[d,a,A]} = \beta_5$	.2137	.0622	$[.2137 \pm 2.069(.0622)] = [.0851, .3423]$
$\mu_{[d,a,C]} - \mu_{[d,a,A]} = \beta_6$	.3818	.0613	$[.3818 \pm 2.069(.0613)] = [.2550, .5085]$
$\mu_{[d,a,C]} - \mu_{[d,a,B]} = \beta_6 - \beta_5$	.1681	.0637	$[.1618 \pm 2.069(.0637)] = [.0363, .2999]$
$\mu_0 = \beta_0 + \beta_1(.10) + \beta_2(6.80) + \beta_3(6.80)^2 + \beta_4(.10)(6.80) + \beta_5(0) + \beta_6(1)$	8.6825	.0597	$[8.6825 \pm 2.069(.0597)] = [8.559, 8.806]$

Thus,

$$[\boldsymbol{\lambda}'\mathbf{b} \pm \sqrt{qF_{[\alpha]}^{(q,n-k)}}s\sqrt{\boldsymbol{\lambda}'(\mathbf{X}'\mathbf{X})^{-1}\boldsymbol{\lambda}}] = [\boldsymbol{\lambda}'\mathbf{b} \pm (2.6153)s\sqrt{\boldsymbol{\lambda}'(\mathbf{X}'\mathbf{X})^{-1}\boldsymbol{\lambda}}]$$

is a Scheffé simultaneous 95% confidence interval for any linear combination $\boldsymbol{\lambda}'\boldsymbol{\beta}$ in Set I. Note that this simultaneous interval is longer (and thus less precise) than

$$[\boldsymbol{\lambda}'\mathbf{b} \pm t_{[\alpha/2]}^{(n-k)}s\sqrt{\boldsymbol{\lambda}'(\mathbf{X}'\mathbf{X})^{-1}\boldsymbol{\lambda}}] = [\boldsymbol{\lambda}'\mathbf{b} \pm (2.069)s\sqrt{\boldsymbol{\lambda}'(\mathbf{X}'\mathbf{X})^{-1}\boldsymbol{\lambda}}]$$

which is the individual 95% confidence interval for $\boldsymbol{\lambda}'\boldsymbol{\beta}$. Therefore, we are "paying for" simultaneous confidence by obtaining less precise intervals.

In Table 8.5 we calculate both individual 95% confidence intervals and Scheffé simultaneous 95% confidence intervals for the three linear combinations in Set I. Whereas we are 95% confident that each of the individual 95% confidence intervals contains the linear combination it is meant to contain, we are less than 95% confident that all three of the individual 95% confidence intervals contain the three linear combinations that they are meant to contain. However, we are 95% confident that all three of the longer Scheffé simultaneous 95% confidence intervals contain the three linear combinations that they are meant to contain. To illustrate that the longer simultaneous confidence intervals are less precise than the individual confidence intervals, note that [.0363, .2999],

Scheffé Simultaneous 95% Confidence Interval for $\boldsymbol{\lambda}'\boldsymbol{\beta}$ (in Set I) $([\boldsymbol{\lambda}'\mathbf{b} \pm (2.6153)s\sqrt{\boldsymbol{\lambda}'(\mathbf{X}'\mathbf{X})^{-1}\boldsymbol{\lambda}}])$	Scheffé Simultaneous 95% Confidence Interval for $\boldsymbol{\lambda}'\boldsymbol{\beta}$ (in Set II) $([\boldsymbol{\lambda}'\mathbf{b} \pm (4.1328)s\sqrt{\boldsymbol{\lambda}'(\mathbf{X}'\mathbf{X})^{-1}\boldsymbol{\lambda}}])$
$[.2137 \pm 2.6153(.0622)] = [.051, .3764]$	$[.2137 \pm 4.1328(.0622)] = [-.0434, .4708]$
$[.3818 \pm 2.6153(.0613)] = [.2215, .5421]$	$[.3818 \pm 4.1328(.0613)] = [.1285, .6351]$
$[.1618 \pm 2.6153(.0637)] = [-.0048, .3284]$	$[.1618 \pm 4.1328(.0637)] = [-.1015, .4251]$
	$[8.6825 \pm 4.1328(.0597)] = [8.4358, 8.9292]$

the individual 95% confidence interval for $\mu_{[d,a,C]} - \mu_{[d,a,B]}$, makes us 95% (individually) confident that $\mu_{[d,a,C]} - \mu_{[d,a,B]}$ does not equal zero, whereas $[-.0048, .3284]$, the simultaneous 95% confidence interval for $\mu_{[d,a,C]} - \mu_{[d,a,B]}$, does not make us 95% (simultaneously) confident that $\mu_{[d,a,C]} - \mu_{[d,a,B]}$ does not equal zero.

We next demonstrate how Enterprise Industries can calculate Scheffé simultaneous 95% confidence intervals for all the linear combinations of regression parameters in the following set (Set II).

Set II

$$\mu_{[d,a,B]} - \mu_{[d,a,A]} = \beta_5$$

$$\mu_{[d,a,C]} - \mu_{[d,a,A]} = \beta_6$$

$$\mu_{[d,a,C]} - \mu_{[d,a,B]} = \beta_6 - \beta_5$$

$$\begin{aligned}\mu_0 &= \beta_0 + \beta_1 x_{04} + \beta_2 x_{03} + \beta_3 x_{03}^2 + \beta_4 x_{04} x_{03} + \beta_5 D_{0,B} + \beta_6 D_{0,C} \\ &= \beta_0 + \beta_1(.10) + \beta_2(6.80) + \beta_3(6.80)^2 + \beta_4(.10)(6.80) \\ &\quad + \beta_5(0) + \beta_6(1)\end{aligned}$$

Since all seven parameters from the regression model

$$y_i = \beta_0 + \beta_1 x_{i4} + \beta_2 x_{i3} + \beta_3 x_{i3}^2 + \beta_4 x_{i4} x_{i3} + \beta_5 D_{i,B} + \beta_6 D_{i,C} + \epsilon_i$$

are involved nontrivially in at least one of the linear combinations in Set II, it follows that $q = 7$ and that

$$\begin{aligned}\sqrt{qF_{[\alpha]}^{(q,n-k)}} &= \sqrt{7F_{[.05]}^{(7,30-7)}} \\ &= \sqrt{7F_{[.05]}^{(7,23)}} \\ &= \sqrt{7(2.44)} \\ &= 4.1328\end{aligned}$$

Thus

$$[\boldsymbol{\lambda}'\mathbf{b} \pm \sqrt{qF_{[\alpha]}^{(q,n-k)}}s\sqrt{\boldsymbol{\lambda}'(\mathbf{X}'\mathbf{X})^{-1}\boldsymbol{\lambda}}] = [\boldsymbol{\lambda}'\mathbf{b} \pm (4.1328)s\sqrt{\boldsymbol{\lambda}'(\mathbf{X}'\mathbf{X})^{-1}\boldsymbol{\lambda}}]$$

is a Scheffé simultaneous 95% confidence interval for any linear combination $\boldsymbol{\lambda}'\boldsymbol{\beta}$ in Set II. Since $q = 7$ for Set II, while $q = 2$ for Set I, this simultaneous 95% interval is longer (less precise) than

$$[\boldsymbol{\lambda}'\mathbf{b} \pm (2.6153)s\sqrt{\boldsymbol{\lambda}'(\mathbf{X}'\mathbf{X})^{-1}\boldsymbol{\lambda}}]$$

the Scheffé simultaneous 95% confidence interval for any linear combination in Set I. In Table 8.5 we calculate the Scheffé simultaneous 95% confidence intervals for the linear combinations in Set II.

Finally, note that *the Scheffé simultaneous confidence interval formula applies to any set of linear combinations of regression parameters. However, there are other simultaneous confidence interval formulas that apply to special sets of linear combinations of regression parameters and that can yield shorter intervals* than those given by the Scheffé simultaneous confidence interval formula. For a discussion of such formulas, see Neter and Wasserman (1974).

8.4 ONE QUALITATIVE INDEPENDENT VARIABLE

8.4.1 The Regression Approach

EXAMPLE 8.5

Part 1. *The Problem and Data.*

North American Oil Company is attempting to develop a reasonably priced unleaded gasoline that will deliver higher gasoline mileages than can be attained by its current gasolines. As part of its development process, North American Oil would like to compare the effects of three basic types of unleaded gasoline (A, B, and C) on gasoline mileage. Since the chemical compositions of these three gasoline types differ, and since these differences in chemical composition cannot be defined to be a function of one or more *quantitative* independent variables, North American Oil defines "gasoline type" to be a *qualitative* independent variable.

For testing purposes, North American Oil will compare the effects of gasoline types A, B, and C on the gasoline mileage obtained by an automobile called the Fire-Hawk, which is produced by National Motors. In particular, North American Oil wishes to compare the effects of gasoline types A, B, and C on the gasoline mileage obtained by the Fire-Hawks in the infinite population of all Fire-Hawks that could potentially be produced this year and that would be driven under normal driving conditions.

Corresponding to each potential Fire-Hawk, there is a gasoline mileage (measured to the nearest tenth of a mile per gallon) that would be obtained if this Fire-Hawk were driven under normal driving conditions *using gasoline type l* (l = A, B, and C). Thus, corresponding to the infinite population of all Fire-Hawks, there is an *infinite population of all Fire-Hawk mileages that would be obtained by using gasoline type l.* We define μ_l to be the *mean Fire-Hawk mileage that would be obtained by using gasoline type l.* Thus, one way that North American Oil can compare the effects of gasoline types A, B, and C on the gasoline mileage obtained by the Fire-Hawk is to find point estimates of μ_A, μ_B, and μ_C. To do this, North American Oil will randomly select

1. A sample of $n_A = 4$ Fire-Hawk mileages from the infinite population of all Fire-Hawk mileages that would be obtained by using gasoline type A.
2. A sample of $n_B = 5$ Fire-Hawk mileages from the infinite population of all Fire-Hawk mileages that would be obtained by using gasoline type B.
3. A sample of $n_C = 3$ Fire-Hawk mileages from the infinite population of all Fire-Hawk mileages that would be obtained by using gasoline type C.

Before describing how North American Oil can randomly select these samples, we note that the sizes of these samples are not equal. Generally, *it is best* (in order to obtain the most information for a given amount of experimental effort) *to make the sizes of the samples drawn from different populations equal.* However, sometimes data are lost, making the sample sizes unequal, or sometimes an analyst must use available data with unequal sample sizes. Because of this, we demonstrate the application of regression analysis to analyzing data with unequal sample sizes.

North American Oil cannot *directly* randomly select a sample from a population of potential Fire-Hawk mileages that would exist if all the Fire-Hawks that could potentially be produced were driven under normal driving conditions using gasoline type l. However, North American Oil can "for all practical purposes" randomly select a sample from such a population by employing a **completely randomized experimental design**. Suppose that National Motors has given North American Oil access to a subpopulation of 1,000 Fire-Hawks and that this finite subpopulation is representative of the infinite population of all

Fire-Hawks that could potentially be produced. Then, in order to employ a completely randomized experimental design, which for all practical purposes is equivalent to randomly selecting the desired samples of $n_A = 4$, $n_B = 5$, and $n_C = 3$ Fire-Hawk mileages, North American Oil should use the following procedure. For l = A, B, and C, randomly select a sample of n_l Fire-Hawks from the subpopulation of 1,000 Fire-Hawks and use gasoline type l to test-drive each of these Fire-Hawks under normal driving conditions for a specified distance. The gasoline used to test-drive each Fire-Hawk should be randomly selected from the potentially infinite supply of gasoline type l. Here, the sample of $n_A = 4$ Fire-Hawks driven using gasoline type A, the sample of $n_B = 5$ Fire-Hawks driven using gasoline type B, and the sample of $n_C = 3$ Fire-Hawks driven using gasoline type C are *different* samples that are randomly selected from the subpopulation of 1,000 Fire-Hawks. Consequently, North American Oil will randomly select a total sample of

$$\begin{aligned} n &= n_A + n_B + n_C \\ &= 4 + 5 + 3 \\ &= 12 \end{aligned}$$

Fire-Hawks. Then, letting $OFH_{l,k}$ denote the kth randomly selected Fire-Hawk that is test-driven using gasoline type l (where $k = 1, 2, \ldots, n_l$), North American Oil should measure

$$y_{l,k} = \text{the gasoline mileage obtained by } OFH_{l,k}$$

which we call *the kth Fire-Hawk mileage observed when using gasoline type l.*† When North American Oil has employed this completely randomized experimental design, we assume that

$$y_{l,1}, y_{l,2}, \ldots, y_{l,n_l}$$

is a sample of n_l Fire-Hawk mileages that North American Oil has randomly selected from the infinite population of all Fire-Hawk mileages that would be obtained by using gasoline type l.

Part 2. *Developing a Regression Model.*

Suppose that when North American Oil has employed the completely randomized experimental design, it has randomly selected three samples and has calculated the means of these samples as shown in Table 8.6. In order to use regression analysis to compare μ_A, μ_B, and μ_C, we first need to express $y_{l,k}$ in terms of a regression model. We define the (l, k)th error term

$$\epsilon_{l,k} = y_{l,k} - \mu_l$$

† The reader should not confuse the use of the subscript k to describe the kth randomly selected Fire-Hawk with the use of the symbol k to denote the number of parameters (that is, β's) in a regression model.

TABLE 8.6 Three Samples of Fire-Hawk Mileages Obtained Using Gasoline Types A, B, and C

Sample of $n_A = 4$ Fire-Hawk Mileages	Sample of $n_B = 5$ Fire-Hawk Mileages	Sample of $n_C = 3$ Fire-Hawk Mileages
$y_{A,1} = 24.0$	$y_{B,1} = 25.3$	$y_{C,1} = 23.3$
$y_{A,2} = 25.0$	$y_{B,2} = 26.5$	$y_{C,2} = 24.0$
$y_{A,3} = 24.3$	$y_{B,3} = 26.4$	$y_{C,3} = 24.7$
$y_{A,4} = 25.5$	$y_{B,4} = 27.0$	
	$y_{B,5} = 27.6$	
$\bar{y}_A = \frac{\sum_{k=1}^{4} y_{A,k}}{4} = \frac{98.8}{4} = 24.7$	$\bar{y}_B = \frac{\sum_{k=1}^{5} y_{B,k}}{5} = \frac{132.8}{5} = 26.56$	$\bar{y}_C = \frac{\sum_{k=1}^{3} y_{C,k}}{3} = \frac{72}{3} = 24.0$

to be the difference between $y_{l,k}$, the kth Fire-Hawk mileage observed when using gasoline type l, and μ_l, the mean Fire-Hawk mileage that would be obtained by using gasoline type l. We can express $y_{l,k}$ in the form

$$y_{l,k} = \mu_l + \epsilon_{l,k}$$

This equation implies that μ_l, the mean Fire-Hawk mileage that would be obtained by using gasoline type l, describes the effects on $y_{l,k}$ of the fact that $OFK_{l,k}$ is a Fire-Hawk that has been test-driven under normal driving conditions using gasoline type l; and (2) $\epsilon_{l,k}$, the (l, k)th error term, describes the effects on $y_{l,k}$ of all *other factors*.

In order to express $y_{l,k} = \mu_l + \epsilon_{l,k}$ in terms of a regression model, we need to write a single equation which, for l = A, B, and C, expresses μ_l as a function of a set of independent variables. An appropriate equation is

$$\mu_l = \beta_A + \beta_B D_{l,B} + \beta_C D_{l,C}$$

where β_A, β_B, and β_C are unknown parameters, and where $D_{l,B}$ and $D_{l,C}$ are independent variables that are called **dummy variables** and that are defined as follows:

$$D_{l,B} = \begin{cases} 1 & \text{if } l = \text{B; that is, if we are using gasoline type B} \\ 0 & \text{otherwise} \end{cases}$$

$$D_{l,C} = \begin{cases} 1 & \text{if } l = \text{C; that is, if we are using gasoline type C} \\ 0 & \text{otherwise} \end{cases}$$

To see how these dummy variables work, we use the equation

$$\mu_l = \beta_A + \beta_B D_{l,B} + \beta_C D_{l,C}$$

to express μ_A, μ_B, and μ_C in terms of the parameters β_A, β_B, and β_C. If $l = A$, then the definitions of $D_{l,B}$ and $D_{l,C}$ state that $D_{A,B} = 0$ and $D_{A,C} = 0$, which implies that

$$\begin{aligned}\mu_A &= \beta_A + \beta_B D_{A,B} + \beta_C D_{A,C} \\ &= \beta_A + \beta_B(0) + \beta_C(0) \\ &= \beta_A\end{aligned}$$

Thus, we can interpret the parameter β_A as

$\beta_A = \mu_A =$ the mean Fire-Hawk mileage obtained by using gasoline type A

Next, if $l = B$, then the definitions of $D_{l,B}$ and $D_{l,C}$ state that $D_{B,B} = 1$ and $D_{B,C} = 0$, which implies that

$$\begin{aligned}\mu_B &= \beta_A + \beta_B D_{B,B} + \beta_C D_{B,C} \\ &= \beta_A + \beta_B(1) + \beta_C(0) \\ &= \beta_A + \beta_B\end{aligned}$$

This implies (since $\beta_A = \mu_A$) that

$$\mu_B = \mu_A + \beta_B$$

Thus, we can interpret the parameter β_B as

$\beta_B = \mu_B - \mu_A =$ the difference between μ_B, the mean Fire-Hawk mileage obtained by using gasoline type B, and μ_A, the mean Fire-Hawk mileage obtained by using gasoline type A

Finally, if $l = C$, then the definitions of $D_{l,B}$ and $D_{l,C}$ state that $D_{C,B} = 0$ and $D_{C,C} = 1$, which implies that

$$\begin{aligned}\mu_C &= \beta_A + \beta_B D_{C,B} + \beta_C D_{C,C} \\ &= \beta_A + \beta_B(0) + \beta_C(1) \\ &= \beta_A + \beta_C\end{aligned}$$

This implies (since $\beta_A = \mu_A$) that

$$\mu_C = \mu_A + \beta_C$$

Thus, we can interpret the parameter β_C as

$\beta_C = \mu_C - \mu_A =$ the difference between μ_C, the mean Fire-Hawk mileage obtained by using gasoline type C, and μ_A, the mean Fire-Hawk mileage obtained by using gasoline type A

To summarize our discussion, the equation

$$\mu_l = \beta_A + \beta_B D_{l,B} + \beta_C D_{l,C}$$

implies that

$$\mu_A = \beta_A$$

$$\mu_B = \beta_A + \beta_B$$

$$\mu_C = \beta_A + \beta_C$$

which in turn implies that

$$\beta_A = \mu_A$$

$$\beta_B = \mu_B - \mu_A$$

$$\beta_C = \mu_C - \mu_A$$

Note that the equation describing μ_l uses the intercept β_A to describe the effect of the mean μ_A and uses a separate dummy variable to describe the effect of each of the other means μ_B and μ_C. For this reason, we sometimes refer to the model using this equation as the **means model**.

Since, for l = A, B, and C, we can express μ_l by the linear equation

$$\mu_l = \beta_A + \beta_B D_{l,B} + \beta_C D_{l,C}$$

it follows that we can describe the kth Fire-Hawk mileage observed when using gasoline type l,

$$y_{l,k} = \mu_l + \epsilon_{l,k}$$

by the linear regression model

$$\begin{aligned} y_{l,k} &= \mu_l + \epsilon_{l,k} \\ &= \beta_A + \beta_B D_{l,B} + \beta_C D_{l,C} + \epsilon_{l,k} \end{aligned}$$

If b_A, b_B, and b_C are the least squares point estimates of the parameters β_A, β_B, and β_C, then

$$\hat{y}_l = b_A + b_B D_{l,B} + b_C D_{l,C}$$

is the point estimate of the mean Fire-Hawk mileage obtained by using gasoline type l,

$$\mu_l = \beta_A + \beta_B D_{l,B} + \beta_C D_{l,C}$$

and the point prediction of the kth Fire-Hawk mileage observed when using gasoline type l,

$$\begin{aligned} y_{l,k} &= \mu_l + \epsilon_{l,k} \\ &= \beta_A + \beta_B D_{l,B} + \beta_C D_{l,C} + \epsilon_{l,k} \end{aligned}$$

Part 3. *Estimating the Regression Parameters.*

In order to calculate the least squares point estimates of β_A, β_B, and β_C, we use the following column vector **y** and matrix **X**:

$$\mathbf{y} = \begin{bmatrix} y_{A,1} \\ y_{A,2} \\ y_{A,3} \\ y_{A,4} \\ y_{B,1} \\ y_{B,2} \\ y_{B,3} \\ y_{B,4} \\ y_{B,5} \\ y_{C,1} \\ y_{C,2} \\ y_{C,3} \end{bmatrix} = \begin{bmatrix} 24.0 \\ 25.0 \\ 24.3 \\ 25.5 \\ 25.3 \\ 26.5 \\ 26.4 \\ 27.0 \\ 27.6 \\ 23.3 \\ 24.0 \\ 24.7 \end{bmatrix} \qquad \mathbf{X} = \begin{array}{c} \begin{matrix} 1 & D_{I,B} & D_{I,C} \end{matrix} \\ \begin{bmatrix} 1 & 0 & 0 \\ 1 & 0 & 0 \\ 1 & 0 & 0 \\ 1 & 0 & 0 \\ 1 & 1 & 0 \\ 1 & 1 & 0 \\ 1 & 1 & 0 \\ 1 & 1 & 0 \\ 1 & 1 & 0 \\ 1 & 0 & 1 \\ 1 & 0 & 1 \\ 1 & 0 & 1 \end{bmatrix} \end{array}$$

Using these quantities, we can calculate $(\mathbf{X}'\mathbf{X})^{-1}$ and $\mathbf{X}'\mathbf{y}$ to be

$$(\mathbf{X}'\mathbf{X})^{-1} = \begin{array}{c} \\ \text{row} \\ A \\ B \\ C \end{array} \begin{array}{c} \text{column} \\ \begin{matrix} A & B & C \end{matrix} \\ \begin{bmatrix} .25 & -.25 & -.25 \\ -.25 & .45 & .25 \\ -.25 & .25 & .58333 \end{bmatrix} \end{array}$$

$$= \begin{bmatrix} c_{AA} & & \\ & c_{BB} & \\ & & c_{CC} \end{bmatrix}$$

and

$$\mathbf{X}'\mathbf{y} = \begin{bmatrix} 303.6 \\ 132.8 \\ 72.0 \end{bmatrix}$$

It follows that the least squares point estimates b_A, b_B, and b_C of the parameters β_A, β_B, and β_C are

$$\begin{bmatrix} b_A \\ b_B \\ b_C \end{bmatrix} = \mathbf{b} = (\mathbf{X}'\mathbf{X})^{-1}\mathbf{X}'\mathbf{y} = \begin{bmatrix} 24.7 \\ 1.86 \\ -.7 \end{bmatrix}$$

These least squares point estimates yield the prediction equation

$$\begin{aligned} \hat{y}_I &= b_A + b_B D_{I,B} + b_C D_{I,C} \\ &= 24.7 + 1.86\, D_{I,B} - .7\, D_{I,C} \end{aligned}$$

DEP VARIABLE: Y

SOURCE	DF[a]	SUM OF SQUARES	MEAN SQUARE	F VALUE	PROB>F
MODEL	2	14.448000[b]	7.224000[e]	12.379[g]	0.0026[h]
ERROR	9	5.252000[c]	0.583556[f]		
C TOTAL	11	19.700000[d]			

[a] $k - 1 = 2$, $n - k = 9$, $n - 1 = 11$ [b] SS_{model} [c] SSE [d] Total variation

[e] MS_{model} [f] $MSE = s^2$ [g] $F(\text{model})$ [h] Prob-value

Figure 8.3 SAS Output of the Analysis of Variance Table for Testing $H_0: \mu_A = \mu_B = \mu_C$ in the North American Oil Problem

and the following quantities:

$$\text{Total variation} = 19.7$$

$$SS_{\text{model}} = \text{Explained variation} = 14.448$$

$$MS_{\text{model}} = \frac{SS_{\text{model}}}{k-1} = \frac{14.448}{3-1} = 7.224$$

$$SSE = \text{Unexplained variation} = 5.252$$

$$s^2 = MSE = \frac{SSE}{n-k} = \frac{5.252}{12-3} = .583556$$

$$F(\text{model}) = \frac{MS_{\text{model}}}{MSE} = \frac{7.224}{.583556} = 12.379$$

$$\begin{aligned} \text{prob-value} &= A[F(\text{model}); k-1, n-k] \\ &= A[12.379; 3-1=2, 12-3=9] \\ &= .0026 \end{aligned}$$

The SAS output of the analysis of variance table summarizing these quantities is given in Figure 8.3.

Part 4. *Testing for Differences Using an Overall F-Test.*

In order to compare the effects of gasoline types A, B, and C, we use the preceding information to perform on overall F-test. Consider the model

$$\begin{aligned} y_{l,k} &= \mu_l + \epsilon_{l,k} \\ &= \beta_A + \beta_B D_{l,B} + \beta_C D_{l,C} + \epsilon_{l,k} \end{aligned}$$

which has $k = 3$ parameters. Since we have seen that

$$\beta_B = \mu_B - \mu_A \quad \text{and} \quad \beta_C = \mu_C - \mu_A$$

it follows that

$$H_0: \beta_B = \beta_C = 0 \quad \text{or} \quad H_0: \beta_B = 0 \quad \text{and} \quad \beta_C = 0$$

TABLE 8.7 Individual and Scheffé Simultaneous 95% Confidence Intervals for Linear Combinations in the North American Oil Problem

Linear Combination $\boldsymbol{\lambda}'\boldsymbol{\beta}$	Point Estimate $\boldsymbol{\lambda}'\mathbf{b}$	$s\sqrt{\boldsymbol{\lambda}'(\mathbf{X}'\mathbf{X})^{-1}\boldsymbol{\lambda}}$
$\mu_B - \mu_A = \beta_B$	1.86	$(.7639)\sqrt{.4499} = .5124$
$\mu_C - \mu_A = \beta_C$	$-.7$	$(.7639)\sqrt{.5833} = .5834$
$\mu_C - \mu_B = \beta_C - \beta_B$	-2.56	$(.7639)\sqrt{.53333} = .5579$
$\mu_B - \dfrac{\mu_C + \mu_A}{2} = \beta_B - .5\beta_C$	2.21	$(.7639)\sqrt{.3458} = .4492$

is equivalent to

$$H_0: \mu_B - \mu_A = 0 \qquad \text{and} \qquad \mu_C - \mu_A = 0$$

which is equivalent to

$$H_0: \mu_A = \mu_B = \mu_C$$

The null hypothesis

$$H_0: \beta_B = \beta_C = 0 \qquad \text{or} \qquad \mu_A = \mu_B = \mu_C$$

says that gasoline types A, B, and C have the same effects on mean Fire-Hawk mileage. The alternative hypothesis

H_1: At least one of β_B and β_C does not equal zero

or

H_1: At least two of μ_A, μ_B, and μ_C differ from each other

says that at least two of gasoline types A, B, and C have different effects on mean Fire-Hawk mileage. If we wish to use condition (1) in the box concerning the overall F-test in Section 7.1 to determine whether we can reject H_0 in favor of H_1 by setting α equal to .05, we would use the rejection point

$$F_{[\alpha]}^{(k-1,n-k)} = F_{[.05]}^{(2,9)} = 4.26$$

Individual 95% Confidence Interval for $\boldsymbol{\lambda}'\boldsymbol{\beta}$ $[\boldsymbol{\lambda}'\mathbf{b} \pm (2.262)s\sqrt{\boldsymbol{\lambda}'(\mathbf{X}'\mathbf{X})^{-1}\boldsymbol{\lambda}}]$	Scheffé Simultaneous 95% Confidence Interval for $\boldsymbol{\lambda}'\boldsymbol{\beta}$ (in Set I) $[\boldsymbol{\lambda}'\mathbf{b} \pm (2.9189)s\sqrt{\boldsymbol{\lambda}'(\mathbf{X}'\mathbf{X})^{-1}\boldsymbol{\lambda}}]$
$[1.86 \pm 2.262(.5124)]$ $= [.701, 3.019]$	$[1.86 \pm 2.9189(.5124)]$ $= [.3644, 3.3556]$
$[-.70 \pm 2.262(.5834)]$ $= [-2.0197, .6197]$	$[-.70 \pm 2.9189(.5834)]$ $= [-2.4029, 1.0029]$
$[-2.56 \pm 2.262(.5579)]$ $= [-3.822, -1.298]$	$[-2.56 \pm 2.9189(.5579)]$ $= [-4.1885, -.9315]$
$[2.21 \pm 2.262(.4492)]$ $= [1.1939, 3.2261]$	$[2.21 \pm 2.9189(.4492)]$ $= [.8988, 3.5212]$

Since $F(\text{model}) = 12.379 > 4.26 = F^{(2,9)}_{[.05]}$, we can reject H_0 in favor of H_1 by setting α equal to .05. Alternatively, since prob-value = .0026 is less than .05 and .01, it follows by condition (2) in that box that we can reject H_0 in favor of H_1 by setting α equal to .05 or .01.

Part 5. *Testing and Estimating Differences Using Linear Combinations of Regression Parameters.*

Since the overall F-test has indicated that there are differences between μ_A, μ_B, and μ_C, we consider Table 8.7 and Figure 8.4, which present information that we can use to investigate the exact nature of these differences. Specifically, Table 8.7 presents several linear combinations of parameters that allow us to investigate these differences, and also presents the least squares point estimate, the individual 95% confidence interval, and the Scheffé simultaneous 95% confidence interval for each of these linear combinations. Note that the least squares point estimates of β_A, β_B, and β_C were calculated previously, the standard error is $s = \sqrt{s^2} = \sqrt{.583556} = .7639$ (s^2 is given in Figure 8.3), and the individual 95% confidence intervals are computed using the formula

$$[\boldsymbol{\lambda}'\mathbf{b} \pm t^{(n-k)}_{[.025]} s\sqrt{\boldsymbol{\lambda}'(\mathbf{X}'\mathbf{X})^{-1}\boldsymbol{\lambda}}] = [\boldsymbol{\lambda}'\mathbf{b} \pm t^{(12-3)}_{[.025]} s\sqrt{\boldsymbol{\lambda}'(\mathbf{X}'\mathbf{X})^{-1}\boldsymbol{\lambda}}]$$
$$= [\boldsymbol{\lambda}'\mathbf{b} \pm 2.262(.7639)\sqrt{\boldsymbol{\lambda}'(\mathbf{X}'\mathbf{X})^{-1}\boldsymbol{\lambda}}]$$

We compare the effects of gasoline types A, B, and C by utilizing the information in Table 8.7 (with the exception of the simultaneous 95% confidence intervals, which are discussed later) and the information in Figure 8.4, the SAS output of

FIGURE 8.4 Hypothesis Tests Concerning Linear Combinations in the North American Oil Problem

VARIABLE[a]	PARAMETER ESTIMATE[b]	T FOR HO: PARAMETER=0[c]	PROB > |T|[d]	STD ERROR OF ESTIMATE[e]
INTERCEPT	24.70000000	64.67	0.0001	0.38195404
DB	1.86000000	3.63	0.0055	0.51244512
DC	-0.70000000	-1.20	0.2609	0.58344443
P1	-2.56000000	-4.59	0.0013	0.55787958
P2	2.21000000	4.92	0.0008	0.44923598

[a] $\boldsymbol{\lambda}'\boldsymbol{\beta}$: [b] $\boldsymbol{\lambda}'\mathbf{b}$: [c] $t_{\boldsymbol{\lambda}'\mathbf{b}}$ [d] Prob-value [e] $s\sqrt{\boldsymbol{\lambda}'(\mathbf{X}'\mathbf{X})^{-1}\boldsymbol{\lambda}}$

$\boldsymbol{\lambda}'\boldsymbol{\beta}$	$\boldsymbol{\lambda}'\mathbf{b}$
$\mu_A = \beta_A$	b_A
$\mu_B - \mu_A = \beta_B$	b_B
$\mu_C - \mu_A = \beta_C$	b_C
$\mu_C - \mu_B = \beta_C - \beta_B$	$b_C - b_B$
$\mu_B - \left(\frac{\mu_C + \mu_A}{2}\right) = \beta_B - .5\beta_C$	$b_B - .5b_C$

the $t_{\boldsymbol{\lambda}'\mathbf{b}}$ statistic and the prob-value for testing $H_0: \boldsymbol{\lambda}'\boldsymbol{\beta} = 0$ versus $H_1: \boldsymbol{\lambda}'\boldsymbol{\beta} \neq 0$ for each of the linear combinations in Table 8.7.

Part 5.1. *Studying the Difference Between* μ_B *and* μ_A.

The least squares point estimate of

$$\begin{aligned}\mu_B - \mu_A &= [\beta_A + \beta_B] - \beta_A \\ &= \beta_B \\ &= 0 \cdot \beta_A + 1 \cdot \beta_B + 0 \cdot \beta_C \\ &= [0 \quad 1 \quad 0]\begin{bmatrix}\beta_A \\ \beta_B \\ \beta_C\end{bmatrix} \\ &= \boldsymbol{\lambda}'\boldsymbol{\beta}, \qquad \text{where } \boldsymbol{\lambda}' = [0 \quad 1 \quad 0]\end{aligned}$$

is

$$\boldsymbol{\lambda}'\mathbf{b} = b_B = 1.86$$

Since the individual 95% confidence interval for $\mu_B - \mu_A$ is calculated in Table 8.7 to be [.701, 3.019], we are 95% confident that μ_B, the mean Fire-Hawk mileage obtained by using gasoline type B, is between .701 mpg and 3.019 mpg greater than μ_A, the mean Fire-Hawk mileage obtained by using gasoline type A. Moreover, since the prob-value for testing

$$H_0: \mu_B - \mu_A = 0$$

versus

$$H_1: \mu_B - \mu_A \neq 0$$

is calculated in Figure 8.4 to be .0055, we have very substantial evidence that H_0 is false and that H_1 is true. Thus, we have very substantial evidence that μ_B is greater than μ_A.

Part 5.2. *Studying the Difference Between μ_C and μ_A.*

The least squares point estimate of

$$\begin{aligned}\mu_C - \mu_A &= [\beta_A + \beta_C] - \beta_A \\ &= \beta_C \\ &= 0 \cdot \beta_A + 0 \cdot \beta_B + 1 \cdot \beta_C \\ &= [0 \quad 0 \quad 1] \begin{bmatrix} \beta_A \\ \beta_B \\ \beta_C \end{bmatrix} \\ &= \boldsymbol{\lambda}'\boldsymbol{\beta}, \qquad \text{where } \boldsymbol{\lambda}' = [0 \quad 0 \quad 1]\end{aligned}$$

is

$$\boldsymbol{\lambda}'\mathbf{b} = b_C = -.7$$

Since the individual 95% confidence interval for $\mu_C - \mu_A$ is calculated in Table 8.7 to be [−2.0197, .6197], we are 95% confident that μ_C, the mean Fire-Hawk mileage obtained by using gasoline type C, is between 2.0197 mpg less than and .6197 mpg greater than μ_A, the mean Fire-Hawk mileage obtained by using gasoline type A. Moreover, since the prob-value for testing

$$H_0: \mu_C - \mu_A = 0$$

versus

$$H_1: \mu_C - \mu_A \neq 0$$

is calculated in Figure 8.4 to be .2609, we do not have substantial evidence that H_0 is false and that H_1 is true. Thus, there is not substantial evidence that μ_C and μ_A differ.

Part 5.3. *Studying the Difference Between μ_C and μ_B.*

The least squares point estimate of

$$\begin{aligned}\mu_C - \mu_B &= [\beta_A + \beta_C] - [\beta_A + \beta_B] \\ &= \beta_C - \beta_B \\ &= 0 \cdot \beta_A + (-1) \cdot \beta_B + 1 \cdot \beta_C \\ &= [0 \quad -1 \quad 1] \begin{bmatrix} \beta_A \\ \beta_B \\ \beta_C \end{bmatrix} \\ &= \boldsymbol{\lambda}'\boldsymbol{\beta}, \qquad \text{where } \boldsymbol{\lambda}' = [0 \quad -1 \quad 1]\end{aligned}$$

is

$$\boldsymbol{\lambda}'\mathbf{b} = b_C - b_B = -.7 - 1.86 = -2.56$$

Since the individual 95% confidence interval for $\mu_C - \mu_B$ is calculated in Table 8.7 to be $[-3.822, -1.298]$, we are 95% confident that μ_C, the mean Fire-Hawk mileage obtained by using gasoline type C, is between 3.822 mpg and 1.298 mpg less than μ_B, the mean Fire-Hawk mileage obtained by using gasoline type B. Furthermore, since the prob-value for testing

$$H_0: \mu_C - \mu_B = 0$$

versus

$$H_1: \mu_C - \mu_B \neq 0$$

is calculated in Figure 8.4 to be .0013, we have very substantial evidence that H_0 is false and that H_1 is true. Thus, we have very substantial evidence that μ_B is greater than μ_C.

Part 5.4. *Studying a More Complicated Difference.*

Gasoline type B contains a chemical—Chemical XX—that is not contained in gasoline type C or in gasoline type A. In order to assess the effect of Chemical XX on gasoline mileage obtained by the Fire-Hawk, we consider

$$\mu_B - \frac{\mu_C + \mu_A}{2} = \text{the difference between } \mu_B \text{ and } \frac{\mu_C + \mu_A}{2}$$

$$\text{(the average of } \mu_C \text{ and } \mu_A\text{)}$$

The least squares point estimate of

$$\begin{aligned}
\mu_B - \frac{\mu_C + \mu_A}{2} &= [\beta_A + \beta_B] - \frac{[\beta_A + \beta_C] + \beta_A}{2} \\
&= \beta_B - .5\beta_C \\
&= 0 \cdot \beta_A + 1 \cdot \beta_B - .5\beta_C \\
&= [0 \quad 1 \quad -.5] \begin{bmatrix} \beta_A \\ \beta_B \\ \beta_C \end{bmatrix} \\
&= \boldsymbol{\lambda}'\boldsymbol{\beta}, \qquad \text{where } \boldsymbol{\lambda}' = [0 \quad 1 \quad -.5]
\end{aligned}$$

is

$$\begin{aligned}
\boldsymbol{\lambda}'\mathbf{b} = b_B - .5b_C &= 1.86 - .5(-.7) \\
&= 2.21
\end{aligned}$$

Since the individual 95% confidence interval for $\mu_B - (\mu_C + \mu_A)/2$ is calculated in Table 8.7 to be [1.1939, 3.2261], we are 95% confident that μ_B (the mean Fire-Hawk mileage obtained by using gasoline type B, which contains Chemical XX) is between 1.1939 mpg and 3.2261 mpg greater than $(\mu_C + \mu_A)/2$ (the average of the mean Fire-Hawk mileages obtained by gasoline types C and A, which do not contain Chemical XX). Moreover, since the prob-value for testing

$$H_0: \mu_B - \frac{(\mu_C + \mu_A)}{2} = 0$$

versus

$$H_1: \mu_B - \frac{(\mu_C + \mu_A)}{2} \neq 0$$

is calculated in Figure 8.4 to be .0008, we have very substantial evidence that H_0 is false and that H_1 is true. Thus, we have very substantial evidence that μ_B is greater than $(\mu_C + \mu_A)/2$. It should be noted that, although Chemical XX might be a major factor causing μ_B to be greater than $(\mu_C + \mu_A)/2$, other chemical differences between gasoline type B and gasoline types A and C, or the combination of these chemical characteristics with Chemical XX, also might be major factors causing μ_B to be greater than $(\mu_C + \mu_A)/2$. Hence, the chemists at North American Oil should use the comparisons of μ_B and $(\mu_C + \mu_A)/2$, along with their knowledge of the chemical compositions of gasoline types A, B, and C, to assess the effect of Chemical XX on gasoline mileage.

Part 6. *Computing Simultaneous Confidence Intervals.*

We next demonstrate how we calculated the Scheffé simultaneous 95% confidence intervals for all the linear combinations of regression parameters (Set I) in Table 8.7.

Set I

$$\mu_B - \mu_A = \beta_B$$

$$\mu_C - \mu_A = \beta_C$$

$$\mu_C - \mu_B = \beta_C - \beta_B$$

$$\mu_B - \frac{\mu_C + \mu_A}{2} = \beta_B - .5\beta_C$$

Since there are two regression parameters, β_B and β_C, from the regression model

$$y_{l,k} = \beta_A + \beta_B D_{l,B} + \beta_C D_{l,C} + \epsilon_{l,k}$$

involved nontrivially in at least one of the linear combinations in Set I, it follows that $q = 2$ and that

$$\begin{aligned}\sqrt{qF_{[\alpha]}^{(q,n-k)}} &= \sqrt{2F_{[.05]}^{(2,12-3)}} \\ &= \sqrt{2F_{[.05]}^{(2,9)}} \\ &= \sqrt{2(4.26)} \\ &= 2.9189\end{aligned}$$

Thus,

$$[\boldsymbol{\lambda}'\mathbf{b} \pm \sqrt{qF_{[\alpha]}^{(q,n-k)}}\, s\sqrt{\boldsymbol{\lambda}'(\mathbf{X}'\mathbf{X})^{-1}\boldsymbol{\lambda}}] = [\boldsymbol{\lambda}'\mathbf{b} \pm (2.9189)s\sqrt{\boldsymbol{\lambda}'(\mathbf{X}'\mathbf{X})^{-1}\boldsymbol{\lambda}}]$$

is a Scheffé simultaneous 95% confidence interval for any linear combination $\boldsymbol{\lambda}'\boldsymbol{\beta}$ in Set I. Note that this simultaneous interval is longer and thus less precise than

$$[\boldsymbol{\lambda}'\mathbf{b} \pm t_{[\alpha/2]}^{(n-k)}\, s\sqrt{\boldsymbol{\lambda}'(\mathbf{X}'\mathbf{X})^{-1}\boldsymbol{\lambda}}] = [\boldsymbol{\lambda}'\mathbf{b} \pm (2.262)s\sqrt{\boldsymbol{\lambda}'(\mathbf{X}'\mathbf{X})^{-1}\boldsymbol{\lambda}}]$$

which is (see Table 8.7) the individual 95% confidence interval for $\boldsymbol{\lambda}'\boldsymbol{\beta}$. For example, $[-3.822, -1.298]$, the individual 95% confidence interval for $\mu_C - \mu_B$ (see Table 8.7), makes us 95% confident that μ_B is at least 1.298 mpg larger than μ_C, whereas $[-4.1885, -.9315]$, the Scheffé simultaneous 95% confidence interval for $\mu_C - \mu_B$ (see Table 8.7), makes us 95% confident that μ_B is at least .9315 mpg larger than μ_C. Thus, if gasoline type B is more expensive to produce than gasoline type C, and if North American Oil feels that the extra cost is justified only if μ_B is at least 1 mpg larger than μ_C, then the individual 95% confidence interval makes us at least 95% confident that it is worthwhile to produce gasoline type B, whereas the Scheffé simultaneous 95% confidence interval does not. In this case, North American Oil must use these results and any other pertinent facts it has available to decide subjectively whether gasoline type B should be produced.

Next, we calculate Scheffé simultaneous 95% confidence intervals for all the linear combinations of regression parameters in the following set (Set II).

Set II

$$\begin{aligned}\mu_A &= \beta_A \\ \mu_B &= \beta_A + \beta_B \\ \mu_C &= \beta_A + \beta_C \\ \mu_B - \mu_A &= \beta_B \\ \mu_C - \mu_A &= \beta_C \\ \mu_C - \mu_B &= \beta_C - \beta_B \\ \mu_B - \frac{\mu_C + \mu_A}{2} &= \beta_B - .5\beta_C\end{aligned}$$

Since all three regression parameters, β_A, β_B, and β_C, from the regression model

$$y_{l,k} = \beta_A + \beta_B D_{l,B} + \beta_C D_{l,C} + \epsilon_{l,k}$$

are involved nontrivially in at least one of the linear combinations in Set II, it follows that $q = 3$ and that

$$\begin{aligned}\sqrt{qF_{[\alpha]}^{(q,n-k)}} &= \sqrt{3F_{[.05]}^{(3,12-3)}}\\ &= \sqrt{3F_{[.05]}^{(3,9)}}\\ &= \sqrt{3(3.86)}\\ &= 3.4029\end{aligned}$$

Thus

$$[\boldsymbol{\lambda}'\mathbf{b} \pm \sqrt{qF_{[\alpha]}^{(q,n-k)}}\, s\sqrt{\boldsymbol{\lambda}'(\mathbf{X}'\mathbf{X})^{-1}\boldsymbol{\lambda}}] = [\boldsymbol{\lambda}'\mathbf{b} \pm (3.4029)s\sqrt{\boldsymbol{\lambda}'(\mathbf{X}'\mathbf{X})^{-1}\boldsymbol{\lambda}}]$$

is a Scheffé simultaneous 95% confidence interval for any linear combination $\boldsymbol{\lambda}'\boldsymbol{\beta}$ in Set II. Since $q = 3$ for Set II, while $q = 2$ for Set I, this simultaneous 95% interval is longer (less precise) than

$$[\boldsymbol{\lambda}'\mathbf{b} \pm (2.9189)s\sqrt{\boldsymbol{\lambda}'(\mathbf{X}'\mathbf{X})^{-1}\boldsymbol{\lambda}}]$$

the Scheffé simultaneous 95% confidence interval for any linear combination in Set I. In Table 8.8 we calculate the Scheffé simultaneous 95% confidence intervals for the seven linear combinations in Set II.

Part 7. *Using the Model to Estimate and Predict.*

We suppose that based on the information presented in Table 8.7 and Figure 8.4, North American Oil has concluded that μ_B is sufficiently greater than μ_A and μ_C so that it will produce gasoline type B. For this reason, we find a point estimate of and a confidence interval for μ_B. These quantities would be of interest to the federal government, since if many Fire-Hawks were to be driven under normal driving conditions using gasoline type B, the average gasoline mileage obtained by this fleet of Fire-Hawks would be close to μ_B. Before finding a confidence interval for μ_B, we note that since it is reasonable to assume that purchasing a Fire-Hawk is equivalent to randomly selecting a Fire-Hawk from the population of all Fire-Hawks, and since using gasoline type B involves randomly selecting a supply of gasoline type B, it follows that a Fire-Hawk owner who plans to drive his Fire-Hawk using gasoline type B would like to have a prediction of

$$y_{B,0} = \mu_B + \epsilon_{B,0}$$

which we define to be the gasoline mileage that will be obtained by a future Fire-Hawk that will be randomly selected from the population of all Fire-Hawks, when this Fire-Hawk is driven under normal driving conditions using (a randomly selected supply of) gasoline type B.

TABLE 8.8 Scheffé Simultaneous 95% Confidence Intervals for the Linear Combinations in Set II

Linear Combination $\boldsymbol{\lambda}'\boldsymbol{\beta}$	Point Estimate $\boldsymbol{\lambda}'\mathbf{b}$	$s\sqrt{\boldsymbol{\lambda}'(\mathbf{X}'\mathbf{X})^{-1}\boldsymbol{\lambda}}$	Scheffé Simultaneous 95% Confidence Interval for $\boldsymbol{\lambda}'\boldsymbol{\beta}$ (in Set II) $[\boldsymbol{\lambda}'\mathbf{b} \pm (3.4029)s\sqrt{\boldsymbol{\lambda}'(\mathbf{X}'\mathbf{X})^{-1}\boldsymbol{\lambda}}]$
$\mu_A = \beta_A$	24.7	$(.7639)\sqrt{.25} = .382$	$[24.7 \pm 3.4029(.382)] = [23.4, 26]$
$\mu_B = \beta_A + \beta_B$	26.56	$(.7639)\sqrt{.20} = .3416$	$[26.56 \pm 3.4029(.3416)] = [25.3976, 27.7224]$
$\mu_C = \beta_A + \beta_C$	24.0	$(.7639)\sqrt{.3333} = .441$	$[24.0 \pm 3.4029(.441)] = [22.4993, 25.5007]$
$\mu_B - \mu_A = \beta_B$	1.86	$(.7639)\sqrt{.4499} = .5124$	$[1.86 \pm 3.4029(.5124)] = [.1164, 3.6036]$
$\mu_C - \mu_A = \beta_C$	−.7	$(.7639)\sqrt{.5833} = .5834$	$[-.7 \pm 3.4029(.5834)] = [-2.6853, 1.2853]$
$\mu_C - \mu_B = \beta_C - \beta_B$	−2.56	$(.7639)\sqrt{.5334} = .5579$	$[-2.56 \pm 3.4029(.5579)] = [-4.4585, -.6615]$
$\mu_B - \frac{(\mu_C + \mu_A)}{2} = \beta_B - .5\beta_C$	2.21	$(.7639)\sqrt{.3458} = .4492$	$[2.21 \pm 3.4029(.4492)] = [.6814, 3.7386]$

Since we can express μ_B by the equation

$$\begin{aligned}\mu_B &= \beta_A + \beta_B D_{I,B} + \beta_C D_{I,C}\\ &= \beta_A + \beta_B D_{B,B} + \beta_C D_{B,C}\\ &= \beta_A + \beta_B(1) + \beta_C(0)\\ &= \beta_A + \beta_B\end{aligned}$$

it follows that we can express $y_{B,0}$ by the equation

$$\begin{aligned}y_{B,0} &= \mu_B + \epsilon_{B,0}\\ &= \beta_A + \beta_B + \epsilon_{B,0}\end{aligned}$$

Thus

$$\begin{aligned}\hat{y}_B &= b_A + b_B D_{I,B} + b_C D_{I,C}\\ &= b_A + b_B D_{B,B} + b_C D_{B,C}\\ &= b_A + b_B(1) + b_C(0)\end{aligned}$$

$$= b_A + b_B$$
$$= 24.7 + 1.86$$
$$= 26.56$$

is the point estimate of μ_B and is the point prediction of $y_{B,0} = \mu_B + \epsilon_{B,0}$. Since

$$\mathbf{x}'_B = [1 \quad 1 \quad 0]$$

is the row vector containing the numbers multiplied by b_A, b_B, and b_C in the preceding prediction equation, and

$$\mathbf{x}'_B(\mathbf{X}'\mathbf{X})^{-1}\mathbf{x}_B = .20$$

and $s = .7639$ is the standard error, and

$$t_{[.025]}^{(12-3)} = t_{[.025]}^{(9)} = 2.262$$

it follows that

$$\left[\hat{y}_B \pm t_{[.025]}^{(9)} s\sqrt{\mathbf{x}'_B(\mathbf{X}'\mathbf{X})^{-1}\mathbf{x}_B}\right] = [26.56 \pm 2.262(.7639)\sqrt{.20}]$$
$$= [26.56 \pm .7727]$$
$$= [25.787, 27.333]$$

is a 95% confidence interval for μ_B, and

$$\left[\hat{y}_B \pm t_{[.025]}^{(9)} s\sqrt{1 + \mathbf{x}'_B(\mathbf{X}'\mathbf{X})^{-1}\mathbf{x}_B}\right] = [26.56 \pm 2.262(.7639)\sqrt{1.20}]$$
$$= [26.56 \pm 1.8931]$$
$$= [24.667, 28.453]$$

is a 95% prediction interval for $y_{B,0} = \mu_B + \epsilon_{B,0}$. Using the preceding 95% confidence interval for μ_B, the federal government can be 95% confident that μ_B, the average Fire-Hawk mileage obtained by using gasoline type B, is at least 25.787 mpg and no more than 27.333 mpg. Using the 95% prediction interval for $y_{B,0} = \mu_B + \epsilon_{B,0}$, the owner of a Fire-Hawk can be 95% confident that $y_{B,0}$, the gasoline mileage that will be obtained by his Fire-Hawk when driven under normal conditions using gasoline type B, will be at least 24.667 mpg and no more than 28.453 mpg.

To complete this example, note (from Table 8.6) that the least squares point estimate of $\mu_A = \beta_A$ is

$$b_A = \bar{y}_A = 24.7$$

the least squares point estimate of $\mu_B = \beta_A + \beta_B$ is

$$b_A + b_B = 24.7 + 1.86 = \bar{y}_B = 26.56$$

and the least squares point estimate of $\mu_C = \beta_A + \beta_C$ is

$$b_A + b_C = 24.7 + (-.7) = \bar{y}_C = 24.0$$

We conclude that the *least squares point estimates* of the mean Fire-Hawk mileages μ_A, μ_B, and μ_C are the corresponding *sample means* $\bar{y}_A$, $\bar{y}_B$, and $\bar{y}_C$, and thus that the *least squares point estimate* of any difference between mean Fire-Hawk mileages is the corresponding *difference in sample means*. In general, it can be shown that *the least squares point estimate of the population mean μ_l is the sample mean $\bar{y}_l$ when we use the means model* (that is, a regression model using a separate dummy variable to describe each of the means) *to compare an arbitrary number of population means.*

*8.4.2 One-Way Analysis of Variance

In the preceding example we compared the three mean mileages μ_A, μ_B, and μ_C by using samples randomly selected from the three populations of potential Fire-Hawk mileages that would be obtained by using gasoline types A, B, and C. In general, consider comparing v population means $\mu_1, \mu_2, \ldots, \mu_v$, where, for $l = 1, 2, \ldots, v$, μ_l is the mean of *the lth population of element values*. In order to make such comparisons, we assume that for $l = 1, 2, \ldots, v$, we have randomly selected a sample of n_l element values

$$y_{l,1}, y_{l,2}, \ldots, y_{l,n_l}$$

from the lth population. Here, for $k = 1, 2, \ldots, n_l$, $y_{l,k}$ denotes the kth such randomly selected element value. It follows that for $l = 1, 2, \ldots, v$, the mean, variance, and standard deviation of the sample of n_l element values

$$\bar{y}_l = \frac{\sum_{k=1}^{n_l} y_{l,k}}{n_l} \qquad s_l^2 = \frac{\sum_{k=1}^{n_l} (y_{l,k} - \bar{y}_l)^2}{n_l - 1} \qquad s_l = \sqrt{s_l^2}$$

are the point estimates of, respectively, μ_l, σ_l^2, and σ_l, the mean, variance, and standard deviation of the lth population.

It can be shown that we can compare the v population means $\mu_1, \mu_2, \ldots, \mu_v$ by using the regression techniques illustrated in Example 8.5, or equivalently, by using algebraic formulas. The validity of the algebraic formulas presented in this section (as well as the regression techniques we have illustrated) depends upon the inference assumptions being satisfied. Inference assumption 1 (Constant Variance) says that the v populations of element values have equal variances. That is, if, for $l = 1, 2, \ldots, v$, σ_l^2 denotes the variance of the lth population, then inference assumption 1 says that $\sigma_1^2 = \sigma_2^2 = \cdots = \sigma_v^2$. Perhaps the best way to check the validity of this assumption is to compare the previously discussed sample variances $s_1^2, s_2^2, \ldots, s_v^2$. If these sample variances are nearly equal, then it is reasonable to believe that inference assumption 1 holds (approximately). There are various statistical tests that use the sample variances to test the equality of the population variances. Although we do not discuss these tests here, we refer the reader to Miller and

Wichern (1977). In general, studies have shown that if the sizes of the v samples randomly selected from the v populations are *equal*, the validity of the formulas used to compare the v population means is not seriously affected by unequal population variances. However, if the sample sizes are substantially different, the validity of the formulas can be affected by unequal population variances. Thus, it is best to try to randomly select samples of equal sizes. In the North American Oil problem, we saw that the sample sizes are $n_A = 4$, $n_B = 5$, and $n_C = 3$. We can use the data in Table 8.6 to calculate the sample variances to be

$$s_A^2 = .46 \qquad s_B^2 = .723 \qquad s_C^2 = .49$$

and thus the standard deviations are

$$s_A = .6782 \qquad s_B = .8503 \qquad s_C = .7$$

Although the sample sizes are not equal, we might conclude (somewhat arbitrarily) that the sample variances do not differ substantially from each other (note that the sample standard deviations, being the square roots of the sample variances, are more nearly equal). Thus, we might conclude that inference assumption 1 approximately holds in the North American Oil problem.

Inference assumption 2 (Independence) will probably hold if the data have been randomly selected at a specific time and if there is no relationship between the element values of one sample and the element values of another sample (in which case we say that the *samples are independent of each other* and that we have performed an *independent samples experiment*). If the element values in different samples result from different measurements being taken on the *same* elements (for example, if the mileages obtained by using gasoline type A were measured on the same Fire-Hawks utilized to measure the mileages obtained by using gasoline type B), then there would be a connection between the element values in different samples (in which case we would say that the *samples are dependent upon each other*). In such a situation, inference assumption 2 would be badly violated, and we definitely should not use the formulas discussed in this section to compare population means. The procedures that should then be used are discussed in Section 8.6. In the North American Oil problem, since each gasoline type was tested using different Fire-Hawks, the three samples of Fire-Hawk mileages should be considered independent of each other. Hence, we conclude that inference assumption 2 holds in the North American Oil problem.

Inference assumption 3 (Normal Populations) says that each of the v populations is normally distributed. This assumption is not crucial, because it has been shown that we can sometimes use the techniques of this section to compare populations that are not normally distributed. In particular, it has been shown that these techniques are approximately valid for populations described by probability curves that are mound-shaped (even if these curves are somewhat skewed to the right or left).

We now present the algebraic formulas that can be used to compare v population means.

FORMULAS FOR COMPARING POPULATION MEANS:

Assume the inference assumptions are satisfied, and let

$$n = \sum_{l=1}^{v} n_l$$

$$\bar{y} = \frac{\sum_{l=1}^{v} \sum_{k=1}^{n_l} y_{l,k}}{n}$$

$$SS_{\text{model}} = \text{Explained variation} = \sum_{l=1}^{v} n_l(\bar{y}_l - \bar{y})^2$$

$$MS_{\text{model}} = \frac{SS_{\text{model}}}{v-1}$$

$$SSE = \text{Unexplained variation} = \sum_{l=1}^{v} \sum_{k=1}^{n_l} (y_{l,k} - \bar{y}_l)^2$$

$$= \sum_{l=1}^{v} (n_l - 1) \frac{\sum_{k=1}^{n_l} (y_{l,k} - \bar{y}_l)^2}{n_l - 1}$$

$$= \sum_{l=1}^{v} (n_l - 1) s_l^2$$

$$s^2 = MSE = \frac{SSE}{n-v}$$

$$s = \sqrt{s^2}$$

$$F(\text{model}) = \frac{MS_{\text{model}}}{MSE}$$

$$\text{prob-value} = A[F(\text{model}); v-1, n-v]$$

Then

1. We can reject

   $H_0: \mu_1 = \mu_2 = \cdots = \mu_v$

   in favor of

   H_1: At least two of $\mu_1, \mu_2, \ldots, \mu_v$ differ from each other

   by setting the probability of a Type I error equal to α if and only if either of the following equivalent conditions hold: $F(\text{model}) > F_{[\alpha]}^{(v-1, n-v)}$ or prob-value $< \alpha$.

2. A point estimate of the mean μ_l is $\bar{y}_l$, and a $100(1-\alpha)\%$ confidence interval for μ_l is

$$\left[\bar{y}_l \pm t_{[\alpha/2]}^{(n-v)}\left(\frac{s}{\sqrt{n_l}}\right)\right]$$

3. A point prediction of the future (randomly selected) element value

$$y_{l,0} = \mu_l + \epsilon_{l,0}$$

is $\bar{y}_l$, and a $100(1-\alpha)\%$ prediction interval for that element value is

$$\left[\bar{y}_l \pm t_{[\alpha/2]}^{(n-v)} s\sqrt{1+\frac{1}{n_l}}\right]$$

4. A point estimate of the difference $\mu_i - \mu_j$ is $\bar{y}_i - \bar{y}_j$, and a $100(1-\alpha)\%$ confidence interval for $\mu_i - \mu_j$ is

$$\left[(\bar{y}_i - \bar{y}_j) \pm t_{[\alpha/2]}^{(n-v)} s\sqrt{\frac{1}{n_i}+\frac{1}{n_j}}\right]$$

5. A point estimate of the linear combination $\sum_{l=1}^{v} a_l\mu_l$ is

$$\sum_{l=1}^{v} a_l\bar{y}_l$$

and a $100(1-\alpha)\%$ confidence interval for that linear combination is

$$\left[\sum_{l=1}^{v} a_l\bar{y}_l \pm t_{[\alpha/2]}^{(n-v)} s\sqrt{\sum_{l=1}^{v}\frac{a_l^2}{n_l}}\right]$$

6. We define a *contrast* to be any linear combination

$$\sum_{l=1}^{v} a_l\mu_l$$

such that

$$\sum_{l=1}^{v} a_l = 0$$

Then the Scheffé simultaneous $100(1-\alpha)\%$ confidence interval for any contrast

$$\sum_{l=1}^{v} a_l\mu_l$$

in the set of *all possible contrasts* is

$$\left[\sum_{l=1}^{v} a_l\bar{y}_l \pm \sqrt{(v-1)F_{[\alpha]}^{(v-1,n-v)}}\, s\sqrt{\sum_{l=1}^{v}\frac{a_l^2}{n_l}}\right]$$

7. The Scheffé simultaneous $100(1-\alpha)\%$ confidence interval for any linear combination

$$\sum_{l=1}^{v} a_l \mu_l$$

in the set of *all possible linear combinations* is

$$\left[\sum_{l=1}^{v} a_l \bar{y}_l \pm \sqrt{vF_{[\alpha]}^{(v,n-v)}}\, s \sqrt{\sum_{l=1}^{v} \frac{a_l^2}{n_l}}\right]$$

8. Consider testing $H_0: \mu_i - \mu_j = 0$ versus $H_1: \mu_i - \mu_j \neq 0$. Then, defining the $t_{[\bar{y}_i - \bar{y}_j, \mu_i - \mu_j]}$ statistic to be

$$t_{[\bar{y}_i - \bar{y}_j, \mu_i - \mu_j]} = \frac{\bar{y}_i - \bar{y}_j - (\mu_i - \mu_j)}{s\sqrt{(1/n_i) + (1/n_j)}}$$

it follows that the population of all possible $t_{[\bar{y}_i - \bar{y}_j, \mu_i - \mu_j]}$ statistics has a t-distribution with $n - v$ degrees of freedom. In order to test $H_0: \mu_i - \mu_j = 0$ versus $H_1: \mu_i - \mu_j \neq 0$, define the $t_{\bar{y}_i - \bar{y}_j}$ statistic and its related prob-value to be

$$t_{\bar{y}_i - \bar{y}_j} = \frac{\bar{y}_i - \bar{y}_j}{s\sqrt{(1/n_i) + (1/n_j)}}$$

$$\text{prob-value} = 2A[[|t_{\bar{y}_i - \bar{y}_j}|, \infty); n - v]$$

where $A[[|t_{\bar{y}_i - \bar{y}_j}|, \infty); n - v]$ is the area under the curve of t-distribution having $n - v$ degrees of freedom to the right of $|t_{\bar{y}_i - \bar{y}_j}|$, the absolute value of the $t_{\bar{y}_i - \bar{y}_j}$ statistic. Then, we can reject $H_0: \mu_i - \mu_j = 0$ in favor of $H_1: \mu_i - \mu_j \neq 0$ by setting the probability of a Type I error equal to α if and only if any of the following three equivalent conditions hold:

(a) $|t_{\bar{y}_i - \bar{y}_j}| > t_{[\alpha/2]}^{(n-v)}$—that is, if $t_{\bar{y}_i - \bar{y}_j} > t_{[\alpha/2]}^{(n-v)}$ or $t_{\bar{y}_i - \bar{y}_j} < -t_{[\alpha/2]}^{(n-v)}$.

(b) Prob-value $< \alpha$.

(c) The $100(1-\alpha)\%$ confidence interval for $\mu_i - \mu_j$,

$$\left[(\bar{y}_i - \bar{y}_j) \pm t_{[\alpha/2]}^{(n-v)} s \sqrt{\frac{1}{n_i} + \frac{1}{n_j}}\right]$$

does not contain zero.

Moreover,

(d) For any level of confidence, $100(1-\alpha)\%$, less than $100[1 - (\text{prob-value})]\%$, the $100(1-\alpha)\%$ confidence interval for $\mu_i - \mu_j$ does not contain zero, and thus we can, with at least $100(1-\alpha)\%$ confidence, reject $H_0: \mu_i - \mu_j = 0$ in favor of $H_1: \mu_i - \mu_j \neq 0$.

(e) For any level of confidence, $100(1-\alpha)\%$, greater than or equal to $100[1 - (\text{prob-value})]\%$, the $100(1-\alpha)\%$ confidence interval for $\mu_i - \mu_j$ does contain zero, and thus we cannot, with at least $100(1-\alpha)\%$ confidence, reject $H_0: \mu_i - \mu_j = 0$ in favor of $H_1: \mu_i - \mu_j \neq 0$.

Note that the proof of the validity of (8a) through (8e) is exactly the same as the proof given in Section 6.5 of the results in the box entitled "Testing $H_0: \beta_j = 0$ versus $H_1: \beta_j \neq 0$," replacing β_j, b_j, $s\sqrt{c_{jj}}$, $t_{[b_j, \beta_j]}$, and t_{b_j} in Section 6.5 by $\mu_i - \mu_j$, $\bar{y}_i - \bar{y}_j$, $s\sqrt{(1/n_i) + (1/n_j)}$, $t_{[\bar{y}_i - \bar{y}_j, \mu_i - \mu_j]}$, and $t_{\bar{y}_i - \bar{y}_j}$, respectively.

EXAMPLE 8.6

In Table 8.9 we again present the three samples of Fire-Hawk mileages that North American Oil has randomly selected from the populations of Fire-Hawk mileages that would be obtained by using gasoline types A, B, and C, and we also calculate several quantities (including F(model) and the related prob-value) that we now use to implement the formulas in the preceding box.

If we wish to use the first condition in (1) to determine whether we can reject

$$H_0: \mu_A = \mu_B = \mu_C$$

in favor of

$$H_1: \text{At least two of } \mu_A, \mu_B, \text{ and } \mu_C \text{ differ from each other}$$

by setting α equal to .05, then we use the rejection point

$$F_{[\alpha]}^{(v-1, n-v)} = F_{[.05]}^{(3-1, 12-3)} = F_{[.05]}^{(2,9)} = 4.26$$

Since F(model) $= 12.379 > 4.26 = F_{[.05]}^{(2,9)}$, we can reject H_0 in favor of H_1 by setting α equal to .05. Alternatively, since prob-value $= .0026$ is less than .05 and .01, it follows by the second condition in (1) that we can reject H_0 in favor of H_1 by setting α equal to .05 or .01. Note that the quantities calculated in Table 8.9 (SS_{model}, MS_{model}, SSE, MSE, F(model), and prob-value) are the same quantities obtained by regression analysis in Figure 8.3.

Next, we illustrate the use of the formulas in (2) through (7).

Using the formulas in (2), we find a point estimate of, and a 95% confidence interval for, the mean μ_B are

$$\bar{y}_B = 26.56$$

$$\left[\bar{y}_B \pm t_{[.025]}^{(9)}\left(\frac{s}{\sqrt{n_B}}\right)\right] = \left[26.56 \pm 2.262\left(\frac{.7639}{\sqrt{5}}\right)\right]$$

$$= [26.56 \pm .7727]$$

$$= [25.787, 27.333]$$

which were also calculated using regression analysis in Example 8.5.

Using the formulas in (3), a point prediction of, and a 95% prediction interval for, a future Fire-Hawk mileage when using gasoline type B, $y_{B,0} =$

TABLE 8.9 Three Samples of Fire-Hawk Mileages Obtained Using Gasoline Types A, B, and C, and Calculations of *F*(model) and the Related Prob-Value

Sample of $n_A = 4$ Fire-Hawk Mileages	Sample of $n_B = 5$ Fire-Hawk Mileages	Sample of $n_C = 3$ Fire-Hawk Mileages
$y_{A,1} = 24.0$	$y_{B,1} = 25.3$	$y_{C,1} = 23.3$
$y_{A,2} = 25.0$	$y_{B,2} = 26.5$	$y_{C,2} = 24.0$
$y_{A,3} = 24.3$	$y_{B,3} = 26.4$	$y_{C,3} = 24.7$
$y_{A,4} = 25.5$	$y_{B,4} = 27.0$	
	$y_{B,5} = 27.6$	
$\bar{y}_A = 24.7$	$\bar{y}_B = 26.56$	$\bar{y}_C = 24.0$
$s_A^2 = .46$	$s_B^2 = .723$	$s_C^2 = .49$

$$n = \sum_{l=A,B,C} n_l = n_A + n_B + n_C = 4 + 5 + 3 = 12$$

$$\bar{y} = \frac{\sum_{l=A,B,C} \sum_{k=1}^{n_l} y_{l,k}}{n} = \frac{\sum_{k=1}^{n_A} y_{A,k} + \sum_{k=1}^{n_B} y_{B,k} + \sum_{k=1}^{n_C} y_{C,k}}{n} = \frac{303.6}{12} = 25.3$$

$$\begin{aligned} SS_{\text{model}} &= \text{Explained variation} \\ &= \sum_{l=A,B,C} n_l(\bar{y}_l - \bar{y})^2 \\ &= n_A(\bar{y}_A - \bar{y})^2 + n_B(\bar{y}_B - \bar{y})^2 + n_C(\bar{y}_C - \bar{y})^2 \\ &= 4(24.7 - 25.3)^2 + 5(26.56 - 25.3)^2 + 3(24 - 25.3)^2 \\ &= 14.448 \end{aligned}$$

$$MS_{\text{model}} = \frac{SS_{\text{model}}}{v - 1} = \frac{14.448}{3 - 1} = 7.224$$

$$\begin{aligned} SSE &= \text{Unexplained variation} \\ &= \sum_{l=A,B,C} \sum_{k=1}^{n_l} (y_{l,k} - \bar{y}_l)^2 \quad \left(= \sum_{l=A,B,C} (n_l - 1)s_l^2\right) \\ &= \sum_{k=1}^{n_A} (y_{A,k} - \bar{y}_A)^2 + \sum_{k=1}^{n_B} (y_{B,k} - \bar{y}_B)^2 + \sum_{k=1}^{n_C} (y_{C,k} - \bar{y}_C)^2 \\ &= [(24 - 24.7)^2 + (25 - 24.7)^2 + (24.3 - 24.7)^2 + (25.5 - 24.7)^2] \\ &\quad + [(25.3 - 26.56)^2 + (26.5 - 26.56)^2 + (26.4 - 26.56)^2 \\ &\quad + (27 - 26.56)^2 + (27.6 - 26.56)^2] + [(23.3 - 24)^2 \\ &\quad + (24 - 24)^2 + (24.7 - 24)^2] \\ &= 5.252 \end{aligned}$$

$$s^2 = MSE = \frac{SSE}{n - v} = \frac{5.252}{12 - 3} = .583556$$

$$s = \sqrt{s^2} = \sqrt{.583556} = .7639$$

$$F(\text{model}) = \frac{MS_{\text{model}}}{MSE} = \frac{7.224}{.583556} = 12.379$$

$$\text{prob-value} = A[F(\text{model}); v - 1, n - v] = A[12.379; 3 - 1, 12 - 3] = .0026$$

$\mu_B + \epsilon_{B,0}$, are

$$\bar{y}_B = 26.56$$

$$\left[\bar{y}_B \pm t_{[.025]}^{(9)} s \sqrt{1 + \frac{1}{n_B}}\right] = \left[26.56 \pm 2.262(.7639)\sqrt{1 + \frac{1}{5}}\right]$$
$$= [26.56 \pm 1.8931]$$
$$= [24.667, 28.453]$$

which were also calculated using regression analysis in Example 8.5.

Using the formulas in (4), a point estimate of, and a 95% confidence interval for, the difference $\mu_C - \mu_B$ are

$$\bar{y}_C - \bar{y}_B = 24 - 26.56$$
$$= -2.56$$

$$\left[\bar{y}_C - \bar{y}_B \pm t_{[.025]}^{(9)} s \sqrt{\frac{1}{n_C} + \frac{1}{n_B}}\right] = \left[-2.56 \pm 2.262(.7639)\sqrt{\frac{1}{3} + \frac{1}{5}}\right]$$
$$= [-2.56 \pm 1.262]$$
$$= [-3.822, -1.298]$$

which were also calculated using regression analysis in Example 8.5 (see Table 8.7).

Recall that in Example 8.5 we used regression analysis to calculate Scheffé simultaneous 95% confidence intervals for all the linear combinations in Set I:

Set I

$\mu_B - \mu_A$

$\mu_C - \mu_A$

$\mu_C - \mu_B$

$\mu_B - \left(\frac{\mu_C + \mu_A}{2}\right)$

We can also use the formula in (6) to calculate Scheffé simultaneous 95% confidence intervals for the preceding linear combinations, because it can be verified that each of these linear combinations is a contrast, that is, is such that

$$\sum_{l=A,B,C} a_l = 0$$

For example, the linear combination $\mu_B - (\mu_C + \mu_A)/2$ is a contrast, since

$$\mu_B - \frac{\mu_C + \mu_A}{2} = -\tfrac{1}{2} \cdot \mu_A + 1 \cdot \mu_B - \tfrac{1}{2} \cdot \mu_C$$
$$= a_A\mu_A + a_B\mu_B + a_C\mu_C$$

where $a_A = -\frac{1}{2}$, $a_B = 1$, and $a_C = -\frac{1}{2}$, which implies that

$$\begin{aligned}\sum_{l=A,B,C} a_l &= a_A + a_B + a_C \\ &= -\tfrac{1}{2} + 1 - \tfrac{1}{2} \\ &= 0\end{aligned}$$

Therefore, since

$$\begin{aligned}\sum_{l=A,B,C} \frac{a_l^2}{n_l} &= \frac{a_A^2}{n_A} + \frac{a_B^2}{n_B} + \frac{a_C^2}{n_C} \\ &= \frac{(-\frac{1}{2})^2}{4} + \frac{(1)^2}{5} + \frac{(-\frac{1}{2})^2}{3} \\ &= .3458\end{aligned}$$

it follows that a Scheffé simultaneous 95% confidence interval for $\mu_B - (\mu_C + \mu_A)/2$ is (by the formula in (6))

$$\begin{aligned}&\left[\bar{y}_B - \frac{\bar{y}_C + \bar{y}_A}{2} \pm \sqrt{(v-1)F_{[.05]}^{(v-1,n-v)}}\, s \sqrt{\sum_{l=A,B,C} \frac{a_l^2}{n_l}}\right] \\ &= \left[26.56 - \frac{24 + 24.7}{2} \pm \sqrt{(3-1)F_{[.05]}^{(3-1,12-3)}}(.7639)\sqrt{.3458}\right] \\ &= [2.21 \pm \sqrt{2(4.26)}(.7639)\sqrt{.3458}] \\ &= [.8988, 3.5212]\end{aligned}$$

which is the same interval calculated using regression analysis in Example 8.5 (see Table 8.7).

Also, recall that in Example 8.5 we used regression analysis to calculate Scheffé simultaneous 95% confidence intervals for all the linear combinations in Set II:

Set II

μ_A

μ_B

μ_C

$\mu_B - \mu_A$

$\mu_C - \mu_A$

$\mu_C - \mu_B$

$\mu_B - \dfrac{\mu_C + \mu_A}{2}$

Since some of these linear combinations are not contrasts, that is, μ_A, μ_B, and μ_C are not such that

$$\sum_{l=A,B,C} a_l = 0$$

we must use the formula in (7) to calculate Scheffé simultaneous 95% confidence intervals for these linear combinations. For example, it follows that a Scheffé simultaneous 95% confidence interval for $\mu_B - (\mu_C + \mu_A)/2$ is

$$\left[\bar{y}_B - \frac{\bar{y}_C + \bar{y}_A}{2} \pm \sqrt{vF_{[.05]}^{(v,n-v)}}\, s\sqrt{\sum_{l=A,B,C} \frac{a_l^2}{n_l}}\right]$$

$$= \left[26.56 - \left(\frac{24 + 24.7}{2}\right) \pm \sqrt{3F_{[.05]}^{(3,12-3)}}(.7639)\sqrt{.3458}\right]$$

$$= [2.21 \pm \sqrt{3(3.86)}(.7639)\sqrt{.3458}]$$

$$= [.6814, 3.7386]$$

which is the same interval calculated using regression analysis in Example 8.5 (see Table 8.8).

Next, it follows from condition (8a) that we can test $H_0: \mu_C - \mu_B = 0$ versus $H_1: \mu_C - \mu_B \neq 0$ by setting α, the probability of a Type I error, equal to .05 if we calculate

$$t_{\bar{y}_C - \bar{y}_B} = \frac{\bar{y}_C - \bar{y}_B}{s\sqrt{(1/n_C) + (1/n_B)}}$$

$$= \frac{24 - 26.56}{.7639\sqrt{(1/3) + (1/5)}}$$

$$= -4.59$$

and use the rejection points

$$t_{[.05/2]}^{(n-v)} = t_{[.025]}^{(12-3)} = t_{[.025]}^{(9)} = 2.262$$

$$-t_{[.025]}^{(9)} = -2.262$$

Since $t_{\bar{y}_C - \bar{y}_B} = -4.59 < -2.262 = -t_{[.025]}^{(9)}$, we can reject $H_0: \mu_C - \mu_B = 0$ in favor of $H_1: \mu_C - \mu_B \neq 0$ by setting α equal to .05. Moreover, since

$$\text{prob-value} = 2A[[|t_{\bar{y}_C - \bar{y}_B}|, \infty); n - v]$$

$$= 2A[[|-4.59|, \infty); 12 - 3]$$

$$= 2A[[4.59, \infty); 9]$$

$$= 2[.00065]$$

$$= .0013$$

is less than .05 and .01, it follows by condition (8b) that we can reject $H_0: \mu_C - \mu_B = 0$ in favor of $H_1: \mu_C - \mu_B \neq 0$ if we set α equal to .05 or .01.

EXAMPLE 8.7

The Um-Good Bakery Company supplies Um-Good Coffee Cake to a large number of supermarkets in a metropolitan area. It wishes to study the effect of *shelf display height,* which has levels B (Bottom), M (Middle), and T (Top), on monthly demand for Um-Good Coffee Cake in these supermarkets. For accounting purposes, Um-Good Bakery defines a month to be a period of exactly four weeks (which implies that the year is divided into 13 such periods) and measures demand by cases, each of which contains 10 coffee cakes. For l = B, M, and T, we define *the lth population of monthly coffee cake demands* to be the infinite population of all possible monthly coffee cake demands that could be obtained at supermarkets which use display height l. We let μ_l denote the mean of this population. In order to compare μ_B, μ_M, and μ_T (where we call μ_l the *mean monthly coffee cake demand that would be obtained by using display height l*), Um-Good Bakery will employ a completely randomized experimental design. Specifically, for l = B, M, and T, Um-Good Bakery will randomly select a sample of $n_l = 6$ supermarkets from supermarkets in the metropolitan area. These supermarkets will sell Um-Good Coffee Cake for one month by using display height l. Then, letting $OSP_{l,k}$ denote the kth randomly selected supermarket (where $k = 1, 2, \ldots, n_l$) that uses display height l, Um-Good Bakery will measure

$$y_{l,k} = \text{the monthly coffee cake demand (measured in cases) that occurs in } OSP_{l,k}$$

which we call *the kth monthly coffee cake demand observed when using display height l.* When the bakery has employed this completely randomized design, we assume that, for l = B, M, and T,

$$y_{l,1}, y_{l,2}, y_{l,3}, y_{l,4}, y_{l,5}, y_{l,6}$$

TABLE 8.10 Three Samples of Monthly Coffee Cake Demands

Display Height B	M	T
$y_{B,1} = 58.2$	$y_{M,1} = 73.0$	$y_{T,1} = 52.4$
$y_{B,2} = 53.7$	$y_{M,2} = 78.1$	$y_{T,2} = 49.7$
$y_{B,3} = 55.8$	$y_{M,3} = 75.4$	$y_{T,3} = 50.9$
$y_{B,4} = 55.7$	$y_{M,4} = 76.2$	$y_{T,4} = 49.9$
$y_{B,5} = 52.5$	$y_{M,5} = 78.4$	$y_{T,5} = 52.1$
$y_{B,6} = 58.9$	$y_{M,6} = 82.1$	$y_{T,6} = 49.9$
$\bar{y}_B = 55.8$	$\bar{y}_M = 77.2$	$\bar{y}_T = 51.5$

TABLE 8.11 Analysis of Variance Table for Testing $H_0: \mu_B = \mu_M = \mu_T$ Versus H_1: At Least Two of μ_B, μ_M, and μ_T Differ From Each Other ($v = 3$, $n = 18$)

Source	df	Sum of Squares	Mean Square	F Statistic	Prob-Value
Model	$v - 1 = 2$	SS_{model} = Explained variation $= \sum_{l=B,M,T} n_l(\bar{y}_l - \bar{y})^2 = 2273.88$	$MS_{model} = \frac{SS_{model}}{v-1} = \frac{2273.88}{2} = 1136.94$	$F(\text{model}) = \frac{MS_{model}}{MSE} = \frac{1136.94}{6.16} = 184.5682$	$A[F(\text{model}); v-1, n-v] = A[184.5682; 2, 15] = .0001$
Error	$n - v = 15$	SSE = Unexplained variation $= \sum_{l=B,M,T} \sum_{k=1}^{n_l} (y_{l,k} - \bar{y}_l)^2 = 92.4$	$MSE = \frac{SSE}{n-v} = \frac{92.4}{15} = 6.16\ (=s^2)$		
Total	$n - 1 = 17$	Total variation $= \sum_{l=B,M,T} \sum_{k=1}^{n_l} (y_{l,k} - \bar{y})^2 = 2366.28$			

is a sample of $n_l = 6$ monthly coffee cake demands that the bakery has randomly selected from the lth population of monthly coffee cake demands. Suppose that when Um-Good Bakery has employed this completely randomized design, it has randomly selected the three samples summarized in Table 8.10. Table 8.11 presents the analysis of variance table summarizing the quantities used to test

$$H_0: \mu_B = \mu_M = \mu_T$$

versus

H_1: At least two of μ_B, μ_M, and μ_T differ from each other

The reader should, as an exercise, verify the calculations of SS_{model} and SSE. Examining Table 8.11, we find that since prob-value = .0001 is less than .05 and .01, we can reject H_0 in favor of H_1 by setting α equal to .05 or .01. In order to investigate the exact nature of the differences between μ_B, μ_M, and μ_T, we first consider

$\mu_M - \mu_B$ = the difference between μ_M, the mean monthly coffee cake demand that would be obtained by using a middle display height, and μ_B, the mean monthly coffee cake demand that would be obtained by using a bottom display height

Using the formulas in (4) of the preceding box, we find that a point estimate of, and a 95% confidence interval for, the difference $\mu_M - \mu_B$ are

$$\bar{y}_M - \bar{y}_B = 77.2 - 55.8 = 21.4 \qquad \text{(see Table 8.10)}$$

$$\begin{aligned}\left[(\bar{y}_M - \bar{y}_B) \pm t_{[.025]}^{(n-v)} s\sqrt{\frac{1}{n_M} + \frac{1}{n_B}}\right] &= \left[21.4 \pm t_{[.025]}^{(15)}(2.4819)\sqrt{\frac{1}{6} + \frac{1}{6}}\right] \\ &= \left[21.4 \pm 2.131(2.4819)\sqrt{\frac{1}{3}}\right] \\ &= [21.4 \pm 3.0536] \\ &= [18.3464, 24.4536]\end{aligned}$$

where $s = \sqrt{s^2} = \sqrt{6.16} = 2.4819$, since $s^2 = 6.16$ from Table 8.11. This interval says that Um-Good Bakery can be 95% confident that μ_M, the mean monthly coffee cake demand that would be obtained by using a middle display height, is between 18.3464 cases and 24.4536 cases greater than μ_B, the mean monthly coffee cake demand that would be obtained by using a bottom display height. Thus, there is strong evidence that μ_M is substantially greater than μ_B. Furthermore, by calculating the 95% confidence intervals for $\mu_M - \mu_T$ and $\mu_T - \mu_B$, the reader can verify that there is strong evidence that μ_M is substantially greater than μ_T and that there is strong evidence that μ_B is somewhat greater than μ_T. Since there is strong evidence that μ_M is substantially greater than both μ_B and μ_T, we assume that Um-Good Bakery will use a middle display height. Using the formulas in (2), a point estimate of, and a 95% confidence interval for, the mean μ_M are

$$\bar{y}_M = 77.2$$

$$\begin{aligned}\left[\bar{y}_M \pm t_{[.025]}^{(15)}\left(\frac{s}{\sqrt{n_M}}\right)\right] &= \left[77.2 \pm 2.131\left(\frac{2.4819}{\sqrt{6}}\right)\right] \\ &= [77.2 \pm 2.1592] \\ &= [75.0408, 79.3592]\end{aligned}$$

This interval says that Um-Good Bakery can be 95% confident that μ_M, the mean monthly coffee cake demand that would be obtained by using a middle display height, is at least 75.0408 cases and no more than 79.3592 cases. Next, using the formulas in (3), a point prediction of, and a 95% prediction interval for, a future monthly coffee cake demand when using a middle display height, $y_{M,0} = \mu_M + \epsilon_{M,0}$, are

$$\bar{y}_M = 77.2$$

$$\left[\bar{y}_M \pm t_{[.025]}^{(15)}\left(s\sqrt{1+\frac{1}{n_M}}\right)\right] = \left[77.2 \pm 2.131(2.4819)\sqrt{1+\frac{1}{6}}\right]$$

$$= [77.2 \pm 5.7127]$$

$$= [71.4873, 82.9127]$$

This interval says that Um-Good Bakery can be 95% confident that $y_{M,0}$, the coffee cake demand that will occur in a future randomly selected month in a randomly selected supermarket using a middle display height, will be at least 71.4873 cases and no more than 82.9127 cases. Before leaving this example, we should point out that if the data in Table 8.10 were observed in supermarkets where Um-Good Coffee Cake had never previously been sold, then these inferences would only pertain to supermarkets introducing the sale of Um-Good Coffee Cake. That is, once customers become aware of Um-Good Coffee Cake, the display height used might have less of an effect upon monthly coffee cake demand.

8.5 TWO QUALITATIVE INDEPENDENT VARIABLES

8.5.1 No Interaction: The Regression Approach

We begin this section by introducing some new terminology. From now on, we use the term **factor** as another name for *independent variable*. An experiment may involve the study of a single factor or the study of several factors. The various levels of a factor (or combination of factors) are called **treatments**. The purpose of most experiments is to estimate and compare the effects of different treatments on a dependent (or response) variable of interest.

EXAMPLE 8.8

Part 1. *The Problem and Data.*

As in Example 8.5, we assume that North American Oil Company is attempting to develop a reasonably priced unleaded gasoline that will deliver higher gasoline mileages that can be achieved by its current gasolines. As part of its development process, North American Oil would like to study the effects of two qualitative factors—unleaded gasoline type and gasoline additive type—on the gasoline mileage obtained by the Fire-Hawk. The factor "gasoline type" is defined to have three levels (A, B, and C), while the factor "additive type" is defined to have four levels (M, N, O, and P). Since the objective of the North

American Oil Company is to estimate and compare the effects of various gasoline types and additive types on gasoline mileage, the treatments in this experiment are gasoline type and additive type combinations. Since there are three levels of gasoline type (A, B, and C) and four levels of additive type (M, N, O, and P), we have 12 combinations of factor levels. That is, this experiment involves 12 treatments. These treatments are

AM	AN	AO	AP
BM	BN	BO	BP
CM	CN	CO	CP

Here, for example, the notation AM denotes the treatment, or factor combination, of gasoline type A and additive type M.

For i = A, B, and C and j = M, N, O, and P, we define *the (i, j)th population of Fire-Hawk mileages* to be the infinite population of all Fire-Hawk mileages that would be obtained by using gasoline type i and additive type j, and we will let μ_{ij} denote the mean of this population. Thus, μ_{ij} is the *mean Fire-Hawk mileage that would be obtained by using gasoline type i and additive type j*. One way that North American Oil could compare the 12 treatments would be to compare the 12 means corresponding to these treatments. Assuming (as in Example 8.5) that National Motors has given North American Oil access to a subpopulation of 1,000 Fire-Hawks, North American Oil can for all practical purposes randomly select a sample of Fire-Hawk mileages from each population of Fire-Hawk mileages by employing the following completely randomized experimental design. For i = A, B, and C and j = M, N, O, and P, North American Oil should randomly select a sample of n_{ij} Fire-Hawks from the subpopulation of 1,000 Fire-Hawks and should use a randomly selected supply of gasoline type i and a randomly selected supply of additive type j to test-drive each of these Fire-Hawks. Then, letting $OFH_{ij,k}$ denote the kth randomly selected Fire-Hawk that is test-driven using gasoline type i and additive type j (where $k = 1, 2, \ldots, n_{ij}$), North American Oil should measure

$$y_{ij,k} = \text{the gasoline mileage obtained by } OFH_{ij,k}$$

which we call *the kth Fire-Hawk mileage observed when using gasoline type i and additive type j*. When North American Oil has employed this completely randomized experimental design, we assume that for i = A, B, and C and j = M, N, O, and P,

$$y_{ij,1}, y_{ij,2}, \ldots, y_{ij,n_{ij}}$$

is a sample of n_{ij} Fire-Hawk mileages that North American Oil has randomly selected from the (i, j)th population of Fire-Hawk mileages.

Suppose that North American Oil has randomly selected the ten samples summarized in Table 8.12. Note that the sizes of these samples are not equal and that no samples have been randomly selected from the (A, P)th or (C, N)th populations of Fire-Hawk mileages. In general, an experiment in which we ran-

TABLE 8.12 Ten Samples of Fire-Hawk Mileages

Unleaded Gasoline Type (i)	Gasoline Additive Type (j)			
	M	N	O	P
A	$n_{AM} = 3$ $y_{AM,1} = 19.4$ $y_{AM,2} = 20.6$ $y_{AM,3} = 20.0$ $\bar{y}_{AM} = 20.0$	$n_{AN} = 2$ $y_{AN,1} = 25.0$ $y_{AN,2} = 24.0$ $\bar{y}_{AN} = 24.5$	$n_{AO} = 2$ $y_{AO,1} = 24.3$ $y_{AO,2} = 25.5$ $\bar{y}_{AO} = 24.9$	$n_{AP} = 0$
B	$n_{BM} = 2$ $y_{BM,1} = 22.6$ $y_{BM,2} = 21.6$ $\bar{y}_{BM} = 22.1$	$n_{BN} = 2$ $y_{BN,1} = 25.3$ $y_{BN,2} = 26.5$ $\bar{y}_{BN} = 25.9$	$n_{BO} = 3$ $y_{BO,1} = 27.6$ $y_{BO,2} = 26.4$ $y_{BO,3} = 27.0$ $\bar{y}_{BO} = 27.0$	$n_{BP} = 1$ $y_{BP,1} = 25.4$ $\bar{y}_{BP} = 25.4$
C	$n_{CM} = 2$ $y_{CM,1} = 18.6$ $y_{CM,2} = 19.8$ $\bar{y}_{CM} = 19.2$	$n_{CN} = 0$	$n_{CO} = 3$ $y_{CO,1} = 24.0$ $y_{CO,2} = 24.7$ $y_{CO,3} = 23.3$ $\bar{y}_{CO} = 24.0$	$n_{CP} = 2$ $y_{CP,1} = 22.0$ $y_{CP,2} = 23.0$ $\bar{y}_{CP} = 22.5$

domly select samples of element values from all the populations corresponding to all the treatments resulting from combining all levels of a given factor with all levels of another factor is called a **complete factorial experiment**. When the sample size is the same for each treatment, such an experiment is a *balanced complete factorial experiment*. When samples are not selected for some of the possible treatments, the experiment is an *incomplete factorial experiment*. Since Table 8.12 indicates that samples are not selected for two treatments (AP and CN), the experiment carried out by North American Oil is an incomplete factorial experiment. If one wishes to obtain as much information as possible from a complete factorial experiment, it is generally best to conduct a balanced complete factorial experiment. However, sometimes data are lost or are unavailable. For example, although data would probably not be lost in a gasoline experiment, sometimes data are lost in agricultural experiments. To see this, consider Table 8.12, and assume that A, B, and C are three fertilizer types, M, N, O, and P are four wheat types, and the $y_{ij,k}$ values are wheat yields in bushels per plot (one-third of an acre) corresponding to the different combinations of fertilizer type and wheat type. Then it is possible that at harvest time the experimenter might find that some of the plots have been mistakenly destroyed (perhaps ploughed or treated with excessive amounts of herbicide). Thus, some of the $y_{ij,k}$ values have been lost, and only the data in Table 8.12 remain. Thus, in some situations, the analyst must deal with complete factorial experiments that are not balanced or must deal with incomplete factorial experiments.

Part 2. *Developing a Regression Model.*

Defining the (ij, k)th error term, $\epsilon_{ij,k} = y_{ij,k} - \mu_{ij}$, to be the difference between $y_{ij,k}$ and μ_{ij}, we may express $y_{ij,k}$ in the form

$$y_{ij,k} = \mu_{ij} + \epsilon_{ij,k}$$

In order to express $y_{ij,k}$ in terms of a regression model, we need to write a single equation which, for $i = 1, 2, \ldots, a$ and $j = 1, 2, \ldots, b$, expresses μ_{ij} as a function of a set of independent variables. To determine an appropriate equation, we consider Figure 8.5, which presents a graphical analysis of the Fire-Hawk gasoline mileage data in Table 8.12. As indicated by Figure 8.5, graphical analysis implies that there is little interaction between unleaded gasoline type and additive type. Therefore, an appropriate equation expressing μ_{ij} as a function of a set of independent variables is

$$\begin{aligned}\mu_{ij} &= \mu + \alpha_i + \gamma_j \\ &= \mu + \alpha_B D_{i,B} + \alpha_C D_{i,C} + \gamma_N D_{j,N} + \gamma_O D_{j,O} + \gamma_P D_{j,P}\end{aligned}$$

where, for example,

$$D_{i,C} = \begin{cases} 1 & \text{if } i = \text{C; that is, if we are using gasoline type C} \\ 0 & \text{otherwise} \end{cases}$$

$$D_{j,N} = \begin{cases} 1 & \text{if } j = \text{N; that is, if we are using additive type N} \\ 0 & \text{otherwise} \end{cases}$$

and the other dummy variables are defined similarly. Considering this equation for μ_{ij}, note that

$$\alpha_i = \alpha_B D_{i,B} + \alpha_C D_{i,C}$$

which implies that

$$\begin{aligned}\alpha_A &= \alpha_B D_{A,B} + \alpha_C D_{A,C} \\ &= \alpha_B(0) + \alpha_C(0) \\ &= 0\end{aligned}$$

and note that

$$\gamma_j = \gamma_N D_{j,N} + \gamma_O D_{j,O} + \gamma_P D_{j,P}$$

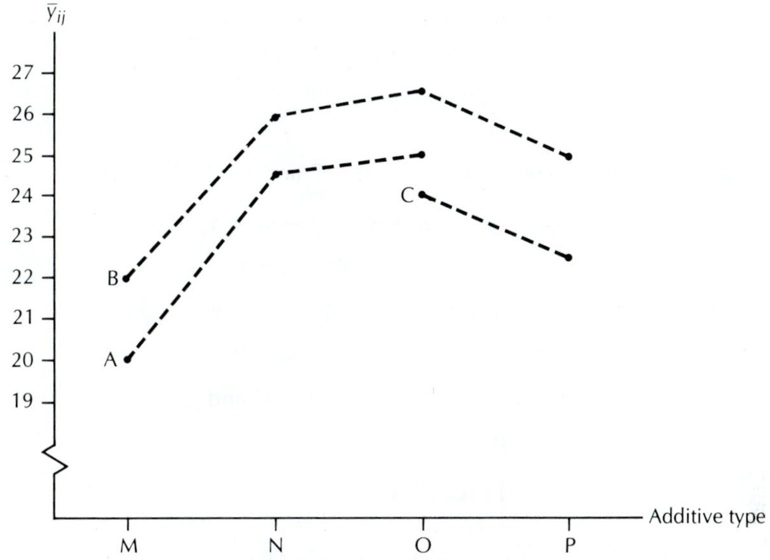

(a) For each unleaded gasoline type (A, B, and C), a plot is made of the change in the $\bar{y}_{ij}$'s associated with changing the additive type (from M to N to O to P). Little interaction means this change in the $\bar{y}_{ij}$'s depends little on the unleaded gasoline type.

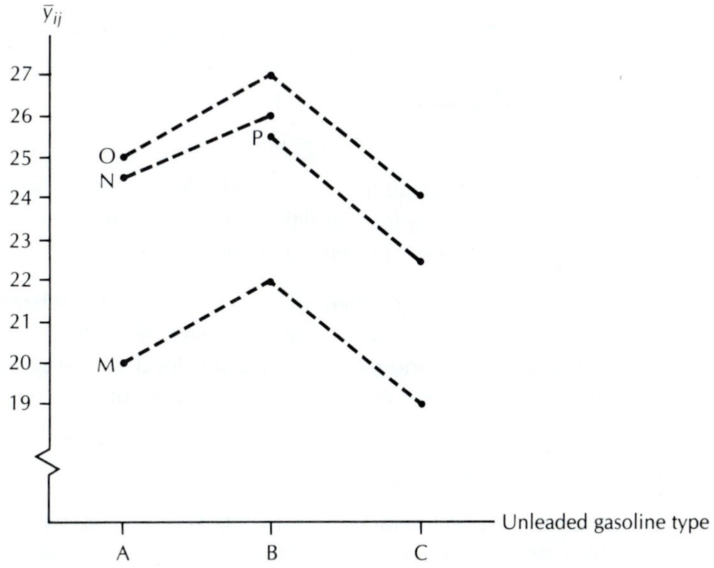

FIGURE 8.5
Graphical Analysis of the Fire-Hawk Mileage Data in Table 8.12: Little Interaction

(b) For each additive type (M, N, O, and P), a plot is made of the change in the $\bar{y}_{ij}$'s associated with changing the unleaded gasoline type (from A to B to C). Little interaction means this change in the $\bar{y}_{ij}$'s depends little on the additive type.

which implies that

$$\begin{aligned}\gamma_M &= \gamma_N D_{M,N} + \gamma_O D_{M,O} + \gamma_P D_{M,P} \\ &= \gamma_N(0) + \gamma_O(0) + \gamma_P(0) \\ &= 0\end{aligned}$$

Therefore, for i = A, B, and C and j = M, N, O, and P

$$\begin{aligned}\mu_{ij} - \mu_{Aj} &= [\mu + \alpha_i + \gamma_j] - [\mu + \alpha_A + \gamma_j] \\ &= \alpha_i \qquad \text{(since } \alpha_A = 0)\end{aligned}$$

$$\begin{aligned}\mu_{ij} - \mu_{iM} &= [\mu + \alpha_i + \gamma_j] - [\mu + \alpha_i + \gamma_M] \\ &= \gamma_j \qquad \text{(since } \gamma_M = 0)\end{aligned}$$

In other words, for i = A, B, and C and j = M, N, O, and P,

$$\begin{aligned}\alpha_i &= \mu_{ij} - \mu_{Aj} \\ &= \begin{cases} \mu_{iM} - \mu_{AM} & \text{if } j = M \\ \mu_{iN} - \mu_{AN} & \text{if } j = N \\ \mu_{iO} - \mu_{AO} & \text{if } j = O \\ \mu_{iP} - \mu_{AP} & \text{if } j = P \end{cases}\end{aligned}$$

= the change in mean Fire-Hawk mileage associated with changing from gasoline type A to gasoline type i while the additive type remains constant

$$\begin{aligned}\gamma_j &= \mu_{ij} - \mu_{iM} \\ &= \begin{cases} \mu_{Aj} - \mu_{AM} & \text{if } i = A \\ \mu_{Bj} - \mu_{BM} & \text{if } i = B \\ \mu_{Cj} - \mu_{CM} & \text{if } i = C \end{cases}\end{aligned}$$

= the change in mean Fire-Hawk mileage associated with changing from additive type M to additive type j while the gasoline type remains constant

Note that since the change $(\mu_{ij} - \mu_{Aj})$ equals α_i, this change is independent of j, the level of the additive type. Moreover, since the change $(\mu_{ij} - \mu_{iM})$ equals γ_j, this change is independent of i, the level of the gasoline type. Since the changes are independent of the level of the "other factor," it follows that the equation

$$\mu_{ij} = \mu + \alpha_i + \gamma_j$$

assumes that no interaction exists between gasoline type and additive type. For this reason, we refer to the model

$$\begin{aligned}y_{ij,k} &= \mu_{ij} + \epsilon_{ij,k} \\ &= \mu + \alpha_i + \gamma_j + \epsilon_{ij,k} \\ &= \mu + \alpha_B D_{i,B} + \alpha_C D_{i,C} + \gamma_N D_{j,N} + \gamma_O D_{j,O} + \gamma_P D_{j,P} + \epsilon_{ij,k}\end{aligned}$$

as the (α, γ) *model*. Here, the notation (α, γ) says that this model assumes that no interaction exists between gasoline type and additive type, and thus we can *separate* the effects of these factors.

Part 3. *Estimating the Regression Parameters.*

In order to use the (α, γ) model to carry out a regression analysis of the data in Table 8.12, we use the following column vector **y** and matrix **X**:

$$\mathbf{y} = \begin{bmatrix} y_{AM,1} \\ y_{AM,2} \\ y_{AM,3} \\ y_{AN,1} \\ y_{AN,2} \\ y_{AO,1} \\ y_{AO,2} \\ y_{BM,1} \\ y_{BM,2} \\ y_{BN,1} \\ y_{BN,2} \\ y_{BO,1} \\ y_{BO,2} \\ y_{BO,3} \\ y_{BP,1} \\ y_{CM,1} \\ y_{CM,2} \\ y_{CO,1} \\ y_{CO,2} \\ y_{CO,3} \\ y_{CP,1} \\ y_{CP,2} \end{bmatrix} = \begin{bmatrix} 19.4 \\ 20.6 \\ 20.0 \\ 25.0 \\ 24.0 \\ 24.3 \\ 25.5 \\ 22.6 \\ 21.6 \\ 25.3 \\ 26.5 \\ 27.6 \\ 26.4 \\ 27.0 \\ 25.4 \\ 18.6 \\ 19.8 \\ 24.0 \\ 24.7 \\ 23.3 \\ 22.0 \\ 23.0 \end{bmatrix} \qquad \mathbf{X} = \begin{array}{c} \begin{matrix} 1 & D_{i,B} & D_{i,C} & D_{j,N} & D_{j,O} & D_{j,P} \end{matrix} \\ \begin{bmatrix} 1 & 0 & 0 & 0 & 0 & 0 \\ 1 & 0 & 0 & 0 & 0 & 0 \\ 1 & 0 & 0 & 0 & 0 & 0 \\ 1 & 0 & 0 & 1 & 0 & 0 \\ 1 & 0 & 0 & 1 & 0 & 0 \\ 1 & 0 & 0 & 0 & 1 & 0 \\ 1 & 0 & 0 & 0 & 1 & 0 \\ 1 & 1 & 0 & 0 & 0 & 0 \\ 1 & 1 & 0 & 0 & 0 & 0 \\ 1 & 1 & 0 & 1 & 0 & 0 \\ 1 & 1 & 0 & 1 & 0 & 0 \\ 1 & 1 & 0 & 0 & 1 & 0 \\ 1 & 1 & 0 & 0 & 1 & 0 \\ 1 & 1 & 0 & 0 & 1 & 0 \\ 1 & 1 & 0 & 0 & 0 & 1 \\ 1 & 0 & 1 & 0 & 0 & 0 \\ 1 & 0 & 1 & 0 & 0 & 0 \\ 1 & 0 & 1 & 0 & 1 & 0 \\ 1 & 0 & 1 & 0 & 1 & 0 \\ 1 & 0 & 1 & 0 & 1 & 0 \\ 1 & 0 & 1 & 0 & 0 & 1 \\ 1 & 0 & 1 & 0 & 0 & 1 \end{bmatrix} \end{array}$$

When the appropriate calculations are carried out, the least squares point estimates of the parameters in the (α, γ) model are found to be

$$\begin{bmatrix} \hat{\mu} \\ \hat{\alpha}_B \\ \hat{\alpha}_C \\ \hat{\gamma}_N \\ \hat{\gamma}_O \\ \hat{\gamma}_P \end{bmatrix} = \mathbf{b} = (\mathbf{X}'\mathbf{X})^{-1}\mathbf{X}'\mathbf{y} = \begin{bmatrix} 20.1029 \\ 1.9038 \\ -.9640 \\ 4.1452 \\ 4.8947 \\ 3.3718 \end{bmatrix}$$

where, for example, $\hat{\gamma}_N = 4.1452$ is the least squares point estimate of γ_N. These least squares point estimates yield the prediction equation

$$\begin{aligned}\hat{y}_{ij} &= \hat{\mu} + \hat{\alpha}_i + \hat{\gamma}_j \\ &= \hat{\mu} + \hat{\alpha}_B D_{i,B} + \hat{\alpha}_C D_{i,C} + \hat{\gamma}_N D_{j,N} + \hat{\gamma}_O D_{j,O} + \hat{\gamma}_P D_{j,P} \\ &= 20.1029 + 1.9038 D_{i,B} - .9640 D_{i,C} + 4.1452 D_{j,N} \\ &\quad + 4.8947 D_{j,O} + 3.3718 D_{j,P}\end{aligned}$$

In addition, the unexplained variation is calculated to be

$$\begin{aligned}SSE_{(\alpha,\gamma)} &= \sum_{i=A,B,C} \sum_{j=M,N,O,P} \sum_{k=1}^{n_{ij}} (y_{ij,k} - \hat{y}_{ij})^2 \\ &= 6.4428\end{aligned}$$

Part 4. *Comparing Effects Using Partial F-Tests.*

To compare the effects of gasoline types A, B, and C and of additive types M, N, O, and P, we first perform two partial F-tests. To perform the first test, consider the following complete (α, γ) model:

$$\begin{aligned}y_{ij,k} &= \mu_{ij} + \epsilon_{ij,k} \\ &= \mu + \alpha_i + \gamma_j + \epsilon_{ij,k} \\ &= \mu + \alpha_B D_{i,B} + \alpha_C D_{i,C} + \gamma_N D_{j,N} + \gamma_O D_{j,O} + \gamma_P D_{j,P} + \epsilon_{ij,k}\end{aligned}$$

which has $k = 6$ parameters and for which the unexplained variation is $SSE_C = SSE_{(\alpha,\gamma)} = 6.4428$. Since we have seen that for i = A, B, and C,

$$\alpha_i = \mu_{ij} - \mu_{Aj}$$

and thus that

$$\alpha_B = \mu_{Bj} - \mu_{Aj}$$

$$\alpha_C = \mu_{Cj} - \mu_{Aj}$$

the null hypothesis

$$H_0: \alpha_B = \alpha_C = 0 \qquad \text{or} \qquad H_0: \alpha_B = 0 \quad \text{and} \quad \alpha_C = 0$$

is equivalent to

$$H_0: \mu_{Bj} - \mu_{Aj} = 0 \qquad \text{and} \qquad \mu_{Cj} - \mu_{Aj} = 0$$

which is equivalent to

$$H_0: \mu_{Aj} = \mu_{Bj} = \mu_{Cj}$$

The null hypothesis,

$$H_0: \alpha_B = \alpha_C = 0 \qquad \text{or} \qquad \mu_{Aj} = \mu_{Bj} = \mu_{Cj}$$

says that gasoline types A, B, and C have the same effects on mean Fire-Hawk mileage. The alternative hypothesis,

H_1: At least one of α_B or α_C does not equal zero

or

H_1: At least two of μ_{Aj}, μ_{Bj}, and μ_{Cj} differ from each other

says that at least two of gasoline types A, B, and C have different effects on mean Fire-Hawk mileage. In order to test H_0 versus H_1, first note that $p - g = 2$, since two parameters (α_B and α_C) are set equal to zero in the statement of the null hypothesis H_0. Also note that under the assumption that H_0 is true, the complete model becomes the following reduced γ model:

$$\begin{aligned} y_{ij,k} &= \mu + \gamma_j + \epsilon_{ij,k} \\ &= \mu + \gamma_N D_{j,N} + \gamma_O D_{j,O} + \gamma_P D_{j,P} + \epsilon_{ij,k} \end{aligned}$$

for which the unexplained variation is $SSE_R = SSE_\gamma = 19.3971$. Thus, to test H_0 versus H_1 we use the following partial F statistic and prob-value:

$$\begin{aligned} F(\alpha|\gamma) &= F(D_{i,B}, D_{i,C}|D_{j,N}, D_{j,O}, D_{j,P}) \\ &= \frac{MS_{drop}}{MSE_C} \\ &= \frac{6.4772}{.4027} \\ &= 16.0844 \end{aligned}$$

since

$$\begin{aligned} MS_{drop} &= \frac{SS_{drop}}{p-g} \\ &= \frac{\{SSE_R - SSE_C\}}{p-g} \\ &= \frac{\{SSE_\gamma - SSE_{(\alpha,\gamma)}\}}{p-g} \\ &= \frac{\{19.3971 - 6.4428\}}{2} \\ &= \frac{12.9543}{2} \\ &= 6.4772 \end{aligned}$$

$$MSE_C = \frac{SSE_C}{n-k}$$

$$= \frac{SSE_{(\alpha,\gamma)}}{n-k}$$

$$= \frac{6.4428}{22-6}$$

$$= \frac{6.4428}{16}$$

$$= .4027$$

and

$$\text{prob-value} = A[F(\alpha|\gamma); p-g, n-k]$$
$$= A[16.0844; 2, 16]$$
$$= .0002$$

If we wish to use condition (1) in the box concerning the partial F-test in Section 7.6 to determine whether we can reject H_0 in favor of H_1 by setting α equal to .05, we would use the rejection point

$$F_{[\alpha]}^{(p-g,n-k)} = F_{[.05]}^{(2,16)} = 3.63$$

Since $F(\alpha|\gamma) = 16.0844 > 3.63 = F_{[.05]}^{(2,16)}$, we can reject H_0 in favor of H_1 by setting α equal to .05. Alternatively, since prob-value = .0002 is less than .05 and .01, it follows by condition (2) that we can reject H_0 in favor of H_1 by setting α equal to .05 or .01. These facts provide substantial evidence indicating that at least two of gasoline types A, B, and C have different effects on mean Fire-Hawk mileage.

To perform the second partial F-test, again consider the complete model to be the (α, γ) model. Since we have seen that for j = M, N, O, and P,

$$\gamma_j = \mu_{ij} - \mu_{iM}$$

and thus that

$$\gamma_N = \mu_{iN} - \mu_{iM}$$

$$\gamma_O = \mu_{iO} - \mu_{iM}$$

$$\gamma_P = \mu_{iP} - \mu_{iM}$$

the null hypothesis

$$H_0: \gamma_N = \gamma_O = \gamma_P = 0 \quad \text{or} \quad H_0: \gamma_N = 0 \text{ and } \gamma_O = 0 \text{ and } \gamma_P = 0$$

is equivalent to

$$H_0: \mu_{iN} - \mu_{iM} = 0 \text{ and } \mu_{iO} - \mu_{iM} = 0 \text{ and } \mu_{iP} - \mu_{iM} = 0$$

which is equivalent to

$$H_0: \mu_{iM} = \mu_{iN} = \mu_{iO} = \mu_{iP}$$

The null hypothesis,

$$H_0: \gamma_N = \gamma_O = \gamma_P = 0 \qquad \text{or} \qquad \mu_{iM} = \mu_{iN} = \mu_{iO} = \mu_{iP}$$

says that additive types M, N, O, and P have the same effects on mean Fire-Hawk mileage. The alternative hypothesis,

H_1: At least one of γ_N, γ_O, or γ_P does not equal zero

or

H_1: At least two of μ_{iM}, μ_{iN}, μ_{iO}, and μ_{iP} differ from each other

says that at least two of additive types M, N, O, and P have different effects on mean Fire-Hawk mileage. To test H_0 versus H_1, first note that $p - g = 3$, since three parameters (γ_N, γ_O, and γ_P) are set equal to zero in the statement of the null hypothesis H_0. Also note that under the assumption that H_0 is true, the complete model becomes the following reduced α model:

$$\begin{aligned} y_{ij,k} &= \mu + \alpha_i + \epsilon_{ij,k} \\ &= \mu + \alpha_B D_{i,B} + \alpha_C D_{i,C} + \epsilon_{ij,k} \end{aligned}$$

for which the unexplained variation is $SSE_R = SSE_\alpha = 101.8886$. Thus, in order to test H_0 versus H_1 we use the following partial F statistic and prob-value:

$$\begin{aligned} F(\gamma|\alpha) &= F(D_{j,N}, D_{j,O}, D_{j,P}|D_{i,B}, D_{i,C}) \\ &= \frac{MS_{\text{drop}}}{MSE_C} \\ &= \frac{31.8153}{.4027} \\ &= 79.005 \end{aligned}$$

since

$$\begin{aligned} MS_{\text{drop}} &= \frac{SS_{\text{drop}}}{p-g} \\ &= \frac{\{SSE_R - SSE_C\}}{p-g} \\ &= \frac{\{SSE_\alpha - SSE_{(\alpha,\gamma)}\}}{p-g} \\ &= \frac{\{101.8886 - 6.4428\}}{3} \\ &= \frac{95.4458}{3} \\ &= 31.8153 \end{aligned}$$

$$MSE_C = \frac{SSE_C}{n-k}$$
$$= \frac{SSE_{(\alpha,\gamma)}}{n-k}$$
$$= \frac{6.4428}{22-6}$$
$$= \frac{6.4428}{16}$$
$$= .4027$$

and

$$\text{prob-value} = A[F(\gamma|\alpha);\, p-g,\, n-k]$$
$$= A[79.005;\, 3,\, 16]$$
$$= .0001 \qquad \text{(that is, } \le .0001\text{)}$$

Since the prob-value is less than .05 and .01, it follows by condition (2) (see the partial F-test in Section 7.6) that we can reject H_0 in favor of H_1 by setting α equal to .05 or .01. Thus, we have substantial evidence indicating that at least two of additive types M, N, O, and P have different effects on mean Fire-Hawk mileage.

Part 5. *Testing and Estimating Differences Using Linear Combinations of Regression Parameters.*

Figure 8.6 presents several linear combinations of parameters that will allow us to investigate the exact nature of the differences between the effects of gasoline types A, B, and C and between the effects of additive types M, N, O, and P. Note that the relationships shown in this figure between the differences in the μ_{ij} values (means) and the parameters α_B, α_C, γ_N, γ_O, and γ_P can be easily established by using the fact that the (α, γ) model assumes that

$$\mu_{ij} = \mu + \alpha_i + \gamma_j$$

For example,

$$\mu_{Cj} - \mu_{Bj} = [\mu + \alpha_C + \gamma_j] - [\mu + \alpha_B + \gamma_j]$$
$$= \alpha_C - \alpha_B$$

= the change in mean Fire-Hawk mileage associated with changing from gasoline type B to gasoline type C while the additive type remains constant

$$\mu_{Bj} - \frac{\mu_{Cj} + \mu_{Aj}}{2} = [\mu + \alpha_B + \gamma_j] - \frac{[\mu + \alpha_C + \gamma_j] + [\mu + \gamma_j]}{2}$$

$$= \alpha_B - .5\alpha_C$$

= the change in mean Fire-Hawk mileage associated with changing from using gasoline type C half the time and gasoline type A half the time to using only gasoline type B while the additive type remains constant

$$\mu_{iO} - \mu_{iN} = [\mu + \alpha_i + \gamma_O] - [\mu + \alpha_i + \gamma_N]$$

$$= \gamma_O - \gamma_N$$

= the change in mean Fire-Hawk mileage associated with changing from additive type N to additive type O while the gasoline type remains constant

In addition, Figure 8.6 presents the SAS output of the least squares point estimate of each of the linear combinations (we previously calculated $\hat{\mu}$, $\hat{\alpha}_B$, $\hat{\alpha}_C$, $\hat{\gamma}_N$, $\hat{\gamma}_O$, and $\hat{\gamma}_P$); the $t_{\boldsymbol{\lambda}'\mathbf{b}}$ statistic and prob-value for testing $H_0: \boldsymbol{\lambda}'\boldsymbol{\beta} = 0$ versus $H_1: \boldsymbol{\lambda}'\boldsymbol{\beta} \neq 0$ for each of the linear combinations; and the individual and Scheffé simultaneous 95% confidence intervals for each of the linear combinations *except* μ. Since we have calculated the unexplained variation for the (α, γ) model to be $SSE_{(\alpha,\gamma)} = 6.4428$, it follows that the standard error is

$$s_{(\alpha,\gamma)} = \sqrt{\frac{SSE_{(\alpha,\gamma)}}{n-k}} = \sqrt{\frac{6.4428}{22-6}} = .6346$$

and thus that the individual 95% confidence intervals are computed using the formula

$$[\boldsymbol{\lambda}'\mathbf{b} \pm t_{[.025]}^{(n-k)} s_{(\alpha,\gamma)} \sqrt{\boldsymbol{\lambda}'(\mathbf{X}'\mathbf{X})^{-1}\boldsymbol{\lambda}}] = [\boldsymbol{\lambda}'\mathbf{b} \pm t_{[.025]}^{(22-6)} s_{(\alpha,\gamma)} \sqrt{\boldsymbol{\lambda}'(\mathbf{X}'\mathbf{X})^{-1}\boldsymbol{\lambda}}]$$
$$= [\boldsymbol{\lambda}'\mathbf{b} \pm (2.12) s_{(\alpha,\gamma)} \sqrt{\boldsymbol{\lambda}'(\mathbf{X}'\mathbf{X})^{-1}\boldsymbol{\lambda}}]$$

where, for example, if we are computing the 95% confidence interval for

$$\mu_{iO} - \mu_{iN} = \gamma_O - \gamma_N$$
$$= 0 \cdot \mu + 0 \cdot \alpha_B + 0 \cdot \alpha_C + (-1) \cdot \gamma_N + 1 \cdot \gamma_O + 0 \cdot \gamma_P$$
$$= [0 \quad 0 \quad 0 \quad -1 \quad 1 \quad 0] \begin{bmatrix} \mu \\ \alpha_B \\ \alpha_C \\ \gamma_N \\ \gamma_O \\ \gamma_P \end{bmatrix}$$
$$= \boldsymbol{\lambda}'\boldsymbol{\beta}$$

FIGURE 8.6 SAS Output of the $t_{\lambda'\mathbf{b}}$ Statistics and Prob-Values for Testing Various Hypotheses When Using the (α, γ) Model to Describe the Fire-Hawk Mileage Data, and Related Individual and Scheffé Simultaneous 95% Confidence Intervals

VARIABLE[a]	PARAMETER ESTIMATE[b]	T FOR HO: PARAMETER=0[c]	PROB > |T|[d]
INTERCEPT	20.10290282	67.28	0.0001
DB	1.90382836	5.66	0.0001
DC	-0.96398822	-2.56	0.0208
DN	4.14518300	10.15	0.0001
DO	4.89465713	14.69	0.0001
DP	3.37181321	7.27	0.0001
P1	-2.86781658	-8.28	0.0001
P2	2.38582247	8.38	0.0001
P3	0.74947413	1.84	0.0848
P4	-0.77336979	-1.45	0.1674
P5	-1.52284392	-3.45	0.0033

[a] $\boldsymbol{\lambda}'\boldsymbol{\beta}$: [b] $\boldsymbol{\lambda}'\mathbf{b}$: [c] $t_{\lambda'\mathbf{b}}$ [d] Prob-value

$\boldsymbol{\lambda}'\boldsymbol{\beta}$	$\boldsymbol{\lambda}'\mathbf{b}$
$\mu_{AM} = \mu$	$\hat{\mu}$
$\mu_{Bj} - \mu_{Aj} = \alpha_B$	$\hat{\alpha}_B$
$\mu_{Cj} - \mu_{Aj} = \alpha_C$	$\hat{\alpha}_C$
$\mu_{iN} - \mu_{iM} = \gamma_N$	$\hat{\gamma}_N$
$\mu_{iO} - \mu_{iM} = \gamma_O$	$\hat{\gamma}_O$
$\mu_{iP} - \mu_{iM} = \gamma_P$	$\hat{\gamma}_P$
$\mu_{Cj} - \mu_{Bj} = \alpha_C - \alpha_B$	$\hat{\alpha}_C - \hat{\alpha}_B$
$\mu_{Bj} - \frac{\mu_{Cj} + \mu_{Aj}}{2} = \alpha_B - .5\alpha_C$	$\hat{\alpha}_B - .5\hat{\alpha}_C$
$\mu_{iO} - \mu_{iN} = \gamma_O - \gamma_N$	$\hat{\gamma}_O - \hat{\gamma}_N$
$\mu_{iP} - \mu_{iN} = \gamma_P - \gamma_N$	$\hat{\gamma}_P - \hat{\gamma}_N$
$\mu_{iP} - \mu_{iO} = \gamma_P - \gamma_O$	$\hat{\gamma}_P - \hat{\gamma}_O$

it follows that $\boldsymbol{\lambda}' = [0 \quad 0 \quad 0 \quad -1 \quad 1 \quad 0]$. Also, note that since there are $q = 5$ parameters—α_B, α_C, γ_N, γ_O, and γ_P—from the (α, γ) model involved nontrivially in at least one of the linear combinations in Figure 8.6, it follows that

$$\begin{aligned}\sqrt{qF_{[\alpha]}^{(q,n-k)}} &= \sqrt{5 \cdot F_{[.05]}^{(5,22-6)}} \\ &= \sqrt{5 \cdot F_{[.05]}^{(5,16)}} \\ &= \sqrt{5(2.85)} \\ &= 3.7749\end{aligned}$$

STD ERROR OF ESTIMATE[e]	Individual 95% Confidence Interval for $\boldsymbol{\lambda}'\boldsymbol{\beta}$ $[\boldsymbol{\lambda}'\mathbf{b} \pm 2.12 s_{(\alpha,\gamma)}\sqrt{\boldsymbol{\lambda}'(\mathbf{X}'\mathbf{X})^{-1}\boldsymbol{\lambda}}]$	Scheffé Simultaneous 95% Confidence Interval for $\boldsymbol{\lambda}'\boldsymbol{\beta}$ $[\boldsymbol{\lambda}'\mathbf{b} \pm 3.7749 s_{(\alpha,\gamma)}\sqrt{\boldsymbol{\lambda}'(\mathbf{X}'\mathbf{X})^{-1}\boldsymbol{\lambda}}]$
0.29879024		
0.33614296	[1.1913, 2.6163]	[.6351, 3.1725]
0.37610087	[−1.7613, −.1667]	[−2.3837, .4557]
0.40828051	[3.2796, 5.0108]	[2.6039, 5.6865]
0.33310546	[4.1894, 5.6008]	[3.6372, 6.152]
0.46365176	[2.389, 4.3546]	[1.6218, 5.1218]
0.34632013	[−3.6019, −2.1337]	[−4.175, −1.5606]
0.28478394	[1.7820, 2.9896]	[1.3107, 3.4609]
0.40786538	[−.1152, 1.6142]	[−.7903, 2.2393]
0.53474573	[−1.907, .3602]	[−2.7918, 1.245]
0.44176097	[−2.4594, −.5862]	[−3.1906, .145]

[e] $s_{(\alpha,\gamma)}\sqrt{\boldsymbol{\lambda}'(\mathbf{X}'\mathbf{X})^{-1}\boldsymbol{\lambda}}$

Thus,

$$[\boldsymbol{\lambda}'\mathbf{b} \pm \sqrt{qF_{[\alpha]}^{(q,n-k)}}\, s_{(\alpha,\gamma)}\sqrt{\boldsymbol{\lambda}'(\mathbf{X}'\mathbf{X})^{-1}\boldsymbol{\lambda}}] = [\boldsymbol{\lambda}'\mathbf{b} \pm (3.7749) s_{(\alpha,\gamma)}\sqrt{\boldsymbol{\lambda}'(\mathbf{X}'\mathbf{X})^{-1}\boldsymbol{\lambda}}]$$

is a Scheffé simultaneous 95% confidence interval for any linear combination $\boldsymbol{\lambda}'\boldsymbol{\beta}$ (except μ) in Figure 8.6.

Examining Figure 8.6, we see that the point estimates of, and the individual and simultaneous 95% confidence intervals for, $\mu_{Bj} - \mu_{Aj}$ and $\mu_{Cj} - \mu_{Bj}$, and the large $t_{\boldsymbol{\lambda}'\mathbf{b}}$ statistics and small prob-values related to the null hypotheses

$$H_0: \alpha_B = \mu_{Bj} - \mu_{Aj} = 0 \qquad \text{(prob-value = .0001)}$$

and

$$H_0: \alpha_C - \alpha_B = \mu_{Cj} - \mu_{Bj} = 0 \qquad \text{(prob-value = .0001)}$$

provide overwhelming evidence indicating that μ_{Bj} is greater than μ_{Aj} and μ_{Cj}. For example, since $\hat{\alpha}_B = 1.9038$ is the least squares point estimate of $\alpha_B = \mu_{Bj} - \mu_{Aj}$, North American Oil estimates that the effect of changing from gasoline type A to gasoline type B (while the additive type remains constant) is to increase mean Fire-Hawk mileage by 1.9038 mpg. Moreover, the individual 95% confidence interval for $\alpha_B = \mu_{Bj} - \mu_{Aj}$, [1.1913, 2.6163], makes North American Oil 95% confident that the effect of changing from gasoline type A to gasoline type B (while the additive type remains constant) is to increase mean Fire-Hawk mileage by between 1.1913 mpg and 2.6163 mpg. Examining Figure 8.6, we also see that the point estimates of, and the individual and simultaneous 95% confidence intervals for, $\mu_{iO} - \mu_{iM}$ and $\mu_{iP} - \mu_{iO}$, and the large $t_{\lambda'b}$ statistics and small prob-values related to the null hypotheses

$$H_0: \gamma_O = \mu_{iO} - \mu_{iM} = 0 \qquad \text{(prob-value = .0001)}$$

and

$$H_0: \gamma_P - \gamma_O = \mu_{iP} - \mu_{iO} = 0 \qquad \text{(prob-value = .0033)}$$

provide overwhelming evidence indicating that μ_{iO} is greater than μ_{iM} and very strong evidence indicating that μ_{iO} is greater than μ_{iP}. To compare μ_{iO} and μ_{iN}, first note that since $\hat{\gamma}_O - \hat{\gamma}_N = .7495$ is the least squares point estimate of $\gamma_O - \gamma_N = \mu_{iO} - \mu_{iN}$, North American Oil estimates that the effect of changing from additive type N to additive type O (while the gasoline type remains constant) is to increase mean Fire-Hawk mileage by .7495 mpg. However, note that the individual 95% confidence interval for $\gamma_O - \gamma_N = \mu_{iO} - \mu_{iN}$, [−.1152, 1.6142], makes North American Oil 95% confident that the effect of changing from additive type N to additive type O (while holding the gasoline type constant) may be to decrease mean Fire-Hawk mileage by .1152 mpg at one extreme and may be to increase mean Fire-Hawk mileage by 1.6142 mpg at the other extreme. Because this confidence interval contains zero, we cannot reject

$$H_0: \gamma_O - \gamma_N = \mu_{iO} - \mu_{iN} = 0$$

in favor of

$$H_1: \gamma_O - \gamma_N = \mu_{iO} - \mu_{iN} \neq 0$$

with at least 95% confidence. However, since the interval is "mostly positive," and the prob-value of .0848 casts a fair amount of doubt on the validity of H_0, we have some evidence indicating that μ_{iO} is greater than μ_{iN}. To summarize, North American Oil has overwhelming evidence indicating μ_{iO} is greater than μ_{iM}, very strong evidence indicating that μ_{iO} is greater than μ_{iP}, and a fair amount of evidence indicating that μ_{iO} is greater than μ_{iN}. Also, (we assume)

North American Oil has theoretical reasons for believing that additive type O is the best additive. Therefore, we assume that North American Oil will produce gasoline type B and use additive type O. Recall that we previously found that gasoline type B maximizes mean Fire-Hawk mileage.

Next, suppose that North American Oil has been producing gasoline type A using additive type N. Because

$$\begin{aligned}
\mu_{BO} - \mu_{AN} &= (\mu + \alpha_B + \gamma_O) - (\mu + \gamma_N) \\
&= \alpha_B + (\gamma_O - \gamma_N) \\
&= 0 \cdot \mu + 1 \cdot \alpha_B + 0 \cdot \alpha_C + (-1) \cdot \gamma_N + 1 \cdot \gamma_O + 0 \cdot \gamma_P \\
&= [0 \quad 1 \quad 0 \quad -1 \quad 1 \quad 0] \begin{bmatrix} \mu \\ \alpha_B \\ \alpha_C \\ \gamma_N \\ \gamma_O \\ \gamma_P \end{bmatrix} \\
&= \boldsymbol{\lambda}'\boldsymbol{\beta}
\end{aligned}$$

where $\boldsymbol{\lambda}' = [0 \quad 1 \quad 0 \quad -1 \quad 1 \quad 0]$, it follows that a point estimate of, and an individual 95% confidence interval for, $\alpha_B + (\gamma_O - \gamma_N) = \mu_{BO} - \mu_{AN}$ are

$$\hat{\alpha}_B + (\hat{\gamma}_O - \hat{\gamma}_N) = 1.9038 + .7495 = 2.6533$$

and

$$\begin{aligned}
[\boldsymbol{\lambda}'\mathbf{b} \pm t_{[.025]}^{(22-6)} s_{(\alpha,\gamma)} \sqrt{\boldsymbol{\lambda}'(\mathbf{X}'\mathbf{X})^{-1}\boldsymbol{\lambda}}] &= [2.6533 \pm 2.12(.6346)\sqrt{.6386}] \\
&= [2.6533 \pm 1.0751] \\
&= [1.5782, 3.7284]
\end{aligned}$$

This interval says that North American Oil can be 95% confident that μ_{BO}, the mean Fire-Hawk mileage obtained by using gasoline type B and additive type O, is between 1.5782 mpg and 3.7284 mpg greater than μ_{AN}, the mean Fire-Hawk mileage obtained by using gasoline type A and additive type N.

Part 6. *Using the Model to Estimate and Predict.*

Finally, since we can describe a future Fire-Hawk mileage that will be observed when using gasoline type B and additive type O by the equation

$$\begin{aligned}
y_{BO,0} &= \mu_{BO} + \epsilon_{BO,0} \\
&= \mu + \alpha_B D_{B,B} + \alpha_C D_{B,C} + \gamma_N D_{O,N} + \gamma_O D_{O,O} + \gamma_P D_{O,P} + \epsilon_{BO,0} \\
&= \mu + \alpha_B(1) + \alpha_C(0) + \gamma_N(0) + \gamma_O(1) + \gamma_P(0) + \epsilon_{BO,0} \\
&= \mu + \alpha_B + \gamma_O + \epsilon_{BO,0}
\end{aligned}$$

it follows that

$$\begin{aligned}\hat{y}_{BO} &= \hat{\mu} + \hat{\alpha}_B + \hat{\gamma}_O \\ &= 20.1029 + 1.9038 + 4.8947 \\ &= 26.9014\end{aligned}$$

is the point estimate of μ_{BO} and is the point prediction of $y_{BO,0}$. Furthermore, the equation describing $y_{BO,0}$ implies that

$$\mathbf{x}'_{BO} = [1 \quad 1 \quad 0 \quad 0 \quad 1 \quad 0]$$

Therefore, a 95% confidence interval for μ_{BO} is

$$\begin{aligned}[\hat{y}_{BO} \pm t_{[.025]}^{(22-6)} s_{(\alpha,\gamma)} \sqrt{\mathbf{x}'_{BO}(\mathbf{X}'\mathbf{X})^{-1}\mathbf{x}_{BO}}] &= [26.9014 \pm 2.12(.6346)\sqrt{.2057}] \\ &= [26.9014 \pm .6102] \\ &= [26.2912, 27.5116]\end{aligned}$$

This interval says that the federal government can be 95% confident that μ_{BO}, the mean Fire-Hawk mileage obtained by using gasoline type B and additive type O, is at least 26.2912 mpg and no more than 27.5116 mpg. Also, a 95% prediction interval for $y_{BO,0}$ is

$$\begin{aligned}[\hat{y}_{BO} \pm t_{[.025]}^{(22-6)} s_{(\alpha,\gamma)} \sqrt{1 + \mathbf{x}'_{BO}(\mathbf{X}'\mathbf{X})^{-1}\mathbf{x}_{BO}}] &= [26.9014 \pm 2.12(.6346)\sqrt{1.2057}] \\ &= [26.9014 \pm 1.4773] \\ &= [25.4241, 28.3787]\end{aligned}$$

This interval says that the owner of a Fire-Hawk can be 95% confident that $y_{BO,0}$, the gasoline mileage that will be obtained by the Fire-Hawk when driven under normal driving conditions using gasoline type B and additive type O, will be at least 25.4241 mpg and no more than 28.3787 mpg.

In general, consider evaluating the effects of two qualitative factors—factor 1 and factor 2—on a dependent variable. As illustrated in the preceding example, if it is reasonable to conclude (by graphical analysis or hypothesis testing) that little or no interaction exists between factor 1 and factor 2, three models can be used to describe two-factor data:

1. The (α, γ) model,

$$y_{ij,k} = \mu + \alpha_i + \gamma_j + \epsilon_{ij,k}$$

which assumes that both factor 1 and factor 2 are significantly related to the dependent variable.

2. The α model,

$$y_{ij,k} = \mu + \alpha_i + \epsilon_{ij,k}$$

which assumes that only factor 1 is significantly related to the dependent variable.

3. The γ model,

$$y_{ij,k} = \mu + \gamma_j + \epsilon_{ij,k}$$

which assumes that only factor 2 is significantly related to the dependent variable.

In addition, a fourth model that can be used to describe two-factor data is the μ model,

$$y_{ij,k} = \mu + \epsilon_{ij,k}$$

which assumes that neither factor 1 nor factor 2 is significantly related to the dependent variable.

Besides the $F(\alpha|\gamma)$ statistic and the $F(\gamma|\alpha)$ statistic (the uses of which are demonstrated in the preceding example), two other statistics that can be used to help determine which of the models best describes a set of two-factor data are the *$F(\alpha)$ statistic*, which we define to be the overall F statistic for the α model, and the *$F(\gamma)$ statistic*, which we define to be the overall F statistic for the γ model. To see how to use these statistics, note that *both the $F(\alpha|\gamma)$ statistic and the $F(\alpha)$ statistic* (and their related prob-values) *measure the importance of factor 1*. However, since the $F(\alpha|\gamma)$ statistic,

$$F(\gamma|\alpha) = \frac{\{SSE_\alpha - SSE_{(\alpha,\gamma)}\}/(p-g)}{SSE_{(\alpha,\gamma)}/(n-k)}$$

compares SSE_γ, the unexplained variation for the γ model, with $SSE_{(\alpha,\gamma)}$, the unexplained variation for the (α, γ) model, *the $F(\alpha|\gamma)$ statistic measures the additional importance of factor 1 over the importance of factor 2*. Also, since the $F(\alpha)$ statistic is the overall F statistic for the α model, *the $F(\alpha)$ statistic measures the importance of factor 1 without considering the importance of factor 2*.

Both the $F(\gamma|\alpha)$ statistic and the $F(\gamma)$ statistic (and their related prob-values) *measure the importance of factor 2*. However, since the $F(\gamma|\alpha)$ statistic,

$$F(\gamma|\alpha) = \frac{\{SSE_\alpha - SSE_{(\alpha,\gamma)}\}/(p-g)}{SSE_{(\alpha,\gamma)}/(n-k)}$$

compares SSE_α, the unexplained variation for the α model, with $SSE_{(\alpha,\gamma)}$, the unexplained variation for the (α, γ) model, *the $F(\gamma|\alpha)$ statistic measures the additional importance of factor 2 over the importance of factor 1*. Also, since the $F(\gamma)$ statistic is the overall F statistic for the γ model, *the $F(\gamma)$ statistic measures the importance of factor 2 without considering the importance of factor 1*.

Now, we say that any of the $F(\alpha|\gamma)$, $F(\alpha)$, $F(\gamma|\alpha)$, and $F(\gamma)$ statistics is *significant* if and only if the prob-value calculated by using this statistic is less than a small (arbitrarily chosen) value (say, .05) of α (which implies that we can reject the null hypothesis tested by this statistic by setting the probability of a Type I error equal to

TABLE 8.13 Models Suggested According to Significance (s) and Nonsignificance (n) of $F(\alpha|\gamma)$, $F(\alpha)$, $F(\gamma|\alpha)$, and $F(\gamma)$ Statistics

	Models Suggested by Cramer and Applebaum (1977)								
	α, γ Model	α Model	γ Model	α Model	γ Model	μ Model	α Model	γ Model	α Model or γ Model
$F(\alpha\|\gamma)$	s	s	n	n	n	n	s	n	n
$F(\alpha)$	—	s	—	s	n	n	n	—	s
$F(\gamma\|\alpha)$	s	n	s	n	n	n	n	s	n
$F(\gamma)$	—	—	s	n	s	n	—	n	s
	α, γ Model	α Model	γ Model	α Model	γ Model	μ Model	α, γ Model	α, γ Model	α, γ Model
							Disagreement		
	Models Suggested by Searle (1971)								

Note: The — means that it does not matter whether the particular F statistic is significant or nonsignificant.

α). It should be noted that multicollinearity between factors 1 and 2 can cause the $F(\alpha)$ statistic to be significant when the $F(\alpha|\gamma)$ statistic is not significant (and vice versa), and can cause the $F(\gamma)$ statistic to be significant when the $F(\gamma|\alpha)$ statistic is not significant (and vice versa). Table 8.13 summarizes the models that are suggested to best describe the two-factor data, according to the significance (s) and nonsignificance (n) of the $F(\alpha|\gamma)$, $F(\alpha)$, $F(\gamma|\alpha)$, and $F(\gamma)$ statistics. Here, we present the suggestions made by Cramer and Applebaum (1977) and the suggestions made by Searle (1971). Note that of the nine cases considered in Table 8.13, the suggestions made by Cramer and Applebaum agree with the suggestions made by Searle in six cases and disagree in three cases. Note that, for example, since we saw in Example 8.8 that the prob-values calculated by using the $F(\alpha|\gamma)$ and $F(\gamma|\alpha)$ statistics are, respectively, .0002 and .0001, it follows that both of these statistics are significant and thus that both Cramer and Applebaum and Searle suggest that the (α, γ) model best describes the Fire-Hawk mileage data concerning unleaded gasoline type and additive type. Next, assume that $F(\alpha|\gamma)$ is significant and that $F(\gamma|\alpha)$ is not significant. In this case, if $F(\alpha)$ is significant, both Cramer and Applebaum and Searle suggest that the α model best describes the two-factor data. However, in this case, if $F(\alpha)$ is not significant, while Cramer and Applebaum still suggest that the α model best describes the two-factor data, Searle believes that factor 2 is important (in helping to clarify the importance of factor 1) and thus that the (α, γ) model best describes the two-factor data. The reader should carefully study Table 8.13. By intelligently using the suggestions made there and the model-building techniques discussed in this book, the reader should be able to determine the model best describing two-factor data in most cases. (Some of the exercises at the end of this chapter offer an opportunity to use Table 8.13.)

Before we continue, it is important to point out that one reason we wish to find the best model is so that we can use the model to investigate the exact nature of the differences between the levels of factor 1 and/or the levels of factor 2 by computing (individual and simultaneous) confidence intervals and $t_{\lambda' b}$ statistics. Some analysts believe, however, that we should not consider determining the best model but rather, because we have performed a *two-factor* experiment, should use the (α, γ) model to compute confidence intervals and $t_{\lambda' b}$ statistics (if little or no interaction exists between the two factors). If one does believe in first determining the best model, then since the α model and the γ model each assumes that only one factor affects the dependent variable, it follows, if we conclude that either of these models is best, that we can use the methods in Section 8.4 to compute the desired confidence intervals and $t_{\lambda' b}$ statistics.

8.5.2 Interaction: The Regression Approach

EXAMPLE 8.9

Part 1. *The Problem and Data.*

To see how graphical analysis indicates that interaction does exist between two factors, suppose that North American Oil is also attempting to develop a reasonably priced, high-mileage *leaded* gasoline and wishes to study the effects of two qualitative factors—leaded gasoline type (which has levels D, E, and F) and gasoline additive type (which has levels Q, R, S, and T)—on the gasoline mileage obtained by an automobile called the Lance. For i = D, E, and F and j = Q, R, S, and T, we define *the* (i, j)*th population of Lance mileages* to be the infinite population of all Lance mileages that would be obtained by using gasoline type i and additive type j. We let μ_{ij} denote the mean of this population.

In order to compare the appropriate means (we call μ_{ij} the *mean Lance mileage obtained by using gasoline type i and additive type j*), North American Oil should, for i = D, E, and F and j = Q, R, S, and T, randomly select a sample of n_{ij} Lances from a subpopulation of Lances and should use a randomly selected supply of gasoline type i and a randomly selected supply of additive type j to test-drive each of these Lances. Let $OLN_{ij,k}$ denote the kth randomly selected Lance that is test-driven using gasoline type i and additive type j ($k = 1, 2, \ldots, n_{ij}$). Then North American Oil should measure

$$y_{ij,k} = \text{the gasoline mileage obtained by } OLN_{ij,k}$$

which we call *the* k*th Lance mileage observed when using gasoline type i and additive type j*. When North American Oil has employed this completely randomized design, we assume that for i = D, E, and F and j = Q, R, S, and T,

$$y_{ij,1}, y_{ij,2}, \ldots, y_{ij,n_{ij}}$$

TABLE 8.14 Ten Samples of Lance Mileages

Leaded Gasoline Type (i)	Gasoline Additive Type (j) Q	R	S	T
D	$n_{DQ} = 3$ $y_{DQ,1} = 18.0$ $y_{DQ,2} = 18.6$ $y_{DQ,3} = 17.4$ $\bar{y}_{DQ} = 18.0$	$n_{DR} = 2$ $y_{DR,1} = 23.0$ $y_{DR,2} = 22.0$ $\bar{y}_{DR} = 22.5$	$n_{DS} = 2$ $y_{DS,1} = 23.5$ $y_{DS,2} = 22.3$ $\bar{y}_{DS} = 22.9$	$n_{DT} = 0$
E	$n_{EQ} = 2$ $y_{EQ,1} = 23.3$ $y_{EQ,2} = 24.5$ $\bar{y}_{EQ} = 23.9$	$n_{ER} = 3$ $y_{ER,1} = 25.6$ $y_{ER,2} = 24.4$ $y_{ER,3} = 25.0$ $\bar{y}_{ER} = 25.0$	$n_{ES} = 1$ $y_{ES,1} = 23.4$ $\bar{y}_{ES} = 23.4$	$n_{ET} = 2$ $y_{ET,1} = 19.6$ $y_{ET,2} = 20.6$ $\bar{y}_{ET} = 20.1$
F	$n_{FQ} = 0$	$n_{FR} = 3$ $y_{FR,1} = 24.0$ $y_{FR,2} = 23.3$ $y_{FR,3} = 24.7$ $\bar{y}_{FR} = 24.0$	$n_{FS} = 2$ $y_{FS,1} = 23.0$ $y_{FS,2} = 22.0$ $\bar{y}_{FS} = 22.5$	$n_{FT} = 2$ $y_{FT,1} = 18.6$ $y_{FT,2} = 19.8$ $\bar{y}_{FT} = 19.2$

is a sample of n_{ij} Lance mileages that North American Oil has randomly selected from the (i, j)th population of Lance mileages.

Suppose that when North American Oil has employed this completely randomized design, it has randomly selected the 10 samples summarized in Table 8.14. In Figure 8.7 we present a graphical analysis of the data in Table 8.14. As indicated by Figure 8.7, graphical analysis implies that there is substantial interaction between leaded gasoline type and additive type. However, the figure also indicates that if we eliminate all sample information concerning gasoline type D, little interaction remains between gasoline type and additive type. That is, little interaction exists between gasoline types E and F and additive types Q, R, S, and T.

Part 2. *Developing a Regression Model.*

Since we are interested in studying the effect of gasoline type D, and since Figure 8.7 indicates that substantial interaction does exist between gasoline types D, E, and F and additive types Q, R, S, and T, we will find that it is probably best to describe $y_{ij,k}$, the kth Lance mileage observed when using

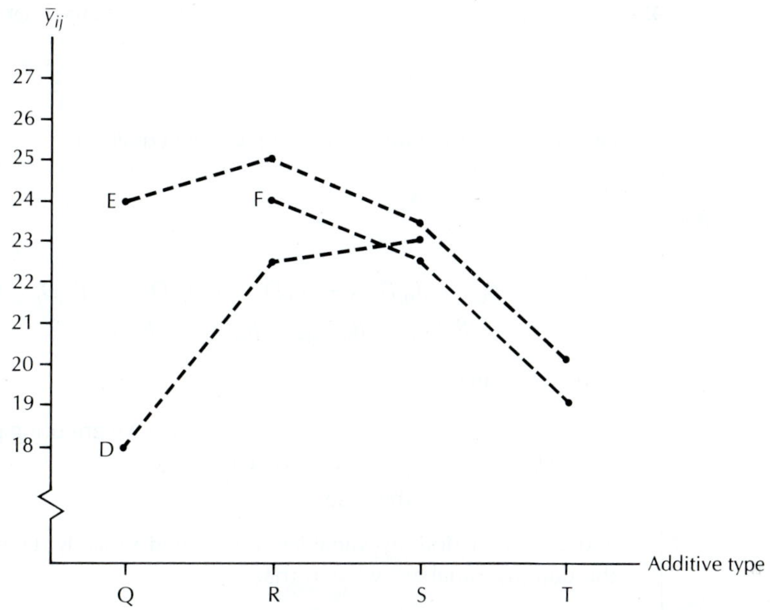

(a) For each leaded gasoline type (D, E, and F), a plot is made of the change in the $\bar{y}_{ij}$'s associated with changing the additive type (from Q to R to S to T). Substantial interaction means this change in the $\bar{y}_{ij}$'s depends substantially on the leaded gasoline type. For example, when changing from additive type R to additive type S, $\bar{y}_{ij}$ increases when using gasoline D but decreases when using gasoline E or F.

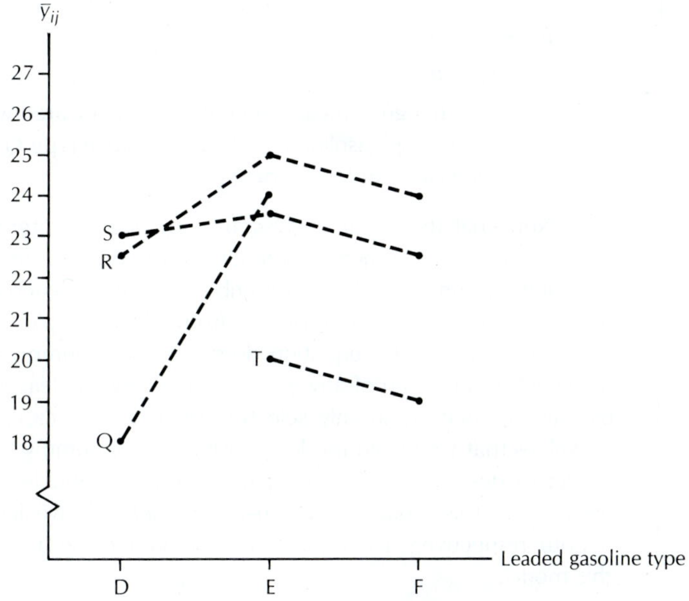

FIGURE 8.7 Graphical Analysis of the Lance Mileage Data in Table 8.14: Substantial Interaction

(b) For each additive type (Q, R, S, and T), a plot is made of the change in the $\bar{y}_{ij}$'s associated with changing the leaded gasoline type (from D to E to F). Substantial interaction means this change in the $\bar{y}_{ij}$'s depends substantially on the additive type. For example, when changing from gasoline type D to gasoline type E, $\bar{y}_{ij}$ increases by much more when using additive type Q than when using additive type S.

gasoline type i and additive type j, by the equation

$$y_{ij,k} = \mu_{ij} + \epsilon_{ij,k}$$

where

$$\begin{aligned}\mu_{ij} = \beta_{DQ} &+ \beta_{DR}D_{ij,DR} + \beta_{DS}D_{ij,DS} + \beta_{EQ}D_{ij,EQ} + \beta_{ER}D_{ij,ER} + \beta_{ES}D_{ij,ES} \\ &+ \beta_{ET}D_{ij,ET} + \beta_{FR}D_{ij,FR} + \beta_{FS}D_{ij,FS} + \beta_{FT}D_{ij,FT}\end{aligned}$$

Here, for example,

$$D_{ij,ER} = \begin{cases} 1 & \text{if } i = \text{E and } j = \text{R; that is, if we are using gasoline type E} \\ & \text{and additive type R} \\ 0 & \text{otherwise} \end{cases}$$

and the other dummy variables are defined similarly. Using the definitions of the dummy variables, we find that

$$\mu_{DQ} = \beta_{DQ}$$

and that for any ij not equal to DQ

$$\mu_{ij} = \beta_{DQ} + \beta_{ij}$$

which implies that for any ij not equal to DQ

$$\begin{aligned}\beta_{ij} &= \mu_{ij} - \beta_{DQ} \\ &= \mu_{ij} - \mu_{DQ} \\ &= \text{the change in mean Lance mileage associated with changing} \\ &\quad \text{from using gasoline type D and additive type Q to using gasoline} \\ &\quad \text{type } i \text{ and additive type } j\end{aligned}$$

Note that the equation describing μ_{ij} uses the intercept β_{DQ} to describe the effect of μ_{DQ}, the mean of the (D, Q)th population of Lance mileages, and a separate dummy variable to describe the effect of each of the means of the other populations of Lance mileages from which samples have been randomly selected. However, this equation does not use a dummy variable to describe the effect of the mean of any population of Lance mileages from which a sample has not been randomly selected (since, in this case, there would be no $y_{ij,k}$ values that we would model by using such a dummy variable). Because the model to describe μ_{ij} uses a separate dummy variable to describe each μ_{ij}, we refer to this model as the *means model*, and we let SS_{means} and SSE_{means} denote, respectively, the explained variation and the unexplained variation for this model.

Part 3. *Estimating the Regression Parameters.*

In order to use the means model to carry out a regression analysis of the data in Table 8.14, we use the following column vector **y** and matrix **X**:

$$
\begin{array}{cccccccccc}
1 & D_{ij,DR} & D_{ij,DS} & D_{ij,EQ} & D_{ij,ER} & D_{ij,ES} & D_{ij,ET} & D_{ij,FR} & D_{ij,FS} & D_{ij,FT}
\end{array}
$$

$$
\mathbf{y} = \begin{bmatrix} y_{DQ,1} \\ y_{DQ,2} \\ y_{DQ,3} \\ y_{DR,1} \\ y_{DR,2} \\ y_{DS,1} \\ y_{DS,2} \\ y_{EQ,1} \\ y_{EQ,2} \\ y_{ER,1} \\ y_{ER,2} \\ y_{ER,3} \\ y_{ES,1} \\ y_{ET,1} \\ y_{ET,2} \\ y_{FR,1} \\ y_{FR,2} \\ y_{FR,3} \\ y_{FS,1} \\ y_{FS,2} \\ y_{FT,1} \\ y_{FT,2} \end{bmatrix} = \begin{bmatrix} 18.0 \\ 18.6 \\ 17.4 \\ 23.0 \\ 22.0 \\ 23.5 \\ 22.3 \\ 23.3 \\ 24.5 \\ 25.6 \\ 24.4 \\ 25.0 \\ 23.4 \\ 19.6 \\ 20.6 \\ 24.0 \\ 23.3 \\ 24.7 \\ 23.0 \\ 22.0 \\ 18.6 \\ 19.8 \end{bmatrix} \qquad \mathbf{X} = \begin{bmatrix}
1 & 0 & 0 & 0 & 0 & 0 & 0 & 0 & 0 & 0 \\
1 & 0 & 0 & 0 & 0 & 0 & 0 & 0 & 0 & 0 \\
1 & 0 & 0 & 0 & 0 & 0 & 0 & 0 & 0 & 0 \\
1 & 1 & 0 & 0 & 0 & 0 & 0 & 0 & 0 & 0 \\
1 & 1 & 0 & 0 & 0 & 0 & 0 & 0 & 0 & 0 \\
1 & 0 & 1 & 0 & 0 & 0 & 0 & 0 & 0 & 0 \\
1 & 0 & 1 & 0 & 0 & 0 & 0 & 0 & 0 & 0 \\
1 & 0 & 0 & 1 & 0 & 0 & 0 & 0 & 0 & 0 \\
1 & 0 & 0 & 1 & 0 & 0 & 0 & 0 & 0 & 0 \\
1 & 0 & 0 & 0 & 1 & 0 & 0 & 0 & 0 & 0 \\
1 & 0 & 0 & 0 & 1 & 0 & 0 & 0 & 0 & 0 \\
1 & 0 & 0 & 0 & 1 & 0 & 0 & 0 & 0 & 0 \\
1 & 0 & 0 & 0 & 0 & 1 & 0 & 0 & 0 & 0 \\
1 & 0 & 0 & 0 & 0 & 0 & 1 & 0 & 0 & 0 \\
1 & 0 & 0 & 0 & 0 & 0 & 1 & 0 & 0 & 0 \\
1 & 0 & 0 & 0 & 0 & 0 & 0 & 1 & 0 & 0 \\
1 & 0 & 0 & 0 & 0 & 0 & 0 & 1 & 0 & 0 \\
1 & 0 & 0 & 0 & 0 & 0 & 0 & 1 & 0 & 0 \\
1 & 0 & 0 & 0 & 0 & 0 & 0 & 0 & 1 & 0 \\
1 & 0 & 0 & 0 & 0 & 0 & 0 & 0 & 1 & 0 \\
1 & 0 & 0 & 0 & 0 & 0 & 0 & 0 & 0 & 1 \\
1 & 0 & 0 & 0 & 0 & 0 & 0 & 0 & 0 & 1
\end{bmatrix}
$$

When the appropriate calculations are carried out, the least squares point estimates of the parameters in the means model are calculated to be

$$
\begin{bmatrix} b_{DQ} \\ b_{DR} \\ b_{DS} \\ b_{EQ} \\ b_{ER} \\ b_{ES} \\ b_{ET} \\ b_{FR} \\ b_{FS} \\ b_{FT} \end{bmatrix} = \mathbf{b} = (\mathbf{X}'\mathbf{X})^{-1}\mathbf{X}'\mathbf{y} = \begin{bmatrix} 18.0 \\ 4.5 \\ 4.9 \\ 5.9 \\ 7.0 \\ 5.4 \\ 2.1 \\ 6.0 \\ 4.5 \\ 1.2 \end{bmatrix}
$$

These least squares point estimates yield the prediction equation

$$\begin{aligned}\hat{y}_{ij} &= b_{DQ} + b_{DR}D_{ij,DR} + b_{DS}D_{ij,DS} + b_{EQ}D_{ij,EQ} + b_{ER}D_{ij,ER} \\ &\quad + b_{ES}D_{ij,ES} + b_{ET}D_{ij,ET} + b_{FR}D_{ij,FR} + b_{FS}D_{ij,FS} + b_{FT}D_{ij,FT} \\ &= 18.0 + 4.5D_{ij,DR} + 4.9D_{ij,DS} + 5.9D_{ij,EQ} + 7.0D_{ij,ER} + 5.4D_{ij,ES} \\ &\quad + 2.1D_{ij,ET} + 6.0D_{ij,FR} + 4.5D_{ij,FS} + 1.2D_{ij,FT}\end{aligned}$$

and the following quantities (note that the means model has $k = 10$ parameters):

$$\text{Total variation} = 127.4727$$

$$SS_{\text{means}} = \text{Explained variation} = 121.3927$$

$$MS_{\text{means}} = \frac{SS_{\text{means}}}{k-1} = \frac{121.3927}{10-1} = 13.4881$$

$$SSE_{\text{means}} = \text{Unexplained variation} = 6.08$$

$$s^2_{\text{means}} = MSE_{\text{means}} = \frac{SSE_{\text{means}}}{n-k} = \frac{6.08}{22-10} = .5067$$

$$s_{\text{means}} = \sqrt{.5067} = .7118$$

$$F(\text{means model}) = \frac{MS_{\text{means}}}{MSE_{\text{means}}} = \frac{13.4881}{.5067} = 26.6212$$

$$\begin{aligned}\text{prob-value} &= A[F(\text{means model}); k-1, n-k] \\ &= A[26.6212; 10-1=9, 22-10=12] \\ &= .0001\end{aligned}$$

Part 4. *Comparing Means Using the Overall F-Test.*

Since we have seen that the means model implies that $\beta_{DQ} = \mu_{DQ}$ and that each of the other parameters is such that $\beta_{ij} = \mu_{ij} - \mu_{DQ}$, it follows that

H_0: All β_{ij} (except β_{DQ}) equal zero

is equivalent to

H_0: All $\{\mu_{ij} - \mu_{DQ}\}$ equal zero

which is equivalent to

H_0: All means μ_{ij} are equal to each other

The alternative hypothesis is

H_1: At least two of the means μ_{ij} differ from each other

If we wish to use condition (2) in the box concerning the overall F-test in Section 7.1 to determine whether we can reject H_0 in favor of H_1, then since prob-value $= .0001$ is less than .05 and .01, we can reject H_0 in favor of H_1 by setting

α equal to .05 or .01. Thus, we have substantial evidence that there are differences in the means.

Part 5. *Estimating a Difference Using a Linear Combination of Regression Parameters.*

We note that $\bar{y}_{ER} = 25$ is the largest sample mean in Table 8.14, which implies that it is reasonable to believe that gasoline type E and additive type R maximize the mean mileage obtained by the Lance. Supposing that North American Oil has been producing gasoline type D with additive type S, we assume that the company wishes to estimate $\mu_{ER} - \mu_{DS}$. Since

$$\begin{aligned}
\mu_{ER} - \mu_{DS} &= (\beta_{DQ} + \beta_{ER}) - (\beta_{DQ} + \beta_{DS}) \\
&= \beta_{ER} - \beta_{DS} \\
&= 0 \cdot \beta_{DQ} + 0 \cdot \beta_{DR} + (-1) \cdot \beta_{DS} + 0 \cdot \beta_{EQ} + 1 \cdot \beta_{ER} \\
&\quad + 0 \cdot \beta_{ES} + 0 \cdot \beta_{ET} + 0 \cdot \beta_{FR} + 0 \cdot \beta_{FS} + 0 \cdot \beta_{FT} \\
&= [0 \quad 0 \quad -1 \quad 0 \quad 1 \quad 0 \quad 0 \quad 0 \quad 0 \quad 0] \begin{bmatrix} \beta_{DQ} \\ \beta_{DR} \\ \beta_{DS} \\ \beta_{EQ} \\ \beta_{ER} \\ \beta_{ES} \\ \beta_{ET} \\ \beta_{FR} \\ \beta_{FS} \\ \beta_{FT} \end{bmatrix} \\
&= \boldsymbol{\lambda}'\boldsymbol{\beta}
\end{aligned}$$

where $\boldsymbol{\lambda}' = [0 \quad 0 \quad -1 \quad 0 \quad 1 \quad 0 \quad 0 \quad 0 \quad 0 \quad 0]$, it follows that a point estimate of, and an individual 95% confidence interval for,

$$\beta_{ER} - \beta_{DS} = \mu_{ER} - \mu_{DS}$$

are

$$b_{ER} - b_{DS} = 7 - 4.9 = 2.1$$

and

$$\begin{aligned}
[\boldsymbol{\lambda}'\mathbf{b} \pm t_{[.025]}^{(22-10)} s_{\text{means}} \sqrt{\boldsymbol{\lambda}'(\mathbf{X}'\mathbf{X})^{-1}\boldsymbol{\lambda}}] &= [2.1 \pm 2.179(.7118)\sqrt{.83333}] \\
&= [2.1 \pm 1.4159] \\
&= [.6841, 3.5159]
\end{aligned}$$

This interval says that North American Oil can be 95% confident that μ_{ER}, the mean Lance mileage obtained by using gasoline type E and additive type R, is between .6841 mpg and 3.5159 mpg greater than μ_{DS}, the mean Lance mileage obtained by using gasoline type D and additive type S.

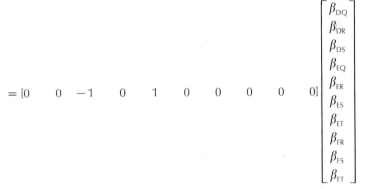

Part 6. *Using the Model to Estimate and Predict.*

Finally, we can describe a future Lance mileage that will be observed when using gasoline type E and additive type R by the equation

$$\begin{aligned} y_{ER,0} &= \mu_{ER} + \epsilon_{ER,0} \\ &= \beta_{DQ} + \beta_{DR}D_{ij,DR} + \beta_{DS}D_{ij,DS} + \beta_{EQ}D_{ij,EQ} + \beta_{ER}D_{ij,ER} + \beta_{ES}D_{ij,ES} \\ &\quad + \beta_{ET}D_{ij,ET} + \beta_{FR}D_{ij,FR} + \beta_{FS}D_{ij,FS} + \beta_{FT}D_{ij,FT} + \epsilon_{ER,0} \\ &= \beta_{DQ} + \beta_{DR}(0) + \beta_{DS}(0) + \beta_{EQ}(0) + \beta_{ER}(1) + \beta_{ES}(0) \\ &\quad + \beta_{ET}(0) + \beta_{FR}(0) + \beta_{FS}(0) + \beta_{FT}(0) + \epsilon_{ER,0} \\ &= \beta_{DQ} + \beta_{ER} + \epsilon_{ER,0} \end{aligned}$$

Therefore,

$$\begin{aligned} \hat{y}_{ER} &= b_{DQ} + b_{ER} \\ &= 18.0 + 7.0 \\ &= 25.0 \end{aligned}$$

is the point estimate of μ_{ER} and the point prediction of $y_{ER,0}$. Moreover, the equation describing $y_{ER,0}$ implies that

$$\mathbf{x}'_{ER} = [1 \quad 0 \quad 0 \quad 0 \quad 1 \quad 0 \quad 0 \quad 0 \quad 0 \quad 0]$$

Therefore, a 95% confidence interval for μ_{ER} is

$$\begin{aligned} [\hat{y}_{ER} \pm t_{[.025]}^{(22-10)} s_{\text{means}} \sqrt{\mathbf{x}'_{ER}(\mathbf{X}'\mathbf{X})^{-1}\mathbf{x}_{ER}}] &= \left[25.0 \pm 2.179(.7118)\sqrt{\frac{1}{3}}\right] \\ &= [25.0 \pm .8955] \\ &= [24.1045, 25.8955] \end{aligned}$$

This interval says that North American Oil can be 95% confident that μ_{ER}, the mean Lance mileage obtained by using gasoline type E and additive type R, is at least 24.1045 mpg and no more than 25.8955 mpg. Also, a 95% prediction interval for $y_{ER,0}$ is

$$\begin{aligned} [\hat{y}_{ER} \pm t_{[.025]}^{(22-10)} s_{\text{means}} \sqrt{1 + \mathbf{x}'_{ER}(\mathbf{X}'\mathbf{X})^{-1}\mathbf{x}_{ER}}] &= \left[25.0 \pm 2.179(.7118)\sqrt{1 + \frac{1}{3}}\right] \\ &= [25.0 \pm 1.7910] \\ &= [23.209, 26.791] \end{aligned}$$

This interval says that the owner of a Lance can be 95% confident that $y_{ER,0}$, the gasoline mileage that will be obtained by the Lance when driven under normal driving conditions using gasoline type E and additive type R, will be at least 23.209 mpg and no more than 26.791 mpg.

We should note that in the preceding example

$$\hat{y}_{ER} = b_{DQ} + b_{ER} \\ = 25.0$$

the least squares point estimate of $\mu_{ER} = \beta_{DQ} + \beta_{ER}$ that is obtained by using the means model, is equal to $\bar{y}_{ER} = 25.0$, the mean of the sample of $n_{ER} = 3$ Lance mileages randomly selected from the (E, R)th population of Lance mileages (see Table 8.14). This is because it can be proved that, in general, *the least squares point estimate of the population mean μ_{ij} is the sample mean $\bar{y}_{ij}$ when we use the means model* (that is, a regression model using a separate dummy variable to describe each of the means μ_{ij}) *to analyze a factorial experiment* (see Example 8.8). In contrast, recall that in Example 8.7

$$\hat{y}_{BO} = \hat{\mu} + \hat{\alpha}_B + \hat{\gamma}_O \\ = 26.9014$$

the least squares point estimate of $\mu_{BO} = \mu + \alpha_B + \gamma_O$, does not equal $\bar{y}_{BO} = 27.0$, the mean of the sample of $n_{BO} = 3$ Fire-Hawk mileages randomly selected from the (B, O)th population of Fire-Hawk mileages (see Table 8.12). This difference is not due to rounding error but rather to the fact that the *least squares point estimate of μ_{ij} usually is not $\bar{y}_{ij}$ when we use the (α, γ) model to analyze a factorial experiment.*

We next discuss why we suggest using the means model to describe data in which substantial interaction exists between the qualitative factors under study. In order to model the effect of interaction between two qualitative factors, one might be tempted to multiply together the appropriate dummy variables in the (α, γ) model. For example, for the Lance mileage data in Table 8.14, this procedure would yield the model

$$\begin{aligned} y_{ij,k} &= \mu_{ij} + \epsilon_{ij,k} \\ &= \mu + \alpha_E D_{i,E} + \alpha_F D_{i,F} + \gamma_R D_{j,R} + \gamma_S D_{j,S} + \gamma_T D_{j,T} + \delta_{ER} D_{i,E} D_{j,R} \\ &\quad + \delta_{ES} D_{i,E} D_{j,S} + \delta_{ET} D_{i,E} D_{j,T} + \delta_{FR} D_{i,F} D_{j,R} + \delta_{FS} D_{i,F} D_{j,S} + \delta_{FT} D_{i,F} D_{j,T} + \epsilon_{ij,k} \end{aligned}$$

In general, if we have carried out a *complete* factorial experiment, we can obtain least squares point estimates of the parameters in the interaction model resulting from multiplying together the appropriate dummy variables in the (α, γ) model, and the unexplained variation (and other results) obtained from the interaction model will be the same as for the means model. However, if we have carried out an *incomplete* factorial experiment, we might not be able to obtain the least squares point estimates of the parameters in the interaction model (because $(\mathbf{X}'\mathbf{X})^{-1}$ might not exist for this model). Therefore, since we are able to obtain the least squares point estimates of the parameters in the means model for both complete and incomplete factorial experiments, we recommend using the means model to estimate and compare the means when substantial interaction exists between the two qualitative factors under study.

We have shown that one way to detect interaction between two qualitative factors is by graphical analysis (see Figures 8.5 and 8.7). Another way to detect interaction is by using the F-test described in the following box.

TESTING FOR INTERACTION BETWEEN TWO QUALITATIVE FACTORS:

Assume the inference assumptions are satisfied, and consider testing

H_0: No interaction exists between two qualitative factors

versus

H_1: Interaction does exist between two qualitative factors

and recall that SSE_{means} and $SSE_{(\alpha,\gamma)}$ denote the unexplained variations for, respectively, the means model and the (α, γ) model. Define n_{means} and $n_{(\alpha,\gamma)}$ to be the number of parameters in, respectively, the means model and the (α, γ) model, and define the F(interaction) statistic and the prob-value to be

$$F(\text{interaction}) = \frac{MS_{\text{interaction}}}{MSE_{\text{means}}}$$

where

$$MS_{\text{interaction}} = \frac{SS_{\text{interaction}}}{n_{\text{means}} - n_{(\alpha,\gamma)}}$$

$$SS_{\text{interaction}} = SSE_{(\alpha,\gamma)} - SSE_{\text{means}}$$

$$MSE_{\text{means}} = \frac{SSE_{\text{means}}}{n - n_{\text{means}}}$$

and

$$\text{prob-value} = A[F(\text{interaction}); n_{\text{means}} - n_{(\alpha,\gamma)}, n - n_{\text{means}}]$$

Then, we can reject H_0 in favor of H_1 by setting the probability of a Type I error equal to α if and only if either of the following equivalent conditions hold:

$$F(\text{interaction}) > F_{[\alpha]}^{(n_{\text{means}} - n_{(\alpha,\gamma)}, n - n_{\text{means}})}$$

or

$$\text{prob-value} < \alpha$$

The first condition, which says that we should reject H_0 in favor of H_1 if F(interaction) is large, is intuitively reasonable, because a large value of F(interaction) would be caused by a large value of

$$MS_{\text{interaction}} = \frac{SS_{\text{interaction}}}{n_{\text{means}} - n_{(\alpha,\gamma)}} = \frac{\{SSE_{(\alpha,\gamma)} - SSE_{\text{means}}\}}{n_{\text{means}} - n_{(\alpha,\gamma)}}$$

which would be caused by a large value of

$$SS_{\text{interaction}} = SSE_{(\alpha,\gamma)} - SSE_{\text{means}}$$

The latter would occur if the means model, which uses a separate dummy variable to describe each of the means μ_{ij}, yields an unexplained variation (SSE_{means}) much smaller than the unexplained variation ($SSE_{(\alpha,\gamma)}$) yielded by the (α, γ) model, which does not describe the effect of interaction. However, it can be shown that although the result in the preceding box does provide an exact test for interaction for complete factorial experiments, it does not provide an exact test for interaction for certain types of incomplete factorial experiments (see Searle (1971)). In spite of this, it is common practice to use this test in both complete and incomplete factorial experiments.

EXAMPLE 8.10

Considering Table 8.14, for i = D, E, and F and j = Q, R, S, and T, the means model and the (α, γ) model describing $y_{ij,k}$, the kth Lance mileage observed when using (leaded) gasoline type i and additive type j, are

$$\begin{aligned}\text{Means model:}\quad y_{ij,k} &= \mu_{ij} + \epsilon_{ij,k}\\ &= \beta_{DQ} + \beta_{DR}D_{ij,DR} + \beta_{DS}D_{ij,DS}\\ &\quad + \beta_{EQ}D_{ij,EQ} + \beta_{ER}D_{ij,ER} + \beta_{ES}D_{ij,ES} + \beta_{ET}D_{ij,ET}\\ &\quad + \beta_{FR}D_{ij,FR} + \beta_{FS}D_{ij,FS} + \beta_{FT}D_{ij,FT} + \epsilon_{ij,k}\end{aligned}$$

which has $n_{\text{means}} = 10$ parameters;

$$\begin{aligned}(\alpha, \gamma)\text{ model:}\quad y_{ij,k} &= \mu_{ij} + \epsilon_{ij,k}\\ &= \mu + \alpha_i + \gamma_j + \epsilon_{ij,k}\\ &= \mu + \alpha_E D_{i,E} + \alpha_F D_{i,F} + \gamma_R D_{j,R} + \gamma_S D_{j,S} + \gamma_T D_{j,T} + \epsilon_{ij,k}\end{aligned}$$

which has $n_{(\alpha,\gamma)} = 6$ parameters. When we use these models to perform regression analyses of the Lance mileage data in Table 8.14, we obtain unexplained variations of

$$SSE_{\text{means}} = 6.08 \qquad \text{and} \qquad SSE_{(\alpha,\gamma)} = 22.4096$$

In order to test

H_0: No interaction exists between gasoline type and additive type

versus

H_1: Interaction does exist between gasoline type and additive type

we use the following F(interaction) statistic and prob-value:

$$F(\text{interaction}) = \frac{MS_{\text{interaction}}}{MSE_{\text{means}}} = \frac{4.0824}{.5067} = 8.0568$$

since

$$\begin{aligned} MS_{\text{interaction}} &= \frac{SS_{\text{interaction}}}{n_{\text{means}} - n_{(\alpha,\gamma)}} \\ &= \frac{\{SSE_{(\alpha,\gamma)} - SSE_{\text{means}}\}}{n_{\text{means}} - n_{(\alpha,\gamma)}} \\ &= \frac{22.4096 - 6.08}{10 - 6} \\ &= \frac{16.3296}{4} \\ &= 4.0824 \end{aligned}$$

$$\begin{aligned} MSE_{\text{means}} &= \frac{SSE_{\text{means}}}{n - n_{\text{means}}} \\ &= \frac{6.08}{22 - 10} \\ &= \frac{6.08}{12} \\ &= .5067 \end{aligned}$$

and

$$\begin{aligned} \text{prob-value} &= A[F(\text{interaction}); n_{\text{means}} - n_{(\alpha,\gamma)}, n - n_{\text{means}}] \\ &= A[8.0568; 4, 12] \\ &= .0021 \end{aligned}$$

Since prob-value = .0021 is less than .05 and .01, it follows by the second condition in the preceding box that we can reject H_0 in favor of H_1 by setting α equal to .05 or .01. This fact and the graphical analysis of Figure 8.7 indicate that substantial interaction exists between (leaded) gasoline type and additive type. Recall, however, that Figure 8.7 also indicates that if we eliminate all sample information concerning gasoline type D, then little interaction remains between gasoline type and additive type. Thus, it would probably be appropriate to use the (α, γ) model (which assumes no interaction) to analyze the data in Table 8.14 concerning gasoline types E and F and additive types Q, R, S, and T. If, however, we wish to include gasoline type D in our study, we should use the means model to analyze these data, as we have done in Example 8.9.

*8.5.3 Balanced Data: Two-Way Analysis of Variance

Suppose we wish to assess the effects of two qualitative factors—factor 1 and factor 2—on a dependent variable. We assume that factor 1 has a levels and factor 2 has b levels. For $i = 1, 2, \ldots, a$ and $j = 1, 2, \ldots, b$, we define *the* (i, j)*th population of element values* to be the infinite population of all possible values of the dependent variable that could be observed when using the ith level of factor 1 and the jth level of factor 2, and we let μ_{ij} denote the mean of this population. In order to estimate and compare the appropriate means (we call μ_{ij} the *mean value of the dependent variable that would be obtained by using level i of factor 1 and level j of factor 2*), we assume that for $i = 1, 2, \ldots, a$ and $j = 1, 2, \ldots, b$,

$$y_{ij,1}, y_{ij,2}, \ldots, y_{ij,m}$$

is a sample of m element values randomly selected from the (i, j)th population of element values. For $k = 1, 2, \ldots, m$, $y_{ij,k}$ denotes the kth such randomly selected element value. It follows that for $i = 1, 2, \ldots, a$ and $j = 1, 2, \ldots, b$,

1. $\bar{y}_{ij} = \dfrac{\sum_{k=1}^{m} y_{ij,k}}{m}$ is the mean of the sample of m element values randomly selected from the (i, j)th population of element values.

2. $\bar{y}_{i\cdot} = \dfrac{\sum_{j=1}^{b}\sum_{k=1}^{m} y_{ij,k}}{bm}$ is the mean of the bm element values observed when using the ith level of factor 1.

3. $\bar{y}_{\cdot j} = \dfrac{\sum_{i=1}^{a}\sum_{k=1}^{m} y_{ij,k}}{am}$ is the mean of the am element values observed when using the jth level of factor 2.

4. $\bar{y} = \dfrac{\sum_{i=1}^{a}\sum_{j=1}^{b}\sum_{k=1}^{m} y_{ij,k}}{abm}$ is the mean of the total of abm element values that we have observed in the experiment.

TABLE 8.15 Analysis of Variance Table Summarizing the *F* Statistics Used to Analyze a Balanced Complete Factorial Experiment

Source	df	Sum of Squares
Means	$ab-1$	$SS_{\text{means}} = m\sum_{i=1}^{a}\sum_{j=1}^{b}(\bar{y}_{ij}-\bar{y})^2$
Factor 1	$a-1$	$SS_{\alpha} = bm\sum_{i=1}^{a}(\bar{y}_{i\cdot}-\bar{y})^2$
Factor 2	$b-1$	$SS_{\gamma} = am\sum_{j=1}^{b}(\bar{y}_{\cdot j}-\bar{y})^2$
Interaction	$(a-1)(b-1)$	$SS_{\text{interaction}} = SS_{\text{means}} - SS_{\alpha} - SS_{\gamma}$
Error (means)	$ab(m-1)$	$SSE_{\text{means}} = \sum_{i=1}^{a}\sum_{j=1}^{b}\sum_{k=1}^{m}(y_{ij,k}-\bar{y}_{ij})^2$
Error (α, γ)	$abm-(a+b-1)$	$SSE_{(\alpha,\gamma)} = SS_{\text{interaction}} + SSE_{\text{means}}$
Error (α)	$a(bm-1)$	$SSE_{\alpha} = SS_{\gamma} + SS_{\text{interaction}} + SSE_{\text{means}}$
Error (γ)	$b(am-1)$	$SSE_{\gamma} = SS_{\alpha} + SS_{\text{interaction}} + SSE_{\text{means}}$

Now, since we assume that the sizes of the samples randomly selected from the *ab* populations of element values are the same (that is, equal to *m*), we assume that we have carried out a balanced complete factorial experiment. Therefore, it can be shown that we can calculate the sums of squares and *F* statistics needed to analyze this experiment. The procedure we recommend (but not the only procedure) for carrying out such calculations is summarized in Table 8.15. Note that besides the *F*

Mean Square	F Statistic	Prob-Value
$MS_{\text{means}} = \dfrac{SS_{\text{means}}}{ab - 1}$	$F(\text{means model}) = \dfrac{MS_{\text{means}}}{MSE_{\text{means}}}$	$A[F(\text{means model}); ab - 1, ab(m - 1)]$
$MS_{\alpha} = \dfrac{SS_{\alpha}}{a - 1}$	$F(\alpha \mid \gamma) = \dfrac{MS_{\alpha}}{MSE_{(\alpha,\gamma)}}$	$A[F(\alpha \mid \gamma); a - 1, abm - (a + b - 1)]$
	$F(\alpha) = \dfrac{MS_{\alpha}}{MSE_{\alpha}}$	$A[F(\alpha); a - 1, a(bm - 1)]$
	$F_{BCF}(\alpha) = \dfrac{MS_{\alpha}}{MSE_{\text{means}}}$	$A[F_{BCF}(\alpha); a - 1, ab(m - 1)]$
$MS_{\gamma} = \dfrac{SS_{\gamma}}{b - 1}$	$F(\gamma \mid \alpha) = \dfrac{MS_{\gamma}}{MSE_{(\alpha,\gamma)}}$	$A[F(\gamma \mid \alpha); b - 1, abm - (a + b - 1)]$
	$F(\gamma) = \dfrac{MS_{\gamma}}{MSE_{\gamma}}$	$A[F(\gamma); b - 1, b(am - 1)]$
	$F_{BCF}(\gamma) = \dfrac{MS_{\gamma}}{MSE_{\text{means}}}$	$A[F_{BCF}(\gamma); b - 1, ab(m - 1)]$
$MS_{\text{interaction}} = \dfrac{SS_{\text{interaction}}}{(a - 1)(b - 1)}$	$F(\text{interaction}) = \dfrac{MS_{\text{interaction}}}{MSE_{\text{means}}}$	$A[F(\text{interaction}); (a - 1)(b - 1), ab(m - 1)]$
$MSE_{\text{means}} = \dfrac{SSE_{\text{means}}}{ab(m - 1)}$		
$MSE_{(\alpha,\gamma)} = \dfrac{SSE_{(\alpha,\gamma)}}{abm - (a + b - 1)}$		
$MSE_{\alpha} = \dfrac{SSE_{\alpha}}{a(bm - 1)}$		
$MSE_{\gamma} = \dfrac{SSE_{\gamma}}{b(am - 1)}$		

statistics we have discussed in previous sections, Table 8.15 presents two additional F statistics:

$$F_{BCF}(\alpha) = \frac{MS_{\alpha}}{MSE_{\text{means}}} \quad \text{and} \quad F_{BCF}(\gamma) = \frac{MS_{\gamma}}{MSE_{\text{means}}}$$

These F statistics (and their related prob-values), which are not (in general) valid for unbalanced complete factorial experiments or for incomplete factorial experiments, are valid (assuming the inference assumptions hold) and generally recommended for use in analyzing a balanced complete factorial experiment (*BCF*). In addition to using these two *F* statistics, we recommend using the $F(\alpha|\gamma)$, $F(\alpha)$, $F(\gamma|\alpha)$, and $F(\gamma)$ statistics to investigate the importance of factors 1 and 2 in a balanced complete factorial experiment.

EXAMPLE 8.11

The Um-Good Bakery Company supplies Um-Good Coffee Cake to a large number of supermarkets in a metropolitan area and wishes to study the effects of two qualitative factors—*shelf display height*, which has levels B(Bottom), M(Middle), and T(Top), and *shelf display width*, which has levels R(Regular) and W(Wide)—on monthly demand for Um-Good Coffee Cake. For accounting purposes, Um-Good Bakery defines a month to be a period of exactly four weeks (which implies that the year is divided into 13 such periods) and measures demand in cases of 10 coffee cakes each. For i = B, M, and T and j = R and W, we define *the* (i, j)*th population of monthly coffee cake demands* to be the infinite population of all possible monthly coffee cake demands that could be obtained at supermarkets which use display height i and display width j. We let μ_{ij} denote the mean of this population. In order to compare the appropriate

TABLE 8.16 Six Samples of Monthly Coffee Cake Demands

Display Height (j)	Display Width (i): R	W	$\bar{y}_{i\cdot}$
B	$y_{BR,1} = 58.2$	$y_{BW,1} = 55.7$	
	$y_{BR,2} = 53.7$	$y_{BW,2} = 52.5$	
	$y_{BR,3} = 55.8$	$y_{BW,3} = 58.9$	
	$\bar{y}_{BR} = 55.9$	$\bar{y}_{BW} = 55.7$	$\bar{y}_{B\cdot} = 55.8$
M	$y_{MR,1} = 73.0$	$y_{MW,1} = 76.2$	
	$y_{MR,2} = 78.1$	$y_{MW,2} = 78.4$	
	$y_{MR,3} = 75.4$	$y_{MW,3} = 82.1$	
	$\bar{y}_{MR} = 75.5$	$\bar{y}_{MW} = 78.9$	$\bar{y}_{M\cdot} = 77.2$
T	$y_{TR,1} = 52.4$	$y_{TW,1} = 54.0$	
	$y_{TR,2} = 49.7$	$y_{TW,2} = 52.1$	
	$y_{TR,3} = 50.9$	$y_{TW,3} = 49.9$	
	$\bar{y}_{TR} = 51.0$	$\bar{y}_{TW} = 52.0$	$\bar{y}_{T\cdot} = 51.5$
$\bar{y}_{\cdot j}$	$\bar{y}_{\cdot R} = 60.8$	$\bar{y}_{\cdot W} = 62.2$	$\bar{y} = 61.5$

means (where we call μ_{ij} the *mean monthly coffee cake demand that would be obtained by using display height i and display width j*), Um-Good Bakery will employ a balanced complete factorial experiment. Specifically, for i = B, M, and T and j = R and W, Um-Good Bakery will randomly select a sample of $m = 3$ metropolitan area supermarkets. Each supermarket will sell Um-Good Coffee Cake for one month using display height i and display width j. Then, letting $OSP_{ij,k}$ denote the kth randomly selected supermarket (where $k = 1, 2, \ldots, m$) that uses display height i and display width j, Um-Good Bakery will measure

$y_{ij,k}$ = the monthly coffee cake demand (measured in cases) that occurs in $OSP_{ij,k}$

which we call *the* k*th monthly coffee cake demand observed when using display height i and display width j*. When the bakery has employed this completely randomized design, we assume that for i = B, M, and T and j = R and W,

$$y_{ij,1}, y_{ij,2}, y_{ij,3}$$

is a sample of $m = 3$ monthly coffee cake demands that Um-Good Bakery has randomly selected from the (i, j)th population of monthly coffee cake demands.

Suppose that when Um-Good Bakery has employed this completely randomized design, it has randomly selected the six samples summarized in Table 8.16. To analyze these data, we construct an analysis of variance table. We first calculate the sums of squares in the order that they appear in Table 8.15 (there are $a = 3$ levels of display height and $b = 2$ levels of display width).

$$\begin{aligned}
SS_{\text{means}} &= m \sum_{i=(B,M,T)} \sum_{j=(R,W)} (\bar{y}_{ij} - \bar{y})^2 \\
&= 3[(\bar{y}_{BR} - \bar{y})^2 + (\bar{y}_{BW} - \bar{y})^2 + (\bar{y}_{MR} - \bar{y})^2 \\
&\quad + (\bar{y}_{MW} - \bar{y})^2 + (\bar{y}_{TR} - \bar{y})^2 + (\bar{y}_{TW} - \bar{y})^2] \\
&= 3[(55.9 - 61.5)^2 + (55.7 - 61.5)^2 + (75.5 - 61.5)^2 \\
&\quad + (78.9 - 61.5)^2 + (51.0 - 61.5)^2 + (52.0 - 61.5)^2] \\
&= 2292.78
\end{aligned}$$

$$\begin{aligned}
SS_{\alpha} &= bm \sum_{i=(B,M,T)} (\bar{y}_{i\cdot} - \bar{y})^2 \\
&= 2 \cdot 3[(\bar{y}_{B\cdot} - \bar{y})^2 + (\bar{y}_{M\cdot} - \bar{y})^2 + (\bar{y}_{T\cdot} - \bar{y})^2] \\
&= 6[(55.8 - 61.5)^2 + (77.2 - 61.5)^2 + (51.5 - 61.5)^2] \\
&= 6[32.49 + 246.49 + 100] \\
&= 2273.88
\end{aligned}$$

$$\begin{aligned}
SS_{\gamma} &= am \sum_{j=(R,W)} (\bar{y}_{\cdot j} - \bar{y})^2 \\
&= 3 \cdot 3[(\bar{y}_{\cdot R} - \bar{y})^2 + (\bar{y}_{\cdot W} - \bar{y})^2] \\
&= 9[(60.8 - 61.5)^2 + (62.2 - 61.5)^2] \\
&= 9[.49 + .49] \\
&= 8.82
\end{aligned}$$

TABLE 8.17 Analysis of Variance Table Summarizing the F Statistics Used to Analyze the Shell Display Experiment

Source	df	Sum of Squares
Means	$ab - 1 = 5$	$SS_{\text{means}} = 2292.78$
Display height	$a - 1 = 2$	$SS_{\alpha} = 2273.88$
Display width	$b - 1 = 1$	$SS_{\gamma} = 8.82$
Interaction	$(a - 1)(b - 1) = 2$	$SS_{\text{interaction}} = SS_{\text{means}} - SS_{\alpha} - SS_{\gamma}$ $= 10.08$
Error (means)	$ab(m - 1) = 12$	$SSE_{\text{means}} = 73.5$
Error (α, γ)	$abm - (a + b - 1) = 14$	$SSE_{(\alpha,\gamma)} = SS_{\text{interaction}} + SSE_{\text{means}}$ $= 10.08 + 73.5 = 83.58$
Error (α)	$a(bm - 1) = 15$	$SSE_{\alpha} = SS_{\gamma} + SS_{\text{interaction}} + SSE_{\text{means}}$ $= 8.82 + 10.08 + 73.5 = 92.4$
Error (γ)	$b(am - 1) = 16$	$SSE_{\gamma} = SS_{\alpha} + SS_{\text{interaction}} + SSE_{\text{means}}$ $= 2273.88 + 10.08 + 73.5 = 2357.46$

Mean Square	F Statistic	Prob-Value
$MS_{\text{means}} = \dfrac{2292.78}{5} = 458.556$	$F(\text{means model}) = \dfrac{MS_{\text{means}}}{MSE_{\text{means}}} = \dfrac{458.556}{6.125} = 74.8663$	$A[74.8663; 5, 12] = .0001$
$MS_{\alpha} = \dfrac{2273.88}{2} = 1136.94$	$F(\alpha \mid \gamma) = \dfrac{MS_{\alpha}}{MSE_{(\alpha,\gamma)}} = \dfrac{1136.94}{5.97} = 190.4422$	$A[190.4422; 2, 14] = .0001$
	$F(\alpha) = \dfrac{MS_{\alpha}}{MSE_{\alpha}} = \dfrac{1136.94}{6.16} = 184.5682$	$A[184.5682; 2, 15] = .0001$
	$F_{BCF}(\alpha) = \dfrac{MS_{\alpha}}{MSE_{\text{means}}} = \dfrac{1136.94}{6.125} = 185.6229$	$A[185.6229; 2, 12] = .0001$
$MS_{\gamma} = \dfrac{8.82}{1} = 8.82$	$F(\gamma \mid \alpha) = \dfrac{MS_{\gamma}}{MSE_{(\alpha,\gamma)}} = \dfrac{8.82}{5.97} = 1.4774$	$A[1.4774; 1, 14] = .2443$
	$F(\gamma) = \dfrac{MS_{\gamma}}{MSE_{\gamma}} = \dfrac{8.82}{147.34125} = .0599$	$A[.0599; 1, 16] = .8098$
	$F_{BCF}(\gamma) = \dfrac{MS_{\gamma}}{MSE_{\text{means}}} = \dfrac{8.82}{6.125} = 1.44$	$A[1.44; 1, 12] = .2533$
$MS_{\text{interaction}} = \dfrac{10.08}{2} = 5.04$	$F(\text{interaction}) = \dfrac{MS_{\text{interaction}}}{MSE_{\text{means}}} = \dfrac{5.04}{6.125} = .8229$	$A[.8229; 2, 12] = .4625$
$MSE_{\text{means}} = \dfrac{73.5}{12} = 6.125$		
$MSE_{(\alpha,\gamma)} = \dfrac{83.58}{14} = 5.97$		
$MSE_{\alpha} = \dfrac{92.4}{15} = 6.16$		
$MSE_{\gamma} = \dfrac{2357.46}{16} = 147.34125$		

$$\begin{aligned} SS_{\text{interaction}} &= SS_{\text{means}} - SS_{\alpha} - SS_{\gamma} \\ &= 2292.78 - 2273.88 - 8.82 \\ &= 10.08 \end{aligned}$$

$$\begin{aligned} SSE_{\text{means}} &= \sum_{i=(B,M,T)} \sum_{j=(R,W)} \sum_{k=1}^{3} (y_{ij,k} - \bar{y}_{ij})^2 \\ &= [(58.2 - 55.9)^2 + (53.7 - 55.9)^2 + (55.8 - 55.9)^2] \\ &\quad + [(55.7 - 55.7)^2 + (52.5 - 55.7)^2 + (58.9 - 55.7)^2] \\ &\quad + [(73.0 - 75.5)^2 + (78.1 - 75.5)^2 + (75.4 - 75.5)^2] \\ &\quad + [(76.2 - 78.9)^2 + (78.4 - 78.9)^2 + (82.1 - 78.9)^2] \\ &\quad + [(52.4 - 51.0)^2 + (49.7 - 51.0)^2 + (50.9 - 51.0)^2] \\ &\quad + [(54.0 - 52.0)^2 + (52.1 - 52.0)^2 + (49.9 - 52.0)^2] \\ &= 73.5 \end{aligned}$$

Using these sums of squares, we calculate $SSE_{(\alpha,\gamma)}$, SSE_{α}, SSE_{γ}, and all the mean squares and F statistics needed to analyze the shelf display experiment as shown in Table 8.17. Examining Table 8.17, we reach the following conclusions:

1. The small prob-value related to F(means model) indicates that display height or display width or the interaction between these factors have important effects upon mean monthly coffee cake demand.
2. The small prob-values related to $F(\alpha|\gamma)$, $F(\alpha)$, and $F_{BCF}(\alpha)$ indicate that at least two of the levels of display height have different effects on mean monthly coffee cake demand. Examining Table 8.16, we see that monthly coffee cake demand was higher when a middle display height was used than when a low or top display height was used.
3. The fairly large prob-values related to $F(\gamma|\alpha)$ and $F_{BCF}(\gamma)$ and the large prob-value related to $F(\gamma)$ do not provide strong evidence indicating that the different levels of display width have different effects on mean monthly coffee cake demand. However, examination of Table 8.16 indicates that, when Um-Good Bakery used a middle display height, a wide display width yielded a sample mean, $\bar{y}_{MW} = 78.9$, which is 3.4 cases higher than the sample mean, $\bar{y}_{MR} = 75.5$, which is yielded by a regular display width.
4. The large prob-value related to F(interaction) indicates that not much interaction exists between display height and display width. However, examination of Table 8.16 indicates that when Um-Good changed from a regular to a wide display width, the sample mean of monthly coffee cake demand *increased from 75.5 to 78.9* for a middle display height; *increased from 51.0 to 52.0* for a top display height; and *decreased from 55.9 to 55.7* for a bottom display height. Thus, we might say that intuitive analysis indicates that perhaps slight interaction existed between display height and display width.

In order to investigate the exact nature of the differences between the levels of factor 1 and the differences between the levels of factor 2 in a balanced complete factorial experiment, we can use the following procedure.

CONFIDENCE INTERVALS IN A BALANCED COMPLETE FACTORIAL EXPERIMENT:

Assume we have performed a balanced complete factorial experiment, assume the inference assumptions are satisfied, and let

$$s_{\text{means}} = \sqrt{MSE_{\text{means}}}$$

where

$$MSE_{\text{means}} = \frac{SSE_{\text{means}}}{ab(m-1)}$$

$$SSE_{\text{means}} = \sum_{i=1}^{a}\sum_{j=1}^{b}\sum_{k=1}^{m}(y_{ij,k} - \bar{y}_{ij})^2$$

Then

1. A point estimate of the mean μ_{ij} is $\bar{y}_{ij}$ and a $100(1-\alpha)\%$ confidence interval for μ_{ij} is

$$\left[\bar{y}_{ij} \pm t_{[\alpha/2]}^{(ab(m-1))}\left(\frac{s_{\text{means}}}{\sqrt{m}}\right)\right]$$

2. A point prediction of the future (randomly selected) element value

$$y_{ij,0} = \mu_{ij} + \epsilon_{ij,0}$$

is $\bar{y}_{ij}$ and a $100(1-\alpha)\%$ prediction interval for that element value is

$$\left[\bar{y}_{ij} \pm t_{[\alpha/2]}^{(ab(m-1))} s_{\text{means}}\sqrt{1 + \frac{1}{m}}\right]$$

3. A point estimate of the difference $\mu_{ij} - \mu_{i'j'}$ is $\bar{y}_{ij} - \bar{y}_{i'j'}$ and a $100(1-\alpha)\%$ confidence interval for $\mu_{ij} - \mu_{i'j'}$ is

$$\left[(\bar{y}_{ij} - \bar{y}_{i'j'}) \pm t_{[\alpha/2]}^{(ab(m-1))} s_{\text{means}}\sqrt{\frac{2}{m}}\right]$$

4. Assume that little or no interaction exists between factors 1 and 2, and let $\mu_{i\cdot} - \mu_{i'\cdot}$ denote the change in the mean value of the dependent variable associated with changing from level i' of factor 1 to level i of factor 1 while the level of factor 2 remains constant. Then a point estimate of $\mu_{i\cdot} - \mu_{i'\cdot}$ is

$\bar{y}_{i\cdot} - \bar{y}_{i'\cdot}$ and a $100(1-\alpha)\%$ confidence interval for $\mu_{i\cdot} - \mu_{i'\cdot}$ is

$$\left[(\bar{y}_{i\cdot}\ \bar{y}_{i'\cdot}) \pm t_{[\alpha/2]}^{(ab(m-1))} s_{\text{means}} \sqrt{\frac{2}{bm}}\right]$$

5. Assume that little or no interaction exists between factors 1 and 2, and let $\mu_{\cdot j} - \mu_{\cdot j'}$ denote the change in the mean value of the dependent variable associated with changing from level j' of factor 2 to level j of factor 2 while the level of factor 1 remains constant. Then a point estimate of $\mu_{\cdot j} - \mu_{\cdot j'}$ is $\bar{y}_{\cdot j} - \bar{y}_{\cdot j'}$ and a $100(1-\alpha)\%$ confidence interval for $\mu_{\cdot j} - \mu_{\cdot j'}$ is

$$\left[(\bar{y}_{\cdot j} - \bar{y}_{\cdot j'}) \pm t_{[\alpha/2]}^{(ab(m-1))} s_{\text{means}} \sqrt{\frac{2}{am}}\right]$$

EXAMPLE 8.12

Consider the Um-Good Coffee Cake example, and recall from Tables 8.16 and 8.17 that $a = 3$, $b = 2$, $m = 3$, $ab(m-1) = 12$, and $MSE_{\text{means}} = 6.125$. Then

$$s_{\text{means}} = \sqrt{MSE_{\text{means}}} = \sqrt{6.125} = 2.4749$$

Since we have concluded that not much interaction exists between display height and display width, we estimate

$\mu_{M\cdot} - \mu_{B\cdot}$ = the change in mean monthly coffee cake demand associated with changing from a bottom display height to a middle display height while the display width remains constant

using the formulas in (4) of the preceding box. We find that a point estimate of, and a 95% confidence interval for, $\mu_{M\cdot} - \mu_{B\cdot}$ are

$$\bar{y}_{M\cdot} - \bar{y}_{B\cdot} = 77.2 - 55.8 = 21.4 \qquad \text{(see Table 8.16)}$$

and

$$\left[(\bar{y}_{M\cdot} - \bar{y}_{B\cdot}) \pm t_{[.025]}^{(ab(m-1))} s_{\text{means}} \sqrt{\frac{2}{bm}}\right] = \left[21.4 \pm t_{[.025]}^{(12)}(2.4749)\sqrt{\frac{2}{2\cdot 3}}\right]$$

$$= \left[21.4 \pm 2.179(2.4749)\sqrt{\frac{1}{3}}\right]$$

$$= [21.4 \pm 3.1135]$$

$$= [18.2865, 24.5135]$$

This interval says that Um-Good Bakery can be 95% confident that the effect of changing from a bottom display height to a middle display height while the display width remains constant is to increase mean monthly coffee cake demand by between 18.2865 cases and 24.5135 cases. Moreover, the reader can verify that similar confidence intervals for $\mu_{T\cdot} - \mu_{M\cdot}$ and $\mu_{T\cdot} - \mu_{B\cdot}$ indicate that there is strong evidence that a middle display height produces greater mean monthly coffee cake demand than a top display height and fairly strong evidence that a bottom display height produces greater mean monthly coffee cake demand than a top display height. We conclude, then, that a middle display height produces the greatest mean monthly coffee cake demand.

Next, consider

$\mu_{\cdot W} - \mu_{\cdot R}$ = the change in mean monthly coffee cake demand associated with changing from a regular display width to a wide display width while the display height remains constant

Using the formulas in (5) of the preceding box, we find that a point estimate of $\mu_{\cdot W} - \mu_{\cdot R}$ is

$$\begin{aligned}\bar{y}_{\cdot W} - \bar{y}_{\cdot R} &= 62.2 - 60.8 \\ &= 1.4\end{aligned}$$

The reader can verify that the 95% confidence interval for $\mu_{\cdot W} - \mu_{\cdot R}$ contains zero. Thus, we cannot reject $H_0: \mu_{\cdot W} - \mu_{\cdot R} = 0$ with at least 95% confidence. However, using the formulas in (3), we find that a point estimate of, and an 80% confidence interval for, $\mu_{MW} - \mu_{MR}$ are

$$\bar{y}_{MW} - \bar{y}_{MR} = 78.9 - 75.5 = 3.4$$

and

$$\begin{aligned}\left[(\bar{y}_{MW} - \bar{y}_{MR}) \pm t_{[.10]}^{(12)} s_{\text{means}} \sqrt{\frac{2}{m}}\right] &= \left[3.4 \pm 1.356(2.4749)\sqrt{\frac{2}{3}}\right] \\ &= [3.4 \pm 2.7401] \\ &= [.6599, 6.1401]\end{aligned}$$

This interval says that Um-Good Bakery can be 80% confident that μ_{MW}, the mean monthly coffee cake demand that would be obtained by using a middle display height and a wide display width, is between .6599 cases (about 7 coffee cakes) and 6.1401 cases (about 61 coffee cakes) greater than μ_{MR}, the mean monthly coffee cake demand that would be obtained by using a middle display height and a regular display width. Thus, we conclude that a middle display height produces the greatest mean monthly coffee cake demand and that Um-Good Bakery will use a middle display height.

We have seen that the point estimate of $\mu_{MW} - \mu_{MR}$ is $\bar{y}_{MW} - \bar{y}_{MR} = 3.4$ cases (34 coffee cakes), and the 80% confidence interval for $\mu_{MW} - \mu_{MR}$, [.6599, 6.1401], makes Um-Good Bakery 80% confident that $\mu_{MW} - \mu_{MR}$ is at least .6599 cases (about 7 coffee cakes). Besides this, suppose that from past experience Um-Good Bakery believes that, when it uses a middle display height, a wide display tends to maximize coffee cake demand. Since wide displays are only slightly more expensive than regular widths, Um-Good Bakery decides to use wide displays. Using the formulas in (1) of the preceding box, we find that a point estimate of, and a 95% confidence interval for, the mean μ_{MW} are

$$\bar{y}_{MW} = 78.9$$

and

$$\left[\bar{y}_{MW} \pm t_{[.025]}^{(12)}\left(\frac{s_{\text{means}}}{\sqrt{m}}\right)\right] = \left[78.9 \pm 2.179\left(\frac{2.4749}{\sqrt{3}}\right)\right]$$
$$= [78.9 \pm 3.1135]$$
$$= [75.7865, 82.0135]$$

This interval says that Um-Good Bakery can be 95% confident that μ_{MW}, the mean monthly coffee cake demand that would be obtained by using a middle display height and a wide display width, is at least 75.7865 cases and no more than 82.0135 cases. Next, using the formulas in (2) of the preceding box, we find that a point prediction of, and a 95% prediction interval for, a future monthly coffee cake demand

$$y_{MW,0} = \mu_{MW} + \epsilon_{MW,0}$$

that will be observed when using a middle display height and a wide display width are

$$\bar{y}_{MW} = 78.9$$

and

$$\left[\bar{y}_{MW} \pm t_{[.025]}^{(12)} s_{\text{means}} \sqrt{1 + \frac{1}{m}}\right] = \left[78.9 \pm 2.179(2.4749)\sqrt{1 + \frac{1}{3}}\right]$$
$$= [78.9 \pm 6.2271]$$
$$= [72.6729, 85.1271]$$

This interval says that Um-Good Bakery can be 95% confident that $y_{MW,0}$, the coffee cake demand that will occur when using a middle display height and a wide display, will be at least 72.6729 cases and no more than 85.1271 cases.

Before concluding this example, we should point out that if the data in Table 8.16 were observed in supermarkets where Um-Good Coffee Cake had never previously been sold, then these inferences would only pertain to supermarkets introducing the sale of Um-Good Coffee Cake. That is, once customers become aware of Um-Good Coffee Cake, the display height and display width used might have less effect on monthly coffee cake demand.

Finally, we must point out that instead of using the methods discussed in this section to analyze a balanced complete factorial experiment, we could use the regression approach discussed in Sections 8.5.1 and 8.5.2 to analyze such an experiment. If the regression approach is used, it will be found that the confidence intervals calculated will be slightly different from the confidence intervals calculated using the formulas in this section.

8.6 BASIC EXPERIMENTAL DESIGN

8.6.1 A Review of the Completely Randomized Design

As we have seen, the purpose of most experiments is to estimate and compare the effects of different treatments on a dependent variable of interest. In order to do this, the various treatments are applied to objects known as **experimental units**. When the treatments are applied to more than one experimental unit, they are said to be **replicated**. The term **randomization** refers to the manner in which experimental units are assigned to the treatments. In Sections 8.4 and 8.5 we presented examples of the *completely randomized experimental design*. In general, suppose that we wish to assign a total of n experimental units (for example, Fire-Hawks) to a total of v treatments (for example, unleaded gasoline types). If we denote the number of experimental units assigned to the lth treatment as n_l, then

$$n = n_1 + n_2 + \cdots + n_v$$

A completely randomized design can be obtained by using the following procedure: Randomly select n_1 experimental units and assign them to the first treatment. Next, randomly select n_2 *different* experimental units and assign them to the second treatment. Then, randomly select n_3 different experimental units (that is, select these units from those not assigned to either the first or second treatment) and assign them to the third treatment. Continue this procedure until the proper number of experimental units have been assigned to each treatment. As we saw in Sections 8.4 and 8.5, when we perform a completely randomized design, we assume that for $l = 1, 2, \ldots, v$, the n_l values of the dependent variable (for example, Fire-Hawk mileages) resulting from assigning the n_l experimental units to the lth treatment

$$y_{l,1}, y_{l,2}, \ldots, y_{l,n_l}$$

are a sample of n_l values of the dependent variable that we have randomly selected from the population of all such values of the dependent variable (for example, the population of all Fire-Hawk mileages that would be obtained by using gasoline type l).

When we perform a completely randomized design, we also assume, in accordance with inference assumption 3 (Independence), that the v samples of values of the dependent variable are independent of each other, which seems reasonable, since the completely randomized design ensures that the values of the dependent variable in different samples result from different measurements being taken on different experimental units.

8.6.2 The Randomized Block Design: The Regression Approach

As the reader might suspect, not all experiments employ a completely randomized design. To see why this is true, consider the following example.

EXAMPLE 8.13

The Black Box Company manufactures cardboard boxes and wishes to perform an experiment to investigate the effects of four production methods (production methods 1, 2, 3, and 4) on the number of defective boxes produced in an hour of production. In order to perform the experiment, the Black Box Company could utilize a completely randomized design, which would be carried out in the following manner. For $l = 1, 2, 3$, and 4, the company would randomly select three machine operators (the number three is chosen arbitrarily) from the pool of all machine operators that it employs, train each operator thoroughly to use production method l, have each operator produce boxes for one hour by using production method l, and record the number of defective boxes produced. The three operators using any one production method would be *different* from the three operators using any other production method (that is, the completely randomized design would utilize a total of 12 machine operators). However, suppose that the abilities of the machine operators employed by the Black Box Company differ substantially. These differences would tend to conceal any real differences between the production methods. For example, if the number of defective boxes obtained by using method 1 were less than the number of defective boxes obtained by using method 2, we might not be able to tell whether the lower number of defective boxes was due to the effectiveness of production method 1 or the (possibly superior) skill of the machine operators using method 1. In order to overcome this disadvantage of the completely randomized design, the Black Box Company will employ a *randomized block experimental design*. This involves randomly selecting three machine operators from the pool of all machine operators, training each operator thoroughly to use all three production methods, having each of the three operators produce boxes for one hour using each of the four production methods (the order in which each operator uses the four methods should be random), and recording the number of defective boxes produced by each operator using each method. The advantage of the randomized block design is that the numbers of defective boxes obtained using the four methods result from employing the *same* operators (a total of three machine operators), and thus any true differences in the effectivenesses of the methods would not be concealed by differences in the abilities of the operators.

In general, a **randomized block design** is an experimental design for comparing v treatments (for example, production methods) by using d blocks (for example, machine operators), where each block is used exactly once to measure the effect of each and every treatment. The advantage of the randomized block design over the completely randomized design is that by comparing the v treatments (for example, production methods) by using the d blocks, we are comparing the treatments by using the same experimental units (for example, the same machine operators), and thus any true differences in the treatments would not be concealed by differences in the experimental units.

As another example, suppose that an insurance company wished to compare the repair costs of moderately damaged automobiles at three garages. An automobile is considered moderately damaged if its repair cost is between \$700 and \$1,400, and thus there are substantial differences in damages to moderately damaged cars. Use of a randomized block design to compare the repair costs at the three garages (the treatments) would involve taking each of d (say, 10) moderately damaged cars (the blocks) to all three garages and obtaining estimates of the repair costs for each car at these garages. The advantage of the randomized block design is that the estimates of the repair costs at the three garages would be obtained on the *same* set of d cars (rather than on *different* sets of cars, as would be the case if we employed a completely randomized design). Thus any true differences in the repair cost estimates would not be concealed by differences in damages to cars.

Finally, note that in some experiments a block consists of *homogeneous* (similar) experimental units. For example, in order to compare the effects of five wheat types (the treatments) on wheat yield, an experimenter might choose four different plots of soil (the blocks) on which to make comparisons. Then, each of the five wheat types would be randomly assigned to a subplot within each of the four different plots of soil. If the experimenter carefully selected the four different plots of soil so that the five subplots within each plot were of roughly the same soil fertility, any true differences in the wheat types would not be concealed by different soil fertility conditions.

EXAMPLE 8.14

Part 1. *The Problem and Data.*

Consider again the Black Box Company problem, and suppose that the company wishes to investigate the effects of machine type, which has levels A and B, and of cardboard grade, which has levels Y and Z, on the number of defective boxes produced in an hour of production, and that production methods 1, 2, 3, and 4 are, respectively, the four combinations AY, AZ, BY, and BZ of machine type and cardboard grade.

Now, suppose that when the Black Box Company has employed the randomized block design discussed in Example 8.13, it has observed the data in

TABLE 8.18 The Number of Defective Cardboard Boxes Obtained by Machine Types A and B, Cardboard Grades Y and Z, and Machine Operators 1, 2, and 3

Production Method (ij)	Machine Operator (h) 1	2	3	$\bar{y}_{ij\cdot}$
AY	$y_{AY1} = 9$	$y_{AY2} = 10$	$y_{AY3} = 12$	$\bar{y}_{AY\cdot} = 10.3333$
AZ	$y_{AZ1} = 8$	$y_{AZ2} = 11$	$y_{AZ3} = 12$	$\bar{y}_{AZ\cdot} = 10.3333$
BY	$y_{BY1} = 3$	$y_{BY2} = 5$	$y_{BY3} = 7$	$\bar{y}_{BY\cdot} = 5.0$
BZ	$y_{BZ1} = 4$	$y_{BZ2} = 5$	$y_{BZ3} = 5$	$\bar{y}_{BZ\cdot} = 4.6667$
$\bar{y}_{\cdot\cdot h}$	$\bar{y}_{\cdot\cdot 1} = 6.0$	$\bar{y}_{\cdot\cdot 2} = 7.75$	$\bar{y}_{\cdot\cdot 3} = 9.0$	$\bar{y} = 7.5833$

Table 8.18. Here, for i = A and B, j = Y and Z, and h = 1, 2, and 3,

y_{ijh} = the number of defective boxes obtained when machine operator h produced boxes for one hour by using machine type i and cardboard grade j

This observation is assumed to have been randomly selected from *the* (i, j, h)*th population of hourly defective box rates,* which we define to be the infinite population of all the possible hourly defective box rates that could be observed when machine operator h uses machine type i and cardboard grade j. We denote the mean of this population by the symbol μ_{ijh}, and we call μ_{ijh} the *mean number of defective boxes* (that would be) *produced per hour by machine operator h using machine type i and cardboard grade j.*

Part 2. *Developing a Regression Model.*

Examining Figure 8.8, we see that graphical analysis of the data in Table 8.18 indicates that there is little interaction between production method and machine operator. Therefore, one possible model describing y_{ijh} is

$$y_{ijh} = \mu_{ijh} + \epsilon_{ijh}$$

where

$$\begin{aligned}\mu_{ijh} &= \mu + \tau_{ij} + \delta_h \\ &= \mu + \tau_{AZ}D_{ij,AZ} + \tau_{BY}D_{ij,BY} + \tau_{BZ}D_{ij,BZ} + \delta_2 D_{h,2} + \delta_3 D_{h,3}\end{aligned}$$

Here, for example,

$$D_{ij,BY} = \begin{cases} 1 & \text{if } i = \text{B and } j = \text{Y; that is, if the machine operator uses machine type B and cardboard grade Y} \\ 0 & \text{otherwise} \end{cases}$$

$$D_{h,3} = \begin{cases} 1 & \text{if } h = 3\text{; that is, if machine operator 3 produces boxes for one hour} \\ 0 & \text{otherwise} \end{cases}$$

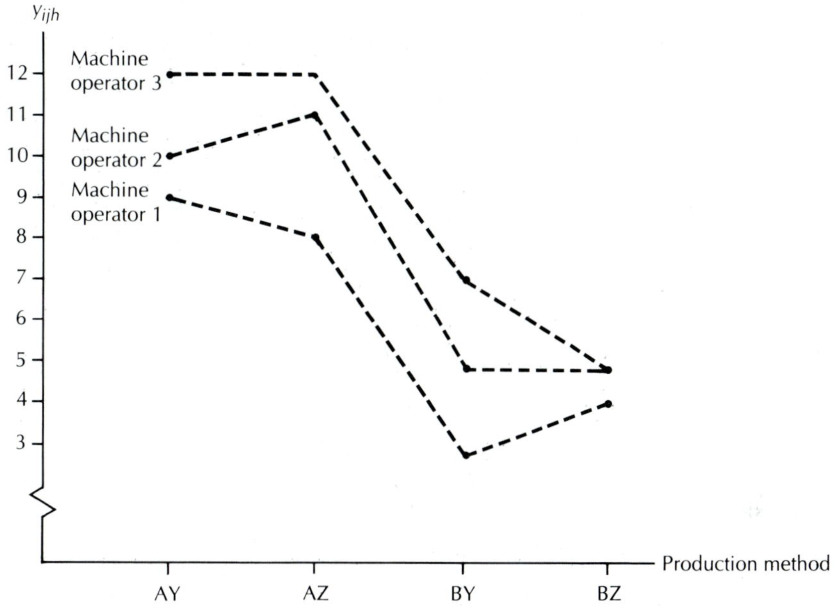

FIGURE 8.8
Graphical Analysis of the Defective Box Data in Table 8.18: Little Interaction

For each machine operator, a plot is made of the change in y_{ijh} associated with changing the production method. Little interaction means this change in the y_{ijh}'s depends little on the machine operator.

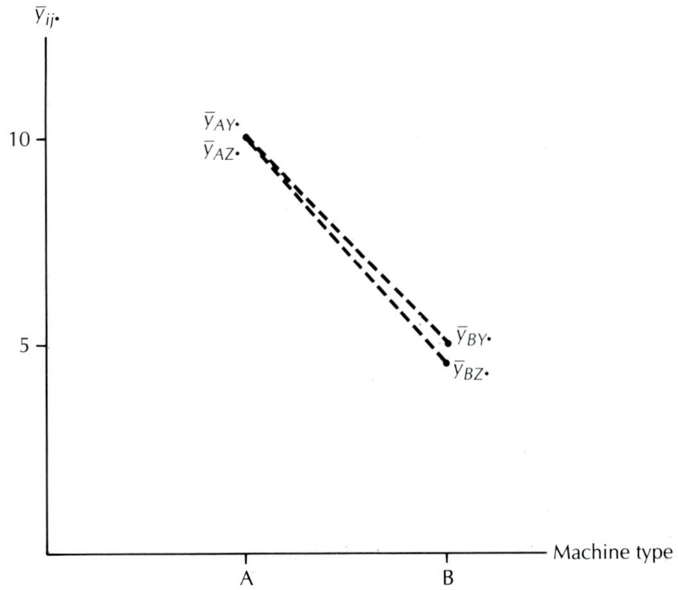

FIGURE 8.9
Graphical Analysis of the Defective Box Data in Table 8.18: Little Interaction

For each cardboard grade, a plot is made of the change in the $\bar{y}_{ij\cdot}$'s associated with changing the machine type. Little interaction means this change in the $\bar{y}_{ij\cdot}$'s depends little on the cardboard grade.

and the other dummy variables are defined similarly. We refer to this model as the (τ, δ) *model*. This model, however, is not the only model we might use to describe the data in Table 8.18. This is because, although the (τ, δ) model correctly assumes that no interaction exists between production method and machine operator (again, see Figure 8.8), the equation describing μ_{ijh} in this model uses a separate dummy variable to describe the effect of each of the combinations of machine type and cardboard grade. Therefore, the (τ, δ) model is the appropriate model to use if interaction exists between machine type and cardboard grade. One way to investigate whether such interaction does exist is to consider the graphical analysis of Figure 8.9.

Since Figure 8.9 indicates that little or no interaction exists between machine type and cardboard grade, we should express μ_{ijh} by the equation

$$\begin{aligned}\mu_{ijh} &= \mu + \alpha_i + \gamma_j + \delta_h \\ &= \mu + \alpha_B D_{i,B} + \gamma_Z D_{j,Z} + \delta_2 D_{h,2} + \delta_3 D_{h,3}\end{aligned}$$

where, for example,

$$D_{i,B} = \begin{cases} 1 & \text{if } i = \text{B; that is, if the machine operator uses machine type B} \\ 0 & \text{otherwise} \end{cases}$$

$$D_{j,Z} = \begin{cases} 1 & \text{if } j = Z\text{; that is, if the machine operator uses cardboard grade } Z \\ 0 & \text{otherwise} \end{cases}$$

$$D_{h,3} = \begin{cases} 1 & \text{if } h = 3\text{; that is, if machine operator 3 produces boxes for one hour} \\ 0 & \text{otherwise} \end{cases}$$

and the other dummy variable is defined similarly. Calling this model the (α, γ, δ) *model*, we note from the equation describing μ_{ijh} that

$$\alpha_i = \alpha_B D_{i,B}$$

which implies that

$$\begin{aligned}\alpha_A &= \alpha_B(0) \\ &= 0\end{aligned}$$

and we note that

$$\gamma_j = \gamma_Z D_{j,Z}$$

which implies that

$$\begin{aligned}\gamma_Y &= \gamma_Z(0) \\ &= 0\end{aligned}$$

and we note that

$$\delta_h = \delta_2 D_{h,2} + \delta_3 D_{h,3}$$

which implies that

$$\begin{aligned}\delta_1 &= \delta_2(0) + \delta_3(0)\\ &= 0\end{aligned}$$

Therefore, the (α, γ, δ) model implies that for $i = \text{A}$ and B, $j = \text{Y}$ and Z, and $h = 1, 2,$ and 3,

$$\begin{aligned}\mu_{\text{B}jh} - \mu_{\text{A}jh} &= [\mu + \alpha_{\text{B}} + \gamma_j + \delta_h] - [\mu + \alpha_{\text{A}} + \gamma_j + \delta_h]\\ &= \alpha_{\text{B}} \qquad (\text{since } \alpha_{\text{A}} = 0)\\ &= \text{the change in the mean number of defective boxes}\\ &\quad\ \text{produced per hour associated with a particular machine}\\ &\quad\ \text{operator changing from machine type A to machine}\\ &\quad\ \text{type B while the cardboard grade remains constant}\end{aligned}$$

$$\begin{aligned}\mu_{i\text{Z}h} - \mu_{i\text{Y}h} &= [\mu + \alpha_i + \gamma_{\text{Z}} + \delta_h] - [\mu + \alpha_i + \gamma_{\text{Y}} + \delta_h]\\ &= \gamma_{\text{Z}} \qquad (\text{since } \gamma_{\text{Y}} = 0)\\ &= \text{the change in the mean number of defective boxes}\\ &\quad\ \text{produced per hour associated with a particular}\\ &\quad\ \text{machine operator changing from cardboard grade Y}\\ &\quad\ \text{to cardboard grade Z while the machine type}\\ &\quad\ \text{remains constant}\end{aligned}$$

and for $h = 2$ and 3,

$$\begin{aligned}\mu_{ijh} - \mu_{ij1} &= [\mu + \alpha_i + \gamma_j + \delta_h] - [\mu + \alpha_i + \gamma_j + \delta_1]\\ &= \delta_h \qquad (\text{since } \delta_1 = 0)\\ &= \text{the change in the mean number of defective boxes}\\ &\quad\ \text{produced per hour associated with changing from}\\ &\quad\ \text{machine operator 1 using a particular machine type and}\\ &\quad\ \text{cardboard grade combination to machine operator } h \text{ using}\\ &\quad\ \text{the same machine type and cardboard grade combination}\end{aligned}$$

Note that since the change $(\mu_{\text{B}jh} - \mu_{\text{A}jh})$ is equal to α_{B} and thus is independent of j and of h; and since the change $(\mu_{i\text{Z}h} - \mu_{i\text{Y}h})$ is equal to γ_{Z} and thus is independent of i and h; and since the change $(\mu_{ijh} - \mu_{ij1})$ is equal to δ_h and thus is independent of i and j, it follows that the (α, γ, δ) model assumes that no interactions exist between machine type, cardboard grade, and machine operator.

Part 3. *Estimating the Regression Parameters.*
In order to use the (α, γ, δ) model

$$\begin{aligned}y_{ijh} &= \mu_{ijh} + \epsilon_{ijh}\\ &= \mu + \alpha_i + \gamma_j + \delta_h + \epsilon_{ijh}\\ &= \mu + \alpha_{\text{B}}D_{i,\text{B}} + \gamma_{\text{Z}}D_{j,\text{Z}} + \delta_2 D_{h,2} + \delta_3 D_{h,3} + \epsilon_{ijh}\end{aligned}$$

to carry out a regression analysis of the defective box data in Table 8.18, we use the following column vector **y** and matrix **X**:

$$\mathbf{y} = \begin{bmatrix} y_{AY1} \\ y_{AY2} \\ y_{AY3} \\ y_{AZ1} \\ y_{AZ2} \\ y_{AZ3} \\ y_{BY1} \\ y_{BY2} \\ y_{BY3} \\ y_{BZ1} \\ y_{BZ2} \\ y_{BZ3} \end{bmatrix} = \begin{bmatrix} 9 \\ 10 \\ 12 \\ 8 \\ 11 \\ 12 \\ 3 \\ 5 \\ 7 \\ 4 \\ 5 \\ 5 \end{bmatrix} \qquad \mathbf{X} = \begin{array}{c} \begin{array}{ccccc} 1 & D_{i,B} & D_{j,Z} & D_{h,2} & D_{h,3} \end{array} \\ \begin{bmatrix} 1 & 0 & 0 & 0 & 0 \\ 1 & 0 & 0 & 1 & 0 \\ 1 & 0 & 0 & 0 & 1 \\ 1 & 0 & 1 & 0 & 0 \\ 1 & 0 & 1 & 1 & 0 \\ 1 & 0 & 1 & 0 & 1 \\ 1 & 1 & 0 & 0 & 0 \\ 1 & 1 & 0 & 1 & 0 \\ 1 & 1 & 0 & 0 & 1 \\ 1 & 1 & 1 & 0 & 0 \\ 1 & 1 & 1 & 1 & 0 \\ 1 & 1 & 1 & 0 & 1 \end{bmatrix} \end{array}$$

When the appropriate calculations are carried out, the least squares point estimates of the parameters in the (α, γ, δ) model are found to be

$$\begin{bmatrix} \hat{\mu} \\ \hat{\alpha}_B \\ \hat{\gamma}_Z \\ \hat{\delta}_2 \\ \hat{\delta}_3 \end{bmatrix} = \mathbf{b} = (\mathbf{X}'\mathbf{X})^{-1}\mathbf{X}'\mathbf{y} = \begin{bmatrix} 8.8333 \\ -5.5 \\ -.1667 \\ 1.75 \\ 3.0 \end{bmatrix}$$

These least squares point estimates yield the prediction equation

$$\begin{aligned} \hat{y}_{ijh} &= \hat{\mu} + \hat{\alpha}_i + \hat{\gamma}_j + \hat{\delta}_h \\ &= \hat{\mu} + \hat{\alpha}_B D_{i,B} + \hat{\gamma}_Z D_{j,Z} + \hat{\delta}_2 D_{h,2} + \hat{\delta}_3 D_{h,3} \\ &= 8.8333 - 5.5D_{i,B} - .1667D_{j,Z} + 1.75D_{h,2} + 3.0D_{h,3} \end{aligned}$$

In addition, the unexplained variation is calculated to be

$$\begin{aligned} SSE_{(\alpha,\gamma,\delta)} &= \sum_{i=A,B} \sum_{j=Y,Z} \sum_{h=1}^{3} (y_{ijh} - \hat{y}_{ijh})^2 \\ &= 3.9167 \end{aligned}$$

Part 4. *Testing for Interaction Between Machine Type and Cardboard Grade.* The unexplained variation for the (τ, δ) model, which is the appropriate model to use if interaction exists between machine type and cardboard grade, can be calculated to be $SSE_{(\tau,\delta)} = 3.8333$. It follows that in order to test

H_0: No interaction exists between machine type and cardboard grade

versus

H_1: Interaction does exist between machine type and cardboard grade

we can intuitively modify the result in the box in Section 8.5.2 entitled "Testing for Interaction Between Two Qualitative Factors." To do this, we let

$n_{(\tau,\delta)} = 6 =$ the number of parameters in the (τ, δ) model

$$y_{ijh} = \mu + \tau_{AZ}D_{ij,AZ} + \tau_{BY}D_{ij,BY} + \tau_{BZ}D_{ij,BZ} + \delta_2 D_{h,2} + \delta_3 D_{h,3} + \epsilon_{ijh}$$

$n_{(\alpha,\gamma,\delta)} = 5 =$ the number of parameters in the (α, γ, δ) model

$$y_{ijh} = \mu + \tau_B D_{i,B} + \gamma_Z D_{j,Z} + \delta_2 D_{h,2} + \delta_3 D_{h,3} + \epsilon_{ijh}$$

and we define the following F(interaction) statistic and prob-value to be

$$F(\text{interaction}) = \frac{MS_{\text{interaction}}}{MSE_{(\tau,\delta)}} = \frac{.0834}{.6389} = .1305$$

since

$$\begin{aligned} MS_{\text{interaction}} &= \frac{SS_{\text{interaction}}}{n_{(\tau,\delta)} - n_{(\alpha,\gamma,\delta)}} \\ &= \frac{SSE_{(\alpha,\gamma,\delta)} - SSE_{(\tau,\delta)}}{n_{(\tau,\delta)} - n_{(\alpha,\gamma,\delta)}} \\ &= \frac{3.9167 - 3.8333}{6 - 5} \\ &= \frac{.0834}{1} \\ &= .0834 \end{aligned}$$

$$\begin{aligned} MSE_{(\tau,\delta)} &= \frac{SSE_{(\tau,\delta)}}{n - n_{(\tau,\delta)}} \\ &= \frac{3.8333}{12 - 6} \\ &= \frac{3.8333}{6} \\ &= .6389 \end{aligned}$$

and

$$\begin{aligned} \text{prob-value} &= A[F(\text{interaction}); n_{(\tau,\delta)} - n_{(\alpha,\gamma,\delta)}, n - n_{(\tau,\delta)}] \\ &= A[.1305; 1, 6] \\ &= .7303 \end{aligned}$$

Since prob-value = .7303 is greater than .10 and .05, we cannot reject H_0 in favor of H_1 by setting α equal to .10 or .05. Thus, we have little evidence of interaction between machine operator and cardboard grade.

Part 5. *Testing for Block Importance.*

Since we have concluded that little or no interaction exists between machine type and cardboard grade, we further analyze the data in Table 8.18 by using the (α, γ, δ) model

$$\begin{aligned} y_{ijh} &= \mu_{ijh} + \epsilon_{ijh} \\ &= \mu + \alpha_B D_{i,B} + \gamma_Z D_{j,Z} + \delta_2 D_{h,2} + \delta_3 D_{h,3} + \epsilon_{ijh} \end{aligned}$$

We have previously seen that for $h = 1$, 2, and 3,

$$\delta_h = \mu_{ijh} - \mu_{ij1}$$

and thus that

$$\delta_2 = \mu_{ij2} - \mu_{ij1}$$

$$\delta_3 = \mu_{ij3} - \mu_{ij1}$$

Therefore, the null hypothesis

$$H_0: \delta_2 = \delta_3 = 0 \qquad \text{or} \qquad H_0: \delta_2 = 0 \quad \text{and} \quad \delta_3 = 0$$

is equivalent to

$$H_0: \mu_{ij2} - \mu_{ij1} = 0 \qquad \text{and} \qquad \mu_{ij3} - \mu_{ij1} = 0$$

which is equivalent to

$$H_0: \mu_{ij1} = \mu_{ij2} = \mu_{ij3}$$

Thus, consider testing

$$H_0: \delta_2 = \delta_3 = 0 \qquad \text{or} \qquad \mu_{ij1} = \mu_{ij2} = \mu_{ij3}$$

which says that the three machine operators have the same effects on the mean number of defective boxes produced per hour, versus

H_1: At least one of δ_2 and δ_3
does not equal zero

or

H_1: At least two of μ_{ij1}, μ_{ij2}, and μ_{ij3} differ from each other

which says that at least two of the machine operators have different effects on the mean number of defective boxes produced per hour. We consider the (α, γ, δ) model to be the complete model (for which $SSE_c = SSE_{(\alpha,\gamma,\delta)} = 3.9167$), and we note that H_0 assumes $p - g = 2$ parameters (δ_2 and δ_3) to be zero. Thus, if H_0 is true, the complete model becomes the following reduced model:

$$y_{ijh} = \mu + \alpha_B D_{i,B} + \gamma_Z D_{j,Z} + \epsilon_{ijh}$$

for which (it can be shown) $SSE_R = 22.0834$. It follows that we can test H_0 versus H_1 by using the following partial F-statistic and prob-value:

$$\begin{aligned} F(D_{h,2}, D_{h,3} \mid D_{i,B}, D_{j,Z}) &= \frac{MS_{\text{drop}}}{MSE_C} \\ &= \frac{9.0833}{.5595} \\ &= 16.2347 \end{aligned}$$

since

$$\begin{aligned} MS_{\text{drop}} &= \frac{SS_{\text{drop}}}{p - g} \\ &= \frac{\{SSE_R - SSE_C\}}{p - g} \\ &= \frac{22.0834 - 3.9167}{2} \\ &= \frac{18.1667}{2} \\ &= 9.0833 \end{aligned}$$

$$\begin{aligned} MSE_C &= \frac{SSE_C}{n - n_{(\alpha,\gamma,\delta)}} \\ &= \frac{3.9167}{12 - 5} \\ &= \frac{3.9167}{7} \\ &= .5595 \end{aligned}$$

and

$$\begin{aligned}\text{prob-value} &= A[F(D_{h,2}, D_{h,3}|D_{i,B}, D_{j,Z}); p - g, n - n_{(\alpha,\gamma,\delta)}] \\ &= A[16.2347; 2, 7] \\ &= .0024\end{aligned}$$

Since prob-value = .0024 is less than .05 and .01, we can reject H_0 in favor of H_1 by setting α equal to .05 or .01. Thus, there is very substantial evidence that at least two of the machine operators have different effects on the mean number of defective boxes produced per hour. This indicates that blocking is important in this problem. In other problems, moreover, even if the test for block importance is inconclusive, it may still be best to use the randomized block design in future experiments if we have theoretical reasons for believing that blocking allows us to compare the treatments under more homogeneous conditions.

Part 6. *Studying Differences Between Means.*

As an exercise, the reader should use the $F(D_{i,B}|D_{j,Z}, D_{h,2}, D_{h,3})$ statistic to test the equality of the machine type effects and the $F(D_{j,Z}|D_{i,B}, D_{h,2}, D_{h,3})$ statistic to test the equality of the cardboard grade effects. Instead of doing this, we now calculate point estimates of, and 95% confidence intervals for, $\mu_{Bjh} - \mu_{Ajh}$ and $\mu_{iZh} - \mu_{iYh}$. Since we have calculated the unexplained variation for the (α, γ, δ) model

$$\begin{aligned}y_{ijh} &= \mu_{ijh} + \epsilon_{ijh} \\ &= \mu + \alpha_B D_{i,B} + \gamma_Z D_{j,Z} + \delta_2 D_{h,2} + \delta_3 D_{h,3} + \epsilon_{ijh}\end{aligned}$$

to be $SSE_{(\alpha,\gamma,\delta)} = 3.9167$, it follows that the standard error is

$$s_{(\alpha,\gamma,\delta)} = \sqrt{\frac{SSE_{(\alpha,\gamma,\delta)}}{n - n_{(\alpha,\gamma,\delta)}}} = \sqrt{\frac{3.9167}{12 - 5}} = .7480$$

Therefore, since the diagonal elements of $(\mathbf{X}'\mathbf{X})^{-1}$ corresponding to α_B and γ_Z in this model can be shown to be $c_{BB} = \frac{1}{3}$ and $c_{ZZ} = \frac{1}{3}$, we have the following results:

A point estimate of, and a 95% confidence interval for, $\mu_{Bjh} - \mu_{Ajh} = \alpha_B$ are

$$\hat{\alpha}_B = -5.5$$

and

$$\begin{aligned}[\hat{\alpha}_B \pm t_{[.025]}^{(n-n(\alpha,\gamma,\delta))} s_{(\alpha,\gamma,\delta)}\sqrt{c_{BB}}] &= \left[-5.5 \pm t_{[.025]}^{(7)}(.748)\sqrt{\frac{1}{3}}\right] \\ &= \left[-5.5 \pm 2.365(.748)\sqrt{\frac{1}{3}}\right] \\ &= [-5.5 \pm 1.0214] \\ &= [-6.5214, -4.4786]\end{aligned}$$

This interval says that the Black Box Company can be 95% confident that the effect of a particular machine operator changing from machine type A to machine type B while the cardboard grade remains constant is to decrease the mean number of defective boxes produced per hour by between 6.5214 and 4.4786 boxes. Thus, there is very strong evidence that machine type B produces a smaller mean number of defective boxes than does machine type A.

A point estimate of, and a 95% confidence interval for, $\mu_{iZh} - \mu_{iYh} = \gamma_Z$ are

$$\hat{\gamma}_Z = -.1667$$

and

$$[\hat{\gamma}_Z \pm t^{(7)}_{[.025]} s_{(\alpha,\gamma,\delta)} \sqrt{c_{ZZ}}] = \left[-.1667 \pm 2.365(.748)\sqrt{\frac{1}{3}}\right]$$
$$= [-.1667 \pm 1.0214]$$
$$= [-1.1881, .8547]$$

This interval says that the Black Box Company can be 95% confident that the effect of a particular machine operator changing from cardboard grade Y to cardboard grade Z while the machine type remains constant may at one extreme be to decrease the mean number of defective boxes produced per hour by 1.1881 boxes and may at the other extreme be to increase the mean number of defective boxes produced per hour by .8547 boxes. Thus, there is little evidence that using one cardboard grade will reduce the mean number of defective boxes more than the other cardboard grade.

Part 7. *Using the Model to Estimate and Predict.*

Since the Black Box Company has concluded that machine type B produces a smaller mean number of defective boxes than does machine type A and that changing cardboard grades has little effect on the mean number of defective boxes produced, and since cardboard grade Z is known to be less expensive and more durable, the Black Box Company decides to produce its boxes by using machine type B and cardboard grade Z. Consider, then, the number of defective boxes that will be produced in a *future* (randomly selected) hour by machine operator 2 using machine type B and cardboard grade Z (here, we arbitrarily consider machine operator 2—we could also consider machine operators 1 and 3). Since the number of defective boxes can be described by the equation

$$y_{BZ2} = \mu_{BZ2} + \epsilon_{BZ2}$$
$$= \mu + \alpha_B D_{B,B} + \gamma_Z D_{Z,Z} + \delta_2 D_{2,2} + \delta_3 D_{2,3} + \epsilon_{BZ2}$$
$$= \mu + \alpha_B(1) + \gamma_Z(1) + \delta_2(1) + \delta_3(0) + \epsilon_{BZ2}$$
$$= \mu + \alpha_B + \gamma_Z + \delta_2 + \epsilon_{BZ2}$$

it follows that

$$\begin{aligned}\hat{y}_{BZ2} &= \hat{\mu} + \hat{\alpha}_B + \hat{\gamma}_Z + \hat{\delta}_2 \\ &= 8.8333 - 5.5 - .1667 + 1.75 \\ &= 4.9166\end{aligned}$$

is the point estimate of μ_{BZ2} and is the point prediction of y_{BZ2}. Furthermore, because the equation describing y_{BZ2} implies that

$$\mathbf{x}'_{BZ2} = [1 \quad 1 \quad 1 \quad 1 \quad 0]$$

it follows that a 95% confidence interval for μ_{BZ2} is

$$\begin{aligned}[\hat{y}_{BZ2} \pm t^{(7)}_{[.025]} s_{(\alpha,\gamma,\delta)} \sqrt{\mathbf{x}'_{BZ2}(\mathbf{X}'\mathbf{X})^{-1}\mathbf{x}_{BZ2}}] &= [4.9166 \pm 2.365(.748)\sqrt{.4166}] \\ &= [4.9166 \pm 1.1418] \\ &= [3.7748, 6.0584]\end{aligned}$$

This interval says that the Black Box Company can be 95% confident that μ_{BZ2}, the mean number of defective boxes (that would be produced) per hour by machine operator 2 using machine type B and cardboard grade Z, is between 3.7748 and 6.0584 boxes.

Moreover, a 95% prediction interval for y_{BZ2} is

$$\begin{aligned}[\hat{y}_{BZ2} \pm t^{(7)}_{[.025]} s_{(\alpha,\gamma,\delta)} \sqrt{1 + \mathbf{x}'_{BZ2}(\mathbf{X}'\mathbf{X})^{-1}\mathbf{x}_{BZ2}}] &= [4.9166 \pm 2.365(.748)\sqrt{1.4166}] \\ &= [4.9166 \pm 2.1055] \\ &= [2.8111, 7.0221]\end{aligned}$$

This interval says that the Black Box Company can be 95% confident that y_{BZ2}, the number of defective boxes that will be produced in a future (randomly selected) hour by machine operator 2 using machine type B and cardboard grade Z, will be between 2.8111 and 7.0221 boxes.

*8.6.3 The Randomized Block Design: An Analysis of Variance

As previously stated, a randomized block design is an experimental design for comparing v treatments (for example, production methods) by using d blocks (for example, machine operators), where each block is used exactly once to measure the effect of each and every treatment. Suppose that when we employ a randomized block design, we observe the data in Table 8.19. Here, for $l = 1, 2, \ldots, v$, and $h = 1, 2, \ldots, d$,

y_{lh} = the value of the dependent variable observed when block h used treatment l

TABLE 8.19
Data Resulting from a Randomized Block Design

	Block				
Treatment	1	2	...	d	$\bar{y}_{l\cdot}$
1	y_{11}	y_{12}	...	y_{1d}	$\bar{y}_{1\cdot}$
2	y_{21}	y_{22}	...	y_{2d}	$\bar{y}_{2\cdot}$
⋮	⋮	⋮		⋮	⋮
v	y_{v1}	y_{v2}	...	y_{vd}	$\bar{y}_{v\cdot}$
$\bar{y}_{\cdot h}$	$\bar{y}_{\cdot 1}$	$\bar{y}_{\cdot 2}$	...	$\bar{y}_{\cdot d}$	$\bar{y}$

and is assumed to have been randomly selected from *the* (l, h)*th population of values of the dependent variable*, which we define to be the infinite population of all possible values of the dependent variable that could be observed when block h uses treatment l. We denote the mean of this population by the symbol μ_{lh}, and we call μ_{lh} the *mean value of the dependent variable that would be obtained by block h using treatment l*.

If it is reasonable to believe that little or no interaction exists between treatments and blocks, then we can analyze the randomized block experiment by using the F statistics and related prob-values presented in Table 8.20. We can check for interaction by using graphical analysis or a statistical test called *Tukey's test for additivity*—see Miller and Wichern (1977). Examining Table 8.20, we see that the $F(\tau|\delta)$ statistic and related prob-value are used to test

H_0: The v treatments have the same effects
on the mean value of the dependent variable

versus

H_1: At least two of the v treatments have different effects
on the mean value of the dependent variable

and the $F(\delta|\tau)$ statistic and related prob-value are used to test

H_0: The d blocks have the same effects
on the mean value of the dependent variable

versus

H_1: At least two of the d blocks have different effects
on the mean value of the dependent variable

In order to investigate the exact nature of any differences between the treatments and any differences between the blocks, we can use the following procedure.

TABLE 8.20 Analysis of Variance Table Summarizing the F Statistics Used to Analyze a Randomized Block Experiment

Source	df	Sum of Squares
Treatments	$v-1$	$SS_\tau = d\sum_{l=1}^{v}(\bar{y}_{l\cdot}-\bar{y})^2$
Blocks	$d-1$	$SS_\delta = v\sum_{h=1}^{d}(\bar{y}_{\cdot h}-\bar{y})^2$
Error	$(v-1)(d-1)$	$SSE_{(\tau,\delta)} = SST - SS_\tau - SS_\delta = \sum_{l=1}^{v}\sum_{h=1}^{d}(y_{lh}-\bar{y})^2 - SS_\tau - SS_\delta$

CONFIDENCE INTERVALS IN A RANDOMIZED BLOCK DESIGN:

Assume that little or no interaction exists between treatments and blocks and let

$$s_{(\tau,\delta)} = \sqrt{MSE_{(\tau,\delta)}}$$

where $MSE_{(\tau,\delta)} = SSE_{(\tau,\delta)}/(v-1)(d-1)$ and $SSE_{(\tau,\delta)} = SST - SS_\tau - SS_\delta$. Then

1. Let $\mu_{l\cdot} - \mu_{l'\cdot}$ denote the change in the mean value of the dependent variable associated with a particular block changing from treatment l' to treatment l. It follows that a point estimate of, and a $100(1-\alpha)\%$ confidence interval for, $\mu_{l\cdot} - \mu_{l'\cdot}$ are

   $$\bar{y}_{l\cdot} - \bar{y}_{l'\cdot}$$

   and

   $$\left[(\bar{y}_{l\cdot} - \bar{y}_{l'\cdot}) \pm t_{[\alpha/2]}^{(v-1)(d-1)} s_{(\tau,\delta)}\sqrt{\frac{2}{d}}\right]$$

2. Let $\mu_{\cdot h} - \mu_{\cdot h'}$ denote the change in the mean value of the dependent variable associated with changing from block h' using a particular treatment to block h using the same treatment. It follows that a point estimate of, and a $100(1-\alpha)\%$ confidence interval for, $\mu_{\cdot h} - \mu_{\cdot h'}$ are

   $$\bar{y}_{\cdot h} - \bar{y}_{\cdot h'}$$

   and

   $$\left[(\bar{y}_{\cdot h} - \bar{y}_{\cdot h'}) \pm t_{[\alpha/2]}^{(v-1)(d-1)} s_{(\tau,\delta)}\sqrt{\frac{2}{v}}\right]$$

Mean Square	F Statistic	Prob-Value
$MS_\tau = \dfrac{SS_\tau}{v-1}$	$F(\tau\|\delta) = \dfrac{MS_\tau}{MSE_{(\tau,\delta)}}$	$A[F(\tau\|\delta); v-1, (v-1)(d-1)]$
$MS_\delta = \dfrac{SS_\delta}{d-1}$	$F(\delta\|\tau) = \dfrac{MS_\delta}{MSE_{(\tau,\delta)}}$	$A[F(\delta\|\tau); d-1, (v-1)(d-1)]$
$MSE_{(\tau,\delta)} = \dfrac{SSE_{(\tau,\delta)}}{(v-1)(d-1)}$		

EXAMPLE 8.15

Consider the Black Box Company problem. If we ignore the fact that the four production methods can be described in terms of the machine type and cardboard grade used, and simply describe these methods as production methods 1, 2, 3, and 4, we can relist the data in Table 8.18 as shown in Table 8.21. We can analyze these data by the methods discussed in this section, since the graphical analysis in Figure 8.8 indicates that there is little interaction between production method and machine operator. Considering Table 8.21, we see that for $l = 1, 2, 3$, and 4, and $h = 1, 2$, and 3,

y_{lh} = the number of defective boxes obtained when machine operator h produced boxes for one hour by using production method l

TABLE 8.21 Numbers of Defective Cardboard Boxes Obtained by Production Methods 1, 2, 3, and 4 and Machine Operators 1, 2, and 3

Production Method	Machine Operator (h)			
(l)	1	2	3	$\bar{y}_{l\cdot}$
1	$y_{11} = 9$	$y_{12} = 10$	$y_{13} = 12$	$\bar{y}_{1\cdot} = 10.3333$
2	$y_{21} = 8$	$y_{22} = 11$	$y_{23} = 12$	$\bar{y}_{2\cdot} = 10.3333$
3	$y_{31} = 3$	$y_{32} = 5$	$y_{33} = 7$	$\bar{y}_{3\cdot} = 5.0$
4	$y_{41} = 4$	$y_{42} = 5$	$y_{43} = 5$	$\bar{y}_{4\cdot} = 4.6667$
$\bar{y}_{\cdot h}$	$\bar{y}_{\cdot 1} = 6.0$	$\bar{y}_{\cdot 2} = 7.75$	$\bar{y}_{\cdot 3} = 9.0$	$\bar{y} = 7.5833$

This value is assumed to have been randomly selected from *the (l, h)th population of hourly defective box rates*, which we define to be the infinite population of all the possible hourly defective box rates that could be produced when machine operator h uses production method l. We denote the mean of this population by the symbol μ_{lh}, and we call μ_{lh} the *mean number of defective boxes* (that would be) *produced per hour by machine operator h using production method l*.

To begin to analyze the data in Table 8.21, we construct an analysis of variance table. We calculate the sums of squares in the order that they are presented in Table 8.20 as follows (note that there are $v = 4$ production methods and $d = 3$ machine operators):

$$\begin{aligned} SS_\tau &= d \sum_{l=1}^{v} (\bar{y}_{l\cdot} - \bar{y})^2 \\ &= 3 \sum_{l=1}^{4} (\bar{y}_{l\cdot} - \bar{y})^2 \\ &= 3[(\bar{y}_{1\cdot} - \bar{y})^2 + (\bar{y}_{2\cdot} - \bar{y})^2 + (\bar{y}_{3\cdot} - \bar{y})^2 + (\bar{y}_{4\cdot} - \bar{y})^2] \\ &= 3[(10.3333 - 7.5833)^2 + (10.3333 - 7.5833)^2 + (5.0 - 7.5833)^2 \\ &\quad + (4.6667 - 7.5833)^2] \\ &= 90.9167 \end{aligned}$$

$$\begin{aligned} SS_\delta &= v \sum_{h=1}^{d} (\bar{y}_{\cdot h} - \bar{y})^2 \\ &= 4[(\bar{y}_{\cdot 1} - \bar{y})^2 + (\bar{y}_{\cdot 2} - \bar{y})^2 + (\bar{y}_{\cdot 3} - \bar{y})^2] \\ &= 4[(6.0 - 7.5833)^2 + (7.75 - 7.5833)^2 + (9.0 - 7.5833)^2] \\ &= 18.1667 \end{aligned}$$

$$\begin{aligned} SSE_{(\tau,\delta)} &= SST - SS_\tau - SS_\delta \\ &= \sum_{l=1}^{v} \sum_{h=1}^{d} (y_{lh} - \bar{y})^2 - SS_\tau - SS_\delta \\ &= \sum_{l=1}^{4} \sum_{h=1}^{3} (y_{lh} - 7.5833)^2 - 90.9167 - 18.1667 \\ &= 3.8333 \end{aligned}$$

Using these sums of squares, we can calculate the F statistics needed to analyze the defective box data as indicated in Table 8.22. Examining that table, we reach the following conclusions:

1. The small prob-value related to $F(\tau|\delta)$ indicates that at least two of the production methods have different effects on the mean number of defective boxes produced per hour.

TABLE 8.22 Analysis of Variance Table Summarizing the *F* Statistics Used to Analyze the Defective Box Data

Source	*df*	Sum of Squares	Mean Square	*F* Statistic	Prob-Value
Production methods	$v - 1 = 3$	$SS_\tau = 90.9167$	$MS_\tau = \frac{90.9167}{3} = 30.3056$	$F(\tau \mid \delta) = \frac{MS_\tau}{MSE_{(\tau,\delta)}} = \frac{30.3056}{.6389} = 47.4348$	$A[47.4348; 3, 6] = .00014$
Machine operators	$d - 1 = 2$	$SS_\delta = 18.1667$	$MS_\delta = \frac{18.1667}{2} = 9.08335$	$F(\delta \mid \tau) = \frac{MS_\delta}{MSE_{(\tau,\delta)}} = \frac{9.08335}{.6389} = 14.2172$	$A[14.2172; 2, 6] = .00529$
Error	$(v - 1)(d - 1) = 6$	$SSE_{(\tau,\delta)} = 3.8333$	$MSE_{(\tau,\delta)} = \frac{3.8333}{6} = .6389$		

2. The small prob-value related to $F(\delta|\tau)$ indicates that at least two of the machine operators have different effects on the mean number of defective boxes produced per hour.

In order to investigate the exact nature of the differences between the production methods and the differences between the machine operators, first note that

$$\begin{aligned} s_{(\tau,\delta)} &= \sqrt{MSE_{(\tau,\delta)}} \\ &= \sqrt{.6389} \\ &= .7993 \end{aligned}$$

Since we have concluded that little interaction exists between production method and machine operator, we estimate

$\mu_{4\cdot} - \mu_{1\cdot}$ = the change in the mean number of defective boxes produced per hour associated with a particular machine operator changing from production method 1 to production method 4

Using the formulas in (1) of the preceding box, we obtain a point estimate of, and a 95% confidence interval for, $\mu_{4\cdot} - \mu_{1\cdot}$ as follows:

$$\bar{y}_{4\cdot} - \bar{y}_{1\cdot} = 4.6667 - 10.3333 = -5.6666 \qquad \text{(see Table 8.21)}$$

$$\begin{aligned} \left[(\bar{y}_{4\cdot} - \bar{y}_{1\cdot}) \pm t_{[.025]}^{(v-1)(d-1)} s_{(\tau,\delta)} \sqrt{\frac{2}{d}}\right] &= \left[-5.6666 \pm t_{[.025]}^{(6)}(.7993)\sqrt{\frac{2}{3}}\right] \\ &= \left[-5.6666 \pm 2.447(.7993)\sqrt{\frac{2}{3}}\right] \\ &= [-5.6666 \pm 1.5970] \\ &= [-7.2636, -4.0696] \end{aligned}$$

This interval says that the Black Box Company can be 95% confident that the effect of a particular machine operator changing from production method 1 to production method 4 is to decrease the mean number of defective boxes produced per hour by between 7.2636 boxes and 4.0696 boxes.

Next, consider estimating

$\mu_{\cdot 1} - \mu_{\cdot 2}$ = the change in the mean number of defective boxes produced per hour associated with changing from machine operator 2 using a particular production method to machine operator 1 using the same production method

Using the formulas in (2) of the preceding box, we obtain a point estimate of, and a 95% confidence interval for, $\mu_{\cdot 1} - \mu_{\cdot 2}$ as follows:

$$\bar{y}_{\cdot 1} - \bar{y}_{\cdot 2} = 6 - 7.75 = -1.75 \qquad \text{(see Table 8.21)}$$

and

$$\left[\bar{y}_{\cdot 1} - \bar{y}_{\cdot 2} \pm t_{[.025]}^{(v-1)(d-1)} s_{(\tau,\delta)} \sqrt{\frac{2}{v}}\right] = \left[-1.75 \pm t_{[.025]}^{(6)}(.7993)\sqrt{\frac{2}{4}}\right]$$

$$= \left[-1.75 \pm 2.447(.7993)\sqrt{\frac{2}{4}}\right]$$

$$= [-1.75 \pm 1.3830]$$

$$= [-3.133, -.367]$$

This interval says that the Black Box Company can be 95% confident that the effect of changing from machine operator 2 using a particular production method to machine operator 1 using the same production method is to decrease the mean number of defective boxes produced per hour by between 3.133 boxes and .367 boxes.

Confidence intervals for other differences can be computed similarly.

*8.7 SOME SPECIAL METHODS FOR COMPARING TWO POPULATION MEANS

In this section (for which the material in optional Sections 3.4 and 6.12 is a prerequisite) we discuss some special methods for comparing two population means, μ_1 and μ_2, where, for $l = 1$ and 2, μ_l is the mean of *the lth population of element values*. In order to compare μ_1 and μ_2, we assume that we have randomly selected the two samples listed in Table 8.23. Specifically, we assume that for $l = 1$ and 2,

$$y_{l,1}, y_{l,2}, \ldots, y_{l,n_l}$$

is a sample of n_l element values that we have randomly selected from the lth population and that the mean, variance, and standard deviation of this sample

$$\bar{y}_l = \frac{\sum_{k=1}^{n_l} y_{l,k}}{n_l} \qquad s_l^2 = \frac{\sum_{k=1}^{n_l} (y_{l,k} - \bar{y}_l)^2}{n_l - 1} \qquad s_l = \sqrt{s_l^2}$$

TABLE 8.23 Two Samples Randomly Selected From Two Populations

Sample of n_1 Element Values From Population 1	Sample of n_2 Element Values From Population 2
$y_{1,1}, y_{1,2}, \ldots, y_{1,n_1}$ Sample mean $= \bar{y}_1$ Sample variance $= s_1^2$	$y_{2,1}, y_{2,2}, \ldots, y_{2,n_2}$ Sample mean $= \bar{y}_2$ Sample variance $= s_2^2$

TABLE 8.24 Formula Sets Based on the Normal Distribution for Making Statistical Inferences About $\mu_1 - \mu_2$

Formula Sets and Conditions for Validity	Parameter θ	Point Estimate $\hat{\theta}$	Standard Deviation (Actual or Estimated) $\sigma_{\hat{\theta}}$ or $s_{\hat{\theta}}$
Formula set 1: Holds approximately under (a) inference assumption 2 (Independent Samples) (b) n_1 large (≥ 30) n_2 large (≥ 30)	$\mu_1 - \mu_2$	$\bar{y}_1 - \bar{y}_2$	$\sigma_{\bar{y}_1-\bar{y}_2} = \sqrt{\frac{\sigma_1^2}{n_1} + \frac{\sigma_2^2}{n_2}}$ or $s_{\bar{y}_1-\bar{y}_2} = \sqrt{\frac{s_1^2}{n_1} + \frac{s_2^2}{n_2}}$
Formula set 2: Holds approximately under (a) inference assumption 2 (b) n_1 large (≥ 30) n_2 large (≥ 30)	$p_1 - p_2$	$\hat{p}_1 - \hat{p}_2$	$\sigma_{\hat{p}_1-\hat{p}_2} = \sqrt{\frac{p_1(1-p_1)}{n_1} + \frac{p_2(1-p_2)}{n_2}}$ or $s_{\hat{p}_1-\hat{p}_2} = \sqrt{\frac{\hat{p}_1(1-\hat{p}_1)}{n_1-1} + \frac{\hat{p}_2(1-\hat{p}_2)}{n_2-1}}$

are the point estimates of, respectively, μ_l, σ_l^2, and σ_l, the mean, variance, and standard deviation of the lth population. It follows that the point estimate of the parameter $\mu_1 - \mu_2$, which is the difference between the *population means* μ_1 and μ_2, is $\bar{y}_1 - \bar{y}_2$, which is the difference between the *sample means* $\bar{y}_1$ and $\bar{y}_2$.

In this section we present six formula sets that can be used to make statistical inferences about $\mu_1 - \mu_2$. These formula sets, which are valid under different sets of assumptions, are summarized in three tables.

There are three assumptions to be considered. The first assumption says that the two *populations* have equal variances (that is, $\sigma_1^2 = \sigma_2^2$). One way to check the validity of this assumption is to compare the sample variances s_1^2 and s_2^2. If these sample variances are nearly equal, then it is reasonable to believe that this "equal variances" assumption holds (or approximately holds). The second assumption is that there is no relationship between the element values of one sample and the element values of the other sample (in which case we say that we have performed an *independent samples experiment*). The last assumption states that each of the two populations is normally distributed.

$z_{[\hat{\theta},\theta]}$ Statistic	$100(1-\alpha)\%$ Confidence Intervals
$z_{[\hat{\theta},\theta]} = \dfrac{\hat{\theta}-\theta}{\sigma_{\hat{\theta}}}$ or $z_{[\hat{\theta},\theta]} = \dfrac{\hat{\theta}-\theta}{s_{\hat{\theta}}}$	$[\hat{\theta} \pm z_{[\alpha/2]}\sigma_{\hat{\theta}}]$ or $[\hat{\theta} \pm z_{[\alpha/2]}s_{\hat{\theta}}]$ $[\hat{\theta} - z_{[\alpha]}\sigma_{\hat{\theta}}, \infty)$ or $[\hat{\theta} - z_{[\alpha]}s_{\hat{\theta}}, \infty)$ $(-\infty, \hat{\theta} + z_{[\alpha]}\sigma_{\hat{\theta}}]$ or $(-\infty, \hat{\theta} + z_{[\alpha]}s_{\hat{\theta}}]$
$z_{[\bar{y}_1-\bar{y}_2,\mu_1-\mu_2]} = \dfrac{\bar{y}_1-\bar{y}_2-(\mu_1-\mu_2)}{\sigma_{\bar{y}_1-\bar{y}_2}}$ or $= \dfrac{\bar{y}_1-\bar{y}_2-(\mu_1-\mu_2)}{s_{\bar{y}_1-\bar{y}_2}}$	$[\bar{y}_1-\bar{y}_2 \pm z_{[\alpha/2]}\sigma_{\bar{y}_1-\bar{y}_2}]$ or $[\bar{y}_1-\bar{y}_2 \pm z_{[\alpha/2]}s_{\bar{y}_1-\bar{y}_2}]$ $[\bar{y}_1-\bar{y}_2 - z_{[\alpha]}\sigma_{\bar{y}_1-\bar{y}_2}, \infty)$ or $[\bar{y}_1-\bar{y}_2 - z_{[\alpha]}s_{\bar{y}_1-\bar{y}_2}, \infty)$ $(-\infty, \bar{y}_1-\bar{y}_2 + z_{[\alpha]}\sigma_{\bar{y}_1-\bar{y}_2}]$ or $(-\infty, \bar{y}_1-\bar{y}_2 + z_{[\alpha]}s_{\bar{y}_1-\bar{y}_2}]$
$z_{[\hat{p}_1-\hat{p}_2,p_1-p_2]} = \dfrac{\hat{p}_1-\hat{p}_2-(p_1-p_2)}{\sigma_{\hat{p}_1-\hat{p}_2}}$ or $= \dfrac{\hat{p}_1-\hat{p}_2-(p_1-p_2)}{s_{\hat{p}_1-\hat{p}_2}}$	$[\hat{p}_1-\hat{p}_2 \pm z_{[\alpha/2]}\sigma_{\hat{p}_1-\hat{p}_2}]$ or $[\hat{p}_1-\hat{p}_2 \pm z_{[\alpha/2]}s_{\hat{p}_1-\hat{p}_2}]$ $[\hat{p}_1-\hat{p}_2 - z_{[\alpha]}\sigma_{\hat{p}_1-\hat{p}_2}, \infty)$ or $[\hat{p}_1-\hat{p}_2 - z_{[\alpha]}s_{\hat{p}_1-\hat{p}_2}, \infty)$ $(-\infty, \hat{p}_1-\hat{p}_2 + z_{[\alpha]}\sigma_{\hat{p}_1-\hat{p}_2}]$ or $(-\infty, \hat{p}_1-\hat{p}_2 + z_{[\alpha]}s_{\hat{p}_1-\hat{p}_2}]$

8.7.1 Large-Sample Formulas

The first two formula sets utilize the normal distribution and are summarized in Table 8.24. Formula set 1 is based on the fact that under the assumptions corresponding to this formula set in Table 8.24, the population of all possible values of $\bar{y}_1 - \bar{y}_2$ (that is, point estimates of $\mu_1 - \mu_2$) approximately has a normal distribution, with mean

$$\mu_{\bar{y}_1-\bar{y}_2} = \mu_1 - \mu_2$$

and standard deviation

$$\sigma_{\bar{y}_1-\bar{y}_2} = \sqrt{\frac{\sigma_1^2}{n_1} + \frac{\sigma_2^2}{n_2}}$$

It follows, if we consider each of the $z_{[\bar{y}_1-\bar{y}_2,\mu_1-\mu_2]}$ statistics in Table 8.24, that the population of all such statistics approximately has a standard normal distribution. We now illustrate the use of formula set 1.

EXAMPLE 8.16

The management of Bargain Town, a large discount chain, wishes to evaluate the performance of its credit managers in Ohio and Illinois by comparing the mean dollar amounts owed by customers with delinquent charge accounts in these two states. Here, a small mean dollar amount owed is desirable, because such a mean would say that bad credit risks are not being extended large amounts of credit. Let μ_O, μ_I, σ_O^2, σ_I^2, σ_O, and σ_I denote the means, variances, and standard deviations of the populations of all possible dollar amounts of delinquent charge accounts in Ohio and Illinois. Suppose that Bargain Town randomly selects two independent samples from these populations in order to make the required comparison. The sample results obtained are as follows:

Sample of Ohio Accounts	Sample of Illinois Accounts
$n_O = 103$	$n_I = 103$
$\bar{y}_O = \$122$	$\bar{y}_I = \$69$
$s_O^2 = \$1521$	$s_I^2 = \$529$

Suppose that before the samples were taken, the management of Bargain Town felt that it should institute new credit policies in Ohio to reduce μ_O, the mean dollar amount of delinquent charge accounts in Ohio, if μ_O is more than \$40 greater than μ_I, the mean dollar amount of delinquent charge accounts in Illinois. Therefore, management has decided to institute such new credit policies if it can use the two samples it has observed to reject

$$H_0: \mu_O - \mu_I \leq 40$$

in favor of

$$H_1: \mu_O - \mu_I > 40$$

by setting α, the probability of a Type I error, equal to .05. Defining the $z_{[\hat{\theta},\theta]}$ statistic to be

$$z_{[\hat{\theta},\theta]} = \frac{\hat{\theta} - \theta}{s_{\hat{\theta}}} = \frac{\bar{y}_O - \bar{y}_I - (\mu_O - \mu_I)}{\sqrt{(s_O^2/n_O) + (s_I^2/n_I)}} = z_{[\bar{y}_O - \bar{y}_I, \mu_O - \mu_I]}$$

we see from formula set 1 in Table 8.24 that since the sample sizes $n_O = 103$ and $n_I = 103$ are large, the population of all possible $z_{[\bar{y}_O - \bar{y}_I, \mu_O - \mu_I]}$ statistics approximately has a standard normal distribution. Testing

$$H_0: \mu_O - \mu_I \leq 40$$

versus

$$H_1: \mu_O - \mu_I > 40$$

is equivalent to testing

$$H_0: \mu_O - \mu_I = 40$$

versus

$$H_1: \mu_O - \mu_I > 40$$

Therefore, it follows from condition (1) in Hypothesis Testing Result IV (see Section 6.12) that we can test

$$H_0: \mu_O - \mu_I = 40 \qquad (\theta = c)$$

versus

$$H_1: \mu_O - \mu_I > 40 \qquad (\theta > c)$$

by setting α equal to .05 if we calculate

$$\begin{aligned} z_{[\hat{\theta},c]} = \frac{\hat{\theta} - c}{s_{\hat{\theta}}} &= \frac{\bar{y}_O - \bar{y}_I - 40}{\sqrt{(s_O^2/n_O) + (s_I^2/n_I)}} = z_{[\bar{y}_O - \bar{y}_I, 40]} \\ &= \frac{122 - 69 - 40}{\sqrt{(1521/103) + (529/103)}} \\ &= \frac{53 - 40}{4.4613} \\ &= 2.91 \end{aligned}$$

and use the rejection point $z_{[\alpha]} = z_{[.05]} = 1.64$. Since

$$z_{[\bar{y}_O - \bar{y}_I, 40]} = 2.91 > 1.64 = z_{[.05]}$$

we can reject

$$H_0: \mu_O - \mu_I = 40$$

in favor of

$$H_1: \mu_O - \mu_I > 40$$

by setting α equal to .05.

Next, we note that

$$\begin{aligned} \text{prob-value} &= A[[z_{[\hat{\theta},c]}, \infty); N(0, 1)] \\ &= A[[z_{[\bar{y}_O - \bar{y}_I, 40]}, \infty); N(0, 1)] \\ &= A[[2.91, \infty); N(0, 1)] \\ &= .0018 \end{aligned}$$

Since this prob-value is less than .05 and less than .01, it follows by condition (2) in Hypothesis Testing Result IV that we can reject

$$H_0: \mu_O - \mu_I = 40$$

in favor of

$$H_1: \mu_O - \mu_I > 40$$

if we set α equal to .05 or .01.

In order to interpret the prob-value as a probability, note that since the population of all possible $z_{[\bar{y}_O - \bar{y}_I, \mu_O - \mu_I]}$ statistics has a standard normal distribution, it follows that if

$$H_0: \mu_O - \mu_I = 40$$

is true, then the population of all possible $z_{[\bar{y}_O - \bar{y}_I, 40]}$ statistics has a standard normal distribution. Therefore, the prob-value equals the proportion of $z_{[\bar{y}_O - \bar{y}_I, 40]}$ statistics in the population of all such statistics which, if the null hypothesis

$$H_0: \mu_O - \mu_I = 40$$

is true, are at least as large as and thus at least as contradictory to

$$H_0: \mu_O - \mu_I = 40$$

as 2.91, the observed $z_{[\bar{y}_O - \bar{y}_I, 40]}$ statistic. Since this prob-value, which equals .0018, says that Bargain Town management must believe that an 18 in 10,000 chance has occurred if it is to believe that

$$H_0: \mu_O - \mu_I = 40$$

is true, management has substantial evidence to conclude that

$$H_0: \mu_O - \mu_I = 40$$

is false and thus that

$$H_0: \mu_O - \mu_I > 40$$

is true.

Finally, we note that it follows from (4) in Hypothesis Testing Result IV that for any level of confidence, 100(1 − α)%, less than

$$\begin{aligned} 100[1 - (\text{prob-value})]\% &= 100[1 - .0018]\% \\ &= 99.82\% \end{aligned}$$

the one-sided 100(1 − α)% confidence interval for $\mu_O - \mu_I$

$$\begin{aligned} &\left[(\bar{y}_O - \bar{y}_I) - z_{[\alpha]}\sqrt{\frac{s_O^2}{n_O} + \frac{s_I^2}{n_I}}, \infty\right) \\ &\quad = \left[(122 - 69) - z_{[\alpha]}\sqrt{\frac{1521}{103} + \frac{529}{103}}, \infty\right) \\ &\quad = [53 - z_{[\alpha]}(4.4613), \infty) \end{aligned}$$

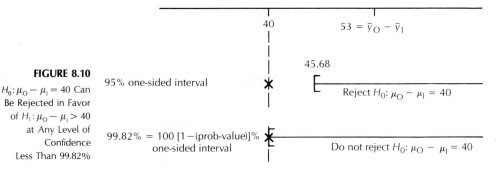

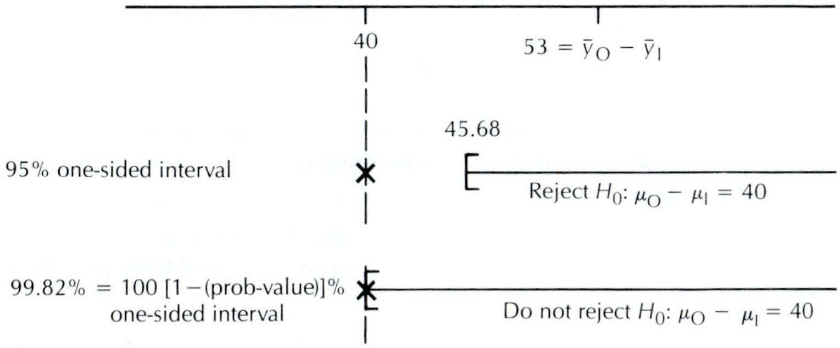

FIGURE 8.10
$H_0: \mu_O - \mu_I = 40$ Can Be Rejected in Favor of $H_1: \mu_O - \mu_I > 40$ at Any Level of Confidence Less Than 99.82%

does not contain 40 (for example, the 95% interval

$$\begin{aligned}[53 - z_{[.05]}(4.4613), \infty) &= [53 - 1.64(4.4613), \infty) \\ &= [53 - 7.32, \infty) \\ &= [45.68, \infty)\end{aligned}$$

does not contain 40—see Figure 8.10), and thus Bargain Town management *can*, with at least $100(1 - \alpha)\%$ confidence, *reject*

$$H_0: \mu_O - \mu_I = 40$$

in favor of

$$H_1: \mu_O - \mu_I > 40$$

It follows from (5) in Hypothesis Testing Result IV that for any level of confidence, $100(1 - \alpha)\%$, greater than or equal to

$$100[1 - (\text{prob-value})]\% = 99.82\%$$

the one-sided $100(1 - \alpha)\%$ confidence interval for $\mu_O - \mu_I$ *does contain 40* (for example, the 99.82% interval does contain 40—see Figure 8.10), and thus Bargain Town management *cannot*, with at least $100(1 - \alpha)\%$ confidence, *reject*

$$H_0: \mu_O - \mu_I = 40$$

in favor of

$$H_1: \mu_O - \mu_I > 40$$

To summarize, management can be up to

$$100[1 - (\text{prob-value})]\% = 99.82\%$$

confident that

$$H_0: \mu_O - \mu_I = 40$$

is false and that

$$H_1: \mu_O - \mu_I > 40$$

is true. Specifically, the 95% confidence interval for $\mu_O - \mu_I$, $[45.68, \infty)$, says that management can be 95% confident that μ_O is at least \$45.68 greater than μ_I.

To conclude this example we note that in addition to the one-sided 95% confidence interval for $\mu_O - \mu_I$, a two-sided 95% confidence interval for $\mu_O - \mu_I$ is (see Table 8.24)

$$\left[(\bar{y}_O - \bar{y}_I) \pm z_{[.025]}\sqrt{\frac{s_O^2}{n_O} + \frac{s_I^2}{n_I}}\right] = \left[(122 - 69) \pm 1.96\sqrt{\frac{1521}{103} + \frac{529}{103}}\right]$$

$$= [53 \pm 8.74]$$

$$= [44.26, 61.74]$$

This interval says that management can be 95% confident that μ_O, the mean dollar amount of delinquent charge accounts in Ohio, is between \$44.26 and \$61.74 more than μ_I, the mean dollar amount of delinquent charge accounts in Illinois.

We next consider formula set 2 in Table 8.24. Let p_1 denote the proportion of elements that have a certain characteristic in a population of elements, and let $\hat{p}_1$ denote the proportion of elements that have the characteristic in a sample of n_1 elements randomly selected from the population. Let p_2 denote the proportion of elements that have a certain characteristic in a second population of elements, and let $\hat{p}_2$ denote the proportion of elements that have the characteristic in a sample of n_2 elements randomly selected from the second population. Then, $\hat{p}_1 - \hat{p}_2$ is the point estimate of $p_1 - p_2$. Moreover, formula set 2, which can be used to make statistical inferences about $p_1 - p_2$, is based on the fact that under the assumptions corresponding to this formula set in Table 8.24, the population of all possible values of $\hat{p}_1 - \hat{p}_2$ approximately has a normal distribution, with mean

$$\mu_{\hat{p}_1 - \hat{p}_2} = p_1 - p_2$$

and standard deviation

$$\sigma_{\hat{p}_1 - \hat{p}_2} = \sqrt{\frac{p_1(1 - p_1)}{n_1} + \frac{p_2(1 - p_2)}{n_2}}$$

It follows, if we consider each of the $z_{[\hat{p}_1 - \hat{p}_2, p_1 - p_2]}$ statistics in Table 8.24, that the population of all such statistics approximately has a standard normal distribution. We now illustrate the use of formula set 2.

EXAMPLE 8.17

A new product was test marketed in the Des Moines, Iowa, and Toledo, Ohio, metropolitan areas. Equal amounts of money were spent on advertising in the two areas. However, different advertising media were employed in the two areas—advertising in the Des Moines area was done entirely on television, while advertising in the Toledo area consisted of a mixture of television, radio, newspaper, and magazine ads. Two months after the advertising campaigns commenced, surveys were taken to estimate consumer awareness of the product. In the Des Moines area 321 out of 510 randomly selected consumers were aware of the product, whereas in the Toledo area 475 out of 601 randomly selected consumers were aware of the product. We define p_1 to be the true proportion of consumers in the Des Moines area who are aware of the product and p_2 to be the true proportion of consumers in the Toledo area who are aware of the product. Then, noting that $\hat{p}_1 = 321/510 = .63$ and $\hat{p}_2 = 475/601 = .79$, we know from formula set 2 in Table 8.24 that a 95% confidence interval for $p_1 - p_2$ is

$$\left[(\hat{p}_1 - \hat{p}_2) \pm z_{[.025]}\sqrt{\frac{\hat{p}_1(1-\hat{p}_1)}{n_1 - 1} + \frac{\hat{p}_2(1-\hat{p}_2)}{n_2 - 1}}\right]$$

$$= \left[(.63 - .79) \pm 1.96\sqrt{\frac{(.63)(.37)}{510 - 1} + \frac{(.79)(.21)}{601 - 1}}\right]$$

$$= [-.16 \pm 1.96(.0271)]$$

$$= [-.16 \pm .0531]$$

$$= [-.2131, -.1069]$$

This interval says that we can be 95% confident that p_1, the proportion of all consumers in the Des Moines area who are aware of the product is between .2131 and .1069 less than p_2, the proportion of all consumers in the Toledo area who are aware of the product. Thus, we have substantial evidence that advertising the new product by using a mixture of television, radio, newspaper, and magazine ads (as in Toledo) is more effective than spending an equal amount of money on television commercials only.

8.7.2 Small-Sample Formulas

Since we sometimes cannot take *large* samples and thus cannot use the formula sets in Table 8.24, we need to discuss the two formula sets in Table 8.25, which utilize the *t*-distribution and which are valid for any (large or small) sample sizes n_1 and n_2. Examining these formula sets, which we call formula sets 3 and 4, we see that formula set 3 holds *exactly* under inference assumption 1 (Equal Variances),

TABLE 8.25 Formula Sets Based on the t-Distribution for Making Statistical Inferences About $\mu_1 - \mu_2$

Formula Sets and Conditions for Validity	Parameter θ	Point Estimate $\hat{\theta}$	Standard Error of the Estimate $s_{\hat{\theta}}$
Formula set 3: Holds exactly under (a) inference assumption 1 (Equal Variances) (b) inference assumption 2 (Independent Samples) (c) inference assumption 3 (Normal Populations)	$\mu_1 - \mu_2$	$\bar{y}_1 - \bar{y}_2$	$s_{\text{equal}} = \sqrt{\frac{(n_1 - 1)s_1^2 + (n_2 - 1)s_2^2}{n_1 + n_2 - 2}\left(\frac{1}{n_1} + \frac{1}{n_2}\right)}$
Formula set 4: Holds approximately under (a) inference assumption 2 (Independent Samples) (b) inference assumption 3 (Normal Populations)	$\mu_1 - \mu_2$	$\bar{y}_1 - \bar{y}_2$	$s_{\text{unequal}} = \sqrt{\frac{s_1^2}{n_1} + \frac{s_2^2}{n_2}}$

[a] If df is not an integer, round down.

inference assumption 2 (Independent Samples), and inference assumption 3 (Normal Populations), whereas formula set 4 holds *approximately* under inference assumptions 2 and 3. It can be shown that formula sets 3 and 4 are completely equivalent (that is, give exactly the same confidence intervals) when the sample sizes n_1 and n_2 are equal. When the sample sizes are not equal and we are choosing between formula sets 3 and 4, since formula set 3 is completely valid and formula set 4 only approximately valid under the appropriate assumptions, it is common practice to use formula set 3 if the sample variances indicate that the population variances are either equal or approximately equal. Finally, it can be shown that inference assumption 3 (Normal Populations) is not crucial to the validity of formula sets 3 and 4. In particular, it has been shown that formula sets 3 and 4 are approximately valid for populations described by probability curves that are mound-shaped (even if these curves are somewhat skewed to the right or left). In the following example we illustrate the use of formula set 3, and in the exercises the reader will have an opportunity to use formula set 4.

$t_{[\hat{\theta},\theta]}$ Statistic $t_{[\hat{\theta},\theta]} = \dfrac{\hat{\theta} - \theta}{s_{\hat{\theta}}}$	Degrees of Freedom[a] df	100(1 − α)% Confidence Intervals $[\hat{\theta} \pm t_{[\alpha/2]}^{(df)} s_{\hat{\theta}}]$ $[\hat{\theta} - t_{[\alpha]}^{(df)} s_{\hat{\theta}}, \infty)$ $(-\infty, \hat{\theta} + t_{[\alpha]}^{(df)} s_{\hat{\theta}}]$
$t_{[\bar{y}_1 - \bar{y}_2, \mu_1 - \mu_2]} = \dfrac{\bar{y}_1 - \bar{y}_2 - (\mu_1 - \mu_2)}{s_{\text{equal}}}$	$n_1 + n_2 - 2$	$[\bar{y}_1 - \bar{y}_2 \pm t_{[\alpha/2]}^{(n_1+n_2-2)} s_{\text{equal}}]$ $[\bar{y}_1 - \bar{y}_2 - t_{[\alpha]}^{(n_1+n_2-2)} s_{\text{equal}}, \infty)$ $(-\infty, \bar{y}_1 - \bar{y}_2 + t_{[\alpha]}^{(n_1+n_2-2)} s_{\text{equal}}]$
$t_{[\bar{y}_1 - \bar{y}_2, \mu_1 - \mu_2]} = \dfrac{\bar{y}_1 - \bar{y}_2 - (\mu_1 - \mu_2)}{s_{\text{unequal}}}$	$df = \dfrac{(n_1 - 1)(n_2 - 1)}{(n_2 - 1)g^2 + (1 - g)^2(n_1 - 1)}$ where $g = \dfrac{s_1^2/n_1}{s_1^2/n_1 + s_2^2/n_2}$	$[\bar{y}_1 - \bar{y}_2 \pm t_{[\alpha/2]}^{(df)} s_{\text{unequal}}]$ $[\bar{y}_1 - \bar{y}_2 - t_{[\alpha]}^{(df)} s_{\text{unequal}}, \infty)$ $(-\infty, \bar{y}_1 - \bar{y}_2 + t_{[\alpha]}^{(df)} s_{\text{unequal}}]$

EXAMPLE 8.18

National Motors wishes to compare the gasoline mileage obtained by the Gas-Mizer with the gasoline mileage obtained by the Gomega (the Gas-Mizer's chief competitor). Letting μ_{GM}, σ_{GM}^2, and σ_{GM} denote the mean, variance, and standard deviation of the population of Gas-Mizer mileages, and letting μ_{GO}, σ_{GO}^2, and σ_{GO} denote the mean, variance, and standard deviation of the population of Gomega mileages, suppose that National Motors obtains the sample results summarized in Table 8.26 (the Gas-Mizers and Gomegas are similarly equipped and are driven under similar conditions).

Since we might conclude (somewhat arbitrarily) that the sample variances $s_{GM}^2 = .723$ and $s_{GO}^2 = .49$ do not differ substantially from each other (note that the sample standard deviations $s_{GM} = .8503$ and $s_{GO} = .7$, being the square roots of the sample variances, are more nearly equal), we might conclude that the equal variances assumption ($\sigma_{GM}^2 = \sigma_{GO}^2$) approximately holds. Therefore, assuming that National Motors has chosen independent random samples and that the populations of Gas-Mizer mileages and Gomega mileages are normally distributed, we see from formula set 3 in Table 8.25 that if we define the $t_{[\hat{\theta},\theta]}$ statistic to be

$$t_{[\hat{\theta},\theta]} = \frac{\hat{\theta} - \theta}{s_{\hat{\theta}}} = \frac{\bar{y}_{GM} - \bar{y}_{GO} - (\mu_{GM} - \mu_{GO})}{\sqrt{\dfrac{(n_{GM} - 1)s_{GM}^2 + (n_{GO} - 1)s_{GO}^2}{n_{GM} + n_{GO} - 2}\left(\dfrac{1}{n_{GM}} + \dfrac{1}{n_{GO}}\right)}}$$

$$= t_{[\bar{y}_{GM} - \bar{y}_{GO}, \mu_{GM} - \mu_{GO}]}$$

then the population of all possible $t_{[\bar{y}_{GM} - \bar{y}_{GO}, \mu_{GM} - \mu_{GO}]}$ statistics has a t-distribution with $df = n_{GM} + n_{GO} - 2 = 5 + 3 - 2 = 6$ degrees of freedom. Therefore, it follows from condition (1) in Hypothesis Testing Result I in Section 6.12 that we can test

$$H_0: \mu_{GM} - \mu_{GO} = 0 \qquad (\theta = c)$$

versus

$$H_1: \mu_{GM} - \mu_{GO} \neq 0 \qquad (\theta \neq c)$$

by setting α, the probability of a Type I error, equal to .05 if we calculate

$$t_{[\hat{\theta},c]} = \frac{\hat{\theta} - c}{s_{\hat{\theta}}} = \frac{(\bar{y}_{GM} - \bar{y}_{GO}) - 0}{\sqrt{\dfrac{(n_{GM} - 1)s_{GM}^2 + (n_{GO} - 1)s_{GO}^2}{n_{GM} + n_{GO} - 2}\left(\dfrac{1}{n_{GM}} + \dfrac{1}{n_{GO}}\right)}}$$

$$= t_{[\bar{y}_{GM} - \bar{y}_{GO}, 0]}$$

$$= \frac{(31.56 - 29) - 0}{\sqrt{\dfrac{(5 - 1)(.723) + (3 - 1)(.49)}{5 + 3 - 2}\left(\dfrac{1}{5} + \dfrac{1}{3}\right)}}$$

$$= \frac{2.56}{.5866}$$

$$= 4.364$$

and use the rejection points

$$t_{[\alpha/2]}^{(df)} = t_{[.05/2]}^{(n_{GM} + n_{GO} - 2)} = t_{[.025]}^{(5+3-2)} = t_{[.025]}^{(6)} = 2.447$$

$$-t_{[\alpha/2]}^{(df)} = -2.447$$

Since

$$t_{[\bar{y}_{GM} - \bar{y}_{GO}, 0]} = 4.364 > 2.447 = t_{[.025]}^{(6)}$$

we can reject

$$H_0: \mu_{GM} - \mu_{GO} = 0$$

TABLE 8.26 Two Samples Randomly Selected from the Populations of Gas-Mizer Mileages and Gomega Mileages

Sample of $n_{GM} = 5$ Gas-Mizer Mileages	Sample of $n_{GO} = 3$ Gomega Mileages
$y_{GM,1} = 30.3$	$y_{GO,1} = 28.3$
$y_{GM,2} = 31.5$	$y_{GO,2} = 29.0$
$y_{GM,3} = 31.4$	$y_{GO,3} = 29.7$
$y_{GM,4} = 32.0$	
$y_{GM,5} = 32.6$	
$\bar{y}_{GM} = 31.56$	$\bar{y}_{GO} = 29.0$
$s^2_{GM} = .723$	$s^2_{GO} = .49$
$s_{GM} = .8503$	$s_{GO} = .7$

in favor of

$$H_1: \mu_{GM} - \mu_{GO} \neq 0$$

by setting α equal to .05.

Next, we note that

$$\begin{aligned}
\text{prob-value} &= 2A[[|t_{[\hat{\theta},c]}|, \infty); df] \\
&= 2A[[|t_{[\bar{y}_{GM}-\bar{y}_{GO},0]}|, \infty); n_{GM} + n_{GO} - 2] \\
&= 2A[[|4.364|, \infty); 5 + 3 - 2] \\
&= 2A[[4.364, \infty); 6] \\
&= 2(.0024) \\
&= .0048
\end{aligned}$$

Since this prob-value is less than .05 and less than .01, it follows by condition (2) in Hypothesis Testing Result I that we can reject

$$H_0: \mu_{GM} - \mu_{GO} = 0$$

in favor of

$$H_1: \mu_{GM} - \mu_{GO} \neq 0$$

if we set α equal to .05 or .01.

In order to interpret the prob-value as a probability, note that since the population of all possible $t_{[\bar{y}_{GM}-\bar{y}_{GO}, \mu_{GM}-\mu_{GO}]}$ statistics has a t-distribution with $n_{GM} + n_{GO} - 2 = 5 + 3 - 2 = 6$ degrees of freedom, it follows that, if

$$H_0: \mu_{GM} - \mu_{GO} = 0$$

is true, then the population of all possible $t_{[\bar{y}_{GM}-\bar{y}_{GO},0]}$ statistics has a t-distribution with $n_{GM} + n_{GO} - 2 = 6$ degrees of freedom. Therefore, this prob-value equals the proportion of $t_{[\bar{y}_{GM}-\bar{y}_{GO},0]}$ statistics in the population of all such statistics which, if the null hypothesis

$$H_0: \mu_{GM} - \mu_{GO} = 0$$

is true, are at least as far away from zero and thus at least as contradictory to

$$H_0: \mu_{GM} - \mu_{GO} = 0$$

as 4.364, the observed $t_{[\bar{y}_{GM}-\bar{y}_{GO},0]}$ statistic. Since this prob-value, which equals .0048, says that National Motors must believe that a 48 in 10,000 chance has occurred if it is to believe that

$$H_0: \mu_{GM} - \mu_{GO} = 0$$

is true, the company has substantial evidence to conclude that

$$H_0: \mu_{GM} - \mu_{GO} = 0$$

is false and thus that

$$H_1: \mu_{GM} - \mu_{GO} \neq 0$$

is true.

Finally, we note that it follows from (4) in Hypothesis Testing Result I that for any level of confidence, 100(1 − α)%, less than

$$\begin{aligned} 100[1 - (\text{prob-value})]\% &= 100[1 - .0048]\% \\ &= 99.52\% \end{aligned}$$

the 100(1 − α)% confidence interval for $\mu_{GM} - \mu_{GO}$

$$\begin{aligned} &\left[(\bar{y}_{GM} - \bar{y}_{GO}) \pm t_{[\alpha/2]}^{(n_{GM}+n_{GO}-2)} \sqrt{\frac{(n_{GM}-1)s_{GM}^2 + (n_{GO}-1)s_{GO}^2}{n_{GM}+n_{GO}-2}\left(\frac{1}{n_{GM}} + \frac{1}{n_{GO}}\right)}\right] \\ &= \left[(31.56 - 29) \pm t_{[\alpha/2]}^{(5+3-2)} \sqrt{\frac{(5-1)(.723) + (3-1)(.49)}{5+3-2}\left(\frac{1}{5} + \frac{1}{3}\right)}\right] \\ &= [2.56 \pm t_{[\alpha/2]}^{(6)}(.5866)] \end{aligned}$$

does not contain zero (for example, the 95% interval

$$\begin{aligned} [2.56 \pm t_{[.025]}^{(6)}(.5866)] &= [2.56 \pm 2.447(.5866)] \\ &= [1.1245, 3.9955] \end{aligned}$$

does not contain zero—see Figure 8.11), and thus National Motors *can*, with at least 100(1 − α)% confidence, *reject*

$$H_0: \mu_{GM} - \mu_{GO} = 0$$

in favor of

$$H_1: \mu_{GM} - \mu_{GO} \neq 0$$

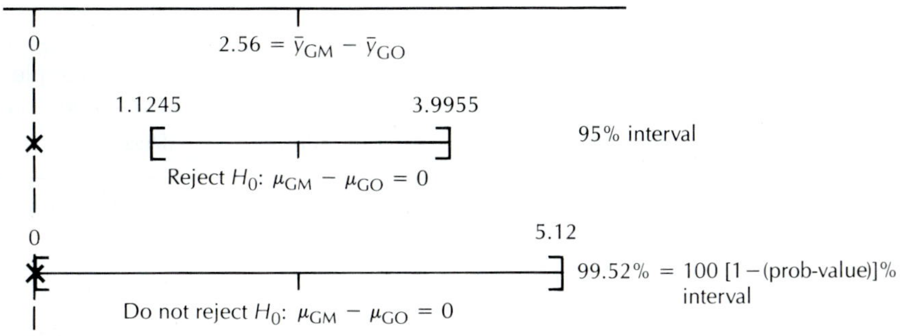

FIGURE 8.11
H_0: $\mu_{GM} - \mu_{GO} = 0$ Can Be Rejected at Any Level of Confidence Less Than 99.52%

It follows from (5) in Hypothesis Testing Result I that for any level of confidence, $100(1 - \alpha)\%$, greater than or equal to

$$100[1 - (\text{prob-value})]\% = 99.52\%$$

the $100(1 - \alpha)\%$ confidence interval for $\mu_{GM} - \mu_{GO}$ *does contain zero* (for example, the 99.52% interval, which can be calculated to be [0, 5.12], does contain zero—see Figure 8.11), and thus National Motors *cannot*, with at least $100(1 - \alpha)\%$ confidence, *reject*

$$H_0: \mu_{GM} - \mu_{GO} = 0$$

in favor of

$$H_1: \mu_{GM} - \mu_{GO} \neq 0$$

To summarize, National Motors can be up to

$$100[1 - (\text{prob-value})]\% = 99.52\%$$

confident that

$$H_0: \mu_{GM} - \mu_{GO} = 0$$

is false and that

$$H_1: \mu_{GM} - \mu_{GO} \neq 0$$

is true. Specifically, the above 95% confidence interval for $\mu_{GM} - \mu_{GO}$, [1.1245, 3.9955], says that National Motors can be 95% confident that μ_{GM} is between 1.1245 mpg and 3.9955 mpg greater than μ_{GO}.

8.7.3 Paired Difference Experiments

All of the previously discussed formulas in this section assume that inference assumption 2 (Independent Samples) holds. However, if the two samples are obtained

TABLE 8.27
Sample of n Paired Differences

First Sample	Second Sample	Sample of n Paired Differences (Assuming $n_1 = n_2 = n$)
$y_{1,1}$	$y_{2,1}$	$d_1 = y_{1,1} - y_{2,1}$
$y_{1,2}$	$y_{2,2}$	$d_2 = y_{1,2} - y_{2,2}$
$\vdots$	$\vdots$	$\vdots$
$y_{1,n}$	$y_{2,n}$	$d_n = y_{1,n} - y_{2,n}$
Sample mean $= \bar{y}_1$	Sample mean $= \bar{y}_2$	$\bar{d} = \bar{y}_1 - \bar{y}_2$ = mean of the sample of paired differences

by taking two different measurements on the same n elements, then the two samples are related to (or depend upon) each other. In such a case we say that we have performed a **paired difference experiment**. The needed formulas in this kind of situation employ what we call the sample of n *paired differences* $d_1, d_2, \ldots, d_n$ summarized in Table 8.27, and the mean, variance, and standard deviations of these paired differences

$$\bar{d} = \frac{\sum_{k=1}^{n} d_k}{n} \qquad s_d^2 = \frac{\sum_{k=1}^{n} (d_k - \bar{d})^2}{n-1} \qquad s_d = \sqrt{s_d^2}$$

which are the point estimates of μ_d, σ_d^2, and σ_d, the mean, variance, and standard deviation of the population of all possible paired differences. Here it is useful to note that

$$\bar{d} = \frac{\sum_{k=1}^{n} d_k}{n} = \frac{\sum_{k=1}^{n} (y_{1,k} - y_{2,k})}{n}$$

$$= \frac{\sum_{k=1}^{n} y_{1,k}}{n} - \frac{\sum_{k=1}^{n} y_{2,k}}{n}$$

$$= \bar{y}_1 - \bar{y}_2$$

That is, $\bar{d}$, the mean of the sample of n paired differences, equals $\bar{y}_1 - \bar{y}_2$, the difference between the sample means. Similarly, it can be shown that μ_d, the mean of the population of all possible paired differences, equals $\mu_1 - \mu_2$, the difference between the population means. Thus, to summarize, $\bar{d} = \bar{y}_1 - \bar{y}_2$ is the point estimate of $\mu_d = \mu_1 - \mu_2$. Moreover, Table 8.28 presents two formula sets—formula sets 5 and 6—that can be used to make statistical inferences about $\mu_d = \mu_1 - \mu_2$. Formula set 5 is based on the fact that if the sample size n is large, then the population of all possible values of $\bar{d} = \bar{y}_1 - \bar{y}_2$ approximately has a normal distribution, with mean $\mu_{\bar{d}} = \mu_d$ and standard deviation $\sigma_{\bar{d}} = \sigma_d/\sqrt{n}$. It follows, if we consider each of the $z_{[\bar{d},\mu_d]}$ statistics in Table 8.28, that the population of all such statistics approximately has a standard normal distribution. Formula set 6 utilizes the t-distribution and is valid for any (large or small) sample size n. Furthermore, the assumption that the population of all possible paired differences has a normal distribution is not crucial to the validity of formula set 6.

EXAMPLE 8.19

Home State Life and Casualty, an insurance company, wishes to compare the repair costs of moderately damaged automobiles at two garages—garage 1 and garage 2. An automobile is moderately damaged if its repair cost is between \$700 and \$1,400. One experimental procedure that could be used in carrying out this study would be to take a sample of $n = 7$ cars that have been recently involved in accidents to garage 1 and obtain repair cost estimates for these cars, and then to take a different sample of $n = 7$ cars that have been recently involved in accidents to garage 2 and obtain repair cost estimates for these cars. This procedure would yield two samples that are independent of each other. However, there are substantial differences in damages to moderately damaged cars. These differences would tend to conceal any real differences between the repair costs at the two garages. For example, if the repair cost estimates for the cars taken to garage 1 were higher than the repair cost estimates for the cars taken to garage 2, we might not be able to tell whether the higher estimates at garage 1 were due to the way garage 1 charges customers for repair work or to the severity of the damage to the cars taken to garage 1. In order to overcome this difficulty, we can perform a paired difference experiment. Here, we would take each of $n = 7$ cars to both garages and obtain repair cost estimates for each car at these two garages. The advantage of this paired difference experiment is that the repair cost estimates at the two garages are obtained by using the same cars, and thus any true differences in these estimates would not be concealed by differences in the damages to the cars.

Suppose that when the paired difference experiment is performed, Home State Life and Casualty obtains the samples, means, variance, and standard deviation in Table 8.29. Defining μ_1 and μ_2 to be the mean repair cost estimates at garages 1 and 2, respectively, it follows that

$$\begin{aligned} \bar{d} &= \bar{y}_1 - \bar{y}_2 \\ &= -.8 \qquad (-\$80) \end{aligned}$$

is the point estimate of

$$\mu_d = \mu_1 - \mu_2$$

TABLE 8.28 Formula Sets Based on a Paired Difference Experiment for Making Statistical Inferences About $\mu_d = \mu_1 - \mu_2$

Formula Sets and Conditions for Validity	Parameter θ	Point Estimate $\hat{\theta}$	Standard Deviation (Actual or Estimated) $\sigma_{\hat{\theta}}$ or $s_{\hat{\theta}}$
Formula set 5: Holds approximately under (a) paired difference experiment (b) n large (≥ 30)	$\mu_d = \mu_1 - \mu_2$	$\bar{d} = \bar{y}_1 - \bar{y}_2$	$\sigma_{\bar{d}} = \frac{\sigma_d}{\sqrt{n}}$ or $s_{\bar{d}} = \frac{s_d}{\sqrt{n}}$
Formula set 6: Holds exactly under (a) paired difference experiment (b) population of all possible paired differences having a normal distribution	$\mu_d = \mu_1 - \mu_2$	$\bar{d} = \bar{y}_1 - \bar{y}_2$	$s_{\bar{d}} = \frac{s_d}{\sqrt{n}}$

TABLE 8.29 Sample of $n = 7$ Paired Differences of the Repair Cost Estimates at Garages 1 and 2 (Cost Estimates in Hundreds of Dollars)

Sample of $n = 7$ Damaged Cars	Sample of $n = 7$ Repair Cost Estimates at Garage 1	Sample of $n = 7$ Repair Cost Estimates at Garage 2	Sample of $n = 7$ Paired Differences
Car 1	$y_{1,1} = 7.1$	$y_{2,1} = 7.9$	$d_1 = -.8$
Car 2	$y_{1,2} = 9.0$	$y_{2,2} = 10.1$	$d_2 = -1.1$
Car 3	$y_{1,3} = 11.0$	$y_{2,3} = 12.2$	$d_3 = -1.2$
Car 4	$y_{1,4} = 8.9$	$y_{2,4} = 8.8$	$d_4 = .1$
Car 5	$y_{1,5} = 9.9$	$y_{2,5} = 10.4$	$d_5 = -.5$
Car 6	$y_{1,6} = 9.1$	$y_{2,6} = 9.8$	$d_6 = -.7$
Car 7	$y_{1,7} = 10.3$	$y_{2,7} = 11.7$	$d_7 = -1.4$
	$\bar{y}_1 = 9.329$	$\bar{y}_2 = 10.129$	$\bar{d} = \bar{y}_1 - \bar{y}_2 = -.8$ $s_d^2 = .2533$ $s_d = .5033$

$z_{[\hat{\theta},\theta]}$ or $t_{[\hat{\theta},\theta]}$ Statistic	$100(1-\alpha)\%$ Confidence Intervals
$z_{[\bar{d},\mu_d]} = \dfrac{\bar{d}-\mu_d}{\sigma_{\bar{d}}}$ or $z_{[\bar{d},\mu_d]} = \dfrac{\bar{d}-\mu_d}{s_{\bar{d}}}$	$[\bar{d} \pm z_{[\alpha/2]}\sigma_{\bar{d}}]$ or $[\bar{d} \pm z_{[\alpha/2]}s_{\bar{d}}]$ $[\bar{d} - z_{[\alpha]}\sigma_{\bar{d}}, \infty)$ or $[\bar{d} - z_{[\alpha]}s_{\bar{d}}, \infty)$ $(-\infty, \bar{d} + z_{[\alpha]}\sigma_{\bar{d}}]$ or $(-\infty, \bar{d} + z_{[\alpha]}s_{\bar{d}}]$
$t_{[\bar{d},\mu_d]} = \dfrac{\bar{d}-\mu_d}{s_{\bar{d}}}$ $df = n-1$	$[\bar{d} \pm t_{[\alpha/2]}^{(n-1)}s_{\bar{d}}]$ $[\bar{d} - t_{[\alpha]}^{(n-1)}s_{\bar{d}}, \infty)$ $(-\infty, \bar{d} + t_{[\alpha]}^{(n-1)}s_{\bar{d}}]$

which is the mean of all possible paired differences of the repair cost estimates at garages 1 and 2, or, the difference between the mean repair cost estimates at garages 1 and 2.

Now, assume that although Home State Life and Casualty has been having its moderately damaged automobiles repaired at garage 2, the insurance company has heard that garage 1 provides less expensive repair service of equal quality. However, because of its past experience with garage 2, Home State has decided that it will have its moderately damaged automobiles repaired at garage 1 only if it is very confident that $\mu_d\,(=\mu_1-\mu_2)$ is less than zero, which says that μ_d, the mean of all possible paired differences of the repair cost estimates at garages 1 and 2, is less than zero, or μ_1, the mean of all possible repair cost estimates at garage 1, is less than μ_2, the mean of all possible repair cost estimates at garage 2. Therefore, Home State has decided to have its moderately damaged automobiles repaired at garage 1 only if it can use the sample of $n = 7$ paired differences of the repair cost estimates at garages 1 and 2 to reject

$$H_0: \mu_d \geq 0 \qquad (\mu_1 - \mu_2 \geq 0)$$

in favor of

$$H_1: \mu_d < 0 \qquad (\mu_1 - \mu_2 < 0)$$

by setting α, the probability of a Type I error, equal to .01. Note that setting α so small (equal to .01) means that there is a very small probability that Home State will conclude that garage 1 is less expensive than garage 2 when garage 1 is not less expensive than garage 2 (that is, reject $H_0: \mu_1 - \mu_2 \geq 0$ when H_0 is true). Assuming that the population of all possible paired differences is approximately normally distributed, and defining the $t_{[\hat{\theta},\theta]}$ statistic to be

$$t_{[\hat{\theta},\theta]} = \frac{\hat{\theta} - \theta}{s_{\hat{\theta}}} = \frac{\bar{d} - \mu_d}{s_d/\sqrt{n}} = t_{[\bar{d},\mu_d]}$$

we find from formula set 6 in Table 8.28 that the population of all possible $t_{[\bar{d},\mu_d]}$ statistics has a t-distribution with $df = n - 1 = 7 - 1 = 6$ degrees of freedom. Testing $H_0: \mu_d \geq 0$ versus $H_1: \mu_d < 0$ is equivalent to testing $H_0: \mu_d = 0$ versus $H_1: \mu_d < 0$, so it follows from condition (1) in Hypothesis Testing Result III in Section 6.12 that we can test $H_0: \mu_d = 0$ versus $H_1: \mu_d < 0$ by setting α equal to .01 if we calculate

$$\begin{aligned} t_{[\hat{\theta},c]} &= \frac{\hat{\theta} - c}{s_{\hat{\theta}}} = \frac{\bar{d} - 0}{s_d/\sqrt{n}} = t_{[\bar{d},0]} \\ &= \frac{\bar{y}_1 - \bar{y}_2 - 0}{s_d/\sqrt{n}} \\ &= \frac{-.8 - 0}{.5033/\sqrt{7}} \\ &= -4.206 \end{aligned}$$

and use the rejection point

$$-t_{[\alpha]}^{(df)} = -t_{[.01]}^{(n-1)} = -t_{[.01]}^{(7-1)} = -t_{[.01]}^{(6)} = -3.143$$

Since

$$t_{[\bar{d},0]} = -4.206 < -3.143 = -t_{[.01]}^{(6)}$$

we can reject H_0 in favor of H_1 by setting α equal to .01.

Next, we note that

$$\begin{aligned} \text{prob-value} &= A[(-\infty, t_{[\hat{\theta},c]}]; df] \\ &= A[(-\infty, t_{[\bar{d},0]}]; n - 1] \\ &= A[(-\infty, -4.206]; 7 - 1 = 6] \\ &= .0014 \end{aligned}$$

Since this prob-value is less than .05 and .01, we can reject H_0 in favor of H_1 if we set α equal to .05 or .01.

In order to interpret the prob-value as a probability, note that since the population of all possible $t_{[\bar{d},\mu_d]}$ statistics has a t-distribution with $n - 1 = 7 - 1 = 6$ degrees of freedom, it follows that, if $H_0: \mu_d = 0$ is true, then the population of all possible $t_{[\bar{d},0]}$ statistics has a t-distribution with $n - 1 = 6$ degrees

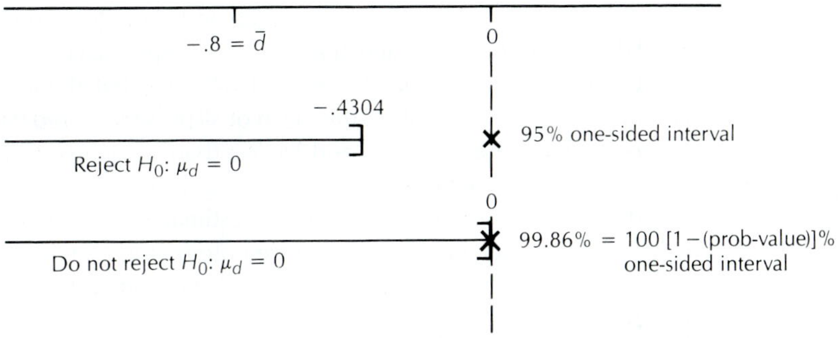

FIGURE 8.12 $H_0: \mu_d (= \mu_1 - \mu_2) = 0$ Can Be Rejected in Favor of $H_1: \mu_d (= \mu_1 - \mu_2) < 0$ at Any Level of Confidence Less Than 99.86%

of freedom. Therefore, this prob-value equals the proportion of $t_{[\bar{d},0]}$ statistics in the population of all such statistics which, if the null hypothesis H_0 is true, are at least as small as and thus at least as contradictory to H_0 as -4.206, the observed $t_{[\bar{d},0]}$ statistic. Since this prob-value, which equals .0014, says that Home State must believe that a 14 in 10,000 chance has occurred if it is to believe that H_0 is true, Home State has substantial evidence to conclude that H_0 is false and that H_1 is true.

Finally, it follows from (4) in Hypothesis Testing Result III that for any level of confidence, $100(1 - \alpha)\%$, less than

$$\begin{aligned} 100[1 - (\text{prob-value})]\% &= 100[1 - .0014]\% \\ &= 99.86\% \end{aligned}$$

the one-sided $100(1 - \alpha)\%$ confidence interval for μ_d,

$$\begin{aligned} \left(-\infty, \bar{d} + t_{[\alpha]}^{(n-1)}\left(\frac{s_d}{\sqrt{n}}\right)\right] &= \left(-\infty, (\bar{y}_1 - \bar{y}_2) + t_{[\alpha]}^{(7-1)}\left(\frac{s_d}{\sqrt{7}}\right)\right] \\ &= \left(-\infty, -.8 + t_{[\alpha]}^{(6)}\left(\frac{.5033}{\sqrt{7}}\right)\right] \\ &= (-\infty, -.8 + t_{[\alpha]}^{(6)}(.1902)] \end{aligned}$$

does not contain zero (for example, the 95% interval

$$\begin{aligned} (-\infty, -.8 + 1.943(.1902)] &= (-\infty, -.8 + .3696] \\ &= (-\infty, -.4304] \end{aligned}$$

does not contain zero—see Figure 8.12), and thus Home State *can*, with at least $100(1 - \alpha)\%$ confidence, *reject* $H_0: \mu_d = 0$ in favor of $H_1: \mu_d < 0$.

It follows from (5) in Hypothesis Testing Result III that for any level of confidence, $100(1 - \alpha)\%$, greater than or equal to

$$100[1 - (\text{prob-value})]\% = 99.86\%$$

the one-sided $100(1 - \alpha)\%$ confidence interval for μ_d *does contain zero* (for example, the 99.86% interval does contain zero—see Figure 8.12), and thus Home State *cannot*, with at least $100(1 - \alpha)\%$ confidence, *reject* $H_0: \mu_d = 0$ in favor of $H_1: \mu_d < 0$.

To summarize, Home State Life and Casualty can be up to 99.86% confident that H_0 is false and that H_1 is true. Specifically, the 95% confidence interval for $\mu_d\,(=\mu_1-\mu_2)$, $(-\infty, -.4304]$, says that the insurance company can be 95% confident that μ_d, the mean of all possible paired differences of the repair cost estimates at garages 1 and 2, is less than or equal to $-\$43.04$, or μ_1, the mean of all possible repair cost estimates at garage 1, is at least \$43.04 less than μ_2, the mean of all possible repair cost estimates at garage 2.

To conclude this example we note that in addition to the one-sided 95% confidence interval for $\mu_d\,(=\mu_1-\mu_2)$, a two-sided 95% confidence interval for $\mu_d\,(=\mu_1-\mu_2)$ is (see Table 8.28)

$$\begin{aligned}\left[\bar{d} \pm t_{[.025]}^{(n-1)}\left(\frac{s_d}{\sqrt{n}}\right)\right] &= \left[(\bar{y}_1-\bar{y}_2) \pm t_{[.025]}^{(7-1)}\left(\frac{s_d}{\sqrt{n}}\right)\right]\\ &= \left[-.8 \pm t_{[.025]}^{(6)}\left(\frac{.5033}{\sqrt{7}}\right)\right]\\ &= [-.8 \pm 2.447(.1902)]\\ &= [-.8 \pm .4654]\\ &= [-1.2654, -.3346]\end{aligned}$$

This interval says that Home State Life and Casualty can be 95% confident that μ_d, the mean of all possible paired differences of the repair cost estimates at garages 1 and 2, is between $-\$126.54$ and $-\$33.46$, or μ_1, the mean of all possible repair cost estimates at garage 1, is between \$126.54 and \$33.46 less than μ_2, the mean of all possible repair cost estimates at garage 2.

8 EXERCISES

8.1 Consider Exercise 7.10 and the following table.

Parameter Number	Parameter $= \boldsymbol{\lambda}'\boldsymbol{\beta}$	$s\sqrt{\boldsymbol{\lambda}'(\mathbf{X}'\mathbf{X})^{-1}\boldsymbol{\lambda}}$	Prob-Value for $H_0: \boldsymbol{\lambda}'\boldsymbol{\beta} = 0$
1	$\mu_{[B,.93]} - \mu_{[A,.93]}$	.5917	.0007
2	$\mu_{[C,.93]} - \mu_{[A,.93]}$	.6587	.0216
3	$\mu_{[C,.93]} - \mu_{[B,.93]}$	.5604	.1360
4	$\dfrac{\mu_{[C,.93]} + \mu_{[B,.93]}}{2} - \mu_{[A,.93]}$	.5598	.0019
5	$\mu_{[B,.93]}$	.3393	—

In parts (n) and (o) of Exercise 7.10 the reader has calculated point estimates of these parameters by assuming that the model

$$\begin{aligned} y_i &= \mu_i + \epsilon_i \\ &= \beta_0 + \beta_1 D_{i,B} + \beta_2 D_{i,C} + \beta_3 x_{i2} + \beta_4 x_{i2}^2 + \beta_5 D_{i,B} x_{i2} \\ &\quad + \beta_6 D_{i,C} x_{i2} + \beta_7 D_{i,B} x_{i2}^2 + \beta_8 D_{i,C} x_{i2}^2 + \epsilon_i \end{aligned}$$

adequately describes the data in Exercise 7.10. Use these point estimates, any other pertinent information from Exercise 7.10, and the information in the table to

a. Find a 95% confidence interval for each parameter in the table by using the formula

$$[\boldsymbol{\lambda}'\mathbf{b} \pm t_{[\alpha/2]}^{(n-k)} s \sqrt{\boldsymbol{\lambda}'(\mathbf{X}'\mathbf{X})^{-1}\boldsymbol{\lambda}}]$$

Also find $\boldsymbol{\lambda}'$ for each parameter.

b. Calculate the $t_{\boldsymbol{\lambda}'\mathbf{b}}$ statistic

$$t_{\boldsymbol{\lambda}'\mathbf{b}} = \frac{\boldsymbol{\lambda}'\mathbf{b}}{s\sqrt{\boldsymbol{\lambda}'(\mathbf{X}'\mathbf{X})^{-1}\boldsymbol{\lambda}}}$$

for testing $H_0: \boldsymbol{\lambda}'\boldsymbol{\beta} = 0$ for each of parameters 1, 2, 3, and 4 in the table. Also show how the prob-values in the table have been calculated, and discuss what these prob-values say about the validity of the appropriate null hypotheses.

8.2 Using the information in the table in Exercise 8.1 and any pertinent information from Exercise 7.10,

a. Calculate Scheffé simultaneous 95% confidence intervals for parameters 1, 2, 3, and 4 in the table.

b. Calculate Scheffé simultaneous 95% confidence intervals for parameters 1, 2, 3, 4, and 5 in the table.

8.3 The Tastee Bakery Company supplies Tastee Bread to a large number of supermarkets in a metropolitan area. The company wishes to study the effect of shelf display height, which has levels B (Bottom), M (Middle), and T (Top), on monthly demand for Tastee Bread (measured in cases of 24 loaves each). For l = B, M, and T, we define *the lth population of monthly bread demands* to be the infinite population of all possible monthly bread demands that could be obtained at supermarkets which use display height l, and we let μ_l denote the mean of this population.

In order to compare μ_B, μ_M, and μ_T (where μ_l is the *mean monthly bread demand that would be obtained by using display height l*), Tastee Bakery will employ a completely randomized experimental design. Specifically, for l = B, M, and T, Tastee Bakery will randomly select a sample of n_l metropolitan area supermarkets. These supermarkets will sell Tastee Bread for one month using display height l. Then, letting $OSP_{l,k}$ denote the kth randomly selected supermarket (where $k = 1, 2, \ldots, n_l$) that uses display height l, Tastee Bakery will measure

$$y_{l,k} = \text{the monthly bread demand (measured in cases) that occurs in } OSP_{l,k}$$

which is *the* k*th monthly bread demand observed when using display height* l. When Tastee Bakery has employed this completely randomized design, we assume that for $l = \text{B, M, and T}$,

$$y_{l,1}, \ldots, y_{l,n_l}$$

is a sample of n_l monthly bread demands that Tastee Bakery has randomly selected from the lth population of monthly bread demands.

Suppose that when Tastee Bakery has employed this completely randomized design, it has randomly selected the three samples summarized in the following table.

Three Samples of Monthly Bread Demands

	Display Height	
B	M	T
$y_{B,1} = 58.2$	$y_{M,1} = 73.0$	$y_{T,1} = 52.5$
$y_{B,2} = 53.7$	$y_{M,2} = 78.1$	$y_{T,2} = 49.8$
$y_{B,3} = 55.6$	$y_{M,3} = 76.2$	$y_{T,3} = 56.0$
	$y_{M,4} = 82.0$	$y_{T,4} = 51.9$
	$y_{M,5} = 78.4$	$y_{T,5} = 53.3$

In order to analyze these data, we use the following model to describe $y_{l,k}$:

$$\begin{aligned} y_{l,k} &= \mu_l + \epsilon_{l,k} \\ &= \beta_B + \beta_M D_{l,M} + \beta_T D_{l,T} + \epsilon_{l,k} \end{aligned}$$

Here, $D_{l,M} = 1$ if a middle display height was used to obtain $y_{l,k}$ and $D_{l,M} = 0$ otherwise, and $D_{l,T} = 1$ if a top display height was used to obtain $y_{l,k}$ and $D_{l,T} = 0$ otherwise.

a. Express the following means and differences in means in terms of β_B, β_M, and β_T:

$$\mu_B \quad \mu_M \quad \mu_T \quad \mu_M - \mu_B \quad \mu_T - \mu_B \quad \mu_T - \mu_M$$

Then show that $H_0\colon \beta_M = \beta_T = 0$ is equivalent to $H_0\colon \mu_B = \mu_M = \mu_T$.

b. Specify the vector **y** and the matrix **X** used to calculate the least squares point estimates of the parameters in the preceding model.

c. When we use this model to perform a regression analysis of these data, we find that the least squares point estimates of the parameters are

$$b_B = 55.8333 \qquad b_M = 21.7067 \qquad b_T = -3.1333$$

and we also obtain the following quantities

$$SS_{model} = \text{Explained variation} = 1741.5844$$

$$SSE = \text{Unexplained variation} = 73.8987$$

By using these quantities to calculate the overall F statistic and by using the appropriate rejection point, determine whether we can, by setting α equal to .05, reject

$$H_0: \beta_M = \beta_T = 0 \qquad \text{or} \qquad \mu_B = \mu_M = \mu_T$$

in favor of

H_1: At least one of β_M and β_T does not equal zero

or

H_1: At least two of μ_B, μ_M, and μ_T differ from each other

The prob-value for testing H_0 versus H_1 can be calculated to be .0001. Demonstrate how this prob-value has been calculated, and use this prob-value to determine whether we can reject H_0 in favor of H_1 by setting α equal to .05 or .01.

d. Consider the following table.

Parameter Number	Parameter = $\boldsymbol{\lambda}'\boldsymbol{\beta}$	$s\sqrt{\boldsymbol{\lambda}'(\mathbf{X}'\mathbf{X})^{-1}\boldsymbol{\lambda}}$	Prob-Value for $H_0: \boldsymbol{\lambda}'\boldsymbol{\beta} = 0$
1	$\mu_M - \mu_B$	1.9853	.0001
2	$\mu_T - \mu_B$	1.9853	.1456
3	$\mu_T - \mu_M$	1.7193	.0001
4	μ_M	1.2157	—

In part (a) the reader has expressed each of these parameters in terms of β_B, β_M, and β_T. By using the least squares point estimates of β_B, β_M, and β_T given in part (c), calculate a point estimate of each parameter in the table, and find a 95% confidence interval for each parameter by using the formula

$$[\boldsymbol{\lambda}'\mathbf{b} \pm t_{[\alpha/2]}^{(n-k)} s\sqrt{\boldsymbol{\lambda}'(\mathbf{X}'\mathbf{X})^{-1}\boldsymbol{\lambda}}]$$

Also find $\boldsymbol{\lambda}'$ for each parameter.

e. Calculate the $t_{\lambda' b}$ statistic for testing $H_0 : \boldsymbol{\lambda}'\boldsymbol{\beta} = 0$ for each of parameters 1, 2, and 3 in the table in part (d). Also, show how the prob-values in the table have been calculated, and discuss what these prob-values say about the validity of the appropriate null hypotheses.

f. Using any pertinent information in this problem, calculate Scheffé simultaneous 95% confidence intervals for parameters 1, 2, and 3 in the table in part (d), and calculate Scheffé simultaneous 95% confidence intervals for parameters 1, 2, 3, and 4 in this table.

g. Calculate a point prediction of, and a 95% prediction interval for, $y_{M,0}$, a future (individual) bread demand that will be observed when using a middle display height.

8.4 Perform a complete analysis of the data in Exercise 8.3 by using one-way analysis of variance (as discussed in Section 8.4.2).

8.5 In order to compare the durability of four different brands of golf balls, the National Golf Association randomly selects five balls of each brand and places each ball into a machine that exerts the force produced by a 250-yard drive. The number of simulated drives needed to crack or chip each ball is recorded, with the following results.

Brand			
A	B	C	D
281	270	218	364
220	334	244	302
274	307	225	325
242	290	273	337
251	331	249	355

Perform a complete analysis of these data by (1) using regression analysis (as discussed in Section 8.4.1) and (2) using one-way analysis of variance (as discussed in Section 8.4.2).

8.6 Again consider the Tastee Bakery Company, and suppose that this company wishes to study the effects of two qualitative factors—shelf display height, which has levels B (Bottom), M (Middle), and T (Top); and shelf display width, which has levels R (Regular) and W (Wide)—on monthly demand for Tastee Bread (measured in cases of 24 loaves each). For i = B, M, and T and j = R and W, we define *the* (i, j)*th population of monthly bread demands* to be the infinite population of all possible monthly bread demands that could be obtained at supermarkets which use display height i and display width j, and we let μ_{ij} denote the mean of this population.

In order to compare the appropriate means (where μ_{ij} is the *mean monthly bread demand that would be obtained by using display height i and display width j*), Tastee Bakery will, for i = B, M, and T, and j = R and W, randomly select a sample of n_{ij} metropolitan area supermarkets. These supermarkets will sell Tastee Bread for one month using display height i and display width j. Then, letting $OSP_{ij,k}$ denote the kth randomly selected supermarket (where $k = 1, 2, \ldots, n_{ij}$) that uses display height i and display width j, Tastee Bakery will measure

$$y_{ij,k} = \text{the monthly bread demand (measured in cases) that occurs in } OSP_{ij,k}$$

which is the kth *monthly bread demand observed when using display height i and display width j*. When Tastee Bakery has employed this completely randomized design, we assume that, for i = B, M, and T and j = R and W,

$$y_{ij,1}, y_{ij,2}, \ldots, y_{ij,n_{ij}}$$

is a sample of n_{ij} monthly bread demands that Tastee Bakery has randomly selected from the (i, j)th population of monthly bread demands.

Suppose that when Tastee Bakery has employed this completely randomized design, it has randomly selected the six samples summarized in the following table.

Six Samples of Monthly Bread Demands

Display Height (i)	Display Width (j): R	Display Width (j): W
B	$y_{BR,1} = 58.2$ $y_{BR,2} = 53.7$	$y_{BW,1} = 55.6$
M	$y_{MR,1} = 73.0$ $y_{MR,2} = 78.1$	$y_{MW,1} = 76.2$ $y_{MW,2} = 82.0$ $y_{MW,3} = 78.4$
T	$y_{TR,1} = 52.5$ $y_{TR,2} = 49.8$	$y_{TW,1} = 56.0$ $y_{TW,2} = 51.9$ $y_{TW,3} = 53.3$

a. Calculate the means of the six samples, and using these sample means, perform a graphical analysis that indicates whether or not interaction exists between display height and display width.

b. Since the graphical analysis performed in (a) implies that little or no interaction exists between display height and display width, we analyze these data by considering the (α, γ) model:

$$\begin{aligned} y_{ij,k} &= \mu_{ij} + \epsilon_{ij,k} \\ &= \mu + \alpha_i + \gamma_j + \epsilon_{ij,k} \\ &= \mu + \alpha_M D_{i,M} + \alpha_T D_{i,T} + \gamma_W D_{j,W} + \epsilon_{ij,k} \end{aligned}$$

where $D_{i,M} = 1$ if $i = M$ and $D_{i,M} = 0$ otherwise, where $D_{i,T} = 1$ if $i = T$ and $D_{i,T} = 0$ otherwise, and where $D_{j,W} = 1$ if $j = W$ and $D_{j,W} = 0$ otherwise. Using the fact that the (α, γ) model implies that

$$\mu_{ij} = \mu + \alpha_i + \gamma_j$$

$$\alpha_i = \alpha_M D_{i,M} + \alpha_T D_{i,T}$$

$$\gamma_j = \gamma_W D_{j,W}$$

show that $\alpha_B = 0$ and $\gamma_R = 0$, and express the following means and differences in means in terms of μ, α_M, α_T, and γ_W:

$$\mu_{MW} \quad \mu_{Bj} \quad \mu_{Mj} \quad \mu_{Tj} \quad \mu_{Mj} - \mu_{Bj} \quad \mu_{Tj} - \mu_{Bj}$$

$$\mu_{Tj} - \mu_{Mj} \quad \mu_{iR} - \mu_{iW} \quad \mu_{iW} - \mu_{iR}$$

Then show that $H_0: \alpha_M = \alpha_T = 0$ is equivalent to $H_0: \mu_{Bj} = \mu_{Mj} = \mu_{Tj}$, and show that $H_0: \gamma_W = 0$ is equivalent to $H_0: \mu_{iR} = \mu_{iW}$.

c. Specify the vector **y** and the matrix **X** used to calculate the least squares point estimates of the parameters in the (α, γ) model.

d. When we use the (α, γ) model to perform a regression analysis of these data, we find that the least squares point estimates of the parameters in this model are

$$\hat{\mu} = 55.0891 \qquad \hat{\alpha}_M = 21.1113 \qquad \hat{\alpha}_T = -3.7287 \qquad \hat{\gamma}_W = 2.2326$$

and we find that the unexplained variation is $SSE_{(\alpha,\gamma)} = 58.6127$. Furthermore, when we use the α model

$$\begin{aligned} y_{ij,k} &= \mu_{ij} + \epsilon_{ij,k} \\ &= \mu + \alpha_i + \epsilon_{ij,k} \\ &= \mu + \alpha_M D_{i,M} + \alpha_T D_{i,T} + \epsilon_{ij,k} \end{aligned}$$

and the γ model

$$\begin{aligned} y_{ij,k} &= \mu_{ij} + \epsilon_{ij,k} \\ &= \mu + \gamma_j + \epsilon_{ij,k} \\ &= \mu + \gamma_W D_{j,W} + \epsilon_{ij,k} \end{aligned}$$

to perform regression analyses of these data, we obtain unexplained variations of, respectively, $SSE_\alpha = 73.8987$ and $SSE_\gamma = 1766.6426$. Using these unexplained variations, calculate the $F(\alpha|\gamma)$ statistic used to test

$$H_0: \alpha_M = \alpha_T = 0 \qquad \text{or} \qquad \mu_{Bj} = \mu_{Mj} = \mu_{Tj}$$

versus

H_1: At least one of α_M or α_T does not equal zero

or

H_1: At least two of μ_{Bj}, μ_{Mj}, and μ_{Tj} differ from each other

and calculate the $F(\gamma|\alpha)$ statistic used to test

$H_0: \gamma_W = 0$ or $\mu_{iR} = \mu_{iW}$

versus

$H_1: \gamma_W \neq 0$ or $\mu_{iR} \neq \mu_{iW}$

Then use these partial F statistics and the appropriate rejection points to determine whether we can reject these null hypotheses by setting α equal to .05. The prob-values related to these null hypotheses can be calculated to be, respectively, .0001 and .1599. Demonstrate how these prob-values have been calculated, and use these prob-values to determine whether we can reject the above null hypotheses by setting α equal to .05 or .01.

e. Consider the following table.

Parameter Number	Parameter $= \boldsymbol{\lambda}'\boldsymbol{\beta}$	$s\sqrt{\boldsymbol{\lambda}'(\mathbf{X}'\mathbf{X})^{-1}\boldsymbol{\lambda}}$	Prob-Value for $H_0: \boldsymbol{\lambda}'\boldsymbol{\beta} = 0$
1	$\mu_{Mj} - \mu_{Bj}$	1.9038	.0001
2	$\mu_{Tj} - \mu_{Bj}$	1.9038	.0818
3	$\mu_{Tj} - \mu_{Mj}$	1.614	.0001
4	$\mu_{iW} - \mu_{iR}$	1.4573	.1599
5	μ_{MW}	1.2815	—

In part (b) the reader has expressed each of these parameters in terms of μ, α_M, α_T, and γ_W. By using the least squares point estimates of μ, α_M, α_T, and γ_W given in part (d), calculate a point estimate of each parameter in the table, and find a 95% confidence interval for each parameter by using the formula

$$[\boldsymbol{\lambda}'\mathbf{b} \pm t_{[\alpha/2]}^{(n-k)} s\sqrt{\boldsymbol{\lambda}'(\mathbf{X}'\mathbf{X})^{-1}\boldsymbol{\lambda}}]$$

Also, find $\boldsymbol{\lambda}'$ for each parameter.

f. Calculate the $t_{\boldsymbol{\lambda}'\mathbf{b}}$ statistic for testing $H_0: \boldsymbol{\lambda}'\boldsymbol{\beta} = 0$ for each of parameters 1, 2, 3, and 4 in the table in part (e). Also, show how the prob-values in the table have been calculated, and discuss what these prob-values say about the validity of the appropriate null hypotheses.

g. Using any pertinent information in this problem, calculate Scheffé simultaneous 95% confidence intervals for parameters 1, 2, 3, and 4 in the table in part (e), calculate Scheffé simultaneous 95% confidence intervals for parameters 1, 2, 3, 4, and 5 in this table, and calculate Scheffé simultaneous 95% confidence intervals for parameters 1, 2, and 3 in this table.

h. Calculate a point prediction of, and a 95% prediction interval for, $y_{MW,0}$, a future (individual) bread demand that will be observed when using a middle display height and a wide display.

i. By using the fact that the total variation can be calculated to be 1815.4831 and that the unexplained variations for the α model and for the γ model are, respectively, $SSE_\alpha = 73.8987$ and $SSE_\gamma = 1766.6426$, calculate the $F(\alpha)$ statistic and the $F(\gamma)$ statistic. Then, use $F(\alpha)$, $F(\gamma)$, $F(\alpha|\gamma)$, and $F(\gamma|\alpha)$ and Table 8.13 to determine which model Cramer and Applebaum (1977) and Searle (1971) suggest best describes the data.

j. Although the model suggested by Cramer and Applebaum and Searle is the α model, it does not necessarily follow that the α model is the best model to use to calculate confidence intervals for differences in means. To see this, note that, if we remove the subscript j from the data in this exercise, these data become the data in Exercise 8.3. Therefore, the individual and simultaneous confidence intervals for $\mu_M - \mu_B$, $\mu_T - \mu_B$, and $\mu_T - \mu_M$ calculated in Exercise 8.3 are the confidence intervals for $\mu_{Mj} - \mu_{Bj}$, $\mu_{Tj} - \mu_{Bj}$, and $\mu_{Tj} - \mu_{Mj}$ that would be calculated by using the α model in this exercise. Compare the lengths of the confidence intervals for $\mu_{Mj} - \mu_{Bj}$, $\mu_{Tj} - \mu_{Bj}$, and $\mu_{Tj} - \mu_{Mj}$ calculated by using the α model with the lengths of the corresponding confidence intervals calculated by using the (α, γ) model. Why do you think that the (α, γ) model yields slightly shorter intervals, even though Cramer and Applebaum and Searle suggest that the α model is best?

8.7 In this exercise consider analyzing the data in Exercise 8.6 by using the means model:

$$\begin{aligned} y_{ij,k} &= \mu_{ij} + \epsilon_{ij,k} \\ &= \beta_{BR} + \beta_{BW}D_{ij,BW} + \beta_{MR}D_{ij,MR} + \beta_{MW}D_{ij,MW} + \beta_{TR}D_{ij,TR} + \beta_{TW}D_{ij,TW} + \epsilon_{ij,k} \end{aligned}$$

where, for example, $D_{ij,MW} = 1$ if $i = M$ and $j = W$, and $D_{ij,MW} = 0$ otherwise.

a. Express the following means and differences in means in terms of β_{BR}, β_{BW}, β_{MR}, β_{MW}, β_{TR}, and β_{TW}:

μ_{BR} μ_{BW} μ_{MR} μ_{MW} μ_{TR} μ_{TW} $\mu_{BW} - \mu_{BR}$ $\mu_{MR} - \mu_{BR}$

$\mu_{MW} - \mu_{BR}$ $\mu_{TR} - \mu_{BR}$ $\mu_{TW} - \mu_{BR}$

Then, show that

$$H_0: \beta_{BW} = \beta_{MR} = \beta_{MW} = \beta_{TR} = \beta_{TW} = 0$$

is equivalent to

$$H_0: \mu_{BR} = \mu_{BW} = \mu_{MR} = \mu_{MW} = \mu_{TR} = \mu_{TW}$$

b. Specify the vector **y** and the matrix **X** used to calculate the least squares point estimates of the parameters in the means model.

c. When we use the means model to perform a regression analysis of the data in Exercise 8.6, we find that the least squares point estimates of the parameters in this model are

$$b_{BR} = 55.95 \quad b_{BW} = -.35 \quad b_{MR} = 19.6 \quad b_{MW} = 22.9167$$

$$b_{TR} = -4.8 \quad b_{TW} = -2.2167$$

and we also obtain the following quantities:

$$SS_{\text{means}} = \text{Explained variation} = 1762.8747$$

$$SSE_{\text{means}} = \text{Unexplained variation} = 52.6083$$

By using these quantities to calculate the F(means model) statistic and by using the appropriate rejection point, determine whether we can, by setting α equal to .05, reject

$$H_0: \beta_{BW} = \beta_{MR} = \beta_{MW} = \beta_{TR} = \beta_{TW} = 0$$

or

$$H_0: \mu_{BR} = \mu_{BW} = \mu_{MR} = \mu_{MW} = \mu_{TR} = \mu_{TW}$$

in favor of

H_1: At least one of β_{BW}, β_{MR}, β_{MW}, β_{TR}, and β_{TW} does not equal zero

or

H_1: At least two of μ_{BR}, μ_{BW}, μ_{MR}, μ_{MW}, μ_{TR}, and μ_{TW} differ from each other

The prob-value for testing H_0 versus H_1 can be calculated to be .0001. Demonstrate how this prob-value has been calculated, and use this prob-value to determine whether we can reject H_0 in favor of H_1 by setting α equal to .05 or .01.

d. Consider the following table.

Parameter Number	Parameter $= \boldsymbol{\lambda}'\boldsymbol{\beta}$	$s\sqrt{\boldsymbol{\lambda}'(\mathbf{X}'\mathbf{X})^{-1}\boldsymbol{\lambda}}$
1	$\mu_{MW} - \mu_{BR}$	2.5026
2	μ_{MW}	1.5827

In part (a) the reader has expressed each of these parameters in terms of β_{BR}, β_{BW}, β_{MR}, β_{MW}, β_{TR}, and β_{TW}. By using the least squares point estimates of these parameters given in part (c), calculate a point estimate of each parameter in the table, and find a 95% confidence interval for each parameter by using the formula

$$[\boldsymbol{\lambda}'\mathbf{b} \pm t_{[\alpha/2]}^{(n-k)} s\sqrt{\boldsymbol{\lambda}'(\mathbf{X}'\mathbf{X})^{-1}\boldsymbol{\lambda}}]$$

Also, find $\boldsymbol{\lambda}'$ for each parameter.

e. Calculate a point prediction of, and a 95% prediction interval for, $y_{MW,0}$, a future (individual) bread demand that will be observed when using a middle display height and a wide display.

f. By using any needed information from this exercise and from Exercise 8.6, calculate the F(interaction) statistic. Then, by using this statistic and the appropriate rejection point, determine whether we can, by setting α equal to .05, reject

H_0: No interaction exists between display height and display width

in favor of

H_1: Interaction does exist between display height and display width

Next, compare the length of the 95% confidence interval for μ_{MW} calculated by using the means model in this exercise with the length of the 95% confidence interval for μ_{MW} calculated by using the (α, γ) model in Exercise 8.6. Which model seems to best describe the data in Exercise 8.6—the (α, γ) model or the means model?

8.8 Consider the following table, and suppose that D, E, and F are three fertilizer types, that Q, R, S, and T are four wheat types, and that the $y_{ij,k}$ values are wheat yields in bushels per plot (where a plot is one-third of an acre) corresponding to the different combinations of fertilizer type and wheat type.

Fertilizer Type (i)	Wheat Type (j) Q	R	S	T
D	$y_{DQ,1} = 17.4$ $y_{DQ,2} = 18.6$	$y_{DR,1} = 23.0$ $y_{DR,2} = 22.0$	$y_{DS,1} = 23.5$ $y_{DS,2} = 22.3$	$y_{DT,1} = 20.8$ $y_{DT,2} = 19.7$
E	$y_{EQ,1} = 23.3$ $y_{EQ,2} = 24.5$	$y_{ER,1} = 25.6$ $y_{ER,2} = 24.4$	$y_{ES,1} = 23.4$ $y_{ES,2} = 23.1$	$y_{ET,1} = 19.6$ $y_{ET,2} = 20.6$
F	$y_{FQ,1} = 23.0$ $y_{FQ,2} = 23.5$	$y_{FR,1} = 24.7$ $y_{FR,2} = 23.3$	$y_{FS,1} = 23.0$ $y_{FS,2} = 22.0$	$y_{FT,1} = 18.6$ $y_{FT,2} = 19.8$

Assuming that these data were obtained by using a completely randomized design, perform a complete analysis of the data by (1) using regression analysis (as discussed in Sections 8.5.1 and 8.5.2) and (2) using two-way analysis of variance (as discussed in Section 8.5.3).

8.9 A consumer preference study involving three different bottle designs (A, B, and C) for the jumbo size of a new liquid laundry detergent was carried out using a randomized block experimental design, with supermarkets as blocks. Specifically, four supermarkets were supplied with all three bottle designs, which were priced the same. The data shown below represent the number of bottles sold in a 24-hour period.

	Bottle Design		
Supermarket	A	B	C
1	16	33	23
2	14	30	21
3	1	19	8
4	6	23	12

Perform a complete analysis of the data by (1) using regression analysis (as discussed in Section 8.6.2) and (2) using analysis of variance (as discussed in Section 8.6.3).

8.10 In order to compare three brands of typewriters (A, B, and C), four typists were randomly selected, each typist used all three typewriters (in a random order) to type the same material for ten minutes, and the numbers of words typed per minute were recorded. The data are as follows.

	Typewriter Brand		
Typist	A	B	C
1	77	67	63
2	71	62	59
3	74	63	59
4	67	57	54

a. Perform a complete analysis of these data by (1) using regression analysis (as discussed in Section 8.6.2) and (2) using analysis of variance (as discussed in Section 8.6.3).

b. Suppose that typist 3 became ill and was not able to use typewriter brand B, and thus we obtained the following data.

	Typewriter Brand		
Typist	A	B	C
1	77	67	63
2	71	62	59
3	74	—	59
4	67	57	54

Perform a complete analysis of the data by using regression analysis (as discussed in Section 8.6.2).

8.11 A sample of 81 60-watt lightbulbs produced by Dynamics, Inc., obtained a mean lifetime of 1,347 hours with a variance of 729 hours. A sample of 50 60-watt lightbulbs produced by National Electronics Corporation obtained a mean lifetime of 1,282 hours with a variance of 800 hours. Calculate a 99% confidence interval for $\mu_1 - \mu_2$, the difference between the true mean lifetimes of the two types of lightbulbs. Interpret the results.

8.12 A medical laboratory is conducting research aimed at finding a cure for the common cold. In a sample of 2,500 patients who have a cold and are treated with Drug XB, 1,500 patients are cured within a two-day period, while in a sample of 2,500 patients who have a cold and are treated with Drug YL, 1,125 patients are cured within a two-day period. Calculate a 95% confidence interval for $p_{XB} - p_{YL}$, the difference between the true proportions of all patients whose colds will be cured within a two-day period by Drugs XB and YL.

8.13 American International Paper Corporation must purchase a new papermaking machine. The company must choose between two machines, machine 1 and machine 2. Since both machines produce paper of equal quality, the company will choose the machine that produces the most paper in a one-hour period (measured in rolls of paper per hour). In order to compare the two machines, five randomly selected machine operators produce paper for one hour using machine 1, and another five randomly selected machine operators produce paper for one hour using machine 2. The results obtained are as follows.

	Hourly Production of Paper (no. of rolls)				
Machine 1 (Sample 1)	53	60	58	48	46
Machine 2 (Sample 2)	56	44	55	50	45

a. Calculate the means and variances of these samples.
b. Discuss why it is reasonable to believe that these samples are independent of each other.
c. Assuming that inference assumption 3 (Normal Populations) holds, select an appropriate formula and calculate a 95% confidence interval for $\mu_1 - \mu_2$, the difference between the true mean numbers of rolls of paper produced per hour by the two machines. Interpret the results.

8.14 Consider the situation in Example 8.16 and suppose that the management of Bargain Town has randomly selected the following two small samples.

Sample of Ohio Accounts	Sample of Illinois Accounts
$n_O = 10$	$n_I = 5$
$\bar{y}_O = \$124$	$\bar{y}_I = \$68$
$s_O^2 = \$1681$	$s_I^2 = \$484$

Calculate a 95% confidence interval for $\mu_O - \mu_I$ using formula set 4 in Table 8.25.

8.15 Since the 95% confidence interval calculated in Exercise 8.13 contains zero, the American International Paper Corporation cannot be at least 95% confident that there is a difference between μ_1 and μ_2. Since the length of this 95% confidence interval might have been caused by differences in the abilities of the machine operators, the company has decided to compare the machines using a paired difference experiment. Suppose, then, that five randomly selected machine operators produce paper for one hour using machine 1 and for one hour using machine 2, with the following results.

Hourly Production of Paper (no. of rolls)	Machine Operator				
	1	2	3	4	5
Machine 1	53	60	58	48	46
Machine 2	50	55	56	44	45

Assuming that the population of all possible paired differences has a normal distribution, calculate a 95% confidence interval for $\mu_d = \mu_1 - \mu_2$. Interpret the results.

The remaining problems in this chapter pertain to testing $H_0: \theta = c$ versus one of the alternative hypotheses $H_1: \theta \neq c$, $H_1: \theta > c$, and $H_1: \theta < c$. For each problem,

a. Define an appropriate $t_{[\hat{\theta},c]}$ statistic or $z_{[\hat{\theta},c]}$ statistic, and use an appropriate rejection point (or appropriate rejection points) to determine whether we would reject $H_0: \theta = c$ by setting α, the probability of a Type I error, equal to .05 and .01.
b. Calculate the appropriate prob-value for testing $H_0: \theta = c$.
c. Using the prob-value calculated in (b), determine whether we would reject $H_0: \theta = c$ by setting α equal to .05, .03, and .01.
d. Interpret the prob-value calculated in (b) as a probability, and discuss how much doubt this prob-value casts upon $H_0: \theta = c$.
e. Discuss what the prob-value calculated in (b) says about a confidence interval for θ and about how confident we can be that we should reject $H_0: \theta = c$. Illustrate by calculating and depicting the appropriate 95% and 100[1 − (prob-value)]% confidence intervals for θ.

8.16 Use the sample information in Exercise 8.11 to test $H_0: \mu_1 - \mu_2 = 50$ versus $H_1: \mu_1 - \mu_2 > 50$.

8.17 Use the sample information in Example 8.17 to test $H_0: p_1 - p_2 = -.10$ versus $H_1: p_1 - p_2 < -.10$.

8.18 Use the sample information in Exercise 8.13 to test $H_0: \mu_1 - \mu_2 = 0$ versus $H_1: \mu_1 - \mu_2 \neq 0$. Use the fact that $A[[.816, \infty); 8] = .266$.

8.19 Use the sample information in Exercise 8.15 to test $H_0: \mu_1 - \mu_2 = 0$ versus $H_1: \mu_1 - \mu_2 \neq 0$. Use the fact that $A[[4.25, \infty); 4] = .007$.

9

THE ASSUMPTIONS BEHIND REGRESSION ANALYSIS

What are the assumptions behind regression analysis? What do these assumptions mean? How can we tell whether or not these assumptions are met? The goal of this chapter is to discuss the answers to these questions.

In order to use regression analysis properly, there are several conditions that must be met. First, the regression model must be a linear function of the model parameters, and the model must have the correct functional form. Second, the formulas we have presented for confidence intervals, prediction intervals, and hypothesis tests (that is, the formulas that allow us to make *statistical inferences*) depend on the validity of three assumptions we have called the *inference assumptions*. We briefly presented these assumptions in Chapter 6. Here we discuss the meanings of these assumptions more completely, and we discuss techniques that can be used to investigate the validity of these assumptions.

We begin this chapter by discussing *residual plots* in Section 9.1. We use these plots in later sections to investigate the validity of the assumptions behind regression analysis. In Section 9.2 we see how these plots can help us to discover when we are employing a regression model that has an incorrect functional form or that has omitted a variable important in describing and predicting the dependent variable. Next, in Section 9.3, in order to discuss the inference assumptions more fully, we introduce the infinite population of potential error terms and present the properties of this population. Inference assumption 1, the Constant Variance assumption, is presented and fully explained in Section 9.4. In this section we also explain how to use residual plots to decide whether or not this assumption is valid for a regression problem. The second inference assumption, the Independence assumption, is the topic of Section 9.5. When this assumption is violated, the error terms for the regression model are said to be *autocorrelated*. Here we see that we can detect violations of the Independence assumption by using residual plots and a statistical test called the *Durbin-Watson test for autocorrelation*. Inference assumption 3, the Normality assumption, is covered in Section 9.6. Here we look at several ways in which residuals can be used to detect violations of this assumption (including use of a *normal plot*). Section 9.7 discusses how to deal with *outlying and influential observations*.

It is important to reiterate that although the formulas for the confidence intervals, prediction intervals, and hypothesis tests in this book are strictly valid only when the inference assumptions hold, these formulas are still approximately correct even when mild departures from the inference assumptions can be detected. In fact, these assumptions very seldom, if ever, exactly hold in any practical regression problem. Therefore, in practice, only pronounced departures from the inference assumptions are considered to be serious enough to need remedial action. However, if these assumptions are seriously violated, then remedies can and should be employed. A detailed discussion of the available remedies is given in Chapter 10.

9.1 RESIDUAL PLOTS

A technique that can be used to help determine whether the assumptions behind regression analysis are valid is the construction and examination of residual plots.

Recall that the ith residual, e_i, is defined to be

$$e_i = y_i - \hat{y}_i$$

which is the difference between the ith observed value of the dependent variable and the ith predicted value of the dependent variable. In order to construct a residual plot, we compute the residual for each observation: For $y_1, y_2, \ldots, y_n$, we compute

$$e_i = y_i - \hat{y}_i = y_i - (b_0 + b_1x_{i1} + b_2x_{i2} + \cdots + b_px_{ip})$$

where $b_0, b_1, b_2, \ldots, b_p$ are the least squares point estimates of the parameters $\beta_0, \beta_1, \beta_2, \ldots, \beta_p$ in the model

$$y_i = \beta_0 + \beta_1x_{i1} + \beta_2x_{i2} + \cdots + \beta_px_{ip} + \epsilon_i$$

The calculated residuals, denoted $e_1, e_2, \ldots, e_n$, are then plotted against some criterion, such as the increasing values of one of the independent variables or the time order in which the historical data have been collected. The resulting plot is called a **residual plot**. We will see that in order to use residual plots to help determine whether the assumptions behind regression analysis are valid, we should make residual plots against

1. Increasing values of each of the independent variables in the model.
2. Increasing values of $\hat{y}$, the predicted value of the dependent variable.
3. The time order in which the historical data have been observed.

In the following two examples we illustrate how to make these residual plots.

EXAMPLE 9.1

Consider the fuel consumption model

$$y = \beta_0 + \beta_1x_1 + \beta_2x_2 + \epsilon$$

which relates weekly fuel consumption, y, to x_1, average hourly temperature, and x_2, the chill index. Using the historical fuel consumption data, we have obtained the prediction equation

$$\hat{y} = b_0 + b_1x_1 + b_2x_2 = 13.109 - .0900x_1 + .0825x_2$$

The eight predicted fuel consumptions and the eight residuals given by this prediction equation are listed in Table 9.1. Using these residuals, we make residual plots against

1. Increasing values of each of the independent variables x_1 and x_2 (Figures 9.1 and 9.2).
2. Increasing values of $\hat{y}$ (Figure 9.3).
3. The time order in which the historical fuel consumption data have been

TABLE 9.1 The Eight Predicted Fuel Consumptions and the Eight Residuals Given by the Prediction Equation $\hat{y} = b_0 + b_1x_1 + b_2x_2 = 13.109 - .0900x_1 + .0825x_2$

The *i*th Observed Week (OWK_i)	The *i*th Observed Average Hourly Temperature (°F) (x_{i1})	The *i*th Observed Chill Index (x_{i2})	The *i*th Observed Fuel Consumption (Tons) y_i	The *i*th Predicted Fuel Consumption ($\hat{y}_i = b_0 + b_1x_{i1} + b_2x_{i2}$ $= 13.109 - .0900x_{i1} + .0825x_{i2}$)	The *i*th Residual ($e_i = y_i - \hat{y}_i$)
OWK_1 ($t = 1$)	28.0	18	12.4	12.0733	.3267
OWK_2 ($t = 3$)	28.0	14	11.7	11.7433	−.0433
OWK_3 ($t = 2$)	32.5	24	12.4	12.1632	.2368
OWK_4 ($t = 4$)	39.0	22	10.8	11.4131	−.6131
OWK_5 ($t = 6$)	45.9	8	9.4	9.6371	−.2371
OWK_6 ($t = 5$)	57.8	16	9.5	9.2259	.2741
OWK_7 ($t = 7$)	58.1	1	8.0	7.9614	.0386
OWK_8 ($t = 8$)	62.5	0	7.5	7.4829	.0171

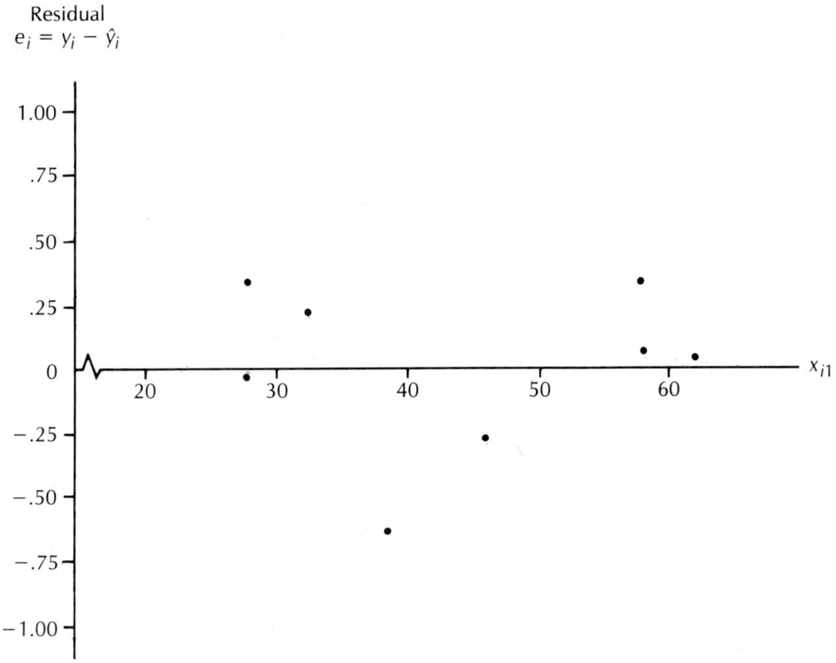

FIGURE 9.1
Residuals for the Two-Variable Fuel Consumption Model Plotted Against Increasing Values of x_1, Average Hourly Temperature

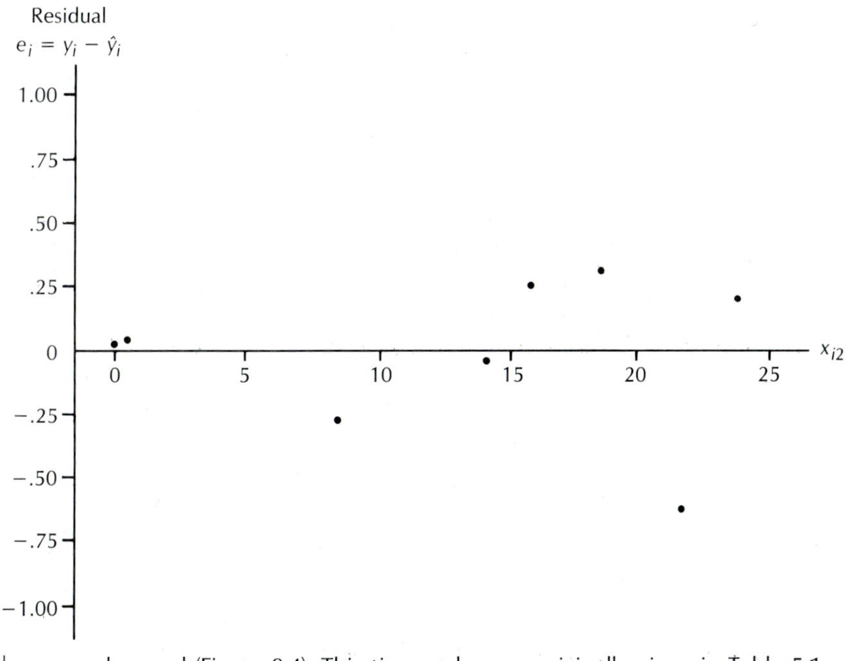

FIGURE 9.2
Residuals for the Two-Variable Fuel Consumption Model Plotted Against Increasing Values of x_2, the Chill Index

observed (Figure 9.4). This time order was originally given in Table 5.1 and is also given in Table 9.1. Here, for example, OWK_3 $(t = 2)$ means that OWK_3 (which is called the third observed week because the third smallest average hourly temperature occurred in this week) is the week that was observed second in the time order.

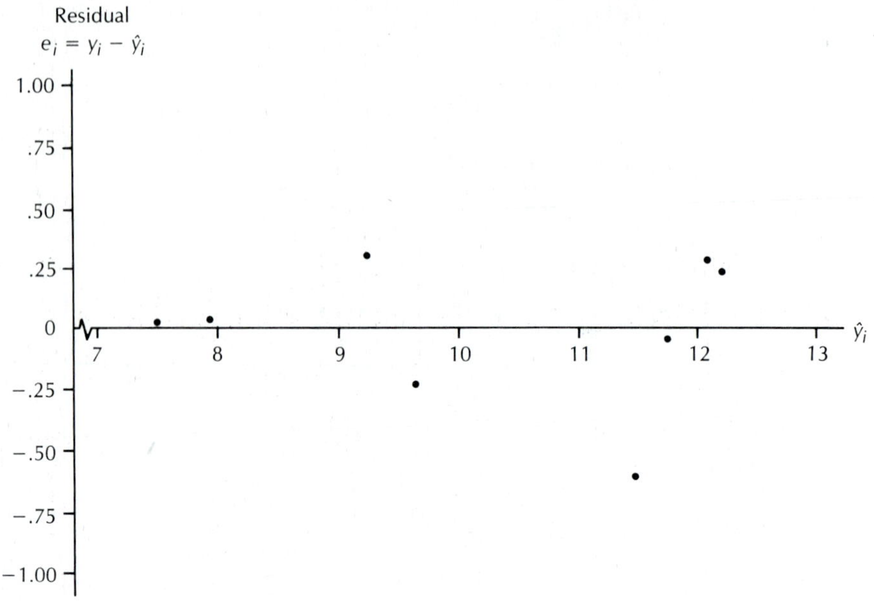

FIGURE 9.3

Residuals for the Two-Variable Fuel Consumption Model Plotted Against Increasing Values of $\hat{y}$, Predicted Fuel Consumption

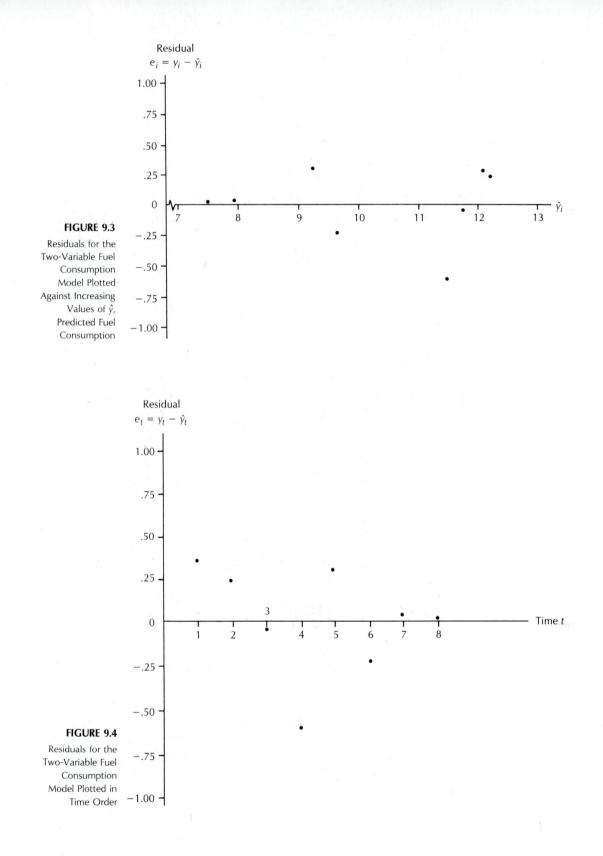

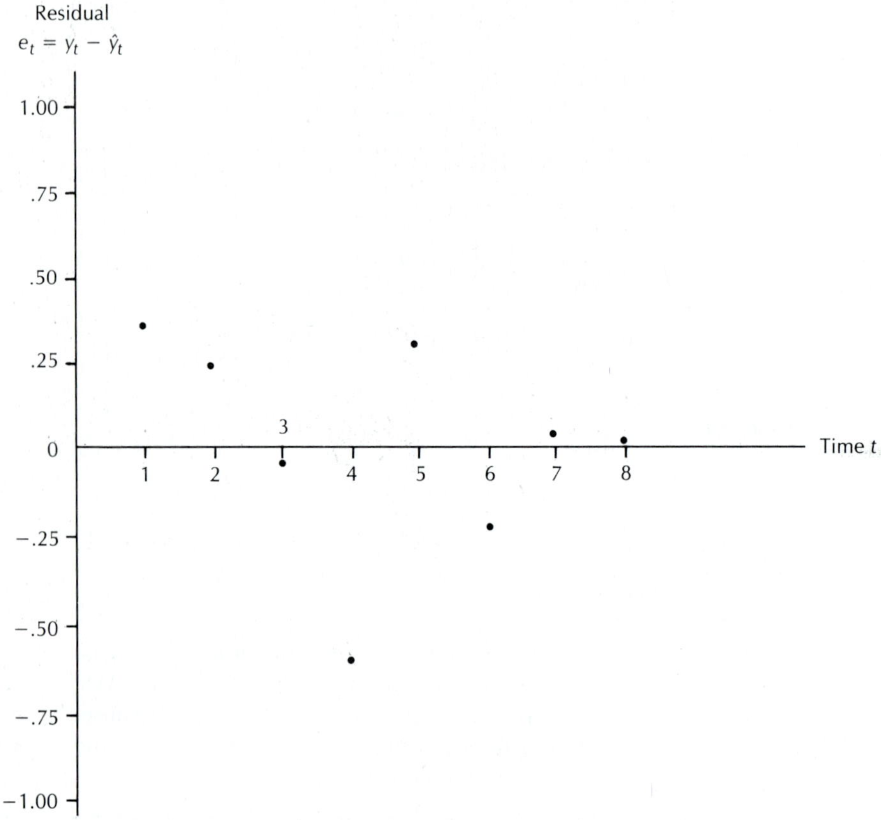

FIGURE 9.4

Residuals for the Two-Variable Fuel Consumption Model Plotted in Time Order

EXAMPLE 9.2

Consider the Fresh Detergent model

$$y = \beta_0 + \beta_1 x_4 + \beta_2 x_3 + \beta_3 x_3^2 + \beta_4 x_4 x_3 + \epsilon$$

which relates y, demand for Fresh, to x_4, the price difference, and x_3, advertising expenditure for Fresh. Using the historical Fresh Detergent data, we have obtained the prediction equation

$$\begin{aligned}\hat{y} &= b_0 + b_1 x_4 + b_2 x_3 + b_3 x_3^2 + b_4 x_4 x_3 \\ &= 29.1133 + 11.1342 x_4 - 7.6080 x_3 + .6712 x_3^2 - 1.4777 x_4 x_3\end{aligned}$$

The residuals given by this prediction equation are calculated, for $i = 1, 2, \ldots, 30$, as

$$\begin{aligned}e_i &= y_i - \hat{y}_i \\ &= y_i - (29.1133 + 11.1342 x_{i4} - 7.6080 x_{i3} + .6712 x_{i3}^2 - 1.4777 x_{i4} x_{i3})\end{aligned}$$

and are listed (in increasing order) in Table 9.2. For future reference, the *standardized residuals* for this model are also given (in increasing order) in this table

TABLE 9.2 Ordered Residuals and Ordered Standardized Residuals for the Fresh Detergent Model $y = \beta_0 + \beta_1 x_4 + \beta_2 x_3 + \beta_3 x_3^2 + \beta_4 x_4 x_3 + \epsilon$

Ordered Residuals (e_i)	Ordered Standardized Residuals $\left(\frac{e_i}{s}\right)$	Ordered Residuals (e_i)	Ordered Standardized Residuals $\left(\frac{e_i}{s}\right)$
−.437250	−2.11948	.005016	.02431
−.327373	−1.58687	.028687	.13905
−.327250	−1.58628	.029666	.14380
−.210185	−1.01883	.029866	.14476
−.143455	−.69537	.035566	.17239
−.138209	−.66994	.102677	.49770
−.133353	−.64640	.109641	.53146
−.122850	−.59549	.126964	.61543
−.101611	−.49254	.141240	.68463
−.074478	−.36102	.176312	.85463
−.066071	−.32026	.233113	1.12997
−.047250	−.22903	.234223	1.13535
−.044139	−.21395	.245527	1.19014
−.038914	−.18862	.325016	1.57545
.004773	.02313	.384097	1.86184

(the standardized residuals are computed by dividing the residuals by the standard error $s = .2063$). Using the residuals, we make residual plots against

1. Increasing values of each of the independent variables x_4 and x_3 (Figures 9.5 and 9.6).

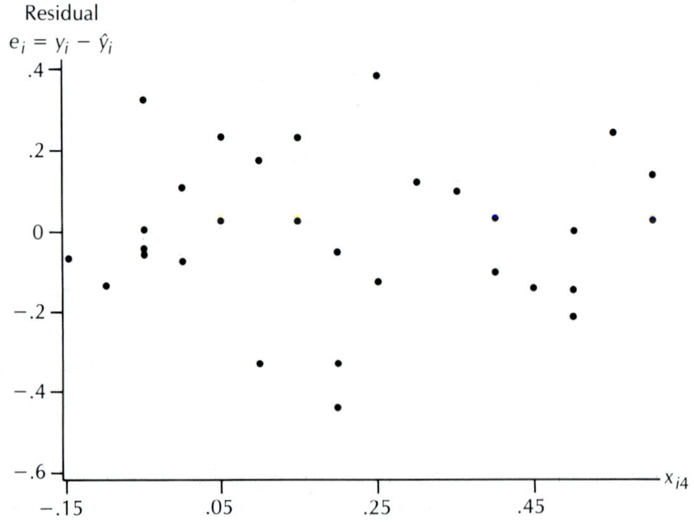

FIGURE 9.5
Residuals for the Fresh Detergent Model Plotted Against Increasing Values of x_4, the Price Difference

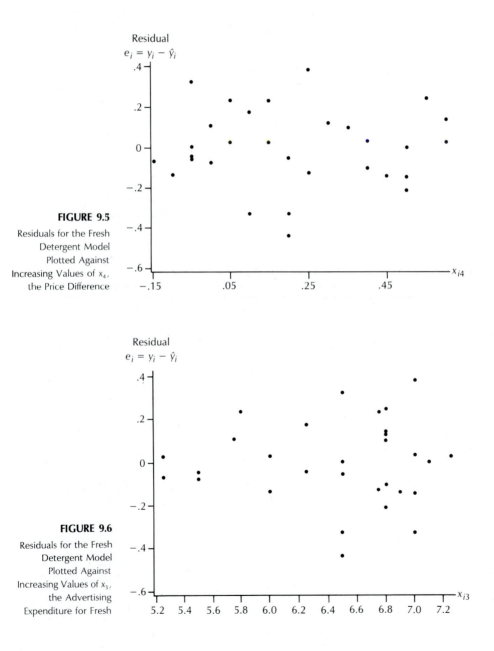

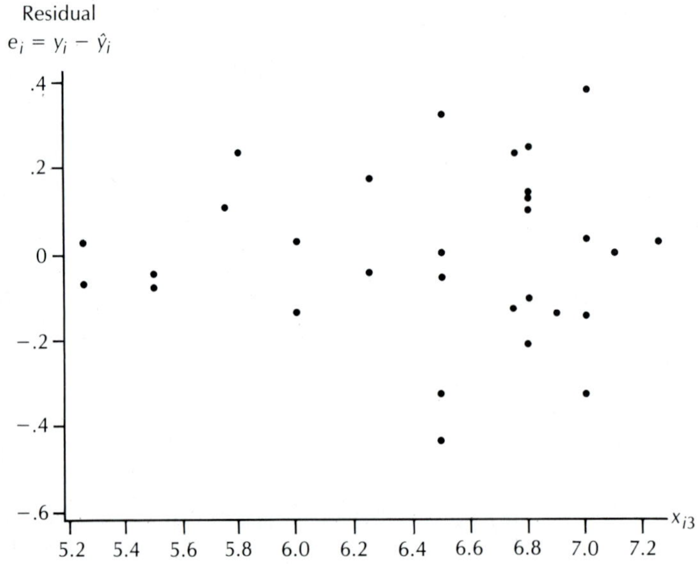

FIGURE 9.6
Residuals for the Fresh Detergent Model Plotted Against Increasing Values of x_3, the Advertising Expenditure for Fresh

2. Increasing values of $\hat{y}$ (Figure 9.7).
3. The time order in which the historical Fresh Detergent data of Table 5.12 were observed (Figure 9.8).

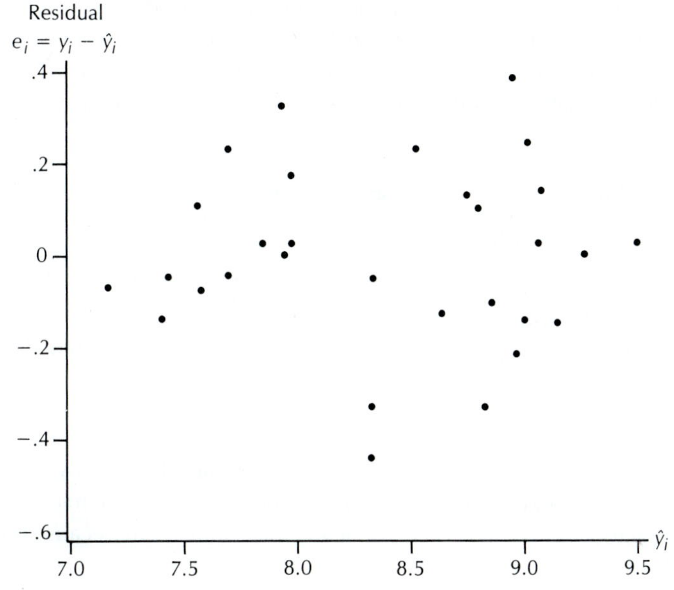

FIGURE 9.7

Residuals for the Fresh Detergent Model Plotted Against Increasing Values of $\hat{y}$, Predicted Demand for Fresh

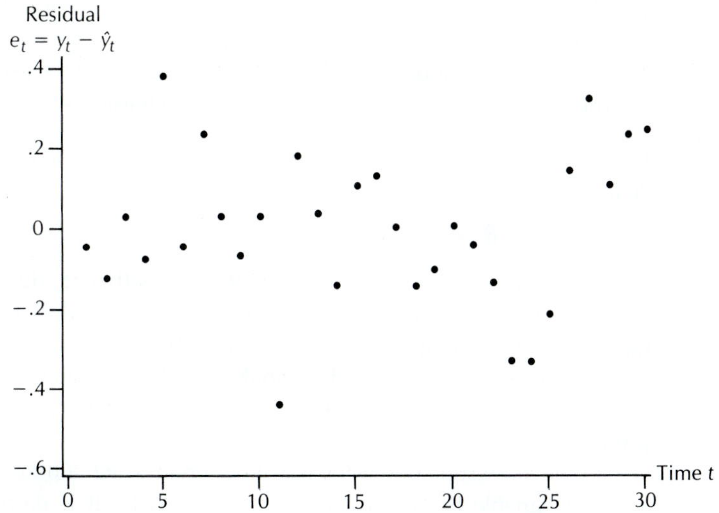

FIGURE 9.8

Residuals for the Fresh Detergent Model Plotted in Time Order

9.2 THE ASSUMPTION OF CORRECT FUNCTIONAL FORM

If the functional form of the relationship between the dependent variable and the independent variables is not properly accounted for by a regression model (and the resulting prediction equation), the residual plots constructed using the model will often display a pattern. This pattern can then be used to determine a more appropriate model. As an illustration of this idea, consider Figure 9.9. In Figure 9.9a we plot the values of a dependent variable y against the increasing values of an independent variable x. These plotted data points (the dots in the figure) indicate that there is a linear relationship between y and x. If we mistakenly use the regression model

$$y = \beta_0 + \epsilon$$

and the resulting prediction equation

$$\hat{y} = b_0$$

which does not properly account for the linear relationship between y and x (because the independent variable x is not included in the model), the predicted values of the dependent variable would be the squares that are plotted along the horizontal line that is defined by the prediction equation

$$\hat{y} = b_0$$

The resulting residuals, whose magnitudes are denoted by the dashed lines in Figure 9.9a, are the observed values of the dependent variable minus the corresponding predicted values of the dependent variable. Notice that because the prediction equation

$$\hat{y} = b_0$$

does not account for the linear relationship between y and x

1. The residuals for low values of x are negative and the residuals for high values of x are positive.
2. When these residuals are plotted against increasing values of x (see Figure 9.9b), they display a straight line pattern with increasing values of x.

This linear pattern in the residuals suggests that a more appropriate model for predicting y is

$$y = \beta_0 + \beta_1 x + \epsilon$$

The example illustrated in Figure 9.9 shows that residual plots against the increasing values of an independent variable not included in a regression model can be useful in indicating that the omission of this variable was a mistake. Of course, this same investigation can be done analytically by including the potentially important independent variable and looking at the appropriate t_{b_j} statistic and associated prob-value.

One interesting possibility is a situation in which a potentially important independent variable is initially included in a model but is dropped from the model

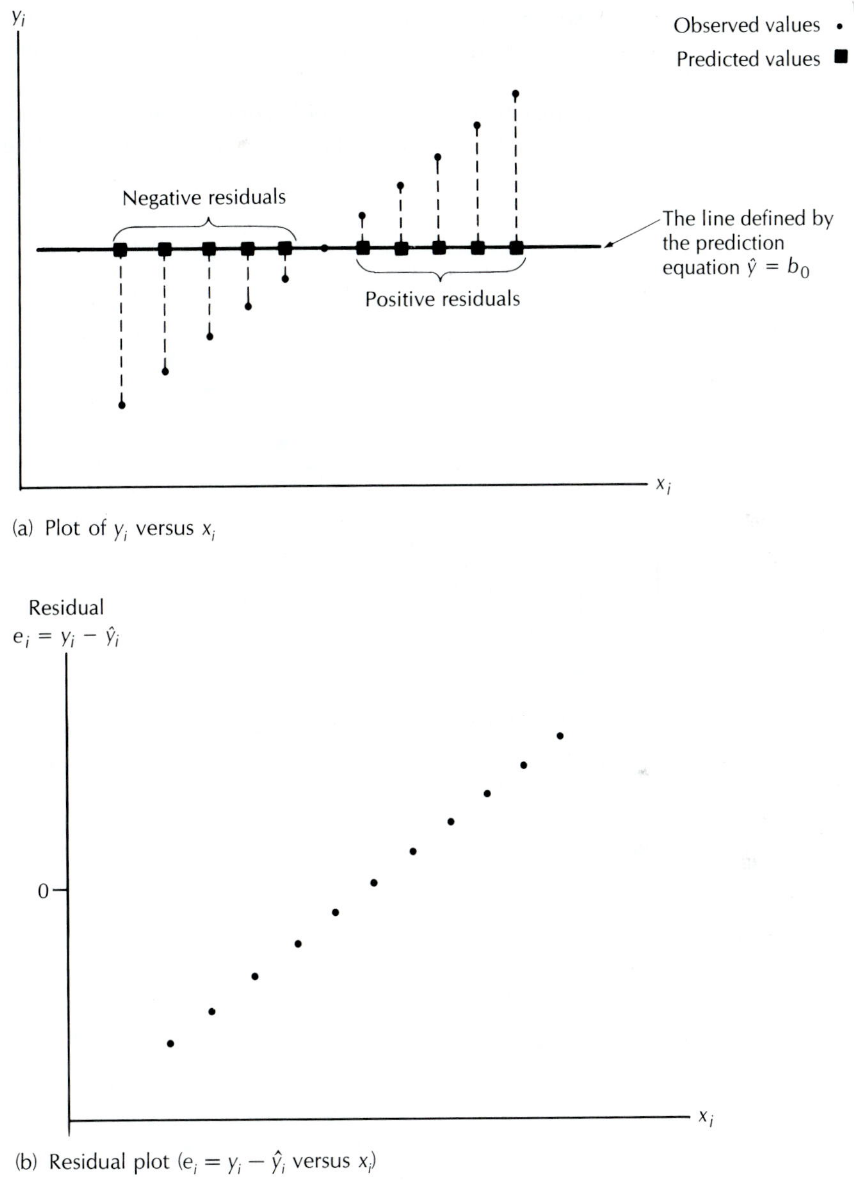

FIGURE 9.9
When the Regression Model Does Not Account for the Linear Relationship Between y and x, the Residual Plot Displays a Linear Trend

because the appropriate t_{b_j} statistic (and associated prob-value) do not indicate that the independent variable significantly affects the dependent variable. It is entirely possible that the independent variable should have been included in the model in a quadratic form. When the variable is included in a linear form, the t_{b_j} statistic and associated prob-value do not indicate that it significantly affects the dependent variable. However, a plot of the residuals against increasing values of the independent variable should show the importance of the independent variable and should indicate the quadratic nature of the relationship between the dependent and independent variables.

In order to illustrate this idea, consider Figure 9.10. In Figure 9.10a we plot the values of a dependent variable y against the increasing values of an independent

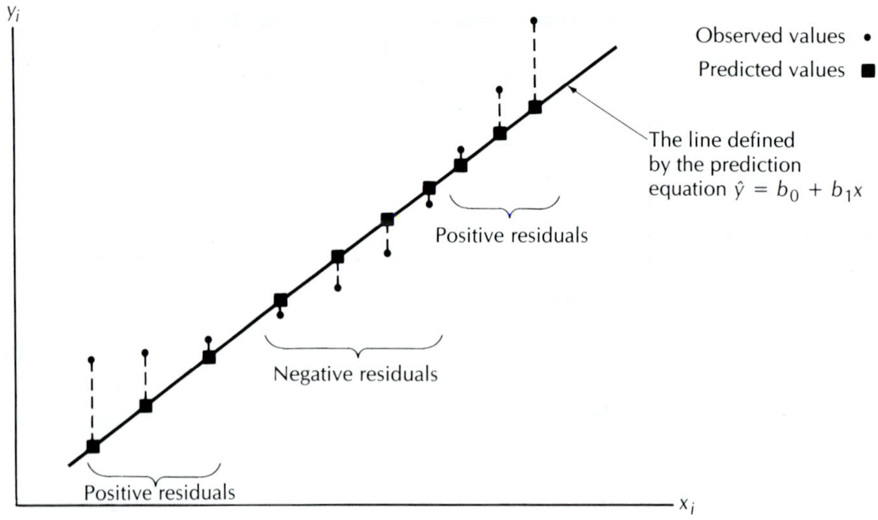

(a) Plot of y_i versus x_i

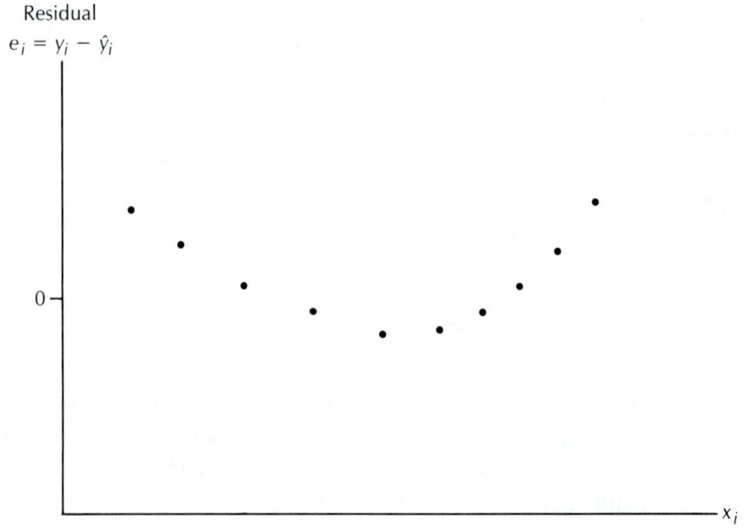

FIGURE 9.10 When the Regression Model Does Not Account for the Quadratic Relationship Between y and x, the Residual Plot Displays a Curved (or Parabolic) Pattern

(b) Residual plot ($e_i = y_i - \hat{y}_i$ versus x_i)

variable x. These plotted data points (the dots in the figure) indicate that there is a quadratic relationship between y and x. If we mistakenly use the regression model

$$y = \beta_0 + \beta_1 x + \epsilon$$

and the resulting prediction equation

$$\hat{y} = b_0 + b_1 x$$

which does not properly account for the quadratic relationship between y and x (because x^2 is not included in the model), the predicted values of the dependent variable would be the squares that are plotted along the line defined by the prediction equation

$$\hat{y} = b_0 + b_1 x$$

The resulting residuals, whose magnitudes are denoted by the dashed lines in Figure 9.10a, are the observed values of the dependent variable minus the corresponding predicted values of the dependent variable. Notice that because the prediction equation

$$\hat{y} = b_0 + b_1 x$$

does not account for the quadratic relationship between y and x

1. The residuals for very low values of x and for very high values of x are positive, while the residuals for intermediate values of x are negative.
2. When these residuals are plotted against increasing values of x (see Figure 9.10b), they display a curved or parabolic pattern with increasing values of x.

This curved pattern in the residuals suggests that a more appropriate model for predicting y is

$$y = \beta_0 + \beta_1 x + \beta_2 x^2 + \epsilon$$

Up to this point, we have illustrated patterns in residual plots that indicate an incorrect functional form. If little or no pattern exists in these plots, this indicates that the relationship between the dependent variable and the independent variables is (probably) properly accounted for. For example, since there is little or no pattern in either Figure 9.1 or Figure 9.2, which are plots of the residuals obtained from the two-variable fuel consumption model

$$y = \beta_0 + \beta_1 x_1 + \beta_2 x_2 + \epsilon$$

against increasing values of x_1 and x_2, we are led to conclude that this two-variable fuel consumption model properly accounts for the relationship between y and x_1 and x_2.

We conclude this section with a more detailed example of the use of residual plots.

EXAMPLE 9.3

Consider the Fresh Detergent problem. We illustrate the ideas of this section by considering the model

$$y = \beta_0 + \beta_1 x_3 + \epsilon$$

Using this model and the Fresh Detergent data of Table 5.12, we obtain a regression analysis yielding the prediction equation

$$\begin{aligned}\hat{y} &= b_0 + b_1 x_3 \\ &= 1.6490 + 1.0434x_3\end{aligned}$$

The residuals are calculated, for $i = 1, 2, \ldots, 30$, as

$$e_i = y_i - \hat{y}_i = y_i - 1.6490 - 1.0434x_{i3}$$

In this model the independent variable x_4 has been omitted, as has x_3^2. Both of these variables (as well as the interaction term x_4x_3) were thought to be necessary in the final Fresh Detergent model. A plot of the residuals against the increasing values of x_3 is shown in Figure 9.11. Notice the pattern of positive residuals associated with high or low values of x_3 and negative residuals associated with intermediate values of x_3. This indicates a quadratic relationship between y and x_3 and hence suggests that x_3^2 be included in the model. In Figure 9.12 a plot of the residuals against the increasing values of x_4 gives a slight indication of a linear relationship between y and x_4 (we see a predominance of positive residuals on the right side of the plot). This suggests that x_4 be included

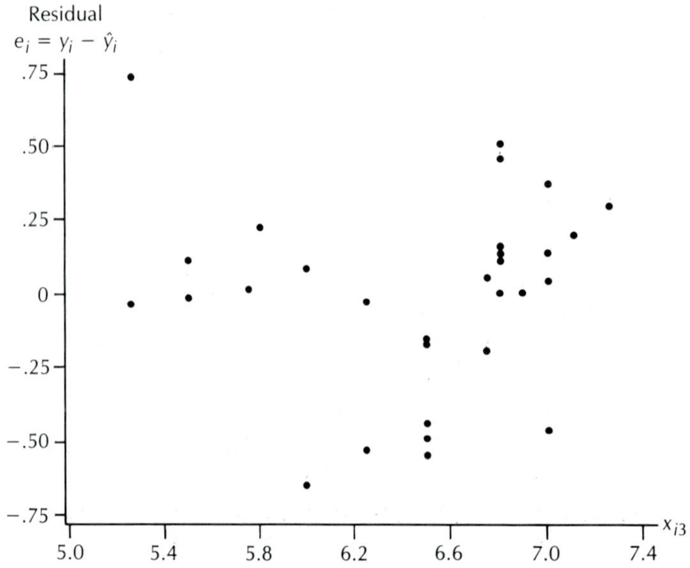

FIGURE 9.11
Residuals for the Fresh Detergent Model $y = \beta_0 + \beta_1 x_3 + \epsilon$ Plotted Against the Increasing Values of x_3

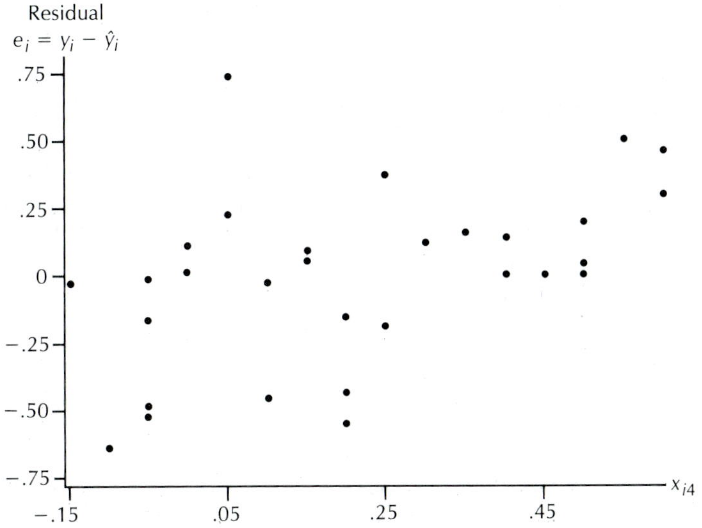

FIGURE 9.12
Residuals for the Fresh Detergent Model $y = \beta_0 + \beta_1 x_3 + \epsilon$ Plotted Against the Increasing Values of x_4

in the model. Since previous discussions also indicate that the interaction term x_4x_3 might be useful, we would be led to try the model

$$y = \beta_0 + \beta_1 x_4 + \beta_2 x_3 + \beta_3 x_3^2 + \beta_4 x_4 x_3 + \epsilon$$

As previously shown in Table 6.6, the t_{b_j} statistics and prob-values related to the hypotheses $H_0: \beta_1 = 0$, $H_0: \beta_2 = 0$, $H_0: \beta_3 = 0$, and $H_0: \beta_4 = 0$ indicate that x_4, x_3, x_3^2, and x_4x_3 should all probably be included in this model. Moreover, the appropriateness of the model can be further verified by examining the residual plots obtained for this model. Since there is little or no pattern in Figure 9.5, which is a plot of these residuals against the increasing values of x_4, and in Figure 9.6, which is a plot of these residuals against the increasing values of x_3, we are led to conclude that the regression model

$$y = \beta_0 + \beta_1 x_4 + \beta_2 x_3 + \beta_3 x_3^2 + \beta_4 x_4 x_3 + \epsilon$$

properly accounts for the relationship between y and x_4 and x_3.

9.3 THE POPULATION OF ERROR TERMS

In order for the regression techniques we have presented in previous chapters to be valid, a regression model

$$y = \mu + \epsilon = \beta_0 + \beta_1 x_1 + \beta_2 x_2 + \cdots + \beta_p x_p + \epsilon$$

must, in addition to having the correct functional form, at least approximately satisfy the inference assumptions. To fully discuss these assumptions, we need to introduce what we call the **population of potential error terms**. Recall that the ith error term,

$$\epsilon_i = y_i - \mu_i$$

is the difference between y_i, the ith observed value of the dependent variable, and μ_i, the ith mean value of the dependent variable. Also, we assume that y_i has been randomly selected from the ith historical population of potential values of the dependent variable. Therefore, since (before selection) y_i could potentially be any one of an infinite number of values, and since an error term could be defined for each of these potential y_i values, we see that ϵ_i could potentially be any one of an infinite number of values. This population of potential ϵ_i values is *the* i*th population of potential error terms*.

This population of potential error terms has two important properties. The first property is as follows (here, if $i = 0$, we are referring to a future population).

> For $i = 1, 2, \ldots, n$, and for $i = 0$, the mean of the ith population of potential error terms is zero.

This property follows because the mean of the potential y_i values in the ith population of potential values of the dependent variable is μ_i, and because the potential ϵ_i values in the ith population of potential error terms are obtained by simply subtracting μ_i from the potential y_i values. Thus, since we have said that the error term ϵ_i is sometimes positive (if y_i is greater than μ_i) and is sometimes negative (if y_i is less than μ_i), it seems intuitively reasonable that the positive potential ϵ_i values might exactly cancel out the negative potential ϵ_i values. This is in fact what happens.

The second property of the population of potential error terms is as follows.

> For $i = 1, 2, \ldots, n$, and for $i = 0$, the variance of the ith population of potential error terms is σ_i^2, where σ_i^2 is the variance of the ith population of potential values of the dependent variable.

This property follows because it is clear that the amount of variability in the potential ϵ_i values in the ith population of potential error terms is the same as the amount of variability in the potential y_i values in the ith population of potential values of the dependent variable (since the potential ϵ_i values are simply the potential y_i values minus the constant μ_i).

Finally, we note that since we can assume that y_i has been randomly selected from the *i*th population of potential values of the dependent variable, we can also assume that ϵ_i has been randomly selected from the *i*th population of potential error terms.

9.4 INFERENCE ASSUMPTION 1: CONSTANT VARIANCE

We are now ready to further discuss the first inference assumption.

INFERENCE ASSUMPTION 1 (CONSTANT VARIANCE):

The *n* historical populations, and any future population, of potential values of the dependent variable have *equal variances*. That is,

$$\sigma_1^2 = \sigma_2^2 = \cdots = \sigma_n^2 = \sigma_0^2$$

Said equivalently, the populations of potential error terms have equal variances.

This assumption essentially says that σ_i^2, the variance of the *i*th population of potential values of the dependent variable, is independent of $x_{i1}, x_{i2}, \ldots, x_{ip}$, the values of the *p* independent variables. When inference assumption 1 holds for a regression model, we say that the model displays a *constant error variance*, and we denote each of the equal variances $\sigma_1^2, \sigma_2^2, \ldots, \sigma_n^2$, and σ_0^2 by the symbol σ^2.

We now present an example of a regression problem in which the Constant Variance assumption does not hold.

EXAMPLE 9.4

The National Association of Retail Hardware Stores (NARHS), a nationally known trade association, wishes to investigate the relationship between *x*, home value (in thousands of dollars), and *y*, yearly expenditure (in dollars) on upkeep (such as lawn care, painting, repairs). A random sample of 40 homeowners is taken, and the results are given in Table 9.3. Figure 9.13 gives a plot of *y* versus *x*. From this plot it appears that *y* is increasing (probably in a quadratic fashion) as *x* increases, and because the *y* values seem to fan out as *x* increases, the variance of the population of potential yearly upkeep expenditures for houses worth *x* (thousand dollars) appears to increase as *x* increases. For example, the variance of the population of potential yearly upkeep expenditures for houses worth \$100,000 would be larger than the variance of the population of potential yearly upkeep expenditures for houses worth \$50,000. Increasing variance makes some

TABLE 9.3 NARHS Upkeep Expenditure Data

House	Value of House (thousands of dollars) (x)	Expenditure on Upkeep (dollars) (y)
1	118.50	706.04
2	76.54	398.60
3	92.43	436.24
4	111.03	501.71
5	80.34	426.45
6	49.84	144.24
7	114.52	644.23
8	50.89	211.54
9	128.93	675.87
10	48.14	189.02
11	85.50	459.04
12	115.51	813.62
13	114.16	602.39
14	102.95	428.52
15	92.86	387.50
16	84.39	434.63
17	123.53	698.00
18	77.77	355.75
19	112.10	737.59
20	101.02	706.66
21	76.52	424.57
22	116.09	656.92
23	62.72	301.03
24	84.91	321.07
25	88.64	519.40
26	81.41	348.50
27	60.22	162.17
28	95.55	482.55
29	79.39	460.07
30	89.25	475.45
31	136.10	835.16
32	24.45	62.70
33	52.28	239.89
34	143.09	1005.32
35	41.86	184.18
36	43.10	212.80
37	66.79	313.45
38	106.43	658.47
39	61.01	195.08
40	99.01	545.42

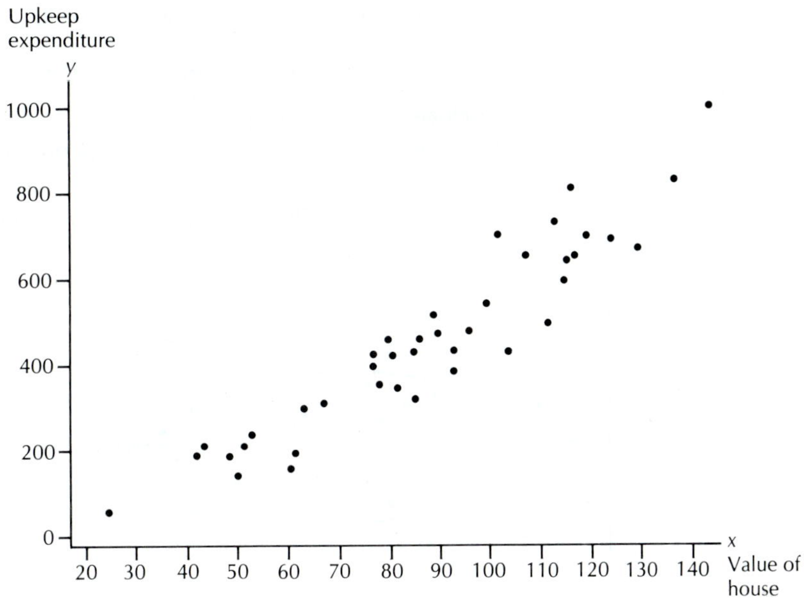

FIGURE 9.13
Upkeep Expenditure Versus Value of House for NARHS Data

intuitive sense here, since people with more expensive homes generally have higher incomes and can afford to pay to have upkeep done by lawn services, painters, and so on if they wish, or can perform upkeep chores themselves if they wish, thus causing a relatively large variation in upkeep expenses.

Residual plots can be used to determine whether or not inference assumption 1 holds for a regression model. Recall that the ith residual is

$$e_i = y_i - \hat{y}_i$$

which is the difference between the ith observed value of the dependent variable and the ith predicted value of the dependent variable. To see how the residuals $e_1, e_2, \ldots, e_n$ relate to inference assumption 1, consider the linear regression model

$$y_i = \mu_i + \epsilon_i = \beta_0 + \beta_1 x_{i1} + \beta_2 x_{i2} + \cdots + \beta_p x_{ip} + \epsilon_i$$

which implies that for $i = 1, 2, \ldots, n$,

$$\epsilon_i = y_i - \mu_i = y_i - (\beta_0 + \beta_1 x_{i1} + \beta_2 x_{i2} + \cdots + \beta_p x_{ip})$$

Then, since the point estimate of μ_i is

$$\hat{y}_i = b_0 + b_1 x_{i1} + b_2 x_{i2} + \cdots + b_p x_{ip}$$

the point estimate of the ith error term ϵ_i is the ith residual

$$e_i = y_i - \hat{y}_i = y_i - (b_0 + b_1x_{i1} + b_2x_{i2} + \cdots + b_px_{ip})$$

Therefore, the residuals $e_1, e_2, \ldots, e_n$ provide point estimates of the error terms $\epsilon_1, \epsilon_2, \ldots, \epsilon_n$. Now, since for $i = 1, 2, \ldots, n$, the ith population of potential error terms has mean zero, we would expect the residuals $e_1, e_2, \ldots, e_n$ to fluctuate around zero. It is the pattern in which the residuals fluctuate around zero that helps

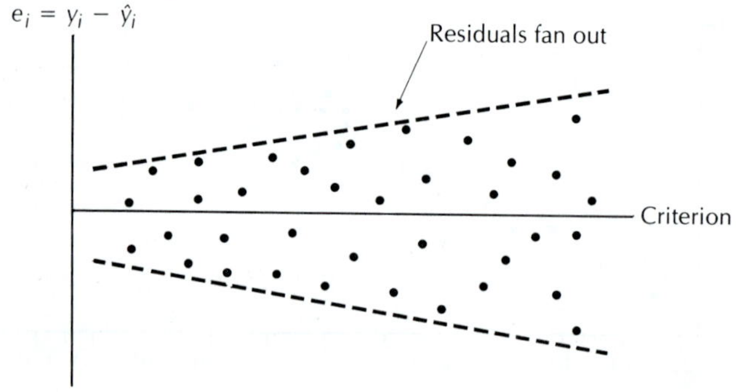

(a) Nonconstant error variance: error variance increases with increasing values of the criterion

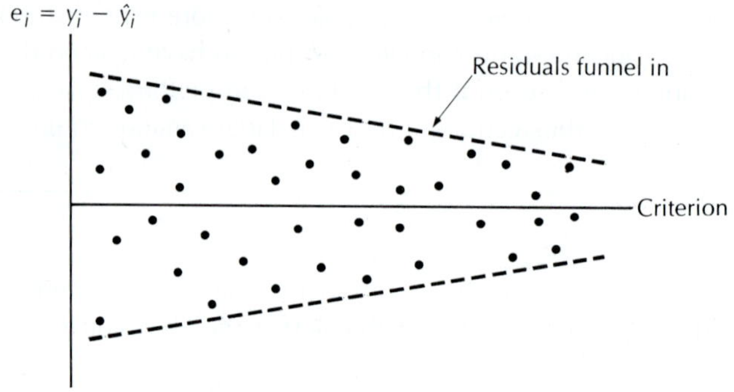

(b) Nonconstant error variance: error variance decreases with increasing values of the criterion

FIGURE 9.14
Possible Implications of a Residual Plot Against Increasing Values of a Particular Criterion (Such as One of the Independent Variables, the Predicted Value of the Dependent Variable, or Time)

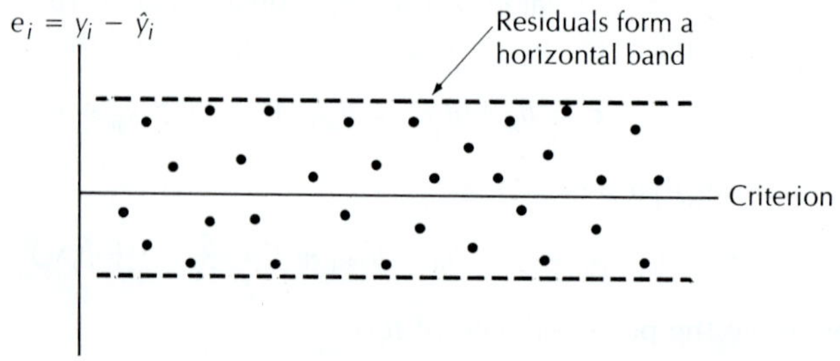

(c) Constant error variance with increasing values of the criterion us determine whether or not a constant error variance exists. If inference assumption 1 holds, the residuals should display a pattern that indicates that σ_i^2, the variance of the ith population of potential error terms, is the same for all i, $i = 1, 2, \ldots, n$. However, if the residuals display a pattern that indicates that σ_i^2 is not the same for all i, $i = 1, 2, \ldots, n$, then there is reason to believe that inference assumption 1 does not hold.

In order to use residual plots to investigate the validity of inference assumption 1, residual plots should be made against the following criteria:

1. Increasing values of each of the independent variables in the model.
2. Increasing values of $\hat{y}$, the predicted value of the dependent variable.
3. The time order in which the historical data have been observed.

A residual plot with the appearance of Figure 9.14a—in which the residuals tend to fan out with increasing values of a particular criterion—indicates that the residuals $e_1, e_2, \ldots, e_n$, and hence the error terms $\epsilon_1, \epsilon_2, \ldots, \epsilon_n$, are increasing in absolute value with increasing values of the criterion. This would suggest that σ_i^2 is increasing with increasing values of the criterion. Hence, a plot with the appearance of Figure 9.14a suggests that a nonconstant error variance exists and that inference assumption 1 does not hold. A plot with the appearance of Figure 9.14b—in which the residuals tend to funnel in with increasing values of a particular criterion—indicates that the residuals $e_1, e_2, \ldots, e_n$, and hence the error terms $\epsilon_1, \epsilon_2, \ldots, \epsilon_n$, are decreasing in absolute value with increasing values of the criterion. Such a plot suggests that σ_i^2 is decreasing with increasing values of the criterion. Hence, a plot with the appearance of Figure 9.14b also suggests that a nonconstant error variance exists and that inference assumption 1 does not hold. However, a plot with the appearance of Figure 9.14c—in which the residuals tend to form a horizontal band—indicates that the residuals $e_1, e_2, \ldots, e_n$, and hence the error terms $\epsilon_1, \epsilon_2, \ldots, \epsilon_n$, are remaining relatively constant in absolute value with increasing values of a particular criterion. This suggests that σ_i^2 is not changing with increasing values of the criterion. Therefore, a plot with the appearance of Figure 9.14c does not provide evidence to suggest that inference assumption 1 is violated.

EXAMPLE 9.5

Consider the plot of the NARHS upkeep expenditure data in Figure 9.13. Since y appears to increase in a quadratic fashion as x increases, we begin by trying the model

$$y_i = \beta_0 + \beta_1 x_i + \beta_2 x_i^2 + \epsilon_i$$

When we estimate β_0, β_1, and β_2 using the data in Table 9.3, we obtain the prediction equation

$$\hat{y}_i = -14.4436 + 3.0585x_i + .0246x_i^2$$

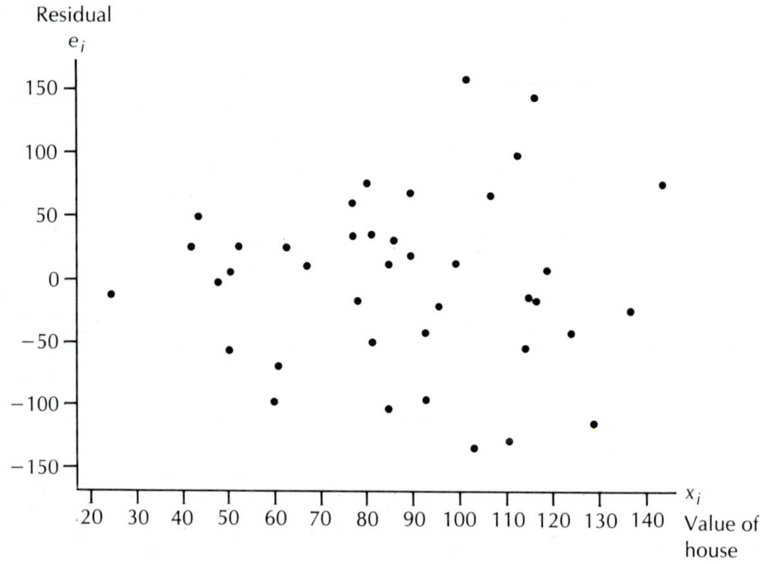

FIGURE 9.15
Residuals for the Quadratic NARHS Model Plotted Against Increasing Values of x_i

and residuals calculated (for $i = 1, 2, \ldots, 40$) as

$$e_i = y_i - \hat{y}_i = y_i - (-14.4436 + 3.0585x_i + .0246x_i^2)$$

Figure 9.15 is a plot of these residuals against increasing values of x. We see that the residuals appear to fan out as x increases, indicating that the variance of the population of potential yearly upkeep expenditures (in dollars) for houses worth x (thousand dollars) increases as x increases. Thus, since a nonconstant error variance seems to exist, we cannot use the formulas of previous chapters to make statistical inferences. In Chapter 10 we discuss how we can make statistical inferences when a nonconstant error variance exists.

Next, note that since each of the residual plots in Figures 9.1, 9.2, 9.3, and 9.4 probably has a horizontal band appearance (admittedly, it is somewhat difficult to tell with only eight residuals), we conclude that none of these residual plots indicate that inference assumption 1 is violated for the fuel consumption model

$$y = \beta_0 + \beta_1 x_1 + \beta_2 x_2 + \epsilon$$

Here we have plotted the residuals against increasing values of $\hat{y}$, predicted fuel consumption (see Figure 9.3). We do this because, although residual plots against increasing values of x_1 or against increasing values of x_2 may not show any evidence of a nonconstant error variance, a residual plot against increasing values of

$$\hat{y} = b_0 + b_1 x_1 + b_2 x_2 = 13.109 - .0900x_1 + .0825x_2$$

could conceivably show evidence of an increasing error variance. (We can think of $\hat{y}$ as a surrogate for "overall coldness," and it is conceivable that the variation in fuel consumption due to factors such as wind velocity (included in the error term for the two-variable fuel consumption model) might increase as it gets colder. In spite of this possibility, Figure 9.3 provides little evidence suggesting that this is the case. Similarly, we have plotted the residuals in time order (see Figure 9.4), because the fuel consumption data have been collected in time order and because the weather has generally been getting warmer over the time period in which the data have been collected. Therefore, it is conceivable that the variation in fuel consumption due to factors included in the error term (such as wind velocity) could decrease over time. Again, in spite of this possibility, Figure 9.4 provides little evidence to suggest that this is the case.

Finally, since each of the residual plots in Figures 9.5, 9.6, 9.7, and 9.8 has a horizontal band appearance, we conclude that none of these residual plots indicate that inference assumption 1 is violated for the Fresh Detergent model

$$y = \beta_0 + \beta_1 x_4 + \beta_2 x_3 + \beta_3 x_3^2 + \beta_4 x_4 x_3 + \epsilon$$

9.5 INFERENCE ASSUMPTION 2: INDEPENDENCE

INFERENCE ASSUMPTION 2 (INDEPENDENCE):

The randomly selected values of the dependent variable

$$y_1, y_2, \ldots, y_n, \quad \text{and} \quad y_0$$

are all *statistically independent,* or equivalently, the randomly selected error terms

$$\epsilon_1, \epsilon_2, \ldots, \epsilon_n, \quad \text{and} \quad \epsilon_0$$

are all statistically independent.

Inference assumption 2 is most likely to be violated when the data being used in a regression problem are *time series data,* that is, data that have been collected in a time sequence. For such data, the error terms

$$\epsilon_1, \epsilon_2, \ldots, \epsilon_n, \quad \text{and} \quad \epsilon_0$$

which we assume are listed here in time order, can be **autocorrelated**. Intuitively, we say that error terms occurring over time have **positive autocorrelation** if a positive error term in time period t tends to produce, or be followed by, another positive error term in time period $t + k$ (a later time period), and if a negative error term in time period t tends to produce, or be followed by, another negative error term in time period $t + k$. In other words, positive autocorrelation exists when positive error terms tend to be followed over time by positive error terms, and negative error terms tend to be followed over time by negative error terms. An example of positive autocorrelation in the error terms is depicted in Figure 9.16a, which illustrates that positive autocorrelation in the error terms can produce a cyclical pattern over time. If we consider μ_i, the mean of the ith historical population of potential values of the dependent variable, we have seen that a positive error term produces a value of the dependent variable that is greater than μ_i, and a negative error term produces a value of the dependent variable that is smaller than μ_i. This says that positive autocorrelation in the error terms means that greater than average values of the dependent variable tend to be followed by greater than average values of the dependent variable, and smaller than average values of the dependent variable tend to be followed by smaller than average values of the dependent variable.

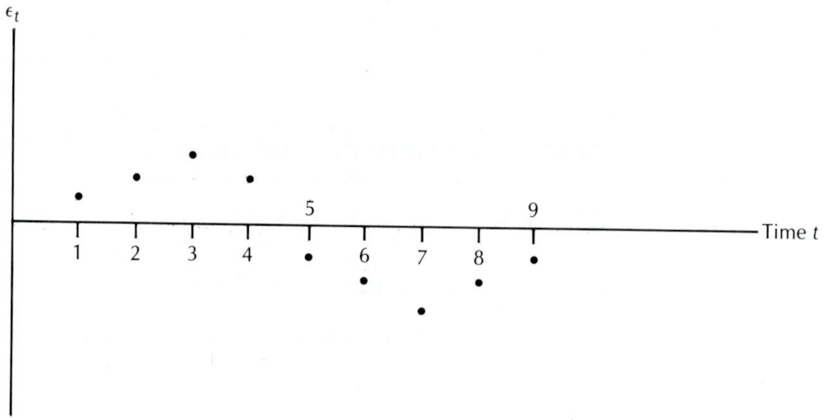

(a) Positive autocorrelation in the error terms: cyclical pattern

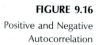

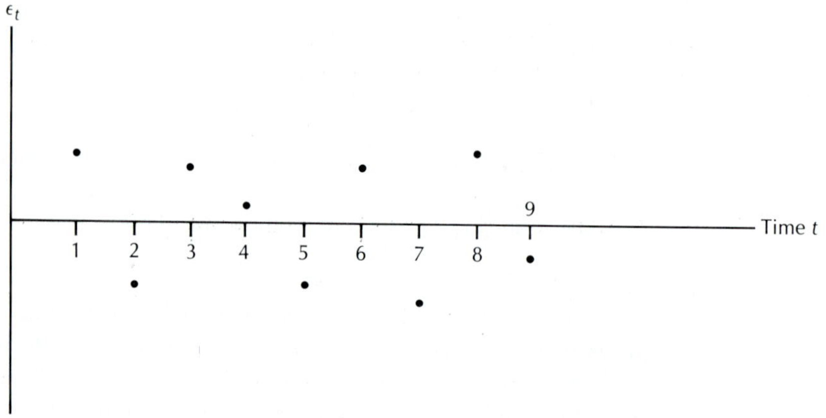

FIGURE 9.16
Positive and Negative Autocorrelation

(b) Negative autocorrelation in the error terms: alternating pattern

EXAMPLE 9.6

In the Fresh Detergent problem, the historical demand data given in Example 5.12 are time series data since they were collected in time sequence over 30 sales periods. If we consider the model

$$\begin{aligned} y &= \mu + \epsilon \\ &= \beta_0 + \beta_1 x_4 + \beta_2 x_3 + \beta_3 x_3^2 + \beta_4 x_4 x_3 + \epsilon \end{aligned}$$

then the effect of the competitors' average advertising expenditure is included in ϵ, the error term. If, for the moment, we assume that the competitors' average advertising expenditure significantly affects the demand for Fresh, then a higher than average competitors' average advertising expenditure probably causes demand for Fresh to be lower than average and hence probably causes a negative error term to occur. On the other hand, a lower than average competitors' average advertising expenditure probably causes demand for Fresh to be higher than average and hence probably causes a positive error term to occur. If, then, Enterprise Industries' competitors tend to spend money on advertising in a cyclical fashion—spending large amounts for several consecutive sales periods (during an advertising campaign) and then spending lesser amounts for several consecutive sales periods—and if the Fresh Detergent demand data are collected in successive sales periods, a negative error term in one sales period will tend to be followed by a negative error term in the next sales period, and a positive error term in one sales period will tend to be followed by a positive error term in the next sales period. In this case, the error terms

$$\epsilon_1, \epsilon_2, \ldots, \epsilon_{30}, \quad \text{and} \quad \epsilon_0$$

would display positive autocorrelation, and thus these error terms, or equivalently, the demand values

$$y_1, y_2, \ldots, y_{30}, \quad \text{and} \quad y_0$$

would be dependent, and inference assumption 2 would be violated. However, in spite of this possibility, we will (in the next example) use a residual plot to verify that this assumption does in fact hold for the Fresh Detergent model.

Intuitively, error terms occurring over time have **negative autocorrelation** if a positive error term in time period t tends to produce, or be followed by, a negative error term in time period $t + k$, and if a negative error term in time period t tends to produce, or be followed by, a positive error term in time period $t + k$. In other words, negative autocorrelation exists when positive error terms tend to be followed over time by negative error terms, and negative error terms tend to be followed over time by positive error terms. An example of negative autocorrelation in the error terms is depicted in Figure 9.16b, which illustrates that negative autocorrelation in the error terms can produce an alternating pattern over time. Since a positive error term produces a greater than average value of the dependent variable, and a negative error term produces a smaller than average value of the dependent variable, negative autocorrelation in the error terms means that greater than average values of the dependent variable tend to be followed by smaller than average values of the dependent variable, and smaller than average values of the dependent variable tend to be followed by greater than average values of the dependent variable. An example of negative autocorrelation might be provided by a retailer's weekly stock orders. Here, a larger than average stock order one week might result in an oversupply and hence a smaller than average order the next week.

The ideas we have presented thus far allow us to give a relatively simple interpretation of inference assumption 2. In essence, the Independence assumption says that the randomly selected error terms

$$\epsilon_1, \epsilon_2, \ldots, \epsilon_n, \quad \text{and} \quad \epsilon_0$$

display no positive autocorrelation and display no negative autocorrelation. This would say that the error terms

$$\epsilon_1, \epsilon_2, \ldots, \epsilon_n, \quad \text{and} \quad \epsilon_0$$

occur in a random pattern over time, as illustrated in Figure 9.17. Such a pattern would imply that these error terms are statistically independent, which would in turn imply that the randomly selected values of the dependent variable

$$y_1, y_2, \ldots, y_n, \quad \text{and} \quad y_0$$

are statistically independent.

If inference assumption 2 holds, we say that we have **independent error terms**. On the other hand, if the error terms

$$\epsilon_1, \epsilon_2, \ldots, \epsilon_n, \quad \text{and} \quad \epsilon_0$$

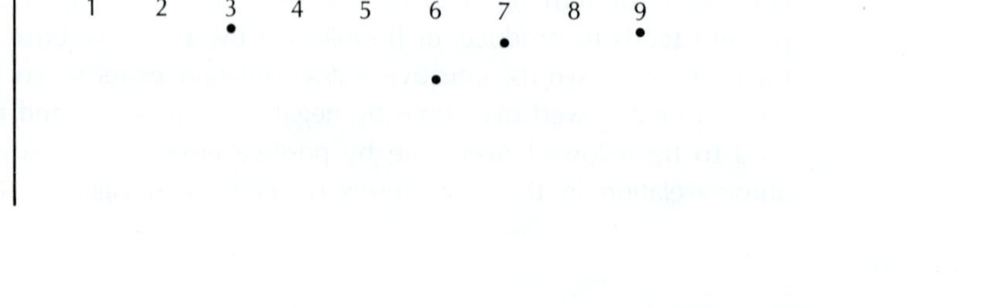

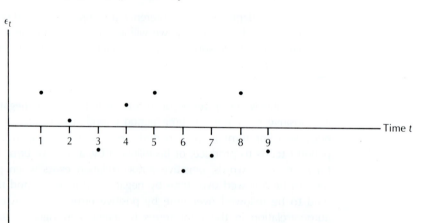

FIGURE 9.17
Little or No Autocorrelation in the Error Terms: Random Pattern

have positive autocorrelation or negative autocorrelation, this would imply that these error terms are related, or **statistically dependent**, which would in turn imply that the values of the dependent variable

$$y_1, y_2, \ldots, y_n, \quad \text{and} \quad y_0$$

are statistically dependent and that inference assumption 2 is violated.

Since the residuals $e_1, e_2, \ldots, e_n$ are point estimates of the error terms $\epsilon_1, \epsilon_2, \ldots, \epsilon_n$, a residual plot against time can be used to detect violations of inference assumption 2. If a residual plot against the time sequence in which the data have been collected has the appearance of Figure 9.16a—that is, if the plot displays a cyclical pattern—the error terms $\epsilon_1, \epsilon_2, \ldots, \epsilon_n$ are positively autocorrelated and the independence assumption does not hold. Another way to detect positive autocorrelation is to look at the signs of the time-ordered residuals. Letting + denote a positive residual and − denote a negative residual, we call a sequence of residuals with the same sign (for instance, + + +) a run. If positive autocorrelation exists, the signs of the residuals should display relatively few runs of fairly long duration. For instance, the pattern + + + + − − − + + + − − − − + + + + − − − in the time-ordered residuals would indicate that positive autocorrelation exists and that the Independence assumption is violated.

If a plot of the time-ordered residuals has the appearance of Figure 9.16b—that is, if the plot displays an alternating pattern—the error terms $\epsilon_1, \epsilon_2, \ldots, \epsilon_n$ are negatively autocorrelated and the Independence assumption does not hold. If we look at the signs of the time-ordered residuals, negative autocorrelation is characterized by many runs of relatively short duration. For example, the pattern + − + − + + − + − + − − + − in the time-ordered residuals would indicate that negative autocorrelation exists and that the Independence assumption is violated.

However, if a plot of the time-ordered residuals displays a random pattern, as illustrated in Figure 9.17, the error terms $\epsilon_1, \epsilon_2, \ldots, \epsilon_n$ have little or no autocorrelation, which suggests that these error terms are independent and that inference assumption 2 holds.

EXAMPLE 9.7

In this example we use residual plots to show that it is reasonable to assume that inference assumption 2 holds for the two-variable fuel consumption model

$$\begin{aligned} y &= \mu + \epsilon \\ &= \beta_0 + \beta_1 x_1 + \beta_2 x_2 + \epsilon \end{aligned}$$

and also for the Fresh Detergent model

$$\begin{aligned} y &= \mu + \epsilon \\ &= \beta_0 + \beta_1 x_4 + \beta_2 x_3 + \beta_3 x_3^2 + \beta_4 x_4 x_3 + \epsilon \end{aligned}$$

The residuals for the two-variable fuel consumption model are plotted in time order in Figure 9.4. Examining this figure, we see that the residuals do not seem to display any well-defined cyclical pattern as in Figure 9.16a, nor do they seem to display a well-defined alternating pattern as in Figure 9.16b. Rather, the residuals in Figure 9.4 probably display a random pattern as in Figure 9.17. This implies that there is probably little or no autocorrelation in the error terms for the two-variable fuel consumption model and that these error terms are probably independent. Hence, it is reasonable to believe that inference assumption 2 holds for this model.

The residuals for the Fresh Detergent model are plotted in time order in Figure 9.8. Looking at this figure, we see that these residuals probably display a random pattern as in Figure 9.17. This implies that there is probably little or no autocorrelation in the error terms for the Fresh Detergent model and that these error terms are probably independent. Hence, it is reasonable to believe that inference assumption 2 holds for this model.

When a residual plot suggests that the error terms may be autocorrelated, one might wish to use a formal statistical test for autocorrelation. Two tests that can be used are the **runs test** and the **Durbin-Watson test**. For a description of the runs test, see Draper and Smith (1981). Here, we discuss the Durbin-Watson test for autocorrelation.

The Durbin-Watson test for autocorrelation is based on the Durbin-Watson statistic.

The **Durbin-Watson statistic** is

$$d = \frac{\sum_{t=2}^{n} (e_t - e_{t-1})^2}{\sum_{t=1}^{n} e_t^2}$$

where $e_1, e_2, \ldots, e_n$ are the time-ordered residuals.

If we consider testing the null hypothesis

H_0: The error terms are not autocorrelated

versus the alternative hypothesis

H_1: The error terms are positively autocorrelated

then Durbin and Watson have shown that there are points (denoted $d_{L,\alpha}$ and $d_{U,\alpha}$) such that, if α is the probability of a Type I error, then

1. If $d < d_{L,\alpha}$, we reject H_0.
2. If $d > d_{U,\alpha}$, we do not reject H_0.
3. If $d_{L,\alpha} \leq d \leq d_{U,\alpha}$, the test is inconclusive.

Here small values of d lead to the conclusion of positive autocorrelation, because if d is small, the differences $(e_t - e_{t-1})$ are small, which indicates that the adjacent residuals e_t and e_{t-1} are of the same magnitude, which in turn says that the adjacent error terms ϵ_t and ϵ_{t-1} are positively correlated.

So that the Durbin-Watson test may be easily done, tables containing the points $d_{L,\alpha}$ and $d_{U,\alpha}$ have been constructed. These tables give the appropriate $d_{L,\alpha}$ and $d_{U,\alpha}$ points for various values of α, the probability of a Type I error; $k - 1$, the number of independent variables in the regression model; and n, the number of observations. Tables E-5 and E-6 in Appendix E give values for $\alpha = .05$ and $\alpha = .01$.

The Durbin-Watson test can also be used to test for negative autocorrelation. If we consider testing the null hypothesis

H_0: The error terms are not autocorrelated

versus the alternative hypothesis

H_1: The error terms are negatively autocorrelated

Durbin and Watson have shown that based on setting the probability of a Type I error equal to α, the points $d_{L,\alpha}$ and $d_{U,\alpha}$ are such that

1. If $(4 - d) < d_{L,\alpha}$, we reject H_0.
2. If $(4 - d) > d_{U,\alpha}$, we do not reject H_0.
3. If $d_{L,\alpha} \leq (4 - d) \leq d_{U,\alpha}$, the test is inconclusive.

Here large values of d (and hence small values of $4 - d$) lead to the conclusion of negative autocorrelation, because if d is large, this indicates that the differences $(e_t - e_{t-1})$ are large, which says that the adjacent error terms ϵ_t and ϵ_{t-1} are negatively autocorrelated.

Finally, if we wish to test the null hypothesis

H_0: The error terms are not autocorrelated

versus the alternative hypothesis

H_1: The error terms are positively or negatively autocorrelated

Durbin and Watson have shown that, based on setting the probability of a Type I error equal to α,

1. If $d < d_{L,\alpha/2}$ or if $(4 - d) < d_{L,\alpha/2}$, we reject H_0.
2. If $d > d_{U,\alpha/2}$ and if $(4 - d) > d_{U,\alpha/2}$, we do not reject H_0.
3. If $d_{L,\alpha/2} \leq d \leq d_{U,\alpha/2}$ and $d_{L,\alpha/2} \leq (4 - d) \leq d_{U,\alpha/2}$, the test is inconclusive.

Before concluding our presentation of the Durbin-Watson test, several comments are relevant. First, the validity of the Durbin-Watson test depends upon the assumption that the population of all possible residuals at any time t has a normal distribution. We will see how to tell whether or not this assumption is reasonable in the next section. Second, positive autocorrelation is found in practice more commonly than negative autocorrelation. Therefore, the first test we have presented (the test for positive autocorrelation) is used more often than the others. Third, most regression computer packages (including SAS) print the Durbin-Watson d statistic. We look at an example in which the Durbin-Watson statistic is used in Chapter 10 (see also Exercise 9.6 at the end of this chapter).

When residual plots or statistical tests indicate that the error terms in a regression problem are autocorrelated, remedies can be employed. We discuss these remedies in Chapter 10.

9.6 INFERENCE ASSUMPTION 3: NORMAL POPULATIONS

INFERENCE ASSUMPTION 3 (NORMAL POPULATIONS):

For $i = 1, 2, \ldots, n$, the ith historical population of potential values of the dependent variable is *normally distributed*. In addition, any future population of potential values of the dependent variable is *normally distributed*.

Or equivalently, for $i = 1, 2, \ldots, n$, the ith population of potential error terms is normally distributed. In addition, any future population of potential error terms is normally distributed.

Validation of the Normality assumption can be accomplished using several methods. First, we can construct a bar chart or histogram of the residuals. If the Normality assumption holds, then each error term ϵ_i (for $i = 1, 2, \ldots, n$) has been randomly selected from a normal distribution with mean zero and (if inference assumption 1 holds) variance σ^2. Thus, if the Normality assumption holds, when a histogram of the residuals (which are point estimates of the error terms) is constructed, the residuals should look like they are values that have been randomly selected from such a normal distribution. That is, the histogram of the residuals should look reasonably bell-shaped and should be reasonably symmetric about zero.

It is important to point out again that mild departures from the inference assumptions do not seriously hinder our ability to make statistical inferences. Hence, in constructing a histogram of the residuals, we are looking for pronounced, rather than subtle, departures from the Normality assumption. We require only that the histogram of the residuals look fairly normal. Also, if one wishes to employ a formal statistical test to determine whether or not the error terms $\epsilon_1, \epsilon_2, \ldots, \epsilon_n$ are normally distributed, two goodness-of-fit tests can be used to analyze the residuals—the chi-square test and the Kolmogorov-Smirnov test. These tests are discussed in many basic statistics texts; for example, see Neter, Wasserman, and Whitmore (1978). One other point worth making here is that violations of inference assumptions 1 or 2, as well as an incorrect functional form, can often cause a histogram of the residuals to look non-normal. Because of this, it is usually a good idea to use residual plots to check for incorrect functional form, nonconstant error variance, and positive or negative autocorrelation before attempting to validate the Normality assumption.

EXAMPLE 9.8

We have seen that inference assumptions 1 and 2 probably hold for the Fresh Detergent model

$$y = \beta_0 + \beta_1 x_4 + \beta_2 x_3 + \beta_3 x_3^2 + \beta_4 x_4 x_3 + \epsilon$$

In order to validate the Normality assumption, we consider Table 9.2, which lists the residuals for this model. Looking at this table, we see that these residuals range between −.5 and .4. We divide this interval into nine subintervals of equal width (the number of subintervals is somewhat arbitrary). We next determine from Table 9.2 how many of the residuals are within each of the subintervals. A frequency distribution of these residuals is given in Table 9.4. This frequency distribution is depicted graphically in the form of a histogram in Figure 9.18. We see that this histogram looks reasonably bell-shaped and symmetric. Since we

TABLE 9.4

Frequency Distribution of the Residuals for the Fresh Detergent Model $y = \beta_0 + \beta_1 x_4 + \beta_2 x_3 + \beta_3 x_3^2 + \beta_4 x_4 x_3 + \epsilon$

Subinterval	Frequency (No. in Subinterval)
−.5 to −.4	1
−.4 to −.3	2
−.3 to −.2	1
−.2 to −.1	5
−.1 to 0	5
0 to .1	6
.1 to .2	5
.2 to .3	3
.3 to .4	2

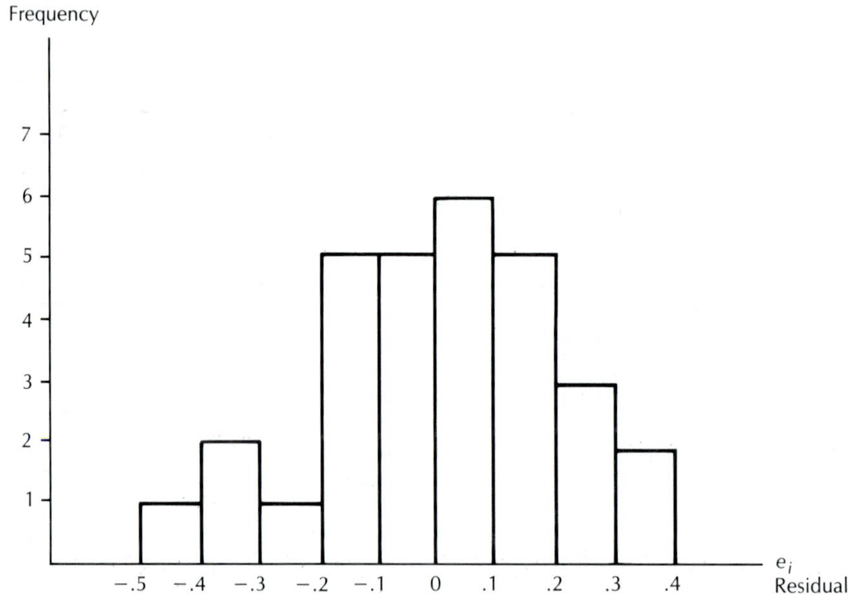

FIGURE 9.18
Histogram of the Residuals for the Fresh Detergent Model

are looking for pronounced departures from the Normality assumption, the histogram in Figure 9.18 does not suggest that any serious violation of inference assumption 3 exists.

Another way to validate the Normality assumption is to use *standardized residuals* (recall that the standardized residuals are computed by dividing the residuals by the standard error s). Since 68.26 percent of the element values in a normally distributed population are within 1 standard deviation of the mean, while 95.44 percent of the element values in a normally distributed population are within 2 standard deviations of the mean, and since inference assumption 3 implies that each error term ϵ_i (for $i = 1, 2, \ldots, n$) has been randomly selected from a normally distributed population of potential error terms having mean zero and (if inference assumption 1 holds) variance σ^2, then inference assumption 3 implies that

$$P(0 - \sigma \leq \epsilon_i \leq 0 + \sigma) = .6826$$
$$P(0 - 2\sigma \leq \epsilon_i \leq 0 + 2\sigma) = .9544$$

Dividing by σ, we obtain the equivalent expressions

$$P\left(-1 \leq \frac{\epsilon_i}{\sigma} \leq 1\right) = .6826$$

$$P\left(-2 \leq \frac{\epsilon_i}{\sigma} \leq 2\right) = .9544$$

Now, since the standardized residual e_i/s is a point estimate of ϵ_i/σ (remember that e_i is the point estimate of ϵ_i and that s is the point estimate of σ), these expressions intuitively say that if inference assumption 3 holds, then about 68 percent of the standardized residuals will be between -1 and 1, while about 95 percent of the standardized residuals will be between -2 and 2.[†] As an illustration, if we consider the Fresh Detergent model

$$y = \beta_0 + \beta_1 x_4 + \beta_2 x_3 + \beta_3 x_3^2 + \beta_4 x_4 x_3 + \epsilon$$

and the standardized residuals for this model (which are given in Table 9.2), we see that 21 of 30 (or 70 percent) of the standardized residuals are between -1 and 1, while 29 of 30 (or 96.7 percent) of the standardized residuals are between -2 and 2. These results are quite close to the 68 percent and 95 percent figures. Therefore, it is reasonable to believe that the Normality assumption holds for this model. If the percentages of standardized residuals between -1 and 1 and between -2 and 2 differed substantially from the 68 percent and 95 percent values, then a violation of inference assumption 3 would be indicated.

Another graphical technique for examining the validity of the Normality assumption is the **normal plot**. This technique (along with several other techniques for examining residuals) is discussed by Anscombe and Tukey (1963). The normal plot technique first requires that the residuals $e_1, e_2, \ldots, e_n$ be arranged in order from smallest to largest. Letting $e_{(i)}$ denote the ith residual in the ordered listing, we plot $e_{(i)}$ on the vertical axis against the point $z_{(i)}$ on the horizontal axis. Here, $z_{(i)}$ is defined to be the point on the scale of the standard normal curve such that

$$A[(-\infty, z_{(i)}]; N(0, 1)] = \frac{3i - 1}{3n + 1}$$

That is, $z_{(i)}$ is defined so that the area under the standard normal curve to the left of $z_{(i)}$ is $(3i - 1)/(3n + 1)$. If the Normality assumption holds, and if the model has the correct functional form, then this plot should appear as a straight line. Substantial departures from a straight line appearance (admittedly a subjective decision) indicate a violation of inference assumption 3.

EXAMPLE 9.9

In this example we construct a normal plot of the residuals for the Fresh Detergent model

$$y = \beta_0 + \beta_1 x_4 + \beta_2 x_3 + \beta_3 x_3^2 + \beta_4 x_4 x_3 + \epsilon$$

[†] For small samples, the corresponding values from the t-distribution with $n - k$ degrees of freedom should be used. For example, about 95 percent of the standardized residuals should be between $-t_{[.025]}^{(n-k)}$ and $t_{[.025]}^{(n-k)}$.

TABLE 9.5

Normal Plot Calculations for the Fresh Detergent Model
$y = \beta_0 + \beta_1 x_4 + \beta_2 x_3 + \beta_3 x_3^2 + \beta_4 x_4 x_3 + \epsilon$

i	$\frac{3i-1}{3n+1}$	$z_{(i)}$	i	$\frac{3i-1}{3n+1}$	$z_{(i)}$
1	$\frac{3(1)-1}{3(30)+1} = \frac{2}{91} = .0220$	−2.01	16	.5165	.04
2	$\frac{3(2)-1}{3(30)+1} = \frac{5}{91} = .0549$	−1.60	17	.5495	.12
3	$\frac{3(3)-1}{3(30)+1} = \frac{8}{91} = .0879$	−1.35	18	.5824	.21
4	.1209	−1.17	19	.6154	.29
5	.1538	−1.02	20	.6484	.38
6	.1868	−.89	21	.6813	.47
7	.2198	−.77	22	.7143	.57
8	.2527	−.67	23	.7473	.67
9	.2857	−.57	24	.7802	.77
10	.3187	−.47	25	.8132	.89
11	.3516	−.38	26	.8462	1.02
12	.3846	−.29	27	.8791	1.17
13	.4176	−.21	28	.9121	1.35
14	.4505	−.12	29	.9451	1.60
15	.4835	−.04	30	.9780	2.02

In order to construct this plot, we must first arrange the residuals in order from smallest to largest. These ordered residuals are given in Table 9.2. Denoting the ith ordered residual as $e_{(i)}$ $(i = 1, 2, \ldots, 30)$, we next compute for each value of i the point $z_{(i)}$ such that

$$A[(-\infty, z_{(i)}]; N(0, 1)] = \frac{3i-1}{3n+1}$$

These computations are summarized in Table 9.5. We now plot the ordered residuals from Table 9.2 on the vertical axis against the values of $z_{(i)}$ from Table 9.5 on the horizontal axis. The resulting normal plot is depicted in Figure 9.19. Looking at this figure, we see that while this normal plot is certainly not exactly a straight line, the plot does have a distinct straight line appearance. This indicates that the Normality assumption is not seriously violated for the Fresh Detergent model.

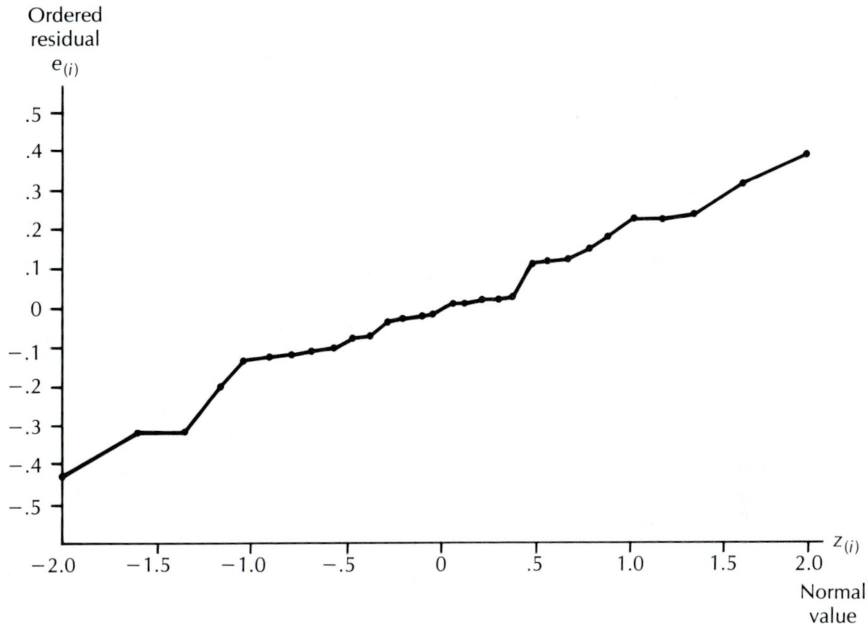

FIGURE 9.19
Normal Plot of the Residuals for the Fresh Detergent Model

If analysis of the residuals indicates that the Normality assumption is seriously violated, then remedies should be employed. We discuss what can be done to remedy violations of the Normality assumption in Chapter 10.

9.7 OUTLYING AND INFLUENTIAL OBSERVATIONS†

An observation that is well separated from the rest of the data is called an **outlier**, and an observation that causes the least squares point estimates to be substantially different from what they would be if the observation were removed from the data set is called **influential**. An observation may be an outlier with respect to its y value and/or its x values, but an outlier may or may not be influential. We illustrate these ideas by considering Figure 9.20, which is a hypothetical plot of the values of a dependent variable y against the increasing values of an independent variable x (we assume that the regression analysis involves only one independent variable). Observation 1 in Figure 9.20 is outlying with respect to its y value but not with respect to its x value (since its x value is near the middle of the other x values). Moreover, observation 1 may not be influential, because there are several observations with

†Much of the discussion of this section is based on a similar discussion by Neter, Wasserman, and Kutner (1985).

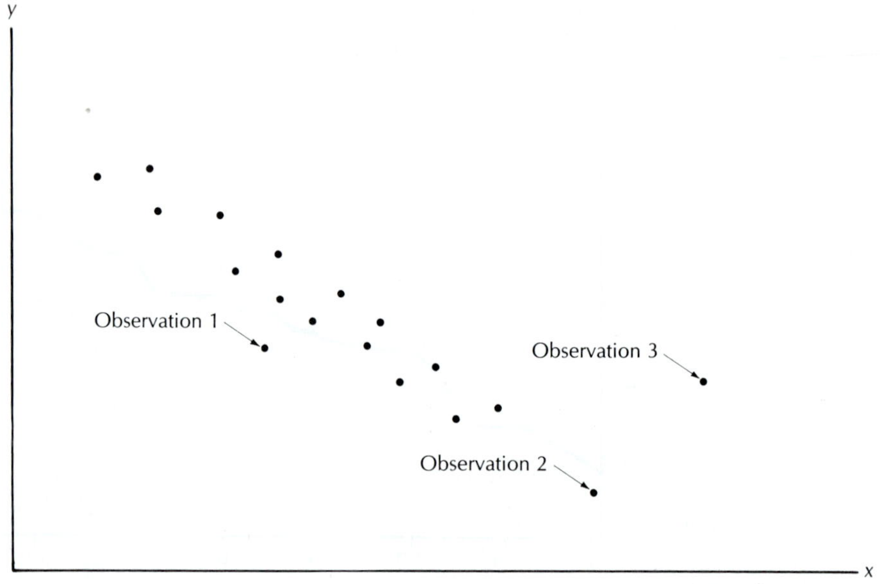

FIGURE 9.20
Data Plot Illustrating Outlying and Influential Observations

similar x values and nonoutlying y values, which will keep the least squares point estimates from being excessively influenced by observation 1. Observation 2 in Figure 9.20 is outlying with respect to its x value, but since its y value is consistent with the regression relationship displayed by the nonoutlying observations, it probably is not influential. Observation 3, however, is probably influential, because it is outlying with respect to its x value, and because its y value is not consistent with the regression relationship displayed by the other observations.

When performing a regression analysis involving one or two independent variables, we can use simple graphical methods, such as data plots and residual plots, to indicate the presence of outliers with respect to their x or y values. This has been illustrated for the case of one independent variable in Figure 9.20. For the case of two independent variables, a plot of the points representing the combinations of the values of x_1 and x_2 will indicate the presence of outliers with respect to their x values. For example, since the plot in Figure 7.4 of the points representing the combinations of the values of x_1 and x_2 in the fuel consumption problem does not indicate that there are any points that are well separated from the remaining points, we conclude that there is little evidence of outliers with respect to their x values. Next, note that residual plots against time, increasing values of the independent variables, or increasing values of the predicted values of the dependent variable can indicate the presence of outliers with respect to their y values. Any residual that is substantially different from the others is suspect. As a rule of thumb, if the absolute value of the standardized residual is greater than 3 or 4, an outlier with respect to its y value is suspected. For example, since Table 9.1 tells us that the largest residual for the fuel consumption model

$$y = \beta_0 + \beta_1 x_1 + \beta_2 x_2 + \epsilon$$

is $e_4 = y_4 - \hat{y}_4 = -.6131$, and since the standard error for this model is $s = .3671$, it follows that the largest standardized residual for the fuel consumption model is

$$\frac{e_4}{s} = \frac{-.6131}{.3671} = -1.6701$$

Since the absolute value of this largest standardized residual is less than 3, we do not have evidence that there are outliers with respect to their y values in the fuel consumption problem.

When more than two independent variables are involved in a regression analysis, identifying outliers by using simple graphical methods is more difficult. There are more sophisticated methods for identifying outliers with respect to their x or y values and for determining whether such outliers are influential. In order to identify outliers with respect to their x values, consider the regression model

$$y_i = \beta_0 + \beta_1 x_{i1} + \beta_2 x_{i2} + \cdots + \beta_p x_{ip} + \epsilon_i$$

which we assume has k parameters, and consider the **hat matrix**

$$\mathbf{H} = \mathbf{X}(\mathbf{X}'\mathbf{X})^{-1}\mathbf{X}'$$

which has n rows and n columns. For $i = 1, 2, \ldots, n$, we let h_{ii} denote the ith diagonal element of $\mathbf{H}$. It can be shown that

$$h_{ii} = \mathbf{x}_i'(\mathbf{X}'\mathbf{X})^{-1}\mathbf{x}_i$$

where

$$\mathbf{x}_i' = [1 \quad x_{i1} \quad x_{i2} \quad \cdots \quad x_{ip}]$$

is a row vector containing the values of the independent variables that have occurred in the ith observed situation. We can also show that

$$0 \le h_{ii} \le 1$$

and

$$\sum_{i=1}^{n} h_{ii} = k$$

The diagonal element h_{ii} is called the *leverage* of the x values

$$x_{i1}, x_{i2}, \ldots, x_{ip}$$

that have occurred in the ith observed situation. It indicates whether these x values are outlying, because it can be shown that h_{ii} is a measure of the distance between these x values and the means

$$\bar{x}_1 = \frac{\sum_{i=1}^{n} x_{i1}}{n}, \bar{x}_2 = \frac{\sum_{i=1}^{n} x_{i2}}{n}, \ldots, \bar{x}_p = \frac{\sum_{i=1}^{n} x_{ip}}{n}$$

of the x values that have occurred in all n observed situations. If the leverage value h_{ii} is large, the ith observation is outlying with respect to its x values, and thus the ith observation will exert substantial leverage in determining the values of the least squares point estimates. (To see why this is true, note that since observation 2 and observation 3 in Figure 9.20 are outlying with respect to their x values, they will exert substantial leverage in determining the values of the least squares point estimates.) A leverage value h_{ii} is generally considered to be large if it is substantially greater than most of the other leverage values or if it is greater than twice the average leverage value—that is, if it is greater than

$$2\bar{h} = 2\frac{\sum_{i=1}^{n} h_{ii}}{n} = 2\frac{k}{n}$$

(where $\sum_{i=1}^{n} h_{ii} = k$). It should be noted, however, that an observation with a large leverage value is not necessarily influential. For example, both observation 2 and observation 3 in Figure 9.20 would have large leverage values, because both observations are outliers with respect to their x values. However, as previously stated, whereas observation 3 probably is very influential (because its y value is not consistent with the regression relationship displayed by the other observations), observation 2 probably is not influential (because its y value is consistent with the regression relationship displayed by the nonoutlying observations).

We have discussed how to determine whether the ith observation is an outlier with respect to its y value by using the ith residual, $e_i = y_i - \hat{y}_i$, and standardized ith residual, e_i/s. We now discuss two modifications of these residuals that will enable us to better identify outliers with respect to their y values. The first modification involves using the **studentized residual**,

$$\frac{e_i}{s\sqrt{1 - h_{ii}}}$$

which has been formed by dividing the ith residual, e_i, by $s\sqrt{1 - h_{ii}}$, which is the point estimate of

$$\sigma_{e_i} = \sigma\sqrt{1 - h_{ii}}$$

which (it can be proved) is the standard deviation of the population of all possible values of the ith residual. The advantage of dividing e_i by $s\sqrt{1-h_{ii}}$ is that doing so guarantees that the n populations of all possible studentized residuals have equal variances (for example, the population of all possible values of $e_1/s\sqrt{1-h_{11}}$ has the same variance as the population of all possible values of $e_2/s\sqrt{1-h_{22}}$). Therefore, dividing e_i by $s\sqrt{1-h_{ii}}$ implies that the n studentized residuals are of the same relative magnitude.

The second modification is to calculate the **deleted residual**,

$$d_i = y_i - \hat{y}_{(i)}$$

where

$$\hat{y}_{(i)} = b_0^{(i)} + b_1^{(i)}x_{i1} + b_2^{(i)}x_{i2} + \cdots + b_p^{(i)}x_{ip}$$

is the point prediction of y_i calculated by using least squares point estimates $b_0^{(i)}$, $b_1^{(i)}$, $b_2^{(i)}$, . . . , $b_p^{(i)}$. These are calculated by using all n observations except for the ith *observation* (that is, the value of the dependent variable and the values of the independent variables that have occurred in the ith observed situation). We do not use the ith observation because, if we do not eliminate the ith observation, and if the ith observation is an outlier with respect to y_i, *its* y value, then the ith observation might cause $e_i = y_i - \hat{y}_i$ to be small, which would falsely imply that the ith observation is not an outlier with respect to its y value. If we let s_{d_i} denote the point estimate of σ_{d_i}, the standard deviation of the population of all possible values of d_i, and if we define the **studentized deleted residual** to be

$$\frac{d_i}{s_{d_i}}$$

then it can be shown that

$$\frac{d_i}{s_{d_i}} = e_i\left[\frac{n-k-1}{SSE(1-h_{ii}) - e_i^2}\right]^{1/2}$$

where $e_i = y_i - \hat{y}_i$, and the population of all possible values of d_i/s_{d_i} has a t-distribution with $n-k-1$ degrees of freedom. If the absolute value of the studentized deleted residual d_i/s_{d_i} is greater than $t_{[.025]}^{(n-k-1)}$, this residual is considered to be large, and the ith observation is considered to be an outlier with respect to its y value.

If we have concluded that the ith observation is an outlier with respect to its x or y value, we can determine whether the ith observation is influential by calculating **Cook's distance measure**,

$$D_i = \frac{(\mathbf{b} - \mathbf{b}^{(i)})'\mathbf{X}'\mathbf{X}(\mathbf{b} - \mathbf{b}^{(i)})}{ks^2}$$

where $s^2 = SSE/(n - k)$ and

$$\mathbf{b} - \mathbf{b}^{(i)} = \begin{bmatrix} b_0 \\ b_1 \\ b_2 \\ \vdots \\ b_p \end{bmatrix} - \begin{bmatrix} b_0^{(i)} \\ b_1^{(i)} \\ b_2^{(i)} \\ \vdots \\ b_p^{(i)} \end{bmatrix} = \begin{bmatrix} b_0 - b_0^{(i)} \\ b_1 - b_1^{(i)} \\ b_2 - b_2^{(i)} \\ \vdots \\ b_p - b_p^{(i)} \end{bmatrix}$$

If D_i is large, the least squares point estimates $b_0, b_1, b_2, \ldots, b_p$ calculated by using all n observations differ substantially from the least squares point estimates $b_0^{(i)}, b_1^{(i)}, b_2^{(i)}, \ldots, b_p^{(i)}$ calculated by using all n observations except for the ith observation, and thus the ith observation is influential. To understand what we mean by a large D_i value, note that although the population of all possible values of D_i does not have an F-distribution, practice has shown that

1. If D_i is less than $F_{[.80]}^{(k,n-k)}$ (the 20th percentile of the F-distribution having k and $n - k$ degrees of freedom), then the ith observation should not be considered influential.
2. If D_i is greater than $F_{[.50]}^{(k,n-k)}$ (the 50th percentile of the F-distribution having k and $n - k$ degrees of freedom), then the ith observation should be considered influential.
3. If $F_{[.80]}^{(k,n-k)} \leq D_i \leq F_{[.50]}^{(k,n-k)}$, then the nearer D_i is to $F_{[.50]}^{(k,n-k)}$, the greater the extent of the influence of the ith observation.

Finally, note that it can be shown that D_i can be calculated by the equation

$$D_i = \frac{e_i^2}{ks^2}\left[\frac{h_{ii}}{(1 - h_{ii})^2}\right]$$

Once we have identified influential outlying observations, we must decide what to do about these observations. If an influential outlying observation has been caused by incorrect measurement (perhaps resulting from a faulty instrument) or erroneous recording (for example, an incorrect decimal point), the observation should be corrected (if it can be corrected), and the regression analysis should be rerun. If the observation cannot be corrected, then it should probably be dropped from the data set. If the influential outlying observation is accurate, it is possible that the regression model under consideration is inadequate in that it does not contain an important independent variable that would explain this observation or does not have the correct functional form. This possibility should be investigated, and if need be, the model should be improved. Finally, if no explanation can be found for an influential outlying observation, it might be appropriate to drop this observation from the data set. As an alternative, instead of calculating point estimates $b_0, b_1, b_2, \ldots, b_p$ that minimize the sum of squared residuals

$$SSE = \sum_{i=1}^{n} e_i^2 = \sum_{i=1}^{n} (y_i - \hat{y}_i)^2$$
$$= \sum_{i=1}^{n} (y_i - (b_0 + b_1x_{i1} + b_2x_{i2} + \cdots + b_px_{ip}))^2$$

we could dampen the effect of the influential outlying observation by calculating point estimates $b_0, b_1, b_2, \ldots, b_p$ that minimize the sum of the absolute values of the residuals

$$\sum_{i=1}^{n} |e_i| = \sum_{i=1}^{n} |y_i - \hat{y}_i|$$
$$= \sum_{i=1}^{n} |y_i - (b_0 + b_1x_{i1} + b_2x_{i2} + \cdots + b_px_{ip})|$$

The reader interested in this approach should see Kennedy and Gentle (1980).

9 EXERCISES

Exercises 9.1 through 9.3 relate to the construction contract data in Exercise 5.4, which are repeated here.

Obs	y_i	x_{i1}	x_{i2}	Model 1 ($e_i = y_i - \hat{y}_i$)	Model 2 ($e_i = y_i - \hat{y}_i$)	Model 3 ($e_i = y_i - \hat{y}_i$)	Model 3 ($\hat{y}_i$)
1	2.0	5.1	4	−.954514	−1.255	.300627	1.699
2	3.5	3.5	4	−1.622	−1.045	.060264	3.440
3	8.5	2.4	2	2.813	3.107	1.597	6.903
4	4.5	4.0	6	−.870052	−.515266	.196875	4.303
5	7.0	1.7	2	.364356	−.043053	−.698810	7.699
6	7.0	2.0	2	.770800	.705576	−.448066	7.448
7	2.0	5.0	4	−1.090	−1.284	.079017	1.921
8	5.0	3.2	2	.396579	1.080	−.091041	5.091
9	8.0	5.2	6	4.256	3.704	1.423	6.577
10	5.0	4.3	6	.036393	.257683	−.074741	5.075
11	6.0	2.9	2	.990134	1.579	.116596	5.883
12	7.5	1.1	2	.051466	−1.226	−.294086	7.794
13	4.0	2.6	4	−2.342	−2.046	1.274	2.726
14	4.0	4.0	6	−1.370	−1.015	−.303125	4.303
15	1.0	5.3	4	−1.684	−2.218	−.211019	1.211
16	5.0	4.9	6	.849282	.617448	−1.212	6.212
17	6.5	5.0	6	2.485	2.153	.151395	6.349
18	1.5	3.9	4	−3.080	−2.557	−1.866	3.366

Consider the following models:

Model 1: $y_i = \beta_0 + \beta_1 x_{i1} + \beta_2 x_{i2} + \epsilon_i$

Model 2: $y_i = \beta_0 + \beta_1 x_{i1} + \beta_2 x_{i2} + \beta_3 x_{i1}^2 + \epsilon_i$

Model 3: $y_i = \beta_0 + \beta_1 x_{i1} + \beta_2 x_{i2} + \beta_3 x_{i1}^2 + \beta_4 x_{i1} x_{i2} + \epsilon_i$

When these models are used to perform regression analyses of these data, we obtain the residuals and predicted values shown in the table.

9.1 For model 1,

a. Plot the residuals against the increasing values of x_{i1}. Does this residual plot indicate the need for the quadratic term x_{i1}^2? Discuss.

b. Plot the residuals against the increasing values of x_{i2} and against the increasing values of $x_{i1}x_{i2}$. Discuss what these residual plots say about the need for the interaction term $x_{i1}x_{i2}$.

9.2 For model 2, plot the residuals against the increasing values of x_{i2} and against the increasing values of $x_{i1}x_{i2}$. Discuss what these residual plots say about the need for the interaction term $x_{i1}x_{i2}$.

9.3 For model 3,

a. Plot the residuals against the increasing values of x_{i1}, against the increasing values of x_{i2}, and against the increasing values of $\hat{y}_i$. Do these residual plots indicate that the functional form of model 3 is incorrect or that inference assumption 1 is violated?

b. Assuming that the data and thus residuals in the table are listed in the time order in which they were observed, plot the residuals against time. Does this residual plot indicate that inference assumption 2 is violated for model 3? Calculate the Durbin-Watson statistic, test for positive autocorrelation at $\alpha = .05$, and test for negative autocorrelation at $\alpha = .05$.

c. Group the residuals into a frequency distribution. Does the histogram constructed using this frequency distribution indicate that inference assumption 3 is violated for model 3?

d. Using the fact that the standard error, s, for model 3 is equal to .9706, calculate the standardized residuals. Do the percentage of these standardized residuals between -1 and 1 or the percentage of these standardized residuals between -2 and 2 indicate that inference assumption 3 is violated for model 3?

e. Construct a normal plot and determine whether this plot indicates that inference assumption 3 is violated.

9.4 Consider again the construction contract data in Exercise 5.4, which are repeated here.

Obs	y_i	x_{i1}	x_{i2}	Model 4 ($e_i = y_i - \hat{y}_i$)	Model 4 ($\hat{y}_i$)
1	2.0	5.1	4	.481948	1.518
2	3.5	3.5	4	.041071	3.459
3	8.5	2.4	2	1.533	6.967
4	4.5	4.0	6	.682848	3.817
5	7.0	1.7	2	− .593227	7.593
6	7.0	2.0	2	− .435951	7.436
7	2.0	5.0	4	.328047	1.672
8	5.0	3.2	2	− .139599	5.140
9	8.0	5.2	6	.664748	7.335
10	5.0	4.3	6	.436063	4.564
11	6.0	2.9	2	.036226	5.964
12	7.5	1.1	2	.092394	7.408
13	4.0	2.6	4	− .061724	4.062
14	4.0	4.0	6	.182848	3.817
15	1.0	5.3	4	− .197210	1.197
16	5.0	4.9	6	−1.323	6.323
17	6.5	5.0	6	− .150576	6.651
18	1.5	3.9	4	−1.578	3.078

When the following model

$$\text{Model 4:}\quad y_i = \beta_0 + \beta_1 x_{i1} + \beta_2 x_{i2} + \beta_3 x_{i1}^2 + \beta_4 x_{i1}x_{i2} + \beta_5 x_{i1}^2 x_{i2} + \epsilon_i$$

is used to perform a regression analysis of these data, we obtain the residuals and predicted values given in the table. Answer Exercise 9.3 (a), (b), (c), (d), and (e) as they relate to model 4, for which the standard error, s, can be calculated to be .8515. Comparing your answers in Exercise 9.3 with your answers in this exercise, which model—model 3 or model 4—seems to best satisfy the inference assumptions? Discuss.

9.5 An economist studied the relationship between x, 1970 yearly income (in thousands of dollars) for a family of four, and y, 1970 yearly clothing expenditure (in hundreds of dollars) for the family of four. Assume that the following data were observed:

x_i:	8	10	12	14	16	18	20
y_i:	6.47	6.17	7.4	10.57	11.93	10.3	14.67

When the regression model

$$y_i = \beta_0 + \beta_1 x_i + \epsilon_i$$

is used to perform a regression analysis of these data, we find that the least squares point estimates of β_0 and β_1 are $b_0 = .2968$ and $b_1 = .6677$ and that the standard error, s, is equal to 1.3434.

a. Calculate the standardized residuals, round them to the nearest .25, and construct a plot of the standardized residuals against increasing values of x_i.
b. By examining the residual plot, determine which inference assumption appears to be violated. Why is it logical that this inference assumption would be violated in this situation? (Note that we discuss in Chapter 10 how to remedy this inference assumption violation.)

9.6 In Example 10.8 of Chapter 10 we use a model containing $k - 1 = 5$ independent variables to perform a regression analysis of a set of data, and when the resulting residuals are plotted against time, we obtain the following residual plot.

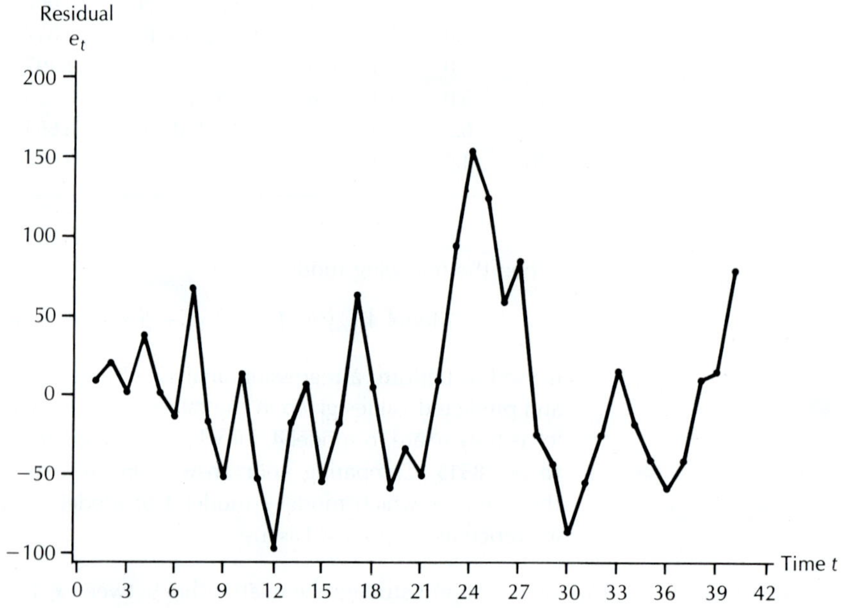

a. Discuss why this residual plot suggests the existence of positive autocorrelation.
b. Assuming that the Durbin-Watson statistic is equal to .8397, test for positive autocorrelation at $\alpha = .05$, and test for positive autocorrelation at $\alpha = .01$. (Note that we discuss in Chapter 10 how to remedy this inference assumption violation.)

10

REMEDIES FOR VIOLATIONS OF THE REGRESSION ASSUMPTIONS

What can be done when the assumptions behind regression analysis are violated? This chapter provides some answers to this question.

We begin Chapter 10 by considering the linearity assumption. We find that some useful regression models are not linear in the regression parameters. However, we can often use data transformations that will produce a linear regression model when applied to a nonlinear model. The use of these *transformations to achieve linearity* is the subject of Section 10.1. In Section 10.2 we turn to a discussion of remedies that can be employed when the Constant Variance assumption is violated. We see that remedies here also involve the use of data transformation techniques. These *variance-equalizing transformations* are explained in Section 10.2.1. Section 10.2.2 presents a related kind of regression problem involving what we call *0–1 dependent variables*. When the Independence assumption is violated, remedies often involve the use of *deterministic time series components* and the use of models employing *autoregressive terms*. These topics are discussed in Sections 10.3 and 10.4. In particular, a technique called the *Cochran-Orcutt procedure*, which can be employed when an autoregressive error structure exists, is explained in detail in Section 10.4. Finally, we conclude this chapter by looking at remedies that can be used when the Normality assumption is violated.

10.1 TRANSFORMATIONS TO ACHIEVE LINEARITY

Up to this point, we have studied models of the form

$$y = \beta_0 + \beta_1 x_1 + \beta_2 x_2 + \cdots + \beta_p x_p + \epsilon$$

Such models are called **linear in the parameters** $\beta_0, \beta_1, \beta_2, \ldots, \beta_p$, because the model contains $p + 1$ terms, each of which is a parameter β_j multiplied by a value determined from the data. Models including squared terms and interaction terms, such as

$$y = \beta_0 + \beta_1 x_1 + \beta_2 x_2 + \beta_3 x_2^2 + \beta_4 x_1 x_2 + \epsilon$$

are also linear in the parameters, since each term in the model is again a parameter β_j multiplied by a value determined from the data.

Sometimes, however, useful models are not linear in the parameters. For instance, the model

$$y = \beta_0(\beta_1^x)\epsilon$$

is not linear in the parameters, since the independent variable x enters as an exponent and β_0 is multiplied by β_1^x. This model further departs from the usual linear model because the error term ϵ is multiplicative rather than additive (that is, multiplied by, rather than added to, $\beta_0(\beta_1^x)$). In order to apply the techniques of estimation and prediction we have presented in previous chapters—for example, to obtain least squares estimates using the formula

$$\mathbf{b} = (\mathbf{X}'\mathbf{X})^{-1}\mathbf{X}'\mathbf{y}$$

we must **transform** such a nonlinear model into a model that is linear in the parameters.

The model

$$y = \beta_0(\beta_1^x)\epsilon$$

can be transformed into a linear model by taking logarithms on both sides. Either base 10 logarithms (denoted log) or natural (base e) logarithms (denoted ln) can be used. Both base 10 and natural logarithms can easily be obtained on many modern pocket calculators or by using computer routines. Two important properties of logarithms are

$$\log AB = \log A + \log B$$

$$\log A^r = r \log A$$

where A and B are positive numbers. These properties allow us to transform some nonlinear models into models that are linear in the parameters.

If $\beta_0 > 0$ and $\beta_1 > 0$, applying a logarithmic transformation to the model

$$y = \beta_0(\beta_1^x)\epsilon$$

yields

$$\log y = \log \beta_0 + x \log \beta_1 + \log \epsilon$$

If we let $\alpha_0 = \log \beta_0$, $\alpha_1 = \log \beta_1$, and $U = \log \epsilon$, the transformed version of the model becomes

$$\log y = \alpha_0 + \alpha_1 x + U$$

Thus, we see that the model with dependent variable log y is linear in the parameters α_0 and α_1. If, in addition, we can assume that the inference assumptions are satisfied, we are justified in using the regression procedures we have presented to estimate α_0 and α_1.

Cases in which a model

$$y = \beta_0(\beta_1^x)\epsilon$$

may be appropriate can be identified by data plots of y versus x. Plots of the expression $\beta_0(\beta_1^x)$ for several combinations of β_0 and β_1 are shown in Figure 10.1. Plots of observed data would have points scattered about such a function. The multiplicative error term would cause more variation around the high parts of the curve and less variation around the low points, because the variation in y depends on the level of $\beta_0(\beta_1^x)$—that is, given the same error term ϵ, the larger $\beta_0(\beta_1^x)$ is, the larger will be the variation in y.

Figure 10.1 shows that the curves described by

$$y = \beta_0(\beta_1^x)$$

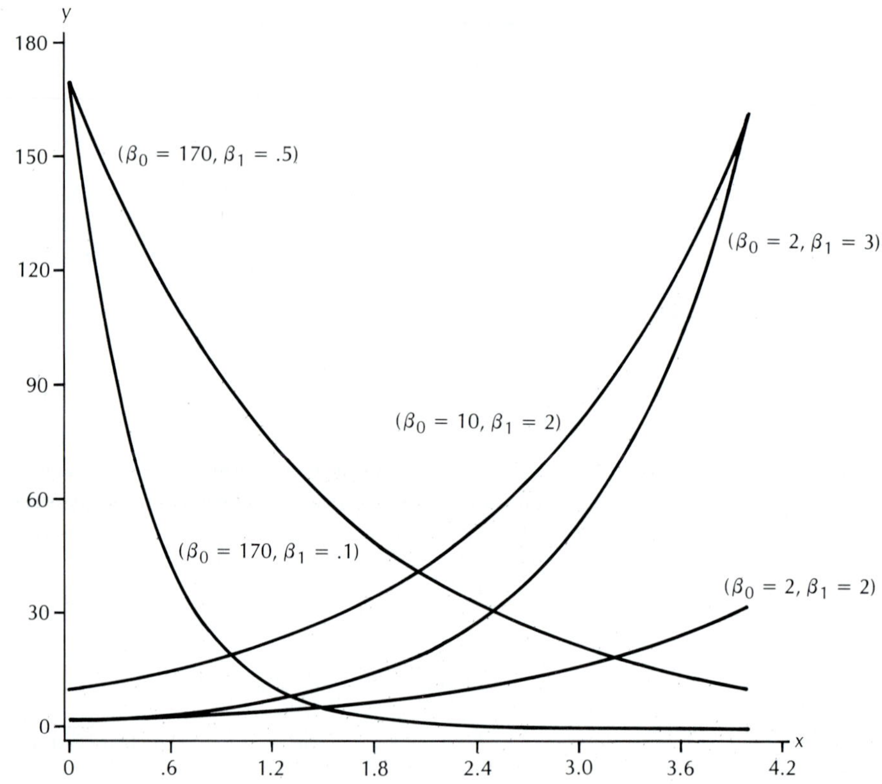

FIGURE 10.1
Exponential Curves of the Form $y = \beta_0(\beta_1^x)$

may be increasing ($\beta_1 > 1$) or decreasing ($0 < \beta_1 < 1$) functions of x. We can see that β_0 is the intercept and that β_1 determines the amount of curvature in the plot. The curvature gets more pronounced as β_1 moves away from 1 in either direction. To illustrate our discussion thus far, we present an example.

EXAMPLE 10.1

Wild Bill's Steakhouses, a fast food chain, opened in 1969. Each year from 1969 to 1983 the number of steakhouses in operation, y_t, is recorded. For convenience, we let $t = 0$ for the year 1969, $t = 1$ for the year 1970, and so on. An analyst for the firm wishes to use these data (presented in Table 10.1) to estimate the growth rate for the company. Here the analyst feels that the model

$$y_t = \beta_0(\beta_1^t)\epsilon_t \qquad \text{for } t = 0, 1, \ldots, 14$$

may be appropriate, in which case $100(\beta_1 - 1)\%$ is the growth rate of the company. Thus, for example, if $\beta_1 = 1.2$, the growth rate is $100(1.2 - 1)\% = 20\%$ per year. The analyst's goal is to obtain both a point estimate and a 95% confidence interval for the growth rate $100(\beta_1 - 1)\%$.

TABLE 10.1 Number of Wild Bill's Steakhouses (y_t) in Operation for the Years 1969–1983

Year	t	y_t	$\ln y_t$	Year	t	y_t	$\ln y_t$
1969	0	11	2.398	1977	8	82	4.407
1970	1	14	2.639	1978	9	99	4.595
1971	2	16	2.773	1979	10	119	4.779
1972	3	22	3.091	1980	11	156	5.050
1973	4	28	3.332	1981	12	257	5.549
1974	5	36	3.584	1982	13	284	5.649
1975	6	46	3.829	1983	14	403	5.999
1976	7	67	4.205				

The growth data are given in Table 10.1, along with the natural (base e) logarithm of y_t for each year. The original steakhouse data (y_t) are plotted against time (t) in Figure 10.2. This plot shows an exponential increase reminiscent of the plots in Figure 10.1 where β_1 is greater than 1. This suggests that the model

$$y_t = \beta_0(\beta_1^t)\epsilon_t$$

where the independent variable is the time period t, may be appropriate. The natural logarithms of the steakhouse data ($\ln y_t$) are plotted in Figure 10.3. This plot suggests that the relationship between $\ln y_t$ and t is linear. Applying the logarithmic transformation to the model

$$y_t = \beta_0(\beta_1^t)\epsilon_t$$

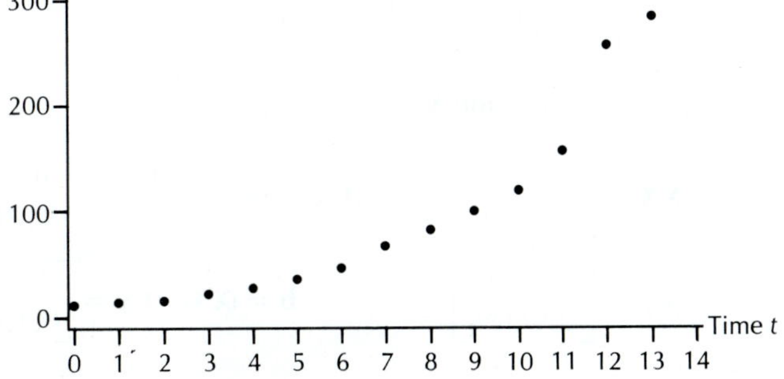

FIGURE 10.2 Number of Wild Bill's Steakhouses in Operation for the Years 1969–1983 Plotted Versus Time

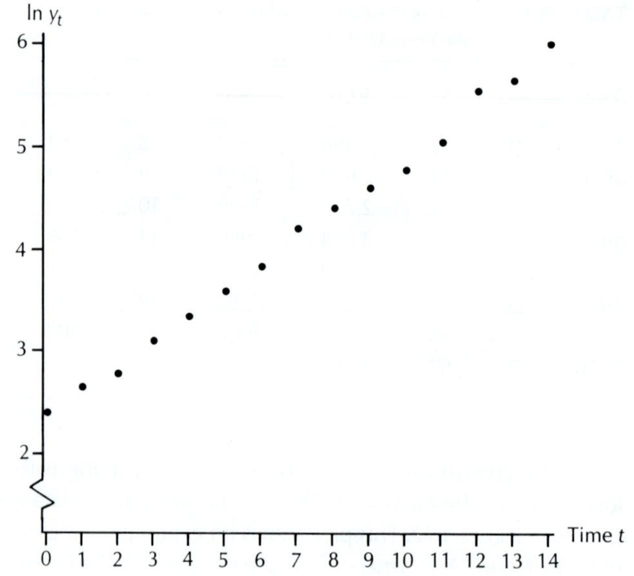

FIGURE 10.3
Natural Logarithms of Steakhouse Data Plotted Versus Time

we obtain

$$\ln y_t = \ln \beta_0 + t \ln \beta_1 + \ln \epsilon_t$$

Defining $\alpha_0 = \ln \beta_0$, $\alpha_1 = \ln \beta_1$, and $U_t = \ln \epsilon_t$, we have

$$\ln y_t = \alpha_0 + \alpha_1 t + U_t$$

We can estimate α_0 and α_1 by letting **y** be a column vector with elements $\ln y_t$ and by letting **X** be a matrix whose first column is a column of 1's and whose second column is the column of t values from zero through 14:

$$\mathbf{y} = \begin{bmatrix} 2.398 \\ 2.639 \\ 2.773 \\ \vdots \\ 5.999 \end{bmatrix} \qquad \mathbf{X} = \begin{bmatrix} 1 & 0 \\ 1 & 1 \\ 1 & 2 \\ \vdots & \vdots \\ 1 & 14 \end{bmatrix}$$

Using this **y** vector and **X** matrix, we obtain

$$\mathbf{X}'\mathbf{X} = \begin{bmatrix} 15 & 105 \\ 105 & 1015 \end{bmatrix} \qquad (\mathbf{X}'\mathbf{X})^{-1} = \begin{bmatrix} .2417 & -.0250 \\ -.0250 & .0036 \end{bmatrix}$$

$$\mathbf{X}'\mathbf{y} = \begin{bmatrix} 61.8774 \\ 505.0686 \end{bmatrix} \qquad \mathbf{b} = (\mathbf{X}'\mathbf{X})^{-1}\mathbf{X}'\mathbf{y} = \begin{bmatrix} 2.3270 \\ .2569 \end{bmatrix}$$

For this model we obtain

$$SSE = .0741 \qquad \text{and} \qquad s = \sqrt{\frac{SSE}{n-k}} = \sqrt{\frac{.0741}{15-2}} = .0754983$$

Note that the actual (computer) computations carried more decimal places than shown here.

To estimate the rate of growth $100(\beta_1 - 1)\%$, Wild Bill's can use the estimate of $\alpha_1 = \ln \beta_1$ in the model

$$\begin{aligned} \ln y_t &= \ln \beta_0 + t \ln \beta_1 + \ln \epsilon_t \\ &= \alpha_0 + \alpha_1 t + U_t \end{aligned}$$

This estimate is .2569. Thus, a point estimate of β_1 is $e^{.2569} = 1.293$, since $e^{\alpha_1} = e^{\ln \beta_1} = \beta_1$. This says that a point estimate of the growth rate $100(\beta_1 - 1)\%$ is

$$100(1.293 - 1)\% = 29.3\%$$

Using the value $t_{[\alpha/2]}^{(n-k)} = t_{[.025]}^{(13)} = 2.160$, a 95% confidence interval for $\alpha_1 = \ln \beta_1$ is

$$\begin{aligned} [.2569 \pm t_{[\alpha/2]}^{(n-k)} s\sqrt{c_{11}}] &= [.2569 \pm 2.160(.0754983)\sqrt{.0036}] \\ &= [.24715, .26665] \end{aligned}$$

Since $e^{.24715} = 1.2804$ and $e^{.26665} = 1.3056$, we are 95% confident that $\beta_1 = e^{\alpha_1}$ is contained in the interval [1.2804, 1.3056] and thus that the growth rate is between

$$100[1.2804 - 1]\% = 28.04\% \quad \text{and} \quad 100[1.3056 - 1]\% = 30.56\%$$

per year. Our confidence interval here is very narrow, because there is little variation in the data points around the estimate of the function $\beta_0(\beta_1^t)$ (see Figure 10.2). While it is probably reasonable to use this model to predict growth of Wild Bill's Steakhouses for the next year, one should be hesitant to use it as a predictor for growth over, say, the next five years. This is because, though the exponential growth model describes the data for 1969–1983 very well, there is no guarantee that growth will continue to be exponential so far outside the experimental region (so far into the future).

Another model that can be linearized by a transformation is the model

$$y = \beta_0 x^{\beta_1} \epsilon$$

This model also employs a multiplicative error term. Taking logarithms (natural or base 10) on both sides, we obtain (with base 10 logs)

$$\log y = \log \beta_0 + \beta_1 \log x + \log \epsilon$$

Letting $\alpha_0 = \log \beta_0$, $\alpha_1 = \beta_1$, and $U = \log \epsilon$, we obtain

$$\log y = \alpha_0 + \alpha_1 \log x + U$$

which is linear in the parameters α_0 and α_1. If we can assume that the inference assumptions are satisfied, we can use our standard regression formulas to estimate $\alpha_0 = \log(\beta_0)$ and $\alpha_1 = \beta_1$. The model

$$y = \beta_0 x^{\beta_1} \epsilon$$

is sometimes used by economists to estimate the relationship between y, the quantity of a good demanded, and x, the per unit price for the good. In this application, $|\beta_1|$ is often referred to as the elasticity of demand for the good.

Plots of the expression $\beta_0 x^{\beta_1}$ are given in Figure 10.4 for several combinations of β_0 and β_1. We can see that if $\beta_0 > 0$ and $\beta_1 < 0$, the graph has the y and x axes as asymptotes. As can be seen by comparing Figures 10.4 and 10.1, the shapes obtainable with this model are similar to those obtainable with our previous model,

$$y = \beta_0(\beta_1^x)\epsilon$$

The choice of a model by visual inspection of a data plot can therefore be difficult (this is especially true when a data plot displays a high degree of variability). Clearly, a statistician would be more comfortable when theoretical considerations (economic theory, for example) suggest a model than when data plot inspection alone is used to find a model. The former situation more often involves regression coefficients with meaningful interpretations (like elasticity). We will illustrate the use of the model

$$y = \beta_0 x^{\beta_1} \epsilon$$

in a demand and price problem at the end of this chapter.

So far we have seen that the regression of $\log y$ on x and the regression of $\log y$ on $\log x$ can both be useful. In addition, although we do not present an example here, we can also regress y on $\log x$ if this seems appropriate for a given data set.

When data plots suggest that the usual linear model may not be appropriate, the reciprocal transformation may be useful. Examples of models involving the reciprocals of y, x, or both, are

$$y = \beta_0 + \beta_1\left(\frac{1}{x}\right) + \epsilon$$

$$\frac{1}{y} = \beta_0 + \beta_1 x + \epsilon$$

$$\frac{1}{y} = \beta_0 + \beta_1\left(\frac{1}{x}\right) + \epsilon$$

Figure 10.5 shows several graphs that are helpful in identifying the general shapes of data plots that suggest these models. As with the logarithmic transformations,

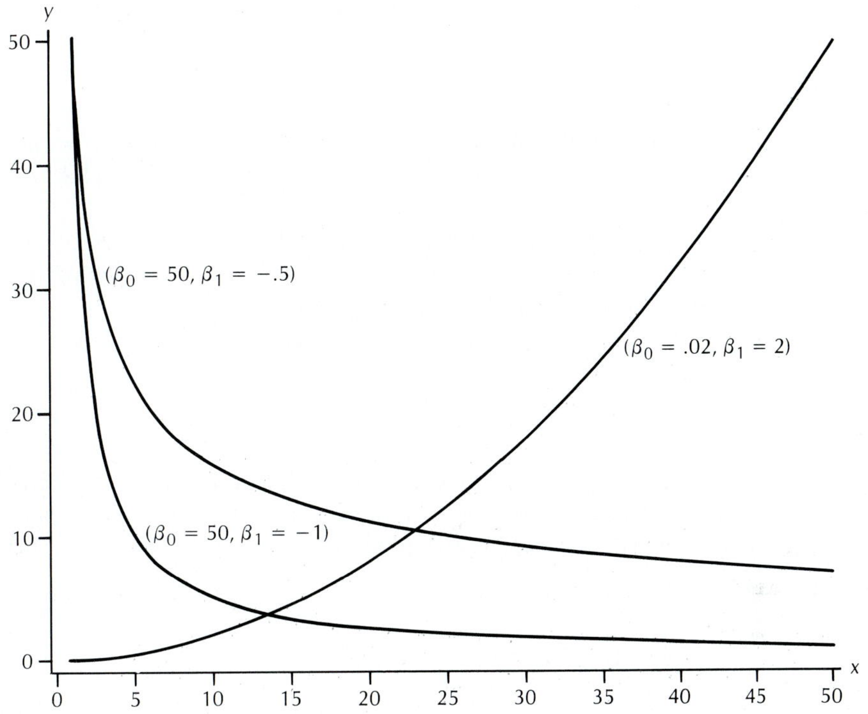

FIGURE 10.4
Plots of $y = \beta_0 x^{\beta_1}$

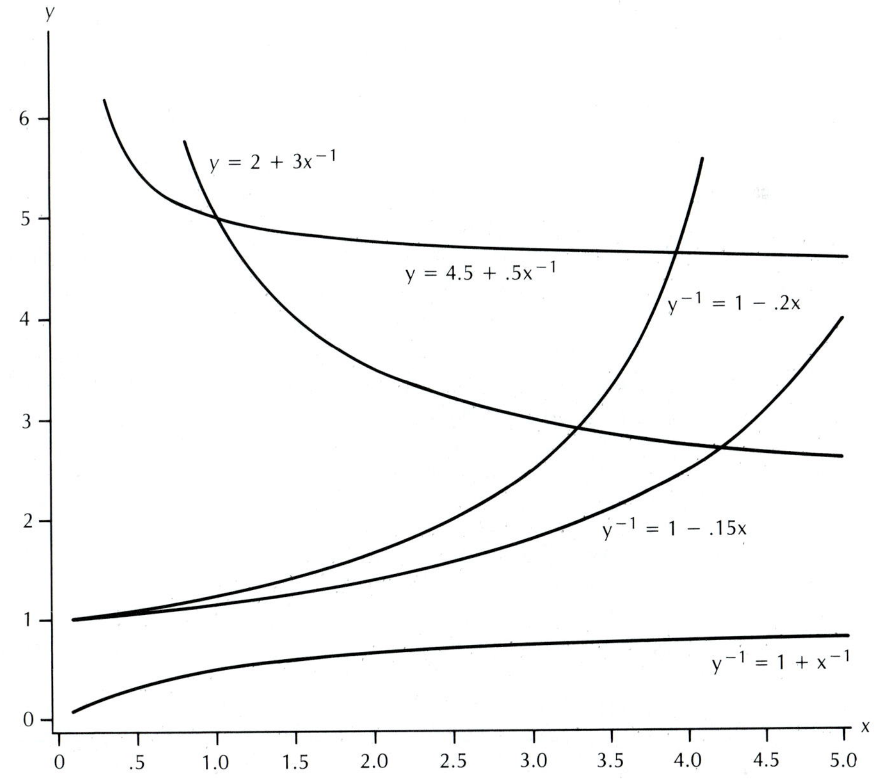

FIGURE 10.5
Plots of Functions Involving Reciprocals of y and x

plots of the transformed data should appear linear. This class of models is easily enlarged by adding higher-order polynomial terms in $1/x$. For example, the model

$$y = \beta_0 + \beta_1\left(\frac{1}{x}\right) + \beta_2\left(\frac{1}{x^2}\right) + \epsilon$$

might be useful in some situations.

Looking at Figure 10.5, we see that in the equation

$$y = \beta_0 + \beta_1\left(\frac{1}{x}\right)$$

y approaches β_0 as x gets larger. This model might be considered in a situation where learning over time (x) causes the time required to perform a task (y) to decrease towards an asymptotic value (β_0). We now present such an example.

EXAMPLE 10.2

The State Department of Taxation wishes to investigate the effect of experience, x, on the amount of time, y, required to fill out Form ST 1040AVG, the state income-averaging form. In order to do this, nine people whose financial status makes income averaging advantageous are chosen at random. Each is asked to fill out Form ST 1040AVG and to report (1) the time y (in hours) required to complete the form and (2) the number of times x (including this one) that he or she has filled out this form. Anticipating a model like

$$y = \beta_0 + \beta_1\left(\frac{1}{x}\right) + \epsilon$$

because learning over time might logically cause the time required to fill out Form ST 1040AVG to decrease towards an asymptotic value β_0, we have defined x in such a way that a zero value for x is impossible.

The goals of this experiment are to see if there is significant evidence to indicate that the mean time to complete Form ST 1040AVG decreases with increasing experience; and to predict the amount of time it will take an individual with a given amount of experience to complete Form ST 1040AVG.

The data obtained in the tax form experiment are given in Table 10.2.

TABLE 10.2 Completion Times for Tax Form ST 1040AVG

Person	1	2	3	4	5	6	7	8	9
Completion time y (in hours)	8.0	4.7	3.7	2.8	8.9	5.8	2.0	1.9	3.3
Experience x	1	8	4	16	1	2	12	5	3

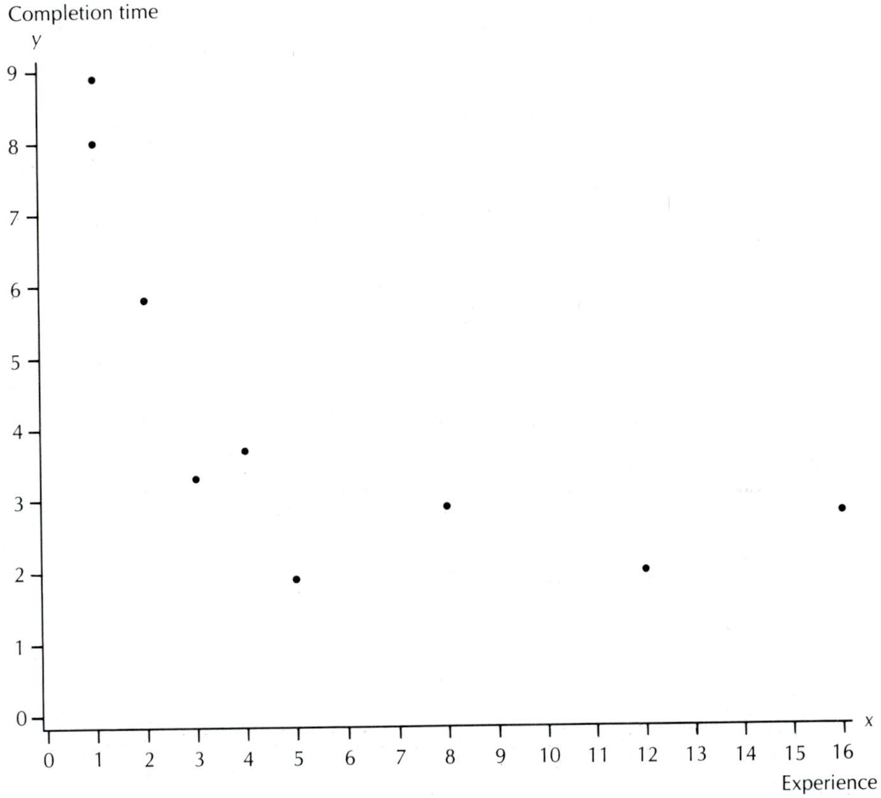

FIGURE 10.6
Completion Time Versus Experience for Tax Form ST 1040AVG

We plot y versus x in Figure 10.6. We see a tendency for a decrease in y with increasing values of x in a pattern consistent with the proposed model

$$y = \beta_0 + \beta_1\left(\frac{1}{x}\right) + \epsilon$$

A plot of y versus $1/x$ is shown in Figure 10.7. Notice that this plot has a linear appearance. We estimate β_0 and β_1 by using the vector $\mathbf{y}$ and the matrix $\mathbf{X}$ as follows:

$$\mathbf{y} = \begin{bmatrix} 8.0 \\ 4.7 \\ 3.7 \\ \vdots \\ 3.3 \end{bmatrix} \qquad \mathbf{X} = \begin{bmatrix} 1 & 1.0000 \\ 1 & .1250 \\ 1 & .2500 \\ \vdots & \vdots \\ 1 & .3333 \end{bmatrix}$$

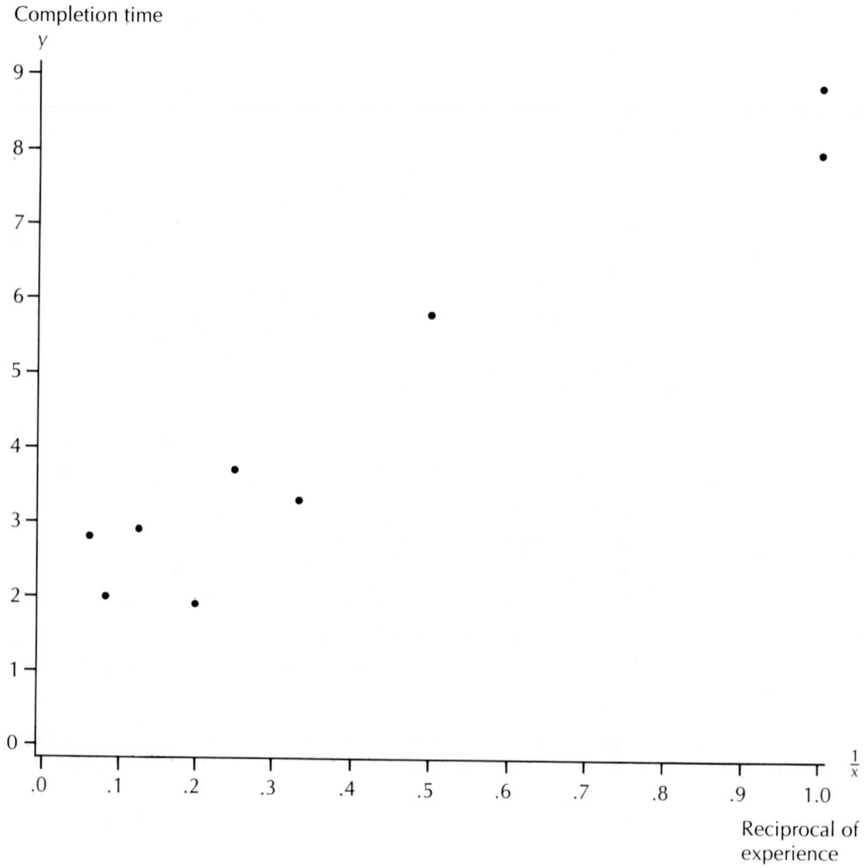

FIGURE 10.7
Completion Time Versus Reciprocal of Experience for Tax Form ST 1040AVG

where the second column of **X** contains the reciprocals of the observations on experience. We find that

$$\mathbf{X}'\mathbf{X} = \begin{bmatrix} 9.0 & 3.5541 \\ 3.5541 & 2.4901 \end{bmatrix} \qquad (\mathbf{X}'\mathbf{X})^{-1} = \begin{bmatrix} .2546 & -.3634 \\ -.3634 & .9203 \end{bmatrix}$$

$$\mathbf{X}'\mathbf{y} = \begin{bmatrix} 41.1000 \\ 23.1340 \end{bmatrix} \qquad \mathbf{b} = (\mathbf{X}'\mathbf{X})^{-1}\mathbf{X}'\mathbf{y} = \begin{bmatrix} 2.0572 \\ 6.3545 \end{bmatrix}$$

Moreover, for this model

$$SSE = 7.4141 \qquad \text{and} \qquad s = \sqrt{\frac{SSE}{n-k}} = \sqrt{\frac{7.4141}{9-2}} = 1.029$$

The first goal of the State Department of Taxation's experiment is to see if there is significant evidence to indicate that the mean completion time decreases with increasing experience. To investigate this, we can test the null hypothesis $H_0: \beta_1 = 0$ against the alternative hypothesis $H_1: \beta_1 \neq 0$. The least squares point estimate of β_1 is $b_1 = 6.3545$, and the t_{b_1} statistic for testing $H_0: \beta_1 = 0$ is

$$t_{b_1} = \frac{b_1}{s\sqrt{c_{11}}} = \frac{6.3545}{1.029\sqrt{.9203}} = 6.44$$

Comparing $t_{b_1} = 6.44$ with the rejection point $t_{[\alpha/2]}^{(n-k)} = t_{[.025]}^{(7)} = 2.365$, since $t_{b_1} > t_{[.025]}^{(7)}$, we reject $H_0: \beta_1 = 0$ with α, the probability of a Type I error, equal to .05. Furthermore, since $b_1 = 6.3545$ is positive, we have significant evidence that $\beta_1 > 0$. Therefore, since if $\beta_1 > 0$, the equation

$$\mu = \beta_0 + \beta_1\left(\frac{1}{x}\right)$$

says that μ, mean completion time, will decrease towards β_0 as x, experience, increases, we have significant evidence to suggest that the mean time to complete Form ST 1040AVG decreases with increasing experience.

The second goal of the State Department of Taxation's experiment is to predict the amount of time it will take an individual with a given amount of experience to complete Form ST 1040AVG. We illustrate the procedure for an individual who has completed four such forms and who is thus filling out the fifth form. A point prediction for an individual completion time when $x = 5$ is

$$\hat{y} = b_0 + b_1\left(\frac{1}{x}\right)$$

$$= 2.0572 + 6.3545\left(\frac{1}{5}\right) = 3.3281 \text{ hours}$$

Since we are predicting an *individual* completion time, we wish to compute a 95% prediction interval when $x = 5$. Using the value $t_{[\alpha/2]}^{(n-k)} = t_{[.025]}^{(7)} = 2.365$ and letting

$$\mathbf{x}_0' = \begin{bmatrix} 1 & \frac{1}{x} \end{bmatrix} = \begin{bmatrix} 1 & \frac{1}{5} \end{bmatrix}$$

we obtain

$$[3.3281 \pm t_{[\alpha/2]}^{(n-k)} s\sqrt{1 + \mathbf{x}_0'(\mathbf{X}'\mathbf{X})^{-1}\mathbf{x}_0}]$$

$$[3.3281 \pm 2.365(1.029)\sqrt{1 + .2138}]$$

$$[.7225, 5.9337]$$

This says that we can be 95% confident that an individual who has completed Form ST 1040AVG four times will require between .7225 and 5.9337 hours to complete the fifth such form.

10.2 HANDLING NONCONSTANT ERROR VARIANCES

10.2.1 Variance-Equalizing Transformations

When the variances of the historical populations of values of the dependent variable (or the variances of the corresponding populations of error terms)—$\sigma_1^2, \sigma_2^2, \ldots, \sigma_n^2$—are unequal, inference assumption 1 is violated. If this assumption is seriously violated, we must equalize these variances. If we do not, the statistical inference formulas (confidence and prediction intervals, hypothesis tests, and so on) that we have presented will not be valid.

In order to equalize variances we use some kind of transformation. For instance, if we consider the general linear regression model

$$y_i = \beta_0 + \beta_1 x_{i1} + \beta_2 x_{i2} + \cdots + \beta_p x_{ip} + \epsilon_i$$

and if the variances $\sigma_1^2, \sigma_2^2, \ldots, \sigma_n^2$ are unequal and *known*, then the variances can be equalized by using the transformed model

$$\frac{y_i}{\sigma_i} = \beta_0\left(\frac{1}{\sigma_i}\right) + \beta_1\left(\frac{x_{i1}}{\sigma_i}\right) + \beta_2\left(\frac{x_{i2}}{\sigma_i}\right) + \cdots + \beta_p\left(\frac{x_{ip}}{\sigma_i}\right) + \eta_i$$

where $\eta_i = \epsilon_i/\sigma_i$. This transformed model has the same parameters $\beta_0, \beta_1, \beta_2, \ldots, \beta_p$ as the original model and also satisfies inference assumption 1, since the properties of the variance tell us that the variance of the potential η_i values in the population of error terms for the transformed model is $\sigma_i^2(1/\sigma_i^2) = 1$. (This follows because, if we divide every element in a population by a constant k, the variance of the new population is k^2 multiplied by the variance of the original population.) The parameters $\beta_0, \beta_1, \beta_2, \ldots, \beta_p$ in the transformed model are estimated using the following **y** vector and **X** matrix:

$$\mathbf{y} = \begin{bmatrix} \frac{y_1}{\sigma_1} \\ \frac{y_2}{\sigma_2} \\ \vdots \\ \frac{y_n}{\sigma_n} \end{bmatrix} \qquad \mathbf{X} = \begin{bmatrix} \frac{1}{\sigma_1} & \frac{x_{11}}{\sigma_1} & \frac{x_{12}}{\sigma_1} & \cdots & \frac{x_{1p}}{\sigma_1} \\ \frac{1}{\sigma_2} & \frac{x_{21}}{\sigma_2} & \frac{x_{22}}{\sigma_2} & \cdots & \frac{x_{2p}}{\sigma_2} \\ \vdots & \vdots & \vdots & & \vdots \\ \frac{1}{\sigma_n} & \frac{x_{n1}}{\sigma_n} & \frac{x_{n2}}{\sigma_n} & \cdots & \frac{x_{np}}{\sigma_n} \end{bmatrix}$$

Notice here that some computer packages by default supply a column of 1's for an intercept term. We would have to override such a default here, since we wish to include a transformed intercept column containing $1/\sigma_i$.

The trouble with this approach is, of course, that the variances $\sigma_1^2, \sigma_2^2, \ldots, \sigma_n^2$ are not precisely known in a practical regression problem. In such a case, it is sometimes possible to estimate these variances. For example, these variances can be estimated when several observations have been randomly selected from each historical population of values of the dependent variable. Defining s_i^2 to be the sample variance of the observations drawn from the ith historical population, the estimates of $\sigma_1^2, \sigma_2^2, \ldots, \sigma_n^2$ are $s_1^2, s_2^2, \ldots, s_n^2$. Then, if the regression model

$$y_i = \beta_0 + \beta_1 x_{i1} + \beta_2 x_{i2} + \cdots + \beta_p x_{ip} + \epsilon_i$$

has a nonconstant error variance, the variances can be approximately equalized by using the transformed model

$$\left(\frac{y_i}{s_i}\right) = \beta_0\left(\frac{1}{s_i}\right) + \beta_1\left(\frac{x_{i1}}{s_i}\right) + \beta_2\left(\frac{x_{i2}}{s_i}\right) + \cdots + \beta_p\left(\frac{x_{ip}}{s_i}\right) + \eta_i$$

where $\eta_i = \epsilon_i/s_i$. The parameters $\beta_0, \beta_1, \beta_2, \ldots, \beta_p$ in this transformed model can be estimated using the **y** vector and **X** matrix specified previously with $\sigma_1, \sigma_2, \ldots, \sigma_n$ replaced by $s_1, s_2, \ldots, s_n$.

When the variances $\sigma_1^2, \sigma_2^2, \ldots, \sigma_n^2$ cannot be estimated (for example, when there is only one observation from each historical population of values of the dependent variable), then other kinds of transformations can often be used to equalize variances. Often the error variance is a function of one of the independent variables. As we have seen, such behavior would be characterized by fanning out or funneling in patterns in a residual plot against the increasing values of the independent variable. When the error variance appears to be related in a simple way to one of the independent variables, we can often use this independent variable to develop a transformed model that satisfies inference assumption 1.

As an example, suppose that the regression model

$$y_i = \beta_0 + \beta_1 x_{i1} + \beta_2 x_{i2} + \epsilon_i$$

is such that $\sigma_i^2 = \sigma^2 x_{i1}$. This says that the variance σ_i^2 is proportional to the value of the first independent variable x_{i1}, where σ^2 is a proportionality constant. In this case, we can obtain a transformed model by dividing all terms in the model by $\sqrt{x_{i1}}$. This transformed model is

$$\left(\frac{y_i}{\sqrt{x_{i1}}}\right) = \beta_0\left(\frac{1}{\sqrt{x_{i1}}}\right) + \beta_1\left(\frac{x_{i1}}{\sqrt{x_{i1}}}\right) + \beta_2\left(\frac{x_{i2}}{\sqrt{x_{i1}}}\right) + \eta_i$$

where $\eta_i = \epsilon_i/\sqrt{x_{i1}}$. For this transformed model, the properties of the variance tell us that for any i, the variance of the potential η_i values in the population of error terms for the transformed model is

$$\sigma^2 x_{i1}\left(\frac{1}{\sqrt{x_{i1}}}\right)^2 = \sigma^2 x_{i1}\left(\frac{1}{x_{i1}}\right) = \sigma^2$$

Therefore, the transformed model satisfies inference assumption 1. The parameters β_0, β_1, and β_2 in the transformed model (which are the same parameters as in the original model) are estimated by using the following **y** vector and **X** matrix:

$$\mathbf{y} = \begin{bmatrix} \frac{y_1}{\sqrt{x_{11}}} \\ \frac{y_2}{\sqrt{x_{21}}} \\ \vdots \\ \frac{y_n}{\sqrt{x_{n1}}} \end{bmatrix} \qquad \mathbf{X} = \begin{bmatrix} \frac{1}{\sqrt{x_{11}}} & \frac{x_{11}}{\sqrt{x_{11}}} & \frac{x_{12}}{\sqrt{x_{11}}} \\ \frac{1}{\sqrt{x_{21}}} & \frac{x_{21}}{\sqrt{x_{21}}} & \frac{x_{22}}{\sqrt{x_{21}}} \\ \vdots & \vdots & \vdots \\ \frac{1}{\sqrt{x_{n1}}} & \frac{x_{n1}}{\sqrt{x_{n1}}} & \frac{x_{n2}}{\sqrt{x_{n1}}} \end{bmatrix}$$

Again, since some computer packages by default supply a column of 1's for an intercept term, we would have to override this default here, since we wish to include an intercept column containing $1/\sqrt{x_{i1}}$ values.

We now (without going into a detailed discussion of each case) list some transformations that can be used to equalize the variances in different regression problems.

1. If the variance σ_i^2 is proportional to x_{ik}, that is, if $\sigma_i^2 = \sigma^2 x_{ik}$, then an appropriate transformed model is obtained by dividing each term in the original model by $\sqrt{x_{ik}}$. We have discussed an example of this model.
2. If the variance σ_i^2 is proportional to x_{ik}^2, or equivalently, if the standard deviation σ_i is proportional to x_{ik}—that is, if $\sigma_i^2 = \sigma^2 x_{ik}^2$ or $\sigma_i = \sigma x_{ik}$—then an appropriate transformed model is obtained by dividing each term in the original model by x_{ik}.
3. If the variance σ_i^2 appears to be an increasing linear function of the predicted values of the dependent variable ($\hat{y}_i$ values), an appropriate transformed model can often be obtained by replacing y by its square root, $\sqrt{y}$, in the regression model.
4. If the standard deviation σ_i appears to be an increasing linear function of the predicted values of the dependent variable ($\hat{y}_i$ values), an appropriate transformed model can often be obtained by replacing y by the logarithm (natural or base 10) of y in the regression model.
5. If the data display a multiplicative error structure, then the variance σ_i^2 will often appear to increase or decrease with increasing values of the dependent variable y. An appropriate transformation that will often equalize the variances is obtained by taking logarithms (natural or base 10) of both sides of the regression equation (for instance, see Example 10.1).
6. When the values of the dependent variable are "count data," inference assumption 1 will often be violated. An appropriate transformation that will often equalize the variances is obtained by replacing y by its square root, $\sqrt{y}$, in the regression model. Here, by count data we mean, for example, the number of occurrences of an event in a given period of time. Such data often can be described by a **Poisson distribution**. For a discussion of the Poisson distribution,

see any basic statistics text (for example, see Mendenhall and Reinmuth (1982)). We present an example of this kind of problem in Example 10.5.

EXAMPLE 10.3

The National Association of Retail Hardware Stores (NARHS), a nationally known trade association, wishes to investigate the relationship between x, home value (in thousands of dollars), and y, yearly expenditure on upkeep like lawn care, painting, repairs (in dollars). A random sample of 40 homeowners is taken; the results are given in Table 10.3. Figure 10.8 gives a plot of y versus x. From this plot it appears that y is increasing (probably in a quadratic fashion) as x increases and that the variance of the y values is also increasing as x increases. Increasing variance makes some intuitive sense here, since people with more expensive homes generally have higher incomes and can afford to pay to have upkeep done or perform upkeep chores themselves if they wish, thus causing a relatively large variation in upkeep expenses.

Since Figure 10.8 indicates the presence of a nonconstant error variance, we consider using a variance-equalizing transformation. It can be difficult to choose the best transformation. For instance, from Figure 10.8 it is not clear whether the variance σ_i^2 or the standard deviation σ_i is proportional to x_i. Our strategy here will be to choose a transformation and then check it using a plot of the residuals from the transformed model. If this plot does not suggest the existence of a nonconstant error variance, then we will be satisfied even though some other transformation may also solve the unequal variances problem for the NARHS data.

Since y appears to increase in a quadratic fashion as x increases, we begin by trying the model

$$y_i = \beta_0 + \beta_1 x_i + \beta_2 x_i^2 + \epsilon_i$$

When we estimate β_0, β_1, and β_2 using the data in Table 10.3, we obtain the prediction equation

$$\hat{y}_i = -14.4436 + 3.0585x_i + .0246x_i^2$$

The t_{b_2} statistic for the x_i^2 term is $t_{b_2} = 1.98$, which has an associated prob-value of .0554. Figure 10.9 shows a plot of the residuals for this model against increasing values of x. We see that the residuals appear to fan out as x increases, indicating the existence of a nonconstant error variance. Therefore, the t-test and prob-value for the x_i^2 term are not valid, since the inference assumptions are not met.

Next, we assume that the standard deviation σ_i is proportional to x_i, which means that the variance σ_i^2 is proportional to x_i^2. Under this assumption, the appropriate transformation is division of each term in the model by x_i. The

TABLE 10.3 NARHS Upkeep Expenditure Data

House	Value of House (thousands of dollars) (x)	Upkeep Expenditure (dollars) (y)
1	118.50	706.04
2	76.54	398.60
3	92.43	436.24
4	111.03	501.71
5	80.34	426.45
6	49.84	144.24
7	114.52	644.23
8	50.89	211.54
9	128.93	675.87
10	48.14	189.02
11	85.50	459.04
12	115.51	813.62
13	114.16	602.39
14	102.95	428.52
15	92.86	387.50
16	84.39	434.63
17	123.53	698.00
18	77.77	355.75
19	112.10	737.59
20	101.02	706.66
21	76.52	424.57
22	116.09	656.92
23	62.72	301.03
24	84.91	321.07
25	88.64	519.40
26	81.41	348.50
27	60.22	162.17
28	95.55	482.55
29	79.39	460.07
30	89.25	475.45
31	136.10	835.16
32	24.45	62.70
33	52.28	239.89
34	143.09	1005.32
35	41.86	184.18
36	43.10	212.80
37	66.79	313.45
38	106.43	658.47
39	61.01	195.08
40	99.01	545.42

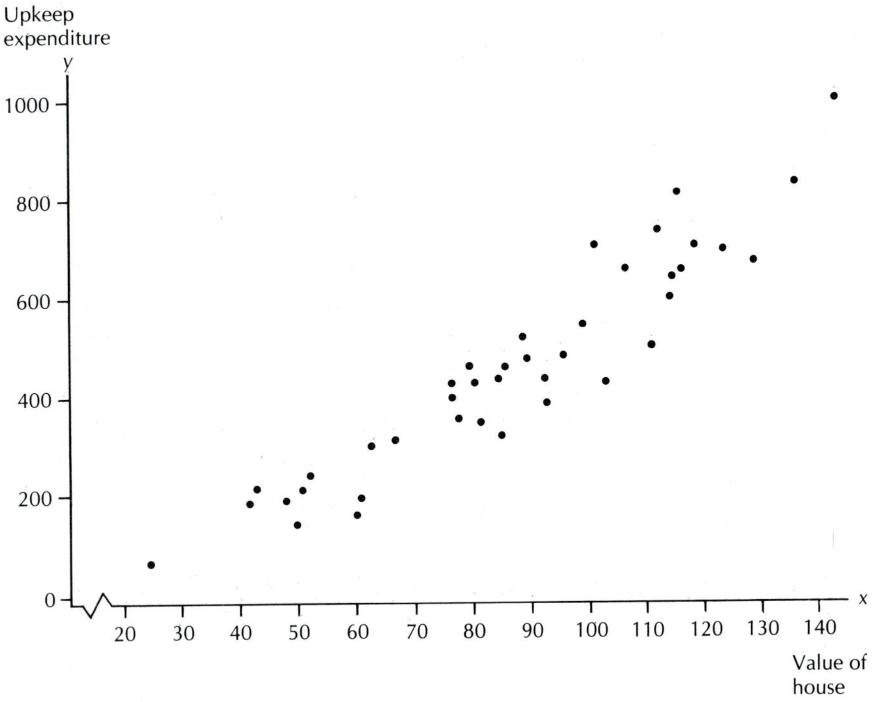

FIGURE 10.8
Upkeep Expenditure Versus Value of House for NARHS Data

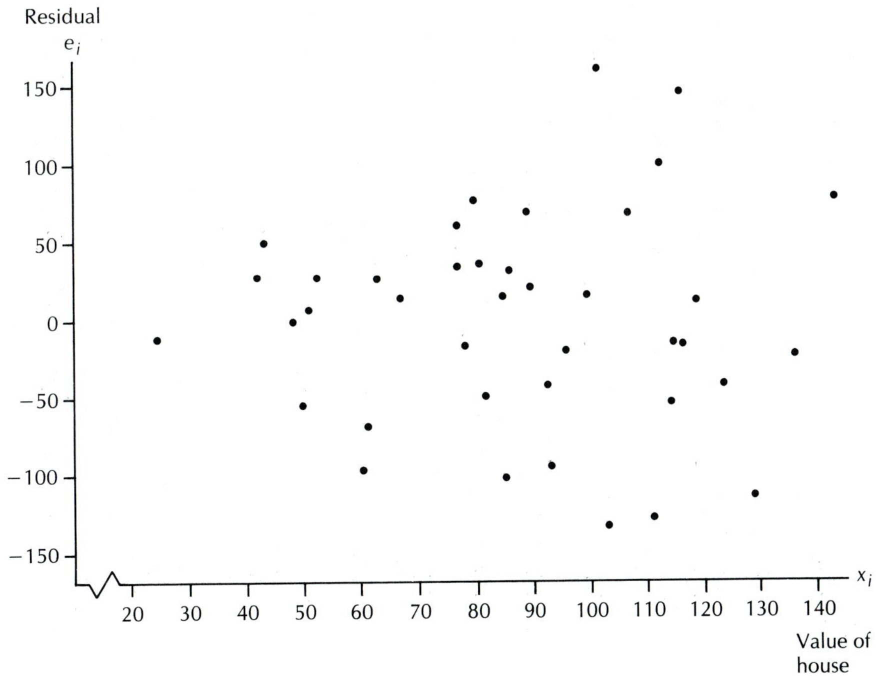

FIGURE 10.9
Residuals for the Quadratic NARHS Model Plotted Against Increasing Values of x_i

transformed model is

$$\frac{y_i}{x_i} = \beta_0\left(\frac{1}{x_i}\right) + \beta_1\left(\frac{x_i}{x_i}\right) + \beta_2\left(\frac{x_i^2}{x_i}\right) + \eta_i$$

$$\frac{y_i}{x_i} = \beta_0\left(\frac{1}{x_i}\right) + \beta_1 + \beta_2 x_i + \eta_i$$

where $\eta_i = \epsilon_i/x_i$. If this transformation remedies the unequal variances problem, a residual plot against increasing values of x should appear as a horizontal band. However, if this transformation is too extreme, a residual plot for the transformed model should show the residuals funneling in as x increases.

We estimate β_0, β_1, and β_2 in the transformed model by using the following **y** vector and **X** matrix (see Table 10.3):

$$\mathbf{y} = \begin{bmatrix} \frac{706.04}{118.50} \\ \frac{398.60}{76.54} \\ \vdots \\ \frac{545.42}{99.01} \end{bmatrix} \qquad \mathbf{X} = \begin{bmatrix} \frac{1}{118.50} & 1 & 118.50 \\ \frac{1}{76.54} & 1 & 76.54 \\ \vdots & \vdots & \vdots \\ \frac{1}{99.01} & 1 & 99.01 \end{bmatrix}$$

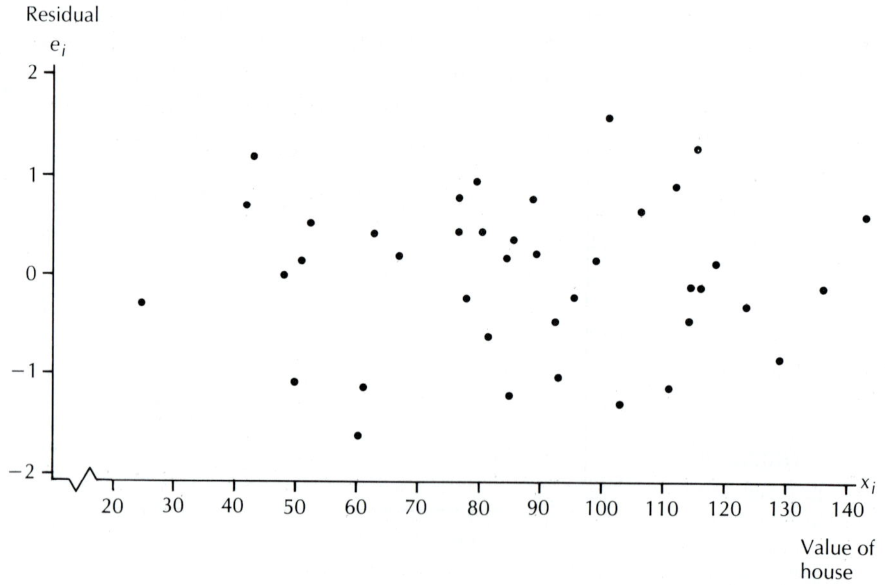

FIGURE 10.10
Residuals for the Transformed NARHS Model Plotted Against Increasing Values of x_i

TABLE 10.4 The t_{b_j} Statistics for the Model $\frac{y_i}{x_i} = \beta_0\left(\frac{1}{x_i}\right) + \beta_1 + \beta_2 x_i + \eta_i$

Parameter (β_j)	Least Squares Point Estimate (b_j)	Standard Error of the Estimate $(s\sqrt{c_{jj}})$	t_{b_j} Statistic $\left(t_{b_j} = \frac{b_j}{s\sqrt{c_{jj}}}\right)$
β_0	$b_0 = -26.7501$	41.5997	$t_{b_0} = -.643$
β_1	$b_1 = 3.4089$	1.3208	$t_{b_1} = 2.581$
β_2	$b_2 = .0224$	.0093	$t_{b_2} = 2.409$

We obtain the prediction equation

$$\frac{\hat{y}_i}{x_i} = \frac{-26.7501}{x_i} + 3.4089 + 0.0224x_i$$

For this transformed model we also obtain

$$SSE = 23.2952 \qquad \text{and} \qquad s = \sqrt{\frac{SSE}{n-k}} = \sqrt{\frac{23.2952}{40-3}} = .7935$$

We can now check by plotting the residuals from the transformed model versus increasing values of x. This plot is given in Figure 10.10. Examining this plot, we see that the residuals tend to form a horizontal band. Thus, we can assume that our transformation has suitably equalized the variances. If the residuals here had fanned out even after division by x, we could have tried a stronger transformation, such as division by x^2. A funneling in of this residual plot would lead us to try a milder transformation such as division by $\sqrt{x}$.

Multiplying the transformed model through by x, we get the following final prediction equation:

$$\hat{y}_i = -26.7501 + 3.4089x_i + .0224x_i^2$$

We would report that the mean square error here is $s^2x^2 = (.7935)^2x^2 = .6296x^2$, emphasizing that the error variance depends on x.

Table 10.4 presents the t_{b_j} statistics for the transformed model

$$\frac{y_i}{x_i} = \beta_0\left(\frac{1}{x_i}\right) + \beta_1 + \beta_2 x_i + \eta_i$$

Although these t_{b_j} statistics have been calculated from the transformed data, they apply no matter how we present the model. Therefore, the values given in Table 10.4 also apply to the prediction equation

$$\hat{y}_i = -26.7501 + 3.4089x_i + .0224x_i^2$$

Notice that the t_{b_2} statistic for the quadratic term is now $t_{b_2} = 2.409$. Unlike the t_{b_2} statistic calculated from the untransformed data, this t_{b_2} statistic can be verified to have a prob-value less than .05. This t_{b_2} statistic and associated prob-value are valid, since the variance has been equalized by our transformation (assuming inference assumptions 2 and 3 also hold). We see here that failure to recognize the nonconstant error variance could cause us to incorrectly omit the quadratic term.

When the error variance σ_i^2 is proportional to one of the independent variables, x_{ik}, or is proportional to x_{ik}^2, procedures for computing confidence intervals and prediction intervals are modifications of the standard procedures. The calculation of confidence intervals for the regression parameters, and thus the calculation of t_{b_j} statistics, present no new problems. We simply note that the transformed data satisfy the inference assumptions and that the transformed model involves the same parameters as does the original model. It follows that a $100(1 - \alpha)\%$ confidence interval for β_j is given by the formula

$$[b_j \pm t_{[\alpha/2]}^{(n-k)} s \sqrt{c_{jj}}]$$

where the standard error, s, is computed using the transformed data and c_{jj} is obtained from the $(\mathbf{X}'\mathbf{X})^{-1}$ *matrix calculated using the transformed data.*

Next, we consider calculating a confidence interval for the mean μ_0 and a prediction interval for the future (individual) value y_0. These intervals are computed as follows.

If the error variance σ_i^2 is proportional to x_{ik}, then

1. An appropriate transformed model is obtained by dividing each term in the original model by $\sqrt{x_{ik}}$.
2. A $100(1 - \alpha)\%$ confidence interval for the mean μ_0 is

   $$[\hat{y}_0 \pm t_{[\alpha/2]}^{(n-k)} s \sqrt{\mathbf{x}_0'(\mathbf{X}'\mathbf{X})^{-1}\mathbf{x}_0}]$$

   where $(\mathbf{X}'\mathbf{X})^{-1}$ and s *are computed from the transformed data,* and $\hat{y}_0$, $t_{[\alpha/2]}^{(n-k)}$, and $\mathbf{x}_0'$ are as previously defined.
3. A $100(1 - \alpha)\%$ prediction interval for the future (individual) value y_0 is

   $$[\hat{y}_0 \pm t_{[\alpha/2]}^{(n-k)} s \sqrt{x_{0k} + \mathbf{x}_0'(\mathbf{X}'\mathbf{X})^{-1}\mathbf{x}_0}]$$

   where $(\mathbf{X}'\mathbf{X})^{-1}$ and s *are computed from the transformed data,* $\hat{y}_0$, $t_{[\alpha/2]}^{(n-k)}$, and $\mathbf{x}_0'$ are as previously defined, and x_{0k} is the value of the independent variable x_{ik} in the future situation.

If the error variance σ_i^2 is proportional to x_{ik}^2, then

1. An appropriate transformed model is obtained by dividing each term in the original model by x_{ik}.
2. A $100(1-\alpha)\%$ confidence interval for the mean μ_0 is

$$[\hat{y}_0 \pm t_{[\alpha/2]}^{(n-k)} s\sqrt{\mathbf{x}_0'(\mathbf{X}'\mathbf{X})^{-1}\mathbf{x}_0}]$$

where $(\mathbf{X}'\mathbf{X})^{-1}$ and s *are computed from the transformed data*, and $\hat{y}_0$, $t_{[\alpha/2]}^{(n-k)}$, and $\mathbf{x}_0'$ are as previously defined.
3. A $100(1-\alpha)\%$ prediction interval for the future (individual) value y_0 is

$$[\hat{y}_0 \pm t_{[\alpha/2]}^{(n-k)} s\sqrt{x_{0k}^2 + \mathbf{x}_0'(\mathbf{X}'\mathbf{X})^{-1}\mathbf{x}_0}]$$

where $(\mathbf{X}'\mathbf{X})^{-1}$ and s *are computed from the transformed data*, $\hat{y}_0$, $t_{[\alpha/2]}^{(n-k)}$, and $\mathbf{x}_0'$ are as previously defined, and x_{0k} is the value of the independent variable x_{ik} in the future situation.

Looking at the preceding formulas for prediction intervals, we see that correct prediction intervals will tend to be wider at larger values of the independent variable x_k and will tend to be narrower at smaller values of x_k. Ignoring the unequal variances problem would produce prediction intervals too narrow for large values of x_k and too wide for small values of x_k.

We now illustrate the calculation of confidence intervals and prediction intervals in the National Association of Retail Hardware Stores problem.

EXAMPLE 10.4

Consider Example 10.3 and the NARHS problem. In this problem, the transformed model

$$\frac{y_i}{x_i} = \beta_0\left(\frac{1}{x_i}\right) + \beta_1 + \beta_2(x_i) + \eta_i$$

involves the same parameters (β_0, β_1, and β_2) as does the original model

$$y_i = \beta_0 + \beta_1 x_i + \beta_2 x_i^2 + \epsilon_i$$

Therefore, if we wish to find a 95% confidence interval for, say, the parameter β_1, we can use the least squares point estimate $b_1 = 3.4089$ and the standard error of the estimate $s\sqrt{c_{11}} = 1.3208$ that we calculated using the transformed model (see Table 10.4). Here c_{11} is obtained from the $(\mathbf{X}'\mathbf{X})^{-1}$ matrix calculated using the transformed data, and s is also computed using the transformed data.

Hence, the 95% confidence interval for β_1 is

$$\begin{aligned}[b_1 \pm t_{[\alpha/2]}^{(n-k)} s\sqrt{c_{11}}] &= [3.4089 \pm t_{[.05/2]}^{(40-3)}(1.3208)] \\ &= [3.4089 \pm 2.026(1.3208)] \\ &= [.7330, 6.0848]\end{aligned}$$

That is, we are 95% confident that β_1 is no less than .7330 and is no more than 6.0848.

Next, we compute a 95% confidence interval for the mean upkeep expenditure for all houses that are worth, say, \$110,000 (call this mean μ_0). Since the original model says that

$$\begin{aligned}y &= \mu + \epsilon \\ &= \beta_0 + \beta_1 x + \beta_2 x^2 + \epsilon\end{aligned}$$

a point estimate of $\mu_0 = \beta_0 + \beta_1(110) + \beta_2(110)^2$ is

$$\begin{aligned}\hat{y}_0 &= b_0 + b_1(110) + b_2(110)^2 \\ &= -26.7501 + 3.4089(110) + .0224(110)^2 = \$619.27\end{aligned}$$

Using $(\mathbf{X}'\mathbf{X})^{-1}$ from the transformed data and $\mathbf{x}_0' = [1 \quad 110 \quad (110)^2]$, we find that for this problem

$$\mathbf{x}_0'(\mathbf{X}'\mathbf{X})^{-1}\mathbf{x}_0 = 506.354$$

(we omit the detailed calculations here). Thus, a 95% confidence interval for μ_0 (using s calculated from the transformed data) is

$$\begin{aligned}[\hat{y}_0 \pm t_{[\alpha/2]}^{(n-k)} s\sqrt{\mathbf{x}_0'(\mathbf{X}'\mathbf{X})^{-1}\mathbf{x}_0}] &= [619.27 \pm 2.026(.7935)\sqrt{506.354}] \\ &= [583.10, 655.44]\end{aligned}$$

Therefore, the National Association of Retail Hardware Stores can be 95% confident that the mean yearly upkeep expenditure for all houses worth \$110,000 is no less than \$583.10 and is no more than \$655.44.

Finally, we compute a 95% prediction interval for the individual yearly expenditure y_0 on a house worth \$110,000 ($x_0 = 110$). This interval is

$$\begin{aligned}[\hat{y}_0 \pm t_{[\alpha/2]}^{(n-k)} s\sqrt{x_0^2 + \mathbf{x}_0'(\mathbf{X}'\mathbf{X})^{-1}\mathbf{x}_0}] &= [619.27 \pm 2.026(.7935)\sqrt{(110)^2 + 506.354}] \\ &= [438.77, 799.77]\end{aligned}$$

Here we again emphasize that $(\mathbf{X}'\mathbf{X})^{-1}$ and s are both computed using the transformed data. Also, the "extra" x_0^2 under the radical is employed because we are assuming that (and residual plots verify that) the error variance is proportional to x^2 (see Example 10.3). This interval says that the National Association of Retail Hardware Stores can be 95% confident that the yearly upkeep expenditure for an individual house valued at \$110,000 will be no less than \$438.77 and no more than \$799.77.

EXAMPLE 10.5

Republic Wholesalers, Inc., wants to investigate its telephone orders. The company feels that there is a repeating daily pattern for the volume of phone business. The goal of the analysis is to describe this daily pattern as a mathematical function of the time of day. More specifically, the company wishes to find 95% prediction intervals for the number of orders coming into the company switchboard at various times of day. The results obtained will be used to help in the scheduling of operators who have been trained to process orders.

Republic Wholesalers takes phone orders from 9 a.m. (hour 1 is defined to be 9 a.m. to 10 a.m.) through 5 p.m. (hour 8 is defined to be 4 p.m. to 5 p.m.). For each hour, 1 through 8, the number of telephone orders is recorded over a span of 10 business days. These telephone order data are given in Table 10.5. Considering this table, we let y_t denote the number of telephone orders occurring in time period t (note there are $10 \times 8 = 80$ different time periods and thus 80 different y_t values). We let H_t (for $t = 1, 2, \ldots, 80$) denote the hour number corresponding to time period t, where H_t can take on any of the values 1, 2, 3, 4, 5, 6, 7, or 8. Thus, for example, $H_{10} = 2$ (see Table 10.5). The different values of y_t are plotted against the increasing values of H_t in Figure 10.11. Two things are apparent from this plot. First, the average number of telephone orders is larger in the middle of the day, and second, the variance of the number of telephone orders is larger in the middle of the day. The second fact indicates

TABLE 10.5 Republic Wholesalers Data: Hourly Number of Telephone Orders

	Hour (H_t)							
Day	1	2	3	4	5	6	7	8
1	10 ($=y_1$)	34 ($=y_2$)	29	44	36	17	8	11 ($=y_8$)
2	31 ($=y_9$)	19 ($=y_{10}$)	18	16	28	12	14	8 ($=y_{16}$)
3	8	9	12	14	23	28	15	1
4	0	23	34	53	56	15	7	2
5	8	13	30	69	26	69	22	0
6	2	35	15	6	19	23	7	3
7	12	16	13	19	7	4	18	8
8	16	17	28	73	35	15	6	14
9	1	16	7	28	19	10	10	0
10	12 ($=y_{73}$)	7 ($=y_{74}$)	46	4	13	13	5	10 ($=y_{80}$)
Hour Avg. $\bar{y}_{H_t}$	10.0	18.9	23.2	32.6	26.2	20.6	11.2	5.7
Hour Var. $s^2_{H_t}$	$(9.1)^2$	$(9.4)^2$	$(12.1)^2$	$(25.5)^2$	$(13.8)^2$	$(18.3)^2$	$(5.8)^2$	$(5.1)^2$

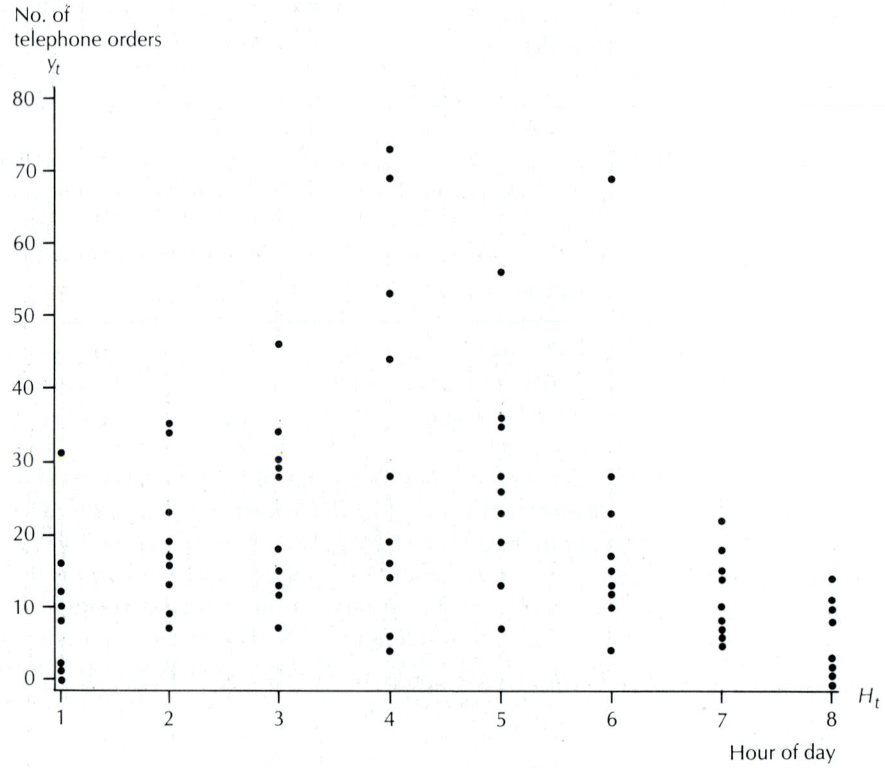

FIGURE 10.11
Number of Telephone Orders Versus Hour of Day

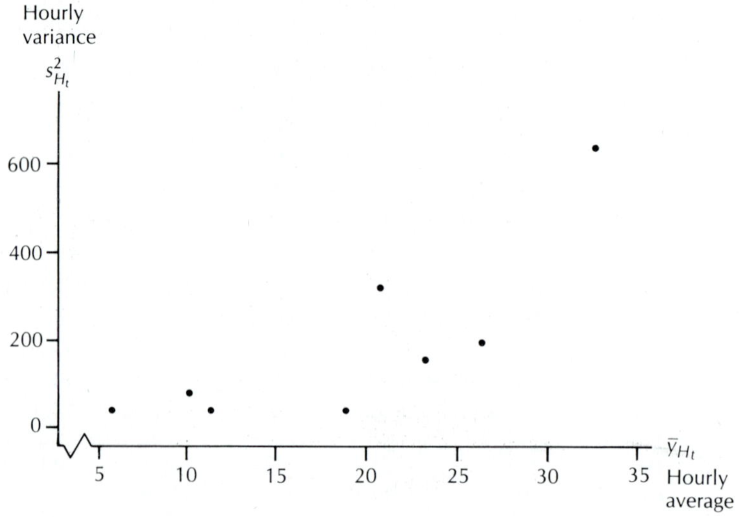

FIGURE 10.12
Plot of $s^2_{H_t}$ Versus $\bar{y}_{H_t}$ for the Republic Wholesalers Data

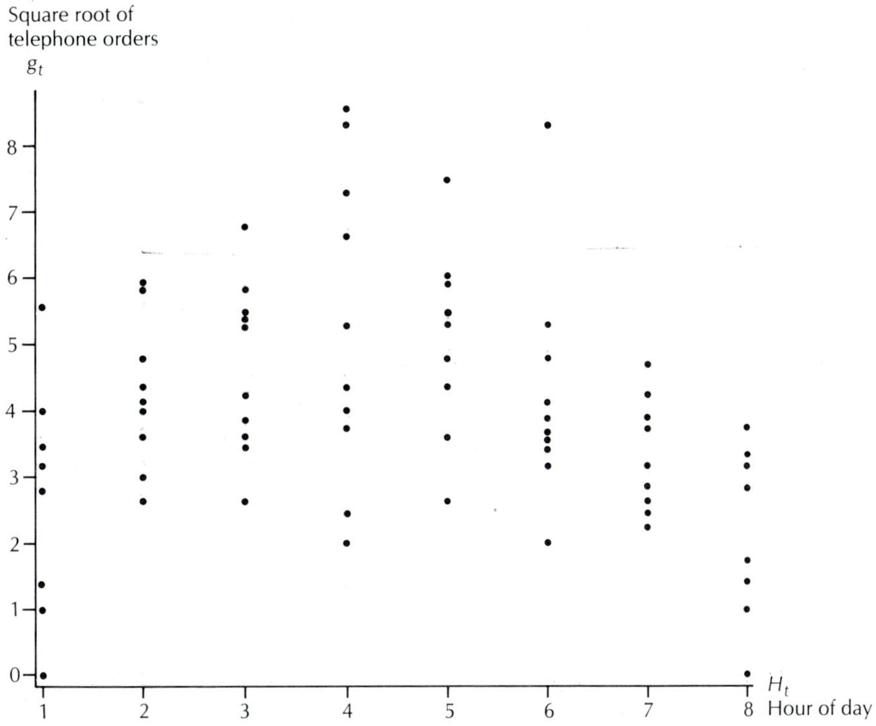

FIGURE 10.13
Square Roots of Telephone Orders Versus Hour of Day

that a nonconstant error variance exists. In fact, letting $\bar{y}_{H_t}$ and $s^2_{H_t}$ denote the mean and variance of the different numbers of telephone orders occurring in hour H_t (for $H_t = 1, 2, \ldots, 8$), noting that these quantities are calculated in Table 10.5, and plotting the different values of $s^2_{H_t}$ against the increasing values of $\bar{y}_{H_t}$ (see Figure 10.12), we see that $s^2_{H_t}$ is probably a linear function of the average level of telephone orders as measured by $\bar{y}_{H_t}$. This indicates that the error variance corresponding to hour H_t is an increasing linear function of the true mean number of telephone orders that could occur in hour H_t. This fact, along with the fact that y represents count data (the number of telephone orders) that might be described by a Poisson distribution, motivates us to use a variance-equalizing transformation in which we replace y by its square root.

We now let g_t denote the square root of the number of telephone orders at time t. A plot of the square root g_t against the hour number H_t is given in Figure 10.13. Notice that the square root transformation seems to have equalized the error variance—the variance of the number of telephone orders is larger in the middle of the day (see Figure 10.11), while the variance of the square roots is not dramatically larger in the middle of the day (see Figure 10.13). To account for the apparent curvature in Figure 10.13 (which is still present after taking

TABLE 10.6 The t_{b_j} Statistics for the Model $g_t = \beta_0 + \beta_1 H_t + \beta_2 H_t^2 + \epsilon_t$

Parameter (β_j)	Least Squares Point Estimate (b_j)	Standard Error of the Estimate ($s\sqrt{c_{jj}}$)	t_{b_j} Statistic $\left(t_{b_j} = \frac{b_j}{s\sqrt{c_{jj}}}\right)$
β_0	$b_0 = 1.2604$	.6434	$t_{b_0} = 1.96$
β_1	$b_1 = 1.8363$	.3280	$t_{b_1} = 5.60$
β_2	$b_2 = -.2197$	.0356	$t_{b_2} = -6.17$

square roots), we try the model

$$g_t = \beta_0 + \beta_1 H_t + \beta_2 H_t^2 + \epsilon_t$$

In order to find the least squares point estimates of the parameters β_0, β_1, and β_2, we use the following **y** vector and **X** matrix (see Table 10.5):

$$\mathbf{y} = \begin{bmatrix} \sqrt{10} \\ \sqrt{34} \\ \sqrt{29} \\ \vdots \\ \sqrt{5} \\ \sqrt{10} \end{bmatrix} \qquad \mathbf{X} = \begin{bmatrix} 1 & 1 & 1 \\ 1 & 2 & 4 \\ 1 & 3 & 9 \\ \vdots & \vdots & \vdots \\ 1 & 7 & 49 \\ 1 & 8 & 64 \end{bmatrix}$$

We obtain the prediction equation

$$\hat{g}_t = 1.2604 + 1.8363H_t - .2197H_t^2$$

For this transformed model we also obtain

$$SSE = 163.7747 \qquad \text{and} \qquad s = \sqrt{\frac{SSE}{n-k}} = \sqrt{\frac{163.7747}{80-3}} = 1.4584$$

Table 10.6 gives the t_{b_j} statistics for this model. Since Figure 10.13 indicates that our transformed model displays a constant error variance, these t_{b_j} statistics are valid (assuming inference assumptions 2 and 3 also hold). We see that the quadratic term is important, as the plot in Figure 10.13 indicated it might be.

We now calculate a 95% prediction interval for the number of telephone orders received on a future day between noon and 1 p.m. (hour 4). First, a point prediction of the square root of the number of telephone orders is

$$\begin{aligned} \hat{g}_0 &= 1.2604 + 1.8363H_0 - .2197H_0^2 \\ &= 1.2604 + 1.8363(4) - .2197(16) = 5.0904 \end{aligned}$$

Therefore, a point prediction of the number of telephone orders received is $(5.0904)^2 = 25.91$. Next, a 95% prediction interval for the square root of the number of telephone orders is

$$\begin{aligned}[\hat{g}_0 \pm t_{[\alpha/2]}^{(n-k)} s\sqrt{1 + \mathbf{x}_0'(\mathbf{X}'\mathbf{X})^{-1}\mathbf{x}_0}] &= [5.0904 \pm t_{[.025]}^{(80-3)}(1.4584)\sqrt{1 + .0287}] \\ &= [5.0904 \pm 1.990(1.4584)\sqrt{1.0287}] \\ &= [2.1468, 8.0340]\end{aligned}$$

Here $\mathbf{x}_0' = [1 \quad 4 \quad 16]$ (we have omitted a detailed calculation of $\mathbf{x}_0'(\mathbf{X}'\mathbf{X})^{-1}\mathbf{x}_0$). It follows that a 95% prediction interval for the number of telephone orders between noon and 1 p.m. is

$$[(2.1468)^2, (8.0340)^2] \qquad \text{or} \qquad [4.61, 64.55]$$

Thus, Republic Wholesalers, Inc., can be 95% confident that between 4.61 and 64.55 (or about 5 to 65) telephone orders will be received between noon and 1 p.m. This interval is quite wide. For this problem we have $\mathbf{x}_0'(\mathbf{X}'\mathbf{X})^{-1}\mathbf{x}_0 = .0287$, which could be reduced by collecting more data. However, even if we were to set $\mathbf{x}_0'(\mathbf{X}'\mathbf{X})^{-1}\mathbf{x}_0$ equal to zero, we would only reduce

$$s\sqrt{1 + \mathbf{x}_0'(\mathbf{X}'\mathbf{X})^{-1}\mathbf{x}_0} = 1.4584\sqrt{1 + .0287} = 1.4792$$

to

$$s\sqrt{1 + 0} = s = 1.4584$$

This implies that the large width of this prediction interval is due mostly to the variability that is inherent in the ordering process. In this case, Republic Wholesalers would have to decide whether or not it is worthwhile to staff their switchboard with enough trained operators to handle the maximum number of calls (65) that might be received. Prediction intervals for the other hours in the day can be found in a similar manner.

10.2.2 0–1 Dependent Variables

We will begin our discussion of 0–1 dependent variables by looking at an example in which the use of a 0–1 dependent variable is appropriate.

EXAMPLE 10.6

Suppose that we wish to investigate the attitudes of people in various income brackets toward a new tax cut. In order to do this, we take a sample survey of n people and ask each person to report his or her income (x) and whether or not a tax cut is favored. We now let $y = 1$ if a tax cut is favored and $y = 0$ otherwise. By defining y in this way, we have set up the problem so that the mean value of y in the population of all people at a particular income level x is simply the proportion p of people with that income who favor a tax cut. Therefore, if we consider a regression in which y is the dependent variable and x is the independent variable, we can try to establish a relationship between income x and the probability p that a person selected at random with income x will favor a tax cut by estimating the mean value of y. That is, we use a regression of y on x to estimate p as a function of x.

When the dependent variable in a regression problem is defined as y is here, we call y a **0–1 dependent variable**. One of the special points about this kind of regression is that the dependent variable y is expressed as a function of x, say $p(x)$, which is interpreted as a probability. This being the case, a prediction equation obtained from such a regression only makes sense when it produces probabilities between zero and 1 (for example, the prediction equation $\hat{y} = .01 + .00002x$ would not be sensible for an income of \$50,000, since this equation would then produce a probability larger than 1, since $.01 + .00002(50{,}000) = 1.01$). Another feature of such a regression is that for a given x, the dependent variable y is 1 with some probability $p(x)$ and is zero with probability $1 - p(x)$. The variance of y at this value of x can be shown to be $p(x)[1 - p(x)]$. The variance $p(x)[1 - p(x)]$ will of course not be the same for all values of x (unless $p(x)$ is a constant function of x, in which case our regression makes little sense in the first place). Thus, a regression of this type displays a nonconstant error variance and hence violates inference assumption 1. Our remedy for this problem will involve a transformation of the data. The effect of unequal variances in this kind of problem is most pronounced when $p(x)$ takes on values near zero or 1. For example, if $.3 \leq p(x) \leq .7$ for all the values of x we are considering, then $.21 \leq p(x)[1 - p(x)] \leq .25$ (since the maximum value that $p(x)[1 - p(x)]$ attains for values of $p(x)$ between zero and 1 is .25, which occurs when $p(x) = .5$). Since the variance of y is practically constant here, little would be gained by transformation.

EXAMPLE 10.7

Happy Valley Real Estate is a company that sells recreational property. The company wishes to investigate the chances that a randomly selected person with a given income will purchase property. As part of its marketing strategy, Happy Valley Real Estate offers incentives (free gifts and vacation trips) to encourage prospective buyers to visit sites at Happy Valley. After a tour, the prospective buyer is asked to fill out a questionnaire, which, among other things, includes yearly income (denoted x). Happy Valley will use this information to conduct its study as follows. If, within a month of the visit, the customer purchases property from Happy Valley, $y = 1$ is recorded for the customer. If the customer does not purchase property within a month, $y = 0$ is recorded. The ultimate goal of the

TABLE 10.7 Happy Valley Real Estate Purchase Data

Income x (thousands of dollars)	50	25	30	28	42	17	20	25	38	
Purchase Decision y (1 = buy, 0 = not buy)	1	0	0	1	1	0	0	0	1	
Income x	15	30	28	48	16	21	13	36	34	22	20
Purchase Decision y	0	1	0	1	0	1	0	1	0	0	0
Income x	37	45	17	24	28	25	22	18	26	44	
Purchase Decision y	0	1	1	0	1	1	0	0	0	0	

analysis is to compute a 95% confidence interval for the probability $p(x)$ that a randomly selected person with income x will purchase property.

Happy Valley Real Estate collects data as described from 30 prospective buyers. These data are given in Table 10.7 and plotted in Figure 10.14. Looking at the data, we see that eight people in this sample have incomes over $35,000, and of those, six purchased property. On the other hand, of the eight people with incomes less than $21,000, only one purchased property. This leads us to suspect that income does indeed affect the buying decision.

In order to establish a relationship between income x and $p(x)$, first notice that for most incomes x we have only one observation of y. This means that we cannot satisfactorily estimate the probability $p(x)$ that a randomly selected person with income x will purchase Happy Valley property by using only the data at a given x value. Instead, our strategy is to use the regression model

$$y = \mu + \epsilon$$
$$= \beta_0 + \beta_1 x + \epsilon$$

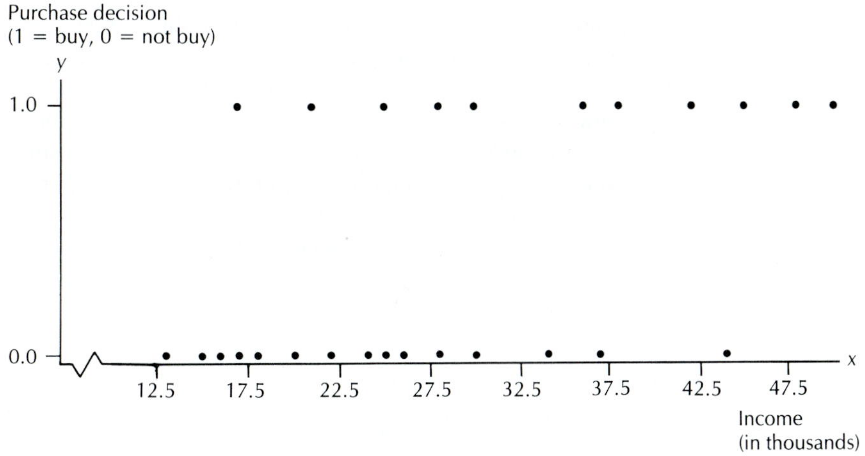

FIGURE 10.14 Purchase Decision Versus Income for the Happy Valley Real Estate Data

which relates the purchase decision y to income x. For this model, the mean value of y in the population of all people at a particular income level x is the proportion $p(x)$ of people with that income who will purchase Happy Valley property. Therefore, we use the Happy Valley data in Table 10.7 to estimate the parameters β_0 and β_1 in the preceding model. Then, since the mean value of y at a particular income level x is

$$\mu = \beta_0 + \beta_1 x$$

our estimate of $p(x)$ is $b_0 + b_1 x$.

However, we are now in a dilemma. We know that the variance of y at a particular value of x is $p(x)[1 - p(x)]$, and we know that this variance varies as x varies, since $p(x)$ changes as x changes. This says that a nonconstant error variance exists and that inference assumption 1 is violated. If we could divide each term in the model

$$y = \beta_0 + \beta_1 x + \epsilon$$

by the quantity $\sqrt{p(x)[1 - p(x)]}$ (call this w), we would have a model

$$\frac{y}{w} = \beta_0\left(\frac{1}{w}\right) + \beta_1\left(\frac{x}{w}\right) + \frac{\epsilon}{w}$$

where ϵ/w has variance 1 at all x values (this is analogous to what we did in Section 10.2.1). The dilemma arises because $p(x)$, and therefore w, are unknown. Of course, if we knew $p(x)$, we would not be trying to estimate it!

Fortunately, the least squares point estimates b_0 and b_1 are not totally unreasonable, even though inference assumption 1 is violated. This means that an iterative scheme can be employed in which b_0 and b_1 are used to estimate $p(x)$, and then this estimate is used to transform the model. We then use the transformed model to re-estimate β_0 and β_1. Following this, a new $p(x)$ estimate can be computed and can be used to obtain a new transformed model, which is in turn used to find yet a third set of estimates of β_0 and β_1. Often, iteration produces only small changes in the estimates of β_0 and β_1, and usually only a few rounds of iteration are necessary. We carry out this iterative procedure for the Happy Valley data in the problems at the end of this chapter. We find that the least squares point estimates of the parameters β_0 and β_1 for our transformed model (these estimates are denoted $\tilde{b}_0$ and $\tilde{b}_1$) are $\tilde{b}_0 = -.2676$ and $\tilde{b}_1 = .0239$. The corresponding t_{b_j} statistics for these estimates are

$$t_{\tilde{b}_0} = \frac{\tilde{b}_0}{s\sqrt{c_{00}}} = \frac{-.2676}{.1861} = -1.438$$

$$t_{\tilde{b}_1} = \frac{\tilde{b}_1}{s\sqrt{c_{11}}} = \frac{.0239}{.0064} = 3.734$$

These t_{b_j} statistics are valid, because the transformed model approximately satisfies inference assumption 1.

Because the transformed model

$$\frac{y}{w} = \beta_0\left(\frac{1}{w}\right) + \beta_1\left(\frac{x}{w}\right) + \frac{\epsilon}{w}$$

has the same parameters β_0 and β_1 as the original model, we use $\tilde{b}_0$ and $\tilde{b}_1$ to estimate $p(x)$. Calling this estimate $\tilde{p}(x)$, we obtain

$$\tilde{p}(x) = -.2676 + .0239x$$

We now use our results to find point and interval estimates of the probability that a randomly selected person with a yearly income of \$23,000 will purchase Happy Valley property within one month of a Happy Valley tour. The point estimate of $p(23)$ is

$$\tilde{p}(23) = -.2676 + .0239(23) = .2821$$

Thus, Happy Valley Real Estate estimates that 28.21% of all people with a yearly income of \$23,000 would purchase property within one month of a Happy Valley tour. In order to compute a 95% confidence interval for $p(x)$, we use the standard error, s, and the $(\mathbf{X}'\mathbf{X})^{-1}$ matrix *computed from the transformed data.* Letting $\mathbf{x}_0' = [1 \quad 23]$, we obtain

$$\mathbf{x}_0'(\mathbf{X}'\mathbf{X})^{-1}\mathbf{x}_0 = .0059217$$

Also, for the transformed model, we have

$$s = \sqrt{\frac{SSE}{n-k}} = \sqrt{\frac{28.8428}{30-2}} = 1.01494$$

It follows that a 95% confidence interval for $p(23)$ is

$$\begin{aligned}[\tilde{p}(23) \pm t_{[\alpha/2]}^{(n-k)} s\sqrt{\mathbf{x}_0'(\mathbf{X}'\mathbf{X})^{-1}\mathbf{x}_0}] &= [\tilde{p}(23) \pm t_{[.025]}^{(30-2)} s\sqrt{\mathbf{x}_0'(\mathbf{X}'\mathbf{X})^{-1}\mathbf{x}_0}] \\ &= [.2821 \pm 2.048(1.01494)\sqrt{.0059217}] \\ &= [.2821 \pm .1600] \\ &= [.1221, .4421]\end{aligned}$$

Notice that we compute a 95% *confidence* interval here rather than a prediction interval. This is because $p(23)$ is the *mean* of the population of y values (0's and 1's) when yearly income is \$23,000 ($x = 23$). The interval says that Happy Valley Real Estate can be 95% confident that between 12.21% and 44.21% of all people with a yearly income of \$23,000 would purchase property within one month of a Happy Valley tour. Point and interval estimates of $p(x)$ at other income levels can be established in a similar fashion.

Before concluding this section, we should point out that in the preceding example we employed a simple linear regression model where $p(x) = \beta_0 + \beta_1 x$. Fortunately, for the range of x values (yearly incomes) in which we wished to estimate $p(x)$, we did not obtain any $\tilde{p}(x)$ values (estimated probabilities) less than zero or greater than 1 (for a verification of this, see Exercise 10.3). Another function that can be used is the **logistic function**. The logistic function, unlike the linear function, can be forced to give estimates of $p(x)$ that are between zero and 1 for all values of x. For a discussion of the logistic function, see Neter, Wasserman, and Kutner (1985).

10.3 MODELING DETERMINISTIC TIME SERIES COMPONENTS

If the appropriate residual plots or statistical tests indicate that the error terms for a regression model are dependent (inference assumption 2 is violated), remedies are available. A common remedy is to use **time series models**. We present some models of this type in this section.

Consider the Fresh Detergent problem. We have analyzed the Fresh residuals and we have found that they appear to be consistent with the inference assumptions. Suppose, however, that Enterprise Industries had just introduced Fresh Detergent prior to the outset of our analysis. In this case, the Fresh sales data might have shown a linear increase over time because of increasing availability of Fresh and increasing consumer awareness of Fresh. This would have shown up in the plot of Fresh residuals versus time as a linear trend and would suggest that time (denoted t) should be included as an independent variable in the model. Such a model would be

$$\begin{aligned} y &= \mu + \epsilon \\ &= \beta_0 + \beta_1 x_4 + \beta_2 x_3 + \beta_3 x_3^2 + \beta_4 x_4 x_3 + \beta_5 t + \epsilon \end{aligned}$$

In general, time series data can be described by *trends, seasonal effects,* and *cyclical effects* as well as *causal variables* (variables like advertising expenditure and the price difference that are related to the dependent variable, demand for Fresh). **Trend** refers to the upward or downward movement that characterizes a time series over a period of time. Thus, trend reflects the long-run growth or decline in the time series. Trends can represent a variety of factors. For example, trends in sales might be determined by (1) technological change, (2) changes in consumer tastes, (3) increases in per capita income, and (4) market growth. Denoting the trend in time period t as TR_t, some common trends are

1. Linear trend, which is modeled as $TR_t = \beta_0 + \beta_1 t$.
2. Quadratic trend, which is modeled as $TR_t = \beta_0 + \beta_1 t + \beta_2 t^2$.
3. Exponential trends (such as the trend in Example 10.1, which we modeled as $TR_t = \beta_0(\beta_1)^t$).

Although these trends are probably most commonly used, other more complicated trends also exist.

Time series data also sometimes show **seasonal effects**. Seasonal variations are periodic patterns in a time series that complete themselves within a calendar year and are then repeated on a yearly basis. Such variations are usually caused by factors like the weather and customs (such as holidays). For instance, air conditioner sales and soft drink sales are seasonal in nature, with highest sales during the summer months. As another example, monthly sales volume for a department store might be seasonal. Here the seasonal variation would be caused by the observance of various holidays, with high sales volumes during December and April, because of shopping activity prior to the Christmas and Easter holidays.

Seasonal patterns can be modeled by using dummy variables. Letting y_t denote the value of the dependent variable in time period t, TR_t denote the trend in time period t, SN_t denote the seasonal factor in time period t, and ϵ_t denote the error term in time period t, we employ the model

$$y_t = TR_t + SN_t + \epsilon_t$$

Moreover, assuming that there are L seasons (months, quarters, and so on) per year, we assume that SN_t is given by the equation

$$SN_t = \beta_{S1}x_{S1,t} + \beta_{S2}x_{S2,t} + \cdots + \beta_{S(L-1)}x_{S(L-1),t}$$

where $x_{S1,t}, x_{S2,t}, \ldots, x_{S(L-1),t}$ are dummy variables, which are defined as follows:

$$x_{S1,t} = \begin{cases} 1 & \text{if time period } t \text{ is season 1} \\ 0 & \text{otherwise} \end{cases}$$

$$x_{S2,t} = \begin{cases} 1 & \text{if time period } t \text{ is season 2} \\ 0 & \text{otherwise} \end{cases}$$

$$\vdots \qquad\qquad \vdots$$

$$x_{S(L-1),t} = \begin{cases} 1 & \text{if time period } t \text{ is season } L-1 \\ 0 & \text{otherwise} \end{cases}$$

For example, if $L = 12$ (monthly data) and period t is season 2 (February), we have

$$\begin{aligned} y_t &= TR_t + SN_t + \epsilon_t \\ &= TR_t + \beta_{S1}x_{S1,t} + \beta_{S2}x_{S2,t} + \beta_{S3}x_{S3,t} + \cdots + \beta_{S11}x_{S11,t} + \epsilon_t \\ &= TR_t + \beta_{S1}(0) + \beta_{S2}(1) + \beta_{S3}(0) + \cdots + \beta_{S11}(0) + \epsilon_t \\ &= TR_t + \beta_{S2} + \epsilon_t \end{aligned}$$

The use of the dummy variables ensures that a seasonal parameter for season 2 is added to the trend in each time period that is season 2. This seasonal parameter, β_{S2}, accounts for the seasonality of the time series in season 2.

In general, the purpose of the dummy variables is to ensure that the appropriate seasonal parameter is included in the regression model in each time period. The use of the dummy variables also assures us that SN_t is a linear function of the parameters $\beta_{S1}, \beta_{S2}, \beta_{S3}, \ldots, \beta_{S(L-1)}$. Thus, if the trend is also a linear function of its parameters, the model

$$y_t = TR_t + SN_t + \epsilon_t$$

is a regression model for which least squares point estimates of the parameters can be computed. Letting tr_t be the estimate of the trend, TR_t, we have, for example,

$$tr_t = b_0 + b_1 t$$

if there is a linear trend, or

$$tr_t = b_0 + b_1 t + b_2 t^2$$

if there is a quadratic trend. Moreover, sn_t, the estimate of SN_t, is

$$sn_t = b_{S1}x_{S1,t} + b_{S2}x_{S2,t} + \cdots + b_{S(L-1)}x_{S(L-1),t}$$

where $b_{S1}, b_{S2}, \ldots, b_{S(L-1)}$ are the least squares point estimates of the parameters $\beta_{S1}, \beta_{S2}, \ldots, \beta_{S(L-1)}$. Therefore, the point prediction of y_t is

$$\hat{y}_t = tr_t + sn_t$$

For example, if period t is season 1, then

$$y_t = TR_t + \beta_{S1} + \epsilon_t$$

and a point prediction of y_t is

$$\hat{y}_t = tr_t + b_{S1}$$

If period t is season 2, then

$$y_t = TR_t + \beta_{S2} + \epsilon_t$$

and a point prediction of y_t is

$$\hat{y}_t = tr_t + b_{S2}$$

If period t is season $L - 1$, then

$$y_t = TR_t + \beta_{S(L-1)} + \epsilon_t$$

and a point prediction of y_t is

$$\hat{y}_t = tr_t + b_{S(L-1)}$$

If period t is season L, then

$$y_t = TR_t + \epsilon_t$$

and a point prediction of y_t is

$$\hat{y}_t = tr_t$$

We see that we have, quite arbitrarily, set the seasonal parameter for season L equal to zero. Thus, the other seasonal parameters—$\beta_{S1}, \beta_{S2}, \ldots, \beta_{S(L-1)}$—are defined with respect to season L. Intuitively, β_{Sj} is the difference, excluding trend, between the level of the time series in season j and the level of the time series in season L. If β_{Sj} is positive, this implies that, excluding trend, the value of the time series in season j can be expected to be greater than the value of the time series in season L. If β_{Sj} is negative, this implies that, excluding trend, the value of the time series in season j can be expected to be smaller than the value of the time series in season L. We do not have to set the seasonal parameter for season L equal to zero. We can set the seasonal parameter for any particular season equal to zero and thus define the other seasonal parameters with respect to that particular season. However, we must arbitrarily set one of the seasonal parameters equal to zero, for if we do not, the columns of the **X** matrix can be shown to be linearly dependent, and the least squares point estimates of the model parameters cannot be computed using the methods we have presented in this book.

EXAMPLE 10.8

Farmers' Bureau Coop, a small agricultural cooperative, would like to predict its propane gas bills for the four quarters of next year. In addition, the coop would like to estimate the trend in its propane gas bills after adjusting for seasonal effects. In order to do this, Farmers' Bureau has compiled its propane gas bills for the last 10 years. The quarterly propane gas bills for this 10-year period (40 quarters) are given in Table 10.8.

A plot of the gas bill data is displayed in Figure 10.15. Looking at this data plot, there appears to be an upward trend in the gas bills. Large seasonal (quarterly) fluctuations, however, make it difficult to ascertain the exact nature of this

TABLE 10.8 Quarterly Propane Gas Bills for Farmers' Bureau Coop

Year	Quarter 1	Quarter 2	Quarter 3	Quarter 4
1	344.39 $(=y_1)$	246.63 $(=y_2)$	131.53 $(=y_3)$	288.87 $(=y_4)$
2	313.45 $(=y_5)$	189.76 $(=y_6)$	179.10 $(=y_7)$	221.10 $(=y_8)$
3	246.84	209.00	51.21	133.89
4	277.01	197.98	50.68	218.08
5	365.10	207.51	54.63	214.09
6	267.00	230.28	230.32	426.41
7	467.06	306.03	253.23	279.46
8	336.56	196.67	152.15	319.67
9	440.00	315.04	216.42	339.78
10	434.66 $(=y_{37})$	399.66 $(=y_{38})$	330.80 $(=y_{39})$	539.78 $(=y_{40})$

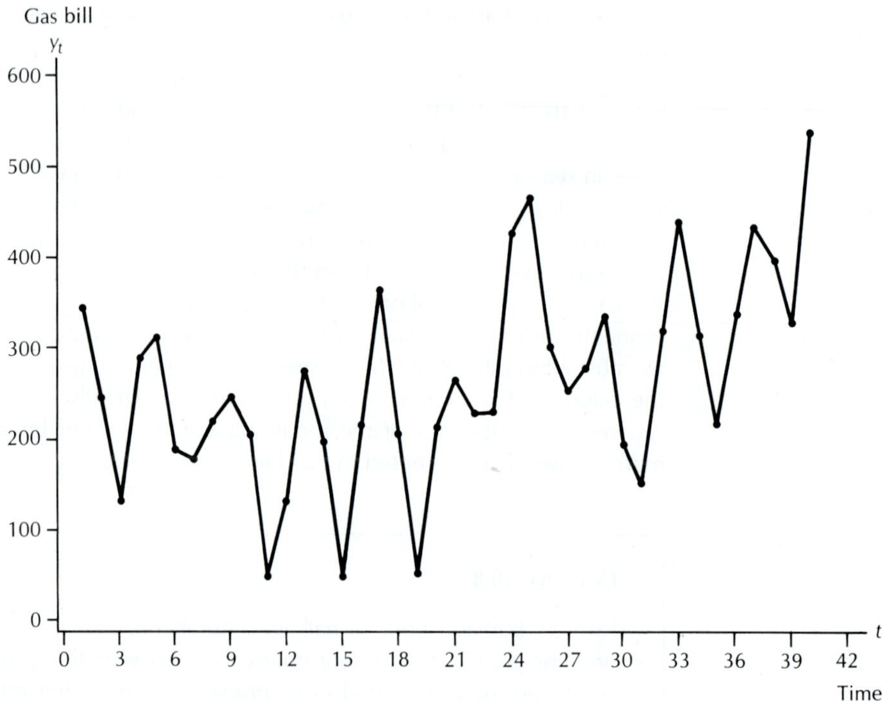

FIGURE 10.15
Quarterly Propane Gas Bill Versus Time for Farmers' Bureau Coop

trend. In Figure 10.16 we plot the yearly mean propane gas bills against the year number (1–10). We do this in order to average out the quarterly effects in the gas bill data. Looking at Figure 10.16, we see that the trend is possibly increasing at an increasing rate. On the basis of this plot, we assume that the trend here is quadratic. That is, the trend, TR_t, is modeled as

$$TR_t = \beta_0 + \beta_1 t + \beta_2 t^2$$

In order to account for the seasonal nature of the gas bill data, we use dummy variables $x_{S1,t}$, $x_{S2,t}$, and $x_{S3,t}$, which are defined as follows:

$$x_{S1,t} = \begin{cases} 1 & \text{if time period } t \text{ is quarter 1} \\ 0 & \text{otherwise} \end{cases}$$

$$x_{S2,t} = \begin{cases} 1 & \text{if time period } t \text{ is quarter 2} \\ 0 & \text{otherwise} \end{cases}$$

$$x_{S3,t} = \begin{cases} 1 & \text{if time period } t \text{ is quarter 3} \\ 0 & \text{otherwise} \end{cases}$$

Our regression model is

$$\begin{aligned} y_t &= TR_t + SN_t + \epsilon_t \\ &= \beta_0 + \beta_1 t + \beta_2 t^2 + \beta_3 x_{S1,t} + \beta_4 x_{S2,t} + \beta_5 x_{S3,t} + \epsilon_t \end{aligned}$$

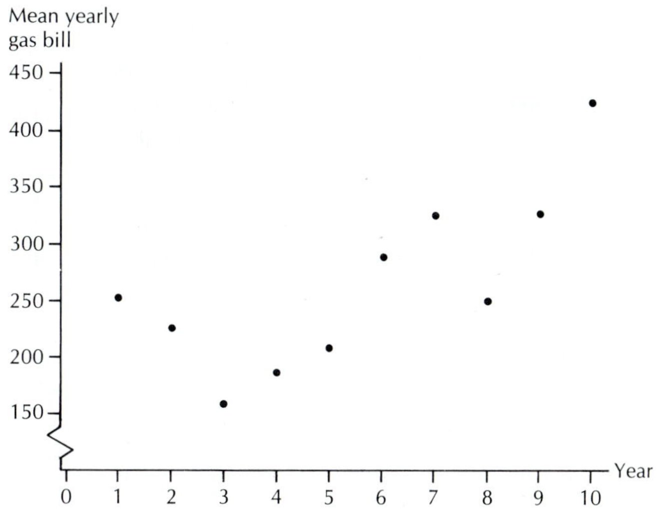

FIGURE 10.16
Mean Yearly Propane Gas Bill Versus Year for Farmers' Bureau Coop

The least squares point estimates of the parameters for this model are calculated using the following **y** vector and **X** matrix:

$$\mathbf{y} = \begin{bmatrix} 344.39 \\ 246.63 \\ 131.53 \\ 288.87 \\ \vdots \\ 434.66 \\ 399.66 \\ 330.80 \\ 539.78 \end{bmatrix} \qquad \mathbf{X} = \begin{bmatrix} 1 & 1 & 1 & 1 & 0 & 0 \\ 1 & 2 & 4 & 0 & 1 & 0 \\ 1 & 3 & 9 & 0 & 0 & 1 \\ 1 & 4 & 16 & 0 & 0 & 0 \\ \vdots & \vdots & \vdots & \vdots & \vdots & \vdots \\ 1 & 37 & 1369 & 1 & 0 & 0 \\ 1 & 38 & 1444 & 0 & 1 & 0 \\ 1 & 39 & 1521 & 0 & 0 & 1 \\ 1 & 40 & 1600 & 0 & 0 & 0 \end{bmatrix}$$

We obtain the prediction equation

$$\hat{y}_t = 276.64 - 7.46t + .30t^2 + 65.77x_{S1,t} - 37.87x_{S2,t} - 127.61x_{S3,t}$$

The t_{b_j} statistics for this model are given in Table 10.9. Using the preceding prediction equation, we obtain point predictions of the quarterly propane gas bills for the four quarters of next year (time periods 41, 42, 43, and 44).

For quarter 1 ($x_{S1,41} = 1$, $x_{S2,41} = 0$, and $x_{S3,41} = 0$), a point forecast of y_{41} (the propane gas bill in time period 41) is

$$\begin{aligned} \hat{y}_{41} &= 276.64 - 7.46(41) + .30(41)^2 + 65.77(1) - 37.87(0) - 127.61(0) \\ &= \$540.85 \end{aligned}$$

TABLE 10.9 The t_{b_j} Statistics for the Farmers' Bureau Coop Dummy Variable Model

Parameter (β_j)	Least Squares Point Estimate (b_j)	Standard Error of the Estimate $(s\sqrt{c_{jj}})$	t_{b_j} Statistic $\left(t_{b_j} = \frac{b_j}{s\sqrt{c_{jj}}}\right)$
β_0	$b_0 = 276.64$	35.05	$t_{b_0} = 7.89$
β_1	$b_1 = -7.46$	3.40	$t_{b_1} = -2.19$
β_2	$b_2 = .30$	.08	$t_{b_2} = 3.75$
β_3	$b_3 = 65.77$	27.16	$t_{b_3} = 2.42$
β_4	$b_4 = -37.87$	27.10	$t_{b_4} = -1.40$
β_5	$b_5 = -127.61$	27.06	$t_{b_5} = -4.72$

For quarter 2 ($x_{S1,42} = 0$, $x_{S2,42} = 1$, and $x_{S3,42} = 0$), a point forecast of y_{42} is

$$\begin{aligned}\hat{y}_{42} &= 276.64 - 7.46(42) + .30(42)^2 + 65.77(0) - 37.87(1) - 127.61(0)\\ &= \$454.65\end{aligned}$$

For quarter 3 ($x_{S1,43} = 0$, $x_{S2,43} = 0$, and $x_{S3,43} = 1$), a point forecast of y_{43} is

$$\begin{aligned}\hat{y}_{43} &= 276.64 - 7.46(43) + .30(43)^2 + 65.77(0) - 37.87(0) - 127.61(1)\\ &= \$382.95\end{aligned}$$

For quarter 4 ($x_{S1,44} = 0$, $x_{S2,44} = 0$, and $x_{S3,44} = 0$), a point forecast of y_{44} is

$$\begin{aligned}\hat{y}_{44} &= 276.64 - 7.46(44) + .30(44)^2 + 65.77(0) - 37.87(0) - 127.61(0)\\ &= \$529.20\end{aligned}$$

The residuals for this model are plotted in time order in Figure 10.17. The plotted residuals have these signs: + + + + + − + − − + − − − + − − + + − − − + + + + + + − − − − − + − − − − + + +. Here we see a tendency for residuals to be followed by residuals of the same sign. This suggests that the error terms for this model display positive autocorrelation and that this model therefore violates inference assumption 2.

The Durbin-Watson statistic for this model is

$$d = \frac{\sum_{t=2}^{n} (e_t - e_{t-1})^2}{\sum_{t=1}^{n} e_t^2} = .8397$$

Our dummy variable regression model has $k - 1 = 5$ independent variables (we exclude the intercept here), and we have $n = 40$ observations in the Farmers' Bureau Coop data. Therefore, letting the probability of a Type I error be $\alpha = .05$, we obtain $d_{L,.05} = 1.23$ and $d_{U,.05} = 1.79$ (see Table E-5). Testing the

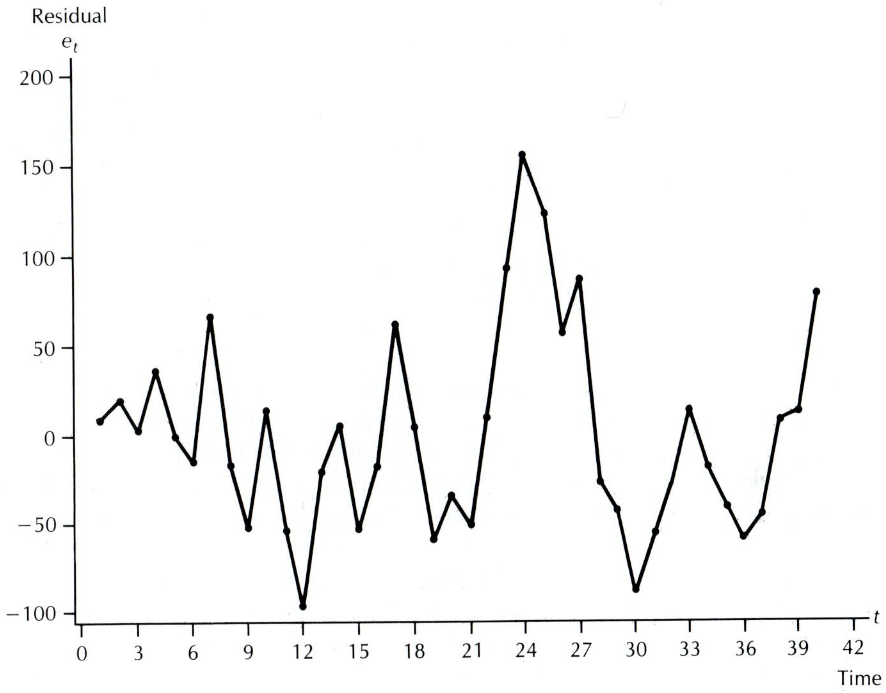

FIGURE 10.17 Residuals Versus Time for Farmers' Bureau Coop Dummy Variable Model

null hypothesis

$$H_0\text{: The error terms are not autocorrelated}$$

versus the alternative hypothesis

$$H_1\text{: The error terms are positively autocorrelated}$$

we see that since $d = .8397 < d_{L,.05} = 1.23$, we reject H_0 and conclude that there is significant evidence that the error terms for the dummy variable regression model are positively autocorrelated. In the next section, we complete the analysis of the Farmers' Bureau Coop data by building a model that will remedy this problem.

Before discussing autocorrelated error structures in detail, we should mention that the dummy variable regression model

$$y_t = TR_t + SN_t + \epsilon_t$$

where

$$SN_t = \beta_{S1}x_{S1,t} + \beta_{S2}x_{S2,t} + \cdots + \beta_{S(L-1)}x_{S(L-1),t}$$

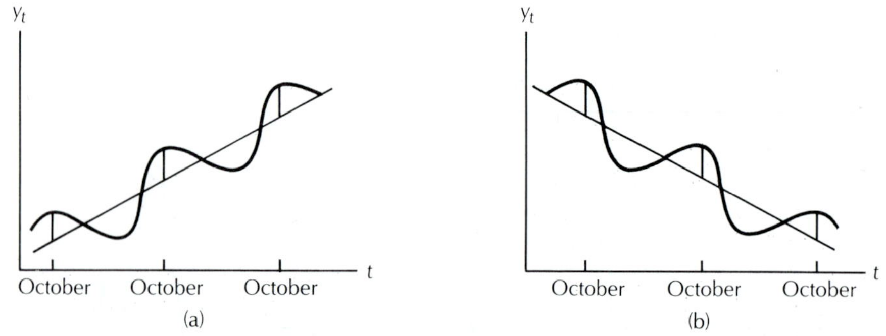

FIGURE 10.18 Additive Seasonal Variation: The Seasonal Swing Is the Same as the Trend Increases or Decreases

assumes that we have *additive seasonal variation*. If a time series displays additive seasonal variation, the magnitude of the seasonal swing is independent of the level of the trend. Additive seasonal variation is illustrated in Figure 10.18. This is not the only kind of seasonal variation, however. Sometimes, seasonal variation is multiplicative. If a time series displays *multiplicative seasonal variation*, the magnitude of the seasonal swing is proportional to the level of the trend. Thus, if the trend of the time series is increasing, so is the magnitude of the seasonal swing, while if the trend of the time series is decreasing, the magnitude of the seasonal swing is also decreasing. Multiplicative seasonal variation is illustrated in Figure 10.19. One should realize that very few actual time series possess seasonal variation that is precisely additive or precisely multiplicative. However, it is useful to try to classify the seasonal variation in a time series as either additive or multiplicative before attempting to find a model that can be used to forecast the series.

As we have stated, the dummy variable regression model is useful for modeling time series with additive seasonal variation. Other regression models can also be used, however. Two models involving *trigonometric terms* that are useful for modeling additive seasonal variation are the following (here L is the number of seasons in a year):

$$y_t = \beta_0 + \beta_1 t + \beta_2 \sin\frac{2\pi t}{L} + \beta_3 \cos\frac{2\pi t}{L} + \epsilon_t$$

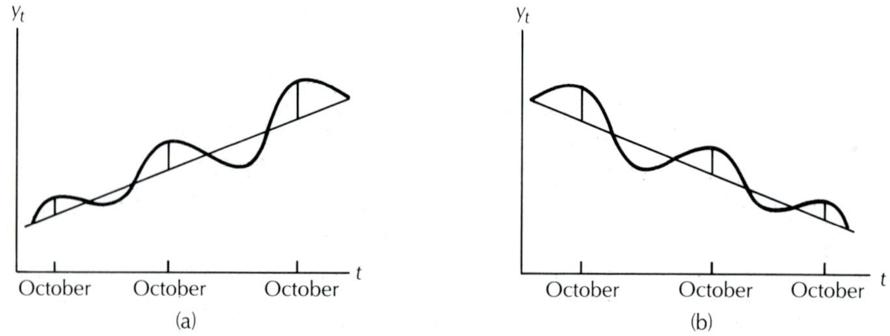

FIGURE 10.19 Multiplicative Seasonal Variation: The Magnitude of the Seasonal Swing Is Proportional to the Trend

and

$$y_t = \beta_0 + \beta_1 t + \beta_2 \sin\frac{2\pi t}{L} + \beta_3 \cos\frac{2\pi t}{L} + \beta_4 \sin\frac{4\pi t}{L} + \beta_5 \cos\frac{4\pi t}{L} + \epsilon_t$$

These models assume a linear trend, but they can be altered to handle other trends. The first of these models is useful in modeling a very regular additive seasonal pattern, while the second model possesses terms that allow modeling of a more complicated additive seasonal pattern.

Regression models can also be employed in modeling multiplicative seasonal variation. Two useful models which involve trigonometric terms are the following:

$$y_t = \beta_0 + \beta_1 t + \beta_2 \sin\frac{2\pi t}{L} + \beta_3 t \sin\frac{2\pi t}{L} + \beta_4 \cos\frac{2\pi t}{L} + \beta_5 t \cos\frac{2\pi t}{L} + \epsilon_t$$

and

$$y_t = \beta_0 + \beta_1 t + \beta_2 \sin\frac{2\pi t}{L} + \beta_3 t \sin\frac{2\pi t}{L} + \beta_4 \cos\frac{2\pi t}{L} + \beta_5 t \cos\frac{2\pi t}{L}$$
$$+ \beta_6 \sin\frac{4\pi t}{L} + \beta_7 t \sin\frac{4\pi t}{L} + \beta_8 \cos\frac{4\pi t}{L} + \beta_9 t \cos\frac{4\pi t}{L} + \epsilon_t$$

Again, these models assume a linear trend but can be altered to handle other trends. The first of these models is useful in modeling a very regular multiplicative seasonal pattern, while the second model possesses terms that allow modeling of a more complicated multiplicative seasonal pattern. For several examples demonstrating the use of these trigonometric regression models, see Bowerman and O'Connell (1979). Besides the regression models we have described in this section, other models (such as *Winters' models* and *decomposition models*) can be useful in modeling time series possessing additive or multiplicative seasonal variation. For a discussion of some of these models, see Bowerman and O'Connell (1979). Included is a discussion of how to handle **cyclical effects**, which are recurring up and down movements around the trend that have a duration of more than a year (for instance, such cycles might be caused by the business cycle or by cyclical weather patterns).

Finally, as indicated at the beginning of this section (when we referred to the introduction of Fresh as a new detergent), we can combine *causal variables* and *time series variables* in the same regression model. As an example of this, consider forecasting monthly sales, y_t, of a fishing lure called the Bass-Grabber. We suppose that the Bass-Grabber has been sold at a constant price for the past several years and that monthly sales of this lure have shown an increasing linear trend because of increasing consumer awareness of the Bass-Grabber. In addition, suppose that monthly sales of this lure are affected by the monthly advertising expenditure, x_t, devoted to promoting the lure and that sales of the Bass-Grabber are seasonal (with highest sales during the summer fishing season). This would suggest using the model

$$y_t = \beta_0 + \beta_1 x_t + \beta_2 t + \beta_3 x_{S1,t} + \beta_4 x_{S2,t} + \cdots + \beta_{13} x_{S11,t} + \epsilon_t$$

where $x_{S1,t}, x_{S2,t}, \ldots, x_{S11,t}$ are monthly dummy variables defined so that

$$x_{Sj,t} = \begin{cases} 1 & \text{if time period } t \text{ is month } j \\ 0 & \text{otherwise} \end{cases}$$

In this model the $\beta_1 x_t$ term (assuming the functional form is correct) accounts for the effect on monthly sales of monthly advertising expenditure, while the $\beta_2 t$ term accounts for the linear trend in sales, and the dummy variables account for the seasonal pattern in Bass-Grabber sales. Thus, in this model we are including a causal variable, x_t, along with the time series variables $t, x_{S1,t}, x_{S2,t}, \ldots,$ and $x_{S11,t}$.

10.4 AUTOREGRESSIVE ERROR STRUCTURES

When the error terms for a regression model are autocorrelated, we can remedy the problem by modeling the autocorrelation. In general, if the error terms are autocorrelated and we ignore this fact, the least squares procedure tends to produce values of $s\sqrt{c_{jj}}$ (the standard error of the estimate b_j) that are too small and, consequently, values of the t_{b_j} statistic that are too large (it should be noted that the opposite behavior is possible but less common). Therefore, if we ignore autocorrelated error terms, we tend to get spurious declarations of significance when variables are really not important. For example, further analysis in this section will reveal that the linear term $-7.46t$ in the Farmers' Bureau Coop dummy variable model is not really significant, even though the t_{b_j} statistic in Table 10.9 ($t_{b_1} = -2.19$) indicates that this term is important. The reason for this misleading t_{b_1} statistic is the autocorrelation in the error terms for the Farmers' Bureau Coop dummy variable model. Another consequence of ignoring autocorrelated error terms is that we pay a penalty in terms of wider prediction intervals. By taking autocorrelation into account, we can achieve more accurate prediction intervals.

One autocorrelated error structure that is frequently encountered is the **first-order autoregressive process**. In order to define this process, we consider the model

$$\epsilon_t = \rho\epsilon_{t-1} + U_t$$

which relates ϵ_t, the error term in time period t, to ϵ_{t-1}, the error term in time period $t - 1$. Here we assume that $U_1, U_2, \ldots, U_n$ each have mean zero and satisfy the inference assumptions, and we define ρ to be the correlation coefficient between ϵ_t and ϵ_{t-1}. That is, ρ is defined to be the correlation coefficient between error terms separated by one time period. If $\rho > 0$, this indicates that the error terms $\epsilon_1, \epsilon_2, \ldots, \epsilon_n$ are positively autocorrelated. (It is easy to see that the equation $\epsilon_t = \rho\epsilon_{t-1} + U_t$ implies that if $\rho > 0$, then a positive error term ϵ_{t-1} will tend to produce another positive error term ϵ_t, and a negative error term ϵ_{t-1} will tend to produce another negative error term ϵ_t.) On the other hand, if $\rho < 0$, the error terms $\epsilon_1, \epsilon_2, \ldots, \epsilon_n$ are negatively autocorrelated. (Again, if $\rho < 0$, then a positive error term ϵ_{t-1} will tend to produce a negative error term ϵ_t, and a negative error term ϵ_{t-1} will tend to produce a positive error term ϵ_t.)

Now consider the general linear regression model

$$y_t = \beta_0 + \beta_1 x_{t1} + \beta_2 x_{t2} + \cdots + \beta_p x_{tp} + \epsilon_t$$

where we assume that

$$\epsilon_t = \rho\epsilon_{t-1} + U_t$$

and that $\rho \neq 0$. This implies that the error terms $\epsilon_1, \epsilon_2, \ldots, \epsilon_n$ for this general model are autocorrelated according to a first-order autoregressive process, which says that the Independence assumption (inference assumption 2) is violated. In order to obtain a transformed model that satisfies the inference assumptions, consider the model

$$\rho y_{t-1} = \rho\beta_0 + \rho\beta_1 x_{t-1,1} + \rho\beta_2 x_{t-1,2} + \cdots + \rho\beta_p x_{t-1,p} + \rho\epsilon_{t-1}$$

and for $t = 2, 3, \ldots, n$ subtract ρy_{t-1} from y_t, which yields the model

$$\begin{aligned} y_t - \rho y_{t-1} = {} & \beta_0(1 - \rho) + \beta_1(x_{t1} - \rho x_{t-1,1}) + \beta_2(x_{t2} - \rho x_{t-1,2}) \\ & + \cdots + \beta_p(x_{tp} - \rho x_{t-1,p}) + [\epsilon_t - \rho\epsilon_{t-1}] \end{aligned}$$

Since $[\epsilon_t - \rho\epsilon_{t-1}] = U_t$, and since $U_1, U_2, \ldots, U_n$ satisfy the inference assumptions, this new transformed model satisfies the inference assumptions.

Notice that when we subtract ρy_{t-1} from y_t we lose information concerning y_1 (that is, $y_t - \rho y_{t-1}$ cannot be computed for $t = 1$). This is not serious if n is large (we have many observations), but this might be serious if n is small. However, we can regain this loss by multiplying both sides of the regression model describing y_1 by $\sqrt{1 - \rho^2}$, which yields

$$\begin{aligned} \sqrt{1 - \rho^2}\, y_1 = {} & \sqrt{1 - \rho^2}\,\beta_0 + \beta_1(\sqrt{1 - \rho^2}\, x_{11}) + \beta_2(\sqrt{1 - \rho^2}\, x_{12}) \\ & + \cdots + \beta_p(\sqrt{1 - \rho^2}\, x_{1p}) + \sqrt{1 - \rho^2}\,\epsilon_1 \end{aligned}$$

Note that it can be shown that this equation, along with the equations

$$\begin{aligned} y_t - \rho y_{t-1} = {} & \beta_0(1 - \rho) + \beta_1(x_{t1} - \rho x_{t-1,1}) + \beta_2(x_{t2} - \rho x_{t-1,2}) \\ & + \cdots + \beta_p(x_{tp} - \rho x_{t-1,p}) + U_t \end{aligned}$$

for $t = 2, 3, \ldots, n$, satisfy the inference assumptions.

We are faced with a dilemma here. In practice, the correlation coefficient ρ must be estimated in order to transform the data and to obtain a model that satisfies the inference assumptions. However, in order to estimate ρ, we must use the untransformed data. Therefore, an iterative procedure is suggested. This procedure is called the **Cochran-Orcutt procedure**, and it works as follows.

First, for the untransformed model

$$y_t = \beta_0 + \beta_1 x_{t1} + \beta_2 x_{t2} + \cdots + \beta_p x_{tp} + \epsilon_t$$

we calculate the least squares point estimates $b_0, b_1, b_2, \ldots, b_p$ using the formula $\mathbf{b} = (\mathbf{X}'\mathbf{X})^{-1}\mathbf{X}'\mathbf{y}$ and the *untransformed data*. Using these least squares point estimates, we compute the residuals $e_1, e_2, \ldots, e_n$.

Second, we perform regression analysis on the residuals using the model

$$e_t = \rho e_{t-1} + U_t$$

Here we let

$$\mathbf{y} = \begin{bmatrix} e_2 \\ e_3 \\ \vdots \\ e_n \end{bmatrix} \qquad \mathbf{X} = \begin{bmatrix} e_1 \\ e_2 \\ \vdots \\ e_{n-1} \end{bmatrix}$$

and we calculate the least squares estimate of ρ (denoted r) to be

$$r = (\mathbf{X}'\mathbf{X})^{-1}\mathbf{X}'\mathbf{y} = \frac{\sum_{t=2}^{n} e_t e_{t-1}}{\sum_{t=2}^{n} e_{t-1}^2}$$

Third, we make the transformation discussed previously, using r as an estimate of the correlation coefficient ρ. Thus, we let

$$\mathbf{y} = \begin{bmatrix} \sqrt{1-r^2}y_1 \\ y_2 - ry_1 \\ y_3 - ry_2 \\ \vdots \\ y_n - ry_{n-1} \end{bmatrix}$$

$$\mathbf{X} = \begin{bmatrix} \sqrt{1-r^2} & \sqrt{1-r^2}x_{11} & \cdots & \sqrt{1-r^2}x_{1j} & \cdots & \sqrt{1-r^2}x_{1p} \\ 1-r & x_{21} - rx_{11} & \cdots & x_{2j} - rx_{1j} & \cdots & x_{2p} - rx_{1p} \\ 1-r & x_{31} - rx_{21} & \cdots & x_{3j} - rx_{2j} & \cdots & x_{3p} - rx_{2p} \\ \vdots & \vdots & & \vdots & & \vdots \\ 1-r & x_{n1} - rx_{n-1,1} & \cdots & x_{nj} - rx_{n-1,j} & \cdots & x_{np} - rx_{n-1,p} \end{bmatrix}$$

and utilizing the new **X** and **y**, we compute new least squares point estimates $b_0, b_1, b_2, \ldots, b_p$ using the formula $\mathbf{b} = (\mathbf{X}'\mathbf{X})^{-1}\mathbf{X}'\mathbf{y}$.

Fourth, using the new least squares estimates $b_0, b_1, b_2, \ldots, b_p$, we recompute the residuals $e_1, e_2, \ldots, e_n$ and return to the second step, where we calculate a revised estimate of ρ using the newest residuals. We then use the revised estimate of ρ to compute newly transformed data (as in the third step), and we use these newly transformed data to calculate revised least squares point estimates $b_0, b_1, b_2, \ldots, b_p$.

The iterative process ends when the new least squares point estimates change little between iterations. Often, one or two iterations are sufficient. When we obtain the final regression, the least squares point estimates $b_0, b_1, b_2, \ldots, b_p$ apply to the untransformed model

$$y_t = \beta_0 + \beta_1 x_{t1} + \beta_2 x_{t2} + \cdots + \beta_p x_{tp} + \epsilon_t$$

where $\epsilon_t = \rho\epsilon_{t-1} + U_t$, since the parameters in this model are the same as the parameters in the transformed model. In addition, the standard error, s, and the t_{b_j} statistics obtained for the transformed model also apply to the untransformed model. Moreover, these values are correct, because the transformed model satisfies the inference assumptions.

Point predictions and prediction intervals are found as follows. First, denoting the predictions of y_t, μ_t, and ϵ_t as $\hat{y}_t$, $\hat{\mu}_t$, and $\hat{\epsilon}_t$, a point prediction of the future value

$$\begin{aligned} y_{n+\tau} &= \mu_{n+\tau} + \epsilon_{n+\tau} \\ &= \beta_0 + \beta_1 x_{n+\tau,1} + \beta_2 x_{n+\tau,2} + \cdots + \beta_p x_{n+\tau,p} + \epsilon_{n+\tau} \end{aligned}$$

where $\epsilon_{n+\tau} = \rho\epsilon_{n+\tau-1} + U_{n+\tau}$, is given by

$$\begin{aligned} \hat{y}_{n+\tau} &= \hat{\mu}_{n+\tau} + \hat{\epsilon}_{n+\tau} = \hat{\mu}_{n+\tau} + r\hat{\epsilon}_{n+\tau-1} + \hat{U}_{n+\tau} \\ &= b_0 + b_1 x_{n+\tau,1} + b_2 x_{n+\tau,2} + \cdots + b_p x_{n+\tau,p} + r\hat{\epsilon}_{n+\tau-1} \end{aligned}$$

where $\hat{\epsilon}_{n+\tau}$ is obtained by substituting the estimate r for ρ and by predicting $U_{n+\tau}$ to be zero. Here, if $\tau = 1$ (the prediction is being made for one time period ahead), then since y_n has been observed,

$$\begin{aligned} \hat{\epsilon}_{n+\tau-1} = \hat{\epsilon}_n &= y_n - \hat{\mu}_n \\ &= y_n - [b_0 + b_1 x_{n1} + b_2 x_{n2} + \cdots + b_p x_{np}] \end{aligned}$$

while, if $\tau > 1$ (the prediction is for more than one time period ahead), then since $y_{n+\tau-1}$ has not been observed,

$$\begin{aligned} \hat{\epsilon}_{n+\tau-1} &= \hat{y}_{n+\tau-1} - \hat{\mu}_{n+\tau-1} \\ &= \hat{y}_{n+\tau-1} - [b_0 + b_1 x_{n+\tau-1,1} + b_2 x_{n+\tau-1,2} + \cdots + b_p x_{n+\tau-1,p}] \end{aligned}$$

Furthermore, approximate $100(1 - \alpha)\%$ bounds on the error of prediction are obtained as follows.

1. If $\tau = 1$, then

$$BP_{\hat{y}_{n+\tau}}[100(1-\alpha)] = BP_{\hat{y}_{n+1}}[100(1-\alpha)] = t_{[\alpha/2]}^{(n-k)}s$$

and an approximate $100(1 - \alpha)\%$ prediction interval for y_{n+1} is

$$[\hat{y}_{n+1} \pm t_{[\alpha/2]}^{(n-k)}s]$$

2. If $\tau = 2$, then

$$BP_{\hat{y}_{n+\tau}}[100(1-\alpha)] = BP_{\hat{y}_{n+2}}[100(1-\alpha)] = t_{[\alpha/2]}^{(n-k)}s\sqrt{1 + r^2}$$

and an approximate $100(1-\alpha)\%$ prediction interval for y_{n+2} is

$$[\hat{y}_{n+2} \pm t_{[\alpha/2]}^{(n-k)} s\sqrt{1+r^2}]$$

3. If $\tau \geq 3$, then

$$BP_{\hat{y}_{n+\tau}}[100(1-\alpha)] = t_{[\alpha/2]}^{(n-k)} s\sqrt{1+r^2+\cdots+r^{2(\tau-1)}}$$

and an approximate $100(1-\alpha)\%$ prediction interval for $y_{n+\tau}$ is

$$[\hat{y}_{n+\tau} \pm t_{[\alpha/2]}^{(n-k)} s\sqrt{1+r^2+\cdots+r^{2(\tau-1)}}]$$

In these formulas, $t_{[\alpha/2]}^{(n-k)}$ is defined as usual, s is the standard error computed from the transformed data, and r is the final estimate of ρ. We demonstrate the use of these formulas in the following example.

EXAMPLE 10.9

Reconsider the Farmers' Bureau Coop problem, in which Farmers' Bureau wishes to predict its quarterly propane gas bill, y_t. We saw in Example 10.8 that there is substantial evidence to suggest that the dummy variable model

$$\begin{aligned} y_t &= TR_t + SN_t + \epsilon_t \\ &= \beta_0 + \beta_1 t + \beta_2 t^2 + \beta_3 x_{S1,t} + \beta_4 x_{S2,t} + \beta_5 x_{S3,t} + \epsilon_t \end{aligned}$$

for this problem has a positively autocorrelated error structure. We now continue to analyze the Farmers' Bureau Coop data by assuming that these data have a first-order autoregressive error structure. That is, we consider the model

$$\begin{aligned} y_t &= \mu_t + \epsilon_t = TR_t + SN_t + \epsilon_t \\ &= \beta_0 + \beta_1 t + \beta_2 t^2 + \beta_3 x_{S1,t} + \beta_4 x_{S2,t} + \beta_5 x_{S3,t} + \epsilon_t \end{aligned}$$

where $\epsilon_t = \rho\epsilon_{t-1} + U_t$.

In order to use the Cochran-Orcutt procedure, we note that we computed (in Example 10.8) least squares point estimates b_0, b_1, b_2, b_3, b_4, and b_5 (see Table 10.9) using the untransformed Farmers' Bureau Coop data, and we obtained the residuals $e_1, e_2, \ldots, e_{40}$ that are plotted in Figure 10.17. Using these residuals and the model

$$e_t = \rho e_{t-1} + U_t$$

we perform a regression analysis on the residuals. Here we let

$$\mathbf{y} = \begin{bmatrix} e_2 \\ e_3 \\ \vdots \\ e_{40} \end{bmatrix} \qquad \mathbf{X} = \begin{bmatrix} e_1 \\ e_2 \\ \vdots \\ e_{39} \end{bmatrix}$$

and we calculate the least squares estimate of ρ to be

$$r = (\mathbf{X}'\mathbf{X})^{-1}\mathbf{X}'\mathbf{y} = .5841$$

Next, we compute transformed data. We find that

$$\mathbf{y} = \begin{bmatrix} \sqrt{1-r^2}y_1 \\ y_2 - ry_1 \\ y_3 - ry_2 \\ \vdots \\ y_{40} - ry_{39} \end{bmatrix} = \begin{bmatrix} \sqrt{1-(.5841)^2}(344.39) \\ 246.63 - .5841(344.39) \\ 131.53 - .5841(246.63) \\ \vdots \\ 539.78 - .5841(330.8) \end{bmatrix}$$

and that

$$\mathbf{X} = \begin{bmatrix} \sqrt{1-r^2} & \sqrt{1-r^2}x_{11} & \cdots & \sqrt{1-r^2}x_{1j} & \cdots & \sqrt{1-r^2}x_{1p} \\ 1-r & x_{21} - rx_{11} & \cdots & x_{2j} - rx_{1j} & \cdots & x_{2p} - rx_{1p} \\ 1-r & x_{31} - rx_{21} & \cdots & x_{3j} - rx_{2j} & \cdots & x_{3p} - rx_{2p} \\ \vdots & \vdots & & \vdots & & \vdots \\ 1-r & x_{n1} - rx_{n-1,1} & \cdots & x_{nj} - rx_{n-1,j} & \cdots & x_{np} - rx_{n-1,p} \end{bmatrix}$$

$$= \begin{bmatrix} & t & t^2 & x_{S1,t} & x_{S2,t} & x_{S3,t} \\ \sqrt{1-(.5841)^2} & \sqrt{1-(.5841)^2}(1) & \sqrt{1-(.5841)^2}(1)^2 & \sqrt{1-(.5841)^2}(1) & \sqrt{1-(.5841)^2}(0) & \sqrt{1-(.5841)^2}(0) \\ 1-.5841 & 2-(.5841)(1) & (2)^2 - .5841(1)^2 & 0-(.5841)(1) & 1-(.5841)(0) & 0-(.5841)(0) \\ 1-.5841 & 3-(.5841)(2) & (3)^2 - .5841(2)^2 & 0-(.5841)(0) & 0-(.5841)(0) & 1-(.5841)(0) \\ \vdots & \vdots & \vdots & \vdots & \vdots & \vdots \\ 1-.5841 & 40-(.5841)(39) & (40)^2 - .5841(39)^2 & 0-(.5841)(0) & 0-(.5841)(0) & 0-(.5841)(1) \end{bmatrix}$$

Using this $\mathbf{y}$ and $\mathbf{X}$ and the formula $\mathbf{b} = (\mathbf{X}'\mathbf{X})^{-1}\mathbf{X}'\mathbf{y}$, we compute updated least squares point estimates b_0, b_1, b_2, b_3, b_4, and b_5. These least squares estimates, which are given in Table 10.10 along with the appropriate t_{b_j} statistics, yield the equation

$$\hat{\mu}_t = 283.9803 - 9.1870t + .3522t^2 + 70.0615x_{S1,t} - 35.4887x_{S2,t} - 126.5643x_{S3,t}$$

For the transformed model we also obtain

$$SSE = 83342.925 \qquad \text{and} \qquad s = \sqrt{\frac{SSE}{n-k}} = \sqrt{\frac{83342.925}{40-6}} = 49.5102$$

(Note that since the transformed model utilizes r, the point estimate of ρ, some computer packages, including SAS, recommend calculating s by the more conservative equation $s = \sqrt{SSE/(n-(k+1))} = \sqrt{83342.925/(40-7)} = 50.2548$.)

TABLE 10.10 The t_{b_j} Statistics for the Farmers' Bureau Coop First-Order Autoregressive Model

Parameter (β_j)	Least Squares Point Estimate (b_j)	Standard Error of the Estimate ($s\sqrt{c_{jj}}$)	t_{b_j} Statistic $\left(t_{b_j} = \frac{b_j}{s\sqrt{c_{jj}}}\right)$
β_0	$b_0 = 283.9803$	51.9057	$t_{b_0} = 5.47$
β_1	$b_1 = -9.1870$	5.6835	$t_{b_1} = -1.62$
β_2	$b_2 = .3522$	.1338	$t_{b_2} = 2.63$
β_3	$b_3 = 70.0615$	17.2703	$t_{b_3} = 4.06$
β_4	$b_4 = -35.4887$	19.2970	$t_{b_4} = -1.84$
β_5	$b_5 = -126.5643$	16.7836	$t_{b_5} = -7.54$

Comparing Tables 10.9 and 10.10, we see that the least squares point estimates obtained from the transformed data do not differ a great deal from those obtained by using the untransformed data. Therefore, further iterations will not be done.

The t_{b_j} statistics in Table 10.10 are valid, because they come from the transformed regression that satisfies the inference assumptions. These results can be compared to the analysis in Example 10.8. Recall that the t_{b_j} statistics reported there are invalid and should be ignored. For example, from Table 10.10 we see that the t_{b_1} statistic ($= -1.62$) indicates that the variable t is not significant in our new correct analysis, while the incorrect t_{b_1} statistic ($= -2.19$) in Table 10.9 had indicated that this variable is significant. (The t term would usually be retained anyway, since the t_{b_2} statistic ($=2.63$) in Table 10.10 indicates that the t^2 term is important.)

Recall that Farmers' Bureau Coop wishes to find out whether the seasonally adjusted trend in the propane gas bills is increasing. One way to remove the seasonality here is to ignore the seasonal dummy variables $x_{S1,t}$, $x_{S2,t}$, and $x_{S3,t}$ and to look only at the estimated trend (denoted tr_t):

$$tr_t = 283.9803 - 9.1870t + .3522t^2$$

This essentially puts all trend estimates on a fourth-quarter basis. Looking at this estimated trend, we see that there is an initial decline because of the negative coefficient on t. However, as t gets large (that is, as time advances), the positive term $.3522t^2$ dominates and the estimated trend tr_t rises. For any polynomial

$$tr_t = a + bt + dt^2$$

where $d > 0$, it can easily be shown using calculus that the minimum value of tr_t occurs at $t = -b/2d$. Therefore, in our case, the estimated propane gas bill trend will increase after

$$t = \frac{-(-9.1870)}{2(.3522)} = 13$$

Since our last observation is at time $t = 40$, the seasonally adjusted propane gas bill is increasing. Also, for the function

$$tr_t = a + bt + dt^2$$

it can easily be shown using calculus that the instantaneous rate of growth is $b + 2dt$. So at $t = 40$ we estimate that the propane gas bill trend is rising at a rate of

$$-9.1870 + 2(.3522)(40) = \$18.99 \text{ per quarter}$$

Farmers' Bureau Coop now wishes to forecast its quarterly propane gas bills for the four quarters of next year (time periods 41, 42, 43, and 44). For time period 41,

$$y_{41} = \mu_{41} + \epsilon_{41} = \mu_{41} + [\rho\epsilon_{40} + U_{41}]$$

Therefore, the predicted propane gas bill for time period 41 is

$$\hat{y}_{41} = \hat{\mu}_{41} + r\hat{\epsilon}_{40}$$

where $\hat{\epsilon}_{40} = y_{40} - \hat{\mu}_{40}$. Here

$$\begin{aligned}\hat{\epsilon}_{40} &= 539.78 - [283.9803 - 9.1870(40) + .3522(40^2) + 70.0615(0) \\ &\quad - 35.4887(0) - 126.5643(0)] \\ &= 539.78 - 480.02 \\ &= 59.76\end{aligned}$$

and

$$\begin{aligned}\hat{y}_{41} &= \hat{\mu}_{41} + r\hat{\epsilon}_{40} \\ &= [283.9803 - 9.1870(41) + .3522(41)^2 + 70.0615(1) \\ &\quad - 35.4887(0) - 126.5643(0)] + .5841(59.76) \\ &= 569.42 + .5841(59.76) = 604.33\end{aligned}$$

Thus, the quarterly propane gas bill for time period 41 is predicted to be \$604.33. Furthermore, an approximate 95% prediction interval for y_{41} is

$$\begin{aligned}[\hat{y}_{41} \pm t_{[\alpha/2]}^{(n-k)}s] &= [604.33 \pm t_{[.025]}^{(40-6)}s] \\ &= [604.33 \pm 2.034(49.5102)] \\ &= [503.63, 705.03]\end{aligned}$$

This interval says that Farmers' Bureau Coop is 95% confident that the propane gas bill for period 41 will be between \$503.63 and \$705.03.

For time period 42,

$$y_{42} = \mu_{42} + \epsilon_{42} = \mu_{42} + [\rho\epsilon_{41} + U_{42}]$$

Therefore, the predicted propane gas bill for time period 42 is

$$\hat{y}_{42} = \hat{\mu}_{42} + r\hat{\epsilon}_{41}$$

where $\hat{\epsilon}_{41} = \hat{y}_{41} - \hat{\mu}_{41}$. Here $\hat{\epsilon}_{41} = \hat{y}_{41} - \hat{\mu}_{41} = 604.33 - 569.42 = 34.91$, and

$$\begin{aligned}\hat{y}_{42} &= \hat{\mu}_{42} + r\hat{\epsilon}_{41}\\ &= [283.9803 - 9.1870(42) + .3522(42)^2 + 70.0615(0) - 35.4887(1)\\ &\quad - 126.5643(0)] + .5841(34.91)\\ &= 483.92 + .5841(34.91) = 504.31\end{aligned}$$

Thus, the quarterly propane gas bill for time period 42 is predicted to be \$504.31. Moreover, an approximate 95% prediction interval for y_{42} is

$$\begin{aligned}[\hat{y}_{42} \pm t_{[.025]}^{(40-6)} s\sqrt{1 + r^2}] &= [504.31 \pm 2.034(49.5102)\sqrt{1 + (.5841)^2}]\\ &= [504.31 \pm 116.62]\\ &= [387.69, 620.93]\end{aligned}$$

This interval says that Farmers' Bureau Coop is 95% confident that the propane gas bill for period 42 will be between \$387.69 and \$620.93.

For time period 43,

$$y_{43} = \mu_{43} + \epsilon_{43} = \mu_{43} + [\rho\epsilon_{42} + U_{43}]$$

Therefore, the predicted propane gas bill for time period 43 is

$$\hat{y}_{43} = \hat{\mu}_{43} + r\hat{\epsilon}_{42}$$

where $\hat{\epsilon}_{42} = \hat{y}_{42} - \hat{\mu}_{42}$. Here $\hat{\epsilon}_{42} = \hat{y}_{42} - \hat{\mu}_{42} = 504.31 - 483.92 = 20.39$, and

$$\begin{aligned}\hat{y}_{43} &= \hat{\mu}_{43} + r\hat{\epsilon}_{42}\\ &= [283.9803 - 9.1870(43) + .3522(43)^2 + 70.0615(0) - 35.4887(0)\\ &\quad - 126.5643(1)] + .5841(20.39)\\ &= 413.59 + .5841(20.39) = 425.50\end{aligned}$$

Therefore, the quarterly propane gas bill for time period 43 is predicted to be \$425.50. Furthermore, an approximate 95% prediction interval for y_{43} is

$$\begin{aligned}&[\hat{y}_{43} \pm t_{[.025]}^{(40-6)} s\sqrt{1 + r^2 + r^4}]\\ &\quad = [425.50 \pm 2.034(49.5102)\sqrt{1 + (.5841)^2 + (.5841)^4}]\\ &\quad = [425.50 \pm 121.58]\\ &\quad = [303.92, 547.08]\end{aligned}$$

This interval says that Farmers' Bureau Coop is 95% confident that the propane gas bill for period 43 will be between \$303.92 and \$547.08.

Finally, for time period 44,

$$y_{44} = \mu_{44} + \epsilon_{44} = \mu_{44} + [\rho\epsilon_{43} + U_{44}]$$

Thus, the predicted propane gas bill for time period 44 is

$$\hat{y}_{44} = \hat{\mu}_{44} + r\hat{\epsilon}_{43}$$

where $\hat{\epsilon}_{43} = \hat{y}_{43} - \hat{\mu}_{43}$. Here $\hat{\epsilon}_{43} = \hat{y}_{43} - \hat{\mu}_{43} = 425.50 - 413.59 = 11.91$, and

$$\begin{aligned}\hat{y}_{44} &= \hat{\mu}_{44} + r\hat{\epsilon}_{43}\\ &= [283.9803 - 9.1870(44) + .3522(44)^2 + 70.0615(0) - 35.4887(0)\\ &\quad - 126.5643(0)] + .5841(11.91)\\ &= 561.61 + 6.96 = 568.57\end{aligned}$$

Therefore, the quarterly propane gas bill for time period 44 is predicted to be \$568.57. Moreover, an approximate 95% prediction interval for y_{44} is

$$\begin{aligned}&[\hat{y}_{44} \pm t_{[.025]}^{(40-6)} s\sqrt{1 + r^2 + r^4 + r^6}]\\ &\quad = [568.57 \pm 2.034(49.5102)\sqrt{1 + (.5841)^2 + (.5841)^4 + (.5841)^6}]\\ &\quad = [568.57 \pm 123.22]\\ &\quad = [445.35, 691.79]\end{aligned}$$

This interval says that Farmers' Bureau Coop is 95% confident that the propane gas bill for period 44 will be between \$445.35 and \$691.79.

In addition to the first-order autoregressive process, other autocorrelated error structures exist. For instance, error terms can follow the **autoregressive process of order *q*.**

$$\epsilon_t = \phi_1\epsilon_{t-1} + \phi_2\epsilon_{t-2} + \cdots + \phi_q\epsilon_{t-q} + U_t$$

which relates ϵ_t, the error term in time period t, to the previous error terms ϵ_{t-1}, $\epsilon_{t-2}, \ldots, \epsilon_{t-q}$. Here $\phi_1, \phi_2, \ldots, \phi_q$ are parameters, and we assume that U_1, $U_2, \ldots, U_n$ each have mean zero and satisfy the inference assumptions. When this model is appropriate, an iterative procedure known as the **Cochran-Orcutt procedure for an autoregressive process of order *q*** can be employed. This procedure works as follows.

First, for the untransformed model

$$y_t = \beta_0 + \beta_1 x_{t1} + \beta_2 x_{t2} + \cdots + \beta_p x_{tp} + \epsilon_t$$

we calculate least squares point estimates $b_0, b_1, b_2, \ldots, b_p$ using the formula $\mathbf{b} = (\mathbf{X}'\mathbf{X})^{-1}\mathbf{X}'\mathbf{y}$ and the *untransformed data*. Using these least squares point estimates, we compute residuals $e_1, e_2, \ldots, e_n$.

Second, we model e_t as an autoregressive process

$$e_t = \phi_1 e_{t-1} + \phi_2 e_{t-2} + \cdots + \phi_q e_{t-q} + U_t$$

and we compute least squares point estimates of the parameters $\phi_1, \phi_2, \ldots, \phi_q$. We denote these estimates as $\hat{\phi}_1, \hat{\phi}_2, \ldots, \hat{\phi}_q$.

Third, we transform y_t to

$$y_t^* = y_t - \hat{\phi}_1 y_{t-1} - \cdots - \hat{\phi}_q y_{t-q}$$

for $t = q + 1, q + 2, \ldots, n$. Similarly, we transform each column of the $\mathbf{X}$ matrix. For example, the intercept column of n 1's becomes a column of $n - q$ entries, each entry being $1 - \hat{\phi}_1 - \hat{\phi}_2 - \cdots - \hat{\phi}_q$. The other columns of the $\mathbf{X}$ matrix are transformed by replacing, for $t = q + 1, q + 2, \ldots, n$, x_{tj} by

$$x_{tj}^* = x_{tj} - \hat{\phi}_1 x_{t-1,j} - \hat{\phi}_2 x_{t-2,j} - \cdots - \hat{\phi}_q x_{t-q,j}$$

We then compute new least squares point estimates $b_0, b_1, b_2, \ldots, b_p$ using the formula $\mathbf{b} = (\mathbf{X}'\mathbf{X})^{-1}\mathbf{X}'\mathbf{y}$ and the *transformed data*.

Fourth, using the new least squares point estimates, we recompute residuals and return to the second step, where we calculate revised estimates of ϕ_1, $\phi_2, \ldots, \phi_q$ using the newest residuals. We then use the revised estimates of ϕ_1, $\phi_2, \ldots, \phi_q$ to compute newly transformed data, which are in turn used to compute revised least squares point estimates $b_0, b_1, b_2, \ldots, b_p$.

Again, the iterative procedure ends when the least squares point estimates $b_0, b_1, b_2, \ldots, b_p$ change little between iterations. Usually, one or two iterations are sufficient. When we obtain the final regression, the final least squares point estimates apply to the untransformed model. In addition, the standard error, s, and the t_{b_j} statistics obtained for the transformed model also apply to the untransformed model. Furthermore, these values are correct, because the transformed model satisfies the inference assumptions.

This procedure loses information from the first q observations. A method similar to that given in the context of the first-order autoregressive process is available to recoup this information. Fortunately, the loss of information is not very severe for large n, and since the method used to recoup this information is somewhat complicated, we do not present it here. The procedure AUTOREG in SAS allows one to run regressions with autoregressive errors using the appropriate transformation.

Another common autoregressive process is the ***p*th-order autoregressive process in the observations** $y_1, y_2, \ldots, y_n$. This process is written as

$$y_t = \beta_0 + \beta_1 y_{t-1} + \beta_2 y_{t-2} + \cdots + \beta_p y_{t-p} + \epsilon_t$$

which expresses y_t in terms of the previous observations $y_{t-1}, y_{t-2}, \ldots, y_{t-p}$ and an error term ϵ_t. Although the Independence assumption does not hold for such a model, it can be shown that for large samples the least squares procedure is appropriate. As an example, the second-order autoregressive process in y_t,

$$y_t = \beta_0 + \beta_1 y_{t-1} + \beta_2 y_{t-2} + \epsilon_t$$

expresses the observation y_t in terms of the previous observations y_{t-1} and y_{t-2}. Here, estimation of the parameters β_0, β_1, and β_2 would be accomplished by using the standard formula $\mathbf{b} = (\mathbf{X}'\mathbf{X})^{-1}\mathbf{X}'\mathbf{y}$, where

$$\mathbf{y} = \begin{bmatrix} y_3 \\ y_4 \\ y_5 \\ \vdots \\ y_n \end{bmatrix} \qquad \mathbf{X} = \begin{bmatrix} 1 & y_2 & y_1 \\ 1 & y_3 & y_2 \\ 1 & y_4 & y_3 \\ \vdots & \vdots & \vdots \\ 1 & y_{n-1} & y_{n-2} \end{bmatrix}$$

Finally, **Box-Jenkins models** express y_t in terms of previous observations $y_{t-1}, y_{t-2}, \ldots, y_{t-p}$ and previous error terms $\epsilon_{t-1}, \epsilon_{t-2}, \ldots, \epsilon_{t-q}$. Such a model can be written as

$$y_t = \beta_0 + \beta_1 y_{t-1} + \beta_2 y_{t-2} + \cdots + \beta_p y_{t-p} + \phi_1 \epsilon_{t-1} + \phi_2 \epsilon_{t-2} + \cdots + \phi_q \epsilon_{t-q} + U_t$$

The use of Box-Jenkins models involves four steps:

1. The *identification* of an appropriate model.
2. The *estimation* of model parameters.
3. *Diagnostic checking* (checking the adequacy of the model).
4. *Forecasting*.

A detailed presentation of the Box-Jenkins methodology can be found in Box and Jenkins (1977) and Bowerman and O'Connell (1979).

10.5 REMEDIES FOR NON-NORMALITY

In this section we discuss what can be done when the Normality assumption is violated. As we have stated, mild departures from inference assumption 3 are not serious. However, if examination of the residuals indicates a pronounced departure from the Normality assumption, there are remedies that can be employed. Generally, these remedies involve transformation of the data. In particular, incorrect functional forms, omitted variables, and violations of the Constant Variance assumption (inference assumption 1) can cause the error terms to look non-normal. Fortunately, remedies for these problems (for example, transformations to achieve equal variances) often correct the non-normality problem as well. We demonstrate this phenomenon in the following example.

EXAMPLE 10.10

Reconsider the Republic Wholesalers, Inc., telephone order data. A frequency distribution and histogram of the residuals for the telephone order data are given in Table 10.11 and Figure 10.20. The residuals are calculated as deviations from the column (hour) means in Table 10.5 (this is reasonable, since we have several observations for each hour). The histogram in Figure 10.20 appears to be skewed to the right, which means that the distribution of residuals appears to have a long tail to the right. Notice that three positive residuals exceed 30, while no negative residuals are more than 30 units away from zero. This suggests a violation of the Normality assumption, although the decision as to whether or not the histogram looks normal is a subjective judgment. The normal plot of these residuals is given in Figure 10.21. This plot was generated on a computer using the SAS procedure PROC RANK. The plot appears to be somewhat nonlinear, although this is again a matter of judgment. Therefore, this normal plot also suggests a violation of the Normality assumption.

As we have said, when data appear to be non-normal, the technique of data transformation is often applied. Recall that in Example 10.5 we saw that the Republic Wholesalers telephone order data appear to violate the Equal Variances assumption (inference assumption 1). In order to remedy this problem, we used a variance-equalizing transformation in which we replaced y (the number of hourly telephone orders) by its square root. Figure 10.22 gives a normal plot of the residuals for the transformed Republic Wholesalers telephone order data. Here the entries in Table 10.5 have been replaced by their square

TABLE 10.11 Frequency Distribution of the Residuals (From Hour Means) for the Republic Wholesalers Telephone Order Data

Subinterval	Frequency (No. in Subinterval)	Subinterval	Frequency (No. in Subinterval)
−29.99 to −25	2	10 to 14.99	3
−24.99 to −20	0	15 to 19.99	2
−19.99 to −15	5	20 to 24.99	3
−14.99 to −10	6	25 to 29.99	1
−9.99 to −5	16	30 to 34.99	0
−4.99 to 0	16	35 to 39.99	1
0 to 4.99	14	40 to 44.99	1
5 to 9.99	9	45 to 49.99	1

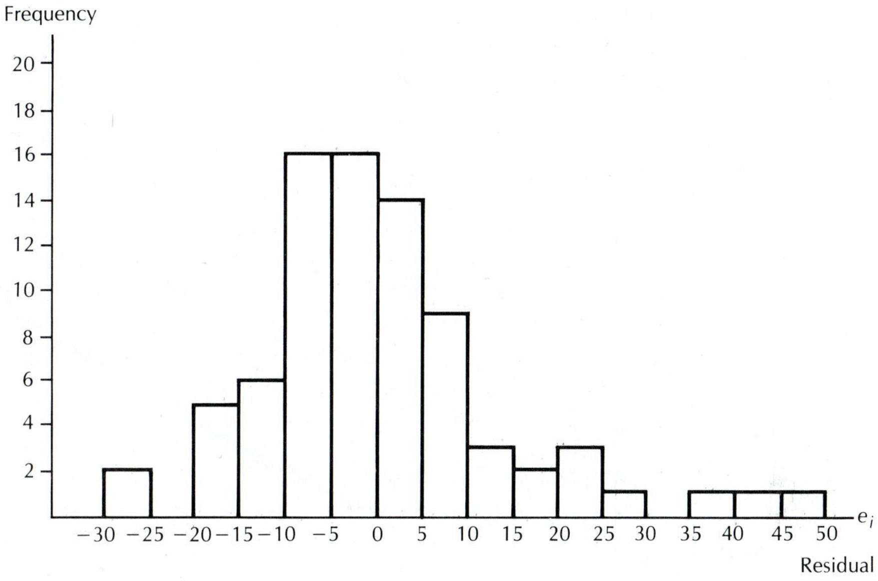

FIGURE 10.20
Histogram of the Residuals (From Hour Means) for the Republic Wholesalers Telephone Order Data

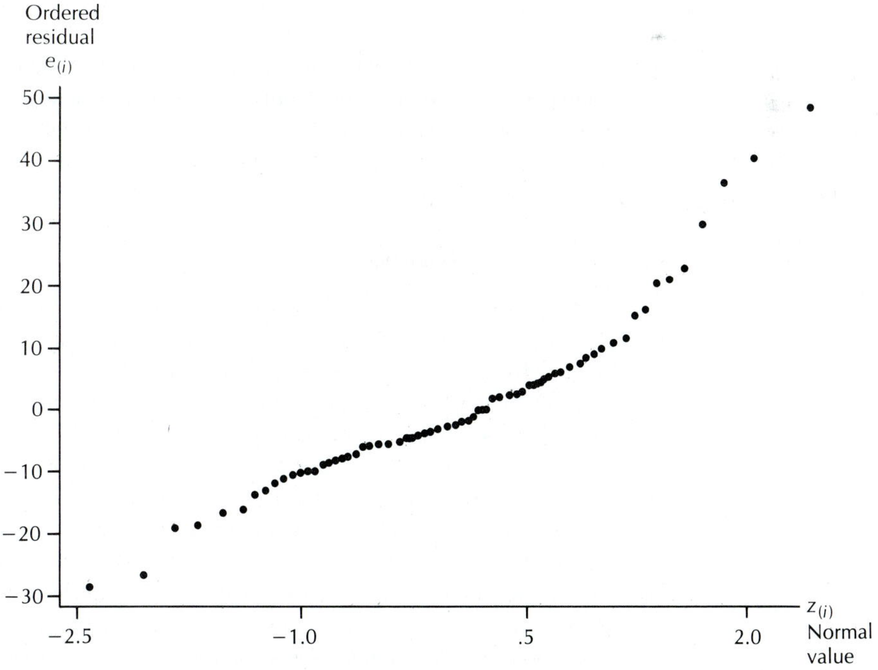

FIGURE 10.21
Normal Plot of the Residuals (From Hour Means) for the Republic Wholesalers Telephone Order Data

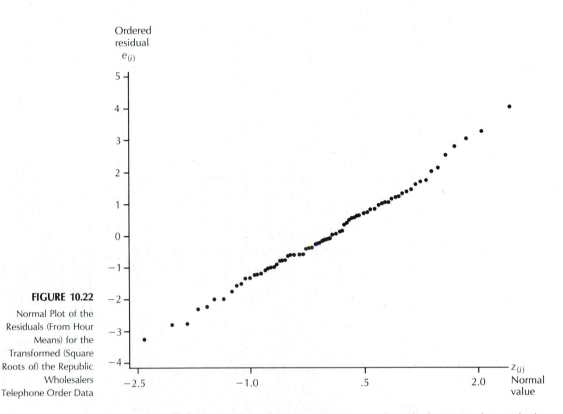

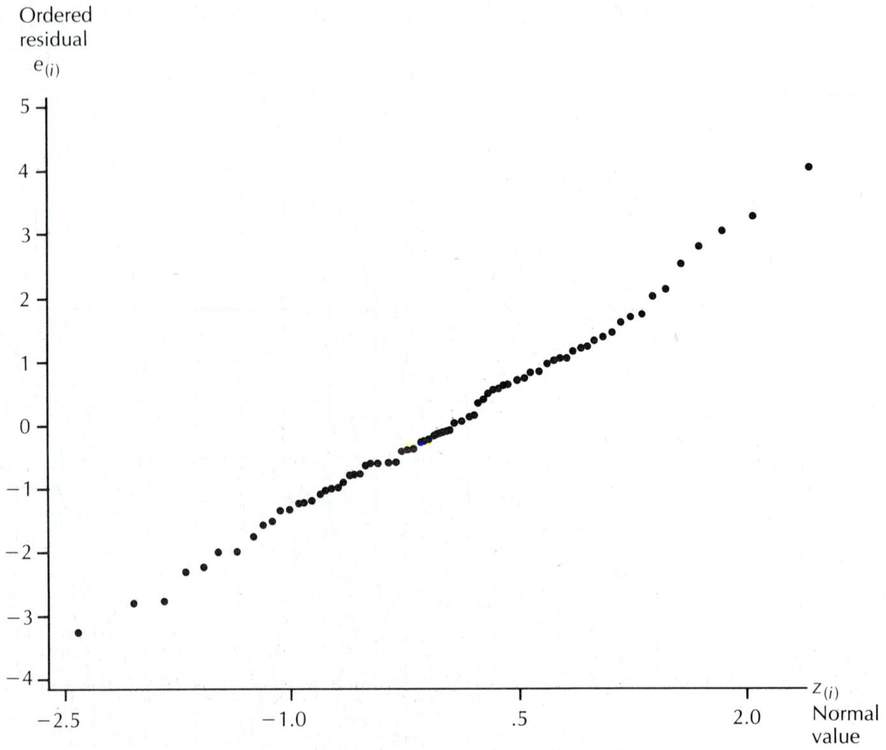

FIGURE 10.22 Normal Plot of the Residuals (From Hour Means) for the Transformed (Square Roots of) the Republic Wholesalers Telephone Order Data

roots before computing column (hour) means and residuals. Again, the residuals have been computed as deviations from column means, since we have several observations for each hour. Looking at Figure 10.22, we see that the normal plot for this transformed data set seems straighter than the normal plot for the original data (see Figure 10.21).

Notice that in Example 10.10 the transformation that appeared to make the variances roughly equal also made the normal plot appear straighter. It is common to observe this phenomenon. In the Republic Wholesalers problem the normal plot of Figure 10.21 and the histogram in Figure 10.20 do not indicate a terribly serious violation of the Normality assumption. Here the most convincing evidence indicating that a transformation is needed is the plot of $s^2_{H_t}$ versus $\bar{y}_{H_t}$ in Figure 10.12, which shows that the error variance tends to increase as $\bar{y}_{H_t}$ increases. However, we see that using an appropriate variance-equalizing transformation simultaneously seems to correct any non-normality problem that might exist. The main point we are making here is that once other problems, such as unequal variances and incorrect functional form, have been remedied, any non-normality problem that might exist will often also be corrected simultaneously. If it is not, then additional data transformation may be necessary.

We emphasize that histograms and normal plots, like checks for autocorrelation, are sensitive to incorrect functional forms. It is crucial that checks for omitted variables and incorrect functional form be made and appropriate changes implemented prior to validation of the Normality assumption.

10 EXERCISES

10.1 The Natugrain Cereal Company wants to set a price for its new cereal, Bran-Nu. In order to do this, Bran-Nu is marketed in a test area at a regular store price of $1.50 per box. To estimate the demand for Bran-Nu at various price levels, the company mails coupons for discounts from 5¢ to $1.10 in 5¢ increments. For each of the 22 discount prices (from $1.45 to 40¢), 500 coupons are mailed. The company will measure demand by the number of coupons, y, (out of 500) redeemed at each discounted price.

The results of the Bran-Nu marketing experiment are given in the following table.

Number of Coupons Redeemed (y)	Discount Price (cents) (x)	ln x	ln y
43	145	4.97673	3.76120
64	140	4.94164	4.15888
62	135	4.90527	4.12713
54	130	4.86753	3.98898
81	125	4.82831	4.39445
93	120	4.78749	4.53260
53	115	4.74493	3.97029
80	110	4.70048	4.38203
87	105	4.65396	4.46591
134	100	4.60517	4.89784
120	95	4.55388	4.78749
166	90	4.49981	5.11199
118	85	4.44265	4.77088
189	80	4.38203	5.24175
174	75	4.31749	5.15906
249	70	4.24850	5.51745
317	65	4.17439	5.75890
217	60	4.09434	5.37990
297	55	4.00733	5.69373
421	50	3.91202	6.04263
488	45	3.80666	6.19032
438	40	3.68888	6.08222

This table presents y, the number of coupons redeemed (out of 500), for each discount price x (expressed in cents). For future reference, the table also gives the natural logarithms of y and x. Using these data, the Natugrain Cereal Company wishes to estimate the elasticity of demand for Bran-Nu and the demand for Bran-Nu at various prices, where demand will be measured as the proportion of the cereal-buying population who would buy Bran-Nu at a given price. The original Bran-Nu data are plotted here:

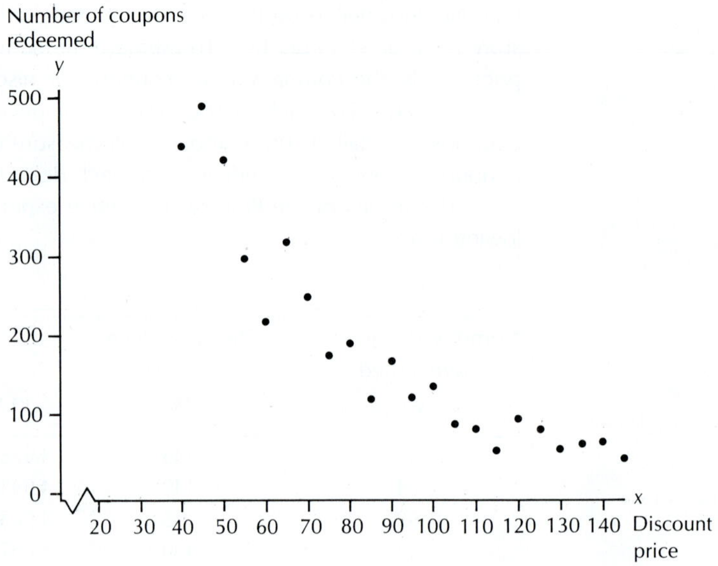

The plot of the Bran-Nu data is consistent with a model of the form

$$y = \beta_0 x^{\beta_1} \epsilon \qquad \text{(see Figure 10.4)}$$

although a model of the form

$$y = \beta_0 \beta_1^x \epsilon \qquad \text{(see Figure 10.1)}$$

might also be appropriate. The choice of a model by visual inspection of a data plot is obviously more difficult in cases with a high degree of variability. We assume here that economists working for the Natugrain Company have theoretical reasons for believing that the model

$$y = \beta_0 x^{\beta_1} \epsilon$$

is appropriate.

a. Show that by taking natural logarithms of both sides of the model

$$y = \beta_0 x^{\beta_1} \epsilon$$

we obtain the model

$$\ln y = \alpha_0 + \alpha_1 \ln x + U$$

where $\alpha_0 = \ln \beta_0$, $\alpha_1 = \beta_1$, and $U = \ln \epsilon$.

b. The natural logarithms of y, plotted against the natural logarithms of x, follow. Noting that the plot is fairly linear, discuss why we would expect this plot to be linear if the model

$$y = \beta_0 x^{\beta_1} \epsilon$$

appropriately describes the original Bran-Nu data.

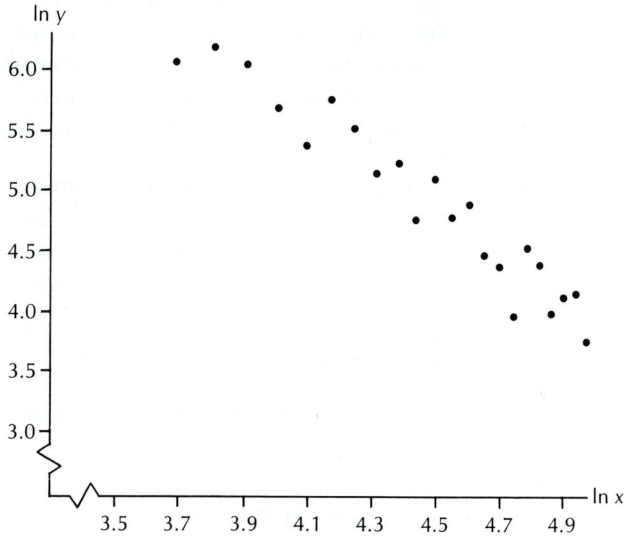

c. Specify the vector **y** and the matrix **X** used to calculate the least squares point estimates of α_0 and α_1 in the model

$$\ln y = \alpha_0 + \alpha_1 \ln x + U$$

d. When we use the model in part (c) to perform a regression analysis of the Bran-Nu data, we obtain

$$(\mathbf{X}'\mathbf{X})^{-1} = \begin{bmatrix} 6.4078 & -1.4262 \\ -1.4262 & 0.3197 \end{bmatrix} \qquad \mathbf{X}'\mathbf{y} = \begin{bmatrix} 108.4154 \\ 477.8322 \end{bmatrix}$$

$$\begin{bmatrix} \alpha_0 \\ \alpha_1 \end{bmatrix} = \mathbf{b} = (\mathbf{X}'\mathbf{X})^{-1}\mathbf{X}'\mathbf{y} = \begin{bmatrix} 13.1956 \\ -1.8543 \end{bmatrix} \qquad SSE = .8300$$

Calculate the standard error, s, for the model in part (c). Using the fact that $\alpha_0 = \ln \beta_0$ and $\alpha_1 = \beta_1$ (see part (a)), calculate point estimates of

β_0 and β_1. Also, noting that $|\beta_1| = |\alpha_1|$ is called the elasticity of demand, calculate a point estimate of $|\beta_1|$, and calculate a 95% confidence interval for $|\beta_1|$.

e. We now turn to the problem of estimating demand (the proportion of the cereal-buying population who would buy Bran-Nu) at a given price. Using the model

$$\ln y = \alpha_0 + \alpha_1 \ln x + U$$

calculate a point estimate of the natural logarithm of the number of people per 500 who would buy Bran-Nu at a \$1.00 price. Then calculate a point estimate of the number of people per 500 who would buy Bran-Nu at the \$1.00 price, and calculate a point estimate of the percentage of the cereal buying population who would buy Bran-Nu at a price of \$1.00. Finally, calculate 95% confidence intervals for these three quantities. *Hint:* First calculate a 95% confidence interval for the natural logarithm of the number of people per 500 who would buy Bran-Nu at the \$1.00 price.

10.2 Recall from Exercise 9.5 that an economist studied the relationship between x, 1970 yearly income (in thousands of dollars) for a family of four, and y, 1970 yearly clothing expenditure (in hundreds of dollars) for the family of four, and recall that the following data were observed:

x_i:	8	10	12	14	16	18	20
y_i:	6.47	6.17	7.4	10.57	11.93	10.3	14.67

Next, recall that when the regression model

$$y_i = \beta_0 + \beta_1 x_i + \epsilon_i$$

is used to perform a regression analysis of these data, we find that the least squares point estimates of β_0 and β_1 are $b_0 = .2968$ and $b_1 = .6677$ and that the standard error, s, is equal to 1.3434. Moreover, recall that the plot of the standardized residuals against the increasing values of x_i fans out, indicating that inference assumption 1 is violated.

a. Assuming that the standard deviation σ_i is proportional to x_i, specify a transformed model that will remedy the unequal variances problem. Then specify the $\mathbf{y}$ vector and $\mathbf{X}$ matrix used to calculate the least squares point estimates b_0 and b_1 of the parameters β_0 and β_1 in the transformed model. Using $\mathbf{y}$ and $\mathbf{X}$, calculate $\mathbf{X'X}$, $(\mathbf{X'X})^{-1}$, $\mathbf{X'y}$, b_0, b_1, and s for the transformed model. Then calculate the standardized residuals, and plot these residuals against the increasing values of x_i to determine whether your transformation has suitably equalized the variances. *Hint:* In order to calculate $(\mathbf{X'X})^{-1}$, it is useful to know that if

$$\mathbf{A} = \begin{bmatrix} a & b \\ c & d \end{bmatrix}$$

is a 2 × 2 matrix, then the inverse of **A** is

$$\mathbf{A}^{-1} = \begin{bmatrix} \dfrac{d}{D} & -\dfrac{b}{D} \\ -\dfrac{c}{D} & \dfrac{a}{D} \end{bmatrix}$$

where $D = ad - bc$.

b. Calculate a 95% confidence interval for β_1.

c. Calculate a 95% confidence interval for the mean 1970 yearly clothing expenditure for all families of four that had a 1970 yearly income of $16,000.

d. Calculate a 95% prediction interval for the 1970 yearly clothing expenditure of an individual family of four that had a 1970 yearly income of $16,000.

10.3 Consider the Happy Valley Real Estate problem in Example 10.7 and the model

$$\begin{aligned} y &= \mu + \epsilon \\ &= p(x) + \epsilon \\ &= \beta_0 + \beta_1 x + \epsilon \end{aligned}$$

where $p(x)$ is the probability that a randomly selected person with income x will purchase Happy Valley property within one month of a Happy Valley tour. We know that the variance of y at a particular value of x is $p(x)[1 - p(x)]$ and that this variance varies as x varies, since $p(x)$ changes as x changes. This says that a nonconstant error variance exists and that inference assumption 1 is violated. If we could divide each term in the model

$$y = \beta_0 + \beta_1 x + \epsilon$$

by the quantity $\sqrt{p(x)[1 - p(x)]}$ (call this w), we would have a model

$$\frac{y}{w} = \beta_0\left(\frac{1}{w}\right) + \beta_1\left(\frac{x}{w}\right) + \frac{\epsilon}{w}$$

where ϵ/w has variance 1 at all x values. However, $p(x)$, and therefore $w = p(x)[1 - p(x)]$, are unknown. Fortunately, the least squares point estimates b_0 and b_1 are not totally unreasonable, even though inference assumption 1 is violated. This means that an iterative scheme can be employed in which b_0 and b_1 are used to estimate $p(x)$, and then this estimate is used to transform the model. We then use the transformed model to reestimate β_0 and β_1. Following this, a new $p(x)$ estimate can be computed and can be used to obtain a new transformed model, which is in turn used to find yet a third set of estimates of β_0 and β_1. Often, iteration produces only small changes in the estimates of β_0 and β_1, and usually only a few rounds of iteration are necessary.

a. In the first step of our procedure, use the data in Table 10.7 to specify the vector **y** and the matrix **X** used to calculate the least squares point estimates b_0 and b_1 of the parameters β_0 and β_1 in the model

$$\begin{aligned} y &= \mu + \epsilon \\ &= p(x) + \epsilon \\ &= \beta_0 + \beta_1 x + \epsilon \end{aligned}$$

b. When we use the model in part (a) to perform a regression analysis of the data in Table 10.7, we obtain $b_0 = -.2411$ and $b_1 = .0228$. It follows that our estimate of the probability $p(x)$ is

$$\hat{p}(x) = -.2411 + .0228x$$

Looking at Table 10.7, we see that the observed range of incomes is from 13 (thousand dollars) to 50 (thousand dollars). This implies that our estimated probabilities at various income levels range from

$$\hat{p}(13) = -.2411 + .0228(13) = .0553$$

to

$$\hat{p}(50) = -.2411 + .0228(50) = .8983$$

Since

$$\hat{p}(13)[1 - \hat{p}(13)] = .0553[1 - .0553] = .0522$$

and

$$\hat{p}(50)[1 - \hat{p}(50)] = .8983[1 - .8983] = .0914$$

the estimated variance $\hat{p}(x)[1 - \hat{p}(x)]$ ranges from .0522 to .25 (recall that $p(x)[1 - p(x)]$ attains a maximum value of $.5[1 - .5] = .25$ when $p(x) = .5$. The table on page 637 shows the original Happy Valley data, the estimated probability $\hat{p}(x)$, and the transformed data $y/\hat{w}$, $1/\hat{w}$, and $x/\hat{w}$ for the transformed model

$$\frac{y}{w} = \beta_0\left(\frac{1}{w}\right) + \beta_1\left(\frac{x}{w}\right) + \frac{\epsilon}{w}$$

Here $\hat{w} = \sqrt{\hat{p}(x)[1 - \hat{p}(x)]}$ is the estimated standard deviation of y computed using the estimated probability $\hat{p}(x)$.

Considering the fourth income from the top of the table, $x = 28$, demonstrate how the values $\hat{w} = \hat{p}(x)$, $y/\hat{w}$, $1/\hat{w}$, and $x/\hat{w}$ corresponding to this income have been calculated. Then, specify the vector **y** and the matrix **X** used to calculate the least squares point estimates $\tilde{b}_0$ and $\tilde{b}_1$ of the parameters β_0 and β_1 in the transformed model.

c. Based on the vector **y** and matrix **X** from part (b), the least squares point estimates of the parameters β_0 and β_1 for the transformed model are found to be $\tilde{b}_0 = -.2676$ and $\tilde{b}_1 = .0239$. Notice that these revised estimates are

Purchase Decision (y)	Income (thousands of dollars) (x)	$\hat{p}(x) = -.2411 + .0228x$	$y/\hat{w} = y/\sqrt{\hat{p}(x)[1 - \hat{p}(x)]}$	$1/\hat{w} = 1/\sqrt{\hat{p}(x)[1 - \hat{p}(x)]}$	$x/\hat{w} = x/\sqrt{\hat{p}(x)[1 - \hat{p}(x)]}$
1	50	0.898276	3.30814	3.30814	165.407
0	25	0.328601	0.00000	2.12900	53.225
0	30	0.442536	0.00000	2.01334	60.400
1	28	0.396962	2.04387	2.04387	57.228
1	42	0.715980	2.21756	2.21756	93.137
0	17	0.146305	0.00000	2.82956	48.103
0	20	0.214666	0.00000	2.43552	48.710
0	25	0.328601	0.00000	2.12900	53.225
1	38	0.624832	2.06541	2.06541	78.485
0	15	0.100731	0.00000	3.32257	49.838
1	30	0.442536	2.01334	2.01334	60.400
0	28	0.396962	0.00000	2.04387	57.228
1	48	0.852702	2.82165	2.82165	135.439
0	16	0.123518	0.00000	3.03923	48.628
1	21	0.237453	2.35005	2.35005	49.351
0	13	0.055157	0.00000	4.38046	56.946
1	36	0.579258	2.02561	2.02561	72.922
0	34	0.533684	0.00000	2.00455	68.155
0	22	0.260240	0.00000	2.27912	50.141
0	20	0.214666	0.00000	2.43552	48.710
0	37	0.602045	0.00000	2.04300	75.591
1	45	0.784341	2.43144	2.43144	109.415
1	17	0.146305	2.82956	2.82956	48.103
0	24	0.305814	0.00000	2.17037	52.089
1	28	0.396962	2.04387	2.04387	57.228
1	25	0.328601	2.12900	2.12900	53.225
0	22	0.260240	0.00000	2.27912	50.141
0	18	0.169092	0.00000	2.66785	48.021
0	26	0.351388	0.00000	2.09466	54.461
0	44	0.761554	0.00000	2.34668	103.254

not very different from the original estimates $b_0 = -.2411$ and $b_1 = .0228$. Therefore, further iteration (yielding a second set of revised estimates) will not be done. Using the estimates $\tilde{b}_0$ and $\tilde{b}_1$, calculate a point estimate of $p(35)$, the proportion of all people with a yearly income of $35,000 who would purchase property within one month of a Happy Valley tour.

10.4 Value City, a small department store, sells the King-Cool air conditioner. The following table contains four years of quarterly sales data for the King-Cool.

Year	Quarter	t	Demand (y_t)	Year	Quarter	t	Demand (y_t)
1	1 (Winter)	1	10	3	1	9	13
	2 (Spring)	2	31		2	10	34
	3 (Summer)	3	43		3	11	48
	4 (Fall)	4	16		4	12	19
2	1	5	11	4	1	13	15
	2	6	33		2	14	37
	3	7	45		3	15	51
	4	8	17		4	16	21

The King-Cool sales data plot follows.

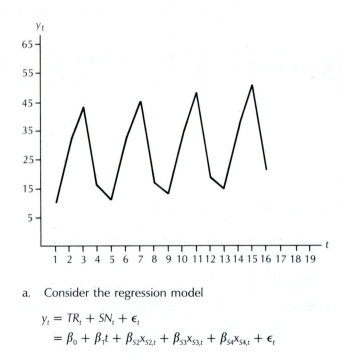

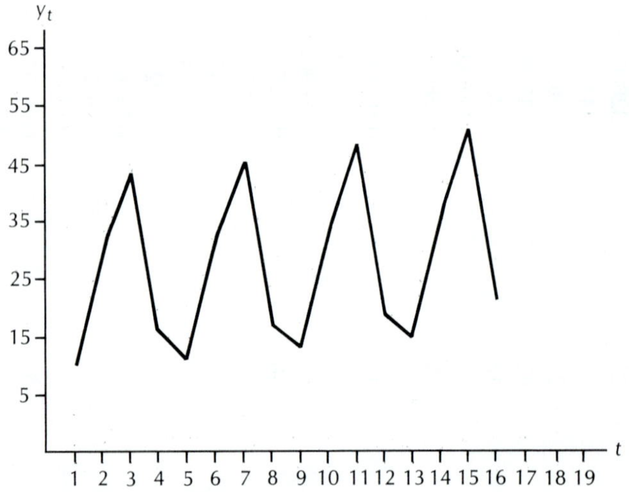

a. Consider the regression model

$$y_t = TR_t + SN_t + \epsilon_t$$
$$= \beta_0 + \beta_1 t + \beta_{S2} x_{S2,t} + \beta_{S3} x_{S3,t} + \beta_{S4} x_{S4,t} + \epsilon_t$$

where $x_{S2,t}$, $x_{S3,t}$, and $x_{S4,t}$ are dummy variables defined as follows:

$$x_{S2,t} = \begin{cases} 1 & \text{if time period } t \text{ is quarter 2} \\ 0 & \text{otherwise} \end{cases}$$

$$x_{S3,t} = \begin{cases} 1 & \text{if time period } t \text{ is quarter 3} \\ 0 & \text{otherwise} \end{cases}$$

$$x_{S4,t} = \begin{cases} 1 & \text{if time period } t \text{ is quarter 4} \\ 0 & \text{otherwise} \end{cases}$$

Discuss the meaning of the seasonal parameters β_{S2}, β_{S3}, and β_{S4}, and discuss why this regression model is a reasonable model to use to describe the data. Also, specify the vector **y** and the matrix **X** used to calculate the least squares point estimates of the parameters in the model.

b. When the regression model in part (a) is used to perform a regression analysis of the data, the least squares point estimates of the parameters in this model are found to be $b_0 = 8.75$, $b_1 = .5$, $b_{S2} = 21$, $b_{S3} = 33.5$, and $b_{S4} = 4.5$, and the squared residuals obtained from the prediction equation

$$\begin{aligned} \hat{y}_t &= tr_t + sn_t \\ &= b_0 + b_1 t + b_{S2}x_{S2,t} + b_{S3}x_{S3,t} + b_{S4}x_{S4,t} \end{aligned}$$

are as follows.

t	y_t	$tr_t = 8.75 + .5t$	sn_t	$\hat{y}_t = tr_t + sn_t$	$(y_t - \hat{y}_t)^2$
1	10	9.25	0	9.25	$(.75)^2$
2	31	9.75	21.0	30.75	$(.25)^2$
3	43	10.25	33.5	43.75	$(-.75)^2$
4	16	10.75	4.5	15.25	$(.75)^2$
5	11	11.25	0	11.25	$(-.25)^2$
6	33	11.75	21.0	32.75	$(.25)^2$
7	45	12.25	33.5	45.75	$(-.75)^2$
8	17	12.75	4.5	17.25	$(-.25)^2$
9	13	13.25	0	13.25	$(-.25)^2$
10	34	13.75	21.0	34.75	$(-.75)^2$
11	48	14.25	33.5	47.75	$(.25)^2$
12	19	14.75	4.5	19.25	$(-.25)^2$
13	15	15.25	0	15.25	$(-.25)^2$
14	37	15.75	21.0	36.75	$(.25)^2$
15	51	16.25	33.5	49.75	$(1.25)^2$
16	21	16.75	4.5	21.25	$(-.25)^2$

$$\sum_{t=1}^{16} (y_t - \hat{y}_t)^2 = 5.00$$

Calculate the standard error, s, and plot the residuals against time to determine whether there is evidence of positive autocorrelation or negative autocorrelation. Also, calculate the Durbin-Watson statistic, test for positive autocorrelation at $\alpha = .05$, and test for negative autocorrelation at $\alpha = .05$.

c. Assuming that the inference assumptions are satisfied, calculate point predictions of, and 95% prediction intervals for, y_{17}, y_{18}, y_{19}, and y_{20}, the demands for the King-Cool air conditioner in periods 17, 18, 19, and 20. *Hint:* If $\mathbf{x}'_{17}$, $\mathbf{x}'_{18}$, $\mathbf{x}'_{19}$, and $\mathbf{x}'_{20}$ denote the row vectors containing the numbers multiplied by b_0, b_1, b_{S2}, b_{S3}, and b_{S4} in, respectively, $\hat{y}_{17}$, $\hat{y}_{18}$, $\hat{y}_{19}$, and $\hat{y}_{20}$, then each of $\mathbf{x}'_{17}(\mathbf{X}'\mathbf{X})^{-1}\mathbf{x}_{17}$, $\mathbf{x}'_{18}(\mathbf{X}'\mathbf{X})^{-1}\mathbf{x}_{18}$, $\mathbf{x}'_{19}(\mathbf{X}'\mathbf{X})^{-1}\mathbf{x}_{19}$, and $\mathbf{x}'_{20}(\mathbf{X}'\mathbf{X})^{-1}\mathbf{x}_{20}$ can be calculated to be .5543. What are $\mathbf{x}'_{17}$, $\mathbf{x}'_{18}$, $\mathbf{x}'_{19}$, and $\mathbf{x}'_{20}$?

10.5 Nite's Rest, Inc., operates four hotels in Central City. The analysts in the operating division of the corporation were asked to develop a model that could be used to obtain short-term forecasts (up to 1 year) of the number of occupied rooms in the hotels. These forecasts were needed by various personnel to assist in decision making with regard to hiring additional help during the summer months, ordering supplies that have long delivery lead times, budgeting of local advertising expenditures, and so on.

The available historical data consisted of the number of occupied rooms during each day for the 15 years from 1973 to 1986. Because it was desired to obtain k-step-ahead monthly forecasts for $k = 1, 2, \ldots, 12$, these data were reduced to monthly averages by dividing each monthly total by the number of days in the month. The monthly room averages for 1973 to 1986, denoted by $y_1, y_2, \ldots, y_{168}$, are given in the following table and are plotted in the figure on page 642.

t	y_t	t	y_t	t	y_t	t	y_t
1	501	43	785	85	645	127	1067
2	488	44	830	86	602	128	1038
3	504	45	645	87	601	129	812
4	578	46	643	88	709	130	790
5	545	47	551	89	706	131	692
6	632	48	606	90	817	132	782
7	728	49	585	91	930	133	758
8	725	50	553	92	983	134	709
9	585	51	576	93	745	135	715
10	542	52	665	94	735	136	788

t	y_t	t	y_t	t	y_t	t	y_t
11	480	53	656	95	620	137	794
12	530	54	720	96	698	138	893
13	518	55	826	97	665	139	1046
14	489	56	838	98	626	140	1075
15	528	57	652	99	649	141	812
16	599	58	661	100	740	142	822
17	572	59	584	101	729	143	714
18	659	60	644	102	824	144	802
19	739	61	623	103	937	145	748
20	758	62	553	104	994	146	731
21	602	63	599	105	781	147	748
22	587	64	657	106	759	148	827
23	497	65	680	107	643	149	788
24	558	66	759	108	728	150	937
25	555	67	878	109	691	151	1076
26	523	68	881	110	649	152	1125
27	532	69	705	111	656	153	840
28	623	70	684	112	735	154	864
29	598	71	577	113	743	155	717
30	683	72	656	114	837	156	813
31	774	73	645	115	995	157	811
32	780	74	593	116	1040	158	732
33	609	75	617	117	809	159	745
34	604	76	686	118	793	160	844
35	531	77	679	119	692	161	833
36	592	78	773	120	763	162	935
37	578	79	906	121	723	163	1110
38	543	80	934	122	655	164	1124
39	565	81	713	123	658	165	868
40	648	82	710	124	761	166	860
41	615	83	600	125	768	167	762
42	697	84	676	126	885	168	877

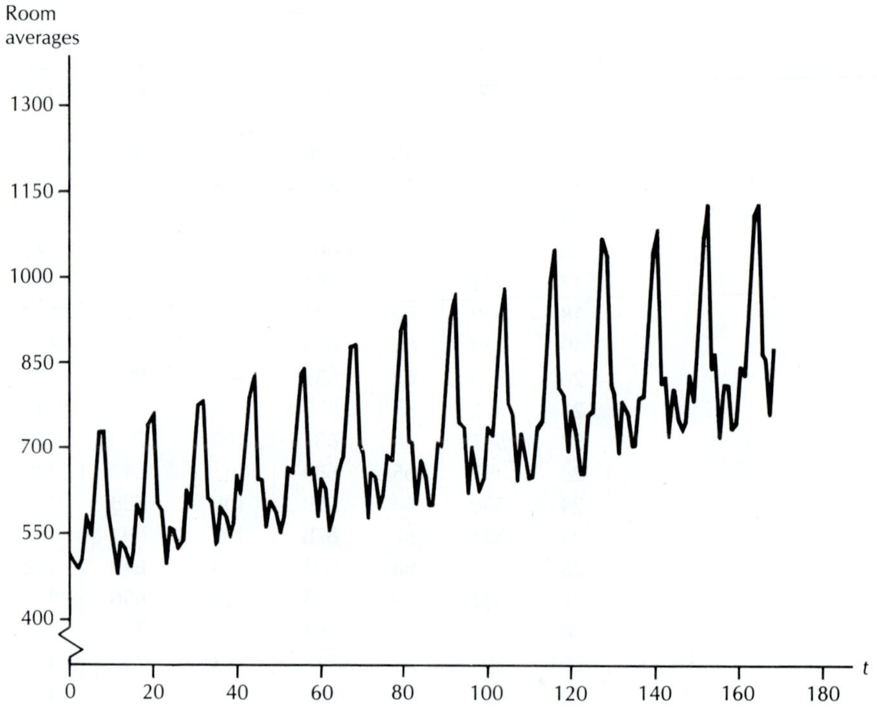

This figure shows that the monthly room averages follow a strong trend and that they have a seasonal pattern with one major and several minor peaks during the year. It also appears that the amount of seasonal variation is increasing with the level of the time series, indicating that the use of a transformation (such as the natural logarithms of the observations) might be warranted. The natural logarithms of the room averages are denoted by $y_1^*, y_2^*, \ldots, y_{168}^*$ and are plotted in the figure on page 643. This figure indicates that the log transformation has equalized the amount of seasonal variation over the range of the data.

a. Consider the regression model

$$\begin{aligned} y_t^* &= \mu_t^* + \epsilon_t^* \\ &= \beta_0 + \beta_1 t + \beta_{S1}x_{S1,t} + \beta_{S2}x_{S2,t} + \beta_{S3}x_{S3,t} + \beta_{S4}x_{S4,t} + \beta_{S5}x_{S5,t} + \beta_{S6}x_{S6,t} \\ &\quad + \beta_{S7}x_{S7,t} + \beta_{S8}x_{S8,t} + \beta_{S9}x_{S9,t} + \beta_{S10}x_{S10,t} + \beta_{S11}x_{S11,t} + \epsilon_t^* \end{aligned}$$

Here, $x_{S1,t}, x_{S2,t}, \ldots, x_{S11,t}$ are dummy variables where

$$x_{S1,t} = \begin{cases} 1 & \text{if time period } t \text{ is season 1 (January)} \\ 0 & \text{otherwise} \end{cases}$$

$$x_{S2,t} = \begin{cases} 1 & \text{if time period } t \text{ is season 2 (February)} \\ 0 & \text{otherwise} \end{cases}$$

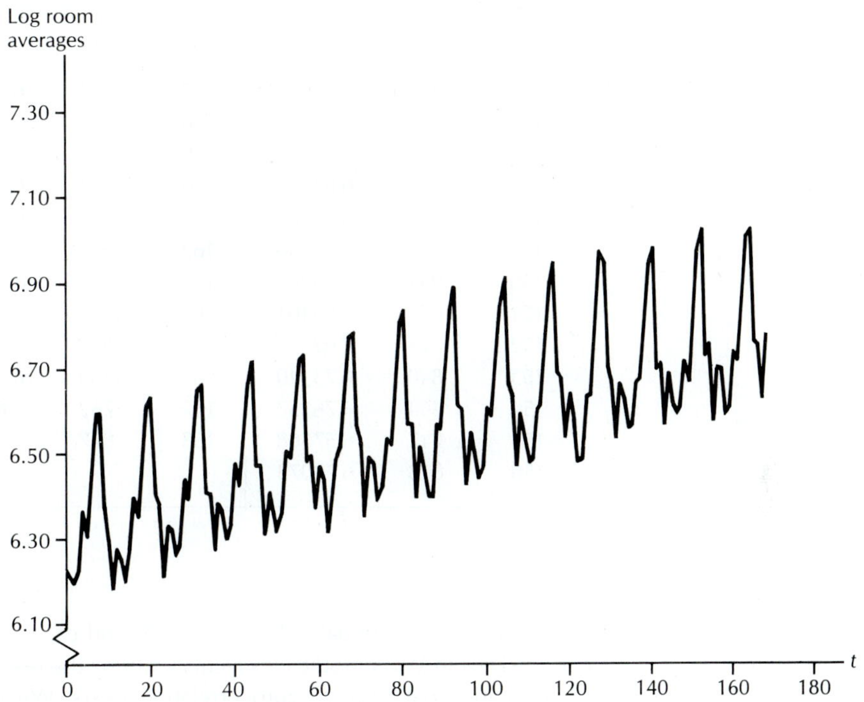

and the other dummy variables are defined similarly. Discuss the meaning of the seasonal parameters $\beta_{S1}, \beta_{S2}, \ldots, \beta_{S11}$, and discuss why this regression model is a reasonable model to use to describe the logarithms of the room averages. Also, specify the first 12 elements of the vector **y** and the first 12 elements of each column of the matrix **X** used to calculate the least squares point estimates of the parameters in the model.

b. When the regression model in part (a) is used to perform a regression analysis of the logarithms of the room averages, the least squares point estimates of the parameters in this model are found to be

$b_0 = 6.287557$ $b_{S6} = .1469094$

$b_1 = .00272528$ $b_{S7} = .2890226$

$b_{S1} = -.0416063$ $b_{S8} = .3111946$

$b_{S2} = -.112079$ $b_{S9} = .05598723$

$b_{S3} = -.084459$ $b_{S10} = .03954382$

$b_{S4} = .03983306$ $b_{S11} = -.112215$

$b_{S5} = .02039515$

Moreover, the observed hotel room averages and the logarithms of these averages for periods 144 through 168 are as follows.

t	y_t	y_t^*	t	y_t	y_t^*
144	802	6.68711	157	811	6.69827
145	748	6.61740	158	732	6.59578
146	731	6.59441	159	745	6.61338
147	748	6.61740	160	844	6.73815
148	827	6.71780	161	833	6.72503
149	788	6.66950	162	935	6.84055
150	937	6.84268	163	1110	7.01212
151	1076	6.98101	164	1124	7.02465
152	1125	7.02554	165	868	6.76619
153	840	6.73340	166	860	6.75693
154	864	6.76157	167	762	6.63595
155	717	6.57508	168	877	6.77651
156	813	6.70073			

Calculate the residuals $e_{167}^* = y_{167}^* - \hat{y}_{167}^*$ and $e_{168}^* = y_{168}^* - \hat{y}_{168}^*$.

c. The Durbin-Watson statistic calculated using the residuals $e_1^*, e_2^*, \ldots, e_{168}^*$ indicates that positive autocorrelation exists. We continue to analyze $y_1^*, y_2^*, \ldots, y_{168}^*$ by assuming that these data have a first-order autoregressive error structure. That is, we consider the model

$$\begin{aligned} y_t^* &= \mu_t^* + \epsilon_t^* \\ &= \beta_0 + \beta_1 t + \beta_{S1}x_{S1,t} + \beta_{S2}x_{S2,t} + \cdots + \beta_{S11}x_{S11,t} + \epsilon_t^* \end{aligned}$$

where $\epsilon_t^* = \rho\epsilon_{t-1}^* + U_t$. Describe the vector **y** and matrix **X** used to calculate r, the least squares point estimate of ρ, by using the model

$$e_t^* = \rho e_{t-1}^* + U_t$$

to perform a regression analysis of the residuals $e_1^*, e_2^*, \ldots, e_{168}^*$ obtained from the model in part (a).

d. When we perform the regression analysis using the model

$$e_t^* = \rho e_{t-1}^* + U_t$$

we find that $r = .390677$. Specify the first element and the last 12 elements of the vector **y** and the first element and the last 12 elements of each column of the matrix **X** used to calculate the revised least squares point estimates of $\beta_0, \beta_1, \beta_{S1}, \beta_{S2}, \ldots, \beta_{S11}$ by using the appropriate transformed model. *Hint:* First specify the appropriate transformed model.

e. When we calculate the revised least squares point estimates by using the equation $\mathbf{b} = (\mathbf{X}'\mathbf{X})^{-1}\mathbf{X}'\mathbf{y}$ (where **y** and **X** are as specified in part (d)), we obtain

$b_0 = 6.28581951250$ $b_{S6} = .14786905558$

$b_1 = .00273459989$ $b_{S7} = .28996311620$

$b_{S1} = -.039732125$ $b_{S8} = .31211113045$

$b_{S2} = -.1107463$ $b_{S9} = .05686103800$

$b_{S3} = -.083343147$ $b_{S10} = .04032441500$

$b_{S4} = .04085815100$ $b_{S11} = -.1116581$

$b_{S5} = .02137845045$

and we find that the unexplained variation, SSE, equals .05860552. Since we are estimating $k = 13$ parameters—$\beta_0, \beta_1, \beta_{S1}, \beta_{S2}, \ldots, \beta_{S11}$—we have suggested in Section 10.4 calculating the standard error, s, by the equation

$$s = \sqrt{\frac{SSE}{n-k}} = \sqrt{\frac{.05860552}{168-13}} = .0194448$$

However, since the appropriate transformed model utilizes r, the point estimate of ρ, some computer packages (including SAS) recommend calculating s by the (more conservative) equation

$$s = \sqrt{\frac{SSE}{n-(k+1)}} = \sqrt{\frac{.05860552}{168-(13+1)}} = .0195078$$

Using the revised least squares point estimates, either of the standard errors, any other pertinent information and the methods discussed in Section 10.4, calculate point predictions of, and 95% prediction intervals for, $y^*_{169}, y_{169}, y^*_{170}, y_{170}, y^*_{171}$, and y_{171}. *Hint:* Since $y^*_t = \ln y_t$, which implies that $y_t = e^{y^*_t}$, it follows that if $\hat{y}^*_t$ is a point prediction of y^*_t, then $e^{\hat{y}^*_t}$ is a point prediction of y_t, and if $[a_t, b_t]$ is a 95% prediction interval for y^*_t, then $[e^{a_t}, e^{b_t}]$ is a 95% prediction interval for y_t.

f. Although we have assumed that $y^*_1, y^*_2, \ldots, y^*_{168}$ has a first-order autoregressive error structure, further analysis shows that the model

$$\begin{aligned} y^*_t &= \mu^*_t + \epsilon^*_t \\ &= \beta_0 + \beta_1 t + \beta_{S1}x_{S1,t} + \beta_{S2}x_{S2,t} + \cdots + \beta_{S11}x_{S11,t} + \epsilon^*_t \end{aligned}$$

where

$$\epsilon^*_t = \phi_1\epsilon^*_{t-1} + \phi_2\epsilon^*_{t-2} + \phi_3\epsilon^*_{t-3} + \phi_{10}\epsilon^*_{t-10} + \phi_{13}\epsilon^*_{t-13} + U_t$$

is perhaps the best model describing $y^*_1, y^*_2, \ldots, y^*_{168}$. Describe the vector $\mathbf{y}$ and the matrix $\mathbf{X}$ used to calculate $\hat{\phi}_1, \hat{\phi}_2, \hat{\phi}_3, \hat{\phi}_{10}$, and $\hat{\phi}_{13}$, the least squares point estimates of ϕ_1, ϕ_2, ϕ_3, ϕ_{10}, and ϕ_{13}, by using the model

$$e_t^* = \phi_1 e_{t-1}^* + \phi_2 e_{t-2}^* + \phi_3 e_{t-3}^* + \phi_{10} e_{t-10}^* + \phi_{13} e_{t-13}^* + U_t$$

to perform a regression analysis of the residuals $e_1^*, e_2^*, \ldots, e_{168}^*$ obtained from the model in part (a). *Note:* This model describing ϵ_t^* has been determined by SAS, which uses a backward elimination technique (see Chapter 7) to determine the best model.

g. When we perform the regression analysis using the model

$$e_t^* = \phi_1 e_{t-1}^* + \phi_2 e_{t-2}^* + \phi_3 e_{t-3}^* + \phi_{10} e_{t-10}^* + \phi_{13} e_{t-13}^* + U_t$$

we obtain

$\hat{\phi}_1 = -.32473663$ $\hat{\phi}_{10} = -.14454908$

$\hat{\phi}_2 = -.17161490$ $\hat{\phi}_{13} = -.18268925$

$\hat{\phi}_3 = .23998988$

Specify the last 12 elements of the vector **y** and the last 12 elements of each column of the matrix **X** used to calculate the revised least squares point estimates of $\beta_0, \beta_1, \beta_{S1}, \beta_{S2}, \ldots, \beta_{S11}$ by using the appropriate transformed model. *Hint:* First specify the appropriate transformed model.

h. When we calculate the revised least squares point estimates by using the equation $\mathbf{b} = (\mathbf{X}'\mathbf{X})^{-1}\mathbf{X}'\mathbf{y}$ (where **y** and **X** are as specified in part (g)), we obtain

$b_0 = 6.28375470060$ $b_{S6} = .14814188966$

$b_1 = .00274053113$ $b_{S7} = .28940666366$

$b_{S1} = -.038092682$ $b_{S8} = .31231859565$

$b_{S2} = -.1089287$ $b_{S9} = .05624999334$

$b_{S3} = -.082511866$ $b_{S10} = .04000362601$

$b_{S4} = .04115035831$ $b_{S11} = -.11246398$

$b_{S5} = .02156220460$

and we find that the unexplained variation, *SSE*, equals .04981187. Since we are estimating $k = 13$ parameters—$\beta_0, \beta_1, \beta_{S1}, \beta_{S2}, \ldots, \beta_{S11}$—and since the appropriate transformed model utilizes $\hat{\phi}_1$, $\hat{\phi}_2$, $\hat{\phi}_3$, $\hat{\phi}_{10}$, and $\hat{\phi}_{13}$, we calculate s by the (conservative) equation

$$s = \sqrt{\frac{SSE}{n - (k + 5)}} = \sqrt{\frac{.04981187}{168 - (13 + 5)}} = .018223$$

Since $s = .018223$, the standard error for the model in part (f), is smaller than $s = .0195078$, the standard error for the model in part (c), we have evidence that the model in part (f) is more appropriate than the model in part (c). Using the revised least squares point estimates and any other pertinent information, calculate point predictions of y^*_{169}, y_{169}, y^*_{170}, y_{170}, y^*_{171}, and y_{171}. (*Hint:* Generalize the methods of Section 10.4.) Using advanced methods, it can be shown that 95% prediction intervals for y^*_{169}, y^*_{170}, and y^*_{171} are

For y^*_{169}: [6.6836, 6.7546]

For y^*_{170}: [6.6117, 6.6861]

For y^*_{171}: [6.6197, 6.6966]

Using these intervals, find 95% prediction intervals for y_{169}, y_{170}, and y_{171}. Then compare the lengths of these intervals with the lengths of the intervals you calculated in part (e) by using the model in part (c). Which model gives shorter intervals—the model in part (f) or the model in part (c)?

APPENDIX A

DERIVATIONS OF THE MEAN AND VARIANCE OF $\bar{y}$

In this appendix we derive the formulas for $\mu_{\bar{y}}$ and $\sigma^2_{\bar{y}}$ given in Chapter 3, Section 3.4, using four properties of means and variances to do the necessary proofs.

Before we randomly select the observed sample

$$SPL = \{y_1, y_2, \ldots, y_n\}$$

from a single population of element values with mean μ and variance σ^2, we note that for $i = 1, 2, \ldots, n$, the ith observed element value, y_i, can potentially be any of the element values in the population. Thus, y_i is a **random variable**, which we define to be a variable whose value is determined by the outcome of an experiment. Moreover, we define the mean of y_i (denoted by μ_{y_i}) and the variance of y_i (denoted by $\sigma^2_{y_i}$) to be the mean and variance of the population of element values. That is, for $i = 1, 2, \ldots, n$,

$$\mu_{y_i} = \mu \qquad \text{(or equivalently, } \mu_{y_1} = \mu_{y_2} = \cdots = \mu_{y_n} = \mu\text{)}$$

$$\sigma^2_{y_i} = \sigma^2 \qquad \text{(or equivalently, } \sigma^2_{y_1} = \sigma^2_{y_2} = \cdots = \sigma^2_{y_n} = \sigma^2\text{)}$$

The following are some important properties of the means and variances of random variables (for proofs of these properties, see Wonnacott and Wonnacott (1977)).

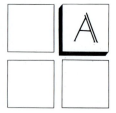

Assume y is a random variable and K is a constant. Then

1. $\mu_{Ky} = K\mu_y$.
2. $\sigma^2_{Ky} = K^2\sigma^2_y$.

Assume $y_1, y_2, \ldots, y_n$ are random variables. Then

3. $\mu_{(y_1+y_2+\cdots+y_n)} = \mu_{y_1} + \mu_{y_2} + \cdots + \mu_{y_n}$.
4. If $y_1, y_2, \ldots, y_n$ are statistically independent, then

$$\sigma^2_{(y_1+y_2+\cdots+y_n)} = \sigma^2_{y_1} + \sigma^2_{y_2} + \cdots + \sigma^2_{y_n}$$

We now use these properties to prove that if we randomly select the observed sample

$$SPL = \{y_1, y_2, \ldots, y_n\}$$

from a single infinite population of element values with mean μ and variance σ^2, and if

$$\bar{y} = \frac{\sum_{i=1}^{n} y_i}{n}$$

then $\mu_{\bar{y}} = \mu$ and $\sigma^2_{\bar{y}} = \sigma^2/n$, which implies that $\sigma_{\bar{y}} = \sigma/\sqrt{n}$.

The first proof is as follows:

$$\mu_{\bar{y}} = \mu\left(\sum_{i=1}^{n} y_i/n\right)$$

$$= \frac{1}{n}\,\mu\left(\sum_{i=1}^{n} y_i\right) \qquad \text{(see Property 1)}$$

$$= \frac{1}{n}(\mu_{y_1} + \mu_{y_2} + \cdots + \mu_{y_n}) \qquad \text{(see Property 3)}$$

$$= \frac{1}{n}(\mu + \mu + \cdots + \mu)$$

$$= \frac{n\mu}{n}$$

$$= \mu$$

since we have previously stated that $\mu_{y_1} = \mu_{y_2} = \cdots = \mu_{y_n} = \mu$.

The second proof is as follows:

$$
\begin{aligned}
\sigma_{\bar{y}}^2 &= \sigma^2_{\left(\sum_{i=1}^{n} y_i/n\right)} \\
&= \left(\frac{1}{n}\right)^2 \sigma^2_{\left(\sum_{i=1}^{n} y_i\right)} && \text{(see Property 2)} \\
&= \frac{1}{n^2}(\sigma^2_{y_1} + \sigma^2_{y_2} + \cdots + \sigma^2_{y_n}) && \text{(see Property 4)} \\
&= \frac{1}{n^2}(\sigma^2 + \sigma^2 + \cdots + \sigma^2) \\
&= \frac{n\sigma^2}{n^2} \\
&= \frac{\sigma^2}{n}
\end{aligned}
$$

since we have previously stated that $\sigma^2_{y_1} = \sigma^2_{y_2} = \cdots = \sigma^2_{y_n} = \sigma^2$. We assume that $y_1, y_2, \ldots, y_n$ are statistically independent, since we assume that the population of element values from which $y_1, y_2, \ldots, y_n$ are randomly selected is infinite.

APPENDIX B

DERIVATION OF THE LEAST SQUARES POINT ESTIMATES

In this appendix we show that the least squares point estimates $b_0, b_1, b_2, \ldots, b_p$ are calculated using the matrix algebra equation

$$\begin{bmatrix} b_0 \\ b_1 \\ b_2 \\ \vdots \\ b_p \end{bmatrix} = \mathbf{b} = (\mathbf{X}'\mathbf{X})^{-1}\mathbf{X}'\mathbf{y}$$

We do this by differentiating

$$\begin{aligned} SSE &= \sum_{i=1}^{n} (y_i - \hat{y}_i)^2 \\ &= \sum_{i=1}^{n} (y_i - (b_0 + b_1 x_{i1} + b_2 x_{i2} + \cdots + b_p x_{ip}))^2 \end{aligned}$$

with respect to $b_0, b_1, b_2, \ldots,$ and b_p as follows.

$$\frac{\partial SSE}{\partial b_0} = -2 \sum_{i=1}^{n} \left(y_i - b_0 - \sum_{j=1}^{p} b_j x_{ij} \right)$$

$$\frac{\partial SSE}{\partial b_k} = -2 \sum_{i=1}^{n} \left(y_i - b_0 - \sum_{j=1}^{p} b_j x_{ij} \right) x_{ik} \qquad \text{for } k = 1, 2, \ldots, p$$

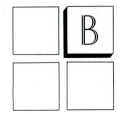

Setting these partial derivatives equal to zero, we obtain

$$\sum_{i=1}^{n}\left(y_i - b_0 - \sum_{j=1}^{p} b_j x_{ij}\right) = 0$$

$$\sum_{i=1}^{n}\left(y_i - b_0 - \sum_{j=1}^{p} b_j x_{ij}\right) x_{ik} = 0 \qquad \text{for } k = 1, 2, \ldots, p$$

which are equivalent to

$$nb_0 + \sum_{j=1}^{p} b_j \sum_{i=1}^{n} x_{ij} = \sum_{i=1}^{n} y_i$$

$$b_0 \sum_{i=1}^{n} x_{ik} + \sum_{j=1}^{p} b_j \sum_{i=1}^{n} x_{ij}x_{ik} = \sum_{i=1}^{n} x_{ik}y_i \qquad \text{for } k = 1, 2, \ldots, p$$

These $p + 1$ equations are called the **normal equations** and can usually be solved simultaneously for $b_0, b_1, b_2, \ldots,$ and b_p. To do this easily, consider the **X** matrix and **y** vector

$$\mathbf{X} = \begin{bmatrix} 1 & x_{11} & \cdots & x_{1p} \\ 1 & x_{21} & \cdots & x_{2p} \\ \vdots & \vdots & & \vdots \\ 1 & x_{n1} & \cdots & x_{np} \end{bmatrix} \qquad \mathbf{y} = \begin{bmatrix} y_1 \\ y_2 \\ \vdots \\ y_n \end{bmatrix}$$

We see that

$$\mathbf{X'X} = \begin{bmatrix} 1 & 1 & \cdots & 1 \\ x_{11} & x_{21} & \cdots & x_{n1} \\ \vdots & \vdots & & \vdots \\ x_{1p} & x_{2p} & \cdots & x_{np} \end{bmatrix} \begin{bmatrix} 1 & x_{11} & \cdots & x_{1p} \\ 1 & x_{21} & \cdots & x_{2p} \\ \vdots & \vdots & & \vdots \\ 1 & x_{n1} & \cdots & x_{np} \end{bmatrix}$$

$$= \begin{bmatrix} n & \sum_{i=1}^{n} x_{i1} & \sum_{i=1}^{n} x_{i2} & \cdots & \sum_{i=1}^{n} x_{ip} \\ \sum_{i=1}^{n} x_{i1} & \sum_{i=1}^{n} x_{i1}x_{i1} & \sum_{i=1}^{n} x_{i2}x_{i1} & \cdots & \sum_{i=1}^{n} x_{ip}x_{i1} \\ \vdots & \vdots & \vdots & & \vdots \\ \sum_{i=1}^{n} x_{ip} & \sum_{i=1}^{n} x_{i1}x_{ip} & \sum_{i=1}^{n} x_{i2}x_{ip} & \cdots & \sum_{i=1}^{n} x_{ip}x_{ip} \end{bmatrix}$$

$$\mathbf{X'y} = \begin{bmatrix} 1 & 1 & \cdots & 1 \\ x_{11} & x_{21} & \cdots & x_{n1} \\ \vdots & \vdots & & \vdots \\ x_{1p} & x_{2p} & \cdots & x_{np} \end{bmatrix} \begin{bmatrix} y_1 \\ y_2 \\ \vdots \\ y_n \end{bmatrix} = \begin{bmatrix} \sum_{i=1}^{n} y_i \\ \sum_{i=1}^{n} x_{i1}y_i \\ \vdots \\ \sum_{i=1}^{n} x_{ip}y_i \end{bmatrix}$$

Thus, letting

$$\mathbf{b} = \begin{bmatrix} b_0 \\ b_1 \\ \vdots \\ b_p \end{bmatrix}$$

we see that the normal equations can be written as

$$\begin{bmatrix} n & \sum_{i=1}^{n} x_{i1} & \sum_{i=1}^{n} x_{i2} & \cdots & \sum_{i=1}^{n} x_{ip} \\ \sum_{i=1}^{n} x_{i1} & \sum_{i=1}^{n} x_{i1}x_{i1} & \sum_{i=1}^{n} x_{i2}x_{i1} & \cdots & \sum_{i=1}^{n} x_{ip}x_{i1} \\ \vdots & \vdots & \vdots & & \vdots \\ \sum_{i=1}^{n} x_{ip} & \sum_{i=1}^{n} x_{i1}x_{ip} & \sum_{i=1}^{n} x_{i2}x_{ip} & \cdots & \sum_{i=1}^{n} x_{ip}x_{ip} \end{bmatrix} \begin{bmatrix} b_0 \\ b_1 \\ \vdots \\ b_p \end{bmatrix} = \begin{bmatrix} \sum_{i=1}^{n} y_i \\ \sum_{i=1}^{n} x_{i1}y_i \\ \vdots \\ \sum_{i=1}^{n} x_{ip}y_i \end{bmatrix}$$

or

$$(\mathbf{X}'\mathbf{X})\mathbf{b} = \mathbf{X}'\mathbf{y}$$

Assuming that the columns of the matrix $\mathbf{X}$ are linearly independent, which implies that the inverse of the matrix $\mathbf{X}'\mathbf{X}$ exists, then $\mathbf{b}$, the vector of least squares point estimates, is given by

$$\mathbf{b} = (\mathbf{X}'\mathbf{X})^{-1}\mathbf{X}'\mathbf{y}$$

Again, using calculus it can be shown that the values of $b_0, b_1, b_2, \ldots, b_p$ given by the preceding matrix algebra equation do in fact minimize

$$SSE = \sum_{i=1}^{n} (y_i - \hat{y}_i)^2$$

(rather than, for example, maximizing SSE).

APPENDIX C

DERIVATION OF THE COMPUTATIONAL FORMULA FOR *SSE*

In this appendix we prove that

$$SSE = \sum_{i=1}^{n} (y_i - \hat{y}_i)^2 = \sum_{i=1}^{n} y_i^2 - \mathbf{b}'\mathbf{X}'\mathbf{y}$$

First recall that

$$\mathbf{b} = \begin{bmatrix} b_0 \\ b_1 \\ \vdots \\ b_p \end{bmatrix} \qquad \mathbf{y} = \begin{bmatrix} y_1 \\ y_2 \\ \vdots \\ y_n \end{bmatrix} \qquad \mathbf{X} = \begin{bmatrix} 1 & x_{11} & \cdots & x_{1p} \\ 1 & x_{21} & \cdots & x_{2p} \\ \vdots & \vdots & & \vdots \\ 1 & x_{n1} & \cdots & x_{np} \end{bmatrix}$$

Then we let

$$\mathbf{Xb} = \begin{bmatrix} 1 & x_{11} & \cdots & x_{1p} \\ 1 & x_{21} & \cdots & x_{2p} \\ \vdots & \vdots & & \vdots \\ 1 & x_{n1} & \cdots & x_{np} \end{bmatrix} \begin{bmatrix} b_0 \\ b_1 \\ \vdots \\ b_p \end{bmatrix} = \begin{bmatrix} b_0 + b_1x_{11} + \cdots + b_px_{1p} \\ b_0 + b_1x_{21} + \cdots + b_px_{2p} \\ \vdots \\ b_0 + b_1x_{n1} + \cdots + b_px_{np} \end{bmatrix}$$

$$= \begin{bmatrix} \hat{y}_1 \\ \hat{y}_2 \\ \vdots \\ \hat{y}_n \end{bmatrix} = \hat{\mathbf{y}}$$

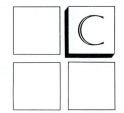

(Recall that $\hat{y}_i = b_0 + b_1 x_{i1} + \cdots + b_p x_{ip}$.) Now, since

$$\mathbf{y} - \hat{\mathbf{y}} = \begin{bmatrix} y_1 \\ y_2 \\ \vdots \\ y_n \end{bmatrix} - \begin{bmatrix} \hat{y}_1 \\ \hat{y}_2 \\ \vdots \\ \hat{y}_n \end{bmatrix} = \begin{bmatrix} y_1 - \hat{y}_1 \\ y_2 - \hat{y}_2 \\ \vdots \\ y_n - \hat{y}_n \end{bmatrix}$$

and

$$(\mathbf{y} - \hat{\mathbf{y}})' = [(y_1 - \hat{y}_1) \quad (y_2 - \hat{y}_2) \quad \cdots \quad (y_n - \hat{y}_n)]$$

it follows that

$$\begin{aligned}(\mathbf{y} - \hat{\mathbf{y}})'(\mathbf{y} - \hat{\mathbf{y}}) &= [(y_1 - \hat{y}_1) \quad (y_2 - \hat{y}_2) \quad \cdots \quad (y_n - \hat{y}_n)] \begin{bmatrix} y_1 - \hat{y}_1 \\ y_2 - \hat{y}_2 \\ \vdots \\ y_n - \hat{y}_n \end{bmatrix} \\ &= (y_1 - \hat{y}_1)^2 + (y_2 - \hat{y}_2)^2 + \cdots + (y_n - \hat{y}_n)^2 \\ &= \sum_{i=1}^{n} (y_i - \hat{y}_i)^2\end{aligned}$$

Thus, since $\hat{\mathbf{y}} = \mathbf{Xb}$, we see that

$$\begin{aligned} SSE &= \sum_{i=1}^{n} (y_i - \hat{y}_i)^2 = (\mathbf{y} - \hat{\mathbf{y}})'(\mathbf{y} - \hat{\mathbf{y}}) \\ &= (\mathbf{y} - \mathbf{Xb})'(\mathbf{y} - \mathbf{Xb}) \end{aligned}$$

Now,

$$\begin{aligned} SSE &= (\mathbf{y} - \mathbf{Xb})'(\mathbf{y} - \mathbf{Xb}) = (\mathbf{y}' - \mathbf{b}'\mathbf{X}')(\mathbf{y} - \mathbf{Xb}) \\ &= \mathbf{y}'\mathbf{y} - \mathbf{b}'\mathbf{X}'\mathbf{y} - \mathbf{y}'\mathbf{Xb} + \mathbf{b}'\mathbf{X}'\mathbf{Xb} \end{aligned}$$

Using the fact that $\mathbf{b} = (\mathbf{X}'\mathbf{X})^{-1}\mathbf{X}'\mathbf{y}$, we see that

$$\mathbf{b}'\mathbf{X}'\mathbf{Xb} = \mathbf{b}'\mathbf{X}'\mathbf{X}(\mathbf{X}'\mathbf{X})^{-1}\mathbf{X}'\mathbf{y} = \mathbf{b}'\mathbf{X}'\mathbf{y}$$

and hence

$$\begin{aligned} SSE &= \mathbf{y}'\mathbf{y} - \mathbf{b}'\mathbf{X}'\mathbf{y} - \mathbf{y}'\mathbf{Xb} + \mathbf{b}'\mathbf{X}'\mathbf{y} \\ &= \mathbf{y}'\mathbf{y} - \mathbf{b}'\mathbf{X}'\mathbf{y} \end{aligned}$$

since

$$\mathbf{y}'\mathbf{Xb} = \mathbf{b}'\mathbf{X}'\mathbf{y}$$

because $\mathbf{b}'\mathbf{X}'\mathbf{y}$ is a number, $(\mathbf{b}'\mathbf{X}'\mathbf{y})' = \mathbf{y}'\mathbf{Xb}$, and the transpose of a number is the number itself.

Finally, noting that

$$\mathbf{y}'\mathbf{y} = [y_1 \quad y_2 \quad \cdots \quad y_n]\begin{bmatrix} y_1 \\ y_2 \\ \vdots \\ y_n \end{bmatrix}$$

$$= y_1^2 + y_2^2 + \cdots + y_n^2 = \sum_{i=1}^{n} y_i^2$$

we have

$$SSE = \mathbf{y}'\mathbf{y} - \mathbf{b}'\mathbf{X}'\mathbf{y} = \sum_{i=1}^{n} y_i^2 - \mathbf{b}'\mathbf{X}'\mathbf{y}$$

APPENDIX D

DERIVATIONS OF THE MEANS AND VARIANCES OF b_j, $\hat{y}_0$, AND $y_0 - \hat{y}_0$

In this appendix we derive the formulas for μ_{b_j}, $\sigma^2_{b_j}$, $\mu_{\hat{y}_0}$, $\sigma^2_{\hat{y}_0}$, $\mu_{(y_0-\hat{y}_0)}$, and $\sigma^2_{(y_0-\hat{y}_0)}$ given in Chapter 6. To do the necessary proofs, which we number (i) through (vi) as we proceed, we use properties of means and variances, which we number (1) through (6) as we proceed. Furthermore, as we present each proof, we indicate which of the six properties implies each equality in the proof by placing the number of the property above the equality.

We now consider regression analysis. Before the observed sample

$$SPL = \{y_1, y_2, \ldots, y_n\}$$

is randomly selected from the n historical populations of potential values of the dependent variable, it follows that for $i = 1, 2, \ldots, n$, the ith observed value of the dependent variable

$$\begin{aligned} y_i &= \mu_i + \epsilon_i \\ &= \beta_0 + \beta_1 x_{i1} + \cdots + \beta_p x_{ip} + \epsilon_i \end{aligned}$$

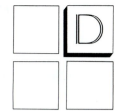

can be any of the potential values of the dependent variable in the ith historical population, which has mean μ_i and (in accordance with inference assumption 1) has variance σ^2. Thus, we define the mean of y_i (denoted by μ_{y_i}) and the variance of y_i (denoted by $\sigma^2_{y_i}$) to be, respectively, the mean and variance of the ith historical population. That is, for $i = 1, 2, \ldots, n$,

$$\mu_{y_i} = \mu_i = \beta_0 + \beta_1 x_{i1} + \cdots + \beta_p x_{ip}$$

$$\sigma^2_{y_i} = \sigma^2 \qquad \text{(or equivalently, } \sigma^2_{y_1} = \sigma^2_{y_2} = \cdots = \sigma^2_{y_n} = \sigma^2\text{)}$$

Moreover, for $i = 1, 2, \ldots, n$, y_i is a *random variable*, which we define to be a variable whose value is determined by the outcome of an experiment.

In general, if $y_1, y_2, \ldots, y_n$ are random variables, and if

$$\mathbf{y} = \begin{bmatrix} y_1 \\ y_2 \\ \vdots \\ y_n \end{bmatrix}$$

then we define the mean of $\mathbf{y}$ (denoted by $\mu_{\mathbf{y}}$) and the variance of $\mathbf{y}$ (denoted by $\sigma^2_{\mathbf{y}}$) as follows:

$$\mu_{\mathbf{y}} = \begin{bmatrix} \mu_{y_1} \\ \mu_{y_2} \\ \vdots \\ \mu_{y_n} \end{bmatrix}$$

$$\sigma^2_{\mathbf{y}} = \begin{bmatrix} \sigma^2_{y_1} & \sigma^2_{y_1y_2} & \cdots & \sigma^2_{y_1y_n} \\ \sigma^2_{y_2y_1} & \sigma^2_{y_2} & \cdots & \sigma^2_{y_2y_n} \\ \vdots & \vdots & & \vdots \\ \sigma^2_{y_ny_1} & \sigma^2_{y_ny_2} & \cdots & \sigma^2_{y_n} \end{bmatrix}$$

We call $\sigma^2_{y_iy_j}$ (for $i = 1, 2, \ldots, n$, and $j = 1, 2, \ldots, n$, and $i \neq j$) the *covariance between y_i and y_j*. For a mathematical definition of $\sigma^2_{y_iy_j}$, see Wonnacott and Wonnacott (1977). For now, suffice it to say that the covariance $\sigma^2_{y_iy_j}$ is a measure of the linear relationship between the random variables y_i and y_j, and if y_i and y_j are statistically independent, then $\sigma^2_{y_iy_j}$ equals zero.

Now, we have seen that in regression analysis, for $i = 1, 2, \ldots, n$,

$$\mu_{y_i} = \mu_i = \beta_0 + \beta_1 x_{i1} + \cdots + \beta_p x_{ip}$$

$$\sigma^2_{y_i} = \sigma^2 \qquad \text{(or equivalently, } \sigma^2_{y_1} = \sigma^2_{y_2} = \cdots = \sigma^2_{y_n} = \sigma^2\text{)}$$

Thus, in regression analysis

$$\mu_{\mathbf{y}} = \begin{bmatrix} \mu_{y_1} \\ \mu_{y_2} \\ \vdots \\ \mu_{y_n} \end{bmatrix} = \begin{bmatrix} \beta_0 + \beta_1 x_{11} + \cdots + \beta_p x_{1p} \\ \beta_0 + \beta_1 x_{21} + \cdots + \beta_p x_{2p} \\ \vdots \\ \beta_0 + \beta_1 x_{n1} + \cdots + \beta_p x_{np} \end{bmatrix}$$

$$= \begin{bmatrix} 1 & x_{11} & \cdots & x_{1p} \\ 1 & x_{21} & \cdots & x_{2p} \\ \vdots & \vdots & & \vdots \\ 1 & x_{n1} & \cdots & x_{np} \end{bmatrix} \begin{bmatrix} \beta_0 \\ \beta_1 \\ \vdots \\ \beta_p \end{bmatrix}$$

$$= \mathbf{X}\boldsymbol{\beta}$$

where $\boldsymbol{\beta}$ is the column vector containing the parameters $\beta_0, \beta_1, \ldots, \beta_p$. Moreover, since inference assumption 2 says that $y_1, y_2, \ldots, y_n$, and y_0 are statistically independent, which implies that for $i = 1, 2, \ldots, n$, and $j = 1, 2, \ldots, n$, and $i \neq j$, $\sigma^2_{y_i y_j}$ equals zero, it follows that

$$\sigma^2_{\mathbf{y}} = \begin{bmatrix} \sigma^2_{y_1} & \sigma^2_{y_1 y_2} & \cdots & \sigma^2_{y_1 y_n} \\ \sigma^2_{y_2 y_1} & \sigma^2_{y_2} & \cdots & \sigma^2_{y_2 y_n} \\ \vdots & \vdots & & \vdots \\ \sigma^2_{y_n y_1} & \sigma^2_{y_n y_2} & \cdots & \sigma^2_{y_n} \end{bmatrix}$$

$$= \begin{bmatrix} \sigma^2 & 0 & \cdots & 0 \\ 0 & \sigma^2 & \cdots & 0 \\ \vdots & \vdots & & \vdots \\ 0 & 0 & \cdots & \sigma^2 \end{bmatrix}$$

$$= \sigma^2 \begin{bmatrix} 1 & 0 & \cdots & 0 \\ 0 & 1 & \cdots & 0 \\ \vdots & \vdots & & \vdots \\ 0 & 0 & \cdots & 1 \end{bmatrix}$$

$$= \sigma^2 \mathbf{I}$$

(Recall $\mathbf{I}$ is the n dimensional identity matrix.) To summarize, in regression analysis

$$\mu_{\mathbf{y}} = \mathbf{X}\boldsymbol{\beta}$$

$$\sigma^2_{\mathbf{y}} = \sigma^2 \mathbf{I}$$

Two important properties concerning $\mu_{\mathbf{y}}$ and $\sigma^2_{\mathbf{y}}$ follow (for proofs of these properties, see Searle (1971)).

Assume $\mathbf{y}$ is a column vector containing the random variables $y_1, y_2, \ldots, y_n$, and assume $\mathbf{A}$ is a matrix of constants such that the matrix product $\mathbf{Ay}$ exists (note that $\mathbf{Ay}$ is a column vector). Then

(1) $\mu_{\mathbf{Ay}} = \mathbf{A}\mu_{\mathbf{y}}$
(2) $\sigma^2_{\mathbf{Ay}} = \mathbf{A}\sigma^2_{\mathbf{y}}\mathbf{A}'$

We first use these properties to prove that if b_j is the least squares point estimate of the parameter β_j in the linear regression model, then

(i) $\mu_{b_j} = \beta_j$
(ii) $\sigma^2_{b_j} = \sigma^2 c_{jj}$

where c_{jj} is the diagonal element of $(\mathbf{X}'\mathbf{X})^{-1}$ corresponding to β_j. The proofs of these results are as follows.

(i) Recalling that $\mathbf{b} = (\mathbf{X}'\mathbf{X})^{-1}\mathbf{X}'\mathbf{y}$ is a column vector containing the least squares point estimates $b_0, b_1, \ldots, b_p$, and recalling that we have previously shown that $\mu_{\mathbf{y}} = \mathbf{X}\boldsymbol{\beta}$, it follows that

$$\begin{bmatrix} \mu_{b_0} \\ \mu_{b_1} \\ \vdots \\ \mu_{b_p} \end{bmatrix} = \mu_{\mathbf{b}} = \mu_{(\mathbf{X}'\mathbf{X})^{-1}\mathbf{X}'\mathbf{y}}$$

$$\begin{aligned} &\overset{(1)}{=} (\mathbf{X}'\mathbf{X})^{-1}\mathbf{X}'\mu_{\mathbf{y}} \\ &= (\mathbf{X}'\mathbf{X})^{-1}\mathbf{X}'[\mathbf{X}\boldsymbol{\beta}] \\ &= (\mathbf{X}'\mathbf{X})^{-1}(\mathbf{X}'\mathbf{X})\boldsymbol{\beta} \\ &= \mathbf{I}\boldsymbol{\beta} \\ &= \boldsymbol{\beta} \\ &= \begin{bmatrix} \beta_0 \\ \beta_1 \\ \vdots \\ \beta_p \end{bmatrix} \end{aligned}$$

Thus, we have shown that for $j = 0, 1, \ldots, p$,

$$\mu_{b_j} = \beta_j$$

or in terms of matrices, we have shown that

$$\mu_{\mathbf{b}} = \boldsymbol{\beta}$$

(ii) In order to prove that $\sigma^2_{b_j} = \sigma^2 c_{jj}$, we first need to perform some matrix manipulations. To do this, we note that it can be proved that if **A** and **B** are matrices such that **A** has an inverse and the matrix product **AB** exists, then

$$(\mathbf{A}')' = \mathbf{A}$$

$$(\mathbf{AB})' = \mathbf{B}'\mathbf{A}'$$

$$(\mathbf{A}^{-1})' = (\mathbf{A}')^{-1}$$

Thus

$$\begin{aligned} [(\mathbf{X}'\mathbf{X})^{-1}\mathbf{X}']' &= (\mathbf{X}')'[(\mathbf{X}'\mathbf{X})^{-1}]' \\ &= \mathbf{X}[(\mathbf{X}'\mathbf{X})']^{-1} \\ &= \mathbf{X}[\mathbf{X}'(\mathbf{X}')']^{-1} \\ &= \mathbf{X}(\mathbf{X}'\mathbf{X})^{-1} \end{aligned}$$

We next recall that we have shown that inference assumptions 1 and 2 imply that $\sigma^2_{\mathbf{y}} = \sigma^2\mathbf{I}$. Hence, if $\sigma^2_{b_ib_j}$ denotes the covariance between b_i and b_j, it follows that

$$\begin{bmatrix} \sigma^2_{b_0} & \sigma^2_{b_0b_1} & \cdots & \sigma^2_{b_0b_p} \\ \sigma^2_{b_1b_0} & \sigma^2_{b_1} & \cdots & \sigma^2_{b_1b_p} \\ \vdots & \vdots & & \vdots \\ \sigma^2_{b_pb_0} & \sigma^2_{b_pb_1} & \cdots & \sigma^2_{b_p} \end{bmatrix} = \sigma^2_{\mathbf{b}}$$

$$\begin{aligned}
&= \sigma^2_{(\mathbf{X}'\mathbf{X})^{-1}\mathbf{X}'\mathbf{y}} \\
&= [(\mathbf{X}'\mathbf{X})^{-1}\mathbf{X}']\sigma^2_{\mathbf{y}}[(\mathbf{X}'\mathbf{X})^{-1}\mathbf{X}']' \\
&= [(\mathbf{X}'\mathbf{X})^{-1}\mathbf{X}']\sigma^2\mathbf{I}[\mathbf{X}(\mathbf{X}'\mathbf{X})^{-1}] \\
&= \sigma^2[(\mathbf{X}'\mathbf{X})^{-1}\mathbf{X}'\mathbf{I}\mathbf{X}(\mathbf{X}'\mathbf{X})^{-1}] \\
&= \sigma^2[(\mathbf{X}'\mathbf{X})^{-1}\mathbf{X}'\mathbf{X}(\mathbf{X}'\mathbf{X})^{-1}] \\
&= \sigma^2[(\mathbf{X}'\mathbf{X})^{-1}\mathbf{I}] \\
&= \sigma^2(\mathbf{X}'\mathbf{X})^{-1} \\
&= \sigma^2 \begin{bmatrix} c_{00} & c_{01} & \cdots & c_{0p} \\ c_{10} & c_{11} & \cdots & c_{1p} \\ \vdots & \vdots & & \vdots \\ c_{p0} & c_{p1} & \cdots & c_{pp} \end{bmatrix}
\end{aligned}$$

Thus, we have shown that for $j = 0, 1, \ldots, p$,

$$\sigma^2_{b_j} = \sigma^2 c_{jj}$$

Moreover, we have shown that for $i = 0, 1, \ldots, p$, and $j = 0, 1, \ldots, p$, and $i \neq j$,

$$\sigma^2_{b_ib_j} = \sigma^2 c_{ij}$$

In terms of matrices, we have shown that

$$\sigma^2_{\mathbf{b}} = \sigma^2(\mathbf{X}'\mathbf{X})^{-1}$$

Finally, we will prove that if

$$\hat{y}_0 = b_0 + b_1x_{01} + \cdots + b_px_{0p}$$

is the point estimate of the future mean value of the dependent variable

$$\mu_0 = \beta_0 + \beta_1x_{01} + \cdots + \beta_px_{0p}$$

and is the point prediction of the future (individual) value of the dependent variable

$$\begin{aligned} y_0 &= \mu_0 + \epsilon_0 \\ &= \beta_0 + \beta_1x_{01} + \cdots + \beta_px_{0p} + \epsilon_0 \end{aligned}$$

and if

$$\mathbf{x}_0' = [1 \quad x_{01} \quad \cdots \quad x_{0p}]$$

is a row vector containing the future values of the independent variables, then

(iii) $\mu_{\hat{y}_0} = \mu_0$.
(iv) $\sigma^2_{\hat{y}_0} = \sigma^2 \mathbf{x}_0'(\mathbf{X}'\mathbf{X})^{-1}\mathbf{x}_0$.
(v) $\mu_{(y_0 - \hat{y}_0)} = 0$.
(vi) $\sigma^2_{(y_0 - \hat{y}_0)} = \sigma^2(1 + \mathbf{x}_0'(\mathbf{X}'\mathbf{X})^{-1}\mathbf{x}_0)$.

To prove these results, we note that

$$\begin{aligned}
\hat{y}_0 &= b_0 + b_1 x_{01} + \cdots + b_p x_{0p} \\
&= [1 \quad x_{01} \quad \cdots \quad x_{0p}] \begin{bmatrix} b_0 \\ b_1 \\ \vdots \\ b_p \end{bmatrix} \\
&= \mathbf{x}_0'\mathbf{b}
\end{aligned}$$

Thus, the proofs for (iii) and (iv) are as follows:

(iii) Since we have shown that $\mu_{\mathbf{b}} = \boldsymbol{\beta}$, we have

$$\begin{aligned}
\mu_{\hat{y}_0} &= \mu_{\mathbf{x}_0'\mathbf{b}} \\
&\overset{(1)}{=} \mathbf{x}_0'\mu_{\mathbf{b}} \\
&= \mathbf{x}_0'\boldsymbol{\beta} \\
&= [1 \quad x_{01} \quad \cdots \quad x_{0p}] \begin{bmatrix} \beta_0 \\ \beta_1 \\ \vdots \\ \beta_p \end{bmatrix} \\
&= \beta_0 + \beta_1 x_{01} + \cdots + \beta_p x_{0p} \\
&= \mu_0
\end{aligned}$$

(iv) Since we have shown that $\sigma^2_{\mathbf{b}} = \sigma^2(\mathbf{X}'\mathbf{X})^{-1}$, we have

$$\begin{aligned}
\sigma^2_{\hat{y}_0} &= \sigma^2_{\mathbf{x}_0'\mathbf{b}} \\
&\overset{(2)}{=} \mathbf{x}_0'\sigma^2_{\mathbf{b}}(\mathbf{x}_0')' \\
&= \mathbf{x}_0'[\sigma^2(\mathbf{X}'\mathbf{X})^{-1}]\mathbf{x}_0 \\
&= \sigma^2\mathbf{x}_0'(\mathbf{X}'\mathbf{X})^{-1}\mathbf{x}_0
\end{aligned}$$

In order to prove (v) and (vi), we first summarize some important properties of the means and variances of random variables. (For proofs of these properties, see Wonnacott and Wonnacott (1977).)

Assume y is a random variable and K is a constant. Then

(3) $\mu_{Ky} = K\mu_y$.
(4) $\sigma^2_{Ky} = K^2\sigma^2_y$.

Assume $y_1, y_2, \ldots, y_n$ are random variables. Then

(5) $\mu_{(y_1+y_2+\cdots+y_n)} = \mu_{y_1} + \mu_{y_2} + \cdots + \mu_{y_n}$.
(6) If $y_1, y_2, \ldots, y_n$ are statistically independent,

$$\sigma^2_{(y_1+y_2+\cdots+y_n)} = \sigma^2_{y_1} + \sigma^2_{y_2} + \cdots + \sigma^2_{y_n}.$$

In order to prove (v) and (vi) we note that before we observe

$$y_0 = \mu_0 + \epsilon_0$$

y_0 can be any of the potential values of the dependent variable in the future population of potential values of the dependent variable, which has mean μ_0 and (in accordance with inference assumption 1) variance σ^2. Thus, we define the mean of y_0 (denoted by μ_{y_0}) and the variance of y_0 (denoted by $\sigma^2_{y_0}$) to be, respectively, the mean and variance of the future population of potential values of the dependent variable. That is,

$$\mu_{y_0} = \mu_0$$
$$\sigma^2_{y_0} = \sigma^2$$

Thus,

(v) $$\mu_{(y_0-\hat{y}_0)} \overset{(5)}{=} \mu_{y_0} + \mu_{(-\hat{y}_0)} \overset{(3)}{=} \mu_{y_0} + (-1)\mu_{\hat{y}_0} = \mu_0 - \mu_0 = 0$$

(vi) Since

$$\hat{y}_0 = b_0 + b_1x_{01} + \cdots + b_px_{0p}$$

is a function of the least squares point estimates $b_0, b_1, \ldots, b_p$, which are in turn functions of $y_1, y_2, \ldots, y_n$, it follows that $\hat{y}_0$ is a function of $y_1, y_2, \ldots, y_n$.

Since inference assumption 2 says that y_0 is statistically independent of y_1, $y_2, \ldots, y_n$, it follows that y_0 and $\hat{y}_0$ are statistically independent. Hence,

$$\begin{aligned}
\sigma^2_{(y_0 - \hat{y}_0)} &\overset{(6)}{=} \sigma^2_{y_0} + \sigma^2_{(-\hat{y}_0)} \\
&\overset{(4)}{=} \sigma^2_{y_0} + (-1)^2 \sigma^2_{\hat{y}_0} \\
&= \sigma^2 + \sigma^2 \mathbf{x}_0'(\mathbf{X}'\mathbf{X})^{-1}\mathbf{x}_0 \\
&= \sigma^2(1 + \mathbf{x}_0'(\mathbf{X}'\mathbf{X})^{-1}\mathbf{x}_0)
\end{aligned}$$

APPENDIX E

STATISTICAL TABLES

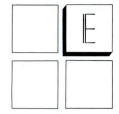

TABLE E-1 Normal Curve Areas

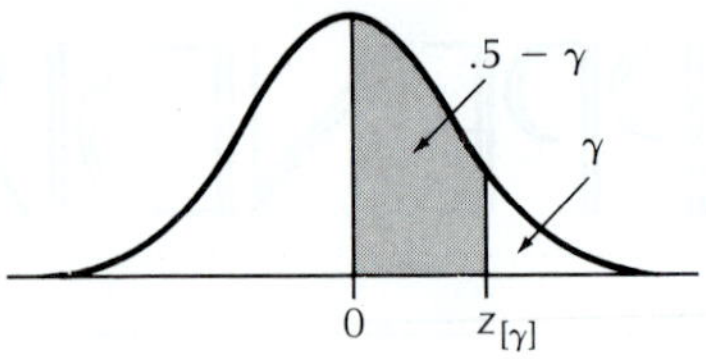

$z_{[\gamma]}$	.00	.01	.02	.03	.04	.05	.06	.07	.08	.09
0.0	.0000	.0040	.0080	.0120	.0160	.0199	.0239	.0279	.0319	.0359
0.1	.0398	.0438	.0478	.0517	.0557	.0596	.0636	.0675	.0714	.0753
0.2	.0793	.0832	.0871	.0910	.0948	.0987	.1026	.1064	.1103	.1141
0.3	.1179	.1217	.1255	.1293	.1331	.1368	.1406	.1443	.1480	.1517
0.4	.1554	.1591	.1628	.1664	.1700	.1736	.1772	.1808	.1844	.1879
0.5	.1915	.1950	.1985	.2019	.2054	.2088	.2123	.2157	.2190	.2224
0.6	.2257	.2291	.2324	.2357	.2389	.2422	.2454	.2486	.2517	.2549
0.7	.2580	.2611	.2642	.2673	.2704	.2734	.2764	.2794	.2823	.2852
0.8	.2881	.2910	.2939	.2967	.2995	.3023	.3051	.3078	.3106	.3133
0.9	.3159	.3186	.3212	.3238	.3264	.3289	.3315	.3340	.3365	.3389
1.0	.3413	.3438	.3461	.3485	.3508	.3531	.3554	.3577	.3599	.3621
1.1	.3643	.3665	.3686	.3708	.3729	.3749	.3770	.3790	.3810	.3830
1.2	.3849	.3869	.3888	.3907	.3925	.3944	.3962	.3980	.3997	.4015
1.3	.4032	.4049	.4066	.4082	.4099	.4115	.4131	.4147	.4162	.4177
1.4	.4192	.4207	.4222	.4236	.4251	.4265	.4279	.4292	.4306	.4319
1.5	.4332	.4345	.4357	.4370	.4382	.4394	.4406	.4418	.4429	.4441
1.6	.4452	.4463	.4474	.4484	.4495	.4505	.4515	.4525	.4535	.4545
1.7	.4554	.4564	.4573	.4582	.4591	.4599	.4608	.4616	.4625	.4633
1.8	.4641	.4649	.4656	.4664	.4671	.4678	.4686	.4693	.4699	.4706
1.9	.4713	.4719	.4726	.4732	.4738	.4744	.4750	.4756	.4761	.4767
2.0	.4772	.4778	.4783	.4788	.4793	.4798	.4803	.4808	.4812	.4817
2.1	.4821	.4826	.4830	.4834	.4838	.4842	.4846	.4850	.4854	.4857
2.2	.4861	.4864	.4868	.4871	.4875	.4878	.4881	.4884	.4887	.4890
2.3	.4893	.4896	.4898	.4901	.4904	.4906	.4909	.4911	.4913	.4916
2.4	.4918	.4920	.4922	.4925	.4927	.4929	.4931	.4932	.4934	.4936
2.5	.4938	.4940	.4941	.4943	.4945	.4946	.4948	.4949	.4951	.4952
2.6	.4953	.4955	.4956	.4957	.4959	.4960	.4961	.4962	.4963	.4964
2.7	.4965	.4966	.4967	.4968	.4969	.4970	.4971	.4972	.4973	.4974
2.8	.4974	.4975	.4976	.4977	.4977	.4978	.4979	.4979	.4980	.4981
2.9	.4981	.4982	.4982	.4983	.4984	.4984	.4985	.4985	.4986	.4986
3.0	.4987	.4987	.4987	.4988	.4988	.4989	.4989	.4989	.4990	.4990

Source: A Hald, *Statistical Tables and Formulas* (New York:Wiley, 1952), abridged from Table 1. Reproduced by permission of the publisher.

TABLE E-2 Critical Values of t

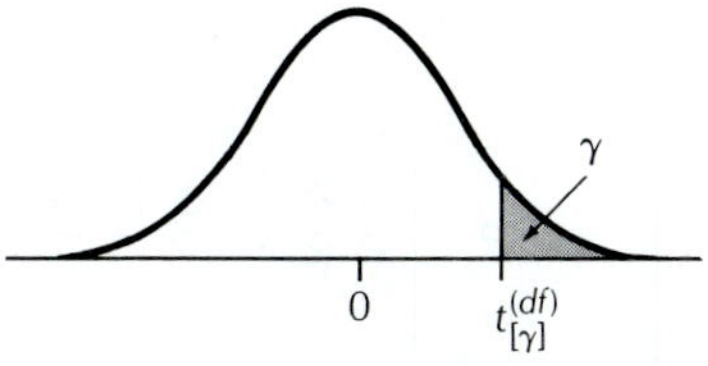

df	$t_{[.10]}^{(df)}$	$t_{[.05]}^{(df)}$	$t_{[.025]}^{(df)}$	$t_{[.01]}^{(df)}$	$t_{[.005]}^{(df)}$
1	3.078	6.314	12.706	31.821	63.657
2	1.886	2.920	4.303	6.965	9.925
3	1.638	2.353	3.182	4.541	5.841
4	1.533	2.132	2.776	3.747	4.604
5	1.476	2.015	2.571	3.365	4.032
6	1.440	1.943	2.447	3.143	3.707
7	1.415	1.895	2.365	2.998	3.499
8	1.397	1.860	2.306	2.896	3.355
9	1.383	1.833	2.262	2.821	3.250
10	1.372	1.812	2.228	2.764	3.169
11	1.363	1.796	2.201	2.718	3.106
12	1.356	1.782	2.179	2.681	3.055
13	1.350	1.771	2.160	2.650	3.012
14	1.345	1.761	2.145	2.624	2.977
15	1.341	1.753	2.131	2.602	2.947
16	1.337	1.746	2.120	2.583	2.921
17	1.333	1.740	2.110	2.567	2.898
18	1.330	1.734	2.101	2.552	2.878
19	1.328	1.729	2.093	2.539	2.861
20	1.325	1.725	2.086	2.528	2.845
21	1.323	1.721	2.080	2.518	2.831
22	1.321	1.717	2.074	2.508	2.819
23	1.319	1.714	2.069	2.500	2.807
24	1.318	1.711	2.064	2.492	2.797
25	1.316	1.708	2.060	2.485	2.787
26	1.315	1.706	2.056	2.479	2.779
27	1.314	1.703	2.052	2.473	2.771
28	1.313	1.701	2.048	2.467	2.763
29	1.311	1.699	2.045	2.462	2.756
inf.	1.282	1.645	1.960	2.326	2.576

Source: From "Table of Percentage Points of the *t*-Distribution," by Maxine Merrington, *Biometrika* 32 (1941), 300. Reproduced by permission of *Biometrika* Trustees.

TABLE E-3 Percentage Points of the *F*-Distribution ($\gamma = .05$)

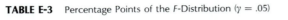

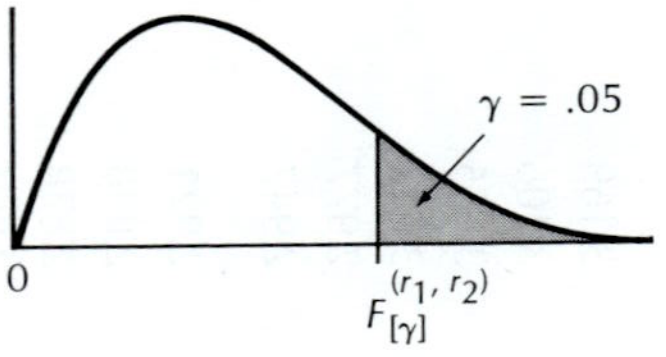

Denominator Degrees of Freedom (r_2)	Numerator Degrees of Freedom (r_1)								
	1	2	3	4	5	6	7	8	9
1	161.4	199.5	215.7	224.6	230.2	234.0	236.8	238.9	240.5
2	18.51	19.00	19.16	19.25	19.30	19.33	19.35	19.37	19.38
3	10.13	9.55	9.28	9.12	9.01	8.94	8.89	8.85	8.81
4	7.71	6.94	6.59	6.39	6.26	6.16	6.09	6.04	6.00
5	6.61	5.79	5.41	5.19	5.05	4.95	4.88	4.82	4.77
6	5.99	5.14	4.76	4.53	4.39	4.28	4.21	4.15	4.10
7	5.59	4.74	4.35	4.12	3.97	3.87	3.79	3.73	3.68
8	5.32	4.46	4.07	3.84	3.69	3.58	3.50	3.44	3.39
9	5.12	4.26	3.86	3.63	3.48	3.37	3.29	3.23	3.18
10	4.96	4.10	3.71	3.48	3.33	3.22	3.14	3.07	3.02
11	4.84	3.98	3.59	3.36	3.20	3.09	3.01	2.95	2.90
12	4.75	3.89	3.49	3.26	3.11	3.00	2.91	2.85	2.80
13	4.67	3.81	3.41	3.18	3.03	2.92	2.83	2.77	2.71
14	4.60	3.74	3.34	3.11	2.96	2.85	2.76	2.70	2.65
15	4.54	3.68	3.29	3.06	2.90	2.79	2.71	2.64	2.59
16	4.49	3.63	3.24	3.01	2.85	2.74	2.66	2.59	2.54
17	4.45	3.59	3.20	2.96	2.81	2.70	2.61	2.55	2.49

18	4.41	3.55	3.16	2.93	2.77	2.66	2.58	2.51	2.46
19	4.38	3.52	3.13	2.90	2.74	2.63	2.54	2.48	2.42
20	4.35	3.49	3.10	2.87	2.71	2.60	2.51	2.45	2.39
21	4.32	3.47	3.07	2.84	2.68	2.57	2.49	2.42	2.37
22	4.30	3.44	3.05	2.82	2.66	2.55	2.46	2.40	2.34
23	4.28	3.42	3.03	2.80	2.64	2.53	2.44	2.37	2.32
24	4.26	3.40	3.01	2.78	2.62	2.51	2.42	2.36	2.30
25	4.24	3.39	2.99	2.76	2.60	2.49	2.40	2.34	2.28
26	4.23	3.37	2.98	2.74	2.59	2.47	2.39	2.32	2.27
27	4.21	3.35	2.96	2.73	2.57	2.46	2.37	2.31	2.25
28	4.20	3.34	2.95	2.71	2.56	2.45	2.36	2.29	2.24
29	4.18	3.33	2.93	2.70	2.55	2.43	2.35	2.28	2.22
30	4.17	3.32	2.92	2.69	2.53	2.42	2.33	2.27	2.21
40	4.08	3.23	2.84	2.61	2.45	2.34	2.25	2.18	2.12
60	4.00	3.15	2.76	2.53	2.37	2.25	2.17	2.10	2.04
120	3.92	3.07	2.68	2.45	2.29	2.17	2.09	2.02	1.96
∞	3.84	3.00	2.60	2.37	2.21	2.10	2.01	1.94	1.88

Source: From "Tables of Percentage Points of the Inverted Beta (*F*)-Distribution," by Maxine Merrington and Catherine M. Thompson, *Biometrika* 33 (1943), 73–88. Reproduced by permission of the *Biometrika* Trustees.

TABLE E-3 Percentage Points of the *F*-Distribution ($\gamma = .05$) (*Continued*)

Denominator Degrees of Freedom (r_2)	Numerator Degrees of Freedom (r_1)									
	10	12	15	20	24	30	40	60	120	∞
1	6056	6106	6157	6209	6235	6261	6287	6313	6339	6366
2	99.40	99.42	99.43	99.45	99.46	99.47	99.47	99.48	99.49	99.50
3	27.23	27.05	26.87	26.69	26.60	26.50	26.41	26.32	26.22	26.13
4	14.55	14.37	14.20	14.02	13.93	13.84	13.75	13.65	13.56	13.46
5	10.05	9.89	9.72	9.55	9.47	9.38	9.29	9.20	9.11	9.02
6	7.87	7.72	7.56	7.40	7.31	7.23	7.14	7.06	6.97	6.88
7	6.62	6.47	6.31	6.16	6.07	5.99	5.91	5.82	5.74	5.65
8	5.81	5.67	5.52	5.36	5.28	5.20	5.12	5.03	4.95	4.86
9	5.26	5.11	4.96	4.81	4.73	4.65	4.57	4.48	4.40	4.31
10	4.85	4.71	4.56	4.41	4.33	4.25	4.17	4.08	4.00	3.91
11	4.54	4.40	4.25	4.10	4.02	3.94	3.86	3.78	3.69	3.60
12	4.30	4.16	4.01	3.86	3.78	3.70	3.62	3.54	3.45	3.36
13	4.10	3.96	3.82	3.66	3.59	3.51	3.43	3.34	3.25	3.17
14	3.94	3.80	3.66	3.51	3.43	3.35	3.27	3.18	3.09	3.00
15	3.80	3.67	3.52	3.37	3.29	3.21	3.13	3.05	2.96	2.87
16	3.69	3.55	3.41	3.26	3.18	3.10	3.02	2.93	2.84	2.75
17	3.59	3.46	3.31	3.16	3.08	3.00	2.92	2.83	2.75	2.65
18	3.51	3.37	3.23	3.08	3.00	2.92	2.84	2.75	2.66	2.57
19	3.43	3.30	3.15	3.00	2.92	2.84	2.76	2.67	2.58	2.49
20	3.37	3.23	3.09	2.94	2.86	2.78	2.69	2.61	2.52	2.42
21	3.31	3.17	3.03	2.88	2.80	2.72	2.64	2.55	2.46	2.36
22	3.26	3.12	2.98	2.83	2.75	2.67	2.58	2.50	2.40	2.31
23	3.21	3.07	2.93	2.78	2.70	2.62	2.54	2.45	2.35	2.26
24	3.17	3.03	2.89	2.74	2.66	2.58	2.49	2.40	2.31	2.21

25	3.13	2.99	2.85	2.70	2.62	2.54	2.45	2.36	2.27	2.17
26	3.09	2.96	2.81	2.66	2.58	2.50	2.42	2.33	2.23	2.13
27	3.06	2.93	2.78	2.63	2.55	2.47	2.38	2.29	2.20	2.10
28	3.03	2.90	2.75	2.60	2.52	2.44	2.35	2.26	2.17	2.06
29	3.00	2.87	2.73	2.57	2.49	2.41	2.33	2.23	2.14	2.03
30	2.98	2.84	2.70	2.55	2.47	2.39	2.30	2.21	2.11	2.01
40	2.80	2.66	2.52	2.37	2.29	2.20	2.11	2.02	1.92	1.80
60	2.63	2.50	2.35	2.20	2.12	2.03	1.94	1.84	1.73	1.60
120	2.47	2.34	2.19	2.03	1.95	1.86	1.76	1.66	1.53	1.38
∞	2.32	2.18	2.04	1.88	1.79	1.70	1.59	1.47	1.32	1.00

TABLE E-4 Percentage Points of the F-Distribution ($\gamma = .01$)

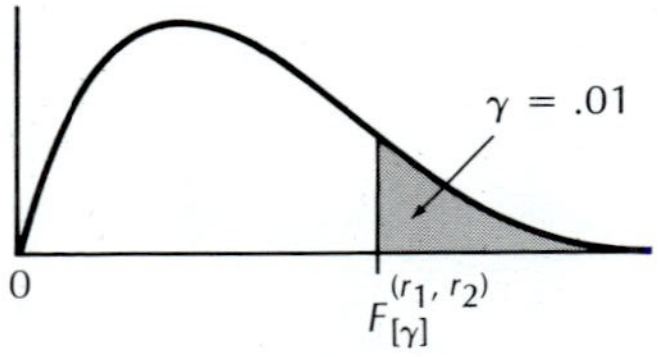

Denominator Degrees of Freedom (r_2)	Numerator Degrees of Freedom (r_1)								
	1	2	3	4	5	6	7	8	9
1	4052	4999.5	5403	5625	5764	5859	5928	5982	6022
2	98.50	99.00	99.17	99.25	99.30	99.33	99.36	99.37	99.39
3	34.12	30.82	29.46	28.71	28.24	27.91	27.67	27.49	27.35
4	21.20	18.00	16.69	15.98	15.52	15.21	14.98	14.80	14.66
5	16.26	13.27	12.06	11.39	10.97	10.67	10.46	10.29	10.16
6	13.75	10.92	9.78	9.15	8.75	8.47	8.26	8.10	7.98
7	12.25	9.55	8.45	7.85	7.46	7.19	6.99	6.84	6.72
8	11.26	8.65	7.59	7.01	6.63	6.37	6.18	6.03	5.91
9	10.56	8.02	6.99	6.42	6.06	5.80	5.61	5.47	5.35
10	10.04	7.56	6.55	5.99	5.64	5.39	5.20	5.06	4.94
11	9.65	7.21	6.22	5.67	5.32	5.07	4.89	4.74	4.63
12	9.33	6.93	5.95	5.41	5.06	4.82	4.64	4.50	4.39
13	9.07	6.70	5.74	5.21	4.86	4.62	4.44	4.30	4.19
14	8.86	6.51	5.56	5.04	4.69	4.46	4.28	4.14	4.03
15	8.68	6.36	5.42	4.89	4.56	4.32	4.14	4.00	3.89
16	8.53	6.23	5.29	4.77	4.44	4.20	4.03	3.89	3.78
17	8.40	6.11	5.18	4.67	4.34	4.10	3.93	3.79	3.68

18	8.29	6.01	5.09	4.58	4.25	4.01	3.84	3.71	3.60
19	8.18	5.93	5.01	4.50	4.17	3.94	3.77	3.63	3.52
20	8.10	5.85	4.94	4.43	4.10	3.87	3.70	3.56	3.46
21	8.02	5.78	4.87	4.37	4.04	3.81	3.64	3.51	3.40
22	7.95	5.72	4.82	4.31	3.99	3.76	3.59	3.45	3.35
23	7.88	5.66	4.76	4.26	3.94	3.71	3.54	3.41	3.30
24	7.82	5.61	4.72	4.22	3.90	3.67	3.50	3.36	3.26
25	7.77	5.57	4.68	4.18	3.85	3.63	3.46	3.32	3.22
26	7.72	5.53	4.64	4.14	3.82	3.59	3.42	3.29	3.18
27	7.68	5.49	4.60	4.11	3.78	3.56	3.39	3.26	3.15
28	7.64	5.45	4.57	4.07	3.75	3.53	3.36	3.23	3.12
29	7.60	5.42	4.54	4.04	3.73	3.50	3.33	3.20	3.09
30	7.56	5.39	4.51	4.02	3.70	3.47	3.30	3.17	3.07
40	7.31	5.18	4.31	3.83	3.51	3.29	3.12	2.99	2.89
60	7.08	4.98	4.13	3.65	3.34	3.12	2.95	2.82	2.72
120	6.85	4.79	3.95	3.48	3.17	2.96	2.79	2.66	2.50
∞	6.63	4.61	3.78	3.32	3.02	2.80	2.64	2.51	2.41

Source: From "Tables of Percentage Points of the Inverted Beta (*F*)-Distribution," by Maxine Merrington and Catherine M. Thompson, *Biometrika* 33 (1943), 73–88. Reproduced by permission of the *Biometrika* Trustees.

TABLE E-4 Percentage Points of the F-Distribution ($\gamma = .01$) (*Continued*)

Denominator Degrees of Freedom (r_2)	Numerator Degrees of Freedom (r_1)									
	10	12	15	20	24	30	40	60	120	∞
1	241.9	243.9	245.9	248.0	249.1	250.1	251.1	252.2	253.3	254.3
2	19.40	19.41	19.43	19.45	19.45	19.46	19.47	19.48	19.49	19.50
3	8.79	8.74	8.70	8.66	8.64	8.62	8.59	8.57	8.55	8.53
4	5.96	5.91	5.86	5.80	5.77	5.75	5.72	5.69	5.66	5.63
5	4.74	4.68	4.62	4.56	4.53	4.50	4.46	4.43	4.40	4.36
6	4.06	4.00	3.94	3.87	3.84	3.81	3.77	3.74	3.70	3.67
7	3.64	3.57	3.51	3.44	3.41	3.38	3.34	3.30	3.27	3.23
8	3.35	3.28	3.22	3.15	3.12	3.08	3.04	3.01	2.97	2.93
9	3.14	3.07	3.01	2.94	2.90	2.86	2.83	2.79	2.75	2.71
10	2.98	2.91	2.85	2.77	2.74	2.70	2.66	2.62	2.58	2.54
11	2.85	2.79	2.72	2.65	2.61	2.57	2.53	2.49	2.45	2.40
12	2.75	2.69	2.62	2.54	2.51	2.47	2.43	2.38	2.34	2.30
13	2.67	2.60	2.53	2.46	2.42	2.38	2.34	2.30	2.25	2.21
14	2.60	2.53	2.46	2.39	2.35	2.31	2.27	2.22	2.18	2.13
15	2.54	2.48	2.40	2.33	2.29	2.25	2.20	2.16	2.11	2.07
16	2.49	2.42	2.35	2.28	2.24	2.19	2.15	2.11	2.06	2.01
17	2.45	2.38	2.31	2.23	2.19	2.15	2.10	2.06	2.01	1.96
18	2.41	2.34	2.27	2.19	2.15	2.11	2.06	2.02	1.97	1.92
19	2.38	2.31	2.23	2.16	2.11	2.07	2.03	1.98	1.93	1.88
20	2.35	2.28	2.20	2.12	2.08	2.04	1.99	1.95	1.90	1.84

21	2.32	2.25	2.18	2.10	2.05	2.01	1.96	1.92	1.87	1.81
22	2.30	2.23	2.15	2.07	2.03	1.98	1.94	1.89	1.84	1.78
23	2.27	2.20	2.13	2.05	2.01	1.96	1.91	1.86	1.81	1.76
24	2.25	2.18	2.11	2.03	1.98	1.94	1.89	1.84	1.79	1.73
25	2.24	2.16	2.09	2.01	1.96	1.92	1.87	1.82	1.77	1.71
26	2.22	2.15	2.07	1.99	1.95	1.90	1.85	1.80	1.75	1.69
27	2.20	2.13	2.06	1.97	1.93	1.88	1.84	1.79	1.73	1.67
28	2.19	2.12	2.04	1.96	1.91	1.87	1.82	1.77	1.71	1.65
29	2.18	2.10	2.03	1.94	1.90	1.85	1.81	1.75	1.70	1.64
30	2.16	2.09	2.01	1.93	1.89	1.84	1.79	1.74	1.68	1.62
40	2.08	2.00	1.92	1.84	1.79	1.74	1.69	1.64	1.58	1.51
60	1.99	1.92	1.84	1.75	1.70	1.65	1.59	1.53	1.47	1.39
120	1.91	1.83	1.75	1.66	1.61	1.55	1.50	1.43	1.35	1.25
∞	1.83	1.75	1.67	1.57	1.52	1.46	1.39	1.32	1.22	1.00

TABLE E-5

Critical Values for the Durbin-Watson d Statistic ($\alpha = .05$)

n	$k-1=1$		$k-1=2$		$k-1=3$		$k-1=4$		$k-1=5$	
	$d_{L,.05}$	$d_{U,.05}$	$d_{L,.05}$	$d_{U,.05}$	$d_{L,.05}$	$d_{U,.05}$	$d_{L,.05}$	$d_{U,.05}$	$d_{L,.05}$	$d_{U,.05}$
15	1.08	1.36	0.95	1.54	0.82	1.75	0.69	1.97	0.56	2.21
16	1.10	1.37	0.98	1.54	0.86	1.73	0.74	1.93	0.62	2.15
17	1.13	1.38	1.02	1.54	0.90	1.71	0.78	1.90	0.67	2.10
18	1.16	1.39	1.05	1.53	0.93	1.69	0.82	1.87	0.71	2.06
19	1.18	1.40	1.08	1.53	0.97	1.68	0.86	1.85	0.75	2.02
20	1.20	1.41	1.10	1.54	1.00	1.68	0.90	1.83	0.79	1.99
21	1.22	1.42	1.13	1.54	1.03	1.67	0.93	1.81	0.83	1.96
22	1.24	1.43	1.15	1.54	1.05	1.66	0.96	1.80	0.86	1.94
23	1.26	1.44	1.17	1.54	1.08	1.66	0.99	1.79	0.90	1.92
24	1.27	1.45	1.19	1.55	1.10	1.66	1.01	1.78	0.93	1.90
25	1.29	1.45	1.21	1.55	1.12	1.66	1.04	1.77	0.95	1.89
26	1.30	1.46	1.22	1.55	1.14	1.65	1.06	1.76	0.98	1.88
27	1.32	1.47	1.24	1.56	1.16	1.65	1.08	1.76	1.01	1.86
28	1.33	1.48	1.26	1.56	1.18	1.65	1.10	1.75	1.03	1.85
29	1.34	1.48	1.27	1.56	1.20	1.65	1.12	1.74	1.05	1.84
30	1.35	1.49	1.28	1.57	1.21	1.65	1.14	1.74	1.07	1.83
31	1.36	1.50	1.30	1.57	1.23	1.65	1.16	1.74	1.09	1.83
32	1.37	1.50	1.31	1.57	1.24	1.65	1.18	1.73	1.11	1.82
33	1.38	1.51	1.32	1.58	1.26	1.65	1.19	1.73	1.13	1.81
34	1.39	1.51	1.33	1.58	1.27	1.65	1.21	1.73	1.15	1.81
35	1.40	1.52	1.34	1.58	1.28	1.65	1.22	1.73	1.16	1.80
36	1.41	1.52	1.35	1.59	1.29	1.65	1.24	1.73	1.18	1.80
37	1.42	1.53	1.36	1.59	1.31	1.66	1.25	1.72	1.19	1.80
38	1.43	1.54	1.37	1.59	1.32	1.66	1.26	1.72	1.21	1.79
39	1.43	1.54	1.38	1.60	1.33	1.66	1.27	1.72	1.22	1.79
40	1.44	1.54	1.39	1.60	1.34	1.66	1.29	1.72	1.23	1.79
45	1.48	1.57	1.43	1.62	1.38	1.67	1.34	1.72	1.29	1.78
50	1.50	1.59	1.46	1.63	1.42	1.67	1.38	1.72	1.34	1.77
55	1.53	1.60	1.49	1.64	1.45	1.68	1.41	1.72	1.38	1.77
60	1.55	1.62	1.51	1.65	1.48	1.69	1.44	1.73	1.41	1.77
65	1.57	1.63	1.54	1.66	1.50	1.70	1.47	1.73	1.44	1.77
70	1.58	1.64	1.55	1.67	1.52	1.70	1.49	1.74	1.46	1.77
75	1.60	1.65	1.57	1.68	1.54	1.71	1.51	1.74	1.49	1.77
80	1.61	1.66	1.59	1.69	1.56	1.72	1.53	1.74	1.51	1.77
85	1.62	1.67	1.60	1.70	1.57	1.72	1.55	1.75	1.52	1.77
90	1.63	1.68	1.61	1.70	1.59	1.73	1.57	1.75	1.54	1.78
95	1.64	1.69	1.62	1.71	1.60	1.73	1.58	1.75	1.56	1.78
100	1.65	1.69	1.63	1.72	1.61	1.74	1.59	1.76	1.57	1.78

Source: From J. Durbin and G. S. Watson, "Testing for Serial Correlation in Least Squares Regression, II," *Biometrika* 30 (1951), 159–178. Reproduced by permission of the *Biometrika* Trustees.

TABLE E-6

Critical Values for the Durbin-Watson d Statistic ($\alpha = .01$)

n	$k-1=1$		$k-1=2$		$k-1=3$		$k-1=4$		$k-1=5$	
	$d_{L,.01}$	$d_{U,.01}$	$d_{L,.01}$	$d_{U,.01}$	$d_{L,.01}$	$d_{U,.01}$	$d_{L,.01}$	$d_{U,.01}$	$d_{L,.01}$	$d_{U,.01}$
15	0.81	1.07	0.70	1.25	0.59	1.46	0.49	1.70	0.39	1.96
16	0.84	1.09	0.74	1.25	0.63	1.44	0.53	1.66	0.44	1.90
17	0.87	1.10	0.77	1.25	0.67	1.43	0.57	1.63	0.48	1.85
18	0.90	1.12	0.80	1.26	0.71	1.42	0.61	1.60	0.52	1.80
19	0.93	1.13	0.83	1.26	0.74	1.41	0.65	1.58	0.56	1.77
20	0.95	1.15	0.86	1.27	0.77	1.41	0.68	1.57	0.60	1.74
21	0.97	1.16	0.89	1.27	0.80	1.41	0.72	1.55	0.63	1.71
22	1.00	1.17	0.91	1.28	0.83	1.40	0.75	1.54	0.66	1.69
23	1.02	1.19	0.94	1.29	0.86	1.40	0.77	1.53	0.70	1.67
24	1.04	1.20	0.96	1.30	0.88	1.41	0.80	1.53	0.72	1.66
25	1.05	1.21	0.98	1.30	0.90	1.41	0.83	1.52	0.75	1.65
26	1.07	1.22	1.00	1.31	0.93	1.41	0.85	1.52	0.78	1.64
27	1.09	1.23	1.02	1.32	0.95	1.41	0.88	1.51	0.81	1.63
28	1.10	1.24	1.04	1.32	0.97	1.41	0.90	1.51	0.83	1.62
29	1.12	1.25	1.05	1.33	0.99	1.42	0.92	1.51	0.85	1.61
30	1.13	1.26	1.07	1.34	1.01	1.42	0.94	1.51	0.88	1.61
31	1.15	1.27	1.08	1.34	1.02	1.42	0.96	1.51	0.90	1.60
32	1.16	1.28	1.10	1.35	1.04	1.43	0.98	1.51	0.92	1.60
33	1.17	1.29	1.11	1.36	1.05	1.43	1.00	1.51	0.94	1.59
34	1.18	1.30	1.13	1.36	1.07	1.43	1.01	1.51	0.95	1.59
35	1.19	1.31	1.14	1.37	1.08	1.44	1.03	1.51	0.97	1.59
36	1.21	1.32	1.15	1.38	1.10	1.44	1.04	1.51	0.99	1.59
37	1.22	1.32	1.16	1.38	1.11	1.45	1.06	1.51	1.00	1.59
38	1.23	1.33	1.18	1.39	1.12	1.45	1.07	1.52	1.02	1,58
39	1.24	1.34	1.19	1.39	1.14	1.45	1.09	1.52	1.03	1.58
40	1.25	1.34	1.20	1.40	1.15	1.46	1.10	1.52	1.05	1.58
45	1.29	1.38	1.24	1.42	1.20	1.48	1.16	1.53	1.11	1.58
50	1.32	1.40	1.28	1.45	1.24	1.49	1.20	1.54	1.16	1.59
55	1.36	1.43	1.32	1.47	1.28	1.51	1.25	1.55	1.21	1.59
60	1.38	1.45	1.35	1.48	1.32	1.52	1.28	1.56	1.25	1.60
65	1.41	1.47	1.38	1.50	1.35	1.53	1.31	1.57	1.28	1.61
70	1.43	1.49	1.40	1.52	1.37	1.55	1.34	1.58	1.31	1.61
75	1.45	1.50	1.42	1.53	1.39	1.56	1.37	1.59	1.34	1.62
80	1.47	1.52	1.44	1.54	1.42	1.57	1.39	1.60	1.36	1.62
85	1.48	1.53	1.46	1.55	1.43	1.58	1.41	1.60	1.39	1.63
90	1.50	1.54	1.47	1.56	1.45	1.59	1.43	1.61	1.41	1.64
95	1.51	1.55	1.49	1.57	1.47	1.60	1.45	1.62	1.42	1.64
100	1.52	1.56	1.50	1.58	1.48	1.60	1.46	1.63	1.44	1.65

Source: From J. Durbin and G. S. Watson, "Testing for Serial Correlation in Least Squares Regression, II," *Biometrika* 30 (1951), 159–178. Reproduced by permission of the *Biometrika* Trustees.

REFERENCES

Anderson, T.W. *The Statistical Analysis of Time Series*. New York: Wiley, 1971.

Anscombe, F.J., and J.W. Tukey. "The Examination and Analysis of Residuals." *Technometrics* 5 (1963): 141–160.

Bowerman, B.L., and R.T. O'Connell. *Time Series and Forecasting: An Applied Approach*. Boston: Duxbury Press, 1979.

Box, G.E.P., and D.R. Cox. "An Analysis of Transformations." *Journal of Royal Statistical Society B* 26 (1964): 211–243.

——— and G.M. Jenkins. *Time Series Analysis: Forecasting and Control*. 2d ed. San Francisco: Holden-Day, 1977.

Cochran, G.W., and G.M. Cox. *Experimental Designs*. 2d ed. New York: Wiley, 1957.

Cramer and Applebaum. "Orthogonal and Nonorthogonal Anova." Speech at 1977 Meeting of the Joint Statistical Societies, Chicago, Ill.

Davis, O.L. *The Design and Analysis of Industrial Experiments*. New York: Hafner, 1956.

Draper, N., and H. Smith. *Applied Regression Analysis*. 2d ed. New York: Wiley, 1981.

Durbin, J., and G.S. Watson. "Testing for Serial Correlation in Least Squares Regression, I." *Biometrika* 37 (1950): 409–428.

———. "Testing for Serial Correlation in Least Squares Regression, II." *Biometrika* 38 (1951): 159–179.

Fuller, W.A. *Introduction to Statistical Time Series*. New York: Wiley, 1976.

Graybill, F.A. *Theory and Application of the Linear Model*. Boston: Duxbury Press, 1976.

Kennedy, W.J., Jr., and J.E. Gentle. *Statistical Computing*. New York: Dekker, 1980.

Kleinbaum, D., and L. Kupper. *Applied Regression Analysis and Other Multivariable Methods*. Boston: Duxbury Press, 1978.

Mendenhall, W. *Introduction to Linear Models and the Design and Analysis of Experiments*. Belmont, Mass.: Wadsworth, 1968.

——— and J. Reinmuth. *Statistics for Management and Economics*. 4th ed. Boston: Duxbury Press, 1982.

Miller, R.B., and D.W. Wichern. *Intermediate Business Statistics: Analysis of Variance, Regression, and Time Series*. New York: Holt, Rinehart, and Winston, 1977.

Nelson, C.R. *Applied Time Series Analysis for Managerial Forecasting*. San Francisco: Holden-Day, 1973.

Neter, J., W. Wasserman, and G.A. Whitmore. *Applied Statistics*. 2d ed. Boston: Allyn and Bacon, 1982.

Neter, J., W. Wasserman, and M.H. Kutner. *Applied Linear Statistical Models*. 2d ed. Homewood, Ill.: Richard Irwin, 1985.

Ott, Lyman. *An Introduction to Statistical Methods and Data Analysis*. 2d ed. Boston: Duxbury Press, 1984.

SAS User's Guide, 1982 Edition. Cary, N. Carolina: SAS Institute, 1982.

Scheffé, H. *The Analysis of Variance*. New York: Wiley, 1959.

Searle, S.R. *Linear Models*. New York: Wiley, 1971.

Winer, B.J. *Statistical Principles in Experimental Design*. New York: McGraw-Hill, 1962.

Wonnacott, T.H., and R.J. Wonnacott. *Introductory Statistics for Business and Economics*, 2d ed. New York: Wiley, 1977.

———. *Regression: A Second Course in Statistics*. New York: Wiley, 1981.

Younger, M.S. *A First Course in Linear Regression*. 2d ed. Boston: Duxbury Press, 1985.

INDEX